T0245199

Principles of Asymmetric Synthesis

Principles of Asymmetric Synthesis

Second Edition

Robert E. Gawley
Department of Chemistry and Biochemistry
University of Arkansas
Fayetteville, AR
USA

and

Jeffrey Aubé
Department of Medicinal Chemistry
University of Kansas
Lawrence, KS
USA

ELSEVIER

AMSTERDAM • BOSTON • HEIDELBERG • LONDON • NEW YORK • OXFORD
PARIS • SAN DIEGO • SAN FRANCISCO • SINGAPORE • SYDNEY • TOKYO

Elsevier
The Boulevard, Langford Lane, Kidlington, Oxford OX5 1GB, UK
Radarweg 29, PO Box 211, 1000 AE Amsterdam, The Netherlands

First edition 1996
Second edition 2012

Notice
No responsibility is assumed by the publisher for any injury and/or damage to persons or
property as a matter of products liability, negligence or otherwise, or from any use or
operation of any methods, products, instructions or ideas contained in the material herein.
Because of rapid advances in the medical sciences, in particular, independent verification of
diagnoses and drug dosages should be made

British Library Cataloguing in Publication Data
A catalogue record for this book is available from the British Library

Library of Congress Cataloging-in-Publication Data
A catalog record for this book is available from the Library of Congress

ISBN: 978-0-08-044860-2

For information on all Elsevier publications
visit our website at www.store.elsevier.com

Printed and bound in UK

12 13 11   10 9 8 7 6 5 4 3 2

Working together to grow
libraries in developing countries

www.elsevier.com | www.bookaid.org | www.sabre.org

ELSEVIER   BOOK AID
            International   Sabre Foundation

Dedicated to Lorraine Gawley and Janet Perkins,
in appreciation for their love and patience.

Contents

Foreword ix
Preface xi

1. Introduction, General Principles, and Glossary of Stereochemical Terms 1
1.1. Why We Do Asymmetric Syntheses 1
1.2. What is an Asymmetric Synthesis? 2
1.3. Stereoselectivity, and What It Takes to Achieve It 4
1.4. Selectivity: Kinetic and Thermodynamic Control 7
1.5. Entropy, the Isoinversion Principle, and the Effect of Temperature on Selectivity 10
1.6. Single and Double Asymmetric Induction 12
1.7. Kinetic Resolution 16
1.8. The Curtin—Hammett Principle 21
1.9. Asymmetric Transformations and Dynamic Resolutions 23
1.10. Asymmetric Catalysis and Nonlinear Effects 28
1.11. Glossary of Stereochemical Terms 34
References 58

2. Practical Aspects of Asymmetric Synthesis 63
2.1. Choosing a Method for Asymmetric Synthesis 63
2.2. How to Get Started 69
2.3. General Considerations for Analysis of Stereoisomers 71
2.4. Chromatography 73
    2.4.1. A Chromatography Primer 73
    2.4.2. Chiral Stationary Phase Chromatography 76
    2.4.3. Achiral Derivatizing Agents 80
2.5. Nuclear Magnetic Resonance 80
    2.5.1. Chiral Derivatizing Agents (CDAs) 81
    2.5.2. Chiral Solvating Agents (CSAs) 85
2.6. Chiroptical Methods 86
2.7. Summary 91
References 92

3. Enolate, Azaenolate, and Organolithium Alkylations 97
3.1. Enolates and Azaenolates 97
    3.1.1. Deprotonation of Carbonyls 100
    3.1.2. The Transition State for Enolate Alkylations 105
    3.1.3. Enolate and Azaenolate Alkylations with Chiral Nucleophiles 110
    3.1.4. Enolate and Azaenolate Alkylations with Chiral Electrophiles 139

3.2. Chiral Organolithiums   151
    3.2.1. Inversion Dynamics of Chiral Organolithiums   153
    3.2.2. Functionalized Organolithiums   157
    3.2.3. Identifying the Stereochemically Defining Step   157
    3.2.4. Asymmetric Deprotonations   161
    3.2.5. Unstabilized Organolithiums   165
References   168

4. 1,2- and 1,4-Additions to C=X Bonds   **179**
4.1. Cram's Rule: Open-Chain Model   180
    4.1.1. The Karabatsos Model   180
    4.1.2. Felkin's Experiments   181
    4.1.3. The Bürgi–Dunitz Trajectory: A Digression   182
    4.1.4. Back to the Cram's Rule Problem (Anh's Analysis)   186
    4.1.5. Further Refinements   187
4.2. Cram's Rule: Rigid, Chelate, or Cyclic Model   191
4.3. Chiral Catalysts and Chiral Auxiliaries   196
    4.3.1. Catalyzed Additions of Organometallic Compounds to Aldehydes   197
    4.3.2. Addition of Organometallics to Azomethines   203
    4.3.3. Additions of Organometallics to Pyridinium Ions   212
    4.3.4. Hydrocyanations of Carbonyls   215
    4.3.5. Hydrocyanations of Azomethines (the Strecker Reaction)   219
4.4. Conjugate Additions   221
    4.4.1. Acyclic Esters and Ketones   222
    4.4.2. Acyclic Amides and Imides   229
    4.4.3. Cyclic Ketones and Lactones   232
References   237

5. Aldol and Michael Additions of Allyls, Enolates, and Enolate Equivalents   **245**
5.1. 1,2-Allylations and Related Reactions   246
    5.1.1. Simple Enantioselectivity   246
    5.1.2. Simple Diastereoselectivity   248
    5.1.3. Single Asymmetric Induction   249
    5.1.4. Double Asymmetric Induction   251
    5.1.5. Other Allyl Metals   254
5.2. Aldol Additions   258
    5.2.1. Simple Diastereoselectivity   259
    5.2.2. Single Asymmetric Induction   263
    5.2.3. Organocatalysis of the Aldol Reaction and its Variants   285
    5.2.4. Double Asymmetric Induction and Beyond: Synthetic
        Applications of the Aldol Reaction   295
5.3. Michael Additions   301
    5.3.1. Simple Diastereoselectivity: Basic Transition State Analysis   301
    5.3.2. Chiral Donors   303
    5.3.3. Chiral Michael Acceptors   315

5.3.4. Interligand Asymmetric Induction and Catalysis   316
5.3.5. Conjugate Addition of Nitrogen Nucleophiles   319
References   326

6. Cycloadditions and Rearrangements   335
6.1. Cycloadditions   335
6.1.1. The Diels−Alder Reaction   335
6.1.2. Hetero Diels−Alder Reaction   360
6.1.3. 1,3-Dipolar Cycloadditions   368
6.1.4. Summary   375
6.1.5. [2 + 1]-Cyclopropanations and Related Processes   375
6.2. Rearrangements   395
6.2.1. [1,3]-Hydrogen Shifts   395
6.2.2. [2,3]-Wittig Rearrangements   401
6.2.3. Other Rearrangements   416
References   422

7. Reductions and Hydroborations   431
7.1. Reduction of Carbon−Heteroatom Double Bonds   431
7.1.1. Modified Lithium Aluminum Hydride   432
7.1.2. Modified Borane   434
7.1.3. Chiral Organoboranes   442
7.1.4. Chiral Transition Metal Catalysts   444
7.2. Reduction of Carbon−Carbon Bonds   463
7.3. Hydroborations   476
References   484

8. Oxidations   491
8.1. Epoxidations and Related Reactions   491
8.1.1. Early Approaches   491
8.1.2. Epoxidations   493
8.1.3. Sharpless Kinetic Resolution   504
8.1.4. Some Applications of Asymmetric Epoxidation and Kinetic
      Resolution Procedures   506
8.1.5. Aziridinations   510
8.2. Asymmetric Dihydroxylation (AD) Reaction   510
8.2.1. Reaction Development   510
8.2.2. Applications of Enantioselective Dihydroxylations   518
8.3. α-Functionalization of Carbonyl Groups and Their Equivalents 521
8.3.1. Hydroxylations   522
8.3.2. Aminations and Halogenations   524
8.4. Miscellaneous Oxidations that Necessitate Differentiation
     of Enantiotopic Groups   529
8.4.1. Oxidation of Sulfides   529
8.4.2. Group-Selective Oxidation of C−H Bonds   531
8.4.3. Group-Selective Oxidative Ring Expansions   531
References   539

Index   545

Chirality in chemistry and asymmetric synthesis (chemical reactions, in which elements of chirality are generated) have developed from a specialty pursued by outsiders ("chiromaniacs") to an art cultured by some learned ones, and now are part of essentially every chemist's daily life. We should, however, not forget that in a multistep synthesis the transition from achiral intermediates or from racemic mixtures to enantiopure intermediates is unique. In all of the other steps, it is functional-group selectivity, regio- and diastereoselectivity that are at stake.

The development of asymmetric synthesis has taken place exponentially in the last three decades, triggered by a number of circumstances. Many more chemists are intrigued and attracted by the phenomenon of chirality and by the origin of homochirality of the molecules of life. Practitioners of organic synthesis and synthetic methodology have annexed transition-metal chemistry (with chiral ligands on the metals) and biological–chemical transformations to achieve enantioselective catalysis. Concomitantly, new chromatographic and spectroscopic methods for determining enantiomer ratios have facilitated the ease and accuracy of analyses of products of enantioselective reactions. Perhaps the strongest driving force for the development was the necessity of producing pharmaceuticals, diagnostics, vitamins, and agrochemicals in enantiopure form (their biological targets are chiral, after all!).

As I stated in the foreword of the first edition in 1996, the authors of *Principles of Asymmetric Synthesis* have managed to cover the subject in a condensed and masterly way; they have chosen well-defined topics for the eight chapters of the book; they have used clear-cut concepts and concise chemical language for the presentations; they have discussed mechanistic considerations with due care; they have included a glossary of stereochemical terms (those to use and those not to use!); they have provided extensive referencing. All of this is still true for the second edition, which has grown from 372 to 556 pages, with an almost doubling of references from *ca.* 1300 to 2400. This is not due to just adding more of the same but mostly due to including the dramatic new developments that have occurred in the past 15 years.

This can be seen as a realization of two dreams, expressed in 1990 by an organic chemist [1]: (i) "The primary center of attention for all synthetic methods will continue to shift towards catalytic and enantioselective variants; indeed, it will not be long before such modifications will be available with every standard reaction for converting achiral educts into chiral products, ... leading to undreamed of efficiency and selectivity" and (ii) "The discovery of truly new reactions is likely to be limited to the realm of transition-metal organic chemistry, which will almost certainly provide us with additional *miracle reagents* in the years to come."

As far as the first dream is concerned, there has been a revival of what was called, in 1935, organocatalysis [2], *i.e.* catalysis of the classical, main-group organic transformations without involvement of metals or metal ions, using *chiral* amino compounds, carbenes, Brønsted acids, counter ions, ureas, or HMPA derivatives as catalysts for essentially all the well-known workhorse reactions of organic synthesis, such as aldol, Michael, Diels−Alder, 1,3-dipolar additions, hydride transfer from Hantzsch ester, Mannich, Strecker, Stetter, Baylis−Hillman reactions, α-alkylation and -functionalization of carbonyl compounds, Friedel−Crafts-type reactions, epoxidations, and aziridinations, to name only a few.

The second dream concerns new types of *miracle reactions* catalyzed by transition metals that enable the synthetic chemist to perform often incredibly complex transformations in one step, which were undreamed of by experts of classical organic reactions; three Nobel prizes have been awarded in this field in the past decade (2001, 2005, and 2010). A plethora of chiral ligands has become available for enantioselective versions of most of these reactions.

Indeed, inclusion of sections covering organocatalysis (in Chapters 3−8), and new transition-metal catalyzed reactions (in Chapters 4−8) have mainly contributed to the increase of the volume of this second edition. But there are also other important additions and changes: there is a scholarly written *highlight box* in each chapter; the presentations of the *formulae*, with some red color for emphasizing steric interactions, is much better; remarkable additions are found in Chapter 1 (entropy, iso-inversion principle, kinetic and dynamic resolution, non-linear effects), in Chapter 2 (enantiomer enrichment during chromatography on achiral column material), in Chapter 4 (sulfinimines and phosphinoyl imines), in Chapter 5 (reductive aldol additions, *N*-, *O*-, and *S*-nucleophiles in Michael additions), in Chapter 6 (reshuffled to emphasize importance of [4 + 2]- and [3 + 2]-cycloadditions), in Chapter 7 ("desymmetrizations"), and in Chapter 8 (dioxiranes, sulfoxidations, Baeyer−Villiger and Beckman−Schmidt type reactions).

The second edition of *Principles of Asymmetric Synthesis* is up-to-date in all aspects of this important part of organic synthesis. Of special note, the first two chapters on "Introduction, General Principles, and Glossary of Stereochemical Terms" and "Practical Aspects of Asymmetric Synthesis" are unique among all the books I have seen on stereochemistry. This book can serve as a textbook for classes, as a monograph on enantio- and diastereoselective synthesis, and as a reference work to find seminal publications, even for the expert in the field.

Dieter Seebach

REFERENCES

[1] Seebach, D. *Angew. Chem.* **1990**, *102*, 1363 Int. Ed. Engl. 1990, 29, 1320.
[2] Langenbeck, W. *Die Organischen Katalysatoren und ihre Beziehung zu den Fermenten*; Springer Verlag: Berlin, 1935.

Preface

The field of asymmetric synthesis continues to grow at an exponential rate. To even address the topic in a significant way is a formidable task. In the end, we have continued the format of the first edition by selecting several reaction categories that comprise many of the most useful synthetic reaction types. As the title implies, the focus is on the principles that govern relative and absolute configurations in transition state assemblies. There are only a few principles, but they recur constantly. For example, organization around a metal atom, $A^{1,3}$ strain, van der Waals interactions, dipolar interactions, etc., are factors affecting transition state energies, and which in turn dictate stereoselectivity *via* transition state theory. One might call these analyses molecular recognition at a saddle point.

The book has 8 chapters, which the publisher will also be making available online on an individual basis to readers interested in only parts of the book. The first chapter provides background, introduces the topic of asymmetric synthesis, outlines principles of transition state theory as applied to stereoselective reactions, and includes the glossary of stereochemical terms. The second chapter begins with a discussion of practical aspects of obtaining an enantiopure compound, and then details methods for analysis of mixtures of stereoisomers. Then follow four chapters on carbon-carbon bond forming reactions, organized by reaction type and presented in order of roughly increasing mechanistic complexity: Chapter 3 discusses enolate and organolithium alkylations, while Chapter 4 covers nucleophilic additions to C=O and C=N bonds; these two chapters cover reactions in which one new stereocenter is formed. Chapter 5 covers aldol and Michael additions that generate at least two new stereocenters, while Chapter 6 covers selected cycloadditions and rearrangements. The last two chapters cover reductions and oxidations.

Transition state analyses are presented to explain - to the current level of understanding - the stereoselectivity of most of the reactions covered. Critical examination of these rationales sometimes exposes the weaknesses of current theories, in that they cannot always explain the experimental observations. These shortcomings provide a challenge for future mechanistic investigations.

Much of the work on the first edition of this book was completed during a sabbatical leave for REG, at the Swiss Federal Institute of Technology (ETH), Zürich, which was funded in part by a Fogarty Senior International Fellowship from the National Institutes of Health, in part by a sabbatical leave from the University of Miami, and in part by the ETH. This financial support is warmly acknowledged, with thanks. Special thanks are also due to Professor Dieter Seebach for his generous hospitality during that sabbatical year, and to his colleagues, Professors Arigoni, Diederich, Dunitz, Prelog, and Vasella, who jointly contributed to making the year in Zürich both enjoyable and memorable.

Many of our friends and colleagues contributed to this work with helpful discussions, or by reading and commenting on various portions of this book. Among these, the late Professor Vladimir Prelog deserves special thanks for his exhaustive critique of an early draft of the glossary. Evidence of his contribution is contained in a highlight box that precedes the glossary in Chapter 1. We are also grateful to the Kansas Book Club, who generously gave their time to assist in literature searching for this edition, discussions about content, and who provided numerous drawings and schemes. They were Thomas Coombs, Erik Fenster, Brooks Maki, Daljit Matharu, Thomas Painter, Digamber Rane, Steven Rogers, and Denise Simpson. We are forever indebted to them for their efforts and for their unwavering good cheer as this project progressed. We also received valuable help from Emily Scott and Jenny Wang in reorganizing the crystal structure in Scheme 6.16. Finally, we thank Christopher Katz of Pequod Book Design for providing us with inspirational cover art.

We are also indebted to our wives, who put up with our numerous weekends in the office and evenings spent staring into our computers. It is to them that this work is dedicated.

Finally, we have been heartened by unsolicited comments from students and teachers who used the first edition of this book in a graduate course at universities around the world, and who found it useful. We hope future readers will feel the same. Comments are welcome.

Robert E. Gawley
Fayetteville, Arkansas

Jeffrey Aubé
Lawrence, Kansas

December 16, 2011

Chapter 1

Introduction, General Principles, and Glossary of Stereochemical Terms

1.1 WHY WE DO ASYMMETRIC SYNTHESES

> *L'univers est dissymétrique*
> Louis Pasteur (1874)

In modern terminology, Pasteur would say "The universe is chiral."[1] We are constantly learning more about the implications of chirality, from weak bosons in nuclear physics to the origins of life on earth and the double helix of DNA [1−5]. Most organic compounds are chiral. Chemists working with perfumes, cosmetics, nutrients, flavors, pesticides, vitamins, and pharmaceuticals, to name a few examples [6−11], require access to enantiomerically pure compounds. Single enantiomer formulations now account for most of the chiral drugs on the market. One estimate suggests that approximately half of the worldwide revenues from chiral products were the result of traditional synthesis from the chiral pool or resolution, whereas less than half result from chemical catalysis [12].

As our ability to produce enantiomerically pure compounds grows, so does our awareness of the differences in pharmacological properties that a chiral compound may have when compared with its enantiomer or racemate [13−19]. We easily recognize that all biological receptors are chiral, and as such can distinguish between the two enantiomers of a ligand or a substrate. Enantiomeric compounds often have different odors or tastes [20−22].[2] Thus, it is obvious that two enantiomers should be considered different compounds when screened for pharmacological activity [10,13,23]. The demand for enantiomerically pure compounds as drug candidates is not likely to let up in the foreseeable future.

How might we obtain enantiomerically pure compounds? Historically, the best answer to that question has been to isolate them from natural sources. Derivatization of natural products or their use as synthetic starting materials has long been a useful tool in the hands of the synthetic chemist, but it has now been raised to an art form by some practitioners, wherein complex molecules are dissected into chiral fragments that may be obtained from natural products [24−34]. Even today, there is no way of obtaining enantiomerically pure compounds without ultimately resorting to Nature, whether for a building block, an auxiliary, or a catalyst.

---

1. Readers may find the glossary at the end of this chapter useful for the definitions of unfamiliar terms.
2. For example, the enantiomers of limonene smell and taste like oranges or lemons, the enantiomers of phenylalanine taste bitter or sweet, the enantiomers of carvone taste like spearmint or caraway.

So if the objective is to obtain an enantiomerically pure compound, one has a choice to make: synthesize the molecule in racemic form and resolve it [35], find a plant or a bacterium that will make it for you, start with a natural product such as a carbohydrate, terpene, or alkaloid (but beware of racemic or partly racemic natural products), or plan an asymmetric synthesis. Among the factors to consider in weighing the alternatives are the amount of material required, the cost of the starting materials, length of synthetic plan, etc., factors that have long been important to synthetic design [36−41]. For the purposes of biological evaluation, it may be *desirable* to include a resolution so that one synthesis will provide both enantiomers. But for the production of a single enantiomer, a classical resolution will have a maximum theoretical yield of 50% unless the unwanted enantiomer can be recovered and recycled, or the process is stereoconvergent *via* an asymmetric transformation or a dynamic resolution. In most cases, starting with a natural product will be restricted to the production of only one enantiomer by a given route, notwithstanding the talent of some investigators to produce both enantiomers of a target from the same chiral starting material. Such practical aspects are discussed more fully in Chapter 2.

1.2 WHAT IS AN ASYMMETRIC SYNTHESIS?

The most quoted definition of an asymmetric synthesis is that published by Marckwald in 1904 [42]:

> "Asymmetrische" Synthesen sind solche, welche aus symmetrisch constituirten Verbindungen unter intermediärer Benutzung optisch-activer Stoffe, aber unter Vermeidung jedes analytischen Vorganges, optisch-activ Substanzen erzeugen.[3]

In modern terminology, the core of Marckwald's definition is the conversion of an achiral substance into a chiral, nonracemic one by the action of a chiral reagent. Marckwald's point of reference, of course, was biochemical processes, so it follows that enzymatic processes [43−45] are included by this definition. Marckwald also asserted that the nature of the reaction was irrelevant, so a self-immolative reaction or sequence[4] such as an intermolecular chirality transfer in a Meerwein−Pondorf−Verley reaction would also be included:

Interestingly, the Marckwald definition is taken from a paper that was rebutting a criticism [46] of Marckwald's claim to have achieved an asymmetric synthesis by a group-selective decarboxylation of the brucine salt of 2-ethyl-2-methylmalonic acid [47,48]:

Thus, from the very beginning, the definition of what an asymmetric synthesis might encompass, or even if one was possible, has been a matter of discussion. On the latter point,

---

3. "Asymmetric" syntheses are those that produce optically active substances from symmetrically constituted compounds with the intermediate use of optically active materials, but with the avoidance of any separations.
4. Self-immolative processes are those that generate a new stereocenter at the expense of an existing one, either in a single reaction or in a sequence whereby the controlling stereocenter is deliberately destroyed in a subsequent step.

the idea that a chemist could synthesize something in optically active form from an achiral precursor was doubted in some circles, even in Marckwald's time. That doubt, expressed in a published lecture in 1898 [49] was one of the last tenets of the vitalism theory to die.[5] This book expands the topic to include many examples of stereoselective synthesis that would not strictly fall under Marckwald's definition.

Several criteria are useful for judging an asymmetric synthesis:

- The synthesis should be highly stereoselective.
- If a chiral catalyst is employed, low catalyst loadings are preferable, and the catalyst must be easily separable from the product.
- If a chiral auxiliary is used, it must be removable without compromising the new stereocenter, and should be recoverable in good yield and without racemization.
- The chiral auxiliary or catalyst should be readily and inexpensively available, preferably as either enantiomer.

Several comments are appropriate regarding these guidelines. The first criterion is obviously the most important, and is universally applicable to all synthetic strategies. It is especially important for reactions that produce mixtures that are difficult to separate. The second criterion can be very important in pharmaceutical processes where trace amounts of impurities, especially toxic metals, are taboo. Point 3 is predicated on the usually high cost of chiral reagents. Condition 4 is less important when a chiral catalyst has a high turnover number, the chiral auxiliary is very inexpensive, or positive nonlinear effects are in force (see Section 1.10).

For the purposes of synthetic planning, the most important variable is the cost of the process relative to the value of the product. Also, the scale of the planned synthesis must be considered: an affordable cost for the preparation of a few grams of product may not be feasible for the production of several hundred kilograms [50] (see also Section 2.1).

The simplicity of the Marckwald definition has been its most enduring feature, but our understanding of structure and mechanism has evolved since Marckwald's time,[6] and chromatographic and spectroscopic techniques have replaced polarimetry as the primary means of measuring stereoisomer ratios. Indeed, many modern methods result in the selective formation of products containing more than one stereocenter, and it is often the case that reagents and catalysts determine the configuration of new stereocenters, even in chiral substrates.

In 1933, two short monographs [51,52] summarized virtually everything known about asymmetric synthesis and asymmetric induction. By 1971, the field was summarized in another short monograph of about 450 pages [53], but by then the art of organic synthesis was ready for a rapid advance: 10 years later, a five-volume treatise [54] was necessary (~1800 pages). The literature continues to grow at such a rate that comprehensive coverage is

---

5. Remnants of the vitalism theory persist in some quarters, such as in advocacy for products containing "all natural ingredients" or that "contain no chemicals."

6. As a point of reference, consider that in Marckwald's time the van't Hoff−le Bel theory of tetrahedral carbon was accepted, but what we now know as an sp^2 or trigonal carbon, was not. It was thought, at least by some, that the fourth site of a carbonyl carbon was an unoccupied site on a tetrahedron. For example, under the term "asymmetric induction" in the first collective index of *Chemical Abstracts*, we find reference to a paper (**8**:3431[1]) entitled "Preparation of *l*-benzaldehyde through asymmetric induction ..." (Erlenmeyer, E.; Landesberger, F.; Hilgendorff, G. *Biochem. Z.* **1914**, *64*, 382−392). The formula for benzaldehyde was PhCHL · OL, where L indicates an unoccupied position.

now impossible. Although not restricted to asymmetric synthesis, the nine-volume treatise *Comprehensive Organic Synthesis*, published in 1991 [55], subtitled "Selectivity, Strategy, and Efficiency in Modern Organic Chemistry" is ~7000 pages. In 1995, a 6000-page treatise entitled *Stereoselective Synthesis* and advertised as "the whole of organic stereochemistry" appeared as volume E21 of the Houben—Weyl series [56]. Beginning with the turn of the twenty-first century, the 48-volume 5th edition of Houben—Weyl, entitled *Science of Synthesis, Houben-Weyl Methods of Molecular Transformations*, is being published in both paper and searchable electronic formats.

It is our primary aim to provide a concise analysis of the stereochemical features of transition states in a variety of reaction types. These features are a manifestation of the intra- and intermolecular forces that govern transition state assemblies, which can sometimes be modeled by computers. The ultimate (attainable?) goal of stereoselective synthesis is clear: the production of any relative and absolute configuration of one or more stereogenic units through the use of chiral catalysts that do not require consideration of chirality elements extant in the substrate.

1.3 STEREOSELECTIVITY, AND WHAT IT TAKES TO ACHIEVE IT

It is the primary goal of this book to analyze the factors that influence stereoselectivity when one stereoisomer predominates over others. For illustrative purposes, consider the metal-mediated addition of a nucleophile to an aldehyde. The faces of aldehydes are heterotopic, either enantiotopic (if there are no stereogenic elements elsewhere in the molecule) or diastereotopic (if there are), as shown in Figure 1.1. In order to achieve a predominance of one stereoisomer (enantiomer or diastereomer) over the other, the transition states resulting from addition of a fourth ligand to the heterotopic *Re* or *Si* faces must be diastereomeric. This will be the case if either the carbonyl compound, the reagent, or a metal catalyst (ML_n) are chiral. More than one stereogenic element, as in R* *and* ML_n^* gives rise to the possibility of matched or mismatched double asymmetric induction. In this chapter, we outline the physical principles that control selectivity, and provide precise definitions of stereochemical terms in the Glossary (Section 1.11).

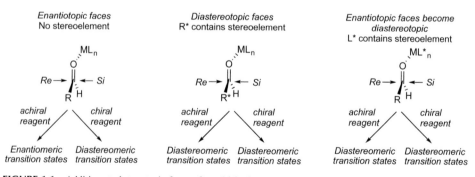

FIGURE 1.1   Additions to heterotopic faces of an aldehyde.

To see how the general principles in Figure 1.1 might apply in fact, consider a reaction from the perspective of the metal. The most important ligand will be the aldehyde substrate; other ligands may simply be solvent, monodentate or bidentate "spectator" ligands, or a ligand that is involved in the reaction (a "player"). In the addition of a Grignard reagent to an aldehyde (Scheme 1.1), the Grignard reagent (ignoring Schlenck processes) has an organic ligand (*e.g.*, Me), a halide (X), and perhaps bound solvent molecules or other ligands (L$_n$) in its coordination sphere. Addition to a carbonyl does not occur, however, until the aldehyde coordinates to the magnesium. Note that four types of ligand are now apparent in the intermediate shown in brackets: the aldehyde substrate, the methyl group that adds to the carbonyl, the halide, and other ligands.

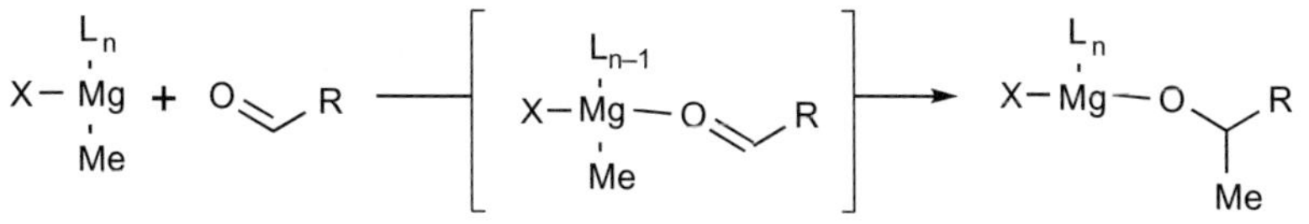

SCHEME 1.1   The addition of a Grignard reagent to an aldehyde.

Several examples of carbonyl additions (see also Chapter 4) exemplify a further ligand classification that has tremendous bearing on stereoselectivity, and which illustrates in a nutshell the history of asymmetric synthesis while also pointing us toward its future. In the 1950s, Cram examined the influence of an adjacent stereocenter on the stereoselectivity of nucleophilic additions to carbonyls [57,58]. In the example illustrated in Scheme 1.2a (taken from Eliel's later work [59]), the aldehyde group has a stereocenter, as in Figure 1.1: R*. One diastereomer is formed to the near exclusion of the other. The chirality sense (*R/S*) of the new stereocenter is determined by the chirality center adjacent to the carbonyl. From the perspective of the metal, both the "old" and the "new" stereocenters are within the same ligand, so the asymmetric induction is *intraligand*. Later, the auxiliary shown in Scheme 1.2b was developed for use in an asymmetric synthesis of α-hydroxy aldehydes. Here, the oxathiane fragment is *removed* after directing the selective formation of one of two possible diastereomers [60]. Note, however, that this example is also a case of *intraligand* asymmetric induction, and that both of these examples have diastereomeric transition states because the carbonyl-containing substrate is chiral (as in R* in Figure 1.1). Scheme 1.2c shows the addition of a Grignard to a ketone, after modifying the reagent with a chiral diol ligand. Here, since the ketone is achiral, the two faces of the carbonyl are enantiotopic, until the transition states are rendered diastereomeric by the chiral diol ligand. Although the details of the reaction are unknown, the only chirality element present in the reactants are in the diol ligand, making this *interligand* asymmetric induction [61]. A metal atom may also be a stereogenic center; a few examples using a complex containing such a center have appeared. The principle has not found widespread application, due in part to issues associated with configurational stability of the metal stereocenter (reviews: [62–64]).

The mental process of "removing" the stereocenter from the substrate and putting it on another ligand of the metal allows the introduction of the element of asymmetric catalysis [65,66], as shown by the addition of a Grignard, through a titanate, as in Scheme 1.2d [67]. Here, the chiral reagent is used in less than stoichiometric quantities, as opposed to the example in Scheme 1.2c, which employs stoichiometric diol ligand. The key to success is the relative rates of the reaction mediated by the chiral catalyst and competing nonselective reactions. Catalytic processes are more cost-effective than stoichiometric processes, and have the added advantage of decreasing the environmental impact of disposing of (or recycling) by-products produced in stoichiometric quantities.

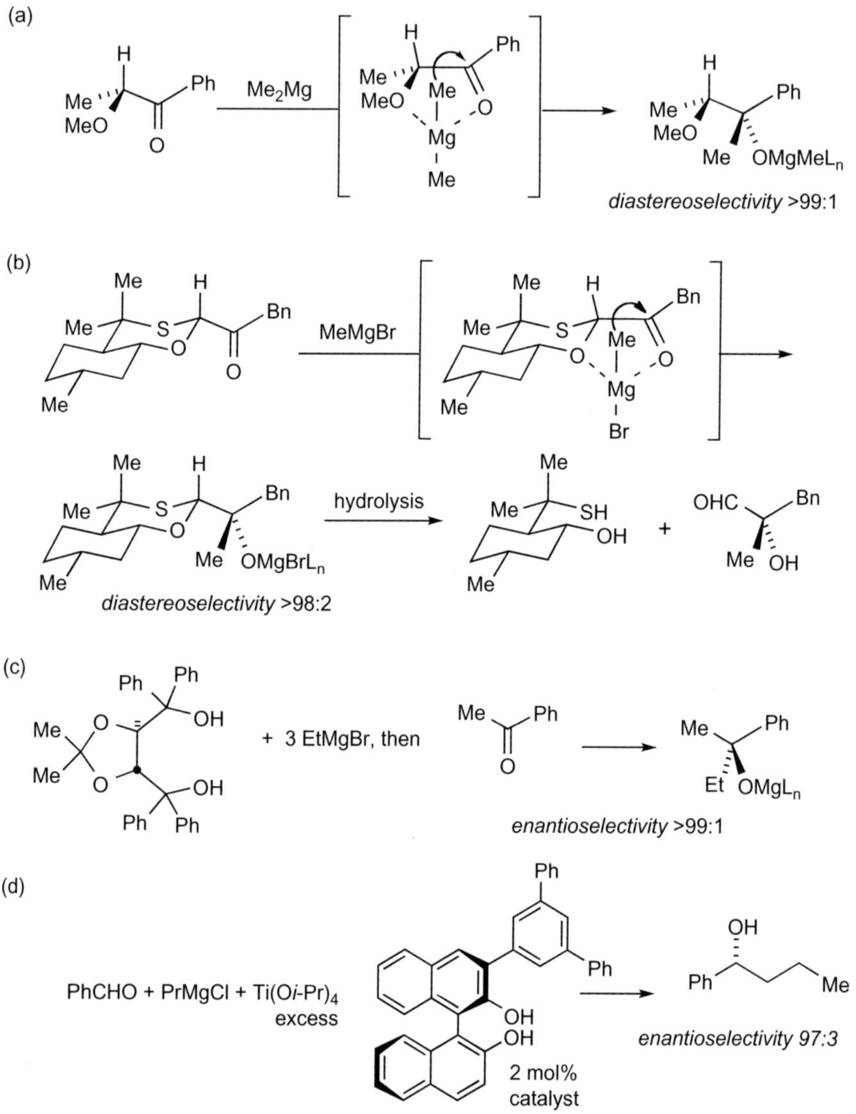

SCHEME 1.2 Intraligand *vs.* interligand asymmetric induction. (a) Diastereoselective addition *via* Cram's cyclic model [59]. (b) Asymmetric synthesis of a pure enantiomer *via* diastereoselective addition to a carbonyl with a chiral auxiliary [60]. (c) Enantioselective addition of ethyl Grignard to an aldehyde using a chiral ligand on magnesium [61]. (d) Catalytic enantioselective addition to aldehyde with a Grignard reagent [67].

The four examples in Scheme 1.2 illustrate the progress made in stereoselective reactions in the last few decades, which has evolved through several distinct phases: (i) diastereoselective synthesis by addition of nucleophiles to carbonyls having a neighboring stereocenter; (ii) the extension of the same notion to the synthesis of a single enantiomer *via* diastereoselection and auxiliary removal; (iii) enantioselective addition to an achiral substrate by a stoichiometric reagent; and (iv) enantioselective addition mediated by a chiral catalyst. Extrapolation of the trend that is apparent in these examples points inexorably toward the goal stated at the end of the previous section: the production of new stereogenic units through the use of chiral catalysts that do not require consideration of existing chirality elements.

1.4 SELECTIVITY: KINETIC AND THERMODYNAMIC CONTROL

The means by which stereoselectivity is achieved in various reactions and processes are widely variable. However, the asymmetric induction that results in any given process must fall into one of only two categories: thermodynamic or kinetic control, the latter being by far the more common.

Consider a starting material, A, that may give two possible products, B and C. Figure 1.2a illustrates how equilibration might occur to afford an equilibrium mixture of B and C by one of two possible routes. The reactions A→B and A→C might be reversible, or B and C could equilibrate by a route that does not involve A. Either way, the product ratio (C/B) is given by

$$C/B = \frac{[C]}{[B]} = K = e^{-\Delta G^{\circ}/RT}, \tag{1.1}$$

where ΔG° is the free energy difference between C and B ($\Delta G^{\circ} = G_C - G_B$), R is the gas constant, and T is the absolute temperature. Processes such as this are under *thermodynamic control*.

Under conditions of *kinetic control* (Figure 1.2b), the conversion of A into either B or C is irreversible, and the relative rates of formation of each product determine the final product ratio. The rates are given by

$$\frac{d[B]}{dt} = k_1[A] \tag{1.2}$$

and

$$\frac{d[C]}{dt} = k_2[A], \tag{1.3}$$

where k_1 and k_2 are the rate constants for the formation of B and C, respectively. From transition state theory,

$$k_1 = \frac{\kappa k_B T}{h} e^{(-\Delta G_B^{\ddagger}/RT)} \tag{1.4}$$

and

$$k_2 = \frac{\kappa k_B T}{h} e^{(-\Delta G_C^{\ddagger}/RT)}, \tag{1.5}$$

where κ is the transmission coefficient (usually taken as unity), k_B is the Boltzmann constant, T is the absolute temperature, h is Planck's constant, and $\Delta G_B^{\ddagger}$ and $\Delta G_C^{\ddagger}$ are the free energies of activation for the formation of B and C, respectively. Assuming equality of the transmission coefficients, at a given temperature,

$$B/C = \frac{k_1}{k_2} = e^{-\Delta\Delta G^{\ddagger}/RT}, \tag{1.6}$$

where $\Delta\Delta G^{\ddagger}$ is the difference in the activation energies for each process:

$$\Delta\Delta G^{\ddagger} = \Delta G_{B}^{\ddagger} - \Delta G_{C}^{\ddagger}. \tag{1.7}$$

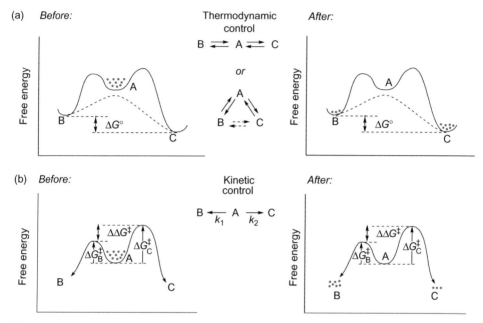

FIGURE 1.2   (a) Conversion of A into a mixture of B and C under *thermodynamic control*. Note that B and C may equilibrate *via* A or by another route (dashed line). (b) Conversion of A into B and C under *kinetic control*. Here, A→B and A→C are irreversible, and the relative free energies of B and C are irrelevant.

Two types of selectivity may be used to establish new stereogenic elements in a molecule: *diastereoselectivity* and *enantioselectivity*. In *diastereoselective* reactions, either kinetic or thermodynamic control is possible, but in *enantioselective* reactions, the products are isoenergetic and only kinetic control is possible.[7]

Equations 1.1 and 1.6 establish the exponential (Boltzmann) dependence of selectivity on free energy and temperature. Figure 1.3 shows plots of these equations at three different temperatures.[8] The Goodman group at Cambridge has made applets available on the web that solve the Boltzmann equation (given any two of the three variables: ratio, $\Delta G^{\ddagger}$, and T) and the Eyring equation to estimate rate constants (given T and $\Delta G^{\ddagger}$): http://www.ch.cam.ac.uk/magnus/ [68].

The curves of Figure 1.3 illustrate a number of points:

- The steepest part of the curves occurs in the region where the selectivity (K or k_1/k_2) is ≤10. Because of the exponential relationship, a doubling of the free energy difference at 10:1 will increase the selectivity to 100:1.

---

7. Unless the reaction is conducted in a chiral solvent (rare).
8. Strictly speaking, these equations and the curves in Figure 1.3 are valid only for a unimolecular reaction in the gas phase, but to a first approximation they serve as useful tools for our purposes.

- The total energy differences that afford 100:1 selectivity are not large. For comparison, recall that ΔG° between the axial and equatorial conformations of methylcyclohexane is $\sim$1700 cal/mol.
- In the "flat" part of the curves, small differences in energy will produce large differences in selectivity. For example at 0 °C, an increase in ΔG° or $\Delta\Delta G^{\ddagger}$ of 873 cal/mol increases selectivity from 20:1 to 100:1. By comparison, ΔG° for the *gauche* and *anti* forms of butane is 900 cal/mol.
- At lower temperatures, the selectivity curves flatten out more quickly. Thus for a given process, subtle changes in the stereochemical control elements will usually have a greater influence if the reaction is carried out at low temperature. Note that this does not necessarily mean that lowering the temperature of a reaction will result in increased selectivity (see Section 1.5).
- Selectivities may be expressed in several ways: as diastereomer ratio (dr), enantiomer ratio (er), percentage of the major enantiomer (% es), or diastereomer (% ds). Usually, dr and er are normalized as a percent, such that 19:1 is written as 95:5.[9] It is worth keeping these parallel scales in mind when evaluating selectivities, as all are used interchangeably in the literature.[10]

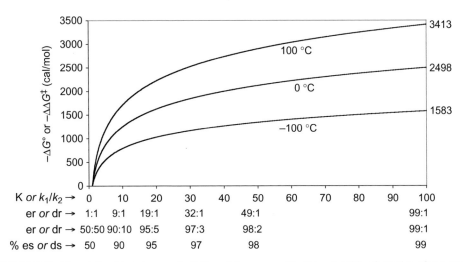

FIGURE 1.3 The relationship between selectivity and free energy (for the competitive formation of two products) at -100 °C, 0 °C, and 100 °C. The free energy values for relative rates of 100:1 are labeled on the right ordinate (cal/mol).

---

9. Normalizing the ratios to 100 has the advantage that, as ratios change from favoring one stereoisomer to another, they pass through a 50:50 mirror point rather than going from a fraction to infinity. For example, $1:99 \rightarrow 50:50 \rightarrow 99:1$ would be $0.01:0.99 \rightarrow 1:1 \rightarrow 99:1$ if not normalized.

10. The term "ee," enantiomeric excess specifies the *difference* between the enantiomeric products. Selectivity is the *ratio* between enantiomeric products. Because of this difference, it is inappropriate to use ee when describing enantioselectivity [69]. See also the highlight box on page 90.

1.5 ENTROPY, THE ISOINVERSION PRINCIPLE, AND THE EFFECT OF TEMPERATURE ON SELECTIVITY

Regarding the effect of temperature on selectivity, total reliance on equations such as Equation 1.6 can be misleading, since free energy itself is temperature dependent:

$$\Delta G = \Delta H - T\Delta S. \tag{1.8}$$

Combination of Equations 1.6 and 1.8 gives

$$B/C = \frac{k_1}{k_2} = (e^{-\Delta\Delta H^{\ddagger}/RT})(e^{\Delta\Delta S^{\ddagger}/R}), \tag{1.9}$$

where $\Delta\Delta H^{\ddagger}$ and $\Delta\Delta S^{\ddagger}$ are the differences in enthalpy and entropy of activation for the formation of B and C, defined as was $\Delta\Delta G^{\ddagger}$ in Equation 1.7.[11] Equation 1.9 suggests that only the enthalpy term is temperature dependent, whereas the entropy term is constant at all temperatures. In many stereoselective reactions, especially when evaluated over a small temperature range, enthalpic contributions tend to dominate.[12] Indeed, many of the sterochemical rationales presented in this book consider only enthalpic effects.

It is important to recognize that this is not always the case. Taking the logarithm of both sides of Equation 1.9 gives a modified form of the Eyring equation:

$$\ln\frac{k_1}{k_2} = -\frac{\Delta\Delta H^{\ddagger}}{RT} + \frac{\Delta\Delta S^{\ddagger}}{R}. \tag{1.10}$$

Plots of $\ln(k_1/k_2)$ vs. $1/T$ are not always linear across a large temperature range [70,71]. The addition of p-anisidine to an α,β-unsaturated N-acyloxazolidinone illustrated in Scheme 1.3 is a particularly striking example [72,73]. At $-38\,°C$, the enantioselectivity reaches a maximum: temperatures above or below this temperature give lower enantioselectivity. The modified Eyring plot of this data is shown in Figure 1.4. The temperature at which maximum selectivity is observed is called the inversion temperature, and is one manifestation of what is known as the isoinversion principle [70]. Note that, both above and below the inversion temperature, the Eyring plot is linear, but leads to two sets of activation parameters ($\Delta H^{\ddagger}$ and $\Delta S^{\ddagger}$) for each temperature range. The S enantiomer is favored over the entire temperature range studied, but to varying degrees.

Using the energy diagram inset in Scheme 1.3 and the generality $S - R$ to calculate differences in activation parameters, the S addition product is favored by negative $\Delta\Delta G^{\ddagger}$ and $\Delta\Delta H^{\ddagger}$ and positive $\Delta\Delta S^{\ddagger}$ (see also Figure 1.2b). From $+25\,°C$ to $-40\,°C$, $\Delta\Delta H^{\ddagger} = -4.61$ kcal/mol (i.e., S is favored by enthalpy) and $\Delta\Delta S^{\ddagger} = 29.0$ cal/mol $\cdot$ K (i.e., S is favored by entropy). Thus, lowering the temperature changes $\Delta\Delta G^{\ddagger}$ from -0.436 to -1.35 kcal/mol, which is reflected in the higher er at $-40\,°C$. From $-40\,°C$ to $-80\,°C$, the slope of the modified Eyring plot is negative, with $\Delta\Delta H^{\ddagger} = +5.47$ kcal/mol (i.e., S is disfavored by enthalpy) and $\Delta\Delta S^{\ddagger} = +28.9$ cal/mol K (i.e., S is favored by entropy). Thus, lowering the temperature from $-40\,°C$ to $-80\,°C$ changes $\Delta\Delta G^{\ddagger}$ from -1.35 to -0.113 kcal/mol, which results in the lower er at $-80\,°C$.

The reason for inversion temperatures in the modified Eyring plot of Figure 1.4 has been interpreted as differences in solvation, such that at different temperatures, the reagents behave

---

11. $\Delta\Delta H^{\ddagger} = \Delta H_B^{\ddagger} - \Delta H_C^{\ddagger}$, negative if B is favored; $\Delta\Delta S^{\ddagger} = \Delta S_B^{\ddagger} - \Delta S_C^{\ddagger}$, positive if B is favored.
12. The derivation of the Eyring equation is predicated on the assumption that the Arrhenius pre-exponential factor (A), enthalpy of activation ($\Delta H^{\ddagger}$), and entropy of activation ($\Delta S^{\ddagger}$) are independent of temperature. This is generally considered to be a safe assumption over small temperature ranges.

like different molecules. In this case (Scheme 1.3), the catalyst may exist in solution as a mixture of monomer, dimer, and oligomers, each of which could be catalytically active but solvated differently. When the reaction was run at −40 °C, but diluted by a factor of 4, lower conversion and lower er were observed, consistent with more monomer in solution. This observation is consistent with the monomer being both less reactive and less enantioselective as a catalyst [73].[13]

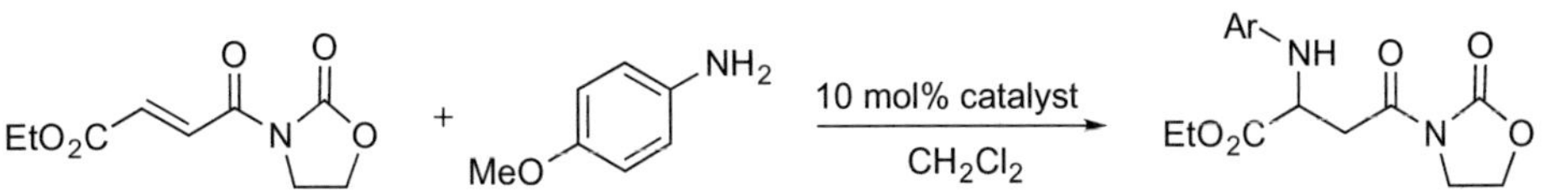

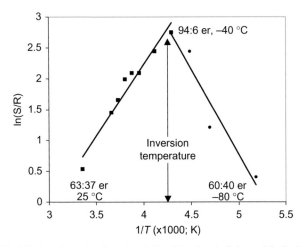

Temperature (°C)	er (S:R)
−80	60:40
−60	77:23
−50	92:8
−40	94:6
−30	92:8
−20	89:11
−15	89:11
−10	88:12
−5	84:16
0	81:19
25	63:37

S is always favored

SCHEME 1.3 Asymmetric addition of *p*-anisidine to an unsaturated oxazolidinone showing a nonlinear temperature effect [72,73].

FIGURE 1.4 Modified Eyring plot, using Equation 1.10, of the data in Scheme 1.3; the inversion temperature for this reaction is −38 °C. [72,73].

13. In addition, a negative nonlinear effect was observed, suggesting that the heterochiral dimer is more active than the homochiral dimer (see Section 1.10).

Often, one of the linear regions of a modified Eyring plot has a zero slope (*i.e.*, $\Delta\Delta H^{\ddagger} = 0$), whereas another linear region may have either a positive or a negative slope. Since most investigators do not do a complete analysis of selectivity as a function of temperature, the general likelihood of encountering an inversion temperature is not known. Interested readers are referred to the available reviews and references cited therein [70,74].

It is intuitively obvious that a change in solvent, for example, going from THF to hexane, could affect the solvation of ground states and transition states, and therefore result in different activation parameters. But it may not be so obvious that a change in the length of a linear hydrocarbon chain can also effect activation parameters, as was observed in the addition of PhLi to 2-phenylpropanal [74,75]. Thus, careful optimization of temperature and solvent is a worthwhile effort, especially in large-scale preparations.

1.6 SINGLE AND DOUBLE ASYMMETRIC INDUCTION

For the purposes of illustration, Figure 1.5 illustrates two reaction types in generic form. Single asymmetric induction occurs if a single chirality element directs the selective formation of one stereoisomer over another by selective reaction at one of the heterotopic (*Re/Si*) faces of a trigonal atom.

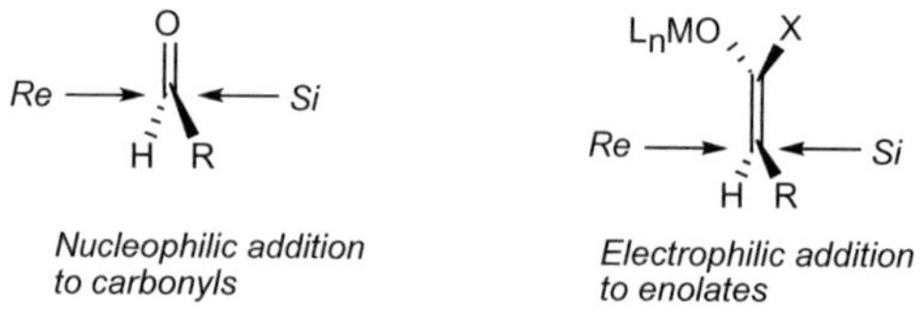

Nucleophilic addition to carbonyls   *Electrophilic addition to enolates*

FIGURE 1.5   Two types of reactions that distinguish heterotopic faces.

An interesting circumstance develops when two of these techniques are combined in the same reaction, such as when the second reactant also contains a chirality element (*e.g.*, when a chiral nucleophile reacts with a chiral electrophile): the chirality elements of each reactant may influence stereoselectivity either in concert or in opposition. This phenomenon is known as double asymmetric induction [76,77]. Two simple illustrations are shown in Scheme 1.4 and involve the reaction of the two enantiomers of a chiral nucleophile [78]. Note the difference in diastereoselectivity observed for the two reactions, clearly resulting from the change in absolute configuration of the remote stereocenter of the nucleophile.

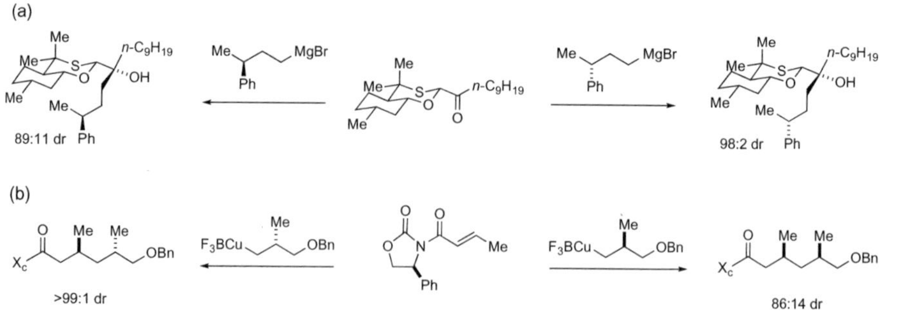

SCHEME 1.4   Double asymmetric induction: changing the absolute configuration of a chiral nucleophile affects the stereoselectivity of addition to (a) a chiral ketone [78]; and (b) a chiral crotonate [79].

To understand the phenomenon of double asymmetric induction, we need to have a clear picture of the inherent selectivities of each of the chiral partners in closely related single asymmetric induction processes. Consider, for example, the kinetically controlled aldol addition reactions shown in Scheme 1.5 [76].[14] The first two illustrated reactions are examples of single asymmetric induction with inherently low selectivities. Scheme 1.5a is the reaction of a chiral *Z(O)*-enolate with an achiral aldehyde [80], and illustrates the *Si*-facial preference of the *S* enantiomer of the enolate of 78:22. In Scheme 1.5b, an achiral enolate that is structurally similar to the chiral enolate of Scheme 1.5a is allowed to react with a chiral aldehyde [81]. The 73:27 product ratio reflects the *Re*-facial preference of the aldehyde.[15] Note that the absolute configuration of the new stereocenters in the major products are the same. Since both chiral reactants, the enolate of the first reaction and the aldehyde of the second, prefer the same absolute configuration in the addition product, we may expect that reaction of the chiral enolate with the chiral aldehyde would afford product having the same absolute configuration.

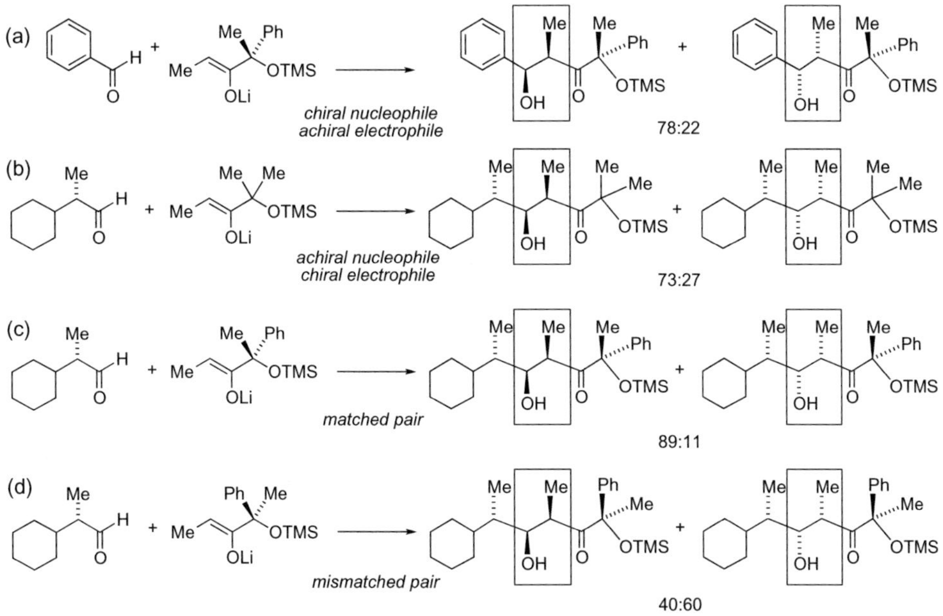

SCHEME 1.5 Examples of single and double asymmetric induction in the aldol addition reaction. (a) Reaction of a chiral enolate and an achiral aldehyde. (b) Reaction of an achiral enolate with a chiral aldehyde. (c) Matched pair double asymmetric induction with a chiral enolate and a chiral aldehyde. (d) Mismatched pair double asymmetric induction with a chiral enolate and the aldehyde enantiomeric to that shown in (a) [76].

---

14. In these examples, two stereocenters are created, but only two of the four possible stereoisomers are observed. As explained in detail in Chapter 5, the two *syn* isomers are produced stereoselectively from the *Z(O)*-enolate (see Glossary and Section 1.11, for definition of this term.)

15. Obviously, if the absolute configuration of either of the chiral reactants in Scheme 1.5a or 1.5b were reversed, the absolute configuration of the new stereocenters would also be reversed, that is, the enantiomers of the illustrated products would be produced in the same ratio.

When the S enolate and the S aldehyde (Scheme 1.5c) were allowed to react, the expected product was indeed formed, but the selectivity was higher (89:11) than in either of the previous examples because the inherent selectivities of the two chiral species are mutually reinforcing [76]. This is an example of *matched pair* double asymmetric induction. A *mismatched* double asymmetric induction would result from reversing the absolute configuration of either of the two chiral reactants. For example, when the S aldehyde and now the R enolate (Scheme 1.5d) were allowed to react, the two products formed in a ratio of 40:60 [76]. The higher selectivity of the enolate (78% ds) over the aldehyde (73% ds) is manifested in the absolute configuration obtained as the major isomer in Scheme 1.5d.

With this example in mind, let us reexamine the principles of selectivity presented earlier and apply them to the case of double asymmetric induction. In Figure 1.2b, two products are possible under kinetic control. This reaction diagram is applicable to the examples of Scheme 1.5a and b, in that *a single chirality element* operates to render the two transition structures diastereomeric. Now imagine what the effect of a second chirality element might be. Figure 1.6a illustrates the case of a matched pair: the second chirality element increases $\Delta\Delta G^{\ddagger}$ by lowering the energy of the already favored transition state ($A \to B$) and/or raising the energy of the disfavored one. The previously favored isomer is formed with increased selectivity. Figure 1.6b illustrates the mismatched case, wherein the second chirality element decreases $\Delta\Delta G^{\ddagger}$ by increasing the energy of the favored transition state and/or decreasing the energy of the disfavored one. In this example, the second chirality element decreases $\Delta\Delta G^{\ddagger}$ but does not change its sign. Obviously, additional perturbation of the transition states could reduce $\Delta\Delta G^{\ddagger}$ to zero, or reverse the selectivity by changing the sign of $\Delta\Delta G^{\ddagger}$.

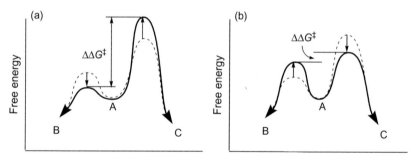

FIGURE 1.6 Double asymmetric induction. The dashed lines represent a hypothetical case of single asymmetric induction. (a) Matched pair: $\Delta\Delta G^{\ddagger}$ is increased by the influence of a second chirality element. (b) Mismatched pair: $\Delta\Delta G^{\ddagger}$ is decreased by the influence of a second chirality element.

There are two important lessons here. The first is that a matched pair will afford higher selectivities than either chiral reactant would afford on its own. The second lesson is more subtle. In considering the two single asymmetric induction reactions, suppose that one of the chiral reagents is much more selective than the other. In this instance, the mismatched pair may still be a highly selective reaction. Figure 1.7 illustrates an energy diagram wherein the stereoselectivity due to the second chirality element completely overwhelms that of the first. The dashed lines indicate a preference, in single asymmetric induction, for product B. Under the influence of a much more highly selective reagent, the double asymmetric induction (bold line) favors C by "changing the sign" of $\Delta\Delta G^{\ddagger}$. Even though this is a mismatched pair, it still may be very selective. In such cases, the chiral reagent is the primary determinant of the absolute configuration of the new stereocenter(s) in the product!

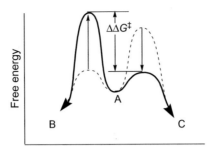

FIGURE 1.7 Reagent-based stereocontrol in double asymmetric induction.

In a general sense, if the species having an overwhelmingly higher inherent selectivity is the chiral auxiliary, chiral reagent or chiral catalyst, and the less selective species is a synthetic intermediate being carried on to a target, then *the reagent can be used to determine the absolute configuration of the product, independent of the chirality sense and bias of the substrate*. This concept is known as *"reagent-based stereocontrol"* [76,77]. As we will see throughout this book, a number of reagents deliver high enough selectivities to achieve this important goal.

Two examples of such processes are shown in Scheme 1.6. One is the titanium TADDOLate-catalyzed addition of diethylzinc to myrtenal (see Section 4.3.1 [82]); the other is the Sharpless asymmetric epoxidation (see Section 8.1.2 [76,83]). In both cases, the diastereoselectivity for the reaction of the substrate with an achiral reagent is low (65−70% ds), whereas the catalysts have enantioselectivities of >95% with achiral substrates. In these cases of double asymmetric induction, the catalyst completely overwhelms the facial bias of the chiral substrate.

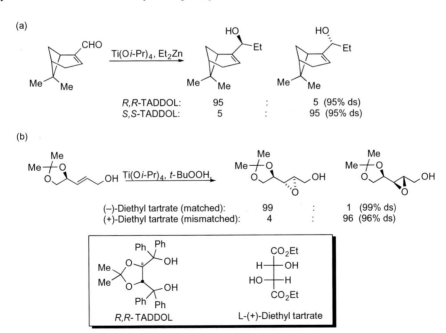

SCHEME 1.6 Matched and mismatched double asymmetric induction demonstrating reagent-based stereocontrol. (a) The diethylzinc addition catalyzed by titanium TADDOLates [82]. (b) The Sharpless asymmetric epoxidation [76,83].

1.7 KINETIC RESOLUTION

Classical resolution involves the separation of enantiomers, often by making diastereomeric derivatives and separating them [35]. Using the principles of double asymmetric induction, it is possible to achieve resolutions by selective reaction of one enantiomer with a chiral reagent, catalyst, or enzyme [84]. The reaction is easily understood in terms of the general reaction and kinetic profile illustrated in Figure 1.8. A pair of enantiomeric substrates, S_S and S_R, that are not interconvertible on the time scale of the reaction (i.e., $k_{ent} = 0$, or k_R, $k_S \gg k_{ent}$), reacts at different rates with chiral reagent R*, or with an achiral reagent, R, in the presence of a chiral catalyst, such that $k_R \neq k_S$. If $k_R > k_S$, the substrate R enantiomer (S_R) will be consumed faster than the $S(S_S)$, leaving the unreacted substrate enriched in S_S.

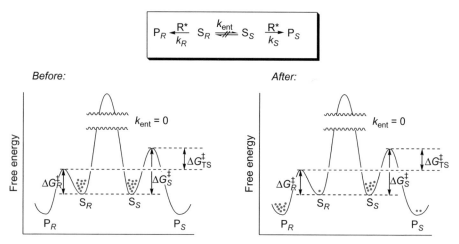

FIGURE 1.8 Energy diagrams and reaction scheme for a kinetic resolution.

The ratio of relative rates of the two enantiomers is known as the selectivity factor, **s**:

$$s = \frac{k_R}{k_S}. \tag{1.11}$$

As explained in detail below, kinetic resolutions can be used to enrich the enantiopurity of unreacted substrate to any desired level, since the enantiopurity of the unreacted starting material continually increases with percent conversion. However, the converse is not true, since the enantiopurity of the product continually decreases with time, meaning that the highest er of product is obtained at low conversion, and, at low conversion, is a direct function of the selectivity factor.[16] Normally, only kinetic resolutions with very high selectivity factors are useful for obtaining products with high er.

New techniques to address this problem include parallel kinetic resolutions, in which two resolving agents are used, one of which reacts faster with S_S, while the other reacts faster

---

16. The rate of product formation is the product of the rate constant times the concentration (i.e., $d[P_S]/dt = k_S[S_S]$ and $d[P_R]/dt = k_R[S_R]$). As the faster-reacting enantiomer is consumed, the relative concentration of the less reactive enantiomer increases, so the rate of its conversion to product also increases, and the er of the product decreases.

with S_R, each affording different products that are easily separable [85−88]. In this strategy, the two resolving agents are often diastereomers that have structures so similar that they are very nearly enantiomeric; they are often termed "quasi-enantiomers." An example from the Vedejs group is shown in Scheme 1.7 [85]. Here, enantiomeric derivatives of dimethylamino-pyridine are acylated with alkyl chloroformates that are structurally very different (Scheme 1.7a). The quasi-enantiomeric pyridiniums are combined in a 1:1 ratio and allowed to react with a racemic alcohol (Scheme 1.7b). Each of the quasi-enantiomeric pyridiniums reacts selectively with only one of the alcohol enantiomers through the principle of double asymmetric induction, and afford structurally different, and separable, carbonate products.

Using chiral catalysts, parallel kinetic resolutions can be made catalytic [89]. Parallel kinetic resolutions work best if the two reactions occur without mutual interference, have similar rates such that $[S_S] \approx [S_R]$ throughout, and have complementary enantioselectivity.

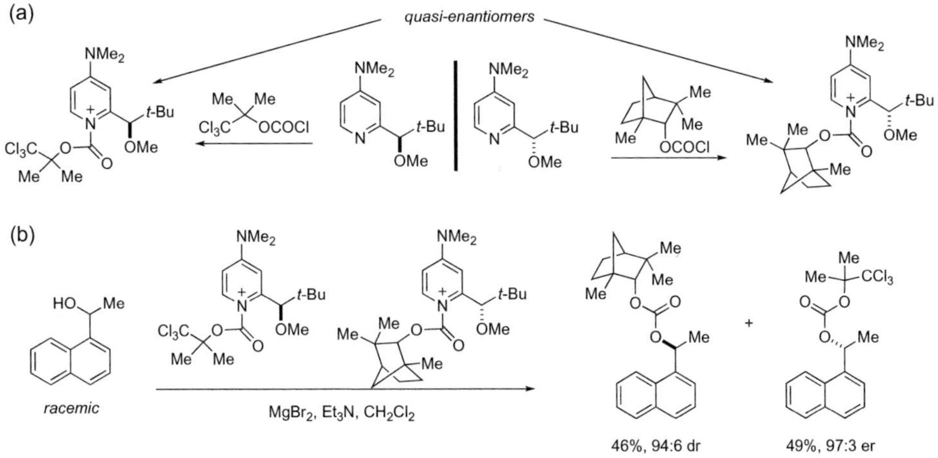

SCHEME 1.7   Parallel kinetic resolution [85]. In the first step (a), enantiomeric pyridines are acylated to give quasi-enantiomeric pyridinium salts. In step (b), the quasi-enantiomeric pyridiniums are allowed to react with the racemic alcohol. Double asymmetric induction renders the reaction of each enantiomer of the alcohol more reactive with only one of the pyridinium ions.

Kinetic resolutions are evaluated by comparing the enantiomer composition as a function of conversion, C, expressed as a fraction. For example if $k_R > k_S$, S_R is consumed faster, and one evaluates the enantiopurity of S_S vs. C. If one begins a kinetic resolution with a racemate, $[S_R]_0 = [S_S]_0 = 0.5$ at t_0. If at time t, conversion is $0 < C < 1$, then at time t,

$$[S_R] + [S_S] + C = 1, \tag{1.12}$$

where $[S_R]$ and $[S_S]$ are the fractions of unreacted substrate enantiomers.

Rearranging,

$$[S_R] = 1 - C - [S_S] \quad \text{and} \quad [S_S] = 1 - C - [S_R]. \tag{1.13}$$

If consumption of S_R and S_S are first order or pseudo-first order in [S], then:

$$\frac{d[S_R]}{dt} = -k_R[S_R] \quad \text{and} \quad \frac{d[S_S]}{dt} = -k_S[S_S]. \tag{1.14}$$

Integration and substitution reveals that the selectivity factor, **s**, is given by

$$\mathbf{s} = \frac{\ln(2[S_R])}{\ln(2[S_S])} = \frac{\ln([S_R] - [S_S] - C + 1)}{\ln([S_S] - [S_R] - C + 1)}. \tag{1.15}$$

Note that after the reaction begins, the $[S_R]$ and $[S_S]$ terms in Equation 1.15 are not the fractions obtained by normalizing enantiomer ratios. If one wishes to use the normalized enantiomer ratio, in which the fractions of each enantiomer, R and S, add to 1 (*e.g.*, 0.83 +0.17, obtained from a normalized er of 83:17), the er fractions should each be multiplied by the quantity $(1 - C)$ to obtain the correct $[S_R]$ and $[S_S]$ terms for Equation 1.15. If one wishes to use the normalized fractions R and S directly, Equation 1.16 should be used [69]:

$$\mathbf{s} = \frac{\ln[(1 - C)(R - S + 1)]}{\ln[(1 - C)(S - R + 1)]}. \tag{1.16}$$

An example illustrates the correct calculation using either equation. Assume that the percent conversion is 45% ($C = 0.45$), and the S/R er normalized as a percent is 83:17. The fraction of original substrate remaining $(1 - C)$ is 0.55, and the calculation of **s** using Equation 1.15 is as follows:

$$\mathbf{s} = \frac{\ln[(0.55)(0.17) - (0.55)(0.83) - 0.45 + 1)]}{\ln[(0.55)(0.83) - (0.55)(0.17) - 0.45 + 1)]} = \frac{\ln[0.187]}{\ln[0.913]} = 18.4.$$

The calculation using Equation 1.16 is as follows:

$$\mathbf{s} = \frac{\ln[(0.55)(0.17 - 0.83 + 1)]}{\ln[(0.55)(0.83 - 0.17 + 1)]} = \frac{\ln[0.187]}{\ln[0.913]} = 18.4.$$

The relationship expressed in Equations 1.15 and 1.16 is valid for all cases where the reaction is first order with respect to substrate, and any order with respect to R* or to an achiral reagent, R, in the presence of a chiral catalyst [69].

By measuring the enantiomer composition at known conversions, C, one may use Equation 1.15 or 1.16 to calculate **s**. Due to potential errors in accurate measurement of er and C, and the exponential nature of Equation 1.15, it is prudent to measure the er at several conversions to determine **s** accurately. Since the product ratio depends on $k_S[S_S]$ and $k_R[S_R]$, accuracy is increased by measuring er at low conversions where $[S_S]$ and $[S_R]$ are in large excess. In situations where the resolution is accomplished by reaction of substrate S with reagent R to form product SR, one may use racemic substrate and racemic reagent to produce $S_S R_S/S_R R_R$ and $S_R R_S/S_S R_R$ diastereomers, whose ratio will equal **s** at any conversion, since $[S_S] = [S_R]$ and $[R_S] = [R_R]$ throughout the course of the reaction [90].

Figure 1.9 plots the relationship of enantiopurity, expressed as fraction of unreacted enantiomer, *vs.* conversion for three values of **s**. For accurate calculations, the Goodman group at Cambridge has developed an applet for kinetic resolution calculations on the web [91]: http://www.ch.cam.ac.uk/magnus/KinRes.html. They note that, at the top of the graph where the lines are near horizontal, a small error in measurement of er results in a large error in **s**. For example, at 95% conversion, a 99:1 er corresponds to **s** = 2.98. If the er is mismeasured as 98:2, **s** = 2.67, a 10% error. Where the slopes are steeper, errors in er are less significant, but errors in measurement of conversion have a greater effect. Thus, calculation of **s** from only one measurement of er and C should be interpreted cautiously.

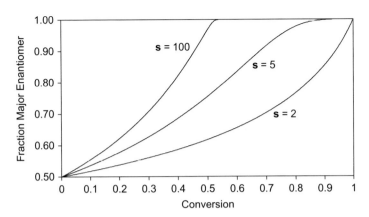

FIGURE 1.9 Relationship between fraction unreacted enantiomer and conversion in a kinetic resolution for three values of **s** [69].

It is possible to prepare a sample of any enantiomeric purity by kinetic resolution, beginning with a racemate or with an enantiomerically enriched substrate, and using the selectivity factor, **s**, to calculate the necessary extent of conversion to achieve the desired er [69,90]. The general equation for determining C in a kinetic resolution, where **s** is known and S_R is the faster reacting enantiomer, is:

$$C = 1 - \left[\left(\frac{[S_R]}{[S_R]_0} \right) \left(\frac{[S_S]_0}{[S_S]} \right)^{\mathbf{s}} \right]^{\frac{1}{(\mathbf{s}-1)}}, \tag{1.17}$$

where $[S_S]$ and $[S_R]$ are the desired fractions of unreacted substrate after the resolution, $[S_S]_0$ and $[S_R]_0$ are their fractions at t_0 [69]. The conversion necessary to achieve the desired er is C, and **s** is the selectivity factor of the kinetic resolution. For example, a kinetic resolution having **s** = 5.0 can be used to enrich a sample from 90:10 er to 99:1 er by carrying the resolution to 50.1% conversion:

$$C = 1 - \left[\left(\frac{0.01}{0.10} \right) \left(\frac{0.90}{0.99} \right)^{5} \right]^{1/4} = 1 - 0.499 = 0.501.$$

If starting with a racemate (50:50 er), Equation 1.17 simplifies to

$$C = 1 - [2[S_R](2[S_S])^{-\mathbf{s}}]^{1/(\mathbf{s}-1)}. \tag{1.18}$$

With a selectivity factor of 5.0, a racemate can be enriched to 99:1 er by carrying the resolution to 84.0% conversion:

$$C = 1 - \left[0.02(2 \times 0.99)^{-5} \right]^{1/4} = 1 - 0.160 = 0.840.$$

As noted above, only kinetic resolutions with high-selectivity factors are capable of producing *products* having high er. Examples of nonenzymatic kinetic resolutions are shown in Scheme 1.8. The first is an acylation that employs a chiral dimethylaminopyridine derivative [92].

The selectivity factor is modest, but the unreacted alcohol is obtained in high enantiopurity by a high conversion; note the low er of the ester product. Scheme 1.8b shows an example of the Sharpless asymmetric epoxidation used as a kinetic resolution [93]. The selectivity factor of 104 affords reasonable yields of both unreacted alcohol and epoxide, both in very high enantiopurity. A special case leading to enantiotopic group selectivity is shown in Scheme 1.8c. The achiral divinyl alcohol shown presents two enantiotopic olefins to the tartrate-derived catalyst. In contrast to the kinetic resolution situation, each molecule now contains two olefins able to react with the chiral reagent system, with the major product taking the same diastereoselective and enantioselective course established through the kinetic resolution studies (Scheme 1.8b). Here, however, when the minor enantiomeric epoxide is formed (along with minor diastereomers that are not shown), this minor product still presents a highly reactive olefin to the catalyst, meaning that carrying out the reaction to higher conversions leads to greater enantiomeric enrichment of the products [94]. See also Section 8.1.4.

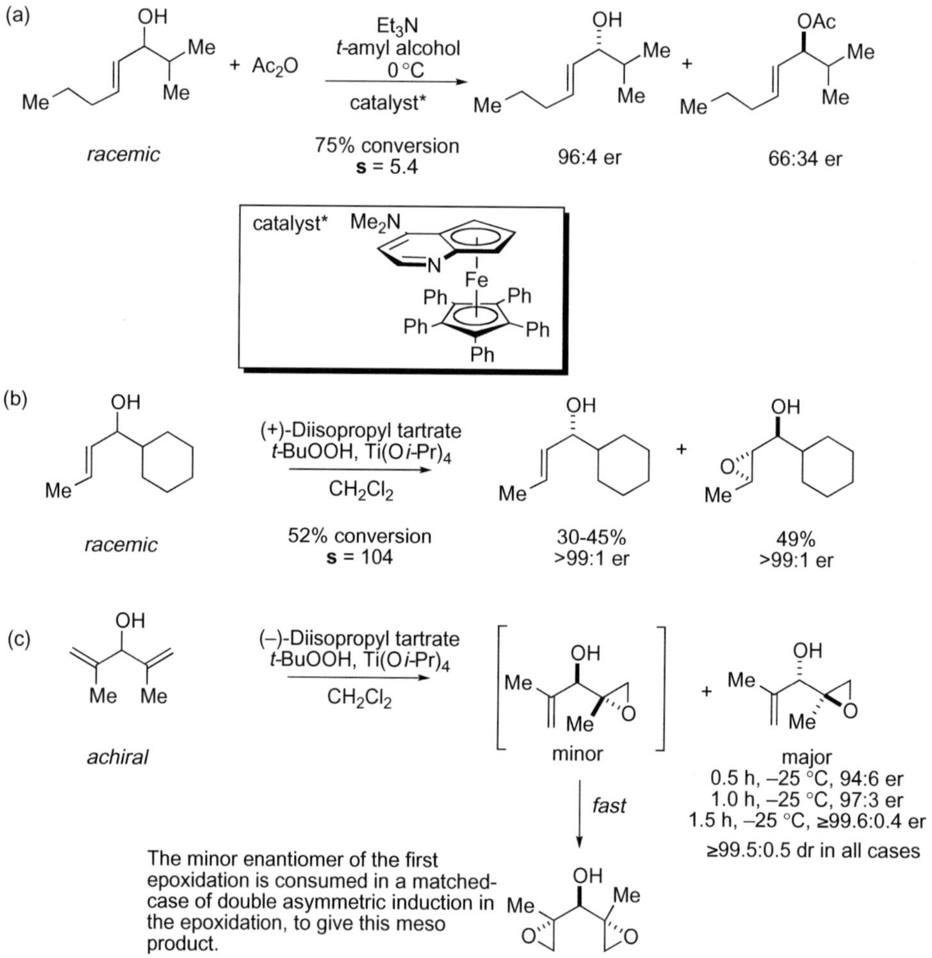

SCHEME 1.8 Kinetic resolutions: (a) by acylation [92]; (b) by epoxidation [93]; and (c) desymmetrization followed by kinetic resolution [94].

1.8 THE CURTIN–HAMMETT PRINCIPLE

It often happens that there is more than one substrate species in solution, each of which produces a different product, but the substrate species are in rapid equilibrium, complicating the analysis of the mechanism. Quite commonly, these are conformational isomers or diastereomeric metal complexes, and in some cases (discussed in the next section), equilibrating enantiomers or diastereomers. When each species reacts to give a different product, and when the barrier to interconversion between substrate species is low compared to the activation energy of conversion to products, a special circumstance ensues, known as Curtin–Hammett kinetics [95–97]. Under such conditions, the product ratio formed from two interconverting species depends on both the difference in transition state energies going to products, $\Delta G_{TS}^{\ddagger}$, and the ratio of the two species.

The general scheme is shown in Figure 1.10a: substrates S_A and S_B interconvert with a low barrier, but reaction of S_A gives P_A and reaction of S_B gives P_B, such that $k_{AB}, k_{BA} \gg k_1, k_2$. We illustrate the principle with two interconverting species, but it is also applicable to more than two. The scheme is general, but for our purposes, consider that the two products are stereoisomers, either enantiomers or diastereomers. Figure 1.10b illustrates the case where S_A is more stable than S_B, and Figure 1.10c illustrates the case where S_B is more stable than S_A. They could also be isoenergetic. These variations in the relative energy of the substrates change the sign of ΔG_{AB}°, but otherwise have no effect on the following equations.

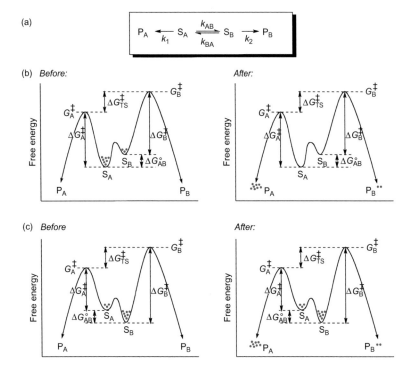

FIGURE 1.10   (a) Curtin–Hammett kinetics applies when two substrate species, S_A and S_B, are in rapid equilibrium relative to their rates of reaction ($k_{AB}, k_{BA} \gg k_1, k_2$), and which react under kinetic control such that $S_A \rightarrow P_A$ and $S_B \rightarrow P_B$. The P_A/P_B product ratio is determined by $\Delta G_{TS}^{\ddagger}$ [95–97]. (b) Situation where S_A is more stable than S_B ($\Delta G_{AB}^{\circ} > 0$). (c) Situation where S_B is more stable than S_A ($\Delta G_{AB}^{\circ} < 0$). Note that in both (b) and (c), the relative energy of the transition states is unchanged, so the product ratio is unchanged.

Under first-order or pseudo-first-order conditions, the following equations describe the process:

$$K_{AB} = \frac{[S_B]}{[S_A]} = e^{-\Delta G^\circ_{AB}/RT} \tag{1.19}$$

$$\frac{d[P_A]}{dt} = k_1[S_A] \tag{1.20}$$

$$\frac{d[P_B]}{dt} = k_2[S_B] = k_2 K_{AB}[S_A], \tag{1.21}$$

where $[S_A]$ and $[S_B]$ are concentrations of substrates A and B, respectively.

Substituting Equations 1.4 and 1.5 into 1.20 and 1.21, then dividing, canceling, and rearranging terms, the ratio of the products is given by

$$\frac{P_A}{P_B} = \frac{d[P_A]/dt}{d[P_B]/dt} = \frac{k_1}{k_2 K_{AB}} = \frac{(e^{(-\Delta G^\ddagger_A/RT)})}{(e^{(-\Delta G^\ddagger_B/RT)})(e^{(-\Delta G^\circ_{AB}/RT)})} = e^{(-\Delta G^\ddagger_A + \Delta G^\ddagger_B + \Delta G^\circ_{AB})/RT}. \tag{1.22}$$

Equation 1.22 reveals that the product ratio depends on both the ratio of the two substrates, K_{AB}, and their relative reactivities, k_1 and k_2. By inspection of Figure 1.10a or b, it is apparent that

$$\Delta G^\ddagger_B + \Delta G^\circ_{AB} - \Delta G^\ddagger_A = G^\ddagger_B - G^\ddagger_A = \Delta G^\ddagger_{TS}. \tag{1.23}$$

Substituting Equation 1.23 into 1.22 reveals that the P_A/P_B product ratio depends on the difference in energy between the transition states:

$$\frac{P_A}{P_B} = e^{\Delta G^\ddagger_{TS}/RT}. \tag{1.24}$$

It is important to emphasize that for the above derivation to hold, the proportions of S_A and S_B ($= K_{AB}$) must remain constant throughout the course of the reaction. Furthermore, even though the product ratio depends only on $\Delta G^\ddagger_{TS}$, one must also recognize that the magnitude of K_{AB}, determined by ΔG°_{AB}, *does* have an effect on the P_A/P_B product ratio, since for the reaction $S_A \rightarrow P_B$, the total energy of activation is $\Delta G^\circ_{AB} + \Delta G^\ddagger_B$. In fact, K_{AB} is inversely proportional to the P_A/P_B product ratio (Equation 1.22).

A few comments on specific cases may provide insight into how this is relevant:

- If A and B are conformational isomers, Curtin–Hammett kinetics are commonly encountered; this is not the case for atropisomers.[17]
- The free energy difference, ΔG°, between S_A and S_B can be zero. This could occur in the case of interconverting enantiomers, for example, in a dynamic kinetic resolution (see Section 1.9).
- On rare occasions $k_1 = k_2$ when $\Delta G^\circ \neq 0$. In this case, the product ratio will reflect the S_A/S_B ratio. For an example, see ref. [98].

For a detailed discussion of the Curtin–Hammett principle and the related Winstein–Holness relationship, the reader is referred to Seeman's excellent reviews [96,97].

A simple example of Curtin–Hammett kinetics as it applies to stereoselectivity is the alkylation of the tropane conformations shown in Scheme 1.9. The rate of pyramidal inversion in tertiary amines is many orders of magnitude faster than alkylation with methyl iodide,

---

17. One must recognize that a molecule that is drawn arbitrarily usually does not depict the reactive conformation.

so Curtin–Hammett kinetics is applicable [99]. The more stable conformation has the
N-methyl in an equatorial configuration relative to the 6-membered ring. However, alkylation
with $^{13}CH_3I$ (*MeI) is faster with the less stable configuration, in which the electrophile
approaches the nitrogen from the equatorial direction.

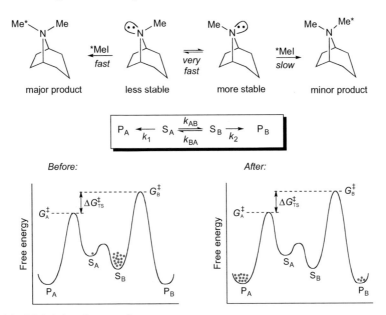

SCHEME 1.9   Methylation of tropane diastereomers.

1.9 ASYMMETRIC TRANSFORMATIONS AND DYNAMIC RESOLUTIONS

When the ratio of a mixture of stereoisomers changes in solution, the phenomenon is called an
asymmetric transformation of the first kind (see Section 1.11 for a precise definition).
Asymmetric transformations of the second kind involve a similar equilibration of stereoisomers,
with concomitant separation of one of them by crystallization or chemical reaction. The equilibrium is therefore drained toward a single (predominant) product. Scheme 1.10 illustrates the two
types of asymmetric transformations using the anomeric equilibria of glucose. Dissolution of
crystalline α-D-glucose in water results in a solution having an initial specific rotation,
$[\alpha]_D = 112$. As the equilibrium ratio of 63:37 β:α is reached, the specific rotation drops to
$[\alpha]_D = 53$; this phenomenon is termed mutarotation (Scheme 1.10a). Because the equilibration
occurs in solution and there is no separation, this is an asymmetric transformation of the first
kind. Scheme 1.10b shows an asymmetric transformation of the second kind. In solution, the
anomers (diastereomers) interconvert rapidly compared to the rate of crystallization, with a β:α
ratio of 63:37. Below 35 °C, the α-anomer selectively crystallizes as the hydrate. Here, equilibration of the stereoisomers in solution is faster than the rate of crystallization. Thus, the α-anomer
crystallizes as the β-anomer epimerizes, such that the glucose in solution remains a 63:37 mixture
of anomers. If one considers the rate of crystallization as a "reaction" that removes one isomer
from solution, we can say that the process follows Curtin–Hammett kinetics. The essential
difference between the two types of asymmetric transformations is that there is a separation and
isolation in the second kind, but not the first kind of asymmetric transformation.

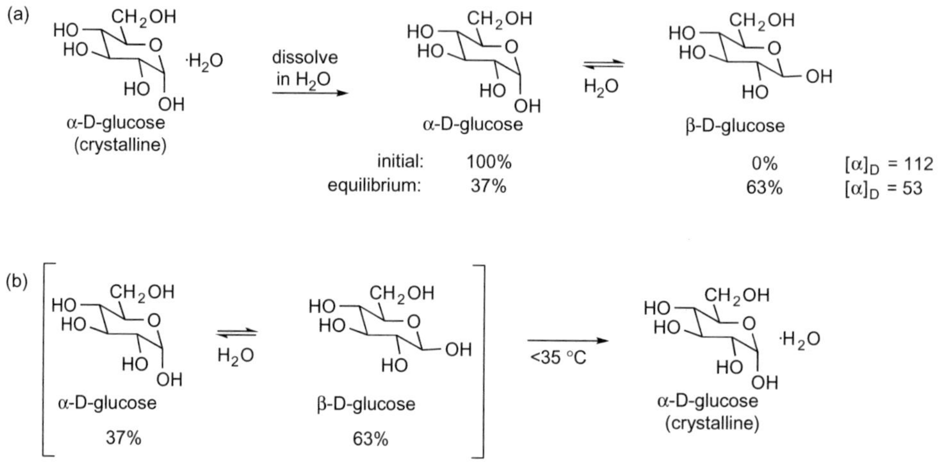

(a)

α-D-glucose (crystalline) · H_2O — dissolve in H_2O → α-D-glucose $\rightleftharpoons_{H_2O}$ β-D-glucose

initial: 100%   0% $[\alpha]_D = 112$
equilibrium: 37%   63% $[\alpha]_D = 53$

(b)

[α-D-glucose (37%) $\rightleftharpoons_{H_2O}$ β-D-glucose (63%)] $\xrightarrow{<35\,°C}$ α-D-glucose (crystalline) · H_2O

SCHEME 1.10 Illustration of asymmetric transformations of the first (a) and second (b) kinds.

In contrast to classical and kinetic resolutions [35,84,100], where the theoretical yield of a resolution is 50%, resolution *via* an asymmetric transformation is stereoconvergent, so the theoretical yield is 100%. For example, as shown in Scheme 1.11a, racemic naproxen methyl ester can be fused with sodium methoxide at 70 °C, supercooled to 67 °C, and seeded with the *S* enantiomer. Under these conditions, crystallization takes place simultaneously with inversion of the *R* ester, affording an 87% yield of the *S* enantiomer [101]. A related process takes advantage of the 13-fold difference in solubilities of the diastereomeric *SR* and *SS* imides shown in Scheme 1.11b. In this example, condensation of racemic naproxen with *S-tert*-butyloxazolidinone gives the naproxen imide as a 44:56 (*SS/RS*) mixture of diastereomers. Diazabicycloundecane (DBU) effects epimerization of a partly soluble slurry of the mixture of diastereomers at 55 °C, taking advantage of the increased solubility of the *RS* naproxen imide diastereomer. After 28 h, the dr is enriched to 98:2 *SS/RS*. Recrystallization and hydrolysis afford enantiopure *S-*(+)-naproxen (99.95:0.05 er) [102].

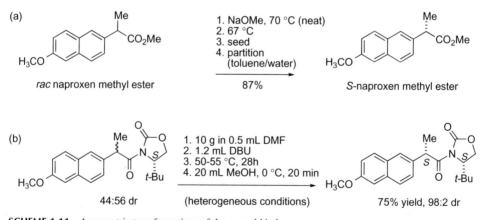

(a) *rac* naproxen methyl ester

1. NaOMe, 70 °C (neat)
2. 67 °C
3. seed
4. partition (toluene/water)

87%

S-naproxen methyl ester

(b) 44:56 dr

1. 10 g in 0.5 mL DMF
2. 1.2 mL DBU
3. 50-55 °C, 28h
4. 20 mL MeOH, 0 °C, 20 min

(heterogeneous conditions)

75% yield, 98:2 dr

SCHEME 1.11 Asymmetric transformations of the second kind.

A subset of asymmetric transformations are dynamic kinetic [103–107] and dynamic thermodynamic [106,107] resolutions. Like kinetic resolutions, dynamic resolutions are examples of double asymmetric induction, since there are two elements of chirality in the key step [108]. For example, as shown in Figure 1.11, a dynamic kinetic resolution occurs when enantiomeric substrates, S, which enantiomerize[18] with rate constant k_{ent}, are allowed to react with a chiral reagent, R*, with rate constants k_R and k_S, acting on the R and S enantiomers, respectively, to give products P_R and P_S. In a dynamic kinetic resolution, if $k_{ent} \gg k_R$, k_S, the process follows Curtin–Hammett kinetics: the P_R/P_S product ratio depends on the difference in transition state energies, $\Delta G_{TS}^{\ddagger}$. Such a dynamic kinetic resolution is stereoconvergent and, unlike classical or kinetic resolutions, has a theoretical yield of 100%. Figure 1.11 illustrates a limiting case, in which $k_{ent} \gg k_R$, k_S. However, there is a continuum of relative rates, and when k_{ent}, k_R, and k_S are closer in value, the mathematical treatment is more complex [107], and product ratio is a function of relative rates as well as percent conversion, just as in a standard kinetic resolution. The product ratio depends on both k_R/k_S and k_{ent}/k_R, where k_R is the rate constant for reaction of the fast-reacting enantiomer [104,108]. That such processes are also asymmetric transformations follows from the facts that: (i) the S_R/S_S ratio may be altered in solution by complexation with R* or L*; (ii) the reaction is stereoconvergent as one enantiomer is selectively removed from reaction mixture while the other inverts; (iii) the theoretical yield of the process is 100%; and (iv) the enriched product is often *not* recovered substrate (as is the case with a classical kinetic resolution). The process differs from the original definition of an asymmetric transformation of the second kind only in the sense that the "separation" required for an asymmetric transformation is now a chemical reaction. Nevertheless, one stereoisomer is selectively removed as it is formed, exactly what happens when one stereoisomer separates from an equilibrium mixture by crystallization.

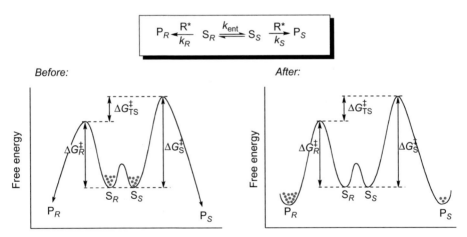

FIGURE 1.11   General scheme of a dynamic kinetic resolution.

---

18. Enantiomerization is the conversion of one enantiomer to another, whereas racemization describes the conversion of an enantioenriched substance to a racemate. As equilibrium is approached, the rate of racemization falls to zero, even if the rate of enantiomerization is fast. The rate constant for racemization, k_{rac}, is double the rate constant for enantiomerization, k_{ent}.

A particularly elegant example of a dynamic kinetic resolution is illustrated in Scheme 1.12a. In this example, where R and S keto ester enantiomers are reduced by a chiral catalyst, the selectivity remained high throughout the reaction (93.8:6.2 er at $t = 0$, and 93.4:6.6 er at 100% conversion). The relative rates were measured as $k_S/k_R = s = 15$ and $k_{ent}/k_S = 6.1$ (here the faster reacting enantiomer is S). A second example is taken from organolithium chemistry, wherein the carbon bearing the metal is often stereogenic. Scheme 1.12b shows an example of a dynamic kinetic resolution of a racemic organolithium [109,110]. Here, tin–lithium exchange produces a racemic organolithium, which is then treated with (−)-sparteine at −78 °C. Under these conditions, the benzylic organolithium · sparteine complex undergoes rapid inversion relative to the rate of reaction with allyl chloride, but the transition states are diastereomeric due to the presence of the sparteine ligand on lithium. Note that, unlike classical kinetic resolutions (Section 1.6), a dynamic kinetic resolution may not involve recovering enantioenriched, unreacted substrate as the major product of the process.

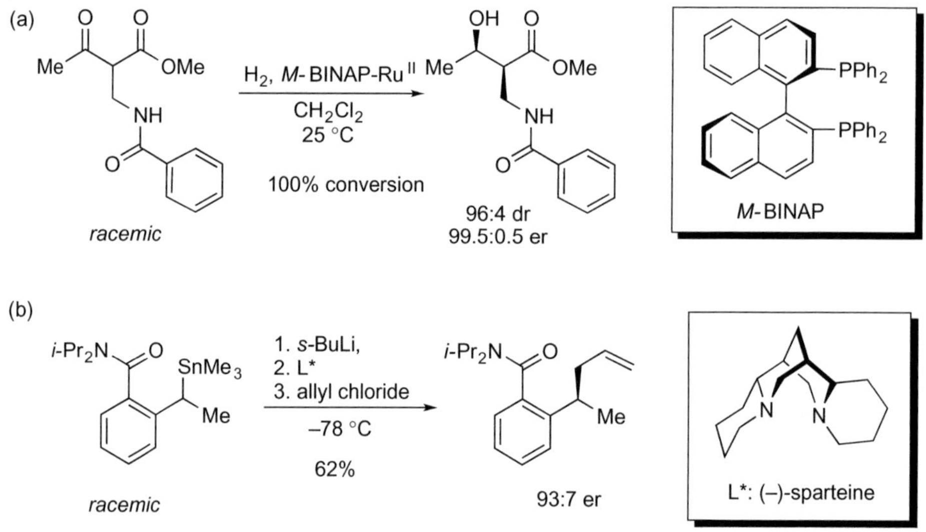

SCHEME 1.12 Dynamic kinetic resolutions that follow Curtin–Hammett kinetics: (a) asymmetric hydrogenation of a chiral β-keto ester [108] and (b) asymmetric electrophilic substitution of a chiral organolithium [109,110].

A second type of dynamic resolution is a dynamic thermodynamic resolution, which applies when an enantiomeric mixture is converted to a mixture of diastereomers by an asymmetric transformation of the first kind [106,107]. Thus, (Figure 1.12), substrate enantiomers, S_R and S_S, react with a chiral reagent, L*, to afford a thermodynamic mixture of two diastereomers, $S_R \cdot L^*$ and $S_S \cdot L^*$. The important feature of this scheme is that $S_R \cdot L^*$ and $S_S \cdot L^*$ interconvert; it does not matter whether this happens by direct epimerization (rate constant k_{inv}) or by dissociation from L* and enantiomerization (rate constant k_{ent}). In dynamic thermodynamic resolutions, k_R, $k_S > k_{inv}$, but the relative rates, k_{inv}, k_{ent}, k_R, k_S, and the percent conversion to products can all influence the product ratio [107]. If the reaction with reagent R is taken to complete conversion, the product ratio will reflect the thermodynamic ratio of the two diastereomeric complexes, $S_R \cdot L^*$ and $S_S \cdot L^*$. However, if k_R, $k_S \gg k_{inv}$, the relative rates of reaction, k_R/k_S, determine the product ratio at partial conversion, just as in kinetic resolutions [107]. In Figure 1.12, the more stable diastereomer has the faster rate of reaction to product, but the opposite situation can also occur. Note that the rates of product formation depends on the quantities $k_R[S_R \cdot L^*]$ and $k_S[S_S \cdot L^*]$, respectively, not just k_R/k_S. Thus, for the less stable diastereomer ($S_S \cdot L^*$ in this example) to provide the major product, the quantity $k_S[S_S \cdot L^*]$ must be greater than $k_R[S_R \cdot L^*]$. Dynamic thermodynamic resolution is distinct from classical kinetic resolutions and from dynamic kinetic resolutions in that equilibration and resolution can be accomplished in separate controllable steps.

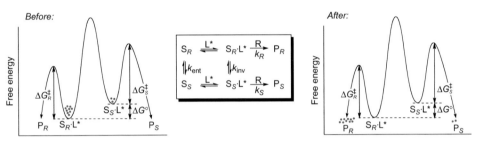

FIGURE 1.12 Energy profile and reaction scheme for a dynamic thermodynamic resolution.

An example of dynamic thermodynamic resolution, and how the dynamic process can be used to advantage, is taken from organolithium chemistry. As shown in Scheme 1.13a, a racemic organolithium dianion is treated with sparteine [110–112]. If the entire process is conducted at −78 °C, the alkylated product is nearly racemic. Note the similarity to the example in Scheme 1.12b. In that case, epimerization of the organolithium · sparteine complex was rapid at low temperature, but here it is slowed by the fact that the organolithium is a dianion, which raises the barrier to inversion. When the organolithium · sparteine complex is allowed to equilibrate at −25 °C, then cooled to −78 °C and quenched, the er is 92:8. Thus, the diastereomeric complexes invert at −25° but not at −78°, and the product ratio reflects the thermodynamic ratio of diastereomeric complexes. The energy profile and equilibria involved in this process are illustrated in Scheme 1.13b [112]. The ratio of organolithium · sparteine complexes is 92:8 (S/R), and the more stable S diastereomer reacts faster, having a free energy of activation that is 0.81 kcal/mol lower than that of the R diastereomer. The fact that the two diastereomers react at different rates can sometimes be used to advantage in dynamic thermodynamic resolutions, as shown by the protocol outlined in Scheme 1.13c. Specifically, use of a substoichiometric amount of electrophile preferentially consumes the more reactive diastereomer. If this is followed by equilibration of the diastereomeric organolithium complexes at −25 °C, the 92:8 S/R ratio is reestablished. Cooling again to −78 °C and quenching gives both an improved yield and an improved enantiomer ratio [112].

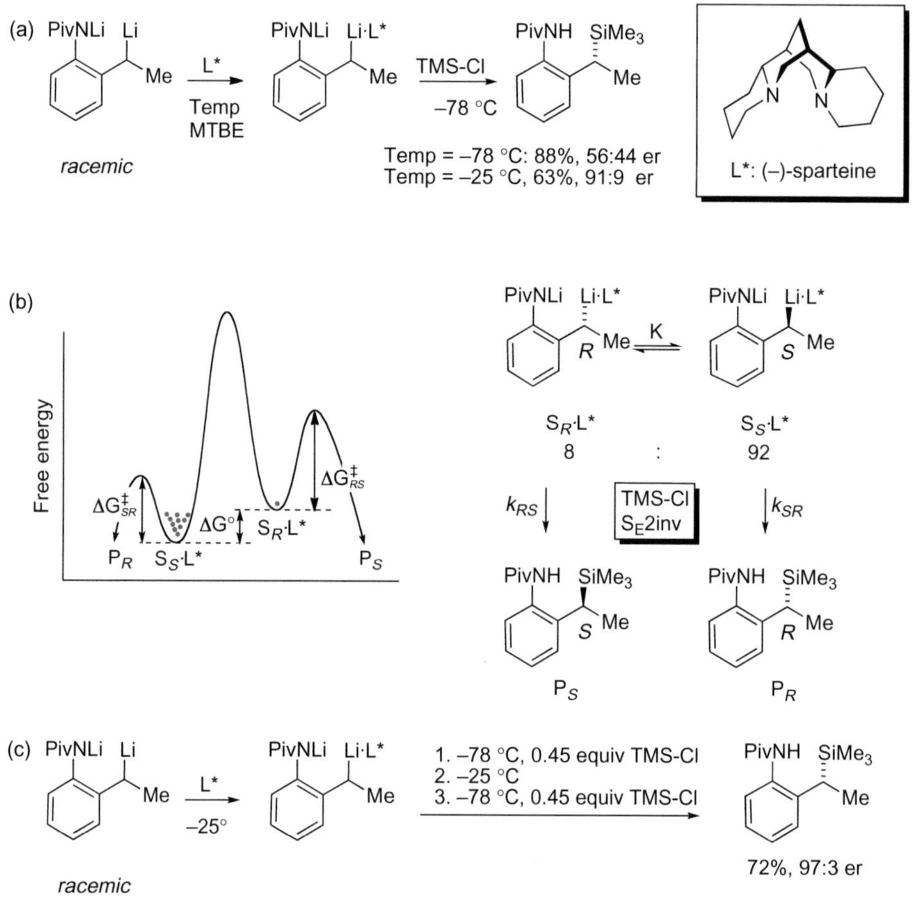

(a)

Temp = −78 °C: 88%, 56:44 er
Temp = −25 °C, 63%, 91:9 er

L*: (−)-sparteine

(b)

$$S_R \cdot L^* \qquad S_S \cdot L^*$$
$$8 \qquad : \qquad 92$$

$$k_{RS} \downarrow \qquad \boxed{\begin{array}{c} \text{TMS-Cl} \\ S_E 2inv \end{array}} \qquad \downarrow k_{SR}$$

$$P_S \qquad P_R$$

(c)

1. −78 °C, 0.45 equiv TMS-Cl
2. −25 °C
3. −78 °C, 0.45 equiv TMS-Cl

72%, 97:3 er

SCHEME 1.13 Dynamic thermodynamic resolution [110]. (a) Asymmetric substitution of a laterally lithiated anilide; (b) equilibria and energy diagram for the system in (a); (c) recycling the less reactive diastereomer by using substoichiometric amounts of electrophile and a warm−cool cycle affords improved stereoconvergence.

The general scheme for dynamic thermodynamic resolution in Figure 1.12 illustrates the formation of diastereomers, $S_R \cdot L^*$ and $S_S \cdot L^*$, by reaction of enantiomers S_R and S_S with a chiral ligand, L^*. In general, asymmetric transformations having rates of epimerization between diastereomers (*e.g.*, Scheme 1.10b) that are comparable to rates of reaction are relatively common. Although these processes are conceptually similar to dynamic resolutions, they are not resolutions since they do not involve equilibrating enantiomers. They are more properly called dynamic asymmetric transformations.

1.10 ASYMMETRIC CATALYSIS AND NONLINEAR EFFECTS

Several examples in the previous sections, and many of the examples described in this book, involve chiral catalysts. In order to evaluate and appreciate asymmetric catalysis, one must think in four dimensions: the x, y, z coordinates of transition state assemblies, as well as the kinetics of the catalyzed reaction [113,114]. In considering such processes, a number of criteria should be considered, including stereoselectivity, substrate scope, and turnover number.[19]

The physical–chemical issues regarding stereoselectivity in the x, y, z dimensions are discussed in the previous sections. In the fourth dimension, one must consider the relative rates of the catalyzed and uncatalyzed reactions. Figure 1.13 illustrates the reaction of an achiral substrate, S, reacting with an achiral reagent, R, with rate constant, k_{cat}, to afford a mixture of enantiomeric products, P_R and P_S. Assuming kinetic control, the enantioselectivity of the catalyzed reaction will be determined by the relative rates of formation of P_R and P_S, determined by k_R/k_S. The uncatalyzed reaction will afford racemic products with rate constant k_{uncat}. Therefore, in order to achieve high enantioselectivity with a chiral catalyst, it is necessary that $k_R \gg k_S$ and that $k_{cat} \gg k_{uncat}$.

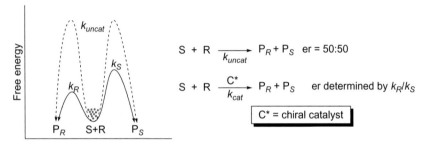

FIGURE 1.13   A stereoselective chiral catalyst, C*, is only effective if $k_R \gg k_S$ and if the catalyzed reaction is significantly faster than the uncatalyzed reaction ($k_{cat} \gg k_{uncat}$). The solid line traces the catalyzed reaction and the dashed line the uncatalyzed reaction. (S = substrate, R = reagent, P = product.)

In most cases, the enantiomeric composition of the products is a linear function of the enantiomeric composition of the catalyst, as shown by the diagonal line of the graph in Figure 1.14. To illustrate, assume $k_R > k_S$. Let L_R be the fraction of the ligand having the R configuration, and define P_R^0 and P_S^0 as the fraction of the product enantiomers that are produced when the ligand is enantiopure. Since $P_R + P_S = 1$, and since $P_R = P_S^0$ when $L_R = 0$, the slope of the line is $(P_R^0 - P_S^0)$ and the y intercept is P_S^0. Note also that the line must pass through the point (0.5, 0.5), since racemic catalyst must afford racemic product. It follows that the equation describing the line is given by [69]:

$$P_R = (P_R^0 - P_S^0)L_R + P_S^0. \qquad (1.25)$$

19. In chemical catalysis, the turnover number is usually defined as the number of moles of substrate consumed per mole of catalyst. In biochemistry, turnover number is defined differently [115]. An enzyme, E, and a substrate, S, are in equilibrium with the enzyme · substrate complex, E · S, which then is converted to product, P. The turnover number is the rate constant, k_2, for the conversion of E · S to P :

$$E + S \underset{k_{-1}}{\overset{k_1}{\rightleftharpoons}} E \cdot S \xrightarrow{k_2} P.$$

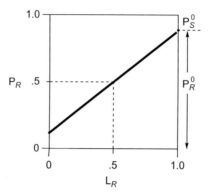

FIGURE 1.14 Linear relationship between enantiomeric composition of ligand and product in asymmetric catalysis [69].

However, there are exceptions, and both positive nonlinear effects (+ NLE) and negative nonlinear effects (− NLE) are known [116−120]. When nonlinearity is observed (and the reaction is homogeneous[20]), it usually implies more than one molecule of the chiral catalyst, ligand, or reagent is involved in the stereodifferentiating step.[21] For example (Scheme 1.14), imagine a metal, M, coordinating with an unequal mixture of R and S ligands to form complexes such as ML, ML_2 or $(ML)_2$. Since both R and S ligands are present, both homochiral and heterochiral dimeric complexes (ML_2 or $(ML)_2$) will be present. For the sake of illustration, assume that monomer, ML_R, or the homochiral complexes containing the R ligand, ML_RL_R or $(ML_R)_2$, produces mainly the R product, and the complexes containing the S ligand produce mainly the S product. The heterochiral complex containing both R and S ligands is achiral and will produce racemic product. The inherent stereoselectivity of the reaction, as expressed by k_R/k_S or k_{RR}/k_{SS}, determines P_R^0, the fraction of R product obtained with enantiopure R ligand.

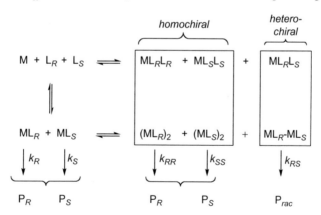

SCHEME 1.14 Kagan's ML_2 model of nonlinear effects (NLE), which relate enantiomer composition of a chiral catalyst with enantiomer composition of products [123].

---

20. If the nonracemic chiral catalyst is not completely dissolved, differential solubilities of conglomerate and racemic compound solids can cause nonlinear effects due to phase behavior [121].
21. A second possibility is a kinetic resolution, in which the minor enantiomer of the product is selectively consumed by the major enantiomer of the catalyst. For an example, see ref. [122].

When L_R is the major enantiomer of the ligand ($L_R > 0.5$), and P_R of the product is greater than expected from Equation 1.25, the situation is as illustrated in Figure 1.15a, and is termed a positive nonlinear effect ($+$ NLE). Conversely, if P_R of the product is less than predicted, it is a negative nonlinear effect ($-$ NLE), as illustrated in Figure 1.15b. If one considers the entire range of L_R from 0 to 1, the positive and negative nonlinear effects plot as shown in Figure 1.15c and d [69].

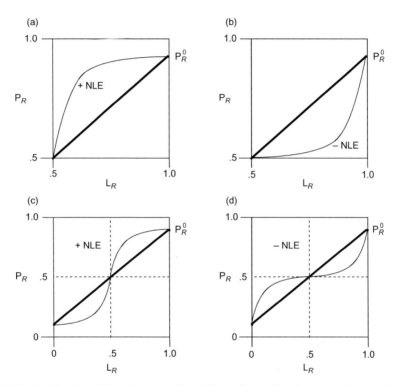

FIGURE 1.15 Positive (a) and (c) and negative (b) and (d) nonlinear effects in asymmetric catalysis.

Three factors, other than k_R/k_S or k_{RR}/k_{SS}, affect the er of the product when the ligand is not enantiopure: the er of the ligand or catalyst, the relative stability of the catalyst complexes having homochiral and heterochiral ligand pairs, and the relative reactivities of the monomeric, homochiral, and heterochiral catalyst systems. The er of the ligand will place upper and lower limits on the statistical distribution between homochiral and heterochiral catalyst complexes. The er of the ligand and the relative stabilities of the monomer (ML) and the homochiral and heterochiral complexes will determine the position of the equilibria between them. The relative reactivities of the monomers, homochiral and heterochiral complexes, will determine whether a positive or negative nonlinear effect is observed.

For example, with reference to Scheme 1.13, consider a case in which the active catalyst is ML_2, and assume that $k_{RR}, k_{SS} \gg k_{RS}$ (i.e., the heterochiral complex is inactive as a catalyst). Since the rate $= k_{RR}[ML_RL_R]$ or $k_{RR}[ML_R]_2$, P_R will be greater than expected from Equation

1.24, since L_S is sequestered in the heterochiral complex. Conversely, if the heterochiral complex is more reactive than the monomer or homochiral complex (*i.e.*, $k_{RS} > k_R$, k_S, k_{SS}, k_{RR}), more racemate will be produced, and a negative nonlinear effect will be observed. Mathematical models of nonlinear behavior have been developed by the Kagan group [123], which cover ML_3 and ML_4 systems as well as the ML_2 systems described here. More recently, Blackmond has amplified on these concepts [117], and separately cautioned that heterogeneous systems, that exhibit linear behavior when homogenous can produce nonlinear effects [121].

A positive nonlinear effect is often called an *asymmetric amplification*. A particularly well-studied example of asymmetric amplification is the amino alcohol catalyzed addition of diethylzinc to benzaldehyde pictured in Scheme 1.15 (see also Section 4.3.1) [124]. In this example, the monomer shown in the inset, formed by the reaction of an amino alcohol with diethylzinc, is the active catalyst. Mechanistic studies revealed that the heterochiral dimer of the catalyst ($ML_R \cdot ML_S$ in Scheme 1.14) is both very stable and inactive as a catalyst. In contrast, the homochiral dimer is not stable, making the ML_S monomer the most abundant species in solution. Thus, the minor enantiomer, ML_R, is sequestered by the major enantiomer as a heterochiral dimer, $ML_R \cdot ML_S$, leaving the ML_S monomer of the major enantiomer free to catalyze the addition with a high degree of selectivity.

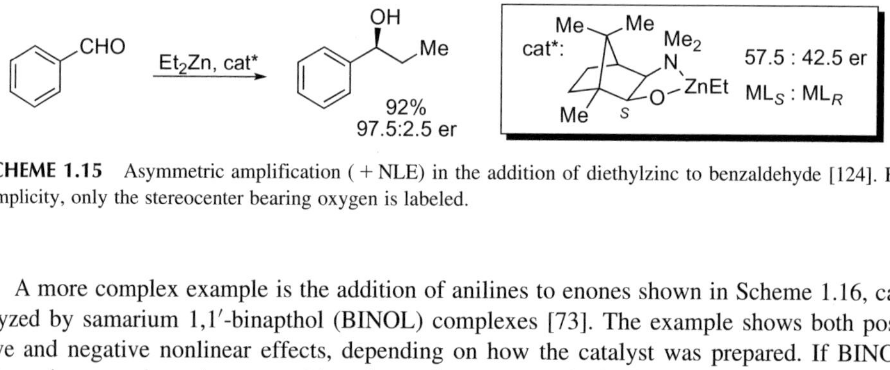

SCHEME 1.15 Asymmetric amplification (+ NLE) in the addition of diethylzinc to benzaldehyde [124]. For simplicity, only the stereocenter bearing oxygen is labeled.

A more complex example is the addition of anilines to enones shown in Scheme 1.16, catalyzed by samarium 1,1′-binapthol (BINOL) complexes [73]. The example shows both positive and negative nonlinear effects, depending on how the catalyst was prepared. If BINOL of varying enantiomeric compositions is used to prepare the iodido(binaphtholato)samarium catalyst (BINOL-SmI), a negative nonlinear effect was observed. In contrast, if enantiopure *P*- and *M*-BINOL were used, a positive nonlinear effect was observed. The authors concluded that the catalyst was forming dimers or oligomers, but no structural information could be obtained.

SCHEME 1.16 The addition of *p*-anisidine shows both positive and negative nonlinear effects depending on how the BINOL-SmI catalyst is prepared [73].

A personal recollection of *Vladimir Prelog*, and his contribution to this book [1].

When writing the first edition of this book, I thought I had a reasonable grasp of the terminology of stereochemistry. So, while on sabbatical leave at the ETH-Zürich in 1993, I showed a first draft of the glossary to my host, Dieter Seebach, and asked for comments. He said it was a good start, and indicated that it would be wise to show it to Professor Prelog and get his input. I agreed and said I would do so as soon as I had put some more work into it. As it turned out, the deed had already been done, and I was shortly summoned to the great man's office for a series of tutorials in the finer points of the nomenclature of stereochemistry [2]. The photo of the two of us during one of these sessions was taken by his longtime friend and colleague, Jack Dunitz [3].

Prelog's critique of my first draft was exacting, precise, constructive, and instructive. Before each of my visits, he made notes in a red pen and pencil to remind him of things he wanted to discuss. See the figure below for an example; the reader is invited to compare the 1993 draft of these terms with the final version that resides in the glossary (Section 1.11).

As can be seen here, it was during these meetings that I first learned the concept of reflection variance; my ignorance is evidenced by the "!!!" notation he wrote beside my definition of *"Enantiotopic."* In the interim, I have noticed that hundreds of fellow chemists, including many journal editors, are equally ignorant, as evidenced by their frequent use of *re* and *si*, when *Re* and *Si* are the correct descriptors. If you, gentle reader, have made it this far, please take careful note, and decide whether the trigonal face you are describing is reflection variant; if it is, then capitalize the *Re* or *Si* descriptor, and don't let anyone tell you it doesn't matter!

In writing this book, we have strived to conform to the stereochemical strictures of the Zürich school; variances are entirely our fault [4].

Bob Gawley

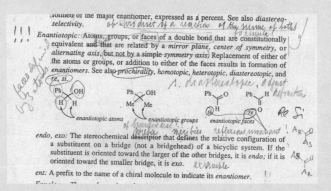

[1] Prelog shared, with John Warcup Cornforth, the Nobel Prize in 1975 "for his work on the stereochemistry of organic molecules and reactions"; he was Professor in the Laboratory for Organic Chemistry, Swiss Federal Institute of Technology (ETH) from 1942 until he "retired" in 1976. His initial "P" identifies him as the third author of the *CIP system* for describing absolute configuration.

[2] In addition to the tutorials, I occasionally joined him for lunch atop of the main building of the ETH (Hauptgebäude), overlooking the city of Zürich. Prelog loved to tell jokes and stories, and I enjoyed hearing as many as he would tell. Some of them are reproduced in his autobiography, which I highly recommend: *My 132 Semesters of Chemistry Studies*, V. Prelog, American Chemical Society, Washington DC, 1991.

[3] For an amusing insight into the man and his relationships with his colleagues, see: *A New Textual Analysis of the Prelog Erlkönig Legend: An Interdisciplinary Approach to Scientific History and Literary Criticism*, published on the occasion of Prelog's 90th birthday, by Heilbronner, E.; Dunitz, J. D. *Helv. Chim. Acta* **1996**, *79*, 1241−1248.

[4] Teachers of stereochemistry may be interested in an article in which Prelog's method of teaching chirality in 1- and 2-dimensions before leaping into the third dimension is explained, with examples: Gawley, R. E. *Chirality made simple: A 1- and 2-dimensional introduction to stereochemistry. J. Chem. Educ.* **2005**, *82*, 1009−1012.

1.11 GLOSSARY OF STEREOCHEMICAL TERMS[22]

A values The free energy difference ($-\Delta G°$) between equatorial and axial conformations of a substituted cyclohexane, positive if equatorial is preferred. The *A* value of several common substituents are (kcal/mol): Cl = 0.53–0.64, MeO = 0.55–0.75, Me = 1.74, Et = 1.79, *i*-Pr = 2.21, Ph = 2.8, and *t*-Bu = 4.7–4.9. For a compilation of values, see ref. [125,126] and references cited therein.

$A^{1,2}$, $A^{1,3}$ *strain* see *allylic strain*.

Absolute asymmetric synthesis A synthesis in which *achiral* reactants are converted to *nonracemic, chiral* products, and where the *enantioselectivity* is induced only by an external force such as circularly polarized light in a photochemical reaction [127].

Absolute configuration The arrangement in space of the ligands of a *stereogenic unit*, which may be specified by a stereochemical descriptor such as *R* or *S*, *D* or *L*, *P* or *M*. See also *chirality sense, chirality element, stereogenic element*.

Achiral See *chiral*.

Achirotopic See *chirotopic*.

Allylic strain The destabilization of a molecule, or an individual conformation, by *van der Waals* repulsion between substituents on a double bond and those in an allylic position [128]. Two types have been identified (see bold bond in the figure below): $A^{1,2}$ strain occurs between substituents on an allylic carbon and the adjacent sp^2 carbon. $A^{1,3}$ strain occurs between substituents on an allylic carbon and the distal sp^2 carbon. The latter effect can be quite strong [129]. Originally [128], the terms were defined in the context of cyclohexane derivatives, but more recently the effects have been recognized as important factors in conformational dynamics of acyclic systems [129].

Alternating symmetry axis (S_n) An axis about which a rotation by an angle of 360/*n*, followed by a reflection across a plane perpendicular to the axis, results in an entity that is indistinguishable from (superimposable on) the original. Also called a rotation−reflection axis. See also *symmetry axis*.

Alpha (α), beta (β) Stereodescriptors used commonly in carbohydrate [130] and steroid [131] nomenclature to describe *relative configuration*. In steroids, "any [substituent] that lies on the same side of the ring plane as the C_3-hydroxyl group of cholesterol [see illustration] is described as β-oriented, and the carbon to which the group is joined has the β-configuration. The opposite orientations and configurations are designated α" [131]. The α, β nomenclature is often extended to other ring systems, but a reference stereocenter must be specified, either explicitly or by convention (see, for example, ref. [132]). Often, reference is made to a two-dimensional drawing in which a reference plane is specified. If the reference plane is horizontal, β is above and α is below the plane, as illustrated below. If the plane is vertical, β is toward the viewer.

In carbohydrates, the β-anomer has the hydroxyl or alkoxyl group at C_1 (circled) on the opposite side of a *Fischer projection* as the substituent (*) that defines D or L (see *Fischer−Rosanoff convention*); this need not be the position at which the ring is closed. The α-anomer has the two on the same side [130].

---

22. Note that the terms defined in this glossary are italicized.

steroids

specified plane within other molecules

carbohydrates, α anomers:

carbohydrates, β anomers:

Angle strain Destabilization of a molecule due to a variation of bond angles from "optimal" values (109 °28′ for a tetrahedral atom). Also called *Baeyer strain*.

Anomeric effect Originally, the unexpected relative stability of a C-1 alkoxy group of a glycopyranoside occupying the *axial* position. This effect is now more generally considered to be a conformational preference of an X−C−Y−C moiety for a *synclinal* (*gauche*) *conformation* (where X and Y are heteroatoms, and at least one is a nitrogen, oxygen, or fluorine). See illustration below [133,134].

synclinal (gauche)

antiperiplanar

Antarafacial, suprafacial In a reaction where a molecule undergoes two changes in bonding (either making or breaking), the relative spatial arrangement is suprafacial if the changes occur on the same face of the molecular fragment and antarafacial if on opposite faces [135].

Anti See *torsion angle; syn, anti.* Also used to describe *antarafacial* addition or elimination reactions [136]. Formerly used to describe the configuration of azomethines such as oximes and hydrazones (see *E, Z*).

Anticlinal See *torsion angle.*

Antiperiplanar See *torsion angle.*

Aracemic Synonym for *nonracemic* [137]. Usage discouraged. See also *scalemic.*

Asymmetric Lacking all symmetry elements, that is, belonging to symmetry point group C_1. Not to be confused with *chiral*. For example, a molecule belonging to symmetry point group C_{2v} is chiral but not asymmetric.

Asymmetric carbon atom van't Hoff's definition for a carbon atom having four different ligands (*i.e.*, Cabcd). See also *stereogenic center, stereogenic element.*

Asymmetric center See *stereogenic center.*

Asymmetric destruction See *kinetic resolution.*

Asymmetric induction The preferential formation of one *enantiomer* or *diastereomer* over another, due to the influence of a *stereogenic element* in the substrate, reagent, catalyst, or environment (such as solvent). Also, the preferential formation of one configuration of a stereogenic element under similar circumstances. When two reactants of a reaction are stereogenic, the stereogenic elements of each reactant may operate either in concert (matched pair) or in opposition (mismatched pair). This phenomenon is known as *double asymmetric induction,* or double diastereoselection [76,77] (see Section 1.5).

Asymmetric synthesis A reaction or reaction sequence that selectively creates one configuration of one or more new *stereogenic elements* by the action of a chiral reagent or auxiliary, acting on *heterotopic* faces, atoms, or groups of a substrate. The stereoselectivity is primarily influenced by the chiral catalyst, reagent, or auxiliary, despite any stereogenic elements that may be present in the substrate (see Section 1.2).

Asymmetric transformation The conversion of a mixture (usually 1:1) of stereoisomers into a single stereoisomer or a mixture in which one isomer predominates. An *"asymmetric transformation of the first kind"* involves such a conversion without separation of the stereoisomers. An *"asymmetric transformation of the second kind"* also involves separation, such as an equilibration accompanied by selective crystallization of one stereoisomer [138]. The terms "first- and second-order asymmetric transformations" to describe these processes are inappropriate. See also *stereoconvergent.*

Atropisomers Stereoisomers arising from *restricted rotation* around a single bond (*i.e., conformers*), with a high enough rotational barrier (16−20 kcal/mol) that the isomers can be isolated at room temperature, such as *ortho*-disubstituted biaryls [139]. The chirality sense of a conformation is best described using the *P, M* system.

Axial, equatorial Bonds or ligands of a cyclohexane (or saturated 6-membered heterocycle) chair conformation. The axial bonds are parallel to the C_3 (S_6) axis of cyclohexane (or the corresponding position of a heterocycle), and each equatorial bond is parallel to two of the ring bonds. In a cyclohexene, the corresponding allylic bonds or ligands are called pseudoaxial (ax') and pseudoequatorial (eq'). In a trigonal bipyramidal structure, the three ligands in a plane with the central atom are also known as equatorial.

Axis of chirality See *chirality element, stereogenic axis, stereogenic element.*

Baeyer strain See *angle strain.*

Bisecting and eclipsing conformations In a structure with the grouping $R_3-C=X$, the conformation in which a torsion angle $R-C-C=X$ is *antiperiplanar*, and the torsion angles to the other two R groups is equal or nearly so is the *bisecting* conformation. The conformation in which a torsion angle $R-C-C=X$ is *synperiplanar* is called *eclipsing.*

Boat See *chair, boat, twist;* and *half-chair, half-boat.*

Bond opposition strain See *eclipsing strain.*

Bowsprit, flagpole In the cyclohexane boat conformation, the ligands on the two carbons that are out of the plane of the other four. Endocyclic ligands are *flagpole*, exocyclic ligands are *bowsprit.*

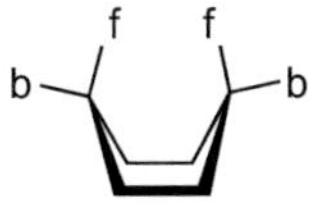

Bürgi-Dunitz trajectory The angle of approach of a nucleophile toward a carbonyl carbon, 107° (probably more accurately $105 \pm 5°$) [140–142] (see Section 4.1.3).

The Bürgi-Dunitz trajectory

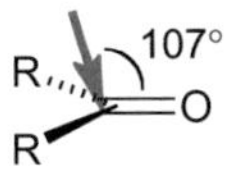

Cahn-Ingold-Prelog method See *CIP method.*

CDA, chiral derivatizing agent A reagent of known enantiomeric purity that is used for derivatization and analysis of enantiomer mixtures by spectroscopic or chromatographic means (see Section 2.5.1).

Center of chirality See *stereogenic center.*

Center of symmetry, center of inversion (i) A point in an object that is the origin of a set of Cartesian axes, such that when all coordinates describing the object (x, y, z) are converted to $(-x, -y, -z)$, an identical entity is obtained. Equivalent to a twofold *alternating axis* (S_2).

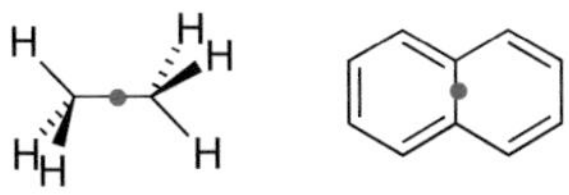

Centers of inversion (•)

Chair, boat, twist-boat The cyclohexane *conformation* (point group D_{3d}) in which carbons 1, 2, 4, and 5 are coplanar and atoms 3 and 6 are on opposite sides of the plane is the *chair*. When atoms 3 and 6 are on the same side of the "1–2–4–5" plane, and also lie in a mirror plane, the conformation (point group C_{2v}) is called a *boat*. If atoms 3 and 6 are moved to either side of the boat's "3–6" mirror plane, the conformation (point group D_2) is the *twist-boat*. The chair and twist-boat conformations are at energy minima ($\Delta G° = 5.6$–8.5 kcal/mol for cyclohexane [143]) while the boat is at a higher energy saddle-point. The *twist-boat* is sometimes called the *skew* conformation (however, see *torsion angle*). These terms are also applied to similar conformations of substituted cyclohexanes and to heterocyclic analogs. See also *axial, equatorial.*

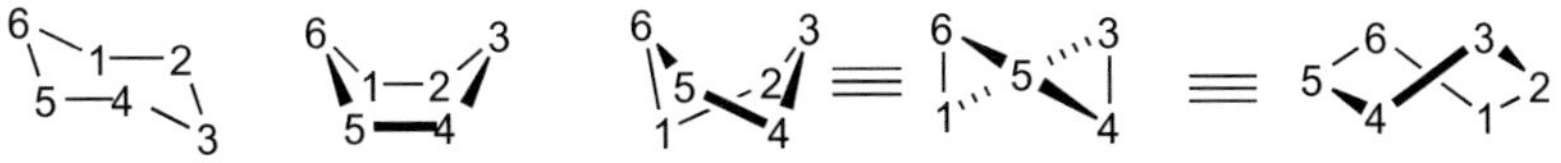

chair   boat   twist-boat

Chair-chair inversion See *ring inversion*

G L O S S A R Y

Chiral A geometric figure, or group of points is chiral if it is nonsuperimposable on its mirror image [144]. A *chiral* object lacks all symmetry elements that include reflections [145]: σ (*mirror plane*), i (*center of symmetry*), and S (*rotation–reflection axis*). In chemistry, the term is (properly) only applied to entire molecules, not to parts of molecules. A chiral compound may be either *racemic* or *nonracemic*. An object that has any of the second-order symmetry elements (*i.e.*, that is superimposable on its mirror image) is *achiral*. It is inappropriate to use the adjective *chiral* to modify an abstract noun: one cannot have a chiral opinion and one cannot execute a chiral resolution or synthesis.

Chiral auxiliary A chiral molecule that is covalently attached to a substrate so as to render *enantiotopic* faces or groups in the substrate *diastereotopic*. After the diastereoselective reaction, the auxiliary should be easily removable (see Sections 1.2 and 2.1).

Chirality The property that is responsible for the nonsuperimposablility of an object, or a group of points, with its mirror image.

Chirality axis See *chirality element*.

Chirality center See *chirality element*.

Chirality element, element of chirality A *stereogenic axis, center,* or *plane* that is *reflection variant.* See also *stereogenic element.*

Chirality plane See *chirality element*.

Chirality sense The property that distinguishes enantiomorphs such as a right- or left-threaded screw. For molecules, the chirality sense may be described by *R, S* or *P, M*. See also *absolute configuration*.

Chirality transfer Asymmetric induction in which one stereogenic element is sacrificed as another is created.

Chiroptic Referring to the optical properties of chiral substances, such as optical rotation, circular dichroism, and optical rotatory dispersion.

Chirotopic The property of "any atom, and, by extension, any point or segment of the molecular model, whether occupied by an atomic nucleus or not, that resides in a chiral environment" [146]. *Achirotopic* is the property of any atom or point that does not reside in a chiral environment (see also ref. [147]). "Chirotopic atoms located in chiral molecules are enantiotopic by external comparison between enantiomers. Chirotopic atoms located in achiral molecules are enantiotopic by internal comparison and therefore also by external comparison. All enantiotopic atoms are chirotopic" [146].

CIP (Cahn, Ingold, Prelog) method, CIP system The CIP sequencing rules establish the conventional ordering of ligands for the unambiguous description of absolute configuration by descriptors such as *R, S; P, M; E, Z.*

There are several steps in the method, which are summarized as follows (for the definitive rules, see refs. [148,149]):

Ligancy complementation All atoms other than hydrogen are complemented to quadriligancy by providing one or two duplicate representations of any ligands that are doubly or triply bonded, respectively, and then adding the necessary number of phantom atoms of atomic number zero [148]. For example, the representation of a carbonyl is expanded as follows:

where 0 denotes a phantom atom of atomic number zero.

Sequence rules:

0. Nearer end of axis or side of plane precedes farther.

1. Higher atomic number precedes lower atomic number.

2. Higher atomic mass precedes lower atomic mass.
3. *Cis* (Z) precedes *trans* (E). Some special cases require the following qualification [149]: when two ligands (indistinguishable by rules 1 and 2) differ by one having the ligand of higher rank in a *cis* position (Z) to the core of the stereogenic unit, and the other in a *trans* position (E), the former takes precedence.
4. Like pair precedes unlike pair. (For a listing of like and unlike pairs, see *l,u* in this glossary).
5. *R* precedes *S*, and *M* precedes *P*

To implement the sequence rules, it is useful to construct a digraph of the ligands to be compared, as shown below. The ligands of the proximal atoms (1 and 2) are placed in the digraph such that 11 has precedence over 12, 12 over 13, 21 over 22, etc. Another layer of ligands, labeled 111, 112, ... 133, could be constructed if necessary (only one such set is shown below). In implementing the sequence rules, the ligands are explored along successive bonds, taking the order of higher precedence at each branch point. For example, if $11 > 12$, branching from 11 is explored. The sequence rules are followed, each to exhaustion, in turn. Therefore, rule 1 is explored in the digraph before proceeding to rule 2, etc. Atom 1 is first compared with 2. If there is no difference, 11 is compared with 21, then 12 with 22, etc., until a decision is reached [148,149].

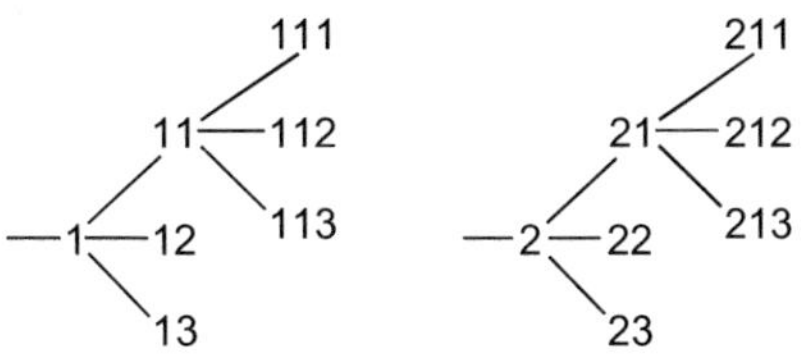

```
        111                    211
       /                      /
   11—112                 21—212
  /      \                /      \
 /        113            /        213
—1—12              —2—22
  \                    \
   13                    23
```

For the vast majority of cases, *CIP* rank can be determined using only ligancy complementation and sequence rule 1. The rules result in the following (descending) sequence of CIP rank for several common functional groups [150]: $COOCH_3$, COOH, COPh, CHO, $CH(OH)_2$, *o*-tolyl, *m*-tolyl, *p*-tolyl, Ph, C≡CH, *t*-Bu, cyclohexyl, vinyl, isopropyl, benzyl, allyl, *n*-pentyl, ethyl, methyl, D, H.

For the assignment of CIP descriptors, see *R, S; P, M;* and *E, Z.*

cis, trans   A stereochemical prefix to describe the relationship between two ligands on a double bond or a ring: *cis* if on the same side, *trans* if on opposite sides. For alkenes, the *cis—trans* nomenclature can be ambiguous and the *E, Z* descriptor is preferred. In a ring, the reference conformation (real or hypothetical) is planar, and approximates a circle or an oval. See *cis—trans isomers*.

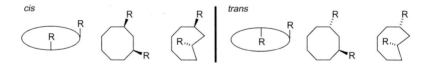

cisoid conformation (usage discouraged)   See *s-cis, s-trans*.
cis—trans isomers   Stereoisomeric alkenes or cycloalkanes (or heterocyclic analogs), that differ in the position of ligands relative to a reference plane: *cis* if on the same side, *trans* if on opposite sides.

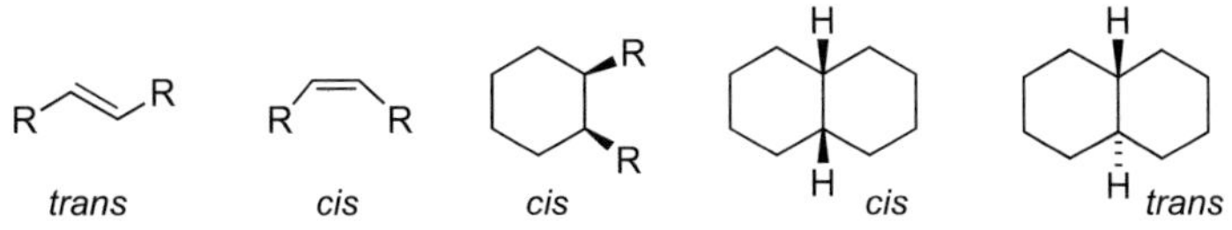

Clinal   See *torsion angle*.

Configuration The arrangement of atoms in space that distinguishes *stereoisomers*, excluding *conformational isomers*. *Atropisomers* are a special case of conformational isomers that, because they are isolable at room temperature, may have an absolute configuration descriptor assigned to the *stereogenic axis*. See also *absolute configuration, chirality sense, relative configuration.*

Conformation In a molecule of a given *constitution* and *configuration*, the spatial array of atoms affording distinction between *stereoisomers* that can be interconverted by rotation around single bonds. The *chirality sense* of conformations may be specified using the *P, M* nomenclature.

Conformational analysis The analysis of the chemical and physical properties of different *conformations* of a molecule.

Conformational isomers (conformers) *Stereoisomers* at potential energy minima (local or global) having identical *constitution* and *configuration*, that differ only in *torsion angles*.

Conglomerate A mechanical mixture of crystals, each of which contains only one of two enantiomers. This occurs when each molecule has a greater affinity for itself than for its enantiomer. The melting point of a conglomerate is always lower than the melting point of the pure enantiomer.

Constitution The description of the number and kind of atoms in a molecule and their bonding (including bond multiplicities, but not *relative* or *absolute configuration*, or *conformation*).

Constitutional isomers Isomers that differ in connectivity, such as CH_3CH_2OH and CH_3OCH_3.

Cornforth model for asymmetric addition to α-heteroatom carbonyls A model for predicting the major stereoisomer in nucleophilic addition to an aldehyde or ketone having an adjacent heteroatom on an adjacent *stereocenter*. This model, an alternative to *Cram's rule (open chain model)*, posits that the ground state conformation places the electronegative substituent *antiperiplanar* to the carbonyl oxygen. The nucleophile then approaches from the side of the smaller substituent [151−153] (see Section 4.1).

Cram's rule (cyclic, or chelate model) A model for predicting the major stereoisomer resulting from nucleophilic addition to an aldehyde or a ketone having an adjacent *stereocenter* that is capable of chelation (especially 5-membered ring chelation). After chelate formation, the nucleophile adds from the side opposite the larger of the remaining substituents on the α-stereocenter [58] (See Section 4.2).

Cram's rule (open chain model) A model to predict the major stereoisomer resulting from nucleophilic addition to a ketone or aldehyde having an adjacent *stereocenter*. The rule originally formulated by Cram in 1952 [57] has evolved into the commonly used Felkin−Anh formulation illustrated below [141,154,155]. In the transition structure, the largest substituent of the stereocenter, or the substituent having the lowest-lying σ* orbital (L), is perpendicular to the carbonyl, and the nucleophile attacks from the opposite side, on a trajectory that places it approximately 107° away from the carbonyl (the *Bürgi−Dunitz trajectory*). The favored transition structure (a) has this trajectory nearly eclipsing the site of the smaller of the two remaining substituents (see Section 4.1).

(a)   (b)

favored TS

CSA (Chiral solvating agent) A diamagnetic additive of known enantiomeric purity used to induce anisochrony in *enantiomers* of a *racemate* for NMR analysis (see Section 2.5.2).

CSP (Chiral stationary phase) A *nonracemic* chiral stationary phase for the chromatographic separation of *enantiomers* (see Section 2.4.2).

D, L See *Fischer-Rosanoff convention.*

d, l, dl Obsolete alternatives for (+)- and (−)- used to designate the sign of rotation of *enantiomers* at 589 nm (the sodium D line), and (±)- for a *racemate*. Sometimes used as arbitrary descriptors for a single enantiomorph.

Diastereoisomers See *diastereomers.*

Diastereomer excess (percent diastereomer excess, % de) In a reaction in which two (and only two) diastereomeric products are possible, the percent diastereomeric excess, % de is given by

$$\% \text{ de} = \frac{|D_1 - D_2|}{D_1 + D_2} \cdot 100 = |\%D_1 - \%D_2|,$$

where D_1 and D_2 are the mole fractions of the two diastereomeric products. Usage is strongly discouraged in favor of *diastereomer ratio, dr* [69]; see also page 90. See also *diastereoselectivity.*

Diastereomer ratio, dr The ratio of two or more diastereomers. The ratio is commonly expressed as percent normalized to 100, as in 95:5 or 99:1, or as a number, as in 19 or 99. The advantage of normalizing the numbers is that, in stereoselective reactions, the stereoisomer ratios pass through a 50:50 mirror point. Use of dr (and er) to express stereoselectivities are preferred over de and ee, because they directly reflect relative rates in kinetically controlled reactions (see Section 1.4 and the highlight box on page 90).

Diastereomers (diastereoisomers) Stereoisomers that are not enantiomers (including alkene *E, Z* isomers).

Diastereofacial selectivity The degree of *diastereoselectivity* observed upon addition to *diastereotopic* faces of a trigonal atom. See also *enantiofacial selectivity.*

Diastereofacial selectivity

diastereomers

Diastereoselectivity (percent diastereoselectivity, % ds) In a reaction in which more than one diastereomer may be formed (with mole fractions $D_1, D_2, \ldots, D_n$ produced), the diastereoselectivity is the mole fraction formed of the major product (or the desired product), expressed as a percent:

$$\% \text{ ds} = \frac{D^*}{D_1 + D_2 + \cdots + D_n} \cdot 100,$$

where D^* is the mole fraction of the desired isomer [156] (see Section 1.4). See also *enantioselectivity, enantiomer ratio,* and *diastereomer ratio.*

GLOSSARY

Diastereotopic The relationship of two ligands of an atom that are constitutionally equivalent, but in positions that are not symmetry related. Replacement of either ligand yields a pair of diastereomers. Also, faces of a trigonal atom that are not symmetry related, such that addition to either face gives a pair of diastereomers. Reflection variant faces may be specified as *Re* or *Si*, and ligands, L, may be specified as or L$_{Si}$, by noting on which face of a triangle the ligand in question sits (see *heterotopic*). Note that addition of a ligand to the *Re* face of a trigonal atom affords a tetrahedral array with the new ligand in the L$_{Re}$ position. Reflection invariant descriptors are *re, si*, as illustrated below [157,158]. See also *Re, Si, homotopic,* and *enantiotopic*.

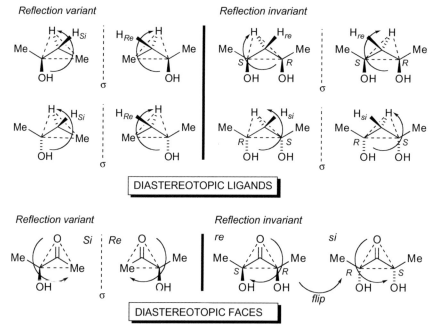

Reflection variant   Reflection invariant

DIASTEREOTOPIC LIGANDS

Reflection variant   Reflection invariant

DIASTEREOTOPIC FACES

Dihedral angle The angle between two defined planes. The term is most commonly applied to vicinal bonds on a *Newman projection*. See also *torsion angle*.

Dissymetric Obsolete synonym for *chiral*. Not equivalent to *asymmetric*, since chiral substances may have symmetry. See also *asymmetric*.

Double asymmetric induction See *asymmetric induction*.

Dunitz angle See *Bürgi−Dunitz trajectory*.

E, Z: Descriptors for the arrangement of ligands around double bonds. On either end of the double bond, the group of highest *CIP* rank is identified. If the two higher ranking groups are on the same side of the double bond, the descriptor of the stereoisomer is *Z* (German zusammen = together); if on opposite sides, *E* (German entgegen = apart). See also *cis, trans* isomers. For enolates, some authors modify this rule such that the OM ligand (anionic oxygen with its metal) takes the highest priority. See *E(O), Z(O)*.

E isomers: Z isomers:

E(O), Z(O) Descriptors for the arrangement of ligands around enolate double bonds. The standard *E/Z* stereochemical descriptor is modified such that the OM group is given priority over the carbonyl substituent, independent of the metal and the other substituent [159]. The priority descriptors for the α-carbon are maintained, as illustrated by the following examples:

Z(O)-enolates:

Eclipsed, Eclipsing Two ligands on adjacent atoms are *eclipsed* if their torsion angle is near 0° (*i.e., synperiplanar*). See also *bisecting conformation, eclipsing strain, torsion angle.*

Eclipsing conformation See *bisecting conformation.*

Eclipsing strain See *torsional strain.*

Element of chirality See *chirality element.*

Enantioconvergent See *stereoconvergent.*

Enantiofacial selectivity The degree of *enantioselectivity* observed upon addition to *enantiotopic* faces of a trigonal atom. See also *diastereofacial selectivity.*

Enantiofacial selectivity

enantiomers

Enantiomer A *stereoisomer* that is not superimposable on its mirror image. See also *enantiomorphous.*

Enantiomer excess, ee (percent enantiomer excess, % ee) For a mixture of a pure *enantiomer* and its *racemate*, the percent excess of the pure enantiomer over the racemate. Percent ee is given by

$$\% \text{ ee} = \frac{|E_1 - E_2|}{E_1 + E_2} \cdot 100 = |\%E_1 - \%E_2|,$$

where E_1 and E_2 are the mole fractions of the two enantiomers. This term was coined by Morrison and Mosher in 1971 as an alternative to the imprecise term *optical purity* (ref. [53], p. 10): "Assuming a linear relationship between rotation and composition and no experimental error, percent 'optical purity' is equated with the percent of one enantiomer over the other, which we shall designate percent enantiomeric excess (% ee)" Since polarimetry is rarely used to determine enantiomer ratios anymore, usage of this term is discouraged [69,160,161]; see also page 90. See *enantiomer purity.*

Enantiomer purity A description of the enantiomer composition of a sample. Preferable quantitative expressions are *enantiomer ratio, er* (expressed as a number or a percentage), *q* (expressed as a whole number), or *enantiomer composition, ec* (expressed as a fraction). For example, a 95:5 mixture of enantiomers could be expressed as 95:5 er, *q* = 19, or ec = 0.95 [160,161]. The obsolete terms *optical purity* and *enantiomer excess (ee)*, both of which are tied to polarimetry as a tool for determining enantiomer composition, are discouraged [69,160,161].

Enantiomer ratio The ratio of two *enantiomers*. When used as an expression of *enantiomer purity*, this ratio is often normalized as a percent (*i.e.*, 99:1, 80:20). The advantage of normalizing the numbers is that, in stereoselective reactions, the enantiomer ratios pass through a 50:50 mirror point. When expressed as a number (or a ratio having a denominator of 1), the symbol q can be used [160]. The latter is convenient for comparison of relative rates in stereoselective reactions (*e.g.*, 100/1, 1000/1) [69] (see Section 1.4).

Enantiomerically enriched (enantioenriched) A sample that has one *enantiomer* in excess.

Enantiomerically pure, enantiopure A sample that contains (within the limits of detection) only one enantiomer. Note that this is not synonymous with *homochiral* [162].

Enantiomerization The conversion of one enantiomer into another. This term is not synonymous with *racemization*.

Enantiomorphous Not superimposable on its mirror image.

Enantioselectivity (percent enantioselectivity, % es) In a reaction or reaction sequence in which one enantiomer (E_1) is produced in excess, the enantioselectivity is the mole fraction formed of the major enantiomer, expressed as a percent:

$$\% \text{ es} = \frac{E_1}{E_1 + E_2} \cdot 100.$$

See also *diastereoselectivity*.

Enantiotopic The relationship of two ligands of an atom, or groups in constitutionally equivalent positions, that are related by a *mirror plane, center of symmetry*, or *alternating axis*, but not by a simple (proper) *symmetry axis*. Replacement of either ligand yields a pair of *enantiomers*. Also, faces of a trigonal atom that are not symmetry related, such that addition to either face gives a pair of enantiomers. Note that addition of a ligand to the *Re* face affords a tetrahedral array with the new ligand in the L_{Re} position. The faces may be specified as *Re* or *Si*, and reflection variant ligands can be specified as L_{Re} or L_{Si}, as illustrated below [157,158]. See also *Re, Si, homotopic, heterotopic*, and *diastereotopic*.

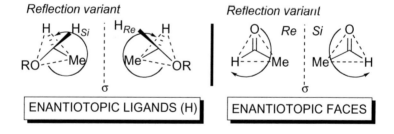

Enantiotopic group selectivity The degree of selectivity encountered by selective reaction of one enantiotopic group over another.

endo, exo The stereochemical prefix that describes the relative configuration of a substituent on a bridge (not a bridgehead) of a bicyclic system. If the substituent is oriented toward the larger of the other bridges, it is *endo*; if it is oriented toward the smaller bridge, it is *exo*.

ent A prefix to the name of a chiral molecule to indicate its *enantiomer.*

Envelope The conformation of a 5-membered ring in which four atoms are coplanar, and the fifth (the flap) is out of the plane.

Epimerization The interconversion of *epimers.*

Epimers *Diastereomers* that differ in *configuration* at one (and only one) of two or more *stereo-genic units.*

Equatorial See *axial, equatorial.*

erythro, threo Terms used to describe relative configuration at adjacent stereocenters. Originally, the term was derived from carbohydrate nomenclature (*cf.* erythrose, threose). In this sense, if the molecule is drawn in a *Fischer projection*, the *erythro* isomer has identical or similar substituents on the same side of the vertical chain and the *threo* isomer has them on opposite sides. In the early 1980s, proposals appeared to redefine these terms based on *zig-zag projections* [163] and *CIP pri-ority* [164], but such usage is now discouraged [165]. See *l, u; syn, anti.*

exo See *endo, exo.*

Felkin-Anh model See *Cram's rule (open chain model).*

Fischer Projection (or Fischer–Tollens projection) A planar projection formula in which the vertical bonds lie behind the plane of the paper and the horizontal bonds lie above the plane. Used commonly in carbohydrate structures, where each carbon in turn is placed in the proper orientation for planar projection.

Fischer projection   hash/wedge view   ball and stick model

Fischer–Rosanoff convention A method for the specification of absolute configuration, still in common use for amino acids and sugars. When drawn in a *Fischer projection* with C_1 at the top, if the functional group of the specified stereocenter is on the right, the absolute configuration is D, if on the left, it is L. For amino acids, the reference stereocenter is C_2; for sugars it is the highest numbered stereocenter [166].

D-glucose   L-glucose   D-serine   L-serine

* Reference stereocenter

Flagpole See *bowsprit, flagpole*

Free rotation, restricted rotation In the context of an experimental observation, *free rotation* is sufficiently fast (*i.e.*, the rotational barrier is sufficiently low) that different *conformations* are not observable. Conversely, *restricted rotation* is sufficiently slow (the barrier is sufficiently high) that conformational isomers can be observed.

Gauche Synonomous with a *synclinal* alignment of groups attached to adjacent atoms (*i.e.*, a torsion angle of near $+60°$ or $-60°$). See *torsion angle.*

Geometric isomers Synonym for *cis-trans* double bond isomers.

Half-boat See *half-chair, half-boat.*

Half-chair, half-boat Terms used most commonly to describe conformations of cyclohexenes in which four contiguous carbon atoms lie in a plane. If the other two atoms lie on opposite sides of the plane, the conformation is a half-chair; if they are on the same side, it is a half-boat. Also used for 5-membered rings, where three adjacent atoms define the plane.

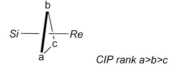

cyclohexene   cyclopentane
half-chair   half-chair

Helicity The chirality sense of a helix. May be specified by *P, M.*

Heterochiral See *homochiral.*

Heterotopic Either *diastereotopic* or *enantiotopic*. Refers to either the *Re* or the *Si* half space of a two-dimensionally chiral triangle, as shown below [157,158]. See also *Re, Si, enantiotopic, diastereotopic.*

b

Si —|— Re

a   c

CIP rank a>b>c

Homochiral A descriptor for objects and molecules having the same *chirality sense* [167]. Thus, "two equal and similar right hands are homochirally similar. Equal and similar right and left hands are heterochirally similar ..." [144]. A set of right shoes, or an assembly of molecules (such as a mixture of amino acids) that have the same *relative configuration* or *chirality sense*, are homochiral [167,168]. This term should not be used to describe enantiomerically pure compounds, since the term homochiral describes a <u>relationship</u>, not a <u>property</u> [162].

Homofacial, heterofacial The *relative configuration* of stereocenters (in different molecules) having three identical ligands and one different is homofacial if the fourth ligand is on the same *heterotopic* face in both, and heterofacial if on opposite faces, as shown below. [157,158,169]. See also *relative configuration.*

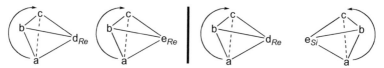

homofacial: *d* and *e* both reside on   heterofacial: *d* is on the *Re* side of the
the *Re* face of the *abc* triangle   *abc* triangle, whereas *e* is on the *Si* face.

Homotopic Ligands that are related by an *n*-fold rotation axis. Similarly, faces of a trigonal atom that are related by an *n*-fold rotation axis. Replacement of any of the ligands or addition to either of the faces gives an identical compound. See also *heterotopic, enantiotopic,* and *diastereotopic.*

Inversion See *Walden inversion, pyramidal inversion,* and *ring inversion.*

Isomers Compounds that have the same molecular formula but which have different *constitutions* (*constitutional isomers*), *configurations* (*enantiomers, diastereomers*), or *conformations* (*conformational isomers*), and may have different chemical, physical, or biological properties.

Kinetic resolution The separation (or partial separation) of *enantiomers* due to a difference in the rate of reaction of the two enantiomers in a *racemic mixture* with an *nonracemic chiral* reagent (see Section 1.6).

l, u Terms for the description of *relative configuration*. A pair of stereogenic units has the relative configuration *l* (for *like*) if the descriptor pairs are *RR, SS, RRe, SSi, ReRe, SiSi, MM, PP, RM, SP, ReM,* or *SiP*. The pair is specified as *u* (*unlike*) if they have descriptor pairs *RS, RSi, ReS, ReSi, MP, RP, SM, ReP,* and *SiM* [149]. Reflection invariant descriptors (*r, s, re, si, p,* and *m*) may be substituted in place of the reflection variant descriptors above. Note the use of lower case *l* and *u* letters, implying a *reflection invariant* relationship (*i.e., SS* and *RR* are both *l*).

lk, ul An extension of the *l, u* nomenclature to describe topicity. If a reagent of configuration *R* (or the *Re* face of a trigonal atom) preferentially approaches the *Re* face of a trigonal atom, the topicity is *lk* (*like*); it is *ul* (*unlike*) if it approaches the *Si* face. Similarly, the approach of an achiral reagent to diastereotopic faces of a trigonal atom is *lk* if the *Re* face is preferred in the *R* enantiomer, and vice versa; the topicity is *ul* if the *Si* face is preferred in the *R* enantiomer, and vice versa. In short, if the first letters of the two *stereochemical descriptors* are the same, the topicity is *lk*. If they are different, it is *ul* [170]. See the more complete listing of like and unlike pairs under *l, u*. Note the use of lower case *l* and *u* letters, implying a *reflection invariant* relationship. *Lk* and *Ul* would be used if the topicity were *reflection variant*, which would occur if one of the components was reflection invariant.

M, P See *P, M*.

meso A stereoisomer that has two or more *stereogenic units*, but which is *achiral* because of a *symmetry plane*. The plane reflects *enantiomorphic* groups.

Newman projection A projection formula that represents the spatial arrangement of the ligands on two adjacent atoms as viewed down the bond joining them.

zig-zag projection — Newman projection of one conformer — ball and stick model

Nonbonded interactions Attractive or repulsive forces between atoms or groups in a molecule (intermolecular or intramolecular) that are not directly bonded to each other. Sometimes called (inappropriately) "through space" interactions.

Nonracemic Not *racemic*. Used as a synonym for enantiomerically enriched.

Optical activity The property of a substance to rotate plane-polarized light (see Section 2.6).

Optical purity (op, % op) The ratio of the observed *specific rotation* of a substance to the maximum possible rotation of the substance, expressed as a percent:

$$\% \ op = \frac{[\alpha]}{[\alpha]_{max}} \ 100.$$

Usually (but not always) it is equal to *enantiomer excess*, or *% ee*. This term is not used any more, since enantiomer ratios are rarely determined by methods that involve polarimetry (see Section 2.6).

Optical yield For a chemical reaction, the *enantiomer excess* of the products relative to that of the starting material, expressed as a percent. In asymmetric synthesis, the denominator may be the ee of the chiral reagent or catalyst. Usage is discouraged.

P, M Descriptors of chirality sense of a helix. Once the axis of the helix is identified, one chooses the ligands of the highest *CIP rank*. If the smallest angle between the ligands (*i.e.*, ≤ 180°) in a projection is clockwise going from front to rear, the chirality sense is *P* (plus), if counterclockwise, it is *M* (minus) [148,171]. Additionally, *P, M* may be used to describe the chirality sense of a helix of any sequence of atoms as long as they are explicitly identified. Note that it does not matter which end of the helix is viewed.

These descriptors can be used to specify enantiomeric *conformers*, such as *gauche* butane, and the *absolute configuration* of *stereogenic axes* and *planes*. Prelog and Helmchen have recommended the use of *P, M* instead of *R, S* for describing the absolute configuration of planes and axes of chirality [149].

P-butane   *M*-2-hydroxyparacyclophane

M-2,3-pentadiene   *P*-binaphthol

Pitzer strain See *torsional strain*.

Planar chirality See *stereogenic element*.

Point group (symmetry point group) The symmetry classification of a molecule based on its symmetry elements (axes, planes, etc.).

Pref, parf Descriptors of relative configuration based on *CIP priority*. The relationship between two chirality centers is pref (priority reflective) if the order of decreasing priority of the three remaining groups at one chirality center is a reflection of the order of decreasing priority of the groups at the other center. When the orders of decreasing priority are not reflective of each other, the relative configuration is parf (priority antireflective). If the chirality centers are not adjacent, the intervening bonds are neglected and the two centers are treated as if they were directly linked. If the two centers are part of a ring, they are treated as if connected by a bond that replaces the shorter path [172]. Not often used.

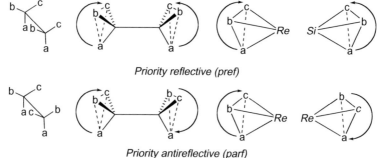

Priority reflective (pref)

Priority antireflective (parf)

Prochiral Tetrahedral atoms having *heterotopic* ligands, or *heterotopic* faces of trigonal atoms, may be described as being prochiral. It is inappropriate to describe an entire molecule as being prochiral [173]. Heterotopic faces are described using *Re, Si* if *reflection variant*, and *re, si* if *reflection invariant* [157]. If the *CIP priority* of the three ligands is clockwise, the face (toward the observer) is *Re*; if counterclockwise, it is *Si* [173]. For heterotopic ligands, two conventions have been used to describe prochirality. Both use the CIP rank of the ligands to specify the "prochirality sense" of each ligand. The broader rule is that of Prelog and Helmchen [157]; see also ref. [158]. In this method, a tetrahedron is constructed of the four ligands around the prochiral center. If the ligand "L" is sitting on the *Re* face of the triangle formed by the other three ligands, it is specified L_{Re} (or L_{re} if *reflection invariant*); similarly, the ligand would be specified L_{Si} or L_{si} if on the *Si* face. See also *enantiotopic, diastereotopic, heterotopic,* and *relative configuration.*

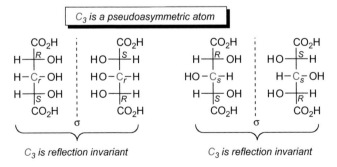

CIP rank a>b>c

PROCHIRAL FACES   PROCHIRAL ATOMS (X), and LIGAND LABELS

Another convention, used in biochemistry to specify the hydrogen atoms of a prochiral methylene, replaces a hydrogen with a deuterium. If such replacement results in the *R* configuration, the ligand position is *pro-R*. If the *S* configuration is obtained, it is *pro-S* [173]. If *reflection invariant*, the descriptors are *pro-r* and *pro-s.*

CIP rank a>b>D>H

Pseudoasymmetric atom A *stereogenic* atom of a stereoisomer that has two *enantiomorphic* ligands and two other different ligands. The absolute configuration of a pseudoasymmetric atom is *reflection invariant.* Exchange of any two ligands generates a *diastereomeric* compound. The *CIP* descriptors for pseudoasymmetric atoms are *r, s.* Use of this term is discouraged in favor of *stereogenic center,* or more explicitly as a *stereogenic, achirotopic* center.

C_3 is a pseudoasymmetric atom

C_3 is reflection invariant C_3 is reflection invariant

Pseudoaxial, pseudoequatorial See *axial, equatorial.*

Pseudorotation Conformational motion that appears to have been produced by rotation of the entire molecule and is superimposable on the initial conformation. In organic chemistry, the term is used by some authors to describe the out-of-plane motion of the ring atoms in cyclopentane during fast conformational interchange of the many envelope and twist conformers.

Pyramidal inversion The change of bond directions in a trivalent central atom having a pyramidal (tripodal) arrangement with the central atom at the apex of the pyramid. The inversion appears to move the central atom to a similar position on the other side of the pyramid. If the central atom is *stereogenic*, pyramidal inversion reverses its *absolute configuration.*

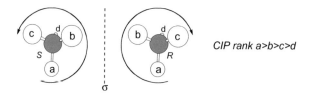

q See *enantiomer purity* and *enantiomer ratio.*

r, s See *pseudoasymmetric atom.*

R, S CIP descriptors for the specification and description of absolute configuration, as follows.

    Stereogenic center After the CIP rank of ligands is determined (see *CIP method*), the tetrahedron is arranged so that the ligand of lowest priority is to the rear. If the order of the other three ligands (highest to lowest) is clockwise, the *absolute configuration* is R (Latin *rectus,* right); if counterclockwise, it is S (Latin *sinister,* left).

CIP rank a>b>c>d

    Stereogenic axis The descriptors may be modified to R_a, S_a (or aR, aS) when applied to a stereogenic axis [148], although it is more convenient (and highly recommended!) to use the P, M system to specify the configuration of stereogenic axes [149]. Taking 2,3-pentadiene as an example, and recalling *CIP* sequence rule 0 (near end of axis precedes farther), the four substituents are prioritized as shown below. It does not matter from which end of the axis the model is viewed. It is general that $M = R_a$ and that $P = S_a$. Note that the correlation between axial and planar R/S descriptors with M/P descriptors are opposite.

M-2,3-pentadiene = R_a-2,3-pentadiene

Stereogenic plane The descriptors may be modified to R_p, S_p (or pR, pS) when applied to a stereogenic plane [148], although it is more convenient (and highly recommended!) to use the P, M system to specify the configuration of stereogenic axes and planes [149]. To describe the absolute configuration of a chirality plane using R_p/S_p, the pilot atom must first be identified; it is the first atom outside the plane at the higher priority end of the plane. Using 2-hydroxy-paracyclophane as an example, the pilot atom (nearer the (C)OH) is indicated. From the vantage point of the pilot atom, the atoms in the plane are then followed in *CIP* sequence: at the branch point (the ipso carbon), one proceeds to the higher priority atom. If counterclockwise, the configuration is S_p. It is general that $M = S_p$ and that $P = R_p$. Note that the correlation between axial and planar *R/S* descriptors with *M/P* descriptors is opposite.

M- 2-hydroxyparacyclophane = S_p- 2-hydroxyparacyclophane

Chiral metallocenes, although formally having a chirality plane, are treated differently when describing *CIP* priority. Each carbon in the ring is considered to be σ-bonded to the metal, and the configuration of the complex is designated by the configuration of the highest priority atom in the arene ring, using the rules for stereocenters. In the ferrocene below, the highest *CIP* priority arene atom is indicated by the red dot. The four ligands are then prioritized as shown; the configuration of the complex is S [148,169,174,175].

If the stereogenic unit is *reflection invariant*, lower case *CIP* descriptors are used (*r, s, m, p*). The symbols $R*$ and $S*$ may be used to describe *relative configuration*. Thus, $R*$, $R*$ describes a racemate of *l* configuration and $R*$, $S*$ describes a racemate of *u* configuration. See *l,u*.

rac A prefix to the name of a chiral molecule to indicate that it is the *racemate*.

Racemate (racemic mixture) An equimolar mixture of two *enantiomers*, whose physical state is unspecified [35]. Some authors restrict the term "racemate" to a crystalline compound whose unit cell contains equal numbers of enantiomeric molecules and "racemic mixture" to a mechanical mixture of two crystals that form a eutectic of two enantiomers. The latter is now referred to as a *conglomerate* [35].

Racemization The conversion of a nonracemic substance into its racemate. See also *enantiomerization* (not a synonym).

Re, Si (re, si) Stereochemical descriptors for *heterotopic* faces. If the *CIP priority* of the three ligands is clockwise, the face (toward the observer) is *Re* (latin *rectus*, right); if counterclockwise, it is *Si* (latin *sinister*, left) [173].

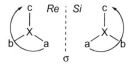

The descriptors may be used to describe the faces of trigonal atoms,

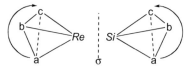

or the ligand position of a tetrahedral stereogenic unit,

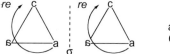

Lower case descriptors *(re, si)* are used for the rare cases that are *reflection invariant* [157,176]:[23]

a and ƨ are enantiomorphous groups
CIP rank: a > ƨ > c

For examples of reflection-invariant stereogenic centers and faces, see *diastereotopic*, and *pseudoasymmetric atom*.

Reflection variant, reflection invariant The terms used to describe an object and its relationship with its mirror image. If the two are identical, the object is reflection invariant. If the object is *enantiomorphous* to its mirror image, it is reflection variant.

Relative configuration The *configuration* of any *stereogenic* element with respect to another. Relative configuration is *reflection invariant*. The relative configuration of pairs of *stereogenic* units in the same molecule may be described as *R*, R** or *l* if they have the same *CIP* descriptor, and *R*, S** or *u* if they are different. (See *l, u* for a complete list of like and unlike descriptors.) The term can also be used in an intermolecular sense as follows: if the two molecules contain stereogenic units abcd and abce, and if e and d both sit on the same *heterotopic face*, the two stereogenic units have the same relative configuration. If not, they have the opposite relative configuration. The term may be applied to starting material and products of a reaction sequence. See also *homofacial, heterofacial.*

Same relative configuration	Opposite relative configuration

---

23. In the highlight box on p. 33, the importance of reflection variance was explained. In his first paper [173], Hanson defined the terms without considering reflection variance. After considering Prelog and Helmchen's suggestion [157], Hanson incorporated the distinction between *Re/Si* and *re/si* into his terminology [176].

Resolution The separation of a *racemic mixture* into (at least one of) its component enantiomers. See also *kinetic resolution.*

Restricted rotation See *free rotation.*

Retention of configuration The product of a chemical reaction has retained its configuration if the product has the same *relative configuration* as the starting material. See also *Walden inversion, relative configuration.*

Ring inversion (ring reversal) The interconversion of cyclohexane *conformations* having similar shapes (chair–chair), accompanied by interchange of the *equatorial* and *axial* substituents. Similarly, the interchange of any such similarly shaped conformations in a cyclic molecule.

Rotamers Stereoisomers of the same constitution and configuration, which differ only by torsion angles.

Rotation angle (α) The rotation of the plane of polarized light (in degree units) after passing through an *optically active* sample. If the angle of rotation is clockwise, the sample is dextrorotatory and the sign of rotation is positive (+). If the angle is counterclockwise, the sample is levorotatory, and the sign of rotation is (−). See also *optical activity, specific rotation,* and Section 2.6.

Rotation–reflection axis See *alternating symmetry axis.*

Rotational barrier The energy barrier between two *conformers.*

s, Selectivity factor In a kinetic resolution, a chiral reagent or catalyst reacts with two enantiomers at different rates; the ratio of the relative rates, k_R/k_S, is the selectivity factor (see Section 1.7).

s-cis, s-trans Conformational descriptors for the single bond linking two double bonds (highlighted red, below). The *synperiplanar* conformation is *s-cis*, and the *antiperiplanar* conformation is *s-trans.* See *torsion angle.*

Sawhorse formula A perspective drawing that indicates the spatial arrangements of the ligands on two adjacent tetrahedral atoms. The bond between the two atoms is a diagonal line, with the nearer atom at the bottom.

sawhorse projection   *ball and stick model*

Scalemic Not *racemic* [177,178]. Synonomous with *aracemic, nonracemic.*

S_E2inv, S_E2ret Abbreviations for the steric course of bimolecular electrophilic substitutions, which may occur with inversion (S_E2inv) or retention (S_E2ret) [179].

Self-disproportion of enantiomers The phenomenon of varying *enantiomer* composition as a function of retention time during chromatography on an *achiral* stationary phase. The phenomenon is only observed during chromatography of a partly racemic compound (er $\neq$ 50:50), and is due to the formation of homochiral and heterochiral dimers during the elution (see Section 2.3).

Sense of chirality See *chirality sense.*

Sequence rules See *CIP method.*

Si, si See *Re, Si.*

Skew See *chair, boat, twist-boat.*

Specific rotation The specific rotation of a sample, [α], is defined as

$$[\alpha]_\lambda^t = \frac{100\alpha}{l \cdot c},$$

where t is temperature, λ is wavelength of the light, α is the observed *rotation*, l is the sample path length (in dm), and c is the concentration (in g/100 mL). [α] is normally reported without units, but the concentration and the solvent are usually specified in parentheses after the value of [α]. The correct units are (degree mL)/(g dm); it is incorrect to report [α] in degrees (see Section 2.6).

Staggered conformation The *conformation* of two tetrahedral carbons is staggered if the *torsion angle* between the ligands is approximately $\pm 60°$.

Stereocenter See *stereogenic center*.

Stereochemical descriptor A letter, symbol, or prefix to specify configuration or conformation, such as *R*, *S*, *E*, *Z*, *P*, *M*, *cis*, *trans*, etc.

Stereoconvergent A reaction or reaction sequence is stereoconvergent if stereoisomerically different starting materials yield the same stereoisomeric product. The sequence may be more specifically labeled either *enantioconvergent* or *diastereoconvergent*.

Stereoelectronic effect An effect on structure and reactivity due to the orientation and alignment of bonded or nonbonded electron pairs [180].

Stereogenic axis A set of two pairs of tetrahedrally arranged bonding positions (D_2 or C_{2v} point symmetry), each occupied by two different ligands. Exchange of the ligands of either pair reverses the *absolute configuration*. Examples include allenes that are unsymmetrically substituted on the terminal carbons and 2,6,2′,6′-tetrasubstituted biphenyls. If the axis is *reflection variant*, it may be called a *chirality axis*. The absolute configuration is best described using the *P*, *M* nomenclature (see *P*, *M*). See also *stereogenic element*.

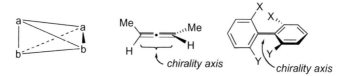

Stereogenic center (stereocenter) An atom in a molecule (or a focal set of atoms) with four equivalent tetrahedral bonding positions (T_d point symmetry), occupied by four different ligands. Exchange of any two ligands reverses the *absolute configuration*. If the center is reflection variant, it may be called a *chirality center*. If it is reflection invariant, it is sometimes called a *pseudoasymmetric atom*, although the usage of this term is discouraged. *Stereogenic center* is thus an extension of the "asymmetric carbon atom" of van't Hoff and LeBel, and now includes species such as N^+abcd and the sulfur atom of unsymmetric sulfoxides (where the fourth "ligand" is a lone pair), as well as tetrahedral arrays of ligands with T_d symmetry. The *absolute configuration* may be described by the *CIP method*. See *R*, *S*.

Stereogenic element, stereogenic unit A *center*, *axis*, or *plane* in a molecule in which exchange of two ligands leads to a new *stereoisomer*. If the stereogenic element is *reflection variant*, the elements are *chirality center*, *chirality axis*, and *chirality plane*. The bonding positions of stereogenic <u>centers</u> have point symmetry T_d; the bonding positions of stereogenic <u>axes</u> have point symmetry D_2 or C_{2v}; and the bonding positions of stereogenic <u>planes</u> have point symmetry C_S. As a result, there must be four different ligands (abcd) on a T_d bonding center to create stereogenicity. On an axis, only the two ligands of each pair need be different (ab/ab), the two pairs may be the same. In a stereogenic plane, only one of the ligands in the plane need be different. See also *stereogenic axis, stereogenic center, and stereogenic plane.*

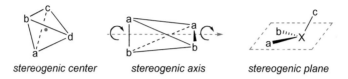

| stereogenic center | stereogenic axis | stereogenic plane |

Stereogenic plane A planar structural fragment that, because of *restricted rotation* or structural requirements, cannot lie in a symmetry plane. If the stereogenic plane is *reflection variant*, the element may be called a *chirality plane*. For example with a monosubstituted paracyclophane, the stereogenic plane includes the plane of the benzene ring. For compounds with two planes, the more highly substituted ring comprises the chirality plane. Note that metal arenes and metal cyclopentadienyl systems are treated as though the metal were sigma bonded to each carbon, and then treated as having stereogenic centers (see *R, S*). The configuration is designated for the highest priority atom in the arene ring. The absolute configuration is best described using the *P, M* nomenclature. For designation of the absolute configuration of metallocenes, see *R, S*. See also *stereogenic element.*

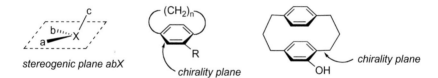

| stereogenic plane abX | chirality plane | chirality plane |

Stereoheterotopic Either *enantiotopic* or *diastereotopic*.

Stereoisomers Isomers of the same *constitution* that differ only in the position of atoms and ligands in space (*i.e., enantiomers* and *diastereomers*).

Stereoselectivity In a reaction, the preferential formation of one *stereoisomer* over another (or others). See also *diastereoselectivity, enantioselectivity*.

Stereospecific A <u>pair of reactions</u> are stereospecific if *stereoisomeric* substrates afford stereoisomeric products. A stereospecific process is necessarily 100% *stereoselective*, but the converse is not necessarily true, even if the stereoselectivity is 100%. Use of the term to describe a reaction that is merely highly stereoselective is incorrect.

Structure The *constitution, configuration,* and *conformation* of a molecule. In the old literature, the term was used as a synonym for *constitution* alone.

Structural isomers Obsolete term for *constitutional isomers*.

Superimposable, superposable Two objects are superimposable if they can be brought into coincidence by translation and rotation. For chemical structures, *free rotation* around single bonds is permissible. Thus, two molecules of *R*-2-butanol are considered superimposable independent of their conformations.

Suprafacial See *antarafacial, suprafacial*.

syn, anti Prefixes that describe the relative configuration of two substituents with respect to a defined plane or ring (*syn* if on the same side, *anti* if opposite). Such planes may be defined arbitrarily, but some that are in common usage are illustrated below. Formerly, these terms were used to describe the configuration of oximes, hydrazones, etc. (see *E, Z*). See also *torsion angle*.

| A/C and B/D rings are anti | methyl and hydroxyl anti hydroxyl and methoxy syn | syn dimethyls | anti dimethyls |

Synclinal See *torsion angle*.

Symmetry axis (C_n) An axis of an object, about which a rotation by an angle of $360/n$ gives an entity that is superimposable on the original. See also *alternating symmetry axis*.

Symmetry elements Axes, centers, or planes of symmetry.

Symmetry plane (σ) A mirror plane that bisects an object, such that reflection of one half produces a fragment that is superimposable on the other half.

Thorpe-Ingold effect The original phenomenon observed by Thorpe and Ingold was an accelerating effect of geminal disubstitution on rates of ring closure [181,182]. Several theories have been advanced to explain the effect, which probably varies among individual examples. For a recent review on the effects of geminal disubstitution, see ref. [183].

Threo See *erythro, threo*.

Torsion angle The angle, in a molecular fragment A−B−C−D (having ABC and BCD bond angles $\le 180°$), between the planes ABC and BCD (see the *Newman projection*, below), always defined such that the absolute value is less than 180°. If (looking from either direction) the turn from A to D or D to A is clockwise, the torsion angle is positive; if it is counterclockwise, the torsion angle is negative (see also *P, M*). If the torsion angle is 0° to $\pm 90°$, the angle is *syn*; if between $\pm 90°$ and 180°, it is *anti*. Similarly, angles from 30° to 150° and $-30°$ to $-150°$ are *clinal*. Combination gives *synperiplanar* for angles between 0° and $\pm 30°$; 30° to 90° and $-30°$ to $-90°$ are *synclinal*; 90° to 150° and $-90°$ to $-150°$ are *anticlinal*; and $\pm 150°$ to 180° are *antiperiplanar* [171]. Often the *synperiplanar* conformation is called eclipsed, the *antiperiplanar* conformation *anti*, and the *synclinal* conformation *gauche* or skew.

Torsional strain Destabilization of a molecule due to a variation of a *torsional angle* from an optimal value (*e.g.*, 60° in a saturated molecule). Also called *Pitzer strain, eclipsing strain.*

Torsional isomers See conformational isomers.

trans See *cis, trans isomers.*

Transannular interaction Literally, cross-ring interactions. Nonbonded interaction between ligands attached to nonadjacent atoms in a ring, for example, in a cyclohexane *boat* or in medium-sized rings.

Transition state, transition structure In a chemical reaction, the transition *state* is the ensemble of molecular structures that are at the free energy saddle point between reactants and products. The transition *structure* corresponds to the single set of atomic coordinates at the saddle-point of the potential energy surface (internal, or enthalpic energy at 0 °K). Thus, coordinates of the transition state vary with temperature, whereas those of the transition structure do not. In a practical sense, a structure that is drawn on a piece of paper (whether derived from a computation or not) should be referred to as a transition structure, since it is static. The transition state is an ensemble of similar structures undergoing translational, vibrational, and rotational motion.

transoid conformation (usage discouraged) See *s-cis, s-trans.*

twist-boat See *chair, boat, twist-boat.*

u See *l, u.*

ul See *lk, ul.*

van der Waals interactions Attractive or repulsive interactions resulting from close approach of two molecules [184−186]. Modern usage (especially in molecular mechanics calculations) also uses the term van der Waals interactions to describe the attractive and repulsive interactions created by intramolecular approach of molecular fragments [187]. See also *nonbonded interactions.*

Walden inversion Conversion of Xabcd into Xabcd (for an identity reaction) or Xabce, of opposite configuration. Synonymous with *inversion of configuration.*

Walden inversion

Z See *E, Z.*

Z(O) See *E(O), Z(O).*

Zig-zag projection A stereochemical projection in which the main chain of an acyclic compound is drawn in the plane of the paper with 180° torsion angles, with substituents above the plane drawn with bold or solid wedges, and hashed lines for substituents behind the plane.

zig-zag projection

G
L
O
S
S
A
R
Y

REFERENCES

[1] Gardner, M. *The New Ambidexterous Universe: Symmetry and Asymmetry from Mirror Reflections to Superstrings*; 3rd ed.; W. H. Freeman: New York, 1990.

[2] *Chirality. From Weak Bosons to the Alpha Helix*; Janoschek, R., Ed.; Springer: New York, 1991.

[3] Hegstrom, R. A.; Kondpudi, D. K. *Scientific American* **1990**, *262*, 108−115.

[4] *Physical Origin of Homochirality in Life*; Cline, D. B., Ed.; AIP Press: Woodbury, NY, 1996; Vol. 379.

[5] Wagnière, G. H. *On Chirality and the Universal Asymmetry. Reflections on Image and Mirror Image*; VHCA: Zürich, 2007.

[6] *Chirality and Biological Activity*; Holmstedt, B.; Frank, H.; Testa, B., Eds.; A. R. Liss: New York, 1990.

[7] *Chirality in Industry: The Commercial Manufacture and Applications of Optically Active Compounds*; Collins, A. N.; Sheldrake, G. N.; Crosby, J., Eds.; Wiley: New York, 1992.

[8] Sheldon, R. A. *Chirotechnology. Industrial Synthesis of Optically Active Compounds*; Marcel Dekker: New York, 1993.

[9] Scott, J. W. *Top. Stereochem.* **1991**, *19*, 209−226.

[10] Crossley, R. *Tetrahedron* **1992**, *48*, 8155−8178.

[11] Brenna, E.; Fuganti, C.; Serra, S. *Tetrahedron: Asymmetry* **2003**, *14*, 1−42.

[12] Rouhi, A. M. *Chem. Eng. News* **2004**, *82*, 47−62.

[13] *Stereochemistry and Biological Activity of Drugs*; Ariëns, E. J.; Soudijin, W.; Timmermans, P. B. M. W. M., Eds.; Blackwell: Oxford, 1983.

[14] Ariëns, E. J. *Trends. Pharm. Sci.* **1986**, *7*, 200−205.

[15] Ariëns, E. J. *Trends Pharm. Sci.* **1993**, *14*, 68−75.

[16] Fassihi, A. R. *Int. J. Pharmaceut.* **1993**, *92*, 1−14.

[17] Triggle, D. J. *Drug Discov. Today* **1997**, *2*, 138−147.

[18] Szelenyi, I.; Geisslinger, G.; Polymeropoulos, E.; Paul, W.; Herbst, M.; Brune, K. *Drug News Persp.* **1998**, *11*, 139−160.

[19] Landoni, M. F.; Soraci, A. *Curr. Drug Metab.* **2001**, *2*, 37−51.

[20] Russel, G. F.; Hills, J. I. *Science* **1971**, *172*, 1043−1044.

[21] Friedman, L.; Miller, J. G. *Science* **1971**, *172*, 1044−1045.

[22] Solms, J.; Vuataz, L.; Egli, R. H. *Experientia* **1965**, *21*, 692−694.

[23] Easson, C. H.; Stedman, E. *Biochem. J.* **1933**, *27*, 1257−1266.

[24] Hanessian, S.; Giroux, S.; Merner, B. L. *Design and Strategy in Organic Synthesis: From the Chiron Approach to Catalysis*; Wiley-VCH: Weinheim, 2012.

[25] Coppola, G. M.; Schuster, H. F. *Asymmetric Synthesis. Construction of Chiral Molecules Using Amino Acids*; Wiley-Interscience: New York, 1987.

[26] *Carbohydrates as Raw Materials*; Lichtenthaler, W., Ed.; VCH: New York, 1991.

[27] Ho, T.-L. *Enantioselective Synthesis: Natural Products from Chiral Terpenes*; Wiley: New York, 1992.

[28] Scott, J. W. In *Asymmetric Synthesis*; Morrison, J. D., Ed.; Academic: Orlando, 1984; Vol. 4, pp. 1−226.

[29] Fleet, G. W. J. *Chem. Br.* **1989**, *25*, 287−292.

[30] Money, T. *Nat. Prod. Rep.* **1985**, 253−289.

[31] Money, T. In *Studies in Natural Products Chemistry*; Atta-Ur-Rahman, Ed.; Elsevier: Amsterdam, 1989; Vol. 4, pp. 625−697.

[32] Hanessian, S.; Franco, J.; Larouche, B. *Pure Appl. Chem.* **1990**, *62*, 1887−1910.

[33] Casiraghi, G.; Zanardi, F.; Rassu, G.; Spanu, P. *Chem. Rev.* **1995**, *95*, 1677−1716.

[34] Boons, G.-J.; Hale, K. *Organic Synthesis with Carbohydrates*; Sheffield Academic Press: Sheffield, 2000.

[35] Jacques, J.; Collet, A.; Wilen, S. H. *Enantiomers, Racemates and Resolutions*; Wiley-Interscience: New York, 1981.

[36] Ireland, R. E. *Organic Synthesis*; Prentice-Hall: Englewood Cliffs, NJ, 1969.

[37] Ho, T.-L. *Tactics of Organic Synthesis*; Wiley-Interscience: New York, 1994.

[38] Hudlicky, T. *Chem. Rev.* **1996**, *96*, 3−30.

[39] Bertz, S. H. *New J. Chem.* **2003**, *27*, 870−879.

[40] Smith, M. B. *Organic Synthesis*; 2nd ed.; McGraw Hill: Boston, 2002.

[41] Carruthers, W.; Coldham, I. *Modern Methods of Organic Synthesis*; Cambridge University Press: Cambridge, 2004.

[42] Marckwald, W. *Chem. Ber.* **1904**, *37*, 1368−1370.
[43] Wong, C.-H.; Whitesides, G. M. *Enzymes in Synthetic Organic Chemistry*; Pergamon: Oxford, 1994.
[44] Holland, H. L. *Organic Synthesis with Oxidative Enzymes*; VCH: New York, 1991.
[45] Pratt, A. J. *Chem. Brit.* **1989**, *25*, 282−286.
[46] Cohen, J. B.; Patterson, T. S. *Chem. Ber.* **1904**, *37*, 1012−1014.
[47] Marckwald, W. *Chem. Ber.* **1904**, *37*, 349−354.
[48] Erlenmeyer, E.; Landesberger, F. *Biochem. Z.* **1914**, *64*, 366−381.
[49] Japp, F. R. *Nature* **1898**, *58*, 452−460.
[50] *Asymmetric Catalysis on Industrial Scale*; Blaser, H. U.; Schmidt, E., Eds.; Wiley-VCH: Weinheim, 2004.
[51] Richie, P. D. *Asymmetric Synthesis and Asymmetric Induction*; Oxford: London, 1933.
[52] Kortum, G. *Sammlung Chimie U. Chemische Technology*; F. Enke: Stuttgart, 1933.
[53] Morrison, J. D.; Mosher, H. S. *Asymmetric Organic Reactions*; Prentice-Hall: Englewood Cliffs, NJ, 1971.
[54] *Asymmetric Synthesis*; Morrison, J. D., Ed.; Academic: Orlando, 1983−1983.
[55] *Comprehensive Organic Synthesis. Selectivity, Strategy, and Efficiency in Modern Organic Chemistry*; Trost, B. M.; Fleming, I., Eds.; Pergamon: Oxford, 1991.
[56] *Stereoselective Synthesis*; Helmchen, G.; Hoffmann, R. W.; Mulzer, J.; Schaumann, E., Eds.; Georg Thieme: Stuttgart, 1995; Vol. E21.
[57] Cram, D. J.; Elhafez, F. A. A. *J. Am. Chem. Soc.* **1952**, *74*, 5828−5835.
[58] Cram, D. J.; Kopecky, K. R. *J. Am. Chem. Soc.* **1959**, *81*, 2748−2755.
[59] Chen, X.; Hortelano, E. R.; Eliel, E. L.; Frye, S. V. *J. Am. Chem. Soc.* **1992**, *114*, 1778−1784.
[60] Frye, S. V.; Eliel, E. L. *Tetrahedron Lett.* **1985**, *26*, 3907−3910.
[61] Weber, B.; Seebach, D. *Angew. Chem., Int. Ed. Engl.* **1992**, *31*, 84−86.
[62] Brunner, H. *Acc. Chem. Res.* **1979**, *12*, 250−257.
[63] Brunner, H. *Angew. Chem., Int. Ed.* **1999**, *38*, 1194−1208.
[64] Ganter, C. *Chem. Soc. Rev.* **2003**, *32*, 130−138.
[65] *Catalytic Asymmetric Synthesis*; 2nd ed.; Ojima, I., Ed.; Wiley-VCH: New York, 2000.
[66] Noyori, R. *Asymmetric Catalysis in Organic Synthesis*; Wiley-Interscience: New York, 1994.
[67] Muramatsu, Y.; Harada, T. *Angew. Chem., Int. Ed.* **2008**, *47*, 1088−1090.
[68] Goodman, J. M.; Kirby, P. D.; Haustedt, L. O. *Tetrahedron Lett.* **2000**, *41*, 9879−9882.
[69] Gawley, R. E. *J. Org. Chem.* **2006**, *71*, 2411−2416; corrigendum *J. Org. Chem.* **2008**, *73*, 6470.
[70] Buschmann, H.; Scharf, H.-D.; Hoffmann, N.; Esser, P. *Angew. Chem., Int. Ed.* **1991**, *30*, 477−515.
[71] Cainelli, G.; Galletti, P.; Giacomini, D. *Chem. Soc. Rev.* **2009**, *38*, 990−1001.
[72] Reboule, I.; Gil, R.; Collin, J. *Tetrahedron: Asymmetry* **2005**, *16*, 3881−3886.
[73] Reboule, I.; Gil, R.; Collin, J. *Eur. J. Org. Chem.* **2008**, 532−539.
[74] Cainelli, G.; Giacomini, d.; Galletti, P. *Chem. Commun.* **1999**, 567−572.
[75] Cainelli, G.; Giacomini, D.; Galletti, P.; Marini, A. *Angew. Chem., Int. Ed.* **1996**, *35*, 2849−2852.
[76] Masamune, S.; Choy, W.; Petersen, J. S.; Sita, L. R. *Angew. Chem., Int. Ed. Engl.* **1985**, *24*, 1−76.
[77] Sharpless, K. B. *Chem. Scr.* **1985**, *25*, 71−77.
[78] Kogure, T.; Eliel, E. L. *J. Org. Chem.* **1984**, *49*, 576−578.
[79] Williams, D. R.; Kissel, W. S.; Li, J. J.; Mullins, R. J. *Tetrahedron Lett.* **2002**, *43*, 3723−3727.
[80] Masamune, S.; Ali, S. A.; Snitman, D. L.; Garvey, D. S. *Angew. Chem., Int. Ed. Engl.* **1980**, *19*, 557−558.
[81] Buse, C. T.; Heathcock, C. H. *J. Am. Chem. Soc.* **1977**, *99*, 8109−8110.
[82] Seebach, D.; Beck, A. K.; Schmidt, B.; Wang, Y. M. *Tetrahedron* **1994**, *50*, 4363−4384.
[83] Katsuki, T.; Lee, A. W. M.; Ma, P.; Martin, V. S.; Masamune, S.; Sharpless, K. B.; Tuddenham, D.; Walker, F. J. *J. Org. Chem.* **1982**, *47*, 1373−1378.
[84] Kagan, H. B.; Fiaud, J. C. *Top. Stereochem.* **1988**, *18*, 249−330.
[85] Vedejs, E.; Chen, X. *J. Am. Chem. Soc.* **1997**, *119*, 2584−2585.
[86] Dehli, J. R.; Gotor, V. *Chem. Soc. Rev.* **2002**, *31*, 365−370.
[87] Eames, J. *Angew. Chem., Int. Ed.* **2000**, *39*, 885−888.
[88] Zhang, Q.; Curran, D. P. *Chem. Eur. J.* **2005**, *11*, 4866−4880.
[89] Duffey, T. A.; MacKay, J. A.; Vedejs, E. *J. Org. Chem.* **2010**, *75*, 4674−4685.
[90] Horeau, A. *Tetrahedron* **1975**, *31*, 1307−1309.
[91] Goodman, J. M.; Köhler, A.-K.; Alderton, S. C. M. *Tetrahedron Lett.* **1999**, *40*, 8715−8718.

[92] Bellemin-Laponnaz, S.; Tweddell, J.; Ruble, J. C.; Breitling, F. M.; Fu, G. C. *Chem. Commun.* **2000**, 1009−1010.

[93] Martin, V. S.; Woodard, S. S.; Katsuki, T.; Yamada, Y.; Ikeda, M.; Sharpless, K. B. *J. Am. Chem. Soc.* **1981**, *103*, 6237−6240.

[94] Schreiber, S. L.; Schreiber, T. S.; Smith, D. B. *J. Am. Chem. Soc.* **1987**, *109*, 1525−1529.

[95] Curtin, D. Y. *Rec. Chem. Progr.* **1954**, *15*, 111−128.

[96] Seeman, J. I. *Chem. Rev.* **1983**, *83*, 83−134.

[97] Seeman, J. I. *J. Chem. Educ.* **1986**, *63*, 42−48.

[98] Hoye, T. R.; Ryba, T. D. *J. Am. Chem. Soc.* **2005**, *127*, 8256−8257.

[99] Seeman, J. I.; Secor, H. V.; Hartung, H.; Galzerano, R. *J. Am. Chem. Soc.* **1980**, *102*, 7741−7747.

[100] Kinbara, K.; Saigo, K. *Top. Stereochem.* **2003**, *23*, 207−265.

[101] Arai, K.; Obara, Y.; Takahashi, Y.; Takakuwa, Y. In *Jpn. Kokai Tokkyo Koho*; Nissan Chemical Industries, Ltd., Japan: Japan, 1986. Vol. JP 61238734, p. 5.

[102] Lopez, F. J.; Ferrino, S. A.; Reyes, M. S.; Roman, R. *Tetrahedron: Asymmetry* **1997**, *8*, 2497−2500.

[103] Ward, R. S. *Tetrahedron: Asymmetry* **1995**, *6*, 1475−1490.

[104] Noyori, R.; Tokunaga, M.; Kitamura, M. *Bull. Chem. Soc. Jpn.* **1995**, *68*, 36−56.

[105] Caddick, S.; Jenkins, K. *Chem. Soc. Rev.* **1996**, *25*, 447−457.

[106] Beak, P.; Basu, A.; Gallagher, D. J.; Park, Y. S.; Thayumanavan, S. *Acc. Chem. Res.* **1996**, *29*, 552−560.

[107] Beak, P.; Anderson, D. R.; Curtis, M. D.; Laumer, J. M.; Pippel, D. J.; Weisenburger, G. A. *Acc. Chem. Res.* **2000**, *33*, 715−727.

[108] Kitamura, M.; Tokunaga, M.; Noyori, R. *J. Am. Chem. Soc.* **1993**, *115*, 144−152.

[109] Thayumanavan, S.; Lee, S.; Liu, C.; Beak, P. *J. Am. Chem. Soc.* **1994**, *116*, 9755−9756.

[110] Thayumanavan, S.; Basu, A.; Beak, P. *J. Am. Chem. Soc.* **1997**, *119*, 8209−8216.

[111] Basu, A.; Beak, P. *J. Am. Chem. Soc.* **1996**, *118*, 1575−1576.

[112] Basu, A.; Gallagher, D. J.; Beak, P. *J. Org. Chem.* **1996**, *61*, 5718−5719.

[113] Noyori, R. In *Asymmetric Catalysis in Organic Synthesis*; Wiley-Interscience: New York, 1994. pp. 124−131.

[114] Walsh, P. J.; Kozlowski, M. C. *Fundamentals of Asymmetric Catalysis*; University Science: Sausalito, 2008.

[115] Espenson, J. H. *Chemical Kinetics and Reaction Mechanisms*; 2nd ed.; McGraw-Hill: New York, 2002.

[116] Fenwick, D. R.; Kagan, H. B. *Top. Stereochem.* **1999**, *22*, 257−296.

[117] Blackmond, D. G. *Acc. Chem. Res.* **2000**, *33*, 402−411.

[118] Kagan, H. B. *Synlett* **2001**, 888−899.

[119] Kagan, H. B. *Adv. Synth. Catal.* **2001**, *343*, 227−233.

[120] Satyanarayana, T.; Abraham, S.; Kagan, H. B. *Angew. Chem., Int. Ed.* **2009**, *48*, 456−494.

[121] Klussmann, M.; Mathew, S. P.; Iwamura, H.; Wells, D. H., Jr.; Armstrong, A.; Blackmond, D. G. *Angew. Chem., Int. Ed.* **2006**, *45*, 7989−7992.

[122] Komatsu, N.; Hashizume, M.; Sugita, T.; Uemura, S. *J. Org. Chem.* **1993**, *58*, 4529−4533.

[123] Guillaneux, D.; Zhao, S.-H.; Samuel, O.; Rainford, D.; Kagan, H. B. *J. Am. Chem. Soc.* **1994**, *116*, 9430−9439.

[124] Kitamura, M.; Okada, S.; Suga, S.; Noyori, R. *J. Am. Chem. Soc.* **1989**, *111*, 4028−4036.

[125] March, J. In *Advanced Organic Chemistry*, 4th Ed.; Wiley-Interscience: New York, 1992. p. 145.

[126] Eliel, E. L.; Wilen, S. H.; Mander, L. N. In *Stereochemistry of Organic Compounds*; Wiley-Interscience: New York, 1994. pp. 696-697.

[127] Bredig, G. *Angew. Chem.,* **1923**, *36*, 456−458.

[128] Johnson, F. *Chem. Rev.* **1968**, *68*, 375−413.

[129] Hoffmann, R. W. *Chem. Rev.* **1989**, *89*, 1841−1860.

[130] Pigman, W.; Horton, D. In *The Carbohydrates. Chemistry and Biochemistry*, 2nd Ed.; Academic: New York, 1972. pp. 1−67.

[131] Fieser, L. F.; Fieser, M. *Natural Products Related to Phenanthrene*, 3rd Ed.; Reinhold: New York, 1949.

[132] Barton, D. H. R. *Experientia* **1950**, *6*, 316−320.

[133] *The Anomeric Effect and Associated Stereoelectronic Effects. ACS Symposium Series 539*; Thatcher, G. R. J., Ed.; American Chemical Society: Washington, 1993.

[134] Kirby, A. J. *The Anomeric Effect and Related Stereoelectronic Effects at Oxygen*; Springer: Berlin, 1983.

[135] Muller, P., Ed. Pure Appl. Chem. **1994**, *66*, 1077−1184.

[136] Gold, V. *Pure Appl. Chem.* **1983**, *55*, 1281−1371.

[137] Eliel, E. L.; Wilen, S. H. *Chem. Eng. News* **1991**, *July 22*, p. 2.

[138] Harris, M. M. In *Progress in Stereochemistry*; Klyne, W., de la Mare, P. B. D., Eds.; Butterworths: London, 1958, pp. 157−195.

[139] Kuhn, R. In *Stereochemie. Eine Zusammenfassung Der Ergebnisse, Grundlagen Und Probleme*; Freudenberg, K., Ed.; Franz Deuticke: Leipzig, 1933, pp. 801−824.

[140] Bürgi, H. B.; Dunitz, J. D.; Schefter, E. *J. Am. Chem. Soc.* **1973**, *95*, 5065−5067.

[141] Anh, N. T.; Eisenstein, O. *Nouv. J. Chimie* **1977**, *1*, 61−70.

[142] Bürgi, H. B.; Dunitz, D.; Lehn, J. M.; Wipff, G. *Tetrahedron* **1974**, *30*, 1563−1572.

[143] Pickett, H. M.; Strauss, H. L. *J. Am. Chem. Soc.* **1970**, *92*, 7281−7290.

[144] Kelvin, L. In *Baltimore Lectures*; C. J. Clay and Sons: London, 1904. p. 436 and 619.

[145] Prelog, V. *Science* **1976**, *193*, 17−24.

[146] Mislow, K.; Siegel, J. *J. Am. Chem. Soc.* **1984**, *106*, 3319−3328.

[147] Dagani, R. *Chem. Eng. News* **1984**, *June 11*, 21−23.

[148] Cahn, R. S.; Ingold, C. K.; Prelog, V. *Angew. Chem., Int. Ed. Engl.* **1966**, *5*, 385−415, 511.

[149] Prelog, V.; Helmchen, G. *Angew. Chem., Int. Ed. Engl.* **1982**, *21*, 567−583.

[150] Cross, L. C.; Klyne, W., collators *Pure Appl. Chem.* **1976**, *45*, 11−30.

[151] Cornforth, J. W.; Cornforth, R. H.; Matthew, K. K. *J. Chem. Soc.* **1959**, 112−127.

[152] Paddon-Row, M. N.; Rondan, N. G.; Houk, K. N. *J. Am. Chem. Soc.* **1982**, *104*, 7162−7166.

[153] Evans, D. A.; Siska, S. J.; Cee, V. J. *Angew. Chem., Int. Ed.* **2003**, *42*, 1761−1765.

[154] Chérest, M.; Felkin, H.; Prudent, N. *Tetrahedron Lett.* **1968**, 2199−2204.

[155] Anh, N. T. *Top. Curr. Chem.* **1980**, *88*, 145−162.

[156] Seebach, D.; Naef, R. *Helv. Chim. Acta* **1981**, *64*, 2704−2708.

[157] Prelog, V.; Helmchen, G. *Helv. Chim. Acta* **1972**, *55*, 2581−2598.

[158] Wintner, C. E. *J. Chem. Educ.* **1983**, *60*, 550−553.

[159] Masamune, S.; Kaiho, T.; Garvey, D. S. *J. Am. Chem. Soc.* **1982**, *104*, 5521−5523.

[160] Kagan, H. B. *Recl. Trav. Chim. Pays-Bas* **1995**, *114*, 203−205.

[161] Schurig, V. *Enantiomer* **1996**, *1*, 139−143.

[162] Eliel, E. L.; Wilen, S. H. *Chem. Eng. News* **1990**, *Sep 19*, p. 2.

[163] Heathcock, C. H.; Buse, C. T.; Kleschick, W. A.; Pirrung, M. C.; Sohn, J. E.; Lampe, J. *J. Org. Chem.* **1980**, *45*, 1066−1081.

[164] Noyori, R.; Nishida, I.; Sakata, J. *J. Am. Chem. Soc.* **1981**, *103*, 2106−2108.

[165] Heathcock, C. H. In *Asymmetric Synthesis*; Morrison, J. D., Ed.; Academic: Orlando, 1984; Vol. 3, pp. 111−212.

[166] Rosanoff, M. A. *J. Am. Chem. Soc.* **1906**, *28*, 114−121.

[167] Ruch, E. *Theor. Chim. Acta* **1968**, *11*, 183−192.

[168] Ruch, E. *Acc. Chem. Res.* **1972**, *5*, 49−56.

[169] Eliel, E. L.; Wilen, S. H.; Mander, L. N. *Stereochemistry of Organic Compounds*; Wiley-Interscience: New York, 1994.

[170] Seebach, D.; Prelog, V. *Angew. Chem., Int. Ed. Engl.* **1982**, *21*, 654−660.

[171] Klyne, W.; Prelog, V. *Experientia* **1960**, *16*, 521−523.

[172] Carey, F. A.; Kuehne, M. E. *J. Org. Chem.* **1982**, *47*, 3811−3815.

[173] Hanson, K. R. *J. Am. Chem. Soc.* **1966**, *88*, 2731−2742.

[174] Schlögl, K. *Top. Stereochem.* **1967**, *1*, 39−91.

[175] Sloan, T. E. *Top. Stereochem.* **1981**, *12*, 1−36.

[176] Hanson, K. R. *Ann. Rev. Biochem.* **1976**, *45*, 307−330.

[177] Heathcock, C. H.; Finkelstein, B. L.; Jarvi, E. T.; Radel, P. A.; Hadley, C. R. *J. Org. Chem.* **1988**, *53*, 1922−1942.

[178] Heathcock, C. H. *Chem. Eng. News* **1991**, *Feb 4*, p. 3.

[179] Gawley, R. E. *Tetrahedron Lett.* **1999**, *40*, 4297−4300.

[180] Deslongchamps, P. *Stereoelectronic Effects in Organic Chemistry*; Pergamon: Oxford, 1983.

[181] Beesley, R. M.; Ingold, C. K.; Thorpe, J. F. *J. Chem. Soc.* **1915**, *107*, 1080−1106.

[182] Ingold, C. K. *J. Chem. Soc.* **1921**, *119*, 305−329.

[183] Jung, M. E.; Piizi, G. *Chem. Rev.* **2005**, *105*, 1735–1766.

[184] Maitland, G. C.; Rigby, M.; Smith, E. B.; Wakeham, W. A. *Intermolecular Forces. Their Origin and Determination*; Clarendon: Oxford, 1981.

[185] Kihara, T. *Intermolecular Forces*; Wiley: New York, 1976.

[186] Margenau, H.; Kestner, N. R. *Theory of Intermolecular Forces*; Pergamon: Oxford, 1969.

[187] Burkert, U.; Allinger, N. L. *Molecular Mechanics. ACS Monograph 177*; American Chemical Society: Washington, DC, 1982.

Chapter 2

Practical Aspects of Asymmetric Synthesis

2.1 CHOOSING A METHOD FOR ASYMMETRIC SYNTHESIS

What is the best way to carry out an asymmetric synthesis of a specific molecule?

The facile answer is simple: The chemist should select an appropriate set of readily available, cheap, nontoxic, and environmentally responsible building blocks, and mix them together with a vanishingly small amount of a chiral catalyst at room temperature (preferably without solvent, but if one *must* use a solvent, it should be readily removable and recycled). Following just enough time for a coffee or tea break (this part is very important), the chemist will return to her or his hood to find that the product has spontaneously crystallized from the reaction mixture to provide the desired product in 100% purity and complete stereochemical integrity.

Of course, this is an idealized situation and most of us live in a decidedly less utopian world.[1] Here, the decisions that one makes depend on numerous factors: the reason for wanting a molecule in enantiomerically pure form (which will usually dictate how pure it needs to be), how much one needs, and how much effort one is willing to go through to get it. In the spirit of offering the inexperienced chemist some help, we are willing to provide some general guidelines with the understanding that there is no single "right" answer to any given scenario.

The decision of how to procure a molecule in enantiomerically pure form requires that the experimentalist balance the pros and cons of the available methods (Table 2.1). We will only consider the most important and widely used methods for this discussion, beginning with purchase and moving in roughly the chronological order that the techniques were adopted by organic synthesis labs. Some important methods such as stereoconvergent transformations or resolution *via* the formation of inclusion complexes are omitted because they are highly situation-dependent and are not as commonly used as those that are in Table 2.1.

A zero-order question that is rarely asked but is worth consideration is "When is it acceptable to work with racemic material?" A purist may well say "never" but most pragmatists would disagree. Racemic materials, even complex ones, often have the advantages of being simpler and less expensive to make relative to enantiopure versions. Even when pursuing a sophisticated total synthesis of a natural product, it is common for chemists to rehearse

---

1. However, the power of visualizing a more perfect world should not be underestimated, and numerous scientists have thoughtfully considered what constitutes more ideal organic syntheses [1−5].

Principles of Asymmetric Synthesis.
© 2012, Professor Robert E. Gawley and Professor Jeffrey Aubé. Elsevier Ltd. All rights reserved. **63**

TABLE 2.1 Advantages and Disadvantages of the Most Important Methods of Obtaining Enantiomerically Pure Molecules

Method	Pros	Cons
Purchase	• *Convenience and time.* If the molecule is available at a reasonable price, it is rarely cost effective to make it oneself.	• *Availability.* Most people reading this book will be interested in rare molecules or those that do not yet exist. • *Cost.* In many cases, obtaining sufficient quantities of a chiral starting material in enantiomerically pure form is beyond a laboratory's means. This most commonly arises when one needs the less-common enantiomer of a naturally occurring compound (*e.g.*, D-amino acids). It is also common that commercially available starting materials are only affordable in small quantities, so that one might buy a sample or three for proof-of-concept work but then commit to making larger quantities if needed for continuing studies (this more commonly occurs in academic research labs than in industry, where cost of goods is less often the major limiting factor). • *Caveat emptor.* Sometimes a sample advertised by the supplier as enantiomerically pure *isn't.* If it's critical, verification of er might be wise.
Classical resolution	• *Practicality and scalability.* Classical resolutions via the formation of diastereomeric salts can be carried out on scales ranging from milligrams to kilograms. • *Reproducibility.* Once conditions for a resolution are worked out, it can more often than not be carried out again and again, with the important proviso that it may be advisable to retain an authentic sample of a seed crystal (as in all crystallizations). In fact, it is rare to see "one-off" resolutions in practical organic synthesis labs.	• *Impracticality and inconvenience.* Not every compound forms separable salts and finding the right conditions for a successful resolution can be tedious. Resolutions can be especially hard to accomplish on some smaller scales. • *Waste.* Many chemists find the idea of discarding the unwanted enantiomer to be abhorrent, meaning that there typically has to be a significant increase in value of the single enantiomer relative to the racemate to make a resolution worthwhile. Of course, this objection vanishes if both enantiomers can be put to good use. • *Inelegance.* Many chemists find the idea of discarding the unwanted enantiomer to be abhorrent, meaning that resolutions rarely appear in total syntheses. In the world of natural product synthesis, this objection only vanishes if the overall route is so elegant that no one notices that a resolution has occurred.

(Continued)

TABLE 2.1 (Continued)

Method	Pros	Cons
Chromatographic separation	• *Effective.* It is often practical to carry out efficient chromatographic separations on a small scale. • *Ability to provide both enantiomers* is especially useful when studying the effect of chirality on a molecule's biological activity.	• *Tedium.* Few organic chemists are enamored with doing HPLC separations and while professional, high-quality separations labs are common in industrial settings, they are rare in academia. The need to validate the method, generally by having a racemic standard in hand, is also unattractive in many cases. • *Scalability.* Large scale chromatographic separations can be done but they are expensive and not available to many involved in asymmetric synthesis.
Kinetic resolution	• *Practicality, reproducibility, and (sometimes) scalability.* Kinetic resolutions, mediated often by enzymes (although increasingly by other catalysts), can be robust synthetic methods suitable for large-scale preparations (but see right).	• *(Sometimes) scalability.* A practical kinetic resolution requires the separation of the product from the unreacted starting material. When this cannot be accomplished by selective crystallization, it may be a scale-limiting consideration. • *Impracticality, inconvenience, waste, and inelegance.* See "Classical Resolutions" above.
Chiral auxiliaries	• *Generality.* For many years, asymmetric synthesis mediated by chiral auxiliaries was the only game in town. For this reason, there are many, many examples of highly selective reactions that are mediated by chiral auxiliaries. • *Separability.* Because the initial products of a stereoselective reaction of a substrate carrying a chiral auxiliary are diastereomers, they can often be separated by crystallization or ordinary column chromatography prior to removal of the auxiliary. This affords a product whose enantiomeric purity is limited only by the resolution of the purification method and the enantiomeric purity of the chiral auxiliary.	• *Inefficiency.* Usually, using a chiral auxiliary adds two additional steps to a synthetic sequence. This always adds time to the process and lowers the yield. • *Cost.* Chiral auxiliaries are often expensive and they need to be used in stoichiometric quantities, not to mention the collateral costs of the attachment and removal steps.

(Continued)

TABLE 2.1 (Continued)

Method	Pros	Cons
Asymmetric catalysis	• *Elegance and efficiency.* A truly effective asymmetric catalyst is hard to beat...	• *Availability.* ...but it can be hard to find. Techniques that have found very wide applicability, such as the Sharpless asymmetric hydroxylation (AD) reaction [6], have succeeded due to broad substrate scope and the commercial availability of the catalysts in either enantiomeric form. If time is of the essence, the need to screen various catalysts for a less well-developed reaction can be a deterrent. • *Enhancement of enantiomeric purity.* The fact that the two products of a catalytic asymmetric process are enantiomers limits one's ability to improve the er of a given product that may not have been formed with sufficiently high selectivity. To use the example above, a secondary reason for the popularity of the Sharpless AD is that the products can often be brought to enantiomeric purity by simple crystallization. The likelihood of being able to do this is greater when the asymmetric step is part of a multistep sequence, since downstream products may well have more favorable crystallization properties.

their chemistry on racemic materials before tackling enantiomerically pure versions.[2,3] However, this is only possible in linear routes wherein each stereocenter is established relative to some existing stereocenter (Scheme 2.1a). For convergent total syntheses that require the joining together of chiral subunits, one is committed to asymmetric syntheses of the building blocks from the very beginning lest one deals with the unattractive possibility of separating diastereomeric products (Scheme 2.1b; mutual kinetic resolutions that avoid this problem are known but rare). In this discussion, we will presume that the decision has been made to prepare enantiomerically enriched products, whatever the stage of the project.

---

2. One has to be careful, though. Heathcock has pointed out that a racemic synthetic intermediate may have different solubility properties relative to a single enantiomer and that "formal syntheses" based on such intermediates may not be valid [7]. This brings to mind R.B. Woodward's statement that examination of "the 'unnatural enantiomer'... is just about the only kind of model study which we regard as wholly reliable!" [8].

3. There is also the rare chiral natural product that is produced in nature as a racemic mixture, such as the endiandric acids [9]. Here, it could be argued that making the natural product in racemic form is simply following the natural example to its logical conclusion!

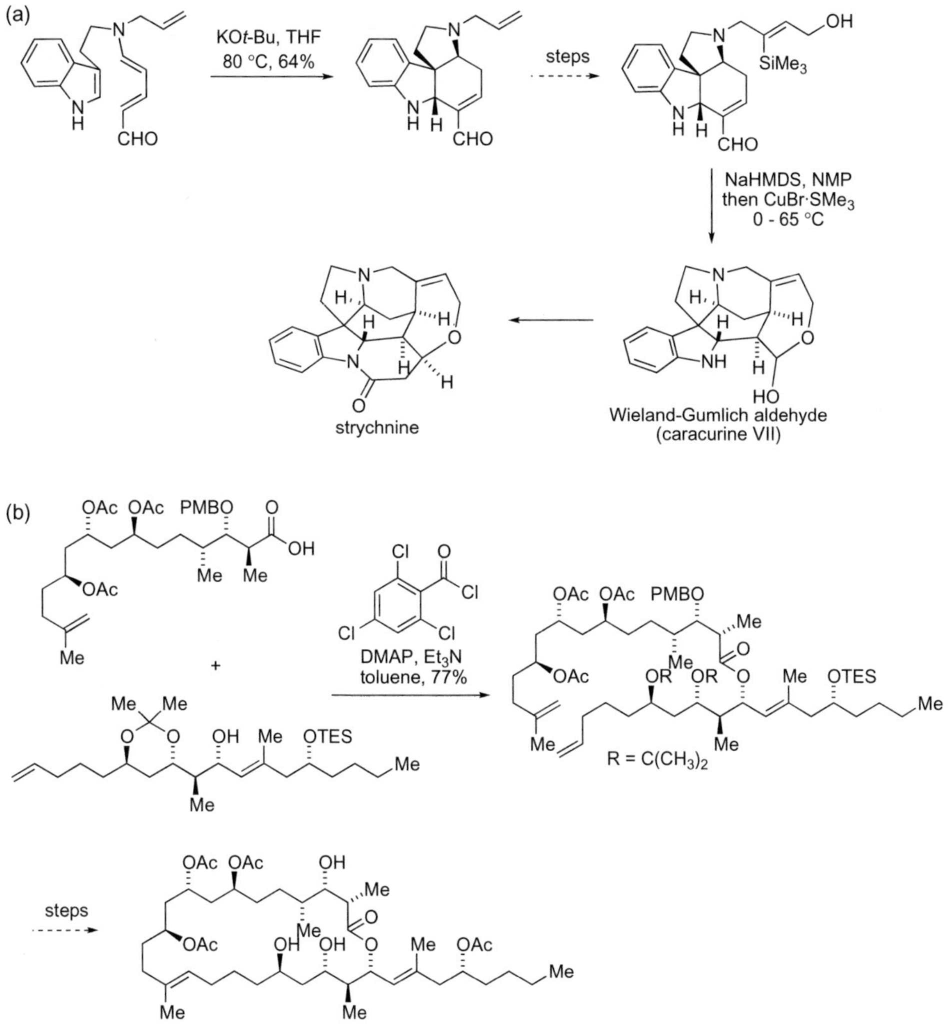

SCHEME 2.1 (a) An example of a linear synthetic scheme of a complex racemic natural product in which all of the stereocenter-generating steps are along a single pathway [10]. (b) Here, two enantiomerically pure building blocks are combined in a convergent synthesis of a macrocyclic lactone [11].

How one proceeds from there will depend on one's goals and limitations. Nearly every road to an asymmetric synthesis starts by examining online databases and the literature for analogous processes. If the precise compound can be found, great, but if not, then one's ability to critically judge a given process and discern how likely it would work on a related system is key.

Situation: One needs a new compound, relatively quickly, for a particular study (laboratory scale). In these situations, the elegance of the chemical route often takes a back seat to concerns about needing the compound sooner rather than later. This is almost a given in medicinal chemistry discovery laboratories, where having a couple of milligrams of high-quality compound in hand is the primary goal.[4] When this is the case, choosing a method that is consistent with convenient separation of the final product can be very attractive, even if the enantiomeric ratio of the initially prepared product is not ideal. In many instances, the efficacy of a chiral auxiliary method combined with the ability to purify the diastereomeric adducts prior to cleavage of the chiral auxiliary proves convenient on scales from a few milligrams to a few grams. In biological chemistry, it is very common for compounds to be initially tested in racemic form and only later separated when found to be biologically active (for an example, see ref. [12]).

Situation: One needs a reliable, long-term source of an enantiomerically pure product. As projects mature, the value of a particular synthetic intermediate to the chemist generally increases. When this happens, an additional investment in time may well be needed to obtain larger quantities — sometimes much larger quantities, as in the specific case of drug candidates. Moreover, the additional cost of materials incurred means that any issues with step count and overall yield become magnified. Accordingly, considerable effort may well be appropriate to provide a process chemistry route to a compound that is as scalable and inexpensive as possible. In industrial settings, reproducibility also becomes paramount in two ways. First, regulatory agencies often require manufacturers to document the amounts and identities of impurities present in a material and to ensure that they are consistent from batch to batch. Moreover, the sheer cost of making materials on large scale (in terms of both materials and time) makes reaction failures unacceptable.

In these settings, chiral auxiliary routes tend to lose favor when compared to preparations that rely on the chiral pool or asymmetric catalysis. For the latter, it also becomes much easier to justify spending the time and effort to select just the right combination of chiral ligand and reaction conditions, even if only to optimize a single specific step. Moreover, a highly effective catalysis route generates chiral material in amounts that are only subject to the limitations of chemical engineering technology. It does not typically matter whether the process uses an organometallic reagent or is biocatalytic as both techniques can be deployed on impressive scales. The historic Monsanto synthesis of L-DOPA *via* asymmetric hydrogenation set the bar early and high in this area [13,14] and multitudinous examples have followed.

It is interesting that the value enhancement of going from racemic to enantiomerically pure materials on large scale is such that workhorse techniques such as classical resolution procedures remain strong contenders. Although not trendy, they can be eminently practical. For example, the resolution shown in Scheme 2.2 was published in 2010 and produced 24 kg of the product [15].

---

4. It's practically a *cliché* that process chemistry divisions in drug discovery companies are often called upon to modify initial routes to important molecules so that the routes can be scaled up.

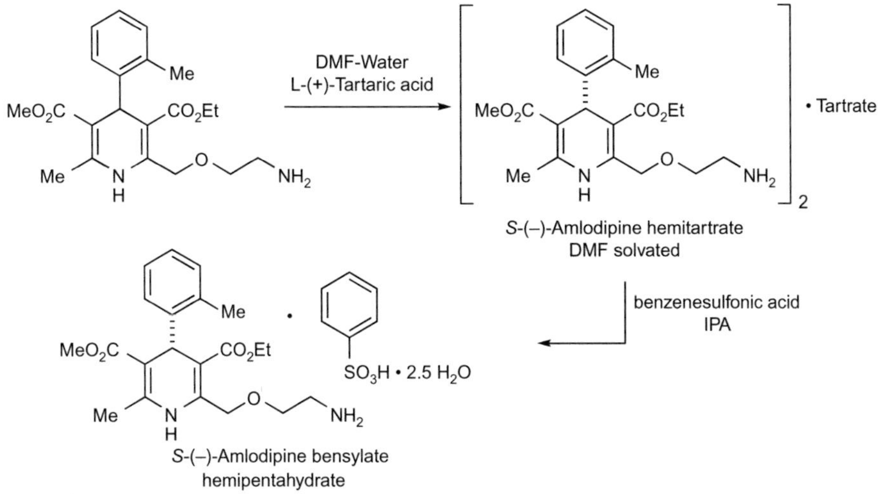

SCHEME 2.2 An example of a classical resolution procedure carried out on large scale [15].

Situation: Natural product total synthesis. The choice of a method for preparing a chiral natural product or a fragment thereof in enantiopure form depends on the stage of the project and the outlook of the scientist. As of this writing, the style points that one accrues from using a sophisticated catalytic method trump those from every other approach. The main practical disadvantages of asymmetric catalysts are the need to carry out separations of enantiomers in the case of imperfect results.[5] Since a total synthesis format often provides numerous opportunities for recrystallization or attachment of two chiral fragments followed by chromatographic purification (which can lead to enantiomeric enrichment of the product), concerns about modest enantioselectivities often evaporate. That being said, it should be recognized that essentially every technique ever invented for procuring a molecule in enantiomerically pure form has appeared in some total synthesis, somewhere.

2.2 HOW TO GET STARTED

If one wishes to simply obtain an enantiomerically pure compound as a research tool or synthetic intermediate, the obvious place to start is with the literature. We are fortunate to live in a time where substructure searching has become automated and routine, enabling the chemist to find routes to the precise desired compound or related ones, often in a matter of minutes. If innovation is not important, then a synthetic route that can be readily adapted from another laboratory is often available.

However, it often becomes necessary to strike out on one's own and develop a new method. A partial list of reasons for this is:

- The starting material is not readily available or is too expensive.
- The compound one wishes to make falls *just* outside the scope of the literature method.

5. For an excellent monograph describing such techniques, see ref. [16].

- The technique in the literature is fine on a small scale, but upon scale-up, the reagent is too expensive or the chemistry does not cooperate.
- The literature route is lengthy and a more streamlined synthesis is desired.
- The catalyst/reagent is great but takes more steps to prepare than one is willing to invest.
- The catalyst/reagent is great but gives the "wrong" enantiomer and the enantiomeric catalyst/reagent is not available (the alkaloid (−)-sparteine is the poster child for this problem [17,18]).

After reviewing the options, one may wish to develop a new chiral auxiliary or resolution scheme to simply get the compound made and/or separated, and then move on to the more interesting part of the project. Or, perhaps, the chemist is more inclined to develop a new reaction or wishes to devote some time to methodology development as an end in itself. In either case, an essential and commonly underappreciated step is to *identify an appropriate analytical method to determine product purity **and validate it using a racemic standard.*** The importance of analytical chemistry to a successful asymmetric synthesis is such that it will constitute the bulk of this chapter.

In designing a new method, it is important to keep in mind what problem one wishes to solve. The old-fashioned virtues of reliable chemical reactions that take place conveniently and in high yields still matter, and without these attributes a new method, no matter what the selectivity, is likely to languish in the literature without being used. Much important work in asymmetric synthesis is directed at developing an asymmetric version of a reaction known only in the racemic world, but new examples that also feature novel kinds of reactivity or design features are especially welcome.

Despite their continued utility, resolution methods or asymmetric reactions that use chiral auxiliaries are no longer at the vanguard of science, and so the bar for developing such methods is particularly high. In general, successful additions to these areas feature some level of practicality that combines a badly needed process with some pragmatic attribute such as easy purification of products *via* crystallization. The Davis and Ellman sulfoximine auxiliaries (Scheme 4.15) for the synthesis of chiral amines are great examples of chiral auxiliaries that enjoy very wide acceptance in the organic synthesis community [19−21].

There is no question that the development of catalytic reactions of all types is of immense current interest. However, it can be challenging work. While the duty of a chiral auxiliary is usually to simply differentiate between diastereofacial attack modes, the catalyst chemist has to also worry about raising the catalytic reaction rates above the nonspecific background. While both sets of scientists typically employ trial and error in the development of their methods,[6] the complexity of many catalytic systems is such that the chemist often screens numerous candidate catalysts before settling on one with useful levels of selectivity and scope. This is why many catalyst systems of real impact involve some element of modularity that allows for easy preparation and examination of the various elements of the system (Jacobsen's salen oxidants provide a nice example [24]).

The old saying that "the proof of the pudding is in the eating" applies to the final stage of catalyst development: adoption and use of asymmetric synthesis methodologies in the nonspecialist laboratory (the chemist's version of natural selection). Clearly, those developments

---

6. We note with some curiosity that validation of potential reagents, auxiliaries, or catalysts using computer modeling has not had more of an impact in the field, although exceptions exist [22,23].

that meet this challenge most readily are those that carry out everyday, essential reactions and that have very wide scope: oxidations, reductions, cycloadditions, and condensations (aldol, etc.) come to mind. Thus, the industrial synthesis of L-DOPA is a justly celebrated landmark event in the coming-of-age of asymmetric synthesis. Another setting for this work is in complex total synthesis. In that context, there are numerous examples of total syntheses that use one or two asymmetric reactions highlighted against a background of more prosaic bond-forming or functional-group modifying reactions or, alternatively, a more substrate-directed synthesis approach (which has its own artistry, of course). There are relatively few examples of multistep syntheses that involve mainly catalytic means of generating stereogenic centers or, for that matter, making all of the C—C bonds in a molecule (for a nice example of the former, see Jacobsen's synthesis of muconin [25]). We do not doubt that there will be more, in time.

To dispel any doubts of the forward progress that has already been achieved in asymmetric synthesis, one need look no further than the specific example of the target erythronolide. In 1981, the Woodward group published three back-to-back communications describing the total synthesis of erythromycin using entirely substrate-based stereocontrol tactics, with the combined authorship of almost 50 chemists [26—28]. Fast forward to 1991, when Mulzer published the synthesis of an auxiliary-based route to erythronolide B, based on only 3 years of work reported in the dissertation of a single student [29]. The latter is a significantly less complex target, true, but the advances in technology over the intervening decade is truly impressive and suggests promisingly of still more exciting developments to come.

2.3 GENERAL CONSIDERATIONS FOR ANALYSIS OF STEREOISOMERS

As explained in Chapter 1 (Section 1.4), analysis of the ratio of stereoisomeric products provides important information about the dynamics of a stereoselective reaction. Before discussing the methods for analysis of stereoisomeric products, it is important to realize the importance of accuracy in determining product ratios, whether diastereomers or enantiomers. If one wants to translate isomer ratios to free energy differences between competing transition states (Equations 1.1 or 1.6), the analysis should reflect the ratio in the crude product without unintended enrichment by chromatographic means during workup and preparation of the sample for analysis.

It is common practice to incorporate a preliminary purification of a crude product mixture into a workup protocol. Such purification is often done chromatographically using an achiral stationary phase such as silica gel. It is well known that such an operation can result in separation of product *diastereomers* as they elute. So, to get an accurate picture of the selectivity of the reaction, care must be taken to ensure that the diastereomer mixture being analyzed is the same as that produced in the reaction. *What is not widely appreciated is the fact that enantiomers in an enriched mixture (≠ 50:50 er) can sometimes be separated by chromatography on an achiral medium.*

In 1983, Cundy and Crooks reported that partial enantiomer separation occurred upon chromatographic purification of a partly racemic nicotine sample [30]. This phenomenon is known as *self-disproportionation of enantiomers* on achiral media. Since then, many examples have been studied, and enantiomer enrichments have been noted. Figure 2.1 reveals the broad structural variety of compounds known to undergo self-disproportionation of enantiomers on an achiral chromatographic column. When the er of these compounds was measured as a function of retention time during chromatography, it was not unusual to observe nearly enantiopure fractions, often eluting before less enantiopure fractions.

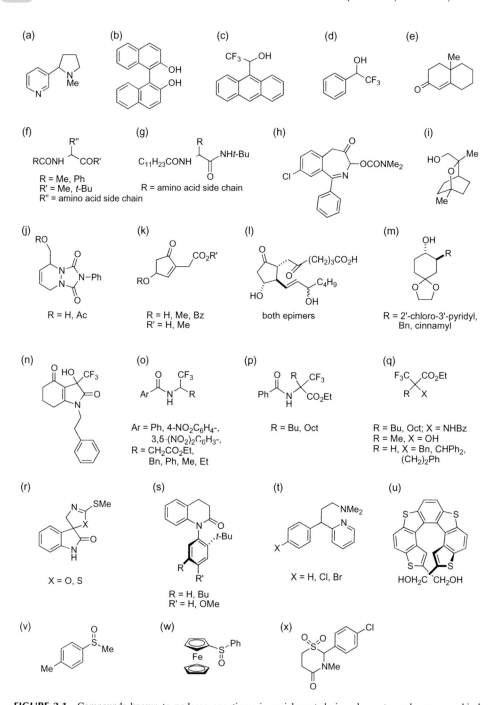

FIGURE 2.1 Compounds known to undergo enantiomeric enrichment during chromatography on an achiral stationary phase such as silica gel. References: (a) [30]; (b) [31–33]; (c,d) [31,34]; (e) [35,36]; (f,g) [31,37,38] (h) [31]; (i) [39]; (j) [40]; (k,l) [41]; (m) [42]; (n) [43]; (o-q) [34,44]; (r) [45,46]; (s) [47]; (t) [48]; (u) [49]; (v,w) [36]; (x) [31].

Self-disproportionation of enantiomers occurs when both homochiral and heterochiral dimers of partly enriched chiral compounds can form. It does not occur with racemic compounds (50:50 er); it only occurs when the sample is partly enriched. Schurig has developed mathematical models to explain the phenomenon in detail [50,51]. Briefly, what is happening during the elution is that there is a rapid and reversible interconversion of monomers to both homochiral and heterochiral dimers. When this dimerization occurs, the chromatographic capacity factors (see below) change during the elution, and the er of the eluting sample is not constant as a function of retention time.

Examination of the structures in Figure 2.1 reveals that there is no obvious common structural feature among these examples. Factors that can influence dimerization include intermolecular forces such as hydrogen bonding and dipole–dipole interactions, and each of the examples in Figure 2.1 possess hydrogen-bonding functional groups, or polar functional groups, or both. Self-disproportionation of enantiomers can only occur when the homochiral and heterochiral dimers differ in stability. *Thus, when preparing a sample for analysis of product ratio, chromatographic fractions should include those eluting before and after the "main band" so as to minimize any adventitious stereoisomer enrichment or depletion.*

Asymmetric synthesis requires the means for the analysis of both enantiomeric and diastereomeric mixtures. Ultimately, the *ratio* of isomers and the *configuration* of each new stereocenter should be determined. In choosing a technique for the analysis of a stereoselective reaction, a number of questions must be addressed:

- What limits of detection are desired?
- Are the stereoisomers enantiomers or diastereomers, and if the latter, how many possible diastereomers are there?
- Do the products have a chromophore that might aid analysis by chromatography?
- Do the products have a functional group available for derivatization, or for interaction with a stationary phase in chromatography or with a chiral agent in solution?
- Once the ratio of stereoisomers is determined, how will the configuration of each new stereocenter be assigned? Can the same method be used for the determination of product ratio *and* the assignment of configuration?

Two types of analysis exist: those that separate the stereoisomers and those that do not. NMR analysis of diastereomers or a mixture of enantiomers with the aid of a chiral solvating agent (CSAs) do not involve a separation. Chiral stationary phase (CSP) chromatographic techniques obviously *do* separate the analyte isomers, and may also facilitate the assignment of absolute configurations. The sections that follow describe the advantages and disadvantages of some of the more popular methods. The following methods of analysis are presented in order of their utility and popularity.[7]

2.4 CHROMATOGRAPHY

2.4.1 A Chromatography Primer

In order to appreciate the forces that are responsible for chromatographic separation, whether gas chromatography (GC), high performance liquid chromatography (HPLC), or supercritical

---

7. A survey of methods for determining enantiomer composition, which were reported in the journal *Tetrahedron: Asymmetry* during the 7-year period of 1995–2002, revealed that 67% of the reports used CSP chromatography, 21% NMR, 8% specific rotation, and 4% other methods [52].

fluid chromatography (SFC), we need to review some of the principles of chromatography.[8]
The elution of a sample through a chromatographic column is accomplished by partitioning
of the sample between a stationary phase and a mobile phase. In the extreme, a sample that
does not interact with the stationary phase is eluted in the amount of time it takes the mobile
phase to travel the column, t_0. Samples that *do* interact with the stationary phase will obviously take longer.

Consider the hypothetical chromatogram in Figure 2.2. The efficiency of a chromatographic column is a measure of peak broadening as the analyte moves through the column,
and is expressed as the number of theoretical plates, n, according to the equation

$$n = 16 \left[\frac{t_R}{w} \right]^2,$$ (2.1)

where t_R is the retention time and w is the peak width at the baseline. Either of these can be
expressed as either volume or time, since n is unitless. Often, this efficiency is expressed as
the height equivalent to one theoretical plate, or plate height, H, which is the length of the
column divided by n:

$$H = L/n.$$ (2.2)

A highly efficient column has a large number of theoretical plates, and is capable of separating a large number of analytes in a single run.

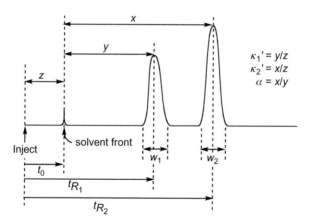

FIGURE 2.2 A hypothetical chromatogram, showing the retention time of an unretained compound (t_0), the
baseline peak widths, the retention times of two analytes, t_{R_1} and t_{R_2}, and the relationship of these quantities to
the capacity ratios, κ_1' and κ_2', and the chromatographic separability factor, α.

Retention of an analyte may be expressed as t_R (retention time), V_R (retention volume), or
κ' (capacity ratio). The latter is defined as

$$\kappa' = \frac{A_s}{A_m},$$ (2.3)

---

8. For comprehensive treatments, see refs. [53,54].

where A_s and A_m represent the amount of solute in the stationary and mobile phases, respectively. Thus, the capacity ratio is the equilibrium constant for the partitioning of the analyte between the mobile and stationary phases. The capacity ratio can also be expressed in terms of retention times:

$$\kappa' = \frac{t_R - t_0}{t_0}, \qquad (2.4)$$

where t_0 is the retention time of an unretained compound, usually visible as the solvent front (see Figure 2.2).

For two peaks to be "resolved" chromatographically, the capacity ratios, κ'_1 and κ'_2, must be different. For analytical purposes, two interdependent chromatographic properties must be considered: the chromatographic separability factor, α, and the resolution, R_S. The chromatographic separability factor is defined as

$$\alpha = \frac{\kappa'_2}{\kappa'_1}, \qquad (2.5)$$

where κ'_1 and κ'_2 are the capacity ratios of the first and second eluting peaks, respectively. Combination of Equations 2.4 and 2.5 gives

$$\alpha = \frac{t_{R_2} - t_0}{t_{R_1} - t_0}, \qquad (2.6)$$

where t_{R_1} and t_{R_2} are the retention times of the first and second eluting peaks, respectively. Using Equations 2.4 and 2.6, capacity ratios and separability factors can be easily obtained from a chromatogram, as shown in Figure 2.2.

Because the capacity ratios reflect the equilibrium between the analytes and the stationary phase, the separability factor, α, is directly related to the free energy difference between the analyte · stationary phase complexes, according to

$$\Delta\Delta G^\circ = -RT \ln \alpha. \qquad (2.7)$$

Rearrangement gives

$$\alpha = e^{-\Delta\Delta G^\circ / RT}. \qquad (2.8)$$

For the interaction of chiral stationary phases with enantiomers, the analyte · CSP complexes are diastereomeric; in order for separation to occur, they must not be isoenergetic. Separability factors of 1.1 are common in CSP chromatography, which translates to a free energy difference, $\Delta\Delta G^\circ$ (at 25 °C), of only 56 cal/mol! It is the *amplification* of this difference during the chromatographic process that accounts for the separation.

Resolution, R_S, of chromatographic peaks is the ratio of the peak separation to the average peak width:

$$R_S = \frac{2(t_{R_2} - t_{R_1})}{w_1 + w_2}, \qquad (2.9)$$

where w_1 and w_2 are the widths of the first and second peaks, respectively (see Figure 2.2). Thus, the resolution is dependent on both the chromatographic separability factor, α, and the column efficiency, expressed as the number of theoretical plates. As shown in Figure 2.3, the same resolution may give rise to two closely spaced narrow peaks or to two broader peaks that are more widely separated.

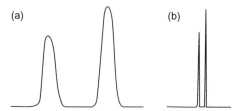

(a)   (b)

FIGURE 2.3   Hypothetical chromatograms of identical resolution. (a) Large α, small n. (b) Smaller α, larger n.

Comparison of Equations 2.1, 2.2, and 2.9 reveals that the resolution of a column is proportional to the square root of the number of theoretical plates ($\sqrt{n}$), and therefore to the square root of the length of the column ($\sqrt{L}$). Thus, doubling the length of a column increases the resolution by $\sqrt{2}$, or ~ 1.4.

Asymmetric reactions and processes give rise to two kinds of stereoisomeric products: diastereomers and enantiomers. The physical separation of these isomers with simultaneous analysis of isomer distribution (peak integration) is an excellent way to determine the selectivity of a reaction when baseline resolution can be achieved.[9] For the analysis of *diastereomers*, achiral stationary phases suffice. In both cases, the chromatographic method should be accompanied by another technique that specifies the configuration of the new centers. Diastereomer analysis also ensues in cases of double asymmetric induction, and the configuration of known centers in the reactants may be used as a point of reference for determination of the new stereocenter(s) by NMR, X-ray diffraction, comparison with known samples, etc.

2.4.2 Chiral Stationary Phase Chromatography

The historical origins of CSP chromatography are early in the twentieth century when it was observed that certain dyes were enantioselectively adsorbed onto biopolymers such as wool [55–57]. Although there were isolated instances of chromatographic resolutions earlier,[10] development of CSP chromatography as a useful tool did not take place until capillary GC and HPLC were popularized in the 1970s and 1980s. The separation of enantiomers on a CSP requires the formation of diastereomeric analyte·CSP "complexes" (for monographs, see refs. [60–71]). CSP chromatography has a number of appealing features:

- The sensitivity of GC or HPLC detectors are such that very small amounts of analyte, as little as a few micrograms under favorable circumstances, may be analyzed.
- If the order of stereoisomer elution for a given class of compounds is known (or can be determined), it may be possible to determine the enantiomer ratio and absolute configuration simultaneously.
- Since integration of chromatographic peaks is more accurate than NMR integration, chromatography is the method of choice when accuracy is important, and is especially applicable to the analysis of samples of high enantiomeric or diastereomeric purity.

---

9. This does entail the usually reasonable and almost always implicit assumption that the isomers have the same response to the detector being used, such as UV extinction coefficients.
10. For more detailed accounts of the early history of CSP chromatography, see refs. [58–63].

- No kinetic resolution arises, as can happen as a result of double asymmetric induction in a derivatization scheme, although care must still be taken to avoid enantiomer enrichment (or depletion) during workup (*cf.* Figure 2.1 and accompanying discussion).
- Scale-up may allow for preparative purification.

For new classes of compounds being studied on a chiral stationary phase, that a mixture of enantiomers are separable must be proven by analyzing a racemate, and the order of elution must be established by correlation with compounds of known configuration. In certain instances, derivatization may be necessary to improve chromatographic behavior and/or detectability.

When using CSP chromatography to separate enantiomers, racemization of either the analyte or the CSP may give rise to peak coalescence, but the two are easily distinguished by their appearance, as shown in Figure 2.4. Over time, racemization of the CSP may occur, and this will reduce the separability factor, α (Figure 2.4a). If racemization of a CSP is possible (such as with an amino acid-derived CSP), it is wise to periodically run a standard to check for peak coalescence. Another type of peak coalescence is due to racemization of the analyte on the column [72,73]. The appearance of such a phenomenon depends on the relative rates of racemization and separation [73]. The two extremes are fast and slow racemization, relative to the separation. Fast racemization of the analyte would yield a single sharp peak, and extremely slow racemization would go undetected. Intermediate cases might appear as a plateau between the peaks, or a hump, as shown in Figure 2.4b and c [73].

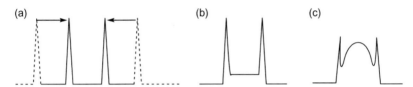

FIGURE 2.4 Hypothetical chromatograms showing peak coalescence due to racemization. (a) Racemization of the chiral stationary phase causes the peaks to move closer together (α decreases, ultimately to unity if the CSP is racemic). (b) and (c) Racemization of the analyte on the column may produce a plateau between the enantiomer peaks, as in (b), which may grow to a hump, as in (c), or even a single peak, depending on the relative rates of racemization and separation [73].

Several structural types of CSPs are available, the most popular and versatile being derivatized polysaccharides and the various categories of Pirkle columns, one example of which is described in detail below. Chiral stationary phases derived from macrocyclic glycopeptides, chiral crown ethers, molecularly imprinted polymers, immobilized proteins, and amino acids have also been developed. Only in a few cases has detailed work been done to rationalize the relative stabilities of the diastereomeric analyte·CSP complexes.

The first rationale for enantiomer discrimination by chiral molecules was the three-point model proposed by Easson and Stedman in 1933 to explain the interaction of racemic drugs with biological receptors [74]. An example of this model is shown in Figure 2.5, which depicts the binding of phenylalanine to a fixed triad of binding sites. A similar model was proposed by Ogston in 1948 to explain the enantioselectivity of enzyme reactions such as kinetic resolutions [75]. These simple models proposed three simultaneous favorable binding interactions to explain enantioselective enzyme reactions. In 1952, Dalgliesh [76] extended the three-point model [74,75] to CSP chromatography, to explain the separation of amino acid enantiomers by paper chromatography. More recently, Pirkle has argued that, although three points are required, all need not be attractive [77]. At least one, however, must be dependent on absolute configuration. Types of favorable intermolecular forces are hydrogen bonding, π-stacking, dipole stacking, etc.

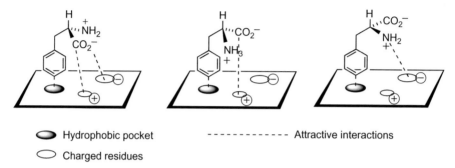

⬬ Hydrophobic pocket   – – – – – – – – – – Attractive interactions
⬭ Charged residues

FIGURE 2.5 The Easson-Stedman three-point binding model. On the left, L-phenylalanine can achieve 3 attractive interactions with the binding site, while in the center and on the right, D-phenylalanine can achieve only two.

Pirkle has designed and studied a number of chiral stationary phases, and formulated detailed binding models for a number of CSP·analyte systems. One of the more popular Pirkle columns is the Whelk-O, which has as a CSP the tetrahydrophenanthrene derivative depicted in Figure 2.6a (which also shows an analyte that has an exceptionally large separability factor, α, of 11.63). This corresponds to a $\Delta\Delta G°$ of 1.44 kcal/mol [78]. The Whelk-O CSP has a hydrogen bond donor and a cleft that is formed by a pseudoaxial dinitrobenzamide substituent. A derivative of the Whelk-O CSP, in which the propylsilane tether is replaced by a methyl group, forms cocrystals with both enantiomers of the analyte. As shown in Figure 2.6b and c, the carbonyl oxygen of the pivalamide analyte is hydrogen bonded to the NH of the benzamide CSP, and also exhibits face to face π-stacking between the *p*-bromophenyl group of the analyte and the dinitrophenyl group of the CSP. However, the binding modes differ in the following ways:

- The orientation of the dinitrophenyl group is twisted significantly in the two complexes.
- The *S*-enantiomer of the analyte is oriented such that the stereocenter resides above the naphthalene ring of the CSP, whereas the *R*-enantiomer of the analyte binds such that the stereocenter is to the rear and the bulky *tert*-butyl group resides above the naphthalene ring.
- The *S*-enantiomer of the analyte has the *p*-bromophenyl group oriented above the naphthalene ring, where it may benefit from edge to face π-stacking with the naphthalene ring.

The cocrystal of the more highly retained enantiomer has a higher melting point than the cocrystal of the faster eluting enantiomer, reflecting the relative stability of the two complexes in the solid state. This fact is consistent with the order of elution, which reflects the relative stability (capacity ratio) of the analyte·CSP complex in solution.

Although detailed models of other CSP separations have not been as extensively studied as the Pirkle systems, it is likely that they also conform to some variant of the three-point rule [77,79]. Indeed, the energy differences that are required for enantiomer separation on an

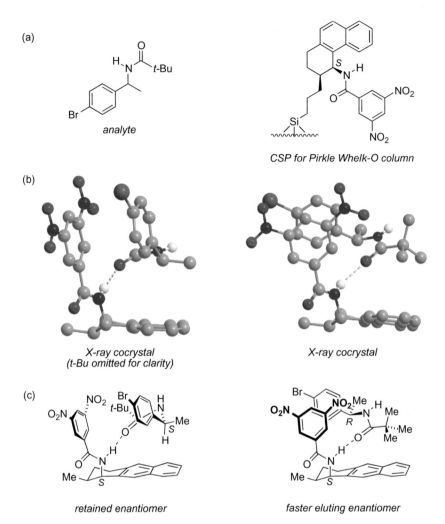

FIGURE 2.6 Model for analyte·CSP pairs of a typical analyte on the Pirkle Whelk-O column [78]. (a) Structures of the analyte and the Whelk-O CSP. (b) Partial X-ray crystal structure of the *S* enantiomer of analyte with *S* enantiomer of the CSP (mp > 200 °C).[11] *(c)* *R* enantiomer of analyte with *S* enantiomer of CSP (mp = 187 °C).

efficient column are so small that caution is advised in overinterpreting enantiomer "recognition" models with small separability factors.[12]

---

11. The published structure is of the *R*-enantiomer of the CSP with the *R*-enantiomer of the analyte, which crystallized as a methylene chloride solvate in a 2:2:1 ratio. Pictured on the left is the inverted structure of one of the two complexes in the unit cell so that the comparison is more clearly representative of a CSP·analyte complex [78].
12. The anthropomorphic notion that a chiral molecule can somehow "recognize" or "discriminate" the chirality sense of another chiral molecule is a convenience that is used commonly, realizing that it is the observer that does the recognizing, not the molecules [80].

2.4.3 Achiral Derivatizing Agents

Imaginative tricks can also be used to analyze enantiomeric mixtures. For example, an achiral, bifunctional derivatizing agent may be used to randomly dimerize a mixture of enantiomers. If a statistical ratio can be proven in control experiments, the ratio of the chiral to meso diastereomers can be used to calculate the enantiomer composition [81]. Figure 2.7 shows several derivatizing agents that are available. In principle, the ratio can be determined by chromatographic or spectroscopic methods. ^{1}H, ^{13}C, and ^{31}P NMR provide a particularly facile method of analysis [81–86]. The following generic reaction illustrates the process:

$$R + S + AX_2 \quad \rightarrow \quad RAR + SAS + RAS + SAR$$
$$\textit{stoichiometry:} \quad 1 \quad x \quad \textit{probabilities:} \quad 1 \cdot 1 \quad x^2 \quad 1 \cdot x \quad 1 \cdot x \quad , \tag{2.10}$$

where R and S indicate the absolute configuration of the analyte, and AX_2 is the "dimerization" reagent bearing an arbitrary leaving group, X. RAR and SAS are a d,l pair, while RAS and SAR are meso. The latter may or may not be identical,[13] but it is necessary to recognize (statistically) that either an "$SAX + R$" or an "$RAX + S$" sequence would produce the meso isomer. If the S/R ratio is x, then the probability for the formation of the d,l pair is $(1 + x^2)$ and $2x$ for the meso isomer(s). Thus the d,l/meso ratio is given by

$$\frac{d,l}{\text{meso}} = y = \frac{1 + x^2}{2x}. \tag{2.11}$$

Solving for x gives

$$x = \text{enantiomer ratio} = \frac{2y \pm \sqrt{4y^2 - 4}}{2}. \tag{2.12}$$

There is an ambiguity in this determination because the mathematics is oblivious to absolute configuration, hence the " $\pm$ " in the quadratic formula. Thus, although x was defined as S/R, the solutions will be S/R and R/S.

It is also of interest to note that this method may be used to enrich the enantiomer ratio of a partially resolved racemate on a preparative scale [81,87–93]. For example, dimerization of a mixture having 90:10 er, followed by separation and then cleavage of the d,l isomer affords an 81:1 (98.8:1.2) mixture of enantiomers with an 82% theoretical yield.

$AX_2 = COCl_2, CH_2(COCl)_2, C_6H_4(COCl)_2, C_6H_4(OH)_2, Me_2SiCl_2, Ph_2SiCl_2, PCl_3, MePOCl_2, B(OH)_3, Ni(NO_3)_2$

FIGURE 2.7 Achiral derivatizing agents for the determination of enantiomer ratio and partial resolution of functional molecules such as an alcohol.

2.5 NUCLEAR MAGNETIC RESONANCE

For the analysis of diastereomeric mixtures, NMR is an obvious choice, and derivatization of enantiomers with a chiral reagent can be an excellent method of analysis (for a monograph

---

13. If "A" is stereogenic in the products, the two will be diastereomers. For example, derivatization of alcohols with PCl_3 affords phosphonates in which the phosphorus is stereogenic and two meso isomers are produced [82].

on this subject, see ref. [94]). In the 1960s, a number of discoveries were made that facilitated the direct observation of diastereomeric and enantiomeric mixtures by NMR. The development of chiral derivatizing agents (CDAs), CSAs, and chiral lanthanide shift reagents (CSRs)[14] made it possible to observe (and integrate) separate signals for enantiomers. Lanthanide shift reagents are not used very much anymore,[15] so the following discussion will focus only on CDAs and CSAs.

2.5.1 Chiral Derivatizing Agents (CDAs)

The derivatization of a mixture of enantiomers with a chiral reagent produces diastereomers that may be analyzed by NMR spectroscopy or by chromatography [102]. In order to be useful, a number of requirements must be met:

- The CDA must be enantiomerically pure or (less satisfactorily) its enantiomeric purity must be known accurately.
- The reaction of the CDA with both enantiomers should go to completion under the reaction conditions, so as to avoid enrichment or depletion of one enantiomer of the analyte by kinetic resolution.
- The CDA must not racemize under the derivatization or analysis conditions, and its attachment should be mild enough so that the substrate does not racemize either.
- If analysis is by HPLC, the CDA should have a chromophore to enhance detectability. If analysis is by NMR, the CDA should have a functional group that gives a singlet and that is remote from other signals for easy integration.

The importance of the first point is evident if we consider the following reactions of an analyte with a CDA:

$$\text{Analyte } (R+S) + CDA \ (R') \rightarrow \text{Diastereomeric derivatives } R\!-\!R' + S\!-\!R' \qquad (2.13)$$

$$\text{Analyte } (R) + CDA \ (R'+S') \rightarrow \text{Diastereomeric derivatives } R\!-\!R' + R\!-\!S' \qquad (2.14)$$

Equation 2.13 illustrates the derivatization of a mixture of R and S enantiomers of analyte with an enantiomerically pure derivatizing agent R'. If the reaction is complete for both enantiomers of analyte, the ratio of diastereomeric derivatives $R\!-\!R'$ and $S\!-\!R'$ will equal the ratio of enantiomers R and S of the analyte. Equation 2.14 shows how a CDA that is not enantiomerically pure can cause problems. The ratio of diastereomeric derivatives $R\!-\!R'$ and $R\!-\!S'$

---

14. Note the distinction between the terms *shift reagent* and *solvating agent*. Because of the differences in the mechanism of binding and induced anisochrony, the former is reserved for lanthanide complexes and the latter for diamagnetic organic compounds.

15. Chiral shift reagents (lanthanide complexes) were very popular for determining enantiomer composition in the 1970s and 1980s because they increased spectral dispersion and permitted accurate integration of peaks arising from diastereotopic complexes [95−97]. However, as high field NMR spectrometers became widely available, the need for increased spectral dispersion was lessened. In addition, many lanthanide shift reagents are paramagnetic and their complexes exhibit broadened lines. Linewidth increases with the square of the field strength. This problem is especially severe when $\Delta\delta$ is large, as is often the case with lanthanide shift reagents. If a paramagnetic CSR must be used, it is wise to conduct the analysis on the lowest field spectrometer available; if line broadening is a problem, warming the sample may help [98−100]. Failing that, spin-echo techniques may be used to eliminate broadened lines [101].

will reflect the enantiomer ratio of the CDA.[16] Note that $S-R'$ (Equation 2.13) and $R-S'$ (Equation 2.14) are enantiomers and standard methods of analysis will not distinguish them. If both the analyte and the CDA have 95:5 er, the four possible diastereomers will have the statistical ratio of .9025/.4075/.4075/.0025 (Equation 2.15). Since the products are two diastereomeric racemates, combination of the enantiomers yields a .9050/.0950 ratio, or 90.5:9.5 er, instead of the correct 95:5 er.

$$.95R/.05S + .95R'/.05S' \rightarrow \underset{.9025}{R-R'} + \underset{.0475}{S-R'} + \underset{.0475}{R-S'} + \underset{.0025}{S-S'} \qquad (2.15)$$
$$\underset{\text{Analyte}}{} \quad \underset{\text{CDA}}{}$$

Over 100 CDAs are detailed in Wenzel's monograph [94],[17] but only a few are discussed here. Several carboxylic acid CDAs are shown in Figure 2.8, along with models that often aid assignment of absolute configuration of derivatized alcohols.

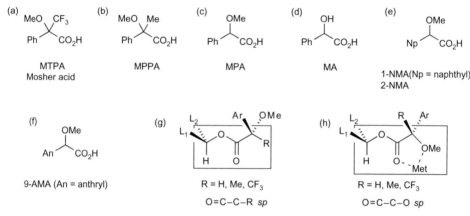

FIGURE 2.8 Carboxylic acid based CDAs: (a) α-Methoxy-α-trifluoromethylphenylacetic acid, MTPA [103−105]; (b) α-Methoxy-α-phenylpropionic acid, MPPA [106]; (c) α-Methoxy-α-phenylacetic acid, MPA [107]; (d) Mandelic acid, MA [104]; (e) α-(1-Naphthyl)-α-methoxyacetic acid, 1-NMA, and α-(2-naphthyl)-α-methoxyacetic acid, 2-NMA [108,109]; (f) α-(9-anthryl)-α-methoxyacetic acid, 9-AMA [108,110]; (g) *Synperiplanar* (*sp*) conformation of C−R and C=O, which predicts that the aryl group shields the nuclei of L_1; (h) Metal chelate in which the aryl group shields the nuclei of L_2.

The first CDA, introduced in 1969, was Mosher's acid, α-methoxy-α-(trifluoromethyl) phenylacetic acid, abbreviated MTPA [103−105]. It is commercially available as either enantiomer, and is used for the derivatization of alcohols and amines. Two reports [111,112] indicate that commercially available material may not be enantiopure, and one might expect similar levels of enantiomeric purity from the original preparation [103]. The enantiomeric purity of MTPA may be determined by esterification of diacetone glucose and examination of the NMR [112], or more accurately by CSP-GC of the MTPA methyl ester [111]. As compared to the other CDAs in Figure 2.8, the quaternary derivatives MTPA and MPPA have the advantage that no racemization is possible during derivatization, although methods that minimize racemization have been reported [113].

---

16. Assuming that there is no kinetic resolution of the CDA by the analyte during derivatization.
17. Wenzel's monograph has 1668 references covering the primary literature into 2006, and readers may wish to consult this reference for more detail than is provided here.

Derivatization of chiral alcohols and amines of general structure RCHZR' (Z=OH or NH$_2$) yields esters and amide derivatives that can be analyzed by NMR. ^{1}H, ^{13}C, or ^{19}F NMR may be used to observe the diastereomeric derivatives [103,114]. Most commonly, the —OCH$_3$ is observed by ^{1}H NMR or the —CF$_3$ by ^{19}F NMR in the MTPA derivatives.[18] In most cases, one or the other of these nuclides will be well enough separated that accurate integration will be possible.

Since the CDA and the analyte are both chiral, care must be taken to ensure complete consumption of the analyte by the CDA. Otherwise, kinetic resolution could enrich one of the products and provide an inaccurate reflection of the enantiomer ratio of the analyte.

Absolute configuration determination. Models have been proposed to correlate chemical shift data with absolute configuration [94]. Most rely on the assumption that the *synperiplanar* (*sp*) conformation of the derivative illustrated in Figure 2.8g predominates in solution. In this conformation, the phenyl, naphthyl, or anthracyl group of the CDA shields the nuclides in L$_1$, shifting them upfield.[19] The *sp* conformation has a larger population when R=H than when R=CH$_3$ or CF$_3$ [110]. The important consequence of this fact is that, for CDAs in which R=H (such as in Figure 2.8c–f), a larger difference in chemical shift difference ($\Delta\delta$) will be observed.

Figure 2.9 elaborates on the model illustrated in Figure 2.8g, in which two situations may arise in practice: determination of absolute configuration of a pure enantiomer, and assignment of configuration of the major enantiomer of a mixture.[20] Assuming that the Cahn–Ingold–Prelog priority of the ligands around the unknown center is X > L$_2$ > L$_1$ > H, derivatization of an enantiopure analyte with racemic MTPA will give *R–R* (relative configuration *l*) and *S–R* diastereomers (relative configuration *u*), illustrated by the top two structures in Figure 2.9. The following empirical rules can be stated:

- The ^{1}H L$_1$ signals of the *l* diastereomer will be upfield of the L$_1$ signals of the *u* isomer.
- The ^{1}H L$_2$ signals of the *l* diastereomer will be downfield of the corresponding signals of the *u* isomer.

If an unequal mixture of enantiomers is present, symmetry considerations dictate that derivatization with only one MTPA enantiomer is necessary, since (see Figure 2.9, two bottom structures) derivatization of a racemate with one enantiomer of MTPA also produces an *l/u* mixture. Thus, either the two left structures or the two right ones could be used to establish configurations. Again assuming the Cahn–Ingold–Prelog priority of the ligands around the unknown center is X > L$_2$ > L$_1$ > H, the following empirical rules apply:

- If *R*-MTPA is used, the *R* configuration at the unknown stereocenter will give L$_1$ signals upfield of the *S* diastereomer and L$_2$ signals downfield of the *S* diastereomer.
- If *S*-MTPA is used, the *S* configuration at the unknown stereocenter will give L$_1$ signals upfield of the *R* diastereomer and L$_2$ signals downfield of the *R* diastereomer.

---

18. Although ^{19}F NMR in an MTPA derivative has the advantage of exhibiting only two peaks, and therefore is an excellent choice for determining the dr of a Mosher ester or amide, differences in the ^{19}F chemical shifts cannot be used reliably for the determination of absolute configuration [115].

19. Modern, 2D methods of peak assignment permit inclusion of several nuclei in L$_1$ and L$_2$, improving the reliability of configurational assignments by this method. Note also that these trends are observed in CDCl$_3$ solution, but not necessarily in C$_6$D$_6$ [115].

20. The analysis presented here is for compounds having a single stereocenter. However, there are many examples where the configuration of a single stereocenter in a complex molecule having many other stereocenters has been determined. See ref. [94].

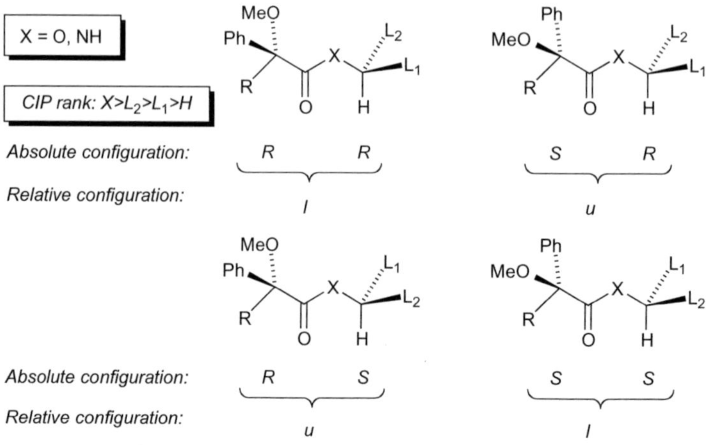

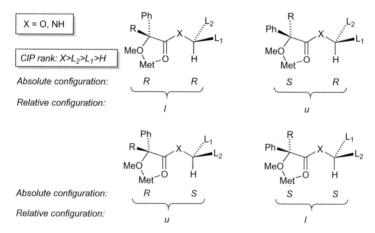

FIGURE 2.9 Diastereomeric CDA derivatives. Note that *l* or *u* isomers may be obtained by derivatization of a single enantiomer with racemic CDA (horizontal pairs) or by derivatization of a racemate with enantiomerically pure CDA (vertical pairs).

In the above analyses, a drawback is the reliance of the model upon the population of only one of several conformations. Contributions of other conformations decreases the $\Delta\delta$ values, thereby lessening the confidence in the assignment. However, if a metal chelates the carbonyl oxygen and the methoxy of the CDA, conformational motion is restricted and $\Delta\delta$ values are increased (Figure 2.10). Mosher esters can be chelated by La(hfa)$_3$ [116], which is diamagnetic and does not suffer the drawbacks of paramagnetic shift reagents. Other esters, such as MPA can sometimes be effectively chelated by Ba(II) [117]. Note that in these chelates, the trends outlined above are reversed, such that L_2 nuclei appear upfield of L_1 in diastereomers having the *l* relative configuration (*R,R* or *S,S*), and L_1 nuclei appear upfield of L_2 nuclei in diastereomers having the *u* relative configuration (*R,S* and *S,R*).

FIGURE 2.10 Diastereomeric CDA derivatives that are chelated by a metal. MTPA derivatives are best chelated by the diamagnetic shift reagent La(hfa)$_3$ (hfa = 1,1,1,5,5,5-hexafluoro-2,4-pentanedione) [116], whereas derivatives such as those from MPA are best chelated by Ba(ClO$_4$)$_2$ [117]. Note that *l* or *u* isomers may be obtained by derivatization of a single enantiomer with racemic CDA (horizontal pairs) or by derivatization of a racemate with enantiomerically pure CDA (vertical pairs).

In conclusion, two points must be emphasized. First, the rationales presented in Figures 2.8–2.10 are models for predicting changes in chemical shift, and do not necessarily represent preferred conformations. Second, it should be restated that in order for the CDA method to be an accurate reflection of isomer ratios, any adventitious kinetic resolution in the derivatization must be quantitated or eliminated. For example, Heathcock has noted that MTPA derivatization of a racemic alcohol (50:50 er) afforded a 1.7:1 mixture of Mosher esters (63:37 dr) and the er determinations had to be corrected accordingly [118]. Svatos used a fivefold excess of the CDA to force a derivatization to completion [119]. If the appropriate control experiments are done, derivatization with CDAs can be a reliable method for determination of enantiomer ratios and absolute configuration of amines and alcohols. For the derivatization of ketones, chiral diols may be used [120], but similar control experiments should be undertaken.

2.5.2 Chiral Solvating Agents (CSAs)

The notion that enantiotopic groups would be anisochronous in a chiral environment was first suggested by Mislow in 1965 [102], and is the basis for the analysis of enantiomer ratios by CSAs. His suggestion was reduced to practice the next year by Pirkle, in the form of a CSAs [121]. In the intervening years, analysis of enantiomer ratios and the assignment of absolute configuration by CSA has become a very useful tool. Virtually any functional group can be analyzed by a CSA [94,122,123]. There are a number of noteworthy features of CSAs:

- CSAs are simple diamagnetic compounds. Since the dynamics of a CSA and its interaction with a solute may be reasonably well understood, deduction of absolute configuration is often possible [123].
- Anisochrony in CSAs is usually induced in enantiotopic groups of a ligand by the presence of an anisotropic moiety in the CSA, such as an aromatic ring.
- Since CSAs are diamagnetic and therefore do not produce line broadening, it is often possible to deduce isomer ratios by comparison of peak *height*, obviating the need for a complete separation and integration of pertinent NMR signals.
- Both enantiomers should be available to ensure the presence of induced anisochrony.[21]

The most studied CSAs are the 1-(aryl)trifluoroethanols and the 1-(aryl)ethyl amines (aryl = phenyl, 1-naphthyl, 9-anthryl) that associate primarily through hydrogen-bonding mechanisms. The equilibria that describe the 1:1 interactions of a CSA and a pair of enantiomeric solutes are shown in Equation 2.16. As a result, the analysis of the geometry of the diastereomeric solvates, $(+)$-CSA $\cdot R$ and $(+)$-CSA $\cdot S$, often allows determination of absolute configuration.

$$(+)\text{-CSA} \cdot R \overset{K_R}{\rightleftharpoons} (+)\text{-CSA} + R, S \overset{K_S}{\rightleftharpoons} \text{CSA} \cdot S \qquad (2.16)$$

Preferential population of one diastereomer over the other ($K_R \neq K_S$) is not a prerequisite for induced anisochrony of enantiotopic groups. Since the CSA is diamagnetic, it may be used in excess over the analyte. A fivefold excess is usually sufficient to drive the equilibria of Equation 2.16 to the "outside," such that the solute is present only as its two diastereomeric

---

21. In principle, induced anisochrony could also be established by studying an enantiomerically pure analyte with racemic CSA (*cf.* Figures 2.9 and 2.10).

solvates. Since the observed spectra are time averages of all the species in solution, this chemical trick simplifies analysis of absolute configuration by focusing on the diastereomeric solvates alone.

Figure 2.11 illustrates the principle with two specific attractive interactions and a third, which provides the anisotropy for enantiomer discrimination.[22] In this example, there are two hydrogen bond donors in the CSA: the hydroxyl and the benzylic hydrogen (which has been rendered acidic by the neighboring trifluoromethyl group). Suppose, for example, that the solute is dibasic and binds preferentially such that OH and B_1 interact and CH and B_2 interact. Note that the solute substituent that is *syn* to the aryl group on the CSA will be shielded relative to the other (R_2 on the left and R_1 on the right). *This is the third point required for discrimination of enantiomers.* Because of the shielding cone above the aromatic ring, the time-averaged spectrum will have R_2 at higher field in the absolute configuration on the left. If the preference of the two attractive interactions between the CSA and the solute are known, then the absolute configuration of the CSA can be used to determine the absolute configuration of the solute. Once again, there need be no difference in stability between the two solvates, since protons that are enantiotopic by internal comparison can also be differentiated by CSAs [124].

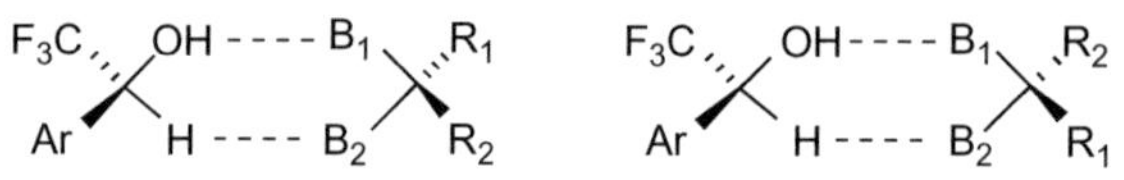

$$F_3C\diagdown OH ---- B_1 \diagdown R_1 \qquad F_3C \diagdown OH ---- B_1 \diagdown R_2$$
$$Ar \diagup H ---- B_2 \diagup R_2 \qquad Ar \diagup H ---- B_2 \diagup R_1$$

FIGURE 2.11   The interaction of a 1-aryltrifluoroethanol chiral solvating agent and the two enantiomers of a dibasic solute.

2.6 CHIROPTICAL METHODS

Although polarimetry was the primary method for determining enantiomer composition for most of the twentieth century, it is rarely used any more for this purpose.[23] Instead, it is now most commonly used to confirm absolute configuration of synthetic compounds when compared to literature data. Modern computational methods, coupled with chiroptical methods, are being developed that permit calculation of specific rotations and determination of absolute configuration [134–138]. The following summary is a brief explanation of chiroptical properties and their use.

Polarimetry measures the rotation of a plane of monochromatic polarized light after having passed through a sample, as shown schematically in Figure 2.12.

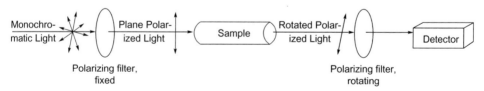

FIGURE 2.12   Schematic representation of a polarimeter.

22. This example illustrates two *attractive* interactions, although one attractive and one repulsive interaction would also suffice, so long as a third is present as well.
23. Monographs: [125–133].

It is not intuitively obvious why a chiral medium should have this effect, until the linearly polarized light beam, represented by the sine wave in Figure 2.13a, is broken down into the two circularly polarized components shown in Figures 2.13b and c.[24]

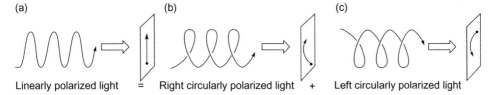

Linearly polarized light   =   Right circularly polarized light   +   Left circularly polarized light

FIGURE 2.13   Representations of the waveforms of polarized light beams passing through a perpendicular plane.

When the linearly polarized beam passes through a perpendicular plane, the point of intersection moves along a line. When the circularly polarized beams pass through the same plane (the helices are moved without being rotated), the point of intersection describes a circle, moving either to the right or the left depending on the chirality sense of the helix. Note that the vector sum of the right- and left-circularly polarized beams equals the linearly polarized beam. The right- and left-circularly polarized beams are refracted equally by an achiral medium; that is, their refractive indices n_R and n_L (which measure change in velocity), are equal. As shown in Figure 2.14a, the vector sum of the two refracted circularly polarized beams remains in the plane of the incident polarized light, that is, the plane is not rotated. In a medium where n_L and n_R are not equal, the two beams are shifted out of phase, and their vector sum is rotated out of plane by an angle, α, as shown in Figure 2.14b.

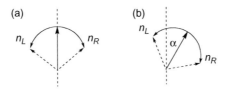

FIGURE 2.14   The effect on the transmitted plane by refraction of circularly polarized light beams relative to the incident plane of polarized light (dashed line). (a) When $n_L = n_R$, the vector sum (solid arrows) of the two circularly polarized beams (dashed arrows) remains in the same plane as the incident beam. (b) When $n_L \neq n_R$, the vector sum of the two waves is rotated $\alpha°$ away from the plane of the incident light.

A medium which produces this effect is *circularly birefringent*. A solution of a pure enantiomer is circularly birefringent. In contrast, an equimolar mixture of two enantiomers will have an equal number of refractions to the right and left, and the net result will be $\alpha = 0$. Thus, a polarimeter cannot distinguish an achiral compound from a racemate. It was Pasteur's discovery of circular birefringence in solutions of enantiomorphous crystals of racemic sodium ammonium tartrate [140] that set the stage for the development of stereochemical theory by establishing the presence of chiral molecules in an optically inactive compound.

To summarize, the differential refraction of right- and left-circularly polarized light by a chiral nonracemic substance results in the rotation of the plane of the sine wave that is the vector sum of the two circularly polarized beams. That the two circularly polarized beams should be refracted differently by a chiral substance is apparent if one considers their helicity and imagines the interaction of a helix with a chiral substance in the context of double

---

24. This analysis is an oversimplification. For a thorough treatment, see ref. [139].

asymmetric induction explained in the previous chapter: any chiral entity will interact differently with the two enantiomeric forms of another chiral entity. On a macroscopic scale, we can easily perceive with our right hand the difference between a right- and a left-handed screw, just as a chiral molecule may detect the difference between right- and left-circularly polarized light. On the molecular scale, whether n_R and n_L differ enough to be measured depends on the system. If the "refractivity"[25] of the various ligands around a stereocenter in a chiral molecule are nearly the same, the difference between n_R and n_L may be too small to detect and no rotation will be observed. From a practical standpoint, it may be possible to change the wavelength to increase the difference in n_R and n_L (see optical rotatory dispersion (ORD), below).

The degree of rotation observed in a polarimeter, α, is dependent on the number of chiral species the light encounters on its passage through the sample chamber, as well as the wavelength of the light. Thus, analytical accuracy dictates strict control of a number of experimental parameters, such as temperature, concentration, light source, and path length. To minimize the effects of these variables and to increase the reproducibility, specific rotation, $[\alpha]$, is defined as

$$[\alpha]_\lambda^T = \frac{100\alpha}{l \cdot c}, \tag{2.17}$$

where T is the temperature, λ is the wavelength of the light (often the D lines of sodium at 589.0 and 589.6 nm and abbreviated simply "D"), α is the observed rotation, l is the sample path length in decimeters, and c is the concentration in grams per 100 milliliters of solution. To ensure reproducibility, it is common practice to report the concentration and solvent along with the specific rotation, and the units are understood.[26,27] For example, if a solution of 0.014 g in 1.0 mL of ethanol solution afforded a measured rotation of +1.38°, the specific rotation would be reported as

$$[\alpha]_D^{25} + 99 \ (c = 1.4, \text{EtOH}).$$

This denotes a specific rotation of +98° mL/g · dm measured at the D line of sodium, temperature 25 °C, at a concentration of 1.4 grams per 100 milliliters of ethanol. This example illustrates the inherent inaccuracy of polarimetry for the accurate determination of enantiomer composition, in that the sample must be weighed accurately[28] and the observed rotation must be determined with extreme accuracy.

For pure liquids (or solids), the equation

$$[\alpha]_\lambda^T = \frac{\alpha}{l \cdot \rho} \tag{2.18}$$

is used, where ρ is the density in grams per milliliter.

---

25. This term is used loosely and is related to the polarizability of each ligand. Interestingly, before the days of IR and NMR spectroscopy attempts were made to quantify the refractive index of individual functional groups as a means of deducing structure. For a summary of "Molar Refraction," see S. Glasstone *Physical Chemistry*, Van Nostrand: Princeton (1946) pp. 528–534, and other texts of the same period.

26. It is incorrect to report specific rotation in "degrees."

27. It is dangerous to assume that the effect of differences in concentration can be ignored, because dimerization or aggregation at higher concentrations will change the refraction properties of the solute. When comparing measured $[\alpha]$ to a literature value, the rotation should be measured at the same concentration reported previously.

28. This is especially problematic nowadays given the small sample quantities often encountered. In the cited example, note that 14 mg (2 significant figures) in a 1 mL volumetric flask limits the accuracy of the rotation to 2 significant figures.

Specific rotation was defined over 150 years ago, which accounts for the unusual units of path length and concentration: decimeters were used because a long path length was needed to get an accurate measurement, and mass was used instead of molecular weight because molecular weights were uncertain in the early nineteenth century. The D line of sodium was chosen because it is easily produced in a flame and is nearly monochromatic. Now that molecular weights are no longer an unknown, molecular rotation, $[\Phi]$, may be used instead of $[\alpha]$:

$$[\Phi]_\lambda^T = \frac{M[\alpha]}{100}, \tag{2.19}$$

where M is the molecular weight. Molecular rotation is commonly used in ORD, which is a plot of rotation as a function of wavelength.

Sign of rotation reflects absolute configuration (and is often used to assign it), and the magnitude of the rotation is used to determine the optical purity, usually expressed as a percent:

$$\% \text{ optical purity} = \frac{100[\alpha]_\lambda^T}{[\alpha_o]_\lambda^T}, \tag{2.20}$$

where $[\alpha]_\lambda^T$ is the observed specific rotation, and $[\alpha_o]_\lambda^T$ is the specific rotation of the pure enantiomer under identical conditions. The optical purity of an enantiomerically pure compound is 100%, and 0% for a racemate. Ideally, the specific rotation of a partly racemic mixture varies linearly with enantiomeric composition. Thus, a sample of 75:25 er, whose $[\alpha_o] = +98$ should exhibit $[\alpha] = +49$, and the optical purity would be 50%.[29]

The optical purity is usually, *but not always,* equal to enantiomer excess. In order for the two to be equal, it is necessary that there be no aggregation. It is possible, for example, that a homochiral or heterochiral dimer (see Glossary, Section 1.11, for definitions) would refract the circularly polarized light differently than the monomer (or each other). In 1968, Krow and Hill showed that the specific rotation of (*S*)-2-ethyl-2-methylsuccinic acid (85% ee) varies markedly with concentration, and even changes from levorotatory to dextrorotatory upon dilution [141]. In 1969, Horeau followed up on Krow and Hill's observation, and showed that the "optical purity" (at constant concentration) and enantiomer excess of (*S*)-2-ethyl-2-methylsuccinic acid were unequal except when enantiomerically pure or completely racemic [142]. This deviation from linearity is known as *the Horeau effect,* and its possible occurrence should be considered when determining enantiomer ratios using polarimetry. For this and many other reasons, the terms "optical purity" and "enantiomer excess" are falling out of favor, and their usage is strongly discouraged [143,144].

For optical purity to accurately reflect enantiomeric purity, it is obvious that the sample must be free of any chiral impurities. It may not be as obvious that achiral impurities can also cause significant error. For example, Yamaguchi and Mosher [145] showed that the specific rotation of enantiopure 1-phenylethanol could be enhanced from $[\alpha]_D^{20} + 43.1$ ($c = 7.19$, cyclopentane) to $[\alpha]_D^{20} + 58.3$ ($c = 2.64$, cyclopentane) by the addition of 10.6 g/100 mL of acetophenone. Presumably, this enhancement is due to an interaction between the alcohol and the ketone, either through hydrogen bonding or hemiacetal formation.

---

29. Specific rotations are notoriously inaccurate as measures of enantiomer composition, which led one pioneer of asymmetric synthesis to say, only partly in jest: "We should all take our polarimeters, put them in a pile, and burn them" (A. I. Meyers).

In their landmark 1971 book [1], Morrison and Mosher coined the term "enantiomeric excess" to describe the relationship of two enantiomers in a mixture: *"Assuming a linear relationship between rotation and composition and no experimental error, percent 'optical purity' [op] is equated with the percent of one enantiomer over the other, which we shall designate percent enantiomeric excess (% e.e.)."* Note that Morrison and Mosher explicitly specified equality between optical rotation

and enantiomer composition, meaning that $op = ee$. As noted in Section 2.6, this is not necessarily true. The presumed equality between ee and the earlier term op was known prior to 1971, but sometimes by other names. Earlier, Horeau used the term "enantiomeric purity" to describe the term $(R - S)/(R + S)$, but then proposed using the fraction of the major enantiomer as a preferable expression of enantiomeric purity: *"$P_E = R/(R + S)$"* [2].

In 2006 [3], the question was posed: *"If chemists had not relied on optical rotation for so many years to determine enantiomer composition, would we ever have defined 'ee' as an expression of enantiomer composition or purity?"*, and answered: *"Certainly not. The association between enantiomer composition and optical 'purity' is unfortunate, as it implies that the 'impurity' is the racemate, and not the minor enantiomer."* From the discussion in Section 1.4, it is obvious that it is product ratio (er or dr) that reflects the relative rates or equilibrium constant, and are the best descriptors of stereoselectivity, not the product excess (ee or de).

As explained in Section 2.6, measurement of specific rotation is often an inaccurate reflection of enantiomer ratios, so few chemists use it any more for that purpose. Equations for the analysis and implementation of kinetic resolutions and nonlinear effects using enantiomer ratios have been published [3], and are also derived and presented in Chapter 1 (Sections 1.7 and 1.10). The usual reason cited for retaining these unwieldy terms is that chemists who do asymmetric synthesis have calibrated themselves to consider an ee of $\geq 90\%$ as a good result and that reporting er values is somehow "cheating". We think this view is readily trumped by the fact that ee is an artifact of a 19th century technique now rarely used for determination of enantiomer composition.

We strongly recommend that the terms ee and (especially) de be abandoned [4]. It may be of interest to teachers of organic chemistry to note that many (most?) introductory textbooks do not mention the concept of enantiomer composition (ee or er). Moreover, ee is no longer on the list of abbreviations in the "Instructions for Authors" of papers in the *Journal of Organic Chemistry* and *Organic Letters*.

References

[1] Morrison, J. D.; Mosher, H. S. *Asymmetric Organic Reactions*; Prentice-Hall: Englewood Cliffs, NJ, 1971.
[2] Horeau, A. *Tetrahedron Lett.* **1969**, 3121–3124.
[3] Gawley, R. E. *J. Org. Chem.* **2006**, *71*, 2411–2416; corrigendum *J. Org. Chem.* **2008**, *73*, 6470.
[4] Gibb, B. C. *Nature Chemistry* **2012**, *4*, 237–238.

The discovery of the anomalies mentioned above are partly responsible for the declining popularity of polarimetry for the determination of enantiomer composition. Even if the experimentalist is alert to these sources of error, the possibility still exists that an early determination of specific rotation, against which a new value must be compared, is itself in error. Thus, caution is advised.

In addition to polarimetry, other chiroptical properties may be useful for the assignment of absolute configuration, although they are rarely used to determine enantiomer ratio

[125–128,131,132,146–148]. Optical rotatory dispersion, ORD, measures the optical rotation of a compound as a function of wavelength, and its theory is the same as for simple polarimetry described above. Circular dichroism, CD, is similar, but differs in that the substrate includes a chromophore that absorbs at the wavelengths employed. In this special case, the molar absorptivities (extinction coefficients) of the right- and left-circularly polarized beams are different. Thus, in addition to being out of phase, the vectors of the transmitted beams are also of unequal magnitude. As a result, the emergent beam no longer traverses a line, but describes an ellipse, and the emergent light is *elliptically polarized.* In the region of such a CD band, the ORD exhibits "anomalous" behavior (a Cotton effect) due to the absorption. The mean wavelength between an ORD peak and trough [125] is close to the λ_{max} of the chromophore absorbing the light. It is not unusual for the ORD curve to change sign in such a region. Because ORD measures a rotation, it is theoretically finite at all wavelengths, but since CD measures a difference in absorption, it only occurs in the vicinity of an absorption band. CD is very commonly used in structural biology for peptide and protein secondary structure analysis, but rarely in organic synthesis labs. An exception to the latter is when there is a very small amount of sample. Because it is very sensitive, CD can be used to determine whether or not two samples have the same configuration.

2.7 SUMMARY

As detailed above, there are a number of issues that must be considered when undertaking the synthesis of an enantiopure compound. A principal consideration is the setting and the amount of compound needed, or whether making the final product is an end in itself (as in many total synthesis projects). The issues that a practitioner will encounter most often are: what is the observed selectivity and what is the absolute configuration of the newly created stereocenter.

Chromatography is most often used to answer the question of selectivity. Chromatography is the method of choice for analysis of stereoisomer ratios, especially for analytes of very high isomeric purity, since it is the only method currently available that can accurately detect and quantify very minor contamination. The goal of asymmetric synthesis is to produce very highly selective reactions, but when this is achieved the job of identifying the chromatographic peaks due to minor stereoisomers becomes more difficult. It is tempting for the analyst to assume that the small peak(s) near the major one is the "other" isomer, but this is a risky assumption. The safest bet is to synthesize the "other" isomer(s) independently. If this is not feasible, the next best thing is to couple the chromatograph to some sort of spectroscopic device such as a mass spectrometer or a diode array UV–VIS detector.[30]

Several of the methods discussed above provide a glimpse of how one might determine absolute configuration. A 2008 issue of *Chirality* was devoted to various methods for determining the absolute configuration of chiral molecules [149], the most reliable of which is anomalous dispersion X-ray diffraction. This is a very powerful tool that has become largely routine and should not be overlooked by the synthetic practitioner [150]. Automated X-ray diffractometers are now available that place the determination of an X-ray crystal structure in the hands of synthetic organic chemists who lack extensive training in the technique.

Chiral or achiral derivatizing agents can facilitate analysis of enantiomers by either chromatography or NMR. For the analysis of compounds that are chiral by virtue of isotopic

---

30. The UV spectra of enantiomers are identical of course, and those of diastereomers are usually superimposable as well.

substitution, NMR is the method of choice, since energetic differences between diastereomeric complexes are not required for induced anisochrony. When it works, NMR is also
one of the simplest and fastest techniques available. For some analytes, CSAs and chiral
derivatizing agents are useful for the assignment of stereoisomer ratios *and* absolute
configuration.

REFERENCES

[1] Hendrickson, J. B. *J. Am. Chem. Soc.* **1975**, *97*, 5784−5800.

[2] Wender, P. A.; Miller, B. L. *Organic Synthesis: Theory and Applications* **1993**, *2*, 27−66.

[3] Wender, P. A.; Handy, S. T.; Wright, D. L. *Chem. Ind. (London)* **1997**, *765*, 767−769.

[4] Wender, P. A.; Bi, F. C.; Gamber, G. G.; Gosselin, F.; Hubbard, R. D.; Scanio, M. J. C.; Sun, R.;
Williams, T. J.; Zhang, L. *Pure Appl. Chem.* **2002**, *74*, 25−31.

[5] Gaich, T.; Baran, P. S. *J. Org. Chem.* **2010**, *75*, 4657−4673.

[6] Kolb, H. C.; VanNieuwenhze, M. S.; Sharpless, K. B. *Chem. Rev.* **1994**, *94*, 2483−2547.

[7] Heathcock, C. H.; Kleinman, E. F.; Binkley, E. S. *J. Am. Chem. Soc.* **1982**, *104*, 1054−1068.

[8] Woodward, R. B. *Pure Appl. Chem.* **1968**, *17*, 519−547.

[9] Nicolaou, K. C.; Petasis, N. A. In *Strategies and Tactics in Organic Synthesis*; Lindberg, T., Ed.;
Academic: Orlando, 1984, pp. 155−173.

[10] Martin, D. B. C.; Vanderwal, C. D. *Chem. Sci.* **2011**, *2*, 649−651.

[11] Hanson, P. R.; Chegondi, R.; Nguyen, J.; Thomas, C. D.; Waetzig, J. D.; Whitehead, A. *J. Org. Chem.*
2011, *76*, 4358−4370.

[12] Ghosh, P.; Aubé, J. *J. Org. Chem.* **2011**, *76*, 4168−4172.

[13] Knowles, W. S.; Noyori, R. *Acc. Chem. Res.* **2007**, *40*, 1238−1239.

[14] Knowles, W. S. *Acc. Chem. Res.* **1983**, *16*, 106−112.

[15] Gotrane, D. M.; Deshmukh, R. D.; Ranade, P. V.; Sonawane, S. P.; Bhawal, B. M.; Gharpure, M. M.;
Gurjar, M. K. *Org. Proc. Res. Dev.* **2010**, *14*, 640−643.

[16] Jacques, J.; Colbert, A.; Wilen, S. H. *Enantiomers, Racemates and Resolutions*; Wiley-Interscience:
New York, 1981.

[17] Kizirian, J.-C. *Top. Stereochem.* **2010**, *26*, 189−251.

[18] Chuzel, O.; Riant, O. *Top. Organomet. Chem.* **2005**, *15*, 59−92.

[19] Davis, F. A.; Zhou, P.; Chen, B.-C. *Chem. Soc. Rev.* **1998**, *27*, 13−18.

[20] Fanelli, D. L.; Szewczyk, J. M.; Zhang, Y.; Reddy, G. V.; Burns, D. M.; Davis, F. A. *Org. Synth.* **2004**,
Coll. Vol. 10, 47−53.

[21] Liu, G.; Cogan, D. A.; Ellman, J. A. *J. Am. Chem. Soc.* **1997**, *119*, 9913−9914.

[22] Denmark, S. E.; Gould, N. D.; Wolf, L. M. *J. Org. Chem.* **2011**, *76*, 4337−4357.

[23] Miller, J. J.; Sigman, M. S. *Angew. Chem., Int. Ed.* **2008**, *47*, 771−774.

[24] Zhang, W.; Loebach, J. L.; Wilson, S. R.; Jacobsen, E. N. *J. Am. Chem. Soc.* **1990**, *112*, 2801−2803.

[25] Schaus, S. E.; Brånalt, J.; Jacobsen, E. N. *J. Org. Chem.* **1998**, *63*, 4876−4877.

[26] Woodward, R. B.; Logusch, E.; Nambiar, K. P.; Sakan, K.; Ward, D. E.; Au-Yeung, B. W.; Balaram, P.;
Browne, L. J.; Card, P. J.; Chen, C. H.; Chênevert, R. B.; Fliri, A.; Frobel, K.; Gais, H.-J.; Garratt, D. G.;
Hayakawa, K.; Heggie, W.; Hesson, D. P.; Hoppe, D.; Hoppe, I.; Hyatt, J. A.; Ikeda, D.; Jacobi, P.; Kim,
K. S.; Kobuke, Y.; Kojima, K.; Krowicki, K.; Lee, V. J.; Leutert, T.; Malchenko, S.; Martens, J.; Matthews,
R. S.; Ong, B. S.; Press, J. B.; Rajan Babu, T. V.; Rousseau, G.; Sauter, H. M.; Suzuki, M.; Tatsuta, K.;
Tolbert, L. M.; Truesdale, E. A.; Uchida, I.; Ueda, Y.; Uyehara, T.; Vasella, A. T.; Vladuchick, W. C.;
Wade, P. A.; Williams, R. M.; Wong, H. N.-C. *J. Am. Chem. Soc.* **1981**, *103*, 3210−3213.

[27] Woodward, R. B.; Logusch, E.; Nambiar, K. P.; Sakan, K.; Ward, D. E.; Au-Yeung, B. W.; Balaram, P.;
Browne, L. J.; Card, P. J.; Chen, C. H.; Chênevert, R. B.; Fliri, A.; Frobel, K.; Gais, H.-J.; Garratt, D. G.;
Hayakawa, K.; Heggie, W.; Hesson, D. P.; Hoppe, D.; Hoppe, I.; Hyatt, J. A.; Ikeda, D.; Jacobi, P.; Kim,
K. S.; Kobuke, Y.; Kojima, K.; Krowicki, K.; Lee, V. J.; Leutert, T.; Malchenko, S.; Martens, J.;
Matthews, R. S.; Ong, B. S.; Press, J. B.; Rajan Babu, T. V.; Rousseau, G.; Sauter, H. M.; Suzuki, M.;
Tatsuta, K.; Tolbert, L. M.; Truesdale, E. A.; Uchida, I.; Ueda, Y.; Uyehara, T.; Vasella, A. T.;
Vladuchick, W. C.; Wade, P. A.; Williams, R. M.; Wong, H. N.-C. *J. Am. Chem. Soc.* **1981**, *103*, 3213−3215.

[28] Woodward, R. B.; Logusch, E.; Nambiar, K. P.; Sakan, K.; Ward, D. E.; Au-Yeung, B. W.; Balaram, P.; Browne, L. J.; Card, P. J.; Chen, C. H.; Chênevert, R. B.; Fliri, A.; Frobel, K.; Gais, H.-J.; Garratt, D. G.; Hayakawa, K.; Heggie, W.; Hesson, D. P.; Hoppe, D.; Hoppe, I.; Hyatt, J. A.; Ikeda, D.; Jacobi, P.; Kim, K. S.; Kobuke, Y.; Kojima, K.; Krowicki, K.; Lee, V. J.; Leutert, T.; Malchenko, S.; Martens, J.; Matthews, R. S.; Ong, B. S.; Press, J. B.; Rajan Babu, T. V.; Rousseau, G.; Sauter, H. M.; Suzuki, M.; Tatsuta, K.; Tolbert, L. M.; Truesdale, E. A.; Uchida, I.; Ueda, Y.; Uyehara, T.; Vasella, A. T.; Vladuchick, W. C.; Wade, P. A.; Williams, R. M.; Wong, H. N.-C. *J. Am. Chem. Soc.* **1981**, *103*, 3215−3217.
[29] Mulzer, J.; Kirstein, H. M.; Buschmann, J.; Lehmann, C.; Luger, P. *J. Am. Chem. Soc.* **1991**, *113*, 910−923.
[30] Cundy, K. C.; Crooks, P. A. *J. Chromatog.* **1983**, *281*, 17−33.
[31] Matusch, R.; Coors, C. *Angew. Chem., Int. Ed. Engl.* **1989**, *28*, 626−627.
[32] Nicoud, R.-M.; Jaubert, J.-N.; Rupprecht, I.; Kinkel, J. *Chirality* **1996**, *8*, 234−243.
[33] Baciocchi, R.; Zenoni, G.; Mazzotti, M.; Morbidelli, M. *J. Chromatog. A* **2002**, *944*, 225−240.
[34] Soloshonok, V. A. *Angew. Chem., Int. Ed.* **2006**, *45*, 766−769.
[35] Tsai, W.-L.; Hermann, K.; Hug, E.; Rohde, B.; Dreiding, A. S. *Helv. Chim. Acta* **1985**, *68*, 2238−2243.
[36] Diter, P.; Daudien, S.; Samuel, O.; Kagan, H. B. *J. Org. Chem.* **1994**, *59*, 370−373.
[37] Dobashi, A.; Motoyama, Y.; Kinoshita, K.; Hara, S. *Anal. Chem.* **1987**, *59*, 2209−2211.
[38] Charles, R.; Gil-Av, E. *J. Chromatog.* **1984**, *298*, 516−520.
[39] Carman, R. M.; Klika, K. D. *Aus. J. Chem.* **1991**, *44*, 895−896.
[40] Ernholt, B. V.; Thomsen, I. B.; Lohse, A.; Plesner, I. W.; Jensen, K. B.; Hazell, R. G.; Liang, X.; Jakobsen, A.; Bols, M. *Chem. Eur. J.* **2000**, *6*, 278−287.
[41] Loza, E.; Lola, D.; Kemme, A.; Freimanis, J. *J. Chromatog. A* **1995**, *708*, 231−243.
[42] Kosugi, H.; Kato, M. *Chem. Commun.* **1997**, 1857−1858.
[43] Ogawa, S.; Nishimine, T.; Tokunaga, E.; Nakamura, S.; Shibata, N. *J. Fluorine Chem.* **2010**, *131*, 521−524.
[44] Soloshonok, V. A.; Berbasov, D. O. *J. Fluorine Chem.* **2006**, *127*, 597−603.
[45] Monde, K.; Harada, N.; Takasugi, M.; Kutschy, P.; Suchy, M.; Dzurilla, M. *J. Nat. Prod.* **2000**, *63*, 1312−1314.
[46] Suchy, M.; Kutschy, P.; Monde, K.; Goto, H.; Harada, N.; Takasugi, M.; Dzurilla, M.; Balentová, E. *J. Org. Chem.* **2001**, *66*, 3940−3947.
[47] Takahashi, M.; Tanabe, H.; Nakamura, T.; Kuribara, D.; Yamazaki, T.; Kitagawa, O. *Tetrahedron* **2010**, *66*, 288−296.
[48] Stephani, R.; Cesare, V. *J. Chromatog. A* **1998**, *813*, 79−84.
[49] Tanaka, K.; Osuga, H.; Suzuki, H.; Shogase, Y.; Kitahara, Y. *J. Chem. Soc. Perkin Trans. 1* **1998**, 935−940.
[50] Gil-Av, E.; Schurig, V. *J. Chromatog. A* **1994**, *666*, 519−525.
[51] Trapp, O.; Schurig, V. *Tetrahedron: Asymmetry* **2010**, *21*, 1334−1340.
[52] Yamamoto, C.; Okamoto, Y. In *Enantiomer Separation. Fundamentals and Practical Methods*; Toda, F., Ed.; Kluwer Academic Publishers: Dordrecht, The Netherlands, 2004, pp. 301−322.
[53] Snyder, L. R.; Kirkland, J. J. *Introduction to Modern Liquid Chromatography*; Wiley-Interscience: New York, 1974.
[54] *Encyclopedia of Chromatography*; Cazes, J., Ed.; CRC Press: Boca Raton, FL, 2010.
[55] Wilstätter, R. *Chem. Ber.* **1904**, *37*, 3758−3760.
[56] Ingersoll, A. W.; Adams, R. *J. Am. Chem. Soc.* **1922**, *44*, 2930−2937.
[57] Konrad, G.; Musso, H. *Chem. Ber.* **1984**, *117*, 423−426.
[58] Feibush, B.; Grinberg, N. In *Chromatographic Chiral Separations*; Zief, M., Crane, L. J., Eds.; Marcel Dekker: New York, 1988, pp. 1−14.
[59] Pryde, A. In *Chiral Liquid Chromotography*; Lough, W. J., Ed.; Blackie: Glasgow, 1989, pp. 23−35.
[60] Souter, R. W. *Chromatographic Separations of Stereoisomers*; CRC: Boca Raton, Fl, 1985.
[61] *Chiral Separations by Liquid Chromatography*; Ahuja, S., Ed.; American Chemical Society: Washington, DC, 1991.
[62] Allenmark, S. *Chromatographic Enantioseparation. Methods and Applications*, 2nd ed.; E. Horwood: New York, 1991.
[63] Beesley, T. E.; Scott, R. P. W. *Chiral Chromatography*; Wiley: Chichester, England, 1998.
[64] *Chiral Liquid Chromotography*; Lough, W. J., Ed.; Blackie: Glasgow and London, 1989.
[65] *Chromatographic Chiral Separations*; Zief, M.; Crane, L. J., Eds.; Marcel Dekker: New York, 1988.
[66] *Chiral Separations*; Stevenson, D.; Wilson, I. D., Eds.; Plenum: New York, 1988.

[67] König, W. A. *The Practice of Enantiomer Separation by Capillary Gas Chromatography*; Huethig: Heidelburg, 1987.

[68] König, W. A. *Gas Chromatographic Enantiomer Separation with Modified Cyclodextrins*; Hüthig Buch Gmbh: Heidelberg, 1992.

[69] *Chiral Separations by Liquid Chromatography. ACS Symposium Series 471*; Satinder, A., Ed.; American Chemical Society: Washington, 1991.

[70] *Enantiomer Separation. Fundamentals and Practical Methods*; Toda, F., Ed.; Kluwer Academic Publishers: Dordrecht, The Netherlands, 2004.

[71] *Chiral Separation Techniques. A Practical Approach*; 3rd ed.; Subramanian, G., Ed.; Wiley-VCH: Weinheim, 2007.

[72] Schurig, V.; Bürkle, W. *J. Am. Chem. Soc.* **1982**, *104*, 7573–7580.

[73] Bürkle, W.; Karfunkle, H.; Schurig, V. *J. Chromatogr.* **1984**, *288*, 1–14.

[74] Easson, C. H.; Stedman, E. *Biochem. J.* **1933**, *27*, 1257–1266.

[75] Ogston, A. G. *Nature* **1948**, *162*, 963.

[76] Dalgliesh, C. E. *J. Chem. Soc.* **1952**, 3940–3942.

[77] Pirkle, W. H.; Pochapsky, T. C. *Chem. Rev.* **1989**, *89*, 347–362.

[78] Koscho, M. E.; Spence, P. L.; Pirkle, W. H. *Tetrahedron: Asymmetry* **2005**, *16*, 3147–3153.

[79] Davankov, V. A. *Chromatographia* **1989**, *27*, 475–482.

[80] Pirkle, W. H.; Hoover, D. J. *Top. Stereochem.* **1982**, *13*, 263–331.

[81] Vigneron, J. P.; Dhaenens, M.; Horeau, A. *Tetrahedron* **1973**, *29*, 1055–1059.

[82] Feringa, B. L.; Smaardijk, A.; Wynberg, H. *J. Am. Chem. Soc.* **1985**, *107*, 4798–4799.

[83] Wang, X. *Tetrahedron Lett.* **1991**, *32*, 3651–3654.

[84] Strijtveen, B.; Feringa, B. L.; Kellogg, R. M. *Tetrahedron* **1987**, *43*, 123–130.

[85] Chan, T. H.; Peng, Q. J.; Wang, D.; Guo, J. A. *J. Chem. Soc. Chem. Commun.* **1987**, 325–326.

[86] Coulbeck, E.; Eames, J. *Tetrahedron: Asymmetry* **2009**, *20*, 635–640.

[87] Pasquier, M. L.; Marty, W. *Angew. Chem., Int. Ed.* **1985**, *24*, 315–316.

[88] D'Arrigo, P.; Feliciotti, L.; Pedrocchi-Fantoni, G.; Servi, S. *J. Org. Chem.* **1997**, *62*, 6394–6396.

[89] Hoye, T. R.; Mayer, M. J.; Vos, T. J.; Ye, Z. *J. Org. Chem.* **1998**, *63*, 8554–8557.

[90] Muci, A. R.; Campos, K. R.; Evans, D. A. *J. Am. Chem. Soc.* **1995**, *117*, 9075–9076.

[91] Fleming, I.; Ghosh, S. K. *J. Chem. Soc. Chem. Commun.* **1994**, 99–100.

[92] Venkatraman, L.; Periasamy, M. *Tetrahedron: Asymmetry* **1996**, *7*, 2471–2474.

[93] Fleming, I.; Ghosh, S. K. *J. Chem. Soc. Perkin Trans. 1* **1998**, 2733– 2748.

[94] Wenzel, T. J. *Discrimination of Chiral Compounds Using NMR Spectroscopy*; Wiley-Interscience: Hoboken, NJ, 2007.

[95] Sievers, R. E. *Nuclear Magnetic Resonance Shift Reagents*; Academic: New York, 1973.

[96] Sullivan, G. R. *Top. Stereochem.* **1978**, *10*, 287–329.

[97] Fraser, R. R. In *Asymmetric Synthesis*; Morrison, J. D., Ed.; Academic: Orlando, 1983; Vol. 1, pp. 173–196.

[98] Grout, D. H. G.; Whitehouse, D. *J. Chem. Soc. Perkin Trans. 1* **1977**, 544–549.

[99] Pasto, D. J.; Borchardt, J. K. *Tetrahedron Lett.* **1973**, 2517–2520.

[100] Wenzel, T. J.; Morin, C. A.; Brechting, A. A. *J. Org. Chem.* **1992**, *57*, 3594–3599.

[101] Bulsing, J. M.; Sanders, J. M. K.; Hall, L. D. *J. Chem. Soc. Chem. Commun.* **1981**, 1201–1203.

[102] Raban, M.; Mislow, K. *Tetrahedron Lett.* **1965**, 4249–4253.

[103] Dale, J. A.; Dull, D. L.; Mosher, H. S. *J. Org. Chem.* **1969**, *34*, 2543 –2549.

[104] Dale, J. A.; Mosher, H. S. *J. Am. Chem. Soc.* **1973**, *95*, 512–519.

[105] Ward, D. E.; Rhee, C. K. *Tetrahedron Lett.* **1991**, *32*, 7165–7166.

[106] Kowalczyk, R.; Skarzewski, J. *Tetrahedron: Asymmetry* **2006**, *17*, 1370–1379.

[107] Dale, J. A.; Mosher, H. S. *J. Am. Chem. Soc.* **1968**, *90*, 3732–3738.

[108] Andersson, T.; Borhan, B.; Berova, N.; Nakanishi, K.; Haugan, J. A.; Liaaen-Jensen, S. *J. Chem. Soc. Perkin Trans. 1* **2000**, 2409–2414.

[109] Duret, P.; Waechter, A.-I.; Figadère, B.; Hocquemiller, R.; Cavé, A. *J. Org. Chem.* **1998**, *63*, 4717–4720.

[110] Latypov, S. K.; Seco, J. M.; Quiñoá, E.; Riguera, R. *J. Org. Chem.* **1996**, *61*, 8569–8577.

[111] König, W. A.; Nippe, K.-S.; Mischnick, P. *Tetrahedron Lett.* **1990**, *31*, 6867–6868.

[112] Roush, W. R.; Hoong, L. K.; Palmer, M. A. J.; Park, J. C. *J. Org. Chem.* **1990**, *55*, 4109–4117.

[113] Trost, B. M.; Belletire, J. L.; Godleski, S.; McDougal, P. G.; Balkovec, J. M.; Baldwin, J. J.; Christy, M. E.; Ponticello, G. S.; Varga, S. L.; Springer, J. P. *J. Org. Chem.* **1986**, *51*, 2370−2374.
[114] Enders, D.; Lohray, B. B. *Angew. Chem., Int. Ed. Engl.* **1988**, *27*, 581−583.
[115] Ohtani, I.; Kusumi, T.; Kashman, Y.; Kakisawa, H. *J. Am. Chem. Soc.* **1991**, *113*, 4092−4096.
[116] Omata, K.; Fujiwara, T.; Kabuto, K. *Tetrahedron: Asymmetry* **2002**, *13*, 1655−1662.
[117] García, R.; Seco, J. M.; Vázquez, S. A.; Quiñoá, E.; Riguera, R. *J. Org. Chem.* **2002**, *67*, 4579−4589.
[118] Dutcher, J. S.; Macmillan, J. G.; Heathcock, C. H. *J. Org. Chem.* **1976**, *41*, 2663−2669.
[119] Svatos, A.; Valterová, J.; Saman, D.; Vrkoc, J. *Collect. Czech. Chem. Commun.* **1990**, *55*, 485−490.
[120] Hiemstra, H.; Wynberg, H. *Tetrahedron Lett.* **1977**, 2183−2186.
[121] Pirkle, W. H. *J. Am. Chem. Soc.* **1966**, *88*, 1837.
[122] Weisman, G. R. In *Asymmetric Synthesis*; Morrison, J. D., Ed.; Academic: Orlando, 1983; Vol. 1, pp. 153−171.
[123] Wenzel, T. J.; Chisholm, C. D. *Chirality* **2011**, *23*, 190−214.
[124] Pirkle, W. H.; Tsipouras, A. *Tetrahedron Lett.* **1985**, *26*, 2989−2992.
[125] Djerassi, C. *Optical Rotary Dispersion*; McGraw-Hill: New York, 1960.
[126] Crabbe, P. *Optical Rotary Dispersion and Circular Dichroism in Organic Chemistry*; Holden-Day: San Francisco, 1965.
[127] Snatzke, G. *Optical Rotary Dispersion and Circular Dichroism in Organic Chemistry*; Heyden and Son: London, 1967.
[128] Mason, S. F. *Modern Optical Activity and the Chiral Discriminations*; Cambridge University: Cambridge, 1982.
[129] Caldwell, D. J.; Eyring, H. *The Theory of Optical Activity*; Wiley: New York, 1971.
[130] Charney, E. *The Molecular Basis of Optical Activity: Optical Rotatory Dispersion and Circular Dichroism*; Wiley: New York, 1979.
[131] *Circular Dichroism: Principles and Applications*; Nakanishi, K.; Berova, N.; Woody, R., Eds.; VCH: New York, 1994.
[132] Lightner, D. A. *Organic Conformational Analysis and Stereochemistry from Circular Dichroism Spectroscopy*; Wiley-VCH: New York, 2000.
[133] Barron, L. D. *Molecular Light Scattering and Optical Activity*, 2nd ed.; Cambridge University: Cambridge, 2004.
[134] Stephens, P. J.; Devlin, F. J.; Cheeseman, J. R.; Frisch, M. J.; Bortolini, O.; Besse, P. *Chirality* **2003**, *15*, S57−S64.
[135] Pecul, M.; Ruud, K. *Adv. Quant. Chem.* **2005**, *50*, 185−212.
[136] Bringmann, G.; Gulder, T. A. M.; Reichert, M.; Gulder, T. *Chirality* **2008**, *20*, 628−642.
[137] Stephens, P. J.; Devlin, F. J.; Pan, J.-J. *Chirality* **2008**, *20*, 643−663.
[138] Polavarapu, P. L. *Chirality* **2008**, *20*, 664−672.
[139] Brewster, J. H. *Top. Stereochem.* **1967**, *2*, 1−72.
[140] Pasteur, L. *Ann. chim. et phys.* **1848**, *24[3]*, 442−459.
[141] Krow, G.; Hill, R. K. *Chem. Commun.* **1968**, 430−431.
[142] Horeau, A. *Tetrahedron Lett.* **1969**, 3121−3124.
[143] Kagan, H. B. *Recl. Trav. Chim. Pays-Bas* **1995**, *114*, 203−205.
[144] Gawley, R. E. *J. Org. Chem.* **2006**, *71*, 2411−2416; corrigendum *J. Org. Chem.* **2008**, *73*, 6470.
[145] Yamaguchi, S.; Mosher, H. S. *J. Org. Chem.* **1973**, *38*, 1870−1877.
[146] Lambert, J. B.; Shurvell, H. F.; Verbit, L.; Cooks, R. G.; Stout, G. H. *Organic Structural Analysis*; Macmillan: New York, 1976.
[147] Legrand, M.; Rougrer, M. J. In *Stereochemistry. Fundamentals and Methods*; Kagan, H. B., Ed.; Georg Thieme: Stuttgart, 1977; Vol. 2, pp. 33−184.
[148] Charney, E. *The Molecular Basis of Optical Activity: Optical Rotatory Dispersion and Circular Dichroism*; Wiley: New York, 1979.
[149] Allenmark, S.; Gawronski, J.; Berova, N. *Chirality* **2008**, *20*, 605.
[150] Flack, H. D.; Bernardinelli, G. *Chirality* **2008**, *20*, 681−690.

Chapter 3

Enolate, Azaenolate, and Organolithium Alkylations

Originally, the term "carbanion" was used to refer to anionic reactive intermediates whose actual structure was rather poorly understood. In recent years, considerable advances have been made in developing the chemistry of carbanionic species and in understanding the structure of "carbanions," especially as regards the involvement of the metal [1−11]. In this chapter, we will focus on three types of intermediate that fall into the category of "carbanion." Our discussion will be further limited to alkylations noted in the chapter title: carbon−carbon bond-forming reactions with electrophiles such as alkyl halides that produce only one stereocenter, that being at the nucleophilic atom. Aldol and Michael additions are covered in Chapter 5, and reactions with heteroatom electrophiles that form carbon−oxygen or carbon−nitrogen bonds are discussed in Chapter 8.

Carbanions that have found use in asymmetric synthesis are stabilized by one or more substituents (Figure 3.1). By far the most common "carbanion stabilizing" functional group is the carbonyl. Although early texts (e.g., [12]) referred to the conjugate base of carbonyl compounds as carbanions, these species are now universally known as enolates (Figure 3.1a). Closely related to enolates are their nitrogen analogs, azaenolates (Figure 3.1b). As we will see, the fact that there is a substituent of the nitrogen is important to asymmetric synthesis because it provides a convenient foothold for attachment of a chiral auxiliary. The third class of "carbanion" comprises organometallic species such as organolithium compounds, in which the metal-bearing carbon is stereogenic (Figure 3.1c and d). Again, the negative charge is stabilized in these species, but not by resonance as is the case with enolates and azaenolates. Instead, heteroatoms on the lithiated carbon provide inductive stabilization; in some instances chelation may be involved.

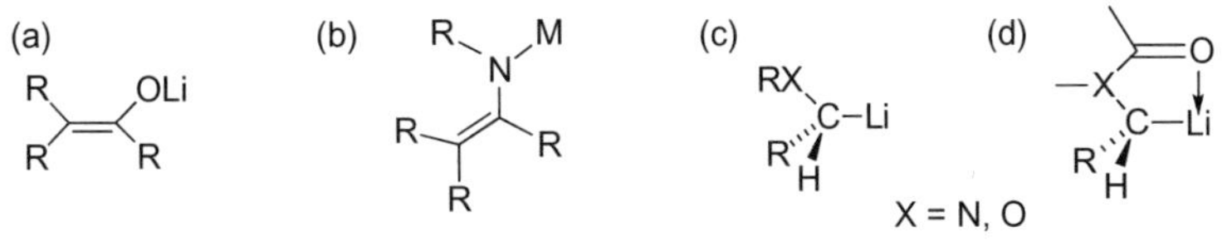

FIGURE 3.1   (a) Enolates, (b) azaenolates, (c) and (d) α-heteroatom organolithiums.

3.1 ENOLATES AND AZAENOLATES[1]

The deprotonation of a carbonyl gives a nucleophilic species that reacts with electrophiles such as alkyl halides to afford products of substitution at the α carbon. Because of this reactivity, the

1. For comprehensive coverage of enolate alkylations, see refs. [13−15].

intermediate species used to be drawn with a negative charge on carbon (Figure 3.2a). Resonance considerations later dictated that the negative charge should be placed on the more electronegative oxygen (Figure 3.2b). When enolate reactions are carried out in aqueous or alcohol solution, the ionic species are separately solvated, and this type of representation is justified. Unfortunately, the same usage has persisted, even when the reactions are conducted in aprotic solvents where solvent-separated ions are not likely to exist. A more appropriate notation is to affix the metal to the oxygen (Figure 3.2c). Spectroscopic and X-ray data have revealed that metal enolates are usually (always?) aggregated, both in the solid state and ethereal or hydrocarbon solution (Figure 3.2d). The illustrations in Figure 3.2 show the historical progression of these notations, which Seebach whimsically called "the route of the sorcerer's apprentice" [5].

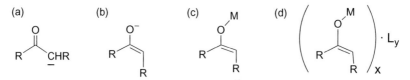

FIGURE 3.2 Various notations for an enolate, from (a) a naked carbanion or (b) an enolate, *via* (c) a metal enolate, to (d) supramolecular aggregates (M, metal) [5].

Enolates may form supramolecular² species such as dimers, tetramers, or hexamers, and these species are often in equilibrium (Scheme 3.1). Enolates may also form mixed aggregates with added salts or with secondary or tertiary amines. The existence of such species has been proven in the solid state by X-ray crystallography, and colligative effects and NMR studies have confirmed their existence in solution (reviews: refs. [5,6,13,18−23]; see also: refs. [24−27]). Interestingly, dimers are even found in crystals of tetrabutylammonium malonates and cyanoacetates [28], indicating that a metal is not necessary for supramolecular organization!

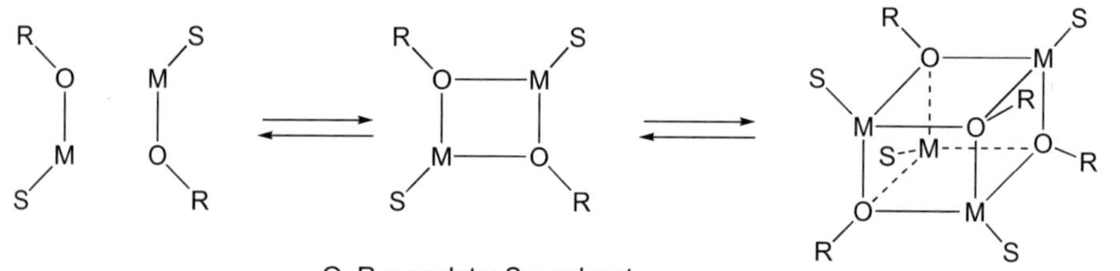

O–R = enolate; S = solvent

SCHEME 3.1 Equilibrating dimeric and tetrameric enolate aggregates. Formal charges are not shown. There is probably more than one solvent molecule coordinated to the monomers (left) and dimer (middle).

Chemical evidence also confirms the presence of supramolecular complexes in enolate reactions. For example, added salts can affect the product ratio of enolate alkylations [29−34]. Evidence of mixed aggregation between enolates and secondary amines includes experiments such as those illustrated in Scheme 3.2, where quenching with D_2O or MeOD produces little or no deuterium incorporation, indicating that the enolate is protonated by the secondary amine from within a supramolecular aggregate [35−40].

---

2. The term supramolecular was coined by Lehn to refer to "organized entities ... that result from the association of two or more chemical species held together by intermolecular forces" [16,17].

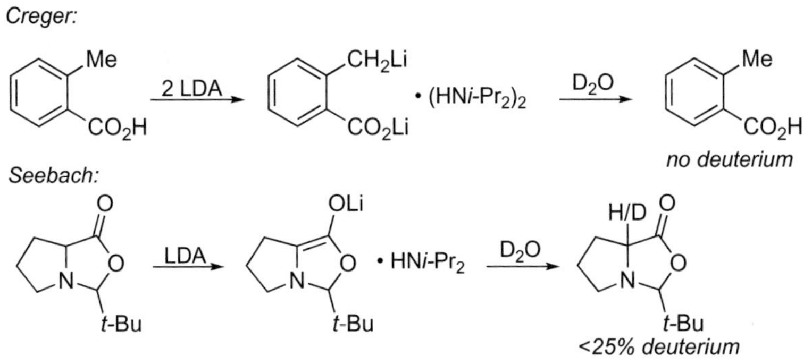

Creger:

Seebach:

SCHEME 3.2 Enolate · diisopropylamine complexes may not incorporate deuterium upon quenching with D$_2$O. (a) Creger demonstrated the phenomenon with *o*-toluic acid [35]. (b) Seebach showed that lactone enolates behave similarly [41].

Early evidence of association of a carbanion with the conjugate acid of a base was obtained by the Cram group in the 1960s. The phrase "conducted tour mechanism" was coined by Cram to describe the removal of a proton by a base and its subsequent return to a different face of the same molecule from which it was removed [1,42−44]. The conducted tour mechanism was postulated to explain the observation that rates of racemization of carbon acids were faster than hydrogen−deuterium exchange in solutions of potassium *tert*-butoxide/ *tert*-butyl alcohol-d_1 (originally termed isoracemization or isoinversion [43,44]). In some systems, this process was envisioned as a rotation of the carbanion within the solvent cage [42]. In others cases, a relay of ROH from one hydrogen bonding site to another on the substrate could occur without ever fully dissociating to a solvent-separated ion pair [44]. Thus, "the basic catalyst takes hydrogen or deuterium on a 'conducted tour' of the substrate from one face of the molecule to the [other]" (ref. [1], p. 101).

The internal return of the proton to the enolate (in preference to deuterium) in the examples of Scheme 3.2 is related to Cram's conducted tour: the secondary amine (conjugate acid of lithium diisopropyl amide (LDA)) forms a mixed aggregate with the metalated anion, such that the reprotonation (and perhaps conformational motion in a conducted tour) is "intrasupramolecular."

A complete understanding of enolate chemistry must include knowledge of the aggregation of the enolate species involved [5]. Consider the reaction of an enolate with an alkyl halide as it may have been depicted over the years (Scheme 3.3), progressing from the simple carbanion alkylation to the reaction of supramolecular aggregates. The mechanism depicted in Scheme 3.3d may be the closest to reality for the reaction of an enolate with an alkyl halide, but this picture is dependent on the individual system under study. For our purposes, we can rationalize most enolate reactions by considering metal enolates as monomers (as in Scheme 3.3c), while realizing that the other coordination sites of the metal may be occupied by ligands that may be solvent molecules, additives such as HMPA, DMPU, or TMEDA,[3] anions of added salts, or another molecule of enolate. The interested reader is referred to Seebach's review to see the types of supramolecular complexes that may arise in the chemistry of lithium enolates [5].

---

3. HMPA, hexamethylphosphoramide. DMPU, *N,N'*-dimethylpropyleneurea. TMEDA, tetramethylethylenediamine. All are additives that coordinate metals and may affect or alter the aggregation state. Note that mechanistic interpretation of the effect of additives, especially TMEDA in THF solvent, is risky [45].

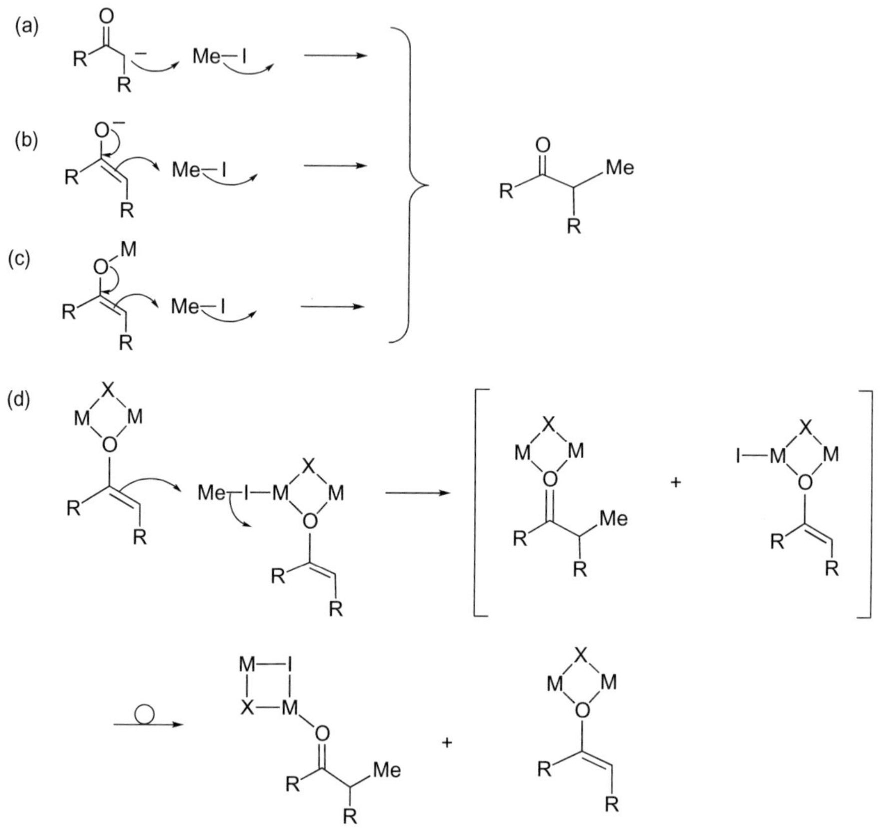

SCHEME 3.3 Alkylation of an "enolate" (a) naked carbanion, (b) naked enolate, (c) metal enolate, and (d) supramolecular alkylation and rearrangement [5].

3.1.1 Deprotonation of Carbonyls[4]

A number of bases may be used for deprotonation, but the most important ones are lithium amide bases such as those illustrated in Figure 3.3.[5] Although other alkali metals may be used with these amides, lithium is the most common. Amide bases efficiently deprotonate virtually all carbonyl compounds, and do so regioselectively with cyclic ketones such as 2-methylcyclohexanone (*i.e.*, C_2 vs. C_6 deprotonation) and stereoselectively with acyclic carbonyls (*i.e.*, *E(O)*- vs. *Z(O)*-[6] enolates). If the carbonyl is added to a solution of the lithium amide, deprotonations are irreversible and kinetically controlled [48–50]. Under such conditions, the configuration of an acyclic enolate is determined during the deprotonation step, and subsequent isomerization is probably not important [50].

---

4. For reviews on enolate formation, see refs. [22,46].
5. LDA is stable in both ether and THF at room temperature for 24 h, but LTMP has a half-life of 12 h in THF and only 4 h in ether at room temperature [47].
6. See glossary (Section 1.11), for the definition of the *E(O)/Z(O)* descriptors of enolate geometry.

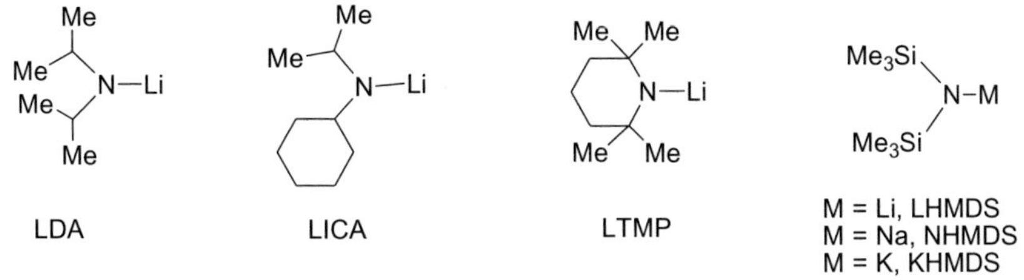

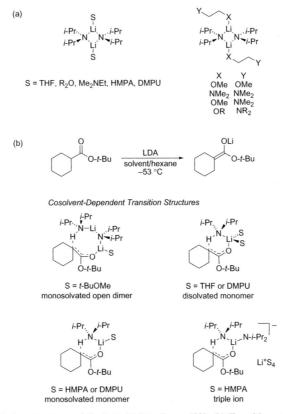

FIGURE 3.3   LDA, lithium diisopropyl amide; LICA, lithium isopropyl cyclohexyl amide; LTMP, lithium tetramethylpiperidide; and LHMDS, lithium hexamethyldisilyl amide.

The deprotonation of a carbonyl by a metal amide base can be very complicated. For example, although LDA is known to be a disolvated dimer at low temperature in quite a few solvents (Figure 3.4a), analysis of kinetics reveals a number of possible stoichiometries of metal, amide, and solvent in the transition state of an ester deprotonation (Figure 3.4b) [22,51–53]. In *tert*-butyl methyl ether (*t*-BuOMe), the deprotonation of *tert*-butyl cyclohexanecarboxylate involves an open LDA dimer in an 8-membered ring transition structure, with one solvent coordinated to one of the lithiums. In THF, a disolvated LDA monomer is involved. In DMPU, transition structures that include disolvated and monosolvated monomers are competitive. In HMPA, monosolvated monomers and triple ions compete.

FIGURE 3.4   (a) Solution structures of disolvated LDA dimers [22]. (b) Transition structures for deprotonation of an ester, depending on solvent [22,52,53].

In 1976, long before these mechanistic insights were realized, Ireland proposed a mechanistic model to rationalize the preferred formation of $E(O)$- or $Z(O)$-enolates of acyclic esters, ketones, and amides [48,54]. In the Ireland model, shown in Scheme 3.4, the deprotonation was proposed to proceed *via* an LDA monomer that is coordinated to the carbonyl oxygen and organized into a chair-like transition structure. Coordination of solvent molecules to the lithium was not specified. In Ireland's rationale, the deprotonation to the $E(O)$- and $Z(O)$-enolates are controlled by a balance of 1,2-eclipsing interactions between an α-methyl group and the carbonyl substituent, R, and 1,3-diaxial interactions between the nitrogen ligand and α-methyl. For esters, the atom attached to the carbonyl is oxygen, and the alkyl group (even one as large as a *tert*-butyl) can rotate away from the α-methyl and have no appreciable influence on the energy of transition structure. In such cases, $E(O)^{\ddagger}$ is more stable than the $Z(O)^{\ddagger}$ due to the destabilizing 1,3-diaxial interaction of the nitrogen ligand and the α-methyl in the latter. As the steric requirements of the carbonyl substituent increase (R = *t*-Bu, Ph, NR$_2$), $A^{1,3}$-strain destabilizes $E(O)^{\ddagger}$ relative to $Z(O)^{\ddagger}$, and the $Z(O)$ enolate is formed exclusively. Molecular mechanics calculations confirm the general validity of these arguments [55], but other analyses have come to different conclusions about the conformation of the transition structure [51,56]. Whether or not the actual transition structure is chair-like, the model has endured due to its simplicity and predictive value.

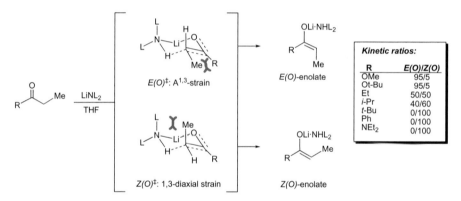

R	E(O)/Z(O)
OMe	95/5
Ot-Bu	95/5
Et	50/50
i-Pr	40/60
t-Bu	0/100
Ph	0/100
NEt$_2$	0/100

SCHEME 3.4 Ireland model transition structures for the deprotonation of acyclic carbonyls [48,54]. The red brackets highlight the sources of strain in the transition states: $A^{1,3}$-strain increasing as the enolate develops in $E(O)^{\ddagger}$ and 1,3-diaxial strain in $Z(O)^{\ddagger}$. For structural information from X-ray data, see ref. [37].

The notion of a highly organized transition state has inspired an elegant solution to the problem of $E(O)/Z(O)$ selectivity in the more difficult situation of deprotonation of α,α'-disubstituted esters [59]. Scheme 3.5a illustrates the problem: deprotonation of an ester having a stereocenter alpha to the carbonyl could be deprotonated to either an $E(O)$ or a $Z(O)$ enolate *via* the two diastereomeric Ireland-transition structures shown. Zakarian noted that similarly sized α-substituents would probably not afford much difference in transition state energies during the formation of the enolate diastereomers, and hypothesized that introduction of chirality into the amide base might have the desired effect. One example is shown in Scheme 3.5b. Lithium diisopropyl amide affords a 67:33 mixture of $Z(O)/E(O)$ diastereomers, whereas the two enantiomers of the chiral base shown, originally developed by Koga [57,58], afford either $Z(O)$ or $E(O)$ enol ethers with 92% diastereoselectivity [59]. A transition state model to rationalize the steric course of the deprotonation is not obvious, but this is a nice example of reagent-based stereoselectivity.

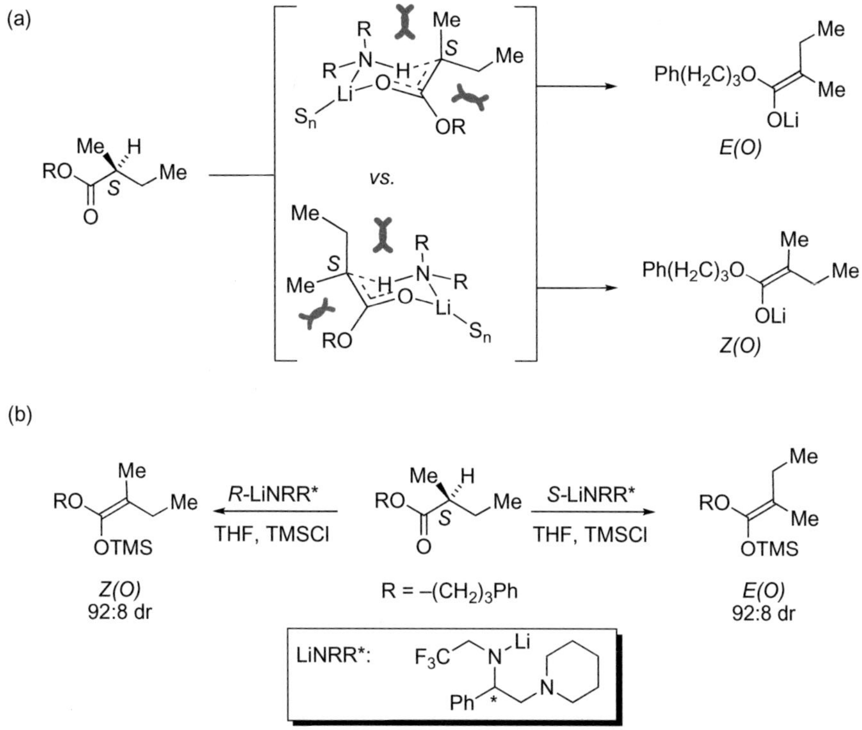

(a)

(b)

SCHEME 3.5 Stereoselective deprotonation of branched esters. (a) Ireland-transition structures for formation of *E(O)* and *Z(O)* enolates with little diastereoselectivity due to the similarity of methyl and ethyl substituents. (b) Stereoselective formation of *E(O)* and *Z(O)* enolates using a chiral base [59].

In the presence of coordinating additives such as HMPA, DMPU, or TMEDA, the trend outlined in Scheme 3.4 may not hold [48,60–62]. As illustrated in Scheme 3.6a, LDA deprotonation of 3-pentanone in the presence of HMPA affords a 5:95 mixture of *E(O)*- and *Z(O)*-enolates under conditions of thermodynamic control (equilibration by reversible aldol addition) [54,60], but a 50:50 mixture under kinetic control [60,61]. For esters (Scheme 3.6b), thermodynamic equilibration of enolates is less likely, but additives can still affect the selectivity. Using LDA in THF for example, deprotonation of ethyl propionate is 94% *E(O)*-selective, but in THF containing 45% DMPU, deprotonation is 93–98% *Z(O)*-selective [48].

For α-silyloxyacetates, Yamamoto reported the selective deprotonations shown in Scheme 3.7, which afford either *E(O)* or *Z(O)* enolates selectively, depending on the base used. These examples are consistent with the trend noted by Ireland, in that HMPA favors formation of the *Z(O)*-enolate, but other factors may also be involved. One possible explanation for formation of the *Z(O)* ester enolate, illustrated in Scheme 3.7b, invokes chelation by the silyloxy group with LHMDS in THF/HMPA. Note, however, that the Ireland argument could also be applied here: as shown in Scheme 3.7a, with LTMP as the base, the *Z(O)*‡ Ireland-transition structure would be destabilized considerably by 1,3-diaxial interactions between the silyloxy and the bulky tetramethylpiperidine. With LHMDS in THF/HMPA, the lithium would be

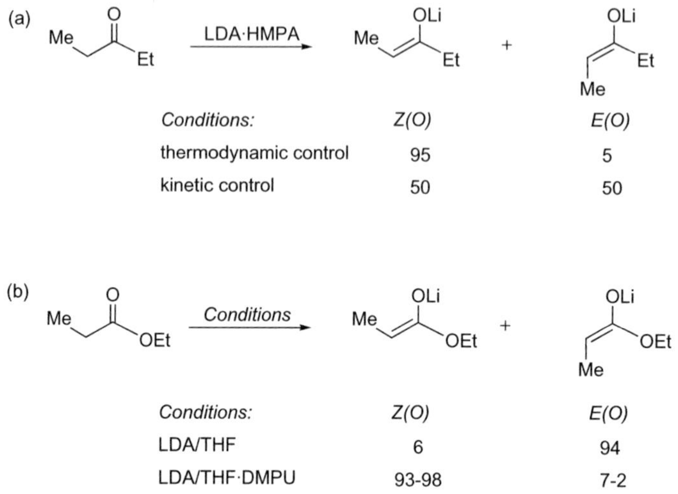

SCHEME 3.6 Selectivity in deprotonations of esters: (a) Equilibration of 3-pentanone enolates through reversible aldol addition (slight excess of ketone) affords high selectivity for the *Z(O)* diastereomer [54,60,61]; (b) Adding DMPU reverses the "normal" *E(O)* selectivity to *Z(O)* for ethyl propionate [48].

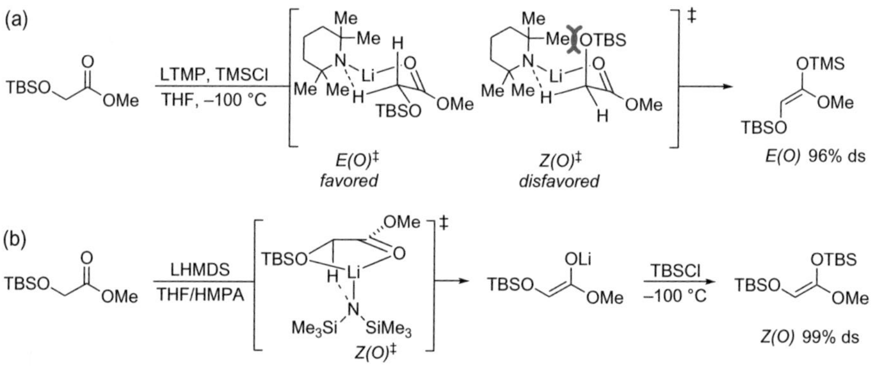

SCHEME 3.7 Selective formation of either *E(O)*- or *Z(O)*-enolates of silyloxyacetates [62]. (a) The authors rationalize the formation of the *E(O)* enolate using the Ireland model. Note the especially severe 1,3-diaxial interaction between the tetramethylpiperidide and the –OTBS group in the disfavored *Z(O)* transition structure. (b) With LHMDS/HMPA, a chelated monomer is postulated to explain the selective formation of the *Z(O)* enolate. However, see the text for an alternative rationale.

solvated by the HMPA, and the 1,3-diaxial interactions would be attenuated as explained above, and they could also be diminished because of the longer Si–N bond distance (compared to C–N) in the amide base. Independent of mechanism, *the bottom line is that both ester E(O)- and Z(O)-enolates can be produced selectively.*

Simple, monomeric transition structures fail to account for phenomena such as changes in selectivity as the reaction proceeds and for the effect of added lithium salts. For example, the deprotonation of 3-pentanone by LTMP affords a 97:3 *E(O)/Z(O)* selectivity at 5% conversion, but <90:10 selectivity at ≥80% conversion [50]. Moreover, the presence of 0.3–0.4 equivalents of lithium chloride or ≥1.0 equivalents of lithium bromide enhances the *E(O)/Z(O)* selectivity (at complete conversion) to 98:2.[7,8] LTMP is one of the most sterically hindered lithium amides known, and there is some evidence that the formation of mixed aggregates is sterically driven: mixed dimerization with sterically unhindered LiX species provides a simple means to alleviate the steric demands of LTMP aggregates (Scheme 3.8). For example, a 50:50 mixture of cyclohexanone enolate and LTMP shows significant heterogeneous aggregation, whereas a similar mixture with LDA shows <5% mixed dimer [63]. The observation of decreased selectivity as enolate accumulates or with added lithium halide [50], as well as the observation of equilibrating mixed aggregates of LTMP, lithium enolates, and lithium halides [63], led to the conclusion that lithium salt dependent selectivities stem from the intervention of mixed aggregates in the product determining transition state(s) [50]. Lithium bis(2-adamantyl) amide, which is even more hindered than LTMP, forms mixed aggregates with ketone enolates but not with lithium halides, and enolizes ketones with a very high degree of $E(O)/Z(O)$ selectivity [26]. The $E(O)/Z(O)$ ratio of ketone enolates is also dependent on the amount of hexane in the THF solvent [64].

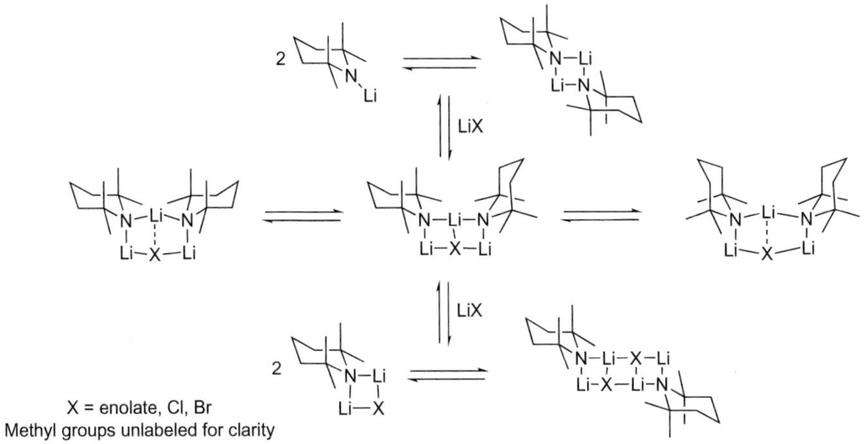

X = enolate, Cl, Br
Methyl groups unlabeled for clarity

SCHEME 3.8 Proposed dynamic equilibria of LTMP and added lithium salts [63].

3.1.2 The Transition State for Enolate Alkylations

The earliest work on the origin of stereoselectivity of enol and enolate reactions was done many years ago in the steroid arena [65,66], at the beginning of the modern era of stereochemistry. More recent efforts have focused on the stereoelectronic effects exerted by the frontier orbitals on the trajectory of electrophilic attack [67,68]. Specifically, Agami suggested that the approach trajectory for the electrophile should be as shown in Figures 3.5a and b [69–71]. Using *ab initio* methods, Houk found a transition structure for the alkylation of acetaldehyde enolate with methyl fluoride, which agreed with Agami's prediction

---

7. For a simple protocol for the preparation of LTMP/LiBr solutions by deprotonation of TMP · HBr with butyllithium, which affords a 50:1 ratio of the $E(O)$ and $Z(O)$ enolates of 3-pentanone, see ref. [50].
8. Curiously, the presence of ≥1.0 equivalents of lithium chloride leaves the $E(O)/Z(O)$ ratio unchanged from the ratio in the absence of lithium halide [50].

of Figure 3.5a [72]. An "out-of-perpendicular" component (as in Figure 3.5b) was not found, but the methyl fluoride transition state is later than that of methyl iodide, and a structure associated with an earlier transition state (less bond making between nucleophile and electrophile) would probably exhibit this feature [72]. Note the pyramidalization of the α-carbon in Houk's transition structure, a feature that crops up in a number of calculated transition structures [73,74] and that appears to be important in other reaction types as well [37,75]. Often, pyramidal sp^2 atoms are found in X-ray crystal structures of ground state reactants such as enones.[9]

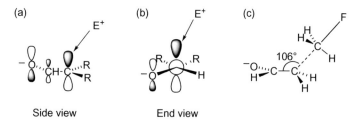

FIGURE 3.5 Theoretical approach trajectories (drawn in the plane of the paper) for electrophilic attack at an enolate carbon. (a) and (b) Agami's trajectory [69–71] and (c) Houk's trajectory [72].

Studies on the stereoselective alkylation of conformationally rigid cyclohexanone enolates (summarized in ref. [20]) indicate that the transition state is early. In these systems, axial attack affords a product in a chair conformation while equatorial attack affords a twist-boat (Scheme 3.9). If the relative stability of these conformers was felt in the transition state, significant selectivity would ensue. However, the low selectivities observed (55:45 for the reaction of the lithium enolate of 4-*tert*-butylcyclohexanone with methyl iodide [76]) suggest an early transition state according to the Hammond postulate. Somewhat higher selectivity for axial deuteration [76] is consistent with a less exothermic reaction. A higher propensity for axial alkylation (70:30) with a tetrabutylammonium cation [77] or when the *tert*-butyl group is in the 3-position (80:20) suggests that other factors (aggregation?) are also at work.

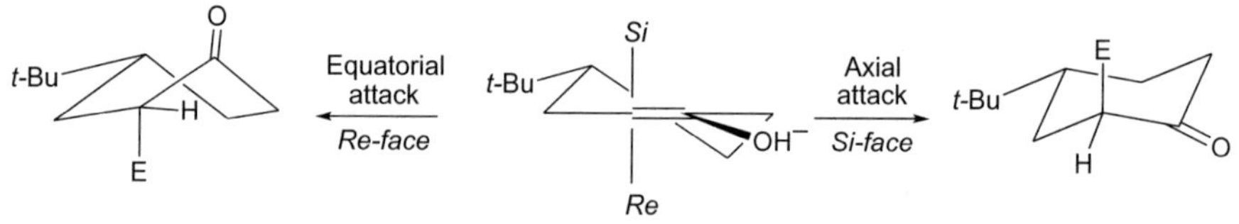

SCHEME 3.9 Equatorial and axial approach of an electrophile to a cyclohexanone enolate.

Stereocenters at the β-position can have an important effect on the differentiation of the enolate faces. The conformation of the 2-3 (allylic) bond of an acyclic enolate is governed

---

9. See Section 4.4.3 (Figure 4.32) for a discussion of the phenomenon of pyramidalization in nucleophilic additions to trigonal atoms.

Chemical Evolution and the Origin of Homochirality

The amino acids in proteins are homochiral. That is, they all have the same L absolute configuration. Homochirality in peptides is responsible for the secondary and tertiary structures of proteins. The direction of the turn of an α-helix and the direction of twist of a β-sheet are the direct result of the L configuration of the amino acid constituents of the peptide chain. Likewise, carbohydrates such as ribose, deoxyribose, glucose, and galactose are homochiral and have the D absolute configuration. The nucleotides that make up DNA and RNA are homochiral, and their absolute configuration is responsible for the helical secondary structure of nucleic acids. *If amino acids and carbohydrates were not homochiral, there would be no uniform secondary structural features in either proteins or nucleic acids, and life could not exist.* Thus, *chemical evolution* that produced a homochiral pool of amino acids and carbohydrates *must have preceded* biological evolution. How this might have occurred remains one of the great mysteries of science.

Virtually all of the reactions in this book that generate a stereogenic center proceed by the addition of a fourth ligand to heterotopic faces of a prochiral trigonal atom. Except in a chiral environment, such an event is random and will occur with equal probability from either face.[1] There is a very small difference in energy between two enantiomers, due to the fact that the weak internuclear force is parity nonconserved [1]. Thus, even an atom is "chiral". For the amino acid alanine, the difference in energy between the enantiomers is 10^{-14} J/mol, resulting in an excess of 1 million molecules of the L enantiomer in a mole of racemate. This corresponds to an equilibrium constant of 1.0000000000000017, that is, equality until you get to the 19th significant figure. It has been suggested that, over geologic time, this tiny excess might be enhanced to enantiopurity [2].

Another possibility was suggested by Sir Frederick Charles Frank, that autocatalysis might provide an answer [3]. That is, enantiomer enrichment could occur if "a chemical substance ... is a catalyst for its own production and an *anti*-catalyst for the production of its [enantiomer]." Although not relevant to the production of amino acids, an example of such an autocatalytic reaction is shown in Scheme 4.11, in which a very small excess of one enantiomer can provide highly enriched product after only a few cycles. No such autocatalytic process has yet been discovered for the production of biomolecules under prebiotic conditions.

Small enrichments of several quaternary amino acids have been found in meteorites on earth [4], and it turns out that homochiral crystals (known as conglomerates) of some amino acids are more soluble than heterochiral crystals that contain equal numbers of both enantiomers in the unit cell (known as racemic compounds) [5]. An interesting theory for the origin of the enantiomer excess is that synchroton radiation from supernovae contains circularly polarized light [6]. Electrons circulating around a neutron star would emit circularly polarized light of opposite chirality on either side of the circulation plane. In 1977, Bonner showed that irradiation of racemic leucine with right circularly polarized light ($\lambda = 212.8$ nm) preferentially photolyzed *R*-leucine, whereas left circularly polarized light preferentially photolyzed *S*-leucine. Enantiomer excesses of 2–2.5% were found.

53.8 : 46.2 er
55.2 : 44.8 er

57.6 : 42.4 er

51.4 : 48.6 er

51.4 : 48.6 er

L:D enantiomer ratios
found on the Murchison meteor

Interesting hypotheses have been offered for how enantiopurity could have evolved from small excesses of one enantiomer in solution. Vlieg showed that irradiation with circularly

---

1. If you flip a coin once, there is an exactly 50:50 chance that it will land on heads. Curiously, the more times you flip, the less likely it is that it will land on heads exactly half the time. However, the chance that the stochastic preference for either heads or tails occurring preferentially is the same.

polarized light, coupled with grinding of a saturated acetonitrile solution of a racemic phenylgly-cine Schiff base, converted the mixture into a single enantiomer. The configuration depended only on the chirality sense of the circularly polarized light [7]. Breslow has shown that quaternary α-amino acids such as those found on the Murchison meteorite can transfer chirality to α-keto acids under conditions that could have occurred on prebiotic earth [5]. He also showed that small enrichment of one enantiomer can be amplified by preferential kinetic dissolution in water, which takes advantage of the higher solubility of conglomerate crystals over racemic compounds.

(L)-α-methylvaline
96:4 er

(L)-phenylalanine
68:72 er

Blackmond has shown that, in a ternary phase system, the enantiomer ratio of the eutectic composition in solution for a number of partially enriched amino acids is quite high [8]. The amino acids that are racemic at their eutectic, threonine and arginine, form conglomerate crystals, whereas the other 20 proteinogenic amino acids form racemic compounds. Remarkably, serine is essentially enantiopure at the eutectic! In an interesting experiment shown below, Viedma and Blackmond showed that a 10% excess of L-aspartic acid conglomerate (homochiral) crystals over D-crystals evolves into enantiopure L-crystals when they are covered by a saturated solution in which racemization is promoted. The process is accelerated by stirring in the presence of glass beads, which serves to keep the size of the crystals small. This results in a faster dissolution of the small crystals and—at the same time—provides a larger surface area for Ostwald ripening[2] of crystals of the major enantiomer. The net result is an exponential growth of the crystals of the major enantiomer [9].

		Racemization in solution		
Conglomerate L-Asp crystals				Conglomerate D-Asp crystals

Start:	55	:		45
End:	>99	:		1

None of these experiments conclusively show how chemical evolution to homochirality *did* occur, but they demonstrate how chemical evolution *could have* occurred. An interesting

---

2. Ostwald ripening is the phenomenon in which molecules on the surface of a small particle dissolve and then crystal-lize on the surface of a larger particle. The rationale is that molecules on the interior of a crystal are surrounded by, for example, six neighbors in the case of a cubic crystal, whereas the molecules on the surface are surrounded by ≤5 neighbors. If the solution becomes supersaturated, the molecules tend to condense on the surface of a larger particle.

question remains: is it coincidence that the major enantiomer of alanine on earth is also the most stable one?

References

[1] Mason, S. F. *Chemical Evolution. Origin of the Elements, Molecules, and Living Systems*; Oxford University Press: Oxford, 1991.

[2] Mason, S. F.; Tranter, G. E. *Proc. R. Soc. London* **1985**, *397*, 45−65.

[3] Frank, F. C. *Biochem. Biophys. Acta* **1953**, *11*, 459−463.

[4] Pizzarello, S. *Acc. Chem. Res.* **2006**, *39*, 231−237.

[5] Levine, M.; Kenesky, C. S.; Mazori, D.; Breslow, R. *Org. Lett.* **2008**, *10*, 2433−2436.

[6] Bonner, W. A. *Origins Life* **1991**, *21*, 59−111.

[7] Noorduin, W. L.; Bode, A. A. C.; van der Meijden, M.; Meekes, H.; van Etteger, A. F.; van Enckevort, W. J. P.; Christianen, P. C. M.; Kaptein, B.; Kellogg, R. M.; Rasing, T.; Vlieg, E. *Nature Chemistry* **2009**, *1*, 729−732.

[8] Klussmann, M.; Iwamura, H.; Mathew, S. P.; Wells, D. H.; Pandya, U.; Armstrong, A.; Blackmond, D. G. *Nature* **2006**, *441*, 621−623.

[9] Viedma, C.; Ortiz, J. E.; Torres, T. d.; Izumi, T.; Blackmond, D. G. *J. Am. Chem. Soc.* **2008**, *130*, 15274−15275.

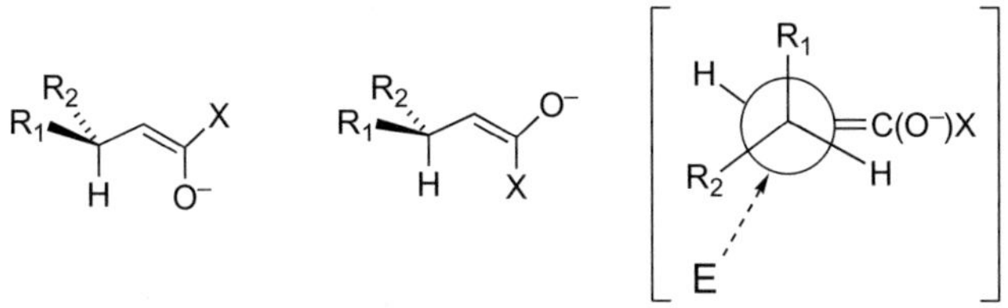

FIGURE 3.6   Ground state (left and center) and transition state conformations of β-substituted enolates.

primarily by $A^{1,3}$-strain (see glossary, section 1.6, and ref. [78]) such that the most stable conformation has the smallest substituent eclipsing the double bond, independent of enolate geometry (Figure 3.6).

In the transition state, the α-carbon is pyramidalized, and the substituents on the β-carbon are rotated such that the substituent that is the better σ-donor (R_1 in Figure 3.6) is perpendicular to the double bond [74,79,80]. The opposite face is then preferred by the approaching electrophile, as shown on the right in Figure 3.6. This *"antiperiplanar* effect" is a phenomenon that occurs quite often in organic chemistry,[10] and arises because of the favorable overlap of an allylic σ-bond with the π-orbital of the enolate. The resulting perturbation raises the energy of the enolate HOMO and renders it more reactive [80]. For substituents that differ only in steric bulk (R_1 and R_2 in Figure 3.6) the selectivity is small (65:35), but the example in Scheme 3.10 illustrates how the electronic effect of an alkoxy substituent can profoundly influence face-selectivity by proper alignment of its lone pairs (Scheme 3.10).

---

10. Another type is the Felkin−Anh descendant of Cram's rule, which is discussed in detail in Chapter 4 (Section 4.1).

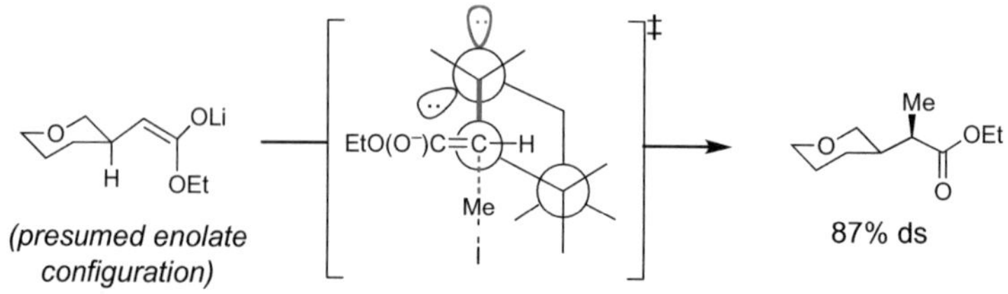

SCHEME 3.10   Stereoselective alkylation of an ester enolate determined solely by stereoelectronic effects. The oxygen lone pair highlighted in red is *antiperiplanar* to the C—C bond (also red) that is itself *antiperiplanar* to the favored approach of the electrophile [80].

3.1.3 Enolate and Azaenolate Alkylations with Chiral Nucleophiles

In Section 1.3, the relationship of extant chirality in a reacting system to any newly created stereocenters was categorized according to the relationship of the former and the latter in a metal complex in the transition state. Thus, *intraligand* asymmetric induction occurs when both the "old" and the "new" stereocenters are on the same ligand of the metal, and *interligand* asymmetric induction occurs when the existing stereocenters are on another ligand. Evans [20] has grouped chiral enolate systems into three categories, based on the location of any existing stereocenter relative to any rings present (intrannular if it is within a ring and extrannular if it is not). In the present context, these categories are subclasses of intraligand asymmetric induction, as shown in Figure 3.7: *intraannular*, in which the existing stereocenter is contained in a ring that is bonded to the enolate at two points; *extraannular*, in which the moiety containing the stereocenter is bonded to the enolate at one point; or *chelate-enforced intraannular*, in which the stereocenter is contained in a chelate ring containing the enolate metal.

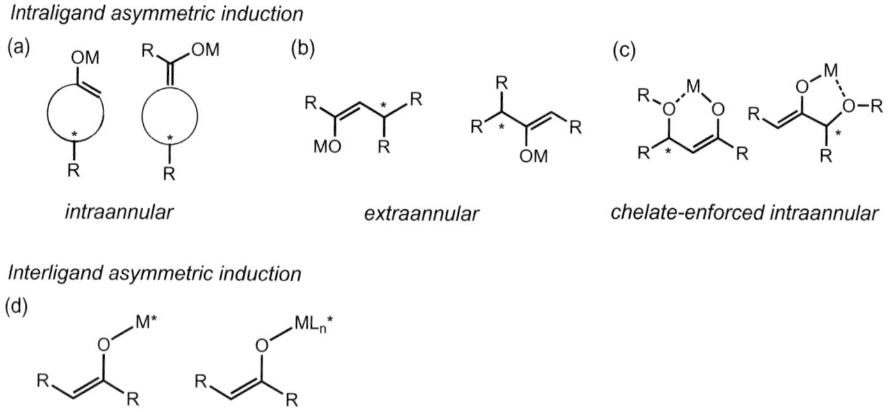

Intraligand asymmetric induction

Interligand asymmetric induction

FIGURE 3.7   Two categories of asymmetric induction are intraligand (a–c) and interligand (d). The former may be subdivided [20] into (a) intraannular, (b) extraannular, and (c) chelate-enforced intraannular.

The widespread use of enolate alkylations for carbon–carbon bond formation has led to the development of a large number of methods for asymmetric synthesis, and the search goes on. The following discussion is intended to highlight enolate alkylation methods that seem to have broad applicability or which illustrate one of the categories mentioned above.

An instructive introduction to intraannular alkylations is the "self-regeneration of chirality centers" concept introduced by Seebach [81–88]. Scheme 3.11 illustrates the concept and Table 3.1 lists several representative examples. A chiral educt, such as an amino acid derivative, is condensed with pivaldehyde or benzaldehyde. This derivatization creates a new stereocenter selectively; whether cis or trans is favored depends on the nature of the heterocycle, as explained below. The new stereocenter at C-2 of the heterocycle then controls the selectivity of the subsequent alkylation by directing the electrophile to the face of the enolate opposite the C-2 acetal substituent, a good example of intraannular 1,3-asymmetric induction. After purification of the alkylation product, hydrolysis affords enantiomerically pure compounds whereby the existing tertiary stereocenter has been transformed to a quaternary one. Whether the replacement of hydrogen by an alkyl group occurs with retention or inversion depends on a number of factors, as explained below.

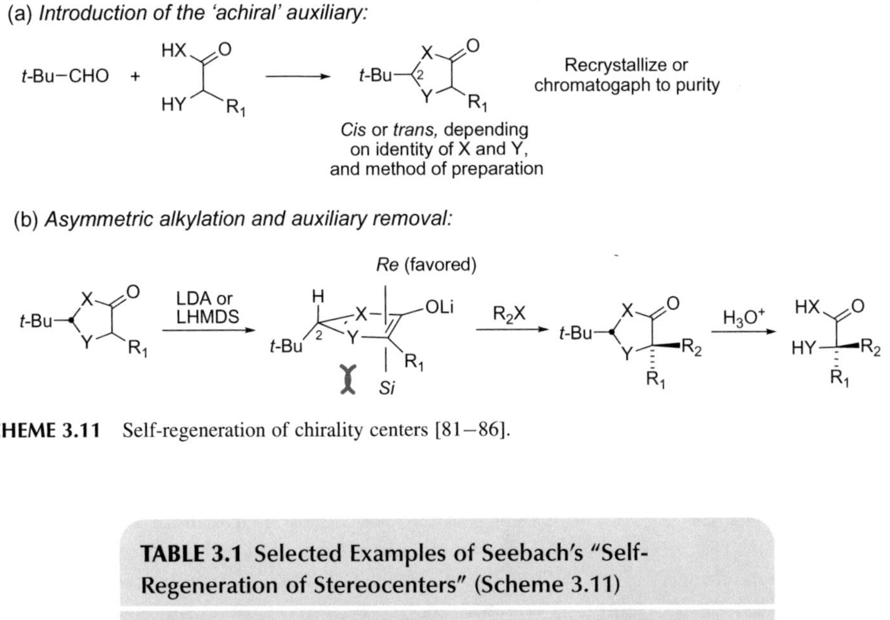

(a) *Introduction of the 'achiral' auxiliary:*

(b) *Asymmetric alkylation and auxiliary removal:*

SCHEME 3.11   Self-regeneration of chirality centers [81–86].

TABLE 3.1 Selected Examples of Seebach's "Self-Regeneration of Stereocenters" (Scheme 3.11)

X/Y	R_1/R_2	dr[a]	% Yield	Ref.
O/O	Me/Et	94:6	82	[81]
O/O	Ph/n-Pr	90:10	84	[81]
O/S	Me/allyl	>96:4	92	[81]
O/NCOPh	Me/Bn	>96:4	93	[83]
O/NCOPh	Bn/Me	>96:4	88	[83]
O/NCOPh	i-Pr/Me	100:0	53	[83]
MeN/NCOPh	Me/Et	>90:10	90	[84]
MeN/NCOPh	Me/Bn	>90:10	73	[84]
MeN/NBoc[b]	H/Bn	100:0	64	[82]
MeN/NBoc[b]	H/allyl	100:0	85	[82]
MeN/Cbz[b]	H/i-Pr	100:0	59	[82]

[a]Diastereomer ratio after purification.
[b]Obtained enantiomerically pure by resolution.

There are two steps in the sequence (Scheme 3.11). The cyclization in the first step can afford either *cis* or *trans* heterocycles, depending on whether the reaction is kinetically or thermodynamically controlled and the nature of X and Y. For example, as shown in Scheme 3.12a, either *cis* or *trans* imidazolidinones are obtained selectively, depending on the method of ring closure. For α-hydroxy acids and α-thiol acids, cyclization with pivaldehyde affords predominantly *cis* heterocycles under thermodynamic control [81]. In the second step of the sequence, electrophile approach is usually favored from the side opposite the *tert*-butyl group. In nitrogen heterocycles such as imidazolidinones and oxazolidinones, when the nitrogen is acylated, $A^{1,3}$-strain forces the *tert*-butyl group into a quasi-axial orientation, whereas in the absence of an *N*-acyl group, the *tert*-butyl occupies a quasi-equatorial position. Subtle structural features in the heterocycles such as nitrogen pyramidalization are caused by ring strain and $A^{1,3}$-strain, and these nuanced effects contribute to the high selectivity observed in the alkylations (Table 3.1) [85,86].

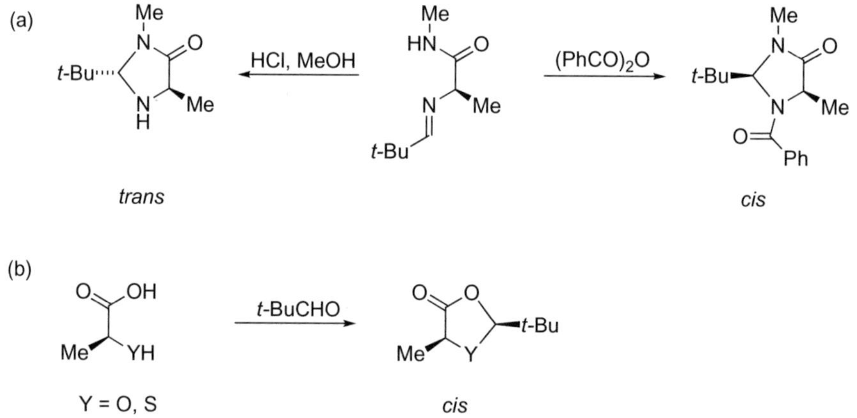

SCHEME 3.12   Stereoselectivity in the cyclizations depends on the conditions and the heteroatoms [81,89].

The concept of self-regeneration has been employed by other groups. For example, Vedejs has shown that condensation of a formamidino amino acid sodium salt with $PhBF_2$ affords a mixture of oxazaborolidine diastereomers that are enriched in one isomer by an asymmetric transformation of the second kind.[11] Recrystallization gives the pure diastereomer shown in Scheme 3.13. Deprotonation and alkylation are highly diastereoselective, with

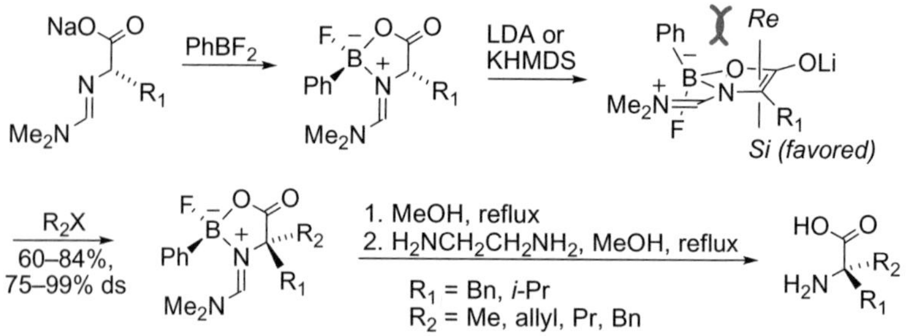

SCHEME 3.13   Extensions of the self-regeneration concept: Vedejs oxazaborolidines [90].

11. See glossary, Section 1.11, for definition of the two types of asymmetric transformation.

alkylation occuring preferentially on the *Si* face, *trans* to the *B*-phenyl group. The major product can be separated from its diastereomer by crystallization and/or chromatography. Removal of the boron and formamidino groups then gives the enantiopure quaternary α-amino acid in good yields [90].

A second example is the 1,2-diacetal methodology developed by Ley and co-workers and illustrated in Scheme 3.14 (reviews: refs. [91–93]). Condensation of lactic acid with the bis-dihydropyran shown gives an 85% yield of a spiro tricyclic acetal with 92% diastereoselectivity. The derivatization creates two spirocyclic ring junctions, each of which is a new stereocenter. The anomeric effect illustrated in the inset, operating in all three rings, renders the illustrated diastereomer the most stable, and effectively freezes conformational motion. Purification by recrystallization and alkylation gives dialkyllactates in excellent diastereoselectivity. The more reactive electrophiles (allyl, benzyl) afford higher diastereoselectivities. The enolate is preferentially alkylated on the *Si* face, *trans* to the 1,3-diaxial acetal oxygen [94].

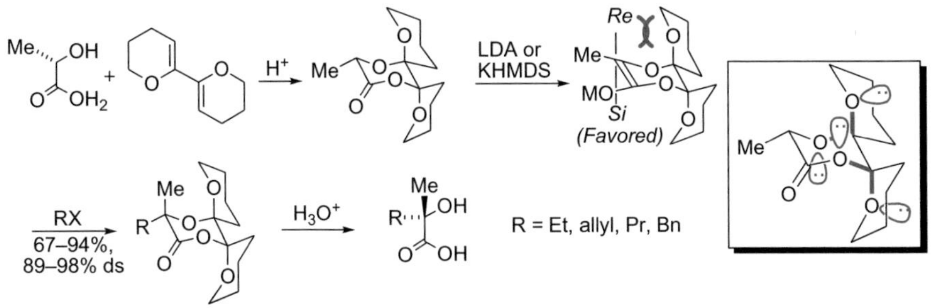

SCHEME 3.14 Ley's "dispoke" acetals for diastereoselective alkylation of lactic acid. Note the overall replacement of hydrogen by an alkyl group with inversion of configuration [94,95].

To prepare a chiral glycolic acid enolate using diacetal methodology, enantiopure 3-bromo-1,2-propanediol is condensed with butane-2,3-dione, as shown in Scheme 3.15. Elimination of HBr and ozonolysis gives the chiral diacetal in 59% overall yield. Here again, as shown in the inset, the anomeric effect directs selective formation of one diastereomer, and then effectively freezes the ring conformation. The enolate formed upon deprotonation can be alkylated in ≥90% diastereoselectivity from the *Re* face opposite the 1,3-diaxial methoxy group on the top face. The authors used substoichiometric amounts of base so that the selectivity of the alkylation could be measured without subsequent equilibration. Excess

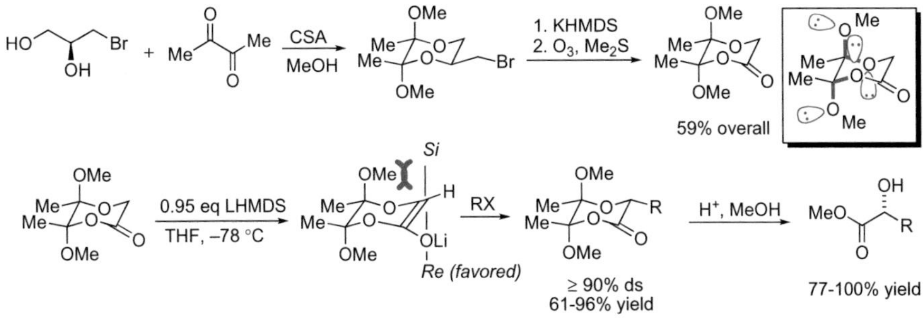

SCHEME 3.15 A self-immolative method for preparation of enantiopure glycolic acid 1,2-diacetals, and their alkylation and hydrolysis to α-hydroxy esters [96].

TABLE 3.2 Ley's Diacetal Alkylation to Make Enantiopure α-Hydroxy Esters (Scheme 3.15)[a]

RX	% ds	% Yield
ClCH$_2$O(CH$_2$)$_2$OMe	91	64
BrCH$_2$CN	91	85
EtI	94	61
t-BuO$_2$CCH$_2$Br	95	92
BuI	96	57
Allyl bromide	98	89
BnBr	>99	96
2-Naphthyl bromide	>99	84
I(CH$_2$)$_3$I	96	62

[a]The diastereoselectivities listed are kinetic; the dr can be increased by base equilibration [96].

base affords even higher diastereomer ratios than those listed in Table 3.2, through equilibration to the more stable equatorial epimer. Subsequent hydrolysis affords enantiopure α-hydroxy esters [96].

In the self-regeneration of stereocenters (SRS) concept, the introduction of additional stereocenters prior to enolate formation guarantees preservation of chirality in subsequent manipulations. Another approach is the development of systems whereby enantioselectivity is achieved by converting one type of chirality element into another [88,97−104].[12] Deprotonation of an enolate destroys the stereogenic center, but if a new stereogenic element is simultaneously introduced, *and if* the barrier to enantiomerization is high enough that racemization is slow, enantioselective alkylation is possible. An example is shown in detail in Scheme 3.16, but first, the relationship between enantiomerization and racemization deserves comment. Enantiomerization is the conversion of one enantiomer to another: for example, $R \to S$, with rate constant k_{RS}. Racemization is the conversion of an enantiopure, or enantioenriched compound, to a 50:50 mixture of enantiomers. The process is reversible:

$$R \underset{k_{SR}}{\overset{k_{RS}}{\rightleftharpoons}} S.$$

If the racemization is first order, the rate constant for racemization is twice the rate constant for enantiomerization (ref. [105], pp 46−49):

$$k_{rac} = k_{RS} + k_{SR} = 2\left(\frac{\kappa k_B T}{h}\right) e^{-\Delta G^{\ddagger}/RT}, \tag{3.1}$$

---

12. Some authors refer to this concept using the anthropomorphic term "memory of chirality." This is unfortunate, since neither chirality nor molecules have any memory. Furthermore, as Carlier has noted, it isn't the *chirality* that is preserved, it is the *configuration* [104]. A more precise term is self-regeneration of stereocenters with stereolabile intermediates.

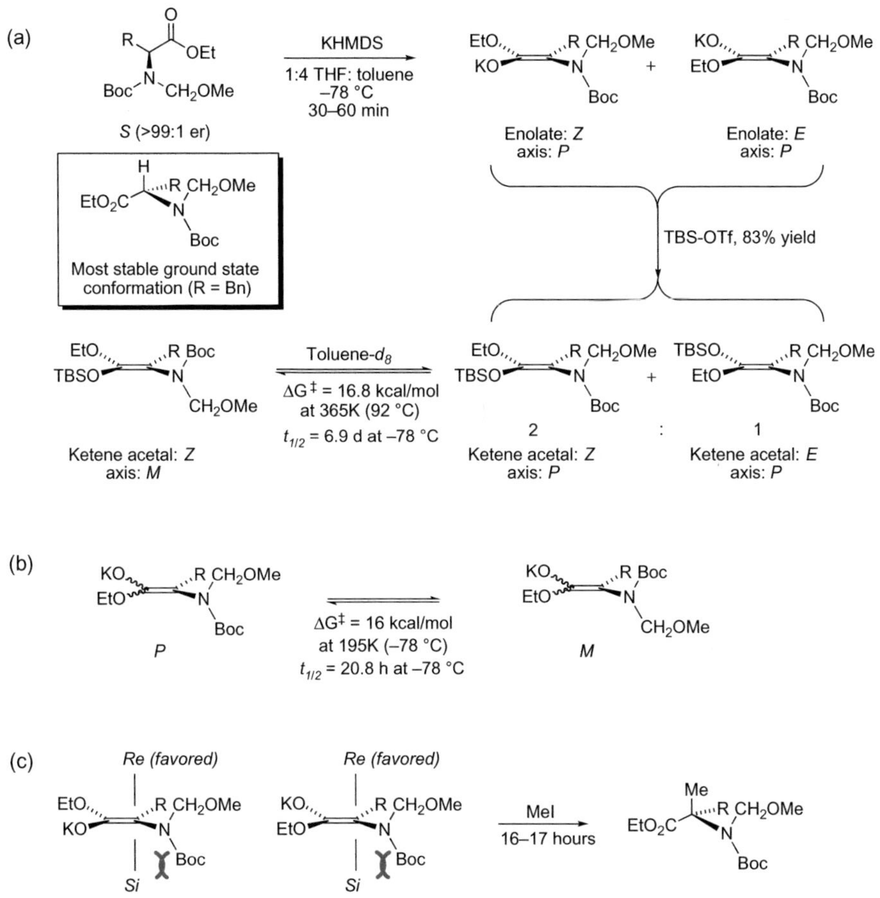

SCHEME 3.16 Auxiliary-free enantioselective α-aminoester alkylations [106]. (a) Deprotonation of the illustrated conformation of the substrate (inset) affords a mixture of E and Z stereoisomers, whose TBS ethers have a free energy barrier to enantiomerization of 16.8 kcal/mol. (b) The barrier to enantiomerization of the enolates was determined to be 16 kcal/mol. (c) Alkylation of either enolate diastereomer occurs preferably from the Re face of the enolate, *anti* to the large Boc group.

where κ is the transmission coefficient (taken as unity), k_B is Boltzmann's constant, h is Planck's constant, R is the gas constant, and T is the absolute temperature. The half-life for racemization is given by

$$t_{1/2} = \frac{\ln 2}{k_{rac}}. \tag{3.2}$$

Substitution and rearrangement give the relationship between the free energy barrier to enantiomerization and the half-life of racemization:

$$\Delta G^{\ddagger} = RT \ln \left(\frac{2k_B T t_{1/2}}{h \ln 2} \right), \tag{3.3}$$

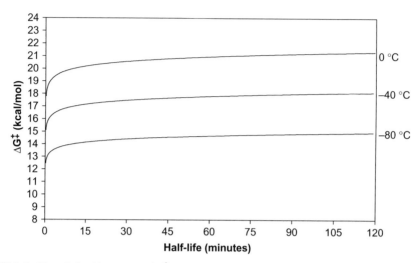

FIGURE 3.8 The relationship between $\Delta G^{\ddagger}$ for enantiomerization and the half-life for racemization at three representative temperatures.

which is plotted graphically in Figure 3.8 for three temperatures. From this graph, it is clear that racemization is very fast for any set of enantiomers with a barrier of ≤ 13 kcal/mol at temperatures of ≥ -80 °C. It is also clear that, in the portions of the curve where the half-life is longer, very small changes in the enantiomerization barrier can produce very large changes in the half-life. Free energy barriers that afford 2 h half-lives at -80 °C, -40 °C, and 0 °C are 14.9, 18.1, and 21.3 kcal/mol, respectively. Another way to look at the curves is this: if $\Delta G^{\ddagger}$ is above the line for a given temperature, racemization will be slow on the laboratory time scale; if $\Delta G^{\ddagger}$ is below the line, racemization will be fast. For reactions run at -80 °C, the critical free energy barrier, $\Delta G^{\ddagger}$, is in the $15-16$ kcal/mol range, which affords half-lives of $1.5-20$ h (see also ref. [102]).

A particularly instructive example of these concepts in asymmetric enolate alkylations, developed by Kawabata, is illustrated in Scheme 3.16 and summarized in Table 3.3 [106]. S-Amino acid esters having N-Boc and N-MOM protecting groups are deprotonated with KHMDS to a mixture of E and Z enolates. The enolate E/Z ratio was established by quenching with *tert*-butyldimethylsilyl triflate, which afforded a 2:1 ratio of the Z and E ketene acetals. Dynamic NMR experiments established that, in toluene-d_8, the barrier to rotation around the chirality axis in the silyl enol ethers is 16.8 kcal/mol at 92 °C. In a series of kinetics experiments, the enolates were allowed to partly racemize before methyl iodide quench. From the enantiomer ratios obtained, the barrier to enolate enantiomerization was found to be 16.0 kcal/mol, corresponding to a half-life of 20.8 h at -78 °C. The enantiomer ratios summarized in Table 3.3 were found after quenching with methyl iodide and stirring $16-17$ h at -78 °C. The steric course of the reaction was rationalized by postulating approach of the electrophile to the Re face of the enolate, *syn* to the N-MOM group, and *anti* to the larger N-Boc substituent, as illustrated in Scheme 3.16c. The loss in enantiopurity could be due to partial racemization of the enolate, less than 100% selective alkylation, or both. Advantages of this method include the elimination of any auxiliaries to maintain chirality in the enolate and the fact that quaternary chirality centers are formed. A limitation is that methyl iodide was the only electrophile reported.

TABLE 3.3 Auxiliary-Free Enantioselective α-Aminoester Alkylation, Which Converts a Chirality Center to a Dynamic Chirality Axis in the Enolate, then Back to a Chirality Center in the Product [106]

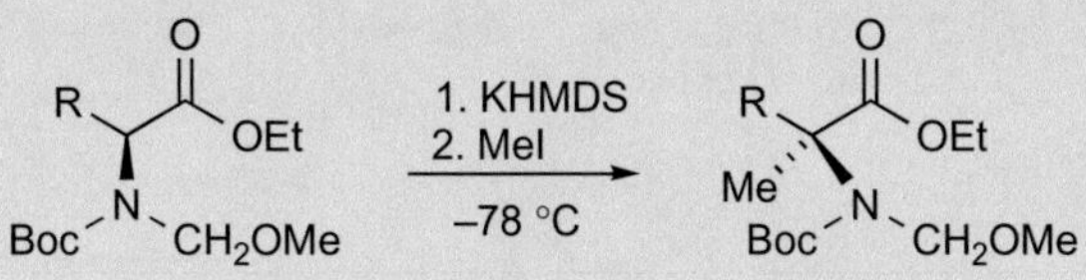

Entry	R	% Yield	% es
1	Bn-	96	90
2	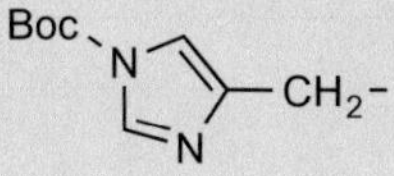 Boc-N-imidazolyl-CH₂-	83	96
3	p-MOMOC$_6$H$_4$CH$_2$-	94	89
4	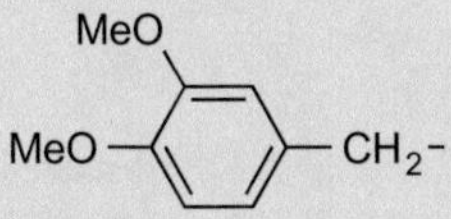 (MeO)(MeO)C$_6$H$_3$-CH$_2$-	95	90
5	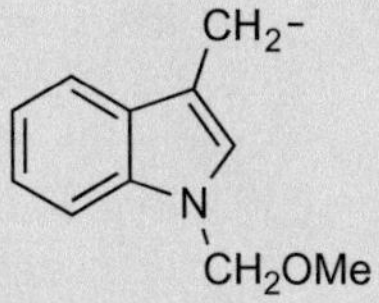 indolyl-CH₂- (N-CH₂OMe)	88	88
6	i-Pr-	81	93
7	i-Bu-	78	89

A method for auxiliary-based amino acid synthesis, which nicely illustrates control of ground state conformations through the avoidance of allylic ($A^{1,3}$) strain, was developed by Williams [107]. As shown in Scheme 3.17, the chiral diphenyloxazinone may be alkylated using LHMDS or NHMDS with excellent diastereoselectivity provided the alkylating agent is activated, such as a benzyl, allyl, or methyl halide. The stereoselectivity is ≥98% and the conformation shown in Scheme 3.17 was postulated to explain the selectivity. In this conformation, $A^{1,3}$-strain due to interactions with the Boc group forces the phenyl group at C-5 into an axial conformation, effectively blocking approach of an electrophile to the *Re* face of the enolate. After the first alkylation, a second alkylation may be executed. After purification of the crystalline oxazinones, reductive cleavage of the benzylic—heteroatom bonds liberates the amino acid. This destruction of the "auxiliary" is a drawback to this strategy because of the high cost of the amino alcohol. Selected examples of this process are listed in Table 3.4.

Another method for asymmetric alkylation of a glycine ester was reported by Yamada and is shown in Scheme 3.18 [108]. In this example of a chiral glycine enolate, the Schiff

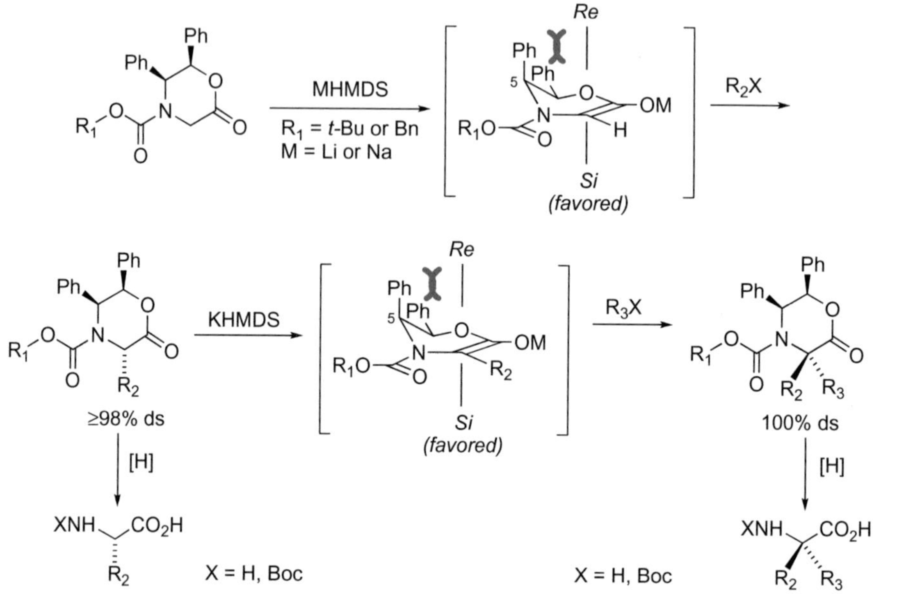

SCHEME 3.17 Williams' oxazinone enolate amino acid synthesis [107]. The conformation shown in the two bracketed structures has the C-5 phenyl in the axial position to avoid $A^{1,3}$ interactions with the adjacent N-Boc group.

TABLE 3.4 Examples of Williams' Amino Acid Synthesis (Scheme 3.17 [107])[a]

R_1	R_2	R_3	Base	% Yield (alkylation)	% Yield (amino acid)	er
t-Bu	allyl	—	LHMDS	86	50-70	99:1
t-Bu	Me	—	NHMDS	91	54	98:2
t-Bu	Bn	—	NHMDS	70	76	99:1
Bn	Bn	—	NHMDS	77	93	>99:1
t-Bu	Me	Allyl	KHMDS[b]	87[b]	70	>99:1
t-Bu	Me	Bn	KHMDS[b]	84[b]	93	>99:1
t-Bu	n-Pr	allyl	KHMDS[b]	90[b]	60	>99:1
Bn	Me	Bn	KHMDS[b]	84[b]	93	>99:1

[a]In all cases, the diastereoselectivity was ≥98%.
[b]Second alkylation.

base of *tert*-butyl glycinate and an α-pinene-derived ketone is dilithiated with two equivalents of LDA. Presumably, the lithium alkoxide is chelated to the nitrogen as shown. This tricyclic chelate is rigid and the dienolate must adopt the *s-trans* conformation in order to avoid severe nonbonding interactions with the α-pinene moiety. Nonbonding interactions that restrict conformational motion are necessary for high selectivity in examples of extraannular asymmetric induction (Figure 3.7) such as these. Approach of the electrophile from the *Re* face gives the configuration shown. Although the authors did not determine the configuration of the enolate, if we assume that the enolate is *E(O)* as shown, then the Agami approach trajectory

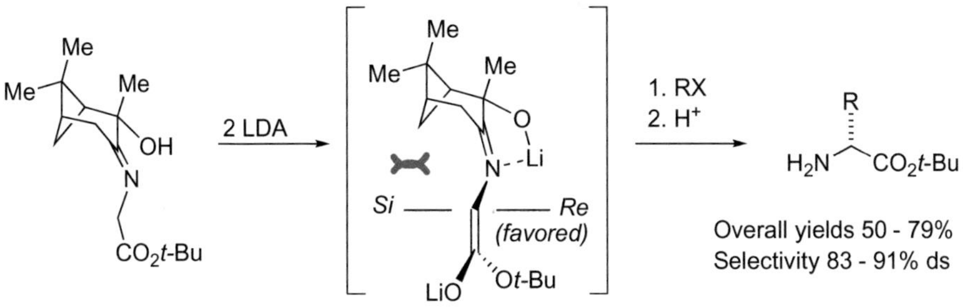

SCHEME 3.18   Yamada's chiral glycine enolate [108].

(Figure 3.5) would be slanted toward the back of the figure, *trans* to the anionic enolate oxygen. Approach from the *Si* face would not only be on the concave face of the structure but would also be slanted back toward the α-pinene moiety. Table 3.5 summarizes the selectivities reported for this asymmetric amino acid synthesis.

TABLE 3.5 Stereoselective Alkylations of Yamada's Glycine Enolate (Scheme 3.18) [108]

R	(Presumed) % ds[a]	% Yield[b]
Me	91	52
i-Bu	91	50
Bn	86	79
3,4−(MeO)$_2$−C$_6$H$_3$CH$_2$	83	62

[a]Calculated from er of the product.
[b]Overall yield of amino acid ester.

During the 1980s, one of the major thrusts of asymmetric synthesis was the development of chiral auxiliaries for the alkylation and aldol addition of propionates. The following paragraphs describe some of the methods that have evolved.

An ester enolate alkylation method, developed by Helmchen uses a camphor nucleus as a chiral auxiliary, and is illustrated in Scheme 3.19 [109−112]. This ester is designed so that one face of the enolate is shielded by a second ligand appended to the camphor nucleus. Scheme 3.19 illustrates the selective formation of either an *E(O)*- or a *Z(O)*-enolate based on the presence or absence of HMPA in the reaction mixture. Thus, deprotonation of the ester with LICA is 98% selective for the *E(O)*-enolate and deprotonation in the presence of HMPA is 96% selective for the *Z(O)*-enolate. Alkylation with benzyl bromide is more selective for the *E(O)*-enolate than for the *Z(O)*, but after diastereomer separation, reduction gives enantiomerically pure *R*- or *S*-2-methyl-3-phenylpropanol, opposite enantiomers from the same auxiliary [111].

The mechanistic rationale for the selectivity of these ester enolate alkylations may be summarized as follows [110]:

- Deprotonation in THF solvent gives *E(O)*-enolates, whereas enolization in THF/HMPA solvent gives *Z(O)*-enolates with a high degree of selectivity.

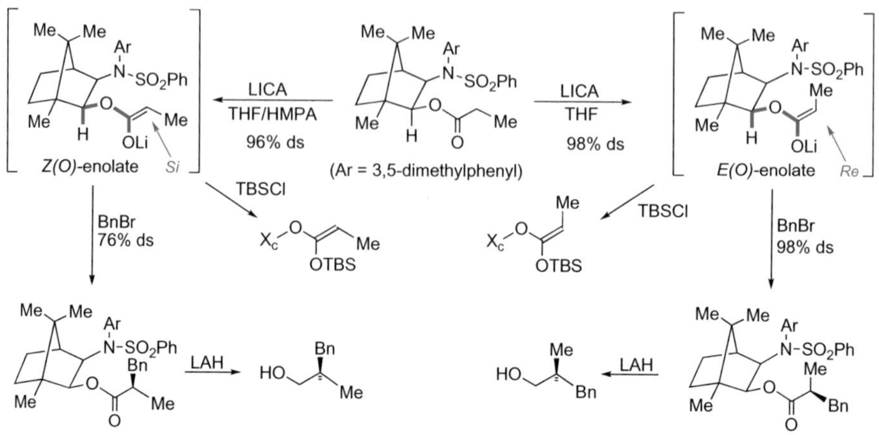

SCHEME 3.19 Controlled stereoselective enolate formation and asymmetric alkylation of a "second generation" camphor ester enolate chiral auxiliary [111].

- After deprotonation, the enolate is oriented as illustrated in the bracketed structures of
  Scheme 3.19 such that the H−C−O−C−OLi moiety (highlighted in red) is coplanar and
  the OLi is *syn* to the *endo* carbinol hydrogen of the camphor.
- Similarly, the H−C−N−S=O sulfonamide moiety is also in the conformation
  illustrated.
- With the rear face of the enolate thus shielded, alkylation occurs from the front face
  (direction of the viewer: *Re* face of the *E(O)*-enolate and *Si* face of the *Z(O)*-enolate).

Although accounting for the gross data, this rationale is not completely satisfactory.
Subsequent studies [111] showed that addition of HMPA *after enolate formation but before
electrophile addition* also had an effect on the selectivity of the alkylation, leading Helmchen
to speculate that the sulfonamide may be chelated to the lithium. Another possibility may be
that the enolates are aggregated, and the effect of HMPA is to disrupt the aggregation.
Additionally, the difference in selectivity between the *E(O)*- and *Z(O)*-enolate alkylations
remains unexplained.

Helmchen has used this methodology in asymmetric synthesis of the three stereoisomers of
the tsetse fly pheromone [109] and the side chain of α-tocopherol [112], illustrated in Figure 3.9.

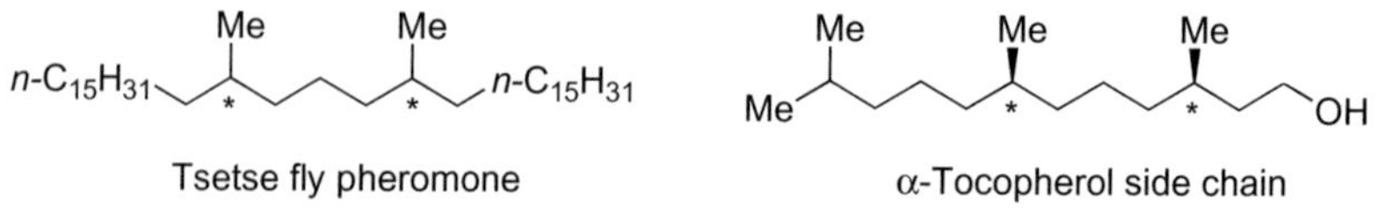

FIGURE 3.9 Natural products synthesized using Helmchen's ester enolates: the tsetse fly pheromone [109]
and the side chain of α-tocopherol [112]. Stereocenters created in the key step are indicated with *.

It is generally true that restrictions on conformational mobility minimize the number of
competing transition states and simplify analysis of the factors that affect selectivity.
Chelation of a metal by a heteroatom often provides such restriction and also often places the
stereocenter of a chiral auxiliary in close proximity to the α-carbon of an enolate. This proximity often results in very high levels of asymmetric induction. A number of auxiliaries have

been developed for the asymmetric alkylation of carboxylic acid derivatives using chelate-enforced intraannular asymmetric induction. The first practical method for asymmetric alkylation of carboxylic acid derivatives utilized oxazolines and was developed by the Meyers group in the 1970s (Scheme 3.20a), whose efforts established the importance and potential for chelation-induced rigidity in asymmetric induction (reviews: refs. [113−115]). In 1980, Sonnet [116] and Evans [117,118] independently reported that the dianions of prolinol amides afford more highly selective asymmetric alkylations (Scheme 3.20b).

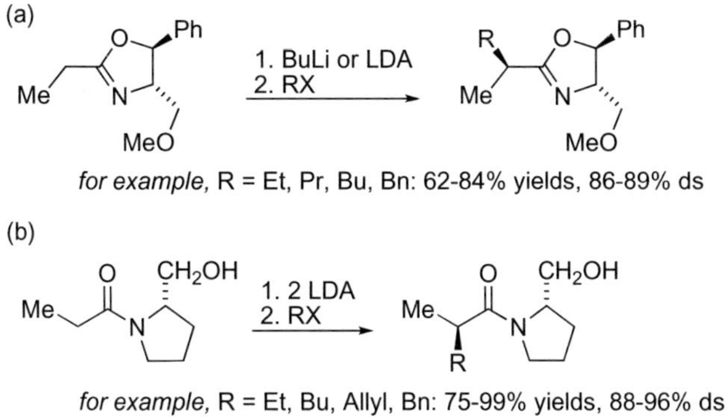

(a)

1. BuLi or LDA
2. RX

for example, R = Et, Pr, Bu, Bn: 62-84% yields, 86-89% ds

(b)

1. 2 LDA
2. RX

for example, R = Et, Bu, Allyl, Bn: 75-99% yields, 88-96% ds

SCHEME 3.20 Early examples of asymmetric enolate alkylations: (a) Meyers' oxazolines [113−115]; (b) Evans' [117,118] and Sonnet's [116] proline amide alkylations.

In 1982, Evans reported that the alkylation of oxazolidinone imides appeared to be superior to either oxazolines or prolinol amides from a practical standpoint, since they are significantly easier to cleave [119]. As shown in Scheme 3.21, enolate formation is at least 99% stereoselective for the Z(O)-enolate, which is chelated to the oxazolidinone carbonyl oxygen as shown. From this intermediate, approach of the electrophile is favored from the *Si* face to give the monoalkylated acyl oxazolidinone as shown. Table 3.6 lists several examples of this process. As can be seen from the last entry in the table, alkylation with unactivated alkyl

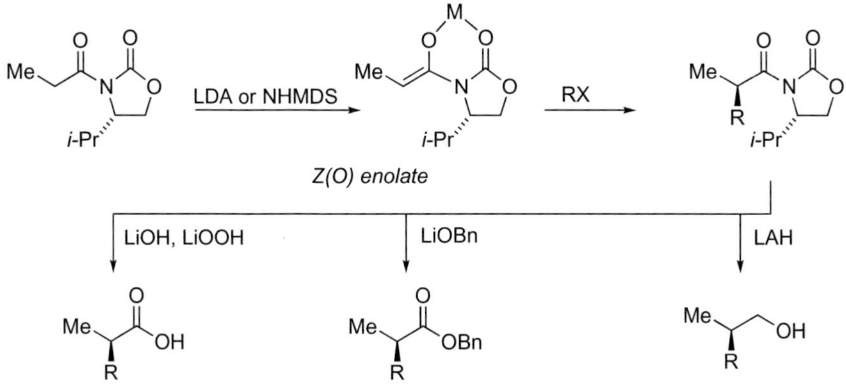

LDA or NHMDS

RX

Z(O) enolate

LiOH, LiOOH

LiOBn

LAH

SCHEME 3.21 Evans' asymmetric alkylation of oxazolidinone imides [119].

TABLE 3.6 Alkylations of Evans' Oxazolidinone Imides
(Scheme 3.21 [119])[a]

R	% ds	% Yield[b]
Bn	>99	92
Methallyl	98	62
Allyl	98	71
Et	94	36

[a]In all cases, the alkylation products were >99% pure after chromatography.
[b]Isolated yields after chromatography.

halides is less efficient, and this low nucleophilicity is the primary weakness of this method. Following alkylation, the chiral auxiliary may be removed by lithium hydroxide or hydroperoxide hydrolysis [120], lithium benzyloxide transesterification, or LAH reduction [121]. Evans has used this methology in numerous total syntheses. Two of the earliest were the Prelog-Djerassi lactone [122] and ionomycin [123]; an industrial application is in the synthesis of metalloproteinase inhibitor Trocade by Hilbert of Hoffmann-La Roche [124] (Figure 3.10).

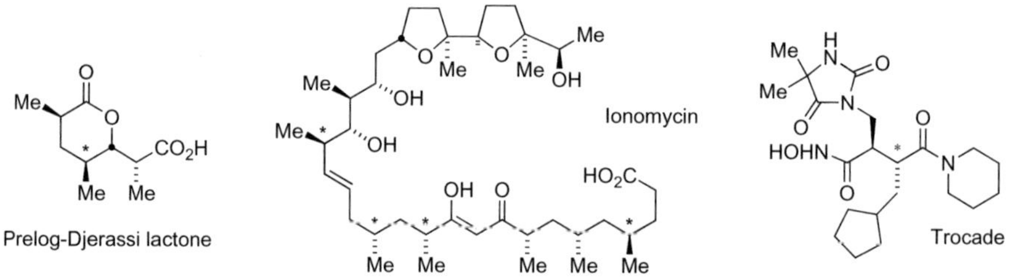

FIGURE 3.10 Syntheses using (in part) asymmetric alkylation of oxazolidinone enolates: Prelog-Djerassi lactone [122], ionomycin [123], and the metalloproteinase inhibitor Trocade [124]. Stereocenters created by alkylation are indicated (*).

In addition to these examples of alkylations that employ chelate-enforced intraannular asymmetric induction, Evans' imides are useful in asymmetric aldol (Section 5.2.2), and Michael additions (Section 5.3.2), Diels−Alder reactions (Section 6.1.1.1), and enolate oxidations (Section 8.3), making this one of the most versatile auxiliaries ever invented.

In 1989, Oppolzer reported that the enolates of N-acyl sultams derived from camphor afford highly diastereoselective alkylation products with a variety of electrophiles *including those which are not allylically activated* [125]. The sultam is deprotonated using either butyllithium with a catalytic amount of cyclohexyl isopropyl amine, butyllithium alone, or sodium hexamethyldisilyl amide.[13] As illustrated in Scheme 3.22, alkylation occurs selectively from the *Re* face of the *Z(O)*-enolate to give monoalkylated sultams, which can be cleaved by LAH reduction or lithium hydroperoxide-catalyzed hydrolysis. Representative examples are listed in Table 3.7.

---

13. These conditions are necessary to avoid competitive deprotonation at C_{10}.

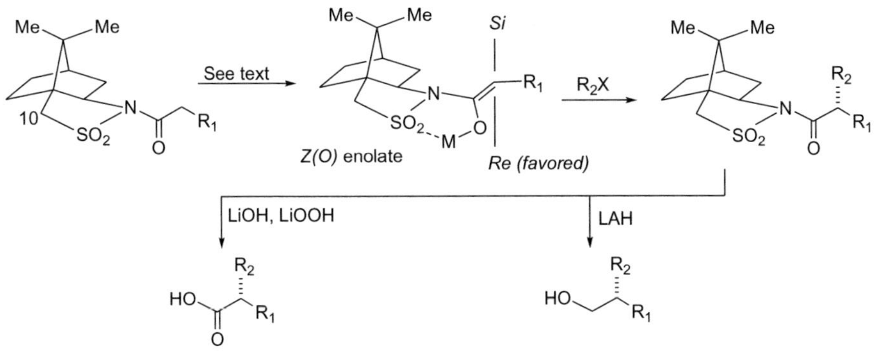

SCHEME 3.22 Oppolzer's asymmetric sultam alkylation [125].

TABLE 3.7 Asymmetric Alkylation of Oppolzer's Sultams (Scheme 3.22) [125]

R_1	R_2	% ds	dr[a]	% Yield
Me	Bn	98	>99:1	89
Bn	Me	97	>99:1	88
Me	allyl	98	>98:2	74
allyl	Me	98	>99:1	NR
Me	methallyl	90	>99:1	70
Me	$n\text{-}C_5H_{11}$	99	98:2	81
$n\text{-}C_5H_{11}$	Me	98	98:2	NR

[a]Diastereomer ratio after recrystallization; NR – not reported.

The selectivity of this reaction is based on the following mechanistic rationale of chelate-enforced intraannular asymmetric induction[14] (Scheme 3.22):

- Following the Ireland model (Scheme 3.4), deprotonation gives the Z(O)-enolate.
- The lithium of the enolate is chelated to the sultam, which also has a pyramidal nitrogen.
- The Si face is shielded by the bridging methyls, and approach is therefore from the Re face, opposite the nitrogen lone pair.

A highly versatile method for the asymmetric alkylation of acyclic amides, using pseudo-ephedrine as the chiral auxiliary, was developed by the Myers group [127–131]. As shown in Scheme 3.23, the dianion of the amide is formed by deprotonation with LDA in THF. The reaction is facilitated by the addition of six molar equivalents of LiCl; the diastereoselectivities are not diminished in the absence of LiCl, but the alkylations are considerably more sluggish. The transition structure that has been proposed to account for the steric course of the reaction is shown in the inset of Scheme 3.23 [128]. Deprotonation of the amide affords the Z(O)-enolate, which lies in the illustrated plane. Note also that the tertiary

---

14. An alternative explanation, based on analogy of the sultam to a trans-2,5-disubstituted pyrrolidine, has also been offered [126].

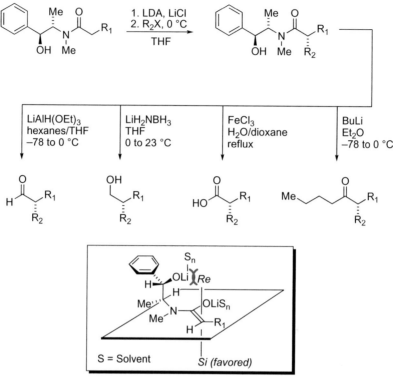

SCHEME 3.23 Myers' asymmetric alkylation of pseudoephedrine amides and their conversion to aldehydes, alcohols, acids, and ketones [128,130,131].

C—H bond α to nitrogen is also in this plane due to $A^{1,3}$-strain considerations. The solvated lithium alkoxide blocks approach of an electrophile to the enolate *Re* face. Lithium chloride suppresses alkylation of the auxiliary alkoxide oxygen, probably by forming a mixed aggregate; the rate-acceleration achieved in the presence of LiCl may be due to the presence of more reactive mixed aggregates of the enolate [128] (see also refs. [132—134]). Selected examples of this procedure are listed in Table 3.8. The reaction is general for primary alkyl halides, including β-branched alkyl halides. The alkylation fails with secondary alkyl halides, and is relatively slow with alkyl halides that are *both* β-branched *and* β-alkoxy substituted. Note that disubstituted products can be obtained with either absolute configuration at the new stereocenter using the same enantiomer of the pseudoephedrine auxiliary (compare entries 1 and 4; 2 and 5). There is a case of possible matched and mismatched double asymmetric induction in β-branched alkyl halides, as shown by entries 8 and 9. The dr of 98.4:1.6 corresponds to 62:1, whereas 98.9:1.1 corresponds to 89:1 diastereoselectivity; it hardly matters since the chemical yield is the same! An advantage of this method is that the pseudoephedrine amides, both substrates and products, are often crystalline, which simplifies isolation of diastereomerically pure products. The practicality of this and related methods is illustrated in the large scale procedures that have been published [130,131]. Scheme 3.23 summarizes the methods that can be used for the conversion of the pseudoephedrine amides to aldehydes, alcohols, acids, and ketones [128,131]. By converting the primary hydroxyl of the reduction products into iodide, and

TABLE 3.8 Selected Examples of Myers' Pseudoephedrine Enolates (Scheme 3.23) [128][a]

Entry	R_1	R_2X	% ds	% Yield
1	Me	BnBr	97	90
2	Me	BuI	99	80
3	Me	Ph(CH$_2$)$_2$I	97	86
4	Bn	MeI	97	99
5	Bu	MeI	97	94
6	i-Pr	BnBr	99	83
7	Me	TBSO(CH$_2$)$_2$I	98	91
8	Me	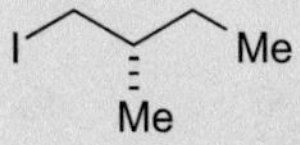	98[b]	94
9	Me	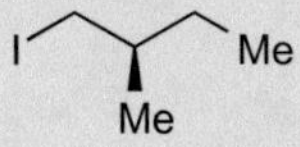	99[b]	94

[a]Alkylations were run at 0 °C unless otherwise noted.
[b]Alkylation conducted at 23 °C.

sequential alkylation with either enantiomer of pseudoephedrine propionamide, 1, 3, *n* (*n* = odd) polymethylated carbon chains of any relative and absolute configuration can be prepared, as illustrated in Figure 3.11 [129] (see also [135]).

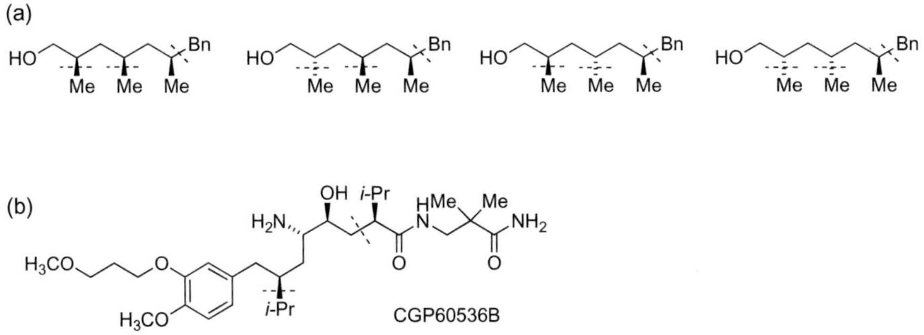

(a)

(b)

CGP60536B

FIGURE 3.11 Applications of Myers' pseudoephedrine alkylations (disconnections at dashed lines): (a) skipped trimethyl compounds prepared by iterative alkylation [129]; (b) Novartis CGP60536B, a renin inhibitor, was prepared using pseudoephedrine-mediated alkylations in two key steps [136].

A versatile method for the synthesis of compounds containing quaternary centers (using an intraannular asymmetric induction strategy) was developed by Meyers and uses the bicyclic lactams illustrated in Scheme 3.24 [137–143]. The bicyclic lactams may be synthesized by condensation of an amino alcohol with a keto acid as illustrated [137,142], or by

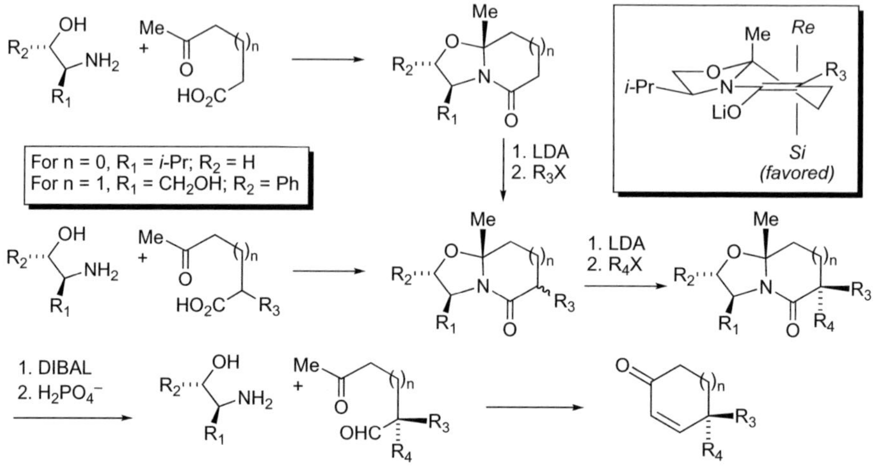

SCHEME 3.24 Meyers' asymmetric alkylations of bicyclic lactams [137–143]. When $n = 0$, $R_1 = i$-Pr and $R_2 = H$; when $n = 1$, $R_1 = CH_2OH$ and $R_2 = Ph$.

condensation of an amino alcohol with an anhydride followed by reductive cyclization [138]. Sequential alkylations proceed with differing degrees of stereoselectivity. The first alkylation is not very selective, but the second is highly so, as shown by the examples listed in Table 3.9. Note that a different auxiliary is used for the two ring systems. Specifically, for the 5,5-bicyclic system ($n = 0$), an auxiliary derived from valinol is used ($R_1 = i$-Pr, $R_2 = H$). For the 5,6-bicyclic system, an auxiliary containing a free hydroxyl group is required in order to enable reduction of the carbonyl by intramolecular delivery of hydride at a later stage [140]. In all of the cases reported to date, the two diastereomeric dialkylated bicyclic lactams have been separable by chromatography, ensuring enantiomerically pure products at the end. Scheme 3.24 details one protocol for the elaboration of these bicyclic lactams into cyclohexenones, and Scheme 3.25 summarizes different ways that have been developed for the elaboration of the alkylated bicyclic lactams into enantiomerically pure cylopentenones and cyclohexenones containing chiral quaternary centers of differing substitution patterns [141] (see also refs. [144,145]).

In all of the examples reported to date, the second alkylation occurs on the Si face as illustrated in Scheme 3.24. The origin of this stereoselectivity is not clear, and there are probably a

TABLE 3.9 Selected Examples of Asymmetric Alkylation of Meyers' Bicyclic Lactams (Scheme 3.24)

n	R_3	R_4	% ds	% Yield	Ref.
0	Allyl	Bn	97	77	[139]
0	Bn	Allyl	95	63	[139]
0	Et	Bn	95	74	[139]
1	Me	Bn	97	48	[137,140]
1	Bn	Me	75	50–80	[140]
1	Bn	Allyl	82	50–80	[140]

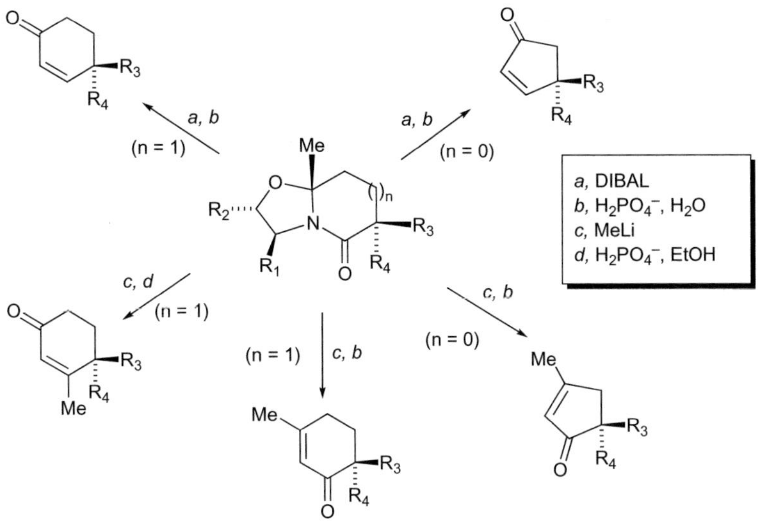

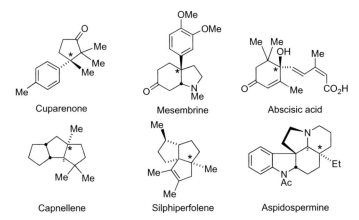

SCHEME 3.25 Protocols for the elaboration of bicyclic lactams into cycloalkenones having different substitution patterns [141] (see also [144,145]).

number of subtle factors that combine to affect the steric course. Among the factors that favor approach from the *Si* face are that this is the axial direction, and it is *anti* to the angular methyl group. On the other hand, the Agami trajectory (Figure 3.5a) for the approach of an electrophile would be slanted toward the β-carbon, *trans* to the enolate oxygen (Figure 3.5b), which would seem to favor the *Re* face (at least in the 6-membered ring), due to the axial hydrogen on the γ-carbon. Meyers has suggested that the selectivity may be due to stereoelectronic factors related to perturbation of the enolate π bond by the nitrogen lone pair [146].

Figure 3.12 illustrates a number of targets that have been synthesized using this methodology. In each case, the stereocenter formed in the asymmetric alkylation is indicated with

FIGURE 3.12 Natural products synthesized using Meyers' bicyclic lactam methodology: cuparenone [147], mesembrine [148], abscisic acid [149], capnellene [149], silphiperfolene [141], and aspidospermine [150]. The stereocenter formed in the key step is indicated with *.

an asterisk. In all cases, the configuration at this center controlled the selective formation of the rest.

For the asymmetric alkylation of ketones and aldehydes, a highly practical method was developed by the Enders group, and this method uses SAMP/RAMP hydrazones (reviews: refs. [151−155]). SAMP and RAMP are acronyms for S- or R-1-amino-2-methoxymethylpyr-rolidine, respectively. This chiral hydrazine is used in an asymmetric version of the dimethyl-hydrazone methodology originally developed by Corey and Enders [156,157]. These auxiliaries are available from either proline or pyroglutamic acid [151,152]. As shown in Scheme 3.26, SAMP hydrazones of aldehydes [158] and ketones [158,159] may be deproto-nated by LDA and alkylated. The diastereoselectivity of the reaction may often be determined by integration of the methoxy singlet in the NMR spectrum. After alkylation, cleavage may be effected with a number of reagents [152,153]. Among these are oxidative cleavage by ozo-nolysis [152], sodium perborate [160], or magnesium peroxyphthalate [161], acidic hydrolysis using methyl iodide and dilute HCl [158,159], or BF_3 and water [162,163]. Table 3.10 lists a few examples of SAMP asymmetric alkylations.

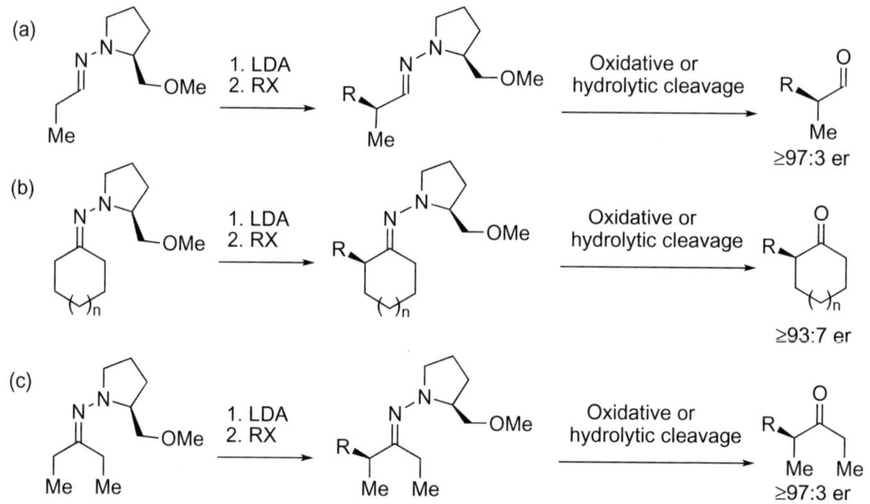

SCHEME 3.26 Asymmetric alkylations of aldehydes and ketones with SAMP hydrazones.

Scheme 3.27 illustrates the mechanistic rationale for this asymmetric alkylation. Deprotonation by the Ireland model (cf. Scheme 3.2) gives the E_{CC},Z_{CN} enolate as shown [164].[15] Cryoscopic and spectroscopic measurements indicate that the lithiated hydrazones are monomeric in THF solution [165], and a crystal structure shows the lithium σ-bonded to the azomethine nitrogen and chelated by the methoxy of the auxiliary [166]. The former azomethine nitrogen (now bonded to lithium) is largely sp²-hybridized, and the nitrogen is pulled 17.5° below the azaallyl plane by the chelating methoxy. If this structural feature is preserved in the transition state, the approach of the electrophile toward the Si face would be hindered by the C-5 methylene of the

---

15. The deprotonation of hydrazones is not regioselective, so the Z_{CN} geometry results from equilibration *after* deprotonation.

TABLE 3.10 Asymmetric Alkylations of SAMP/RAMP Hydrazones (Scheme 3.26)

Product	Electrophile	% Yield	er	Ref.
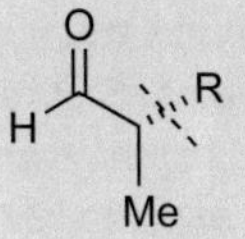	EtI $C_6H_{13}I$	71 52	97:3 ≥97:3	[151,158]
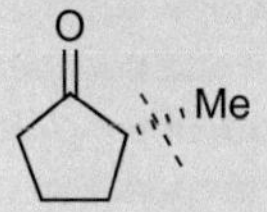	Me_2SO_4	66	93:7	[158]
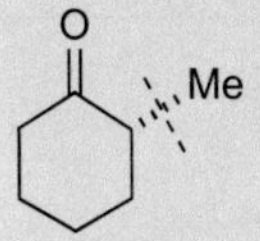	Me_2SO_4	70	>99:1	[158]
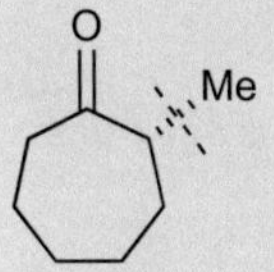	MeI	59	97:3	[158]
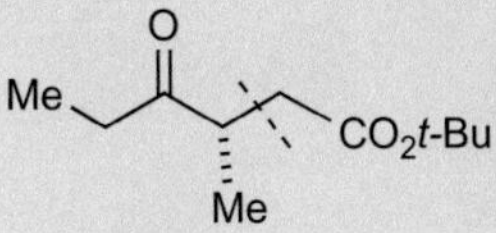	$BrCH_2CO_2t\text{-Bu}$	53	>97:3	[151]
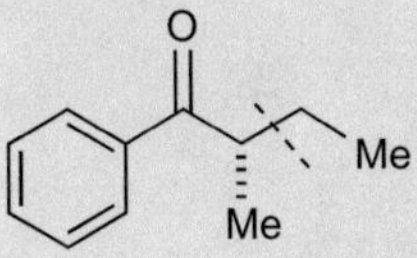	EtI	44	>98:2	[159]

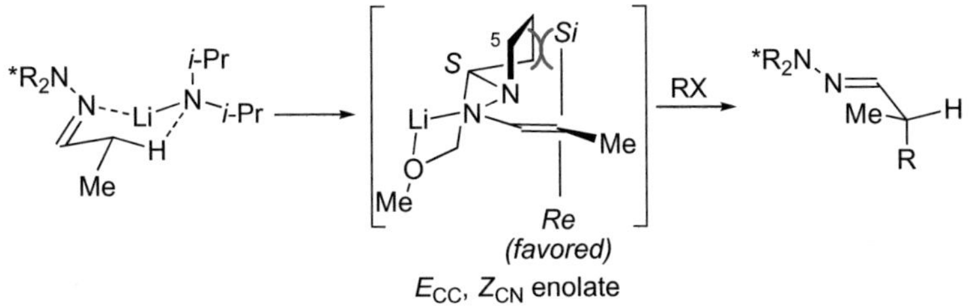

SCHEME 3.27   Mechanistic rationale for face-selectivity of SAMP hydrazone alkylation [166].

pyrrolidine ring and approach toward the *Re* face would be favored, as is observed [166].[16,17] Note that this substructure is equally accessible from both cyclic and acyclic ketones as well as aldehydes. Approach from the *Re* face gives the configuration shown, which is uniformly predictable independent of the ketone or aldehyde educt.

The practicality of the SAMP/RAMP hydrazone method for the asymmetric alkylation of aldehydes and ketones is responsible for widespread applications in synthesis [151,155]. Figure 3.13 illustrates only a few of the natural products that have been synthesized using this methodology. These include a number of insect pheromones as well as the sesquiterpene eremophilenolide and the antibiotic X-14547A. The latter two compounds have multiple stereocenters but the asymmetric alkylation using the SAMP/RAMP hydrazone method produces one stereocenter, which is then used to direct the selective formation of the others. The synthesis of pectinatone highlights the SAMP/RAMP method in the iterative preparation of skipped trimethyl stereocenters in an alkyl chain.

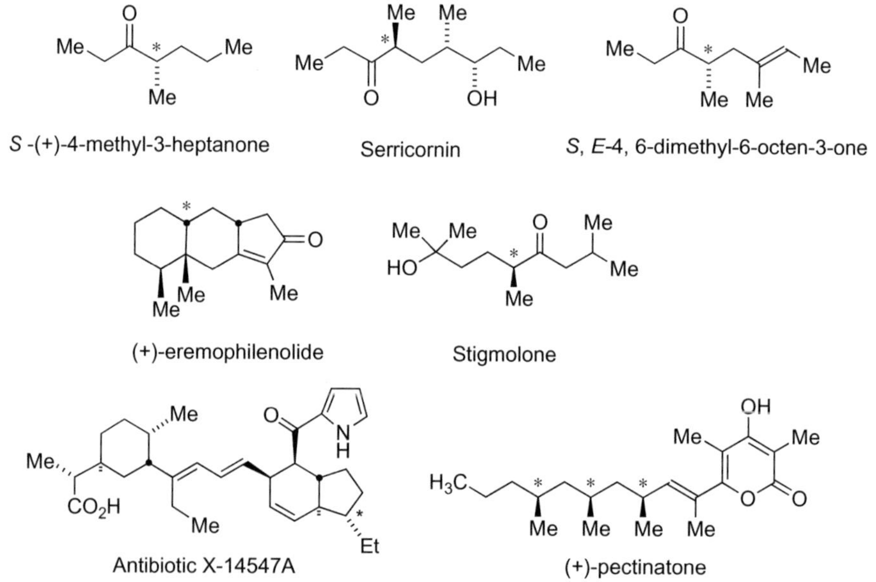

S -(+)-4-methyl-3-heptanone   Serricornin   S, E-4, 6-dimethyl-6-octen-3-one

(+)-eremophilenolide   Stigmolone

Antibiotic X-14547A   (+)-pectinatone

FIGURE 3.13 Natural product syntheses employing SAMP/RAMP hydrazones: S-(+)-4-methyl-3-heptanone, the leaf cutting ant alarm pheromone [152]; serricornin, the sex pheromone of the cigarette beetle [167,168]; S, E-4,6-dimethyl-6-octen-3-one, the defense substance of "daddy longlegs" [167]; (+)-eremophilenolide [167]; antibiotic X-14547A [169]; the antibacterial, antifungal, and cytotoxic pectinatone [170]; and stigmolone, the fruiting body inducing pheromone of the myxobacterium *Stigmatella aurantiaca* [171]. Stereocenters formed by asymmetric alkylation are indicated by *.

Another hydrazone auxiliary that has properties complementary to the SAMP/RAMP hydrazones was developed by Coltart and is illustrated in Scheme 3.28 [172,173]. In this

---

16. The Agami trajectory (Figure 3.5) would seem to suggest an approach trajectory that is slanted *away* from the pyrrolidine (*i.e.,* towards the viewer), decreasing the effect of the auxiliary in directing the approach. On the other hand, *gem*-dimethyls at the 5-position of the auxiliary enhance the selectivity [166].
17. Another rationale, which postulates a chelated lithium that is situated on top of the π-cloud of the azaallyl anion, has also been proposed [14,153].

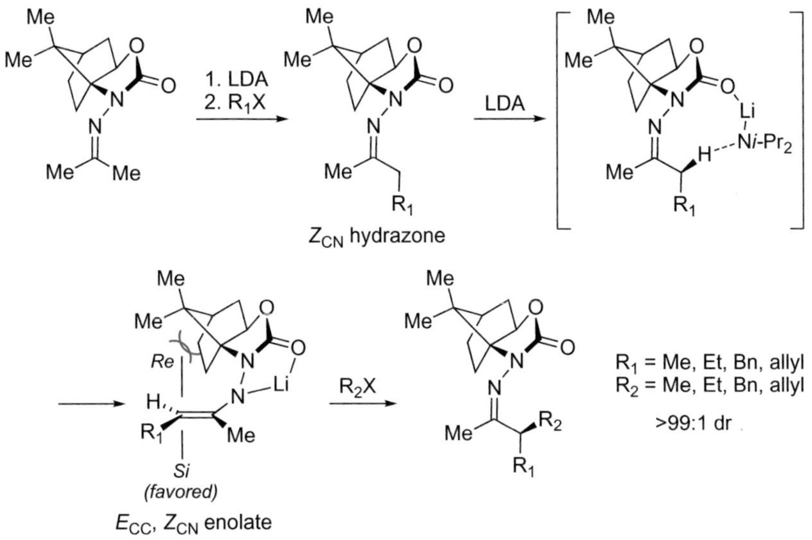

SCHEME 3.28 Coltart's *N*-amino cyclic carbamate hydrazone asymmetric alkylation [172,173].

system, a camphor-derived *N*-amino oxazolidinone is condensed with, for example, acetone. Deprotonation and alkylation furnish exclusively the Z_{CN} hydrazone, which can then be regio-selectively deprotonated *syn* to the carbonyl oxygen by complexation with the lithium amide base as illustrated in the brackets. The lithium cation on the azaenolate is then chelated by the oxazolidinone oxygen and alkylated on the *Si* face by a second alkyl halide. Mild acid hydrolysis produces the corresponding ketone. Several examples illustrate synthesis of either enantiomer in excellent yield and selectivity.

The MacMillan group has introduced a concept in catalytic α-alkylation of aldehydes through oxidation of enamines to radical cations, which they term SOMO-activation (SOMO stands for singly occupied molecular orbital) [174,176]. The general process, illustrated in Scheme 3.29a, is an example of reactivity umpolung, whereby a site which is normally thought of as nucleophilic becomes electrophilic [177,178]. In this case, condensation of an aldehyde with a chiral secondary amine affords an enamine that is oxidized *in situ* by ceric ammonium nitrate (CAN) to a radical cation. An appropriate nucleophile then captures the radical cation, and further one-electron oxidation affords an iminium ion product, which is hydrolyzed to the aldehyde. Nucleophiles appropriate to this process include allyl silanes and *N*-Boc pyrrole [174], silyl enol ethers [175], and vinyl trifluoroborate salts [176]. General conditions are shown in Scheme 3.29b−d. Scheme 3.29e illustrates the probable catalytic cycle. The critical step is oxidation of the enamine to form a radical cation, whose conforma-tion is oriented *anti* to the *tert*-butyl group. In this conformation, the benzyl group shields the upper (*Re*) face of the radical, and the nucleophiles approach preferentially from the bottom (*Si*) face. Nucleophilic addition, accompanied by a second one-electron oxidation, then hydro-lysis affords the product and the catalyst.

These one-pot procedures have several requirements for successful application. First, it is necessary that the enamine intermediate undergoes selective oxidation in the presence of the amine catalyst, the aldehyde, the iminium ion intermedate, and the nucleophile. Table 3.11 lists selected examples to show the variety of electrophiles that have been used in these

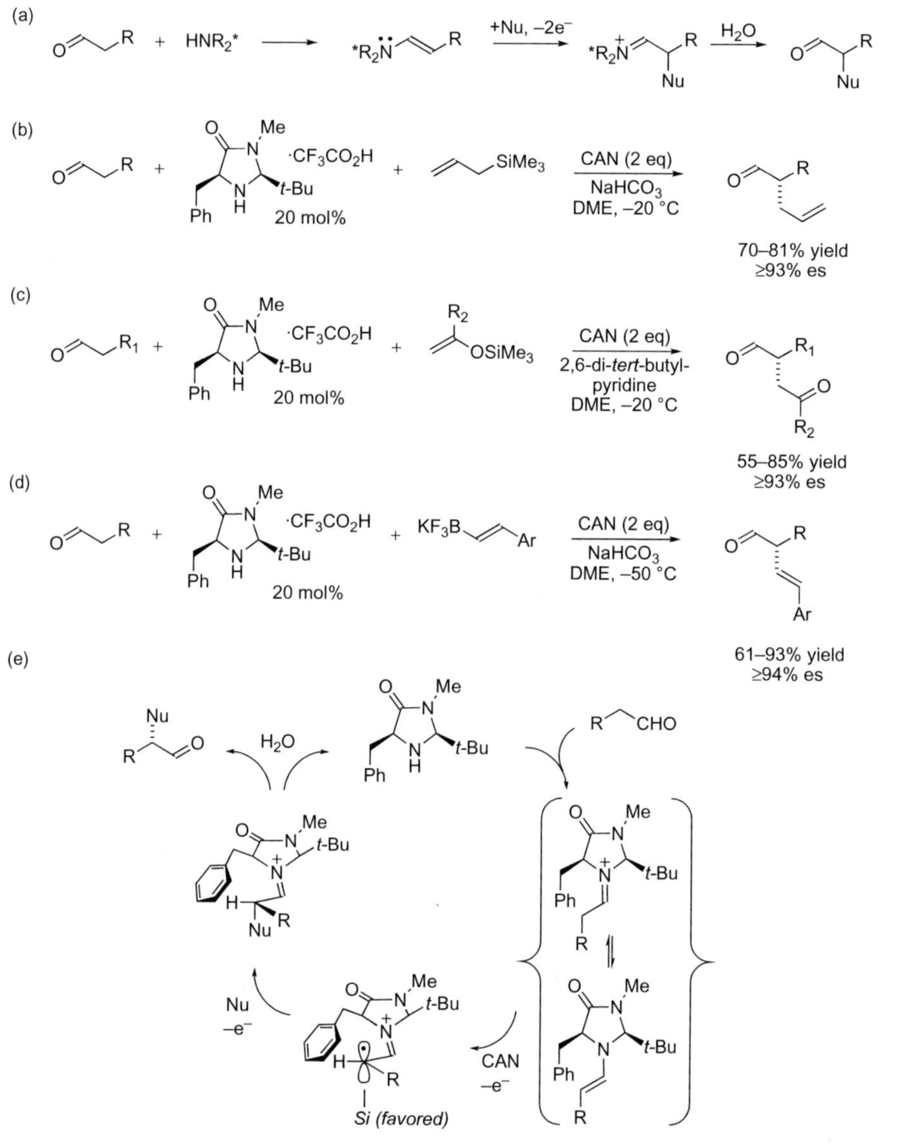

SCHEME 3.29 (a) The concept of SOMO organocatalytic activation of aldehydes [174]; (b) α-allylation [174]; (c) α-enolation [175]; (d) α-vinylation [176]; and (e) catalytic cycle.

reactions, and illustrate the variety of functional groups that are compatible with the process, including esters, ketones, olefins, and carbamates.

Jørgensen and colleagues have developed a related organocatalytic method for α-arylation of aldehydes that is based on enamine conjugate addition to quinones, as illustrated in Scheme 3.30a [179]. The enantioselectivities are excellent, and only 2 of the 13 examples gave yields less than 75%. The product having the bis-phenol in the α-position of the aldehyde exists as a mixture of phenol-aldehyde and hemiacetal, as shown. The catalytic cycle is

TABLE 3.11 Examples of SOMO-Activation of Aldehyde Enamines

Entry	Aldehyde	Nucleophile	Product	% Yield	er	Ref.
1	C$_7$H$_{15}$CHO	CH$_2$=CHCH$_2$SiMe$_3$	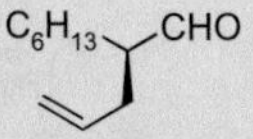 C$_6$H$_{13}$—CHO (allyl)	81	95:5	[174]
2	C$_7$H$_{15}$CHO	 (pyrrole)NBoc	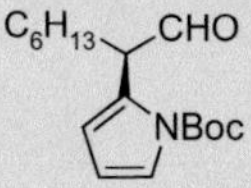 C$_6$H$_{13}$—CHO, NBoc-pyrrole	85	92:8	[174]
3	C$_7$H$_{15}$CHO	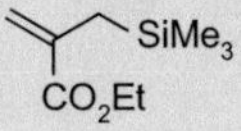 =SiMe$_3$, CO$_2$Et	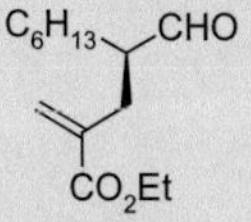 C$_6$H$_{13}$—CHO, =CO$_2$Et	81	95:5	[174]
4	C$_7$H$_{15}$CHO	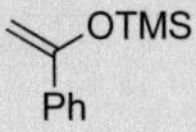 =OTMS, Ph	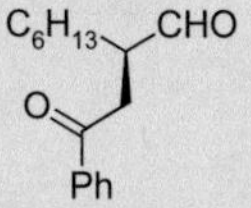 C$_6$H$_{13}$—CHO, O=, Ph	85	95:5	[175]
5	C$_7$H$_{15}$CHO	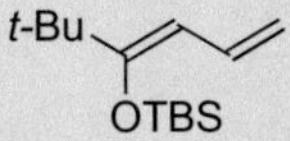 t-Bu, OTBS (diene)	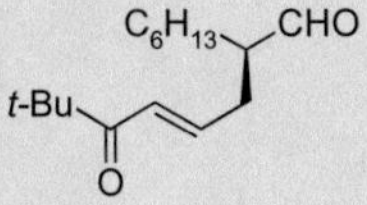 C$_6$H$_{13}$—CHO, t-Bu—O	74	98:2	[175]
6	C$_7$H$_{15}$CHO	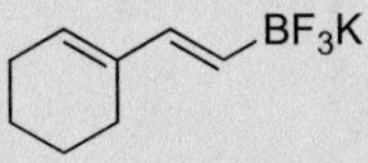 (cyclohexenyl)—BF$_3$K	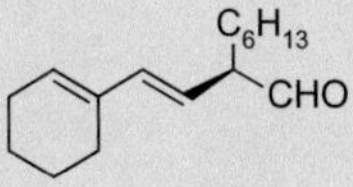 (cyclohexenyl), C$_6$H$_{13}$, CHO	73	96:4	[176]
7	C$_7$H$_{15}$CHO	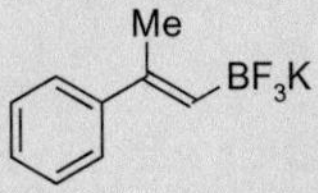 Me, (Ph)—BF$_3$K	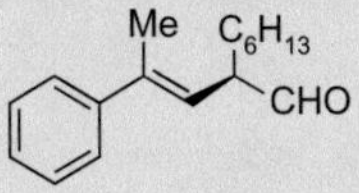 Me, C$_6$H$_{13}$, CHO	93	97:3	[176]
8	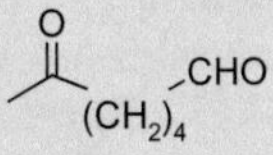 O=, CHO, (CH$_2$)$_4$	CH$_2$=CHCH$_2$SiMe$_3$	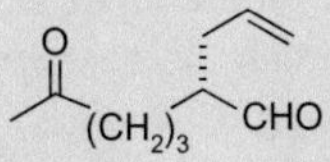 O=, (CH$_2$)$_3$, CHO (allyl)	72	94:6	[174]
9	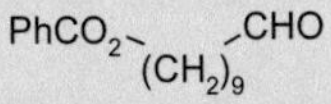 PhCO$_2$—, CHO, (CH$_2$)$_9$	CH$_2$=CHCH$_2$SiMe$_3$	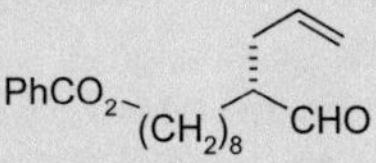 PhCO$_2$—(CH$_2$)$_8$, CHO (allyl)	72	97:3	[174]
10	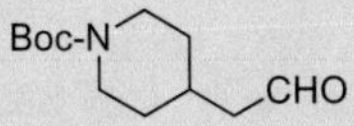 Boc-N (piperidine), CHO	CH$_2$=CHCH$_2$SiMe$_3$	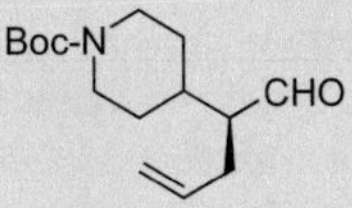 Boc-N (piperidine), CHO (allyl)	70	96:4	[174]
11	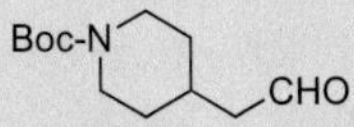 Boc-N (piperidine), CHO	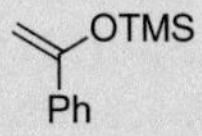 =OTMS, Ph	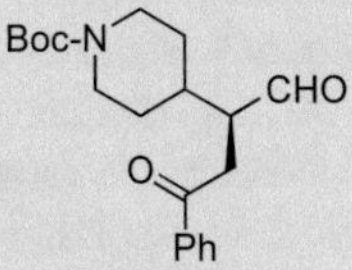 Boc-N (piperidine), CHO, O=, Ph	84	97:3	[175]
12	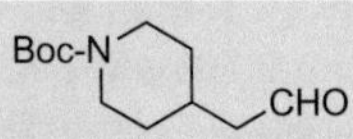 Boc-N (piperidine), CHO	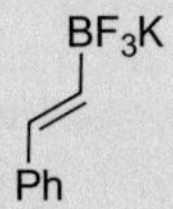 BF$_3$K, Ph	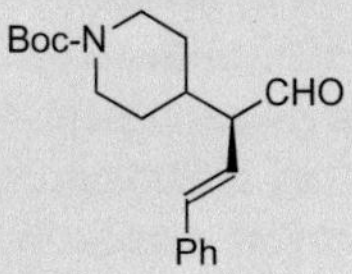 Boc-N (piperidine), CHO, Ph	73	96:4	[176]

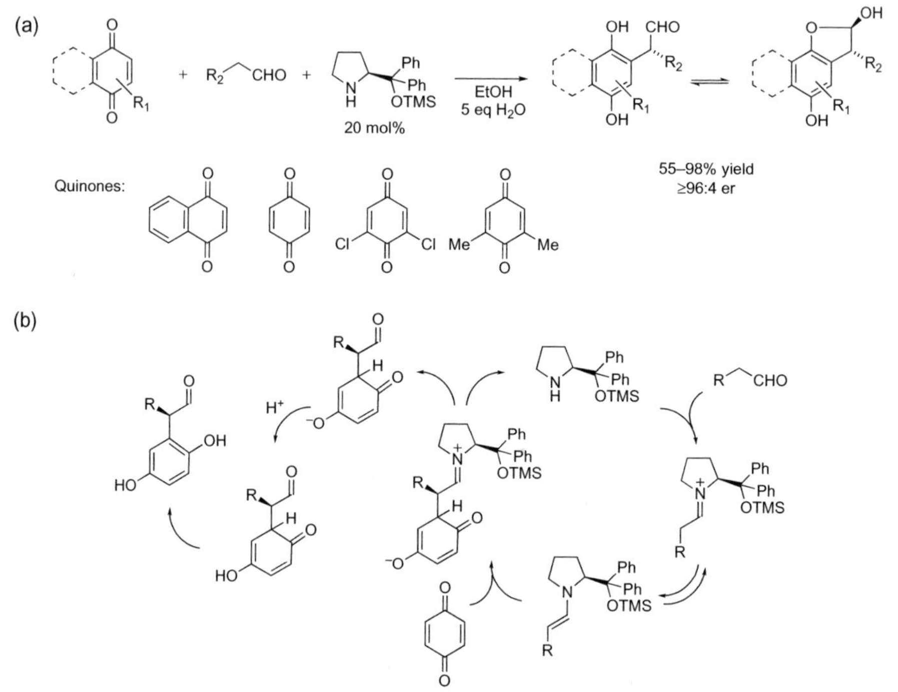

(a)

Quinones:

(b)

SCHEME 3.30 Jørgensen's organocatalytic α-arylation of aldehydes by enamine addition to quinones [179].

shown in Scheme 3.30b. Condensation of the catalyst with the aldehyde affords the enamine *via* the iminium ion shown at the right of the scheme. Conjugate addition and hydrolysis regenerate the catalyst, whereas the quinone adduct tautomerizes to the diphenol shown on the left. Although an equivalent of water is formed in the condensation step, the reaction fails in anhydrous DMSO. The optimal solvent was found to be ethanol containing 5 molar equivalents of water.

Group-selective reactions are ones in which heterotopic ligands are distinguished. Recall from the discussion at the beginning of this chapter that secondary amines form complexes with lithium enolates (Scheme 3.2) and that lithium amides form complexes with carbonyl compounds (Section 3.1.1). So, if the ligands on a carbonyl are enantiotopic, they become diastereotopic on complexation with chiral lithium amides. Thus, deprotonation of certain ketones can be rendered enantioselective by using a chiral lithium amide base [180], as shown in Scheme 3.31 for the deprotonation of cyclohexanones [181–187]. *Cis*-2,6-dimethyl cyclohexanone (Scheme 3.31a) is meso, whereas 4-*tert*-butylcyclohexanone (Scheme 3.31b) and 2,2-disubstituted cyclohexanediones (Scheme 3.31c) are achiral. Nevertheless, the enolates of these ketones are chiral. Alkylation of the enolates affords nonracemic products and *O*-silylation affords a chiral enol ether, which can be further manipulated by a number of means. Although crystallographic and spectroscopic characterizations of chiral lithium amides have been carried out [184], a rationale explaining the relative topicity of these deprotonations has not been offered. Note that any heterotopic protons may, in principle, be distinguished by this concept. Scheme 3.31c illustrates the selective deprotonation of cyclohexanediones, and *in situ* reduction of the unreacted carbonyl [188]. In this example, the Simpkins group used

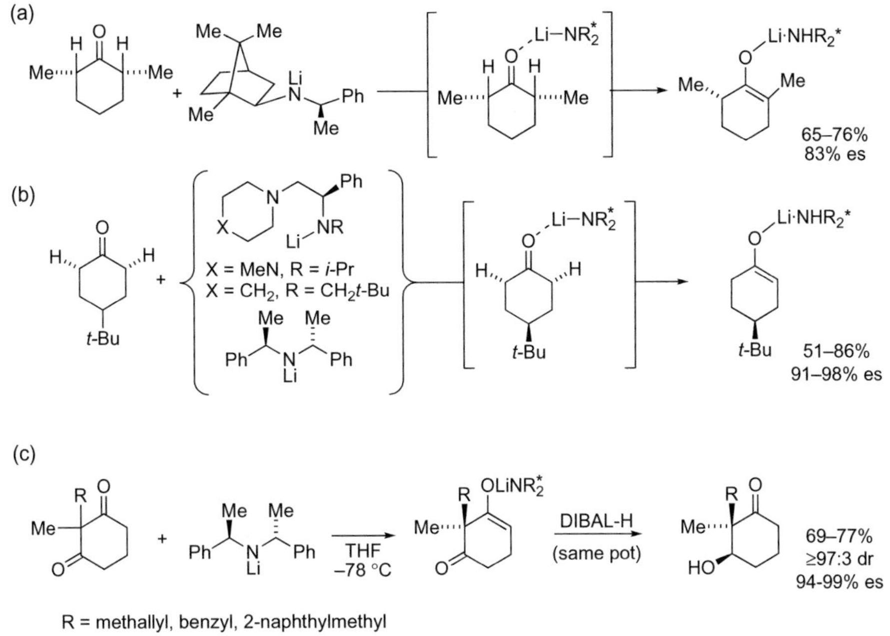

(a)

65–76%
83% es

(b)

X = MeN, R = *i*-Pr
X = CH₂, R = CH₂*t*-Bu

51–86%
91–98% es

(c)

DIBAL-H
(same pot)

69–77%
≥97:3 dr
94–99% es

R = methallyl, benzyl, 2-naphthylmethyl

SCHEME 3.31 Enantioselective deprotonation of achiral ketones with chiral lithium amide bases: (a) [182], (b) [183–185,187], and *(c)* one-pot deprotonation of cyclohexanediones, and *in situ* reduction [188].

enolization to protect one of the carbonyls from reduction. The enantioselectivity of this one-pot process was superior to a two-step process in which the enol silyl ether was isolated.

This concept has been extended to kinetic resolution (selective reaction of protons that are enantiotopic by external comparison) [38,189,190] and to selective reaction at proton pairs that are diastereotopic (double asymmetric induction) [191].

A further extension of these concepts is the enantioselective alkylation of lithium enolate· amine complexes. Following several early observations [192–194], systematic investigations were undertaken by the Koga group [33,34,195–199]. These efforts have resulted in a very selective asymmetric alkylation of cyclohexanone, δ-valerolactone, *N*-alkyl-2-piperidones, and α-tetralone with activated alkyl halides. As summarized in Scheme 3.32 and Table 3.12, alkylation of these carbonyl compounds affords up to 98% enantioselectivity. During the optimization studies in which the enolates were generated by deprotonation, Koga observed an increase in enantioselectivity and chemical yield as the reaction time increased, and ascribed the phenomenon to the formation of a mixed aggregate that includes the lithium bromide formed as the reaction proceeds [200]. Further experiments revealed that generation of the lithium enolate from the silyl enol ether and addition of lithium bromide and a chiral amine afford comparable enantioselectivities [200]. The reactive species is thought to be a lithium enolate/secondary amine/lithium bromide mixed aggregate (Scheme 3.32 inset) [196,200]. As indicated in Scheme 3.32a and tabulated in Table 3.12, the process can be made catalytic in chiral amine in the presence of an excess of tetramethylpropylenediamine (TMPDA). This implies that there is ligand exchange between the chiral triamine and the TMPDA (Scheme 2.35b), and that the enolate · TMPDA complex is less reactive in the alkylation. A possible explanation is that the TMPDA breaks up the enolate · LiBr mixed

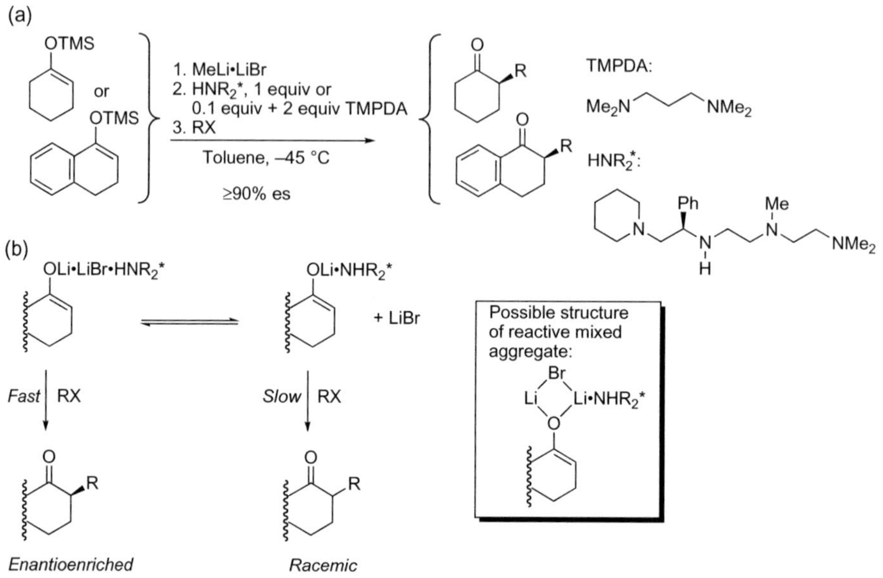

(a)

(b)

SCHEME 3.32 Enantioselective alkylation of lithium enolate/secondary amine/lithium bromide complexes by interligand asymmetric induction [198].

TABLE 3.12 Koga's Asymmetric Alkylation of Ketones (Scheme 3.32) [198]

Silyl enol Ether of	HNR₂*/TMPDA	Electrophile	% Yield	% es
α-tetralone	1.0/0	BnBr	56	98
α-tetralone	0.1/2.0	BnBr	78	97
α-tetralone	0.05/2.0	BnBr	75	98
α-tetralone	0.05/2.0	Allyl bromide	69	98
α-tetralone	0.05/2.0	Prenyl bromide	62	98
α-tetralone	0.05/2.0	Cinnamyl bromide	69	95
α-tetralone	0.05/2.0	MeO_2CCH_2Br	84	90
cyclohexanone	1.0/0	BnBr	43	95
cyclohexanone	0.1/2.0	BnBr	52	95

aggregate, producing an enolate · TMPDA complex that is less reactive. A similar catalytic approach has been used to accomplish enantioselective alkylation of 2-methyl-α-tetralone, affording a quaternary stereocenter with up to 94:6 er [201].

A conceptually different approach to interligand asymmetric induction uses chiral phase-transfer catalysts (reviews: refs. [202–204]). Scheme 3.33 illustrates two examples of such a process using an *N*-benzylcinchonium halide catalyst. O'Donnell reported the asymmetric alkylation of the Schiff base of *tert*-butyl glycinate using *N*-benzylcinchonium chloride (Scheme 3.33a, [205]). This process, which works for methyl, primary alkyl, allyl, and benzyl halides, is noteworthy because the substrate is acyclic and because monoalkylation is achieved without racemization under the reaction conditions. Hughes et al. reported a detailed kinetic study

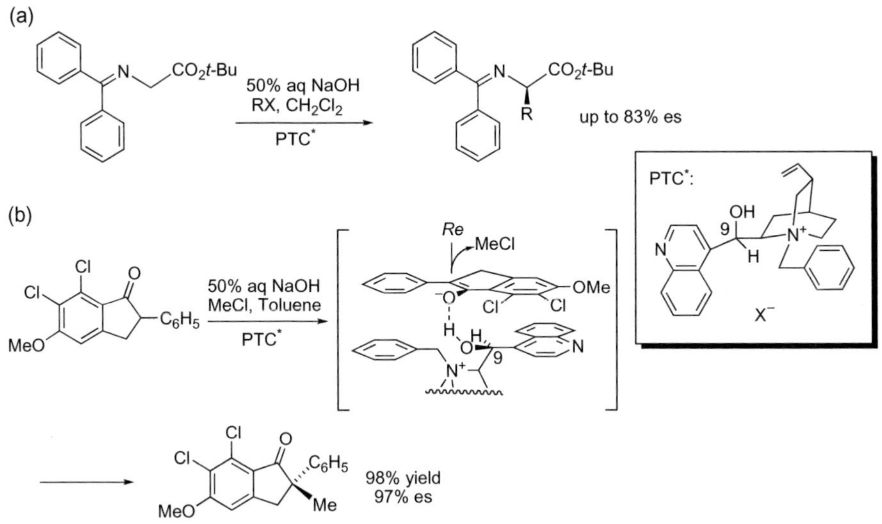

SCHEME 3.33 Enantioselective alkylations using chiral phase-transfer catalysts derived from a cinchona alkaloid. (a) [206] and (b) [205].

of the indanone methylation illustrated in Scheme 3.33b, which revealed a mechanism significantly more complicated than a simple phase-transfer process: the reaction is 0.55 order in catalyst and 0.7 order in methyl chloride, deprotonation of the indanone occurs at the interface, and methylation of the enolate (not deprotonation) is rate determining [206]. Nevertheless, the rationale for the enantioselectivity involves a 1:1 complex of catalyst and enolate, as illustrated in brackets. Molecular modeling studies and an X-ray crystal structure suggest that the most stable conformation of the catalyst has the quinoline ring, the C_9-O bond, and the phenyl of the N-benzyl group nearly coplanar (Scheme 3.33b) [206]. Hydrogen bonding with the enolate, dipole alignment, and π-stacking of the aromatic moieties result in the assembly shown. Methylation then occurs from the *Re* face, opposite the catalyst. For the O'Donnell alkylation in Scheme 3.33a, the observed chirality sense may be rationalized by assuming an *E(O)*-enolate and π-stacking of the benzophenone rings of the enolate above the quinoline ring on the catalyst, and approach of the electrophile as before.

Subsequent to these pioneering investigations, Corey and Lygo independently developed an improved phase-transfer catalyst based on N-anthracylmethyl cinchona alkaloids [207,208]. The rationale for this catalyst design relies on a combination of ion pairing of the enolate with the quaternary ammonium, and attractive van der Waals interactions as illustrated in Figure 3.14. The red lines indicate the tetrahedron surrounding the quaternary nitrogen. The rear face is blocked by the quinuclidine ring, the right face by the anthracene, and the bottom face by the O-allyl group. Only the front-left face is sterically accessible for ion pairing with an anion. Two X-ray crystal structures support this analysis [207].

The anthracylmethyl cinchodiniums afford better selectivity than the N-benzyl analogs. Several examples from the two independent reports are listed in Table 3.13. A rationale for alkylation of the *Si* face of the *E(O)*-enolate is shown in the inset of Scheme 3.34. The N-anthracylmethyl cinchonidium system can be used in organic solvents and in solid phase synthesis [209], on solid supports [210,211], or in water in the presence of Triton X-100 micelles [212]. In the latter instance, catalyst loadings as low as 0.1 mol% are possible.

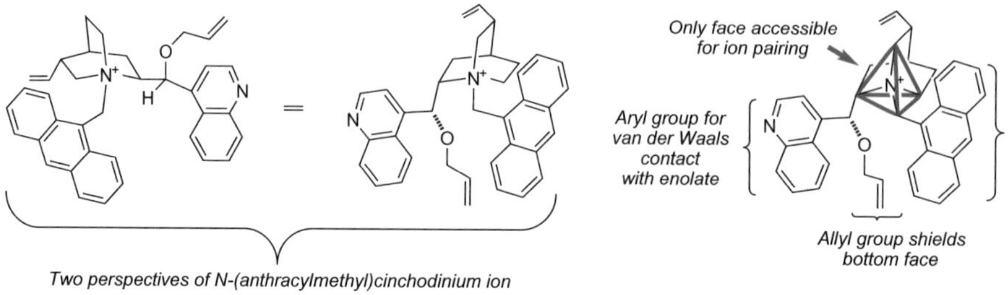

Two perspectives of N-(anthracylmethyl)cinchodinium ion

FIGURE 3.14 Corey's analysis of cinchodinium phase-transfer catalyst [207]. On the left, note the similarity to the mirror image of the cinchona alkaloid in the inset of Scheme 3.33. On the right, the red lines indicate the tetrahedral faces of the ammonium ion, with only one available for ion pairing with the enolate.

TABLE 3.13 Examples of Glyine Enolate Alkylation Using *N*-Anthracylmethyl Cinchona Alkaloids[a]

RX	Conditions	% Yield	% es	Ref.
MeI	A	71	98	[207]
MeI	B	41	94	[208]
Allyl bromide	A	89	98	[207]
Allyl bromide	B	76	94	[208]
Benzyl bromide	A	87	97	[207]
Benzyl bromide	B	68	95	[208]
Hexyl iodide	A	79	>99	[207]
Butyl iodide	B	42	94	[208]

[a]*Conditions are described in the caption to scheme 3.34.*

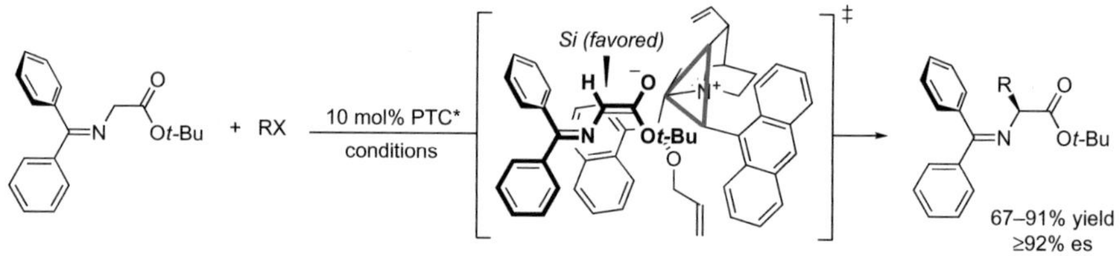

SCHEME 3.34 Improved glycine alkylation using phase-transfer catalysis. For specific examples, see Table 3.13. The inset shows the postulated position of the enolate according to the Corey model in Figure 3.14. Conditions: (*a*) R=allyl, CsOH·H$_2$O, CH$_2$Cl$_2$, −60 °C or −78 °C [207]; (*b*) R=H, 50% aq. KOH, toluene, room temperature [208]. Condition (*a*) gives somewhat higher enantioselectivities and higher yields.

The lack of racemization and dialkylation in the glycine alkylation processes deserves comment. Apparently, the rate of deprotonation of the product is significantly slower than deprotonation of the starting material. The reasons for the reduced acidity of the product become apparent upon examination of models (Figure 3.15).[18] $A^{1,3}$-strain considerations dictate that the α-carbon$-$hydrogen bond (nearly) eclipses the nitrogen$-$carbon double bond and also forces the *syn* phenyl group out of planarity. The three lowest energy conformers[19] are illustrated looking down the α-carbon$-$nitrogen bond. The global minimum (conformation shown in Figure 3.15a) has the α-proton near the nodal plane of the carbonyl π-system, and therefore nonacidic. The other two have the α-proton in better alignment with the carbonyl, but shielded from the approach of the base by the phenyl group at the top. Note that a proton in the position of the α-methyl in the conformation shown in Figure 3.15a (as in the starting material) would be quite acidic due to overlap with *both* π-systems. Additionally, to the extent that the proximal phenyl lies in the plane of the $C=N-C=C$ π system, the substituted enolate is significantly destabilized by $A^{1,3}$-strain as shown in Figure 3.15d.

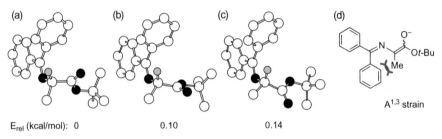

E_{rel} (kcal/mol):   0   0.10   0.14

FIGURE 3.15   (a)-(c) Low-energy conformations of *tert*-butyl alaninate-benzophenone Schiff base. Only the α-hydrogen is shown (hydrogen is shaded, nitrogen and oxygen are black); note that the phenyl that is *cis* to the α-hydrogen is rotated out of plane in all three conformers due to $A^{1,3}$-strain. *(d)* Substituted enolate, showing allylic strain due to the methyl group.

3.1.4 Enolate and Azaenolate Alkylations with Chiral Electrophiles

In all the preceding examples, stereoselectivity has been achieved by creating a chiral environment in the vicinity of the carbonyl α-carbon, such that approach to only one face of the enolate is favored. A complementary approach, with an entirely different set of challenges, is to employ chiral electrophiles. The examples illustrated in Scheme 3.35 are presented according to the hybridization state of the electrophilic carbon. If the leaving group (LG) is on a primary sp^3 carbon, then one or more stereocenters in R* may influence the stereoselectivity of the alkylation (Scheme 3.35a). If the leaving group is on a stereogenic secondary carbon, S_N2 inversion will control one of the stereocenters, and other factors may influence the configuration of the α-carbon (Scheme 3.35b). The most common type of sp^2-hybridized electrophiles are π-allyl complexes, wherein a chiral ligand on the metal may influence the approach of a prochiral electrophile (Schemes 3.35c and d). With unsymmetrical π-allyls, the question of regiochemistry enters the picture (Scheme 3.35e).

---

18. Molecular modeling calculations were done by the author using the MM2* force field as supplied in Macromodel [213].
19. Three other accessible conformers lie ≥ 1.8 kcal/mol above the global minimum.

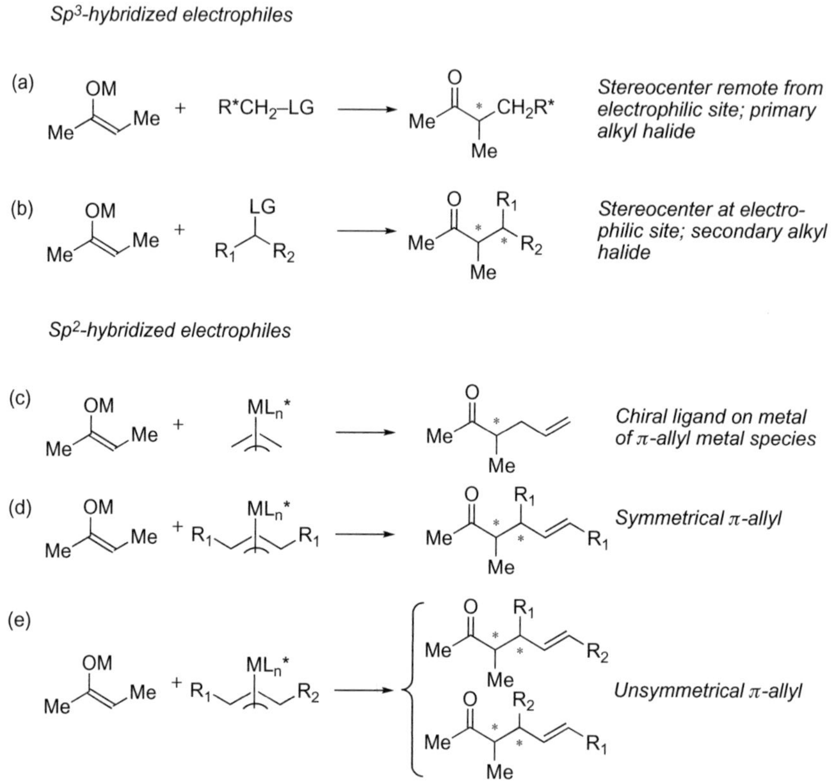

Sp³-hybridized electrophiles

(a) Stereocenter remote from electrophilic site; primary alkyl halide

(b) Stereocenter at electrophilic site; secondary alkyl halide

Sp²-hybridized electrophiles

(c) Chiral ligand on metal of π-allyl metal species

(d) Symmetrical π-allyl

(e) Unsymmetrical π-allyl

SCHEME 3.35 Possible modes of enolate alkylation featuring chiral electrophiles.

The reaction of a prochiral enolate with an electrophile having the leaving group at a primary carbon was exploited by the Overman group in the total synthesis of the pyrrolidinoindoline alkaloids shown in Figure 3.16 [214–216]. The challenge presented in all these compounds is the stereoselective preparation of (sometimes adjacent) quaternary stereocenters, which was met by employing the ditriflate shown.

The key alkylation of the ditriflate with the potassium enolate of the indanone is shown in Scheme 3.36 [215]. The carefully optimized conditions afforded the major C_2-symmetric product with 89% diastereoselectivity; the product was isolated in 70% yield. Lithium, sodium, and potassium enolates were found to afford similar levels of diastereoselectivity, but the potassium enolate was the most reactive, affording complete conversion in 3 h. The major by-product was the C_1-symmetric isomer shown; a third diastereomer, of C_2-symmetry, was formed in less than 1% yield. Acetonide hydrolysis of the major diastereomer, oxidative cleavage, and functional group manipulations afforded (−)-esermethole, (−)-physostigmine, and (−)-phenserine in excellent overall yields.

The origin of the stereoselectivity is interesting. First of all, note that the diastereoselectivity in favor of the major C_2-symmetric isomer is favored on statistical grounds. If the *Si/Re* facial selectivity for both alkylations is $n:1$, then the diastereomer ratio will be $n^2:2n:1$ (C_2 major: $C_1:C_2$ minor). For this alkylation, this corresponds to a facial bias of about 20:1.

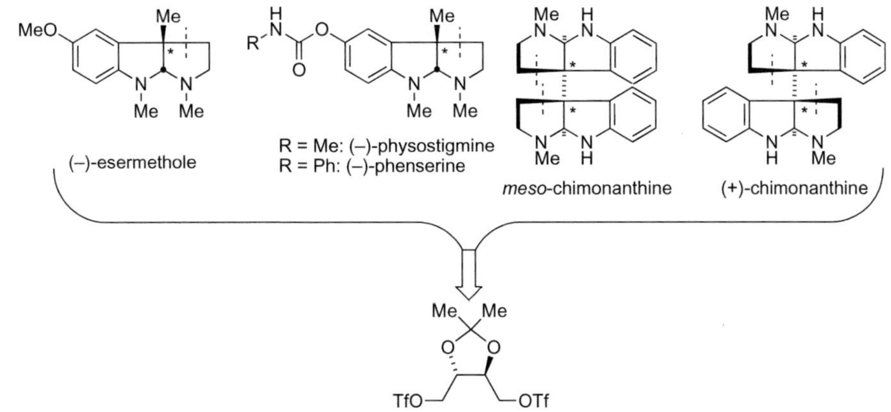

FIGURE 3.16 Overman's synthesis of pyrrolidinoindoline alkaloids employed a tartate-derived ditriflate in alkylations of prochiral enolates [214–216]. The stereocenters formed in the alkylation are indicated (*).

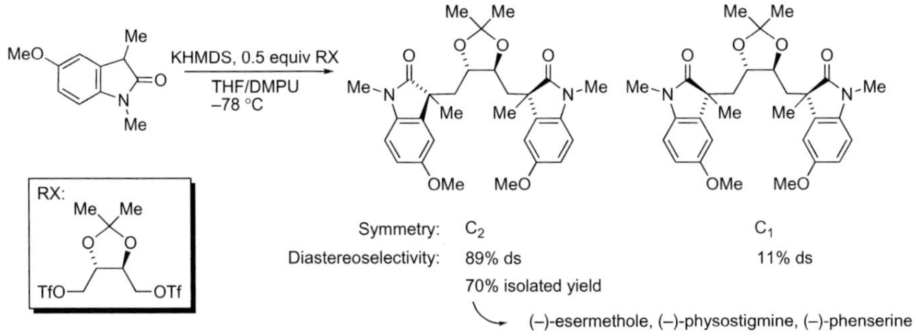

SCHEME 3.36 The key step of Overman's asymmetric alkylation of 2-methylindanones using a chiral ditriflate electrophile [215].

The cation does not appear to affect the diastereoselectivity. The optimal solvent mixture is 2% DMPU in THF, a mixture that would be expected to coordinate the cations strongly. For this reason, the authors postulated open transition structures [215]. Projections illustrating six possible staggered, open transition structures are shown in Scheme 3.37a. These structures are tentative, since the orientation of the dioxolane ring relative to the approaching enolate is not known. The dioxolane is important, since a dimethylcyclopentane ditriflate analog (the two dioxolane oxygens are replaced with methylenes) showed diminished stereoselectivity (69:23:8, C_2-major:C_1:C_2-minor). Two rationales for the observed Si-facial bias were suggested. In the first rationale, structures A, B, D, and E appear to be disfavored due to severe steric interactions, leaving C and F as the most likely based on steric grounds. The authors then suggest that C should have lower energy due to unfavorable electrostatic or dipole interactions between the enolate oxygen and the proximal oxygen in the dioxolane of structure F. In the second rationale, opposite dipole alignment between the enolate and the C_2–O bond of the triflate could be the controlling factor. This would appear to favor A, with D coming in second. Scheme 3.37b illustrates an earlier example of stereocontrol, reported by Tadano in 1995, using a chiral triflate to preferentially alkylate one face of a 5-membered ring enolate,

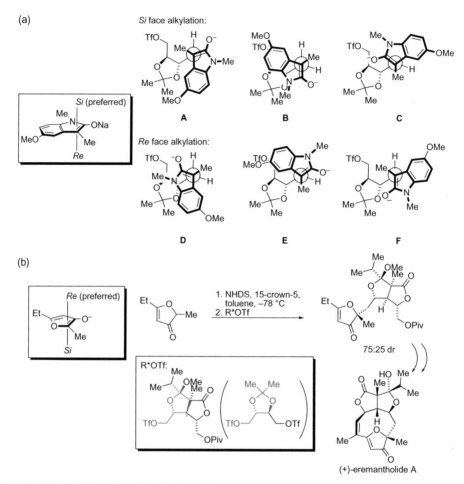

SCHEME 3.37 (a) Projections of possible transition structures for the asymmetric alkylations of Scheme 3.36 [215]. (b) Alkylation of a chiral triflate with structural and stereoelectronic properties similar to Overman's tartrate-derived ditriflate (note highlights in red) [217].

as part of a total synthesis of (+)-eremantholide A [217]. There are striking similarities between the steric course of this alkylation and the Overman examples (highlighted in the insets). Both alkylations are done with a sodium enolate in a 5-membered ring, and both likely proceed through open transition states (note the Na$^+$·15-crown-5 complex). Although the *CIP* sequence is reversed, when drawn as in the insets at the left, the preferred electrophile approach is similar on steric grounds. Examination of the structurally complex triflate shows remarkable similarities to the tartrate-derived ditriflate shown in the inset of Scheme 3.37b (highlighted in red). Thus, the Overman analysis of Scheme 3.37a may be general for structurally rigid β-alkoxy triflates.

The tartrate-derived ditriflate was also used by the Overman group to effect dialkylation of the 1,4-dicarbonyls of 3,3'-dioxindoles, as illustrated in Scheme 3.38 [214,216]. Depending on the conditions used, either the C_1-symmetric product or one of the two C_2-symmetric products could be obtained selectively. Double deprotonation of the dioxindole

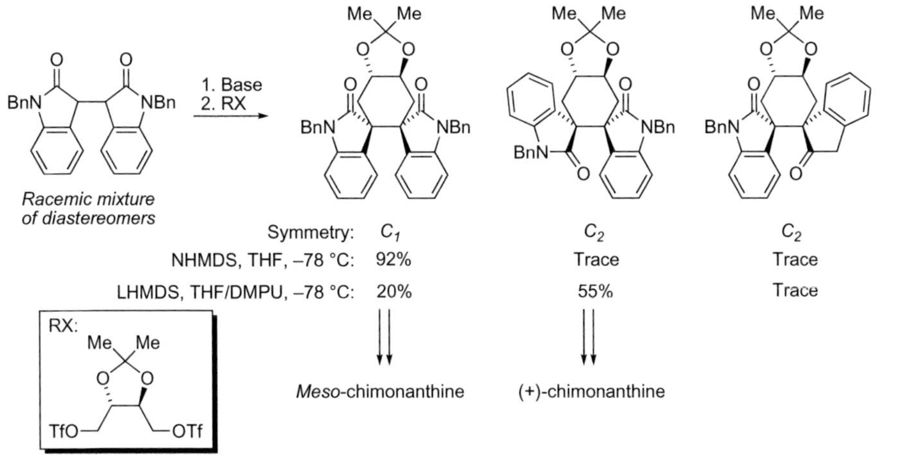

SCHEME 3.38   Overman's synthesis of chimonanthine diastereomers by double alkylation of a 1,4-dicarbonyl with a chiral ditriflate [214,216].

using sodium hexamethyldisilazide (NHMDS) in THF, followed by addition of the ditriflate shown in the inset, afforded the C_1-symmetric dispirocyclohexane in 92% yield. Hydrolysis of the dioxolane ring, oxidative cleavage, and six additional steps afforded *meso*-chimonanthine in a (curiously) asymmetric synthesis of an achiral alkaloid. Changing the cation from sodium to lithium and the solvent to 90:10 THF/DMPU (or later 70:30 THF/HMPA [218]) reversed the selectivity to one of the C_2-symmetric diastereomers. This diastereomer was then converted by a similar route to (+)-chimonanthine. These schemes are notable because they provide a method for preparing adjacent quaternary stereocenters by alkylation of a 1,4-dione with good-to-excellent levels of diastereoselection and yield.

An investigation was undertaken to elucidate the mechanistic features of these dialkylation processes [218]. It was determined that optimal diastereoselectivity was achieved if the dione was doubly deprotonated prior to addition of the electrophile, although the nature of the dienolate (chelated *s-cis* or dipole-aligned *s-trans*) could not be determined. Scheme 3.39a outlines the formation of the C_1-symmetric product. At the outset, note that the two alkylations must occur on opposite faces (one *Re* and one *Si*) in order to produce the C_1-symmetric product. The first (intermolecular) alkylation could take place from either the *Re* or the *Si* face of one of the enolates to give either of the two intermediates shown in the brackets. The second (intramolecular) alkylation occurs through one of the chelated intermediates shown, which exposes the opposite *CIP* face of the second enolate to the triflate. Optimized conditions for the formation of the C_2-symmetric product utilize metal-ligating solvents such as DMPU or HMPA (Scheme 3.39b). A model study revealed that the first step of the alkylation of the dilithium dienolate proceeds with 95% diastereoselectivity, adding to the *Si* face in this solvent system. Dipolar alignment of the amide carbonyl and the enolate is then followed by the second alkylation. Statistical analysis of the product mixtures revealed that the second alkylation occurred with 92% diastereoselectivity. It may be that in THF solvent, a chelated *s-cis* dienolate is the reactive species, whereas in the presence of DMPU or HMPA, chelation is avoided and the conformation of the dienolate is *s-trans*.

(a) Selective preparation of the C_1-symmetric dialkylation product

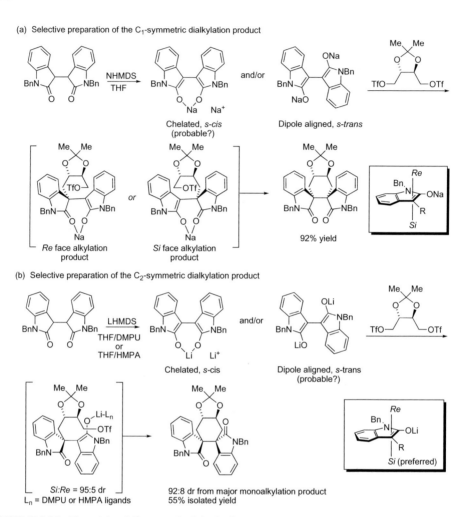

SCHEME 3.39 The origin of diastereoselectivity in the dialkylation of dioxoindoles with a chiral ditriflate [214,216,218]. (a) NHMDS in THF; (b) LHMDS in THF with either DMPU or HMPA.

Conceptually, the simplest example of a chiral electrophile in an enolate alkylation is the reaction of an unsymmetrical secondary alkyl halide. One interesting application, reported by Vedejs, is shown in Scheme 3.40 [219,220]. The lithium enolate of 2,2-dimethylcyclopentanone was alkylated with lactate triflate esters. Of course, the lactate stereocenter is inverted by the S_N2 substitution, but the facial preference of the enolate is influenced by the ester group of the triflate: the *Si* face of the enolate is preferentially alkylated by lactate triflate ethyl ester in THF, but the methoxyethoxyethyl (MEE) ester was preferentially approached by the *Re* face of the enolate when conducted in toluene. The purpose of the MEE ester was to introduce affinity for the lithium cation of the enolate, so as to produce a more ordered transition state in a hydrocarbon solvent. The postulated transition structures for these two examples are shown in brackets. Recall (Figure 3.5) that the optimal approach trajectory of

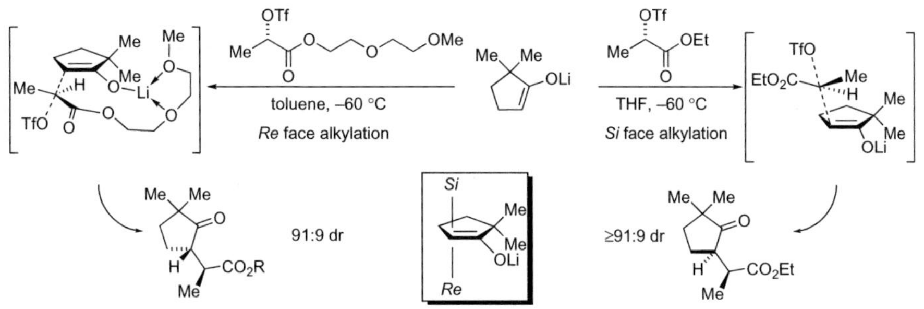

SCHEME 3.40 The steric course of Vedejs' asymmetric alkylation of a cyclopentanone enolate is dependent on the lactate ester group [219,220].

an electrophile to an enolate is tilted away from both the carbonyl carbon and the enolate oxygen. Recall also that the nucleophile must approach the electrophile at an angle of 180° from the leaving group. The favored approach therefore places the electrophile somewhat over the ring. Therefore, in both transition structures, the small hydrogen of the triflate is projected over or under the cyclopentene ring to minimize steric repulsion. With the ethyl ester, the carboxylate is *antiperiplanar* to the enolate. With the MEE ester, the lithium is chelated by two or three oxygens of the MEE group. With this chelation, in order for the hydrogen to project toward the cyclopentene ring, the electrophile must be positioned on the *Re* face of the enolate. Presumably, a similar orientation of the ethyl ester on the *Re* face is disfavored due to unfavorable electrostatic and/or steric interactions with the solvated O−Li moiety.

Palladium-catalyzed allylic allylations were first reported by Tsuji in 1965 [223−226]. The literature on the type of transition metal-mediated asymmetric allylic alkylations (AAAs) shown in Schemes 3.35c−e is vast (recent reviews: refs. [224−238]), and only selected examples are presented here.

The catalytic cycle is illustrated in Scheme 3.41. An allylic substrate coordinates in an η^2 fashion to a metal catalyst (ML_2), then loss of a leaving group gives the cationic π-allyl metal complex. Nucleophilic addition is then followed by decomplexation and regeneration of the catalyst. In this illustration, there are no stereochemical or regiochemical consequences involving the π-allyl metal species. Such issues arise when the allyl group is substituted on either end, or when the nucleophile is prochiral.

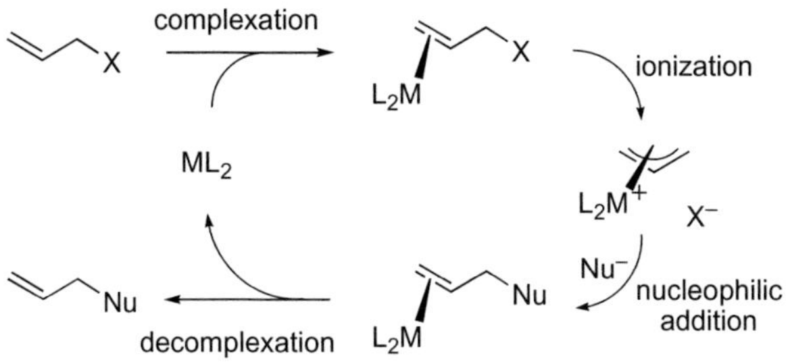

SCHEME 3.41 Catalytic cycle for transition metal-mediated allylic alkylation.

The mechanism of any allylation depends on a number of factors, but an important classification is the basicity of the nucleophile. In these systems, nucleophiles whose conjugate acids have $pK_a \leq 25$ are considered soft nucleophiles, whereas nucleophiles whose conjugate acids have $pK_a \geq 25$ are considered hard [226]. Soft nucleophiles generally approach the face of the allyl group from the side opposite the transition metal, displacing the metal in an S_N2-like fashion. In contrast, hard nucleophiles tend to first coordinate to the metal, and then couple with the allyl group by reductive elimination (Scheme 3.42). These mechanistic differences may result in net retention with soft nucleophiles, and net inversion with hard nucleophiles, but not always.

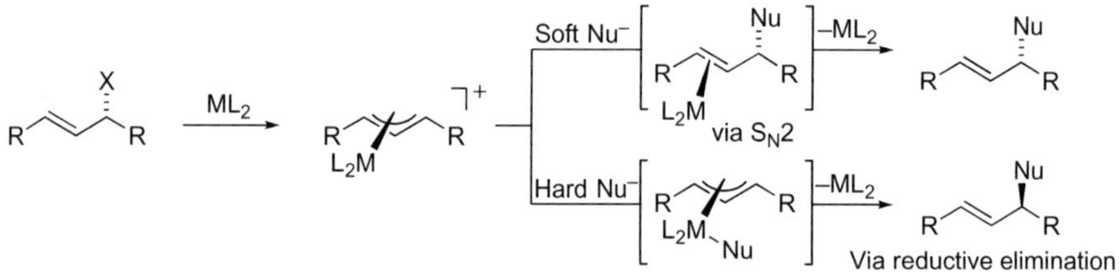

SCHEME 3.42   Soft *vs.* hard nucleophiles in allylic alkylation reactions.

Much of the work in the development of asymmetric allylic alkylations (AAAs) has been done by the Trost group, using the catalysts illustrated in Figure 3.17. This family of catalysts employs chiral diamine scaffolds (Figure 3.17a), which are linked to diphenylphospino benzoic acid (DPPBA) through amide linkages. Structural studies reveal that the π-allyl palladium complexes of these ligands are in equilibrium with oligomeric species, which complicates efforts to understand the steric course of the reaction. Nevertheless, the allyl group resides in a chiral pocket created by the four phenyl groups attached to phosphorous. A predictive model has been formulated for understanding and rationalizing the steric course of the AAA [239], which focuses on the two phenyl groups highlighted in red in Figure 3.17b. These two phenyl groups are thought to block two quadrants of space around the allyl group: in the *R,R* series, the right rear and left-front quadrants, and the left rear and right-front quadrants of the *S,S* series. The cartoon in Figure 3.17c illustrates a simplified model of the DPPBA ligands. The plane of the allyl group is at an angle of 110° relative to the Pd, as shown in Figure 3.17d. Although the *DPPBA ligand is C₂-symmetric, the π-allyl Pd DPPBA complex is not.* Since the ionization and nucleophilic addition steps of the catalytic cycle are S_N2 displacements, the orientation in which the Pd is *antiperiplanar* to the leaving group or nucleophile is favored for both ionization and nucleophilic displacement. Thus, the open rear quadrants of the cartoons in Figure 3.17c, which would require *anticlinal* approach, are disfavored. Using the *R,R*-ligand as a model, the model posits that the leaving group in the ionization step and the nucleophile in the addition step act in the right-front quadrant, as represented by the red double arrows of Figure 3.17c. Similarly, it is the left-front quadrant that is active with the *S,S* series of ligands.

An example of the AAA in which the precursor is a meso diester is shown in Scheme 3.43a [240,241]. Using an *R,R*-DPPBA catalyst and the lithium anion of phenylsulfonyl nitromethane as nucleophile, an isoxazoline *N*-oxide was obtained in excellent yield and enantioselectivity. The steric course of the reaction is rationalized as shown in Scheme 3.43b. Complexation of the meso dibenzoate to the palladium complex is followed by ionization of the benzoate in the right-front quadrant to give the π-allyl complex shown in the center. Nucleophilic approach, again from the front right quadrant, gives the initial adduct (shown in brackets), which then

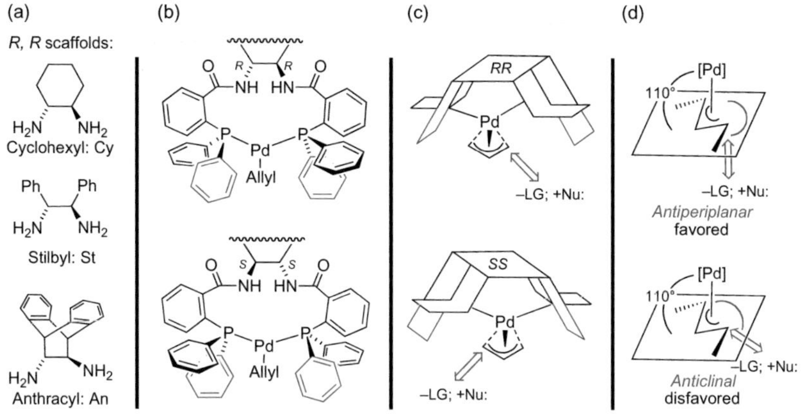

FIGURE 3.17 Models for understanding the steric course of AAA reactions using Trost's DPPBA ligands [239].

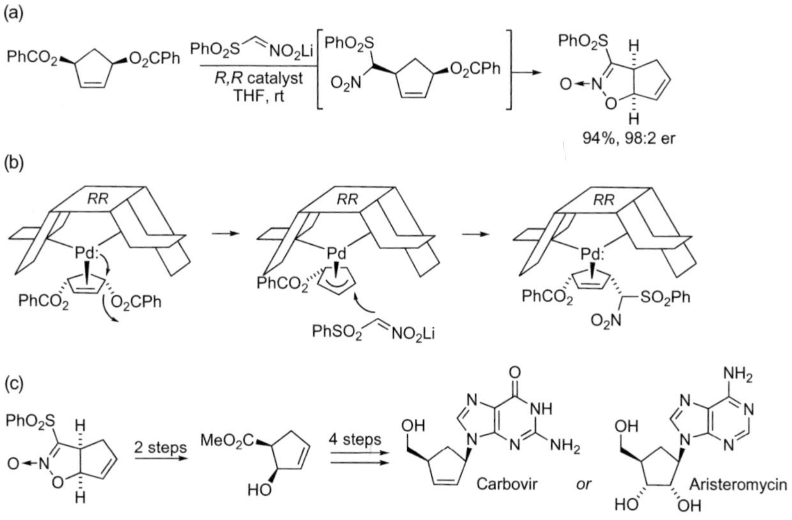

SCHEME 3.43 Asymmetric allylic alkylation in which the ionization step sets the steric course [240,241].

cyclizes to the isoxazoline *N*-oxide. The cyclization is an intramolecular π-allyl palladium alkylation, but in this case, ionization is mismatched. Addition of (Ph₃P)₄Pd at this stage can shorten reaction times [241]. Methanolysis, to replace the phenylsulfonyl group with methoxy, and reduction of the N—O bond afford the hydroxy ester shown in Scheme 3.43c. This intermediate could be converted in four steps to either carbovir or aristeromycin.

A stereoconvergent AAA, which begins with racemic cycloalkenyl acetates and carbonates, is shown in Scheme 3.44a [242]. Recall from Figure 3.17 that the *R,R* catalyst should react more quickly with the *S*-allylic acetate, as shown in Scheme 3.44b; this is the matched pair. The reaction of the catalyst with the *R*-allylic substrate is slower, but both substrate enantiomers afford the same π-allyl palladium intermediate. Approach of the malonate

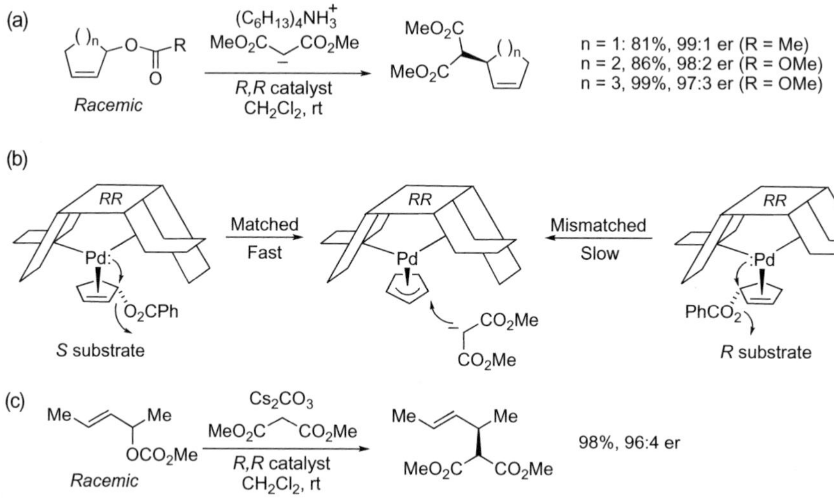

(a)

n = 1: 81%, 99:1 er (R = Me)
n = 2, 86%, 98:2 er (R = OMe)
n = 3, 99%, 97:3 er (R = OMe)

(b)

(c)

98%, 96:4 er

SCHEME 3.44 Stereoconvergent AAA using racemic allylic substrates: (a) cyclic alkenes [242], (b) mechanistic rationale [239], and (c) acyclic alkene [243].

nucleophile is preferred from the front right quadrant, and works best when the cation is tetrahexylammonium. A conceptually similar stereoconvergent AAA, employing an acyclic allylic substrate is shown in Scheme 3.44c [243]. This example worked very well with the cesium salt of dimethyl malonate and other nucleophiles, but the analog in which the two methyls of the pentenyl system are replaced by phenyls afforded poor yield and almost no enantioselectivity. The authors surmised that the phenyls were too large to fit nicely in the chiral pocket of the DPPBA−Pd complex. With reference to Figure 3.17, note that bulky groups could encounter steric hindrance in both the complexation and the ionization steps.

In the examples of AAA shown in Scheme 3.44, the only chirality center generated in the reaction is in the allyl system; the nucleophile is malonate ion; and the nucleophilic enolate carbon is not prochiral. Most enolate anions are prochiral, and employment of such nucleophiles in the AAA reaction imposes a new challenge. The chiral pocket must somehow restrict approach of the enolate in such a way as to favor only one enantiotopic face of the approaching enolate. Scheme 3.45a and b shows the alkylation of β-ketoesters of cyclohexanone and tetralone with allylic acetates, in which the single stereocenter in the product is the carbon α to the carbonyl [244]. Examples of cyclic allylic acetates are illustrated in Scheme 3.45c. In this system, two stereocenters are formed, and the diastereoselectivity is as high as the enantioselectivity. From the absolute configuration of the products, it appears that the *R,R*-DPPBA ligands favor approach of the chiral π-allyl palladium complex to the *Si* face of the nucleophile, as shown in Scheme 3.45d. As before, the *R,R* ligand favors approach of the nucleophile toward the *Si* face of the π-allyl in the front right quadrant (Scheme 3.45d, middle), suggesting a model (Scheme 3.45d, right) combining these two elements. Note, however, that the illustrated orientation of the enolate relative to the DPPBA ligand is entirely speculative [239].

Unstabilized enolates, such as the lithium or magnesium enolates of ketones, can also be employed in AAA reactions, as demonstrated by Braun [245,246] and Trost [247] (review: ref. [237]). Using *M*-BINAP as catalyst, Braun demonstrated the asymmetric allylation of cyclohexanone, as illustrated in Scheme 3.46. The racemic diphenylallyl acetate shown in

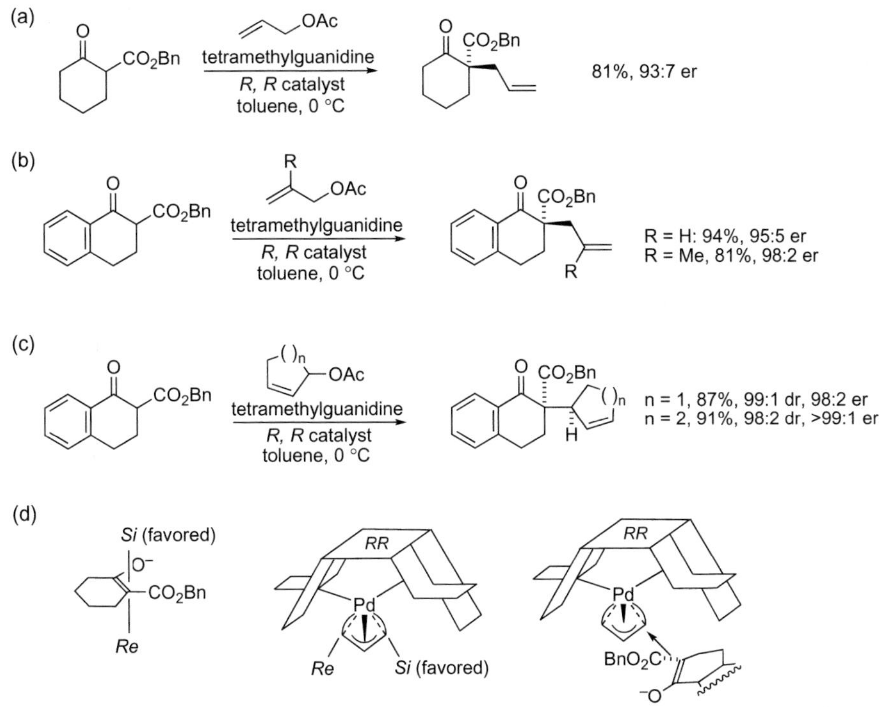

SCHEME 3.45 AAA reactions using prochiral nucleophiles with (a) and (b) achiral electrophiles and using a chiral catalyst [244]; (c) chiral electrophiles. (d) The *Si* face of the enolate is apparently favored with the *R,R* catalyst, as is the *Si* face of the prochiral π-allyl electrophile.

Scheme 3.46a did not behave well with the Trost DPPBA catalysts (see above), but with *M*-BINAP ligand (inset), the palladium catalyst effected a nicely stereoconvergent reaction: the allylated product was obtained in excellent diastereoselectivity and enantioselectivity [245]. Cyclopentanone and cyclohexanone lithium enolates can be allylated using a similar system in good yield; the enantioselectivity is higher with cyclohexanone, however [246]. The lithium enolate allylations benefit from the presence of LiCl in the reaction mixture. Another example of a stereoconvergent alkylation was achieved using racemic dimethyl allyl carbonate, *M*-BINAP-Pd catalyst to achieve a highly diastereo- and enantioselective allylation, albeit in modest yield (Scheme 3.46c). No rationale has been offered to rationalize the steric course of these reactions, but recall that it is generally thought that hard nucleophiles such as these coordinate to the palladium rather than approaching the allyl system from the face opposite the metal. These examples reveal that the *M*-BINAP-Pd-allyl electrophile preferably alkylates the *Si* face of the enolates, whereas the *P* catalyst enantiomer prefers the *Re* face (see inset). Additional complications with nucleophiles such as ketone enolates is the competing reactivity of the π-allyl complex and the carbonyl group in the ketone product, and the possibility of dialkylation *via* enolization of the product. In some cases, the reactivity of hard nucleophile enolates can be attenuated through the use of zinc, boron, or tin counterions [237].

In early work with achiral catalysts, Tsuji showed that enol carbonates could serve as precursors to both nucleophile and electrophile in allylation reactions [221–223]. More recently,

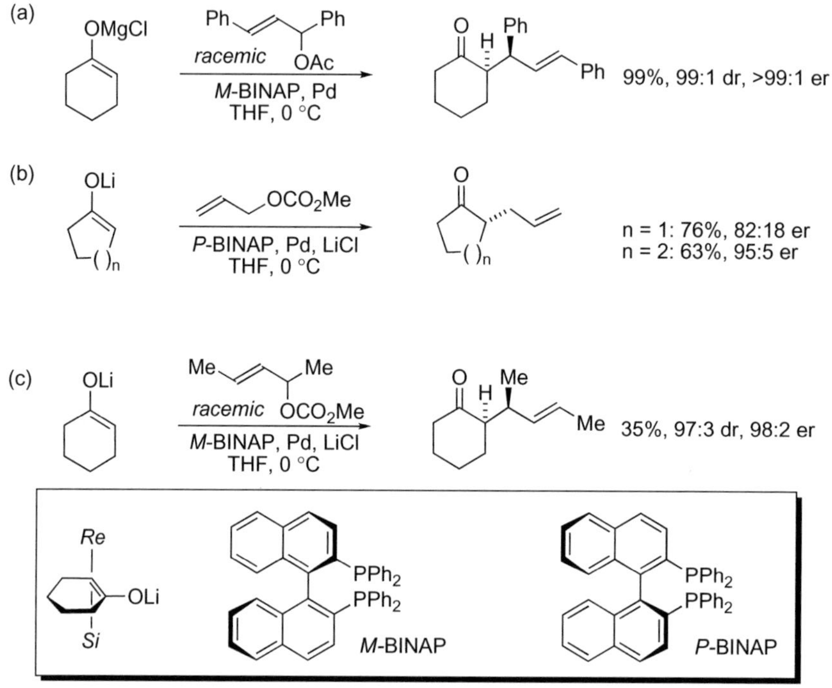

SCHEME 3.46 (a) Cyclohexanone magnesium enolate in a stereoconvergent allylation [245]; (b and c) cyclopentanone and cyclohexanone lithium enolates in AAA reactions [246].

Tunge [248], Stoltz [249], and Trost [250] used DPPBA ligands to demonstrate the application of the Tsuji allylation to asymmetric synthesis.

Scheme 3.47 shows examples of allylations of acetone enolate, reported by Tunge, which was generated by decarboxylation of acetoacetate subsequent to ionization of the allylic ester [248]. In these examples, the only stereocenter is at the allylic position, and the steric course using the R,R-DPPBA ligand is the same as the intermolecular variant illustrated in Scheme 3.44. A control experiment with an acetoacetic ester disubstituted between the carbonyls rearranged at a rate similar to the unsubstituted compounds, suggesting that decarboxylation of acetoacetate-to-acetone enolate precedes nucleophilic attack on the π-allyl palladium intermediate. Crossover experiments revealed that this appears to be an outer sphere process in which the steric course follows that illustrated in Figure 3.17, even though the enolate's counterion is palladium.

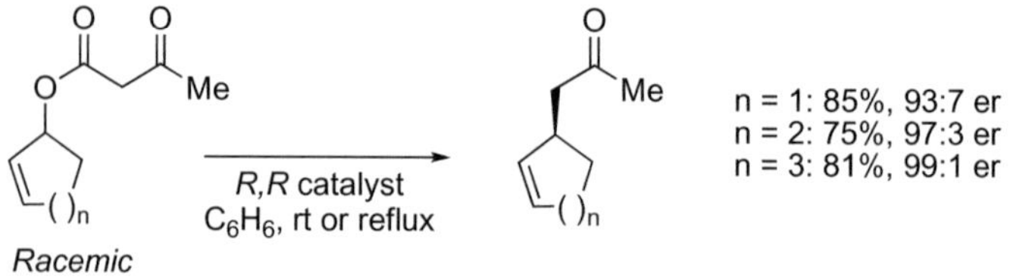

SCHEME 3.47 Stereoconvergent AAA of allylic acetoacetic esters [248].

Nearly simultaneously, Stoltz [249] and Trost [250] reported a similar approach to generate the enolate of 2-methylcyclohexanone. Both groups tested several ligands and solvents; selected results of these studies are listed in Table 3.14. Comparison of entries 1, 2, and 5 reveal a solvent effect, and entries 2−4 reveal that subtle changes in the ligand architecture can have a dramatic effect on the enantioselectivity. With the *R,R*-DPPBA ligands, the steric course favors alkylation of the *Si* face of the enolate, consistent with the model shown earlier in Scheme 3.45d. The *tert*-butyl PHOX and anthracene DPPBA ligands afforded the highest yields and enantioselectivities (entries 4 and 6).

TABLE 3.14 Optimization of the Allylation of 2-Methylcyclohexanone Enolate[a]

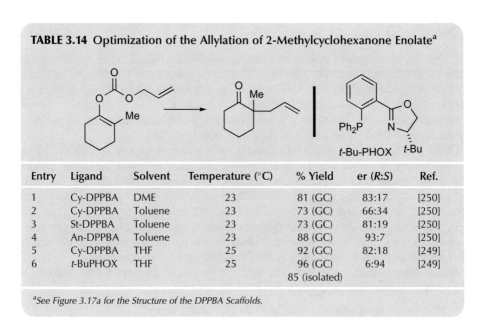

Entry	Ligand	Solvent	Temperature (°C)	% Yield	er (*R:S*)	Ref.
1	Cy-DPPBA	DME	23	81 (GC)	83:17	[250]
2	Cy-DPPBA	Toluene	23	73 (GC)	66:34	[250]
3	St-DPPBA	Toluene	23	73 (GC)	81:19	[250]
4	An-DPPBA	Toluene	23	88 (GC)	93:7	[250]
5	Cy-DPPBA	THF	25	92 (GC)	82:18	[249]
6	*t*-BuPHOX	THF	25	96 (GC) 85 (isolated)	6:94	[249]

[a]*See Figure 3.17a for the Structure of the DPPBA Scaffolds.*

3.2 CHIRAL ORGANOLITHIUMS

Beak has written that "organolithiums are the most widely used organometallics in contemporary organic chemistry" [251]. This statement derived partly from the widespread use of butyllithium bases, but it is also true that functionalized organolithiums in more complex molecules are important species in their own right, as evidenced by the large number of reviews [251−280] and monographs that have appeared since 2000 [8−11]. Lithium is the smallest of the metals, and its bonding is entirely electrostatic. Nevertheless, the structure of organolithiums and their reaction chemistry can be diabolically complex. This is partly due to the tendency of organolithium compounds to aggregate [281−283].

Several strategies may be used to generate an organolithium, including deprotonation, reductive lithiation, and transmetalation (Figure 3.18). Deprotonation is only useful if the conjugate acid is activated in some way, so as to lower the kinetic barrier to deprotonation. For example, a proximal directing group on a heteroatom, a double bond, or an aromatic ring could activate the α-proton(s) toward removal by a strong base such as butyllithium. Recent years have seen numerous examples of butyllithium (BuLi) or *sec*-butyllithium (*s*-BuLi) complexed with a chiral ligand such as sparteine, to create a chiral base that is able to effect enantioselective deprotonations by distinguishing enantiotopic protons [251,260,264,274,284−287].

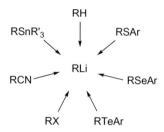

FIGURE 3.18 Some possible sources of organolithium compounds.

Reductive lithiations include reductions of sulfides, selenides, tellurides, or halides with lith-ium metal or lithium arenes [281,288—294]. Tertiary α-aminoorganolithiums can be made by reductive decyanation [295]. The metal most commonly exchanged for lithium is tin, *via* the reaction of trialkylstannyl compounds with BuLi [296—308]. The most popular precursors to configurationally stable α-alkoxy-organolithiums are α-alkoxyorganostannanes, which are now readily available by asymmetric synthesis [301,302,309—311].

Unlike S_N2 reactions, in which only invertive substitution is allowed by a concerted path-way, S_E2 reactions are allowed for both retentive and invertive concerted (polar) pathways [312—314]. Because both are allowed, we distinguish the steric course by adding the suffixes "ret" and "inv" for retentive and invertive substitutions, respectively [315]. Thus, electrophilic, aliphatic, and bimolecular substitutions, abbreviated S_E2 in the Hughes—Ingold terminology [316], may be categorized as $S_E2$ret or $S_E2$inv, depending on the steric course. Single-electron transfer (SET) can occur by oxidation of the carbanion to a radical, which opens new reaction manifolds that can be problematic to a synthesis plan due to lack of selectivity.

When contemplating the use of stereogenic "carbanions" in asymmetric synthesis, one must consider several factors (Scheme 3.48):

- Is the organolithium configurationally stable (Scheme 3.48a)?
- Does the reaction with an electrophile proceed with retention or inversion of configuration at the carbanionic carbon (Scheme 3.48b)?

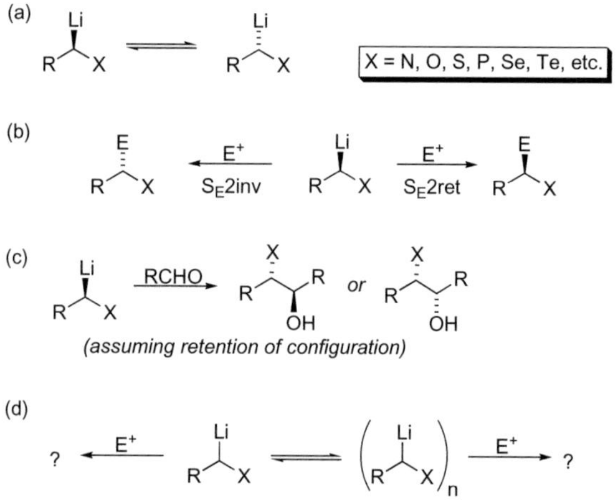

SCHEME 3.48 Factors to consider in evaluating reactions of chiral organolithiums.

- If the electrophile is an aldehyde or an unsymmetrical ketone, does the organometallic add selectively to one of the heterotopic faces (Scheme 3.48c)?
- What is the aggregation state of the organometallic? If there are aggregates, are they homochiral or heterochiral? If there is more than one species present in solution, which one is responsible for the observed behavior (Scheme 3.48d)?

3.2.1 Inversion Dynamics of Chiral Organolithiums

Organolithiums in which the carbanionic carbon is neither allylic nor benzylic are termed η^1 organolithiums, and the metal-bearing carbon may be stereogenic (Figure 3.19, left). Mesomerically stabilized carbanions may be either η^1 or η^3, and may possess a chirality plane in the latter instance (Figure 3.19, right). They may also be equilibrating mixtures of η^1 and η^3. In the absence of additional chirality elements, the organolithiums are enantiomers, but an additional chirality element, either in the carbanion or in a ligand coordinated to lithium, renders the two species diastereomeric. The free energy differences (ΔG^0) between interconverting diastereomers can have a profound effect on their chemistry. The free energy barriers to inversion ($\Delta G^{\ddagger}$) can vary widely, and also have a profound effect on the chemistry.

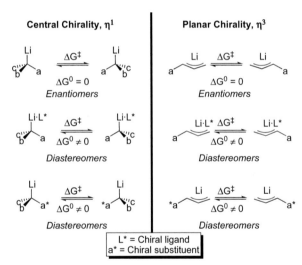

FIGURE 3.19 General structures of η^1 and η^3 organolithium compounds.

Questions of configurational stability are critical to the use of chiral organolithium compounds in asymmetric synthesis. The pertinent issue is the *time frame* of configurational stability. Some organolithiums are configurationally stable for hours, others for only seconds. Figure 3.8 (p. 116) illustrates the relationship between $\Delta G^{\ddagger}$ for enantiomerization and the half-life toward racemization.

Fast Inversion

When the rate of organolithium epimerization is fast relative to electrophilic quench, Curtin–Hammett kinetics [317,318] ensue (see Section 1.8). The energy profile is as shown in Figure 3.20. The relative energies of the enantiomeric organolithiums, *R*-RLi and *S*-RLi,

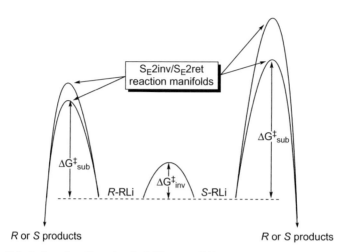

FIGURE 3.20 Substitution of configurationally labile organolithiums.

are not relevant: they may be enantiomers or diastereomers. In Figure 3.20, the barrier to inversion, $\Delta G_{inv}^{\ddagger}$, is low compared to the energy of activation for any substitution manifold, $\Delta G_{sub}^{\ddagger}$, by whichever pathway is favored. Under these conditions, evaluating the steric course of the reaction is not possible, except perhaps computationally.

Slow Inversion

At the other extreme are compounds that are configurationally stable for long periods of time. The reaction profile for this situation is illustrated in Figure 3.21. If the configuration of the organolithium is known, as would be the case with tin–lithium exchange, then the configuration of the product(s) allows determination of the steric course. A mixture of stereoisomeric products could imply competing S_E2inv/S_E2ret reaction pathways, or competition between such polar pathways and single-electron transfer. If the latter can be ruled out, for example by using radical clock electrophiles [319,320], the product ratio can be used to

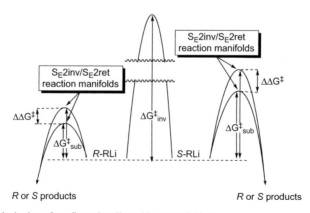

FIGURE 3.21 Substitution of configurationally stable organolithiums.

evaluate the relative rates of invertive *vs.* retentive substitution using transition state theory ($\Delta\Delta G^{\ddagger}$ in Figure 3.21).

Competing Inversion and Substitution

In between the limiting situations of configurational stability and configurational lability are instances where the rates of inversion and substitution are similar. The energy profile for such a situation is shown in Figure 3.22. What is most relevant is whether the organolithium is configurationally stable on the time scale of its reaction with an electrophile. In the continuum of relative rates, when k_{inv}, k_R, and k_S are close in value, the mathematical treatment is complex [321], and the product ratio is a function of both relative rates and percent conversion.

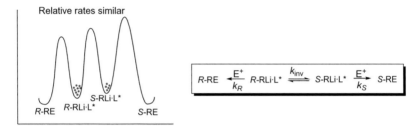

FIGURE 3.22 Energy profile of system where k_{inv}, k_R, and k_S are comparable.

The Hoffmann test – a users manual[20]

It is sometimes possible to use mutual kinetic resolutions to evaluate relative rates of inversion *vs.* electrophilic substitution [322−324]. This test is based on the reaction of a chiral organolithium with a chiral electrophile, and analysis of the diasteromer ratio of the products. The analysis begins with the assumption that there are equal amounts of the two organolithium stereoisomers, which may only be true if there are no other chirality elements present (*i.e.*, the organolithiums are enantiomers, not diastereomers, and there are no chiral ligands such as sparteine present).[21] Although this equivalence of stereoisomer amounts is not a requirement (the isomer ratio must be known, however), the analysis becomes more complicated if this is not the case.

Two experiments are run (Scheme 3.49). In Experiment 1, racemic organolithium is allowed to react with racemic electrophile (*rac*-E*); the selectivity factor, **s**, is evaluated by determining the diastereomer ratio of the products, which reflects the relative rates k_{RR}/k_{RS} for formation of the two products.[22] For reasons of symmetry, $k_{SR} = k_{RS}$ and $k_{SS} = k_{RR}$. For obvious reasons, the ratio must be $\neq 1$; for practical reasons (see point 1 below), it is best if the kinetic resolution has a low selectivity factor, $1.2 < s < 3.0$. In Experiment 2, the same reaction is carried out with enantiopure electrophile; and the reaction carried to 100% conversion. If the organolithium is configurationally stable, the er of *R*-RLi (50:50) will be carried

---

20. For an interesting recollection of how the test was discovered and developed, see ref. [322].

21. A related analysis that uses racemic organolithium and (inexpensive) achiral electrophiles in the presence of chiral ligands such as sparteine has been dubbed "the poor man's Hoffmann test" [325].

22. For discussions of the selectivity factor in kinetic resolutions, see Section 1.7 and refs. [326,327].

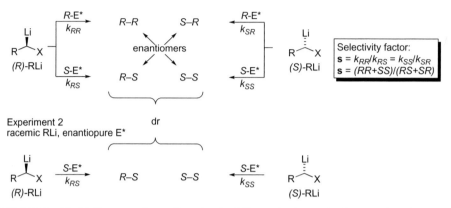

SCHEME 3.49   The Hoffmann test for configurational stability on the time scale of the reaction with an electrophile [323–325].

over to the dr of the product in Experiment 2 (50:50). If the dr for the two experiments is the same (*i.e.*, ≠ 50:50), the organolithium is configurationally labile on the time scale of the reaction, and the organolithium has been kinetically resolved in both experiments. If different, the organolithium is configurationally stable. In other words, comparison of the dr for the two experiments reveals the configurational stability of the organolithium compared to the time scale of the electrophilic substitution.

There are some limitations of the Hoffmann test that should be recognized.

- The dr in Experiment 2 depends on the percent conversion. In order to ensure complete conversion in Experiment 2, an excess of the chiral electrophile should be used. This will require longer reaction times than were required in Experiment 1, and the extra length of time depends on the selectivity factor, **s**. For this reason, it is prudent to employ reactions in which **s** is small (≤3.0).
- If the rate of reaction in Experiment 2 is faster than the rate of the addition of the organolithium to the electrophile, inverse addition should be employed.
- A 50:50 dr in Experiment 2 can only be reached if the electrophile is enantiopure. Good results are still possible if a moderate excess (1–5 equivalents) of electrophile having ≥98.5:1.5 er are used.
- If there are side reactions that consume the organolithium, they should not exceed 10%.
- It is assumed that the steric course of the reaction of the chiral organometal is 100% selective, *i.e.,* that the electrophilic substitution is 100% S$_E$2ret or 100% S$_E$2inv, and that SET is not a competing mechanism. This may not always be true, especially if both homochiral and heterochiral aggregates are reactive.
- It is possible that the rate of racemization changes as a reaction proceeds. Many electrophilic substitutions accumulate lithium halide or lithium alkoxide salts as the reaction proceeds. Lithium salts sometimes facilitate inversion of organolithiums [328].
- An asymmetric synthesis would use an enantiopure organolithium. Since the Hoffmann test uses racemic organolithium, it assumes that the aggregation states of enantiopure and racemic organolithiums are the same, and that the kinetics of reactions of electophiles with homochiral and heterochiral aggregates (or monomers in equilibrium with them) are the same.

3.2.2 Functionalized Organolithiums

sec-Butyllithium is chiral, but it is usually found in racemic form.[23] Indeed, many secondary organolithiums (and Grignard reagents) are chiral, but those used in asymmetric synthesis often employ α-heteroatom organometallics. In addition to possible resonance stabilization, α-heteroatom "carbanions" can be stabilized by inductive and dipole effects (Figure 3.23), or both, and sometimes by chelation [329]. The heteroatom may be a first-row element such as nitrogen or oxygen, or main group elements such as phosphorous, sulfur, selenium, or tellurium. There is ample evidence, both theoretical [330−332] and structural [333,334], that dipole-stabilized organometallics are chelated by the carbonyl oxygen. There is also good evidence that inductively stabilized α-heteroatom organolithiums have the metal bridged across the carbon-heteroatom bond [334−337]. Note that a distinction is made between bridging and chelation, even though the former might be called α-chelation. The simple reason is that there are distinct differences in stability and reactivity between the two types of compounds (*e.g.*, see ref. [259]).

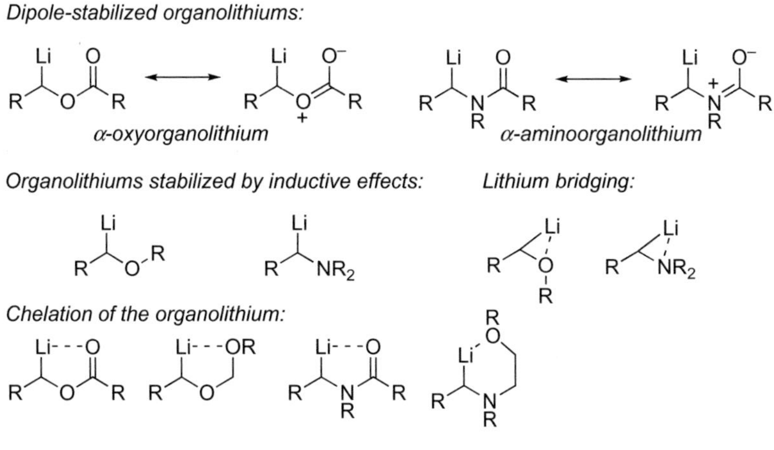

FIGURE 3.23 Classification of α-oxy- and α-aminoorganolithiums as either dipole-stabilized or inductively stabilized. Metal atom bridging and internal chelation may also play a role in both stabilization and chemical properties such as configurational stability.

3.2.3 Identifying the Stereochemically Defining Step

In electrophilic substitutions that begin with a deprotonation, the stereoselectivity of the sequence can be the result of either the deprotonation (if a chiral base is used) or the substitution. The reaction sequence in Scheme 3.50 illustrates an indirect method of establishing the stereochemically defining step by beginning with racemic deuterated substrate [338]. The chiral base complex *tert*-butyllithium · Box was used to deprotonate racemic α-deutero methyl benzyl ether. Carboxylation afforded a product in 81% yield and 91% enantioselectivity, but for which the deuterium content was >96%. This indicates that the deprotonation is governed by the deuterium isotope effect, and is *not* the determinant of organolithium

---

23. Reich has shown that 2° alkyllithiums possess reasonable configurational stability, even in THF [328].

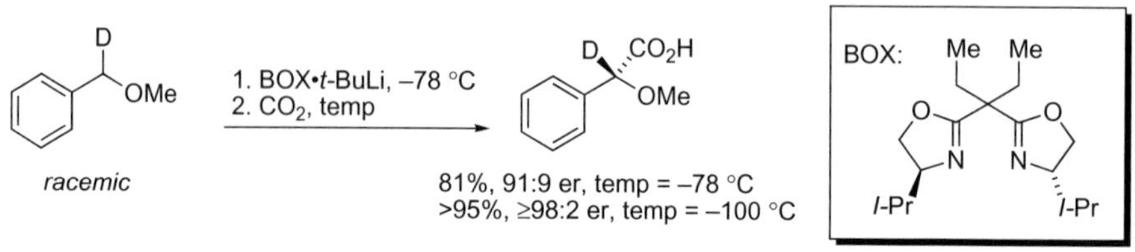

SCHEME 3.50 Dynamic thermodynamic resolution using a bisoxazoline (Box) ligand [338].

configuration or the steric course of the reaction. That a dynamic thermodynamic resolution (see Section 1.9) is operative is suggested by two facts. First, the er is increased when the electrophilic substitution is conducted at lower temperatures. Second, quenching experiments with substoichiometric quantities of electrophile afforded diminished enantiomer ratios (e.g., 73:27 er at 10% conversion), which is consistent with higher reactivity of the less stable organolithium epimer. These facts are best explained by a dynamic thermodynamic resolution in which the er at complete conversion reflects the dr of the organolithium · Box complex, and in which the minor organolithium stereoisomer is more reactive. The steric course of the electrophilic substitution is uncertain because the configuration of the intermediate organolithium is unknown.

Metalated amides and amidines of tetrahydroisoquinolines are useful intermediates in alkaloid synthesis [255,257,339–345], and the details of the stereoselective lithiation and electrophilic substitution are instructive. An example of Curtin–Hammett kinetics in organolithium substitutions is found in the chemistry of tetrahydroisoquinolyl oxazolines shown in Scheme 3.51 [346]. This example also illustrates a method for determining the source of stereoselectivity in an electrophilic substitution. In these oxazolines, the deprotonation is stereoselective, but the stereoselectivity in the deprotonation is not the source of the diastereoselectivity of the overall sequence. This was shown by a pair of deprotonations using diastereomeric deuterium-labeled compounds. With H_{Re} replaced by deuterium, the methylated product is obtained with 50% deuterium incorporation and 92:8 dr. With deuterium in place of H_{Si}, the methylated product is obtained with 97% deuterium incorporation and 91:9 dr. From this data, the relative rate for removal of H_{Re} over H_{Si} was calculated to be 5.8:1. Prior coordination of the butyllithium to the oxazoline nitrogen was postulated to account for the selectivity in the deprotonation. The variable deuterium content in the products, coupled with the consistent diastereoselectivity in the two experiments, clearly implicates equilibration of

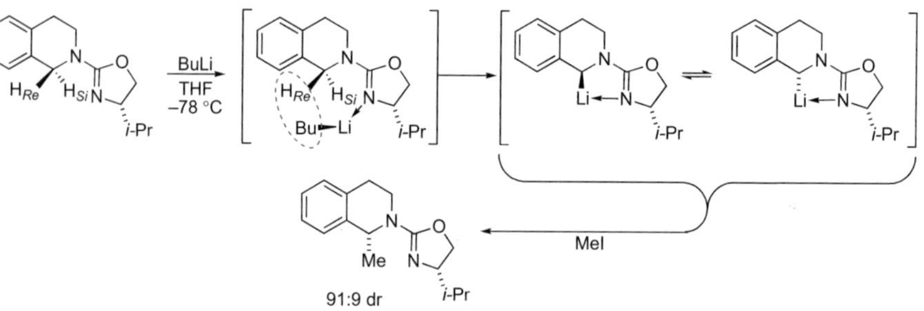

SCHEME 3.51 Lithiation/alkylation of tetrahydroisoquinolyloxazolines [345,346,348].

the organolithium diastereomers shown. The initial conclusion in 1987 [346] was that the stereoselectivity in the alkylation was due to a thermodynamic preference for one organo-lithium diastereomer, as in the formamidines described below [347], but subsequent studies were less conclusive and Curtin–Hammett kinetics were also considered [348]. Later, DNMR studies placed an upper limit of 8.2 kcal/mol on the barrier to inversion, indicating that Curtin–Hammett kinetics (see Figure 3.20) were the most likely explanation for the stereoselectivity in the alkylation [345]. Under Curtin–Hammett conditions, the steric course of the alkylation (retention or inversion) cannot be determined.

The formamidine auxiliary developed in the Meyers laboratory has been applied to the asym-metric synthesis of a number of isoquinoline and indole alkaloids (reviews: refs. [255,257]). When the α-deuteriotetrahydroisoquinoline shown in Scheme 3.52a was lithiated and alkylated with methyl iodide, the product *S* diastereomer was obtained with 98% diastereoselectivity, and

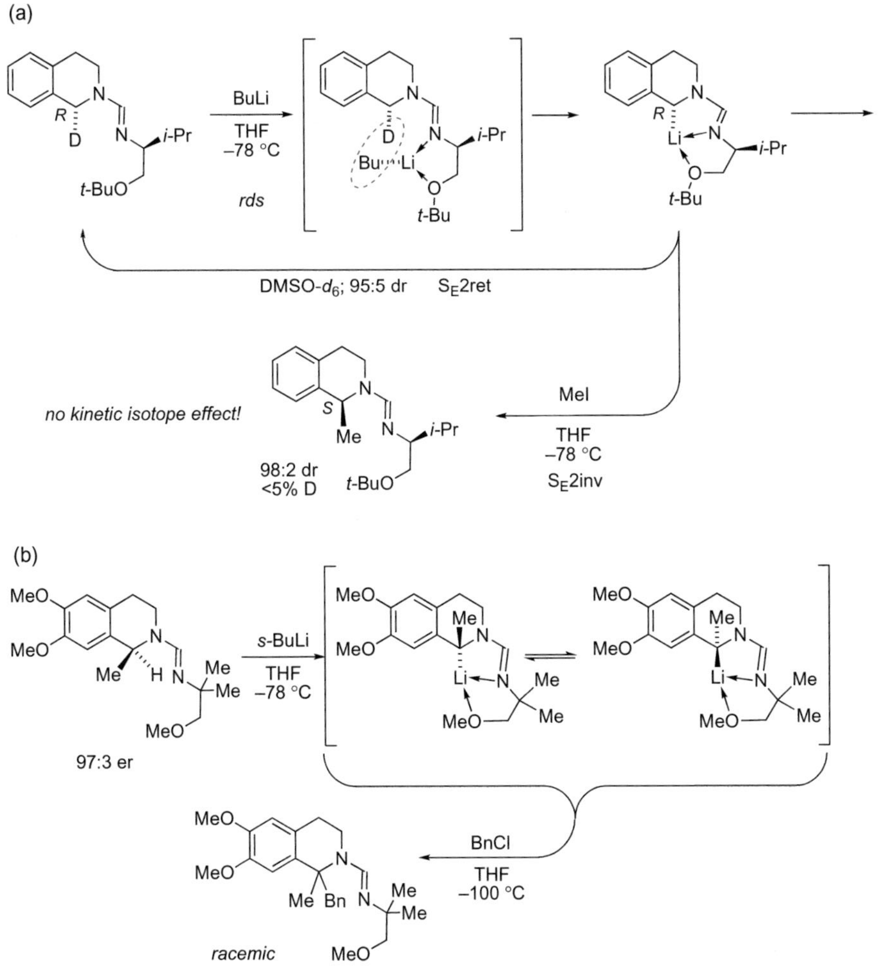

SCHEME 3.52 Formamidines: (a) configurational stability imparted by a chiral auxiliary [347]; (b) configu-rational lability in the absence of a chiral auxiliary [351].

less than 5% deuterium [347]. To account for the lack of a primary deuterium isotope effect, the postulated mechanism has the organization of a bidentate chelate of the butyllithium as the rate-determining step in the metalation. In the 5-membered chelate, the butyl group is thought to be *trans* to the isopropyl, placing it proximal to the H$_{Si}$ proton (or deuterium), resulting in a stereoselective deprotonation [349,350]. In contrast to the oxazolines above, when this proton is replaced by deuterium, there is no isotope effect observed in the deprotonation step, meaning that deprotonation is not rate determining. Scheme 3.52b illustrates an experiment designed to test the inherent configurational stability of a similar organolithium, which lacks a stereocenter in the formamidine auxiliary [351]. In the event, deprotonation of enantioenriched 1-methyltetrahydroisoquinoline at −78 °C, followed by alkylation with benzyl chloride at −100 °C, gave racemic product. These results are consistent with an inherently labile organolithium that can be stabilized in a single configuration (presumably *R*) by the chiral auxiliary.

A general process for asymmetric synthesis of alkaloids was developed using these alkylations, and is illustrated in Scheme 3.53, along with representative examples of tetrahydroisoquinoline [352] and β-carboline alkylations [353]. The authors speculated that, because of the chelation illustrated in brackets, the organolithium is more stable in the configuration shown and that alkylation occurs by inversion of configuration [349,350]. The energy profile for this scheme is shown in Figure 3.21, wherein chelation of the lithium by the chiral auxiliary provides the thermodynamic bias toward one configuration (note the −100 °C temperature of alkylation). Figure 3.24 illustrates several natural products synthesized using the asymmetric alkylation strategy, with the stereocenter formed in the asymmetric alkylation indicated (*).

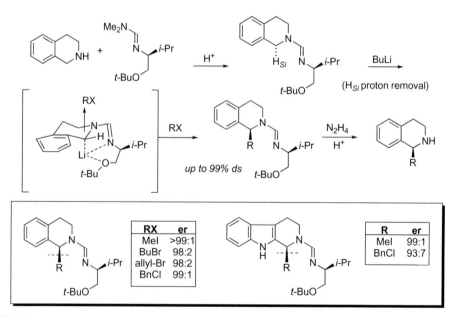

SCHEME 3.53 Formamidine approach to asymmetric alkylation of tetrahydroisoquinolines [352] and β-carbolines [353]. The indicated enantiomeric excesses were determined after auxiliary removal, derivatization, and CSP−HPLC analysis (Chapter 2).

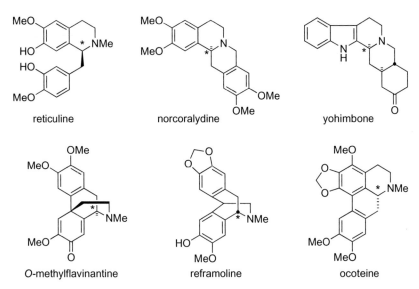

FIGURE 3.24 Natural products synthesized using the asymmetric alkylation strategy: reticuline [354], norcoralydine [339,355], yohimbone [356], O-methylflavinantine [343], reframoline [339], and ocoteine [339,355]. The stereocenter formed by asymmetric alkylation is indicated with *.

3.2.4 Asymmetric Deprotonations

The idea of using the sparteine·sec-BuLi complex as a chiral base was first explored by Nozaki in 1971, with limited success [357]. It was not until Hoppe's 1990 report on enantioselective deprotonations, α to oxygen, of hindered carbamates [358], that real progress began to occur. The growth in applications of sparteine as a chiral ligand for lithium has been explosive, with several reviews documenting progress in the use of sparteine and the development of alternative diamines for similar purposes [251,274,284−287,359,360].

Thus, Hoppe showed that a complex of sec-butyllithium and sparteine deprotonates O-ethyl, O-butyl, O-isobutyl, and O-hexyl (i.e., unactivated, nonallylic or nonbenzylic) carbamates to afford stereogenic organolithiums enantioselectively, as illustrated in Scheme 3.54. Reaction with certain electrophiles affords high yields of product, and the oxazolidine/ carbamate may be cleaved by acid hydrolysis. The rationale for the observed configuration (Scheme 3.54) is based on the X-ray structure of another α-carbamoyloxyorganolithium· sparteine complex [361] and a theoretical study [362]. The source of the enantioselectivity is the deprotonation [358,362]. The carbonyl coordinates to the lithium, and the bulky oxazolidine is oriented away from the sparteine. Note also that conformational motion could present either the H_{Si} or H_{Re} protons to the base. Schemes 3.31 and 3.32, in the previous section, illustrated two examples of group-selective reactions where enantiotopic groups on a carbonyl were distinguished by a chiral base. In this case, the enantiotopic groups are the protons of a prochiral methylene, and the chiral base is the sec-butyllithium·sparteine complex. In all these examples, retentive substitution is observed.

It may be tempting to assume that similar organolithiums would also alkylate with retention of configuration at the metal-bearing carbon. Not so. Unlike S_N2 reactions, transition states for S_E2 electrophilic substitution reactions giving retention and inversion are not

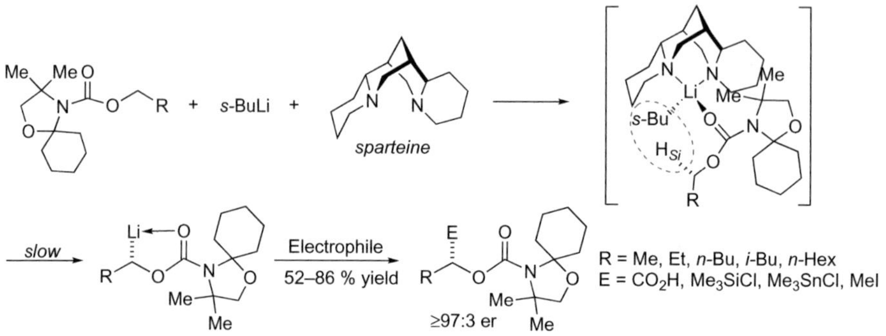
s-BuLi
sparteine
s-Bu
H_{Si}
slow
Electrophile
52–86 % yield
≥97:3 er
R = Me, Et, n-Bu, i-Bu, n-Hex
E = CO_2H, Me_3SiCl, Me_3SnCl, MeI

SCHEME 3.54 Enantioselective deprotonation and alkylation of carbamates [260,363].

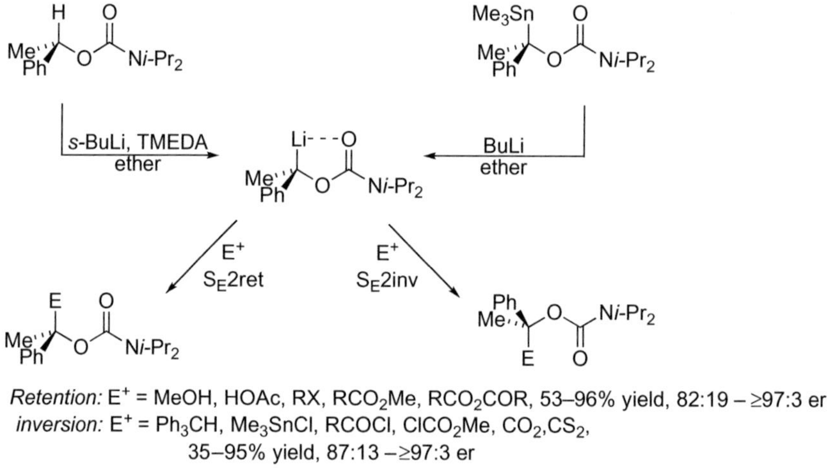
s-BuLi, TMEDA
ether
BuLi
ether
E^+
S_E2ret
E^+
S_E2inv
Retention: E^+ = MeOH, HOAc, RX, RCO_2Me, RCO_2COR, 53–96% yield, 82:19 – ≥97:3 er
inversion: E^+ = Ph_3CH, Me_3SnCl, RCOCl, $ClCO_2Me$, CO_2,CS_2,
35–95% yield, 87:13 –≥97:3 er

SCHEME 3.55 The stereochemical course of the alkylation of chiral organolithiums may depend on the electrophile [363,364].

far apart in energy [314], and both reaction manifolds are common. For example, the carbamate shown in Scheme 3.55, obtained either by deprotonation or by tin–lithium exchange, affords products of either retention or inversion, depending on the electrophile: esters, anhydrides, and alkyl halides afford products of retention, whereas acid chlorides, acyl cyanides, carbon dioxide, carbon disulfide, isocyanates, and tin chlorides afford products of inversion [363,364].

The authors speculate that the stereochemical divergence may be related to the ability of the electrophile to coordinate with the lithium, coupled with the presence or absence of a low-lying LUMO. Hoppe also noted that the enantiomeric purity of the products also depends on the solvent. In THF, the products were nearly racemic, and the enantiomeric purity of several of the other alkylation products was variable in solvents such as ether and pentane. This variability is due, at least in part, to the degree of covalency of the C–Li bond. In donor solvents such as THF, racemization is more facile.

Unlike lithiated tetrahydroisoquinolines, α-lithio derivatives of saturated nitrogen heterocycles exhibit significant configurational stability [365−370] (reviews: refs. [259,371]), and they have a considerably higher kinetic barrier to deprotonation [368,369]. Nevertheless, there have been a number of activating groups developed for the alkylation of α-lithio amines. The most popular activating group for the α-deprotonation of nitrogen heterocycles is the Boc group, introduced for this purpose by Beak in 1989 [372]. In 1991, Beak showed that the complex of (−)-sparteine and *sec*-butyllithium enantioselectively deprotonates Boc-pyrrolidine, and that the derived organolithium is a good nucleophile for the reaction with several electrophiles, as shown in Scheme 3.56a [367,373,374]. Due to the lack of ready access to (+)-sparteine, O'Brien introduced a (+)-sparteine surrogate, which is almost as enantioselective as (−)-sparteine in the deprotonation of *N*-Boc-pyrrolidine [375], and which is readily available by reduction of the alkaloid cytisine [376].

A limitation of these methods is the failure of the lithiated pyrrolidine to react efficiently with alkyl halide electrophiles [251,377]. In 2006, Campos showed that transmetalation of the organolithium compound with $ZnCl_2$ generated a configurationally stable organozinc compound, which could be subjected to Negishi arylation with $Pd(OAc)_2$, $t\text{-}Bu_3P \cdot HBF_4$ to effect an efficient and highly enantioselective arylation of *N*-Boc-pyrrolidine, as shown in Scheme 3.56b [378].

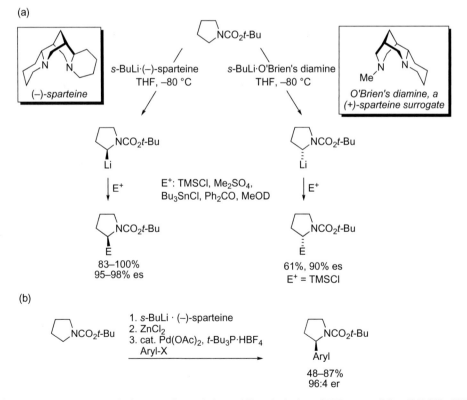

SCHEME 3.56 Asymmetric deprotonation and electrophilic substitution of *N*-Boc-pyrrolidine [367,373−375] (a) Direct electrophilic substitutions of the organolithium compounds; (b) arylations via transmetalations.

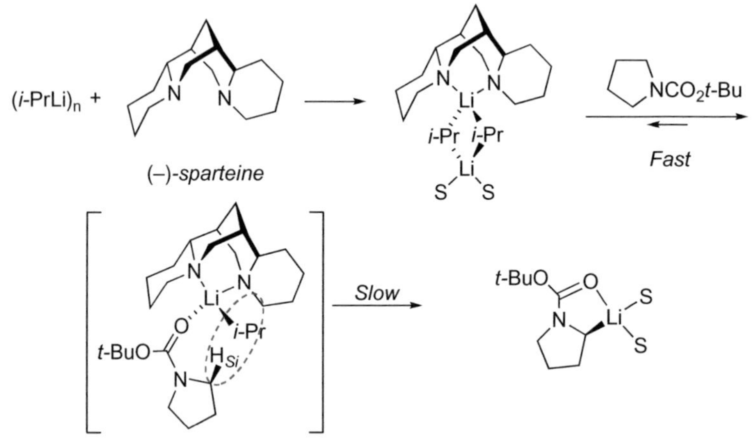

SCHEME 3.57 Mechanism for the asymmetric deprotonation of Boc-pyrrolidine [380–383].

Pertinent aspects of the mechanism are illustrated in Scheme 3.57. NMR studies indicate that sparteine and isopropyllithium form an unsymmetrical complex wherein one of the lithium atoms of the isopropyllithium dimer is chelated by sparteine while the other is not [379]. Kinetic studies indicate that when Boc-pyrrolidine is added to this complex, an equilibrium is established with a ternary complex of isopropyllithium, sparteine, and Boc-pyrrolidine (favoring the ternary complex with an equilibrium constant ≥ 300). The kinetic data further indicate that the deprotonation step is rate determining [380]. Wiberg and Bailey used computational techniques to investigate the structure of the ground state and the transition state and found the structure in the lower left of Scheme 3.57 to be lowest in energy [381–384]. The relative energies of several ground state isomers were investigated, and they were determined largely by steric effects. (Note the similarity and the differences between the bracketed ground state structure and that of the carbamate shown in brackets in Scheme 3.54.) Coincidentally, this low-energy ground state also led to the lowest energy transition structure. Pertinent factors include the placement of the metal-bearing C—H bond *syn* to the sparteine; the lowest energy structure fixes the orientation of the pyrrolidine such that the H_{Si} proton is closest to the carbanionic carbon of the base. Transfer of the H_{Si} proton relieves the repulsive steric interactions present in the low-energy ternary complex.

O'Brien showed that the deprotonation of *N*-Boc-pyrrolidine could be accomplished using substoichiometric quantities of sparteine or his diamine [385,386]. The protocol developed uses a ligand exchange approach, since the reaction requires stoichiometric quantities of *s*-BuLi, which, in the absence of a coordinating ligand, can effect stereorandom deprotonation. The success of the catalytic cycle shown in Scheme 3.58 depends on three things: (i) that ligand exchange occurs; (ii) that the deprotonation of achiral *s*-BuLi · L (L = *N,N*'-diisopropylbispidine) is slower than deprotonation of *s*-BuLi · L* (L* = sparteine); and (iii) that the chiral organolithium complexed to L does not racemize. The dynamics of ligand exchange and the relative rates of deprotonation have not been determined. However, the free energy barrier to inversion of *N*-Boc-2-lithiopyrrolidine at $-78\,°C$ was determined to be approximately 21 kcal/mol in the absence of ligands, 19 kcal/mol in the presence of sparteine, and 21 kcal/mol in the presence of *N,N*'-diisopropylbispidine [369]. So, all the chiral organolithium species are configurationally stable under the reaction conditions ($t_{1/2} \geq 5$ years at $-78\,°C$).

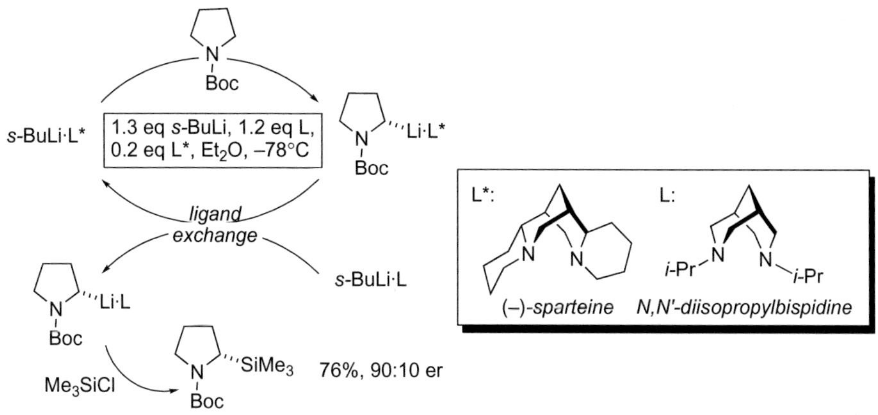

SCHEME 3.58 O'Brien's catalytic asymmetric deprotonation of *N*-Boc-pyrrolidine [385].

The asymmetric deprotonations using (−)-sparteine or O'Brien's diamine fails with *N*-Boc-piperidine [387,388]. However, organolithiums with relatively high barriers to inversion can undergo dynamic resolution in the presence of a chiral ligand on the lithium, thus obviating the need for an asymmetric deprotonation [389]. Dynamic resolutions are the best way to achieve enantioselective substitutions in the piperidine series, and Coldham did the seminal experiments in this regard [370,390]. The ligands that the Coldham group evaluated were tertiary amine/lithium alkoxides derived from amino acids and dipeptides [390]; one of the better ligands is shown in Scheme 3.59a. Later, Gawley tested dilithio derivatives and found significantly higher enantioselectivities (Scheme 3.59a) [391]. Often, the electrophile used in these reactions consumes the resolving ligand, which is a limitation of the method. Following up on a catalytic resolution discovered in the pyrrolidine series [392], Beng and Gawley found that catalytic amounts of the chiral ligand were able to effect the dynamic resolution, achieving er's that equaled those obtained with stoichiometric amounts of resolving ligand, as shown in Scheme 3.59b [391]. They later reported that the Campos procedure − developed for *N*-Boc-pyrrolidine − for arylation *via* the organozinc compound, could be adapted in the piperidine series [393]. When the catalytically resolved organolithium is quenched with dimethyl sulfate, the 2-methyl piperidine is obtained in 96:4 er. This enantioenriched 2-methyl piperidine could then be converted to 2,6-disubstituted piperidines using the diastereoselective alkylations reported by Beak in 1990 [394,395]. Scheme 3.59c shows some simple piperidine alkaloids that were prepared using this catalytic dynamic resolution.

3.2.5 Unstabilized Organolithiums

Many useful organolithiums are not available by deprotonation, and tin−lithium exchange is the most common method to access such compounds. In 1980, Still reported that the α-alkoxyorganolithium reagents derived from tin−lithium exchange of α-alkoxyorganostannanes are configurationally stable [299].[24] The tin−lithium exchange reaction takes place with retention of configuration at the metal-bearing carbon [299,300],[25] so obtaining an

---

24. For a theoretical explanation of this stability, see ref. [396].
25. The stereochemical course at tin depends on the tin ligands and the solvent [304].

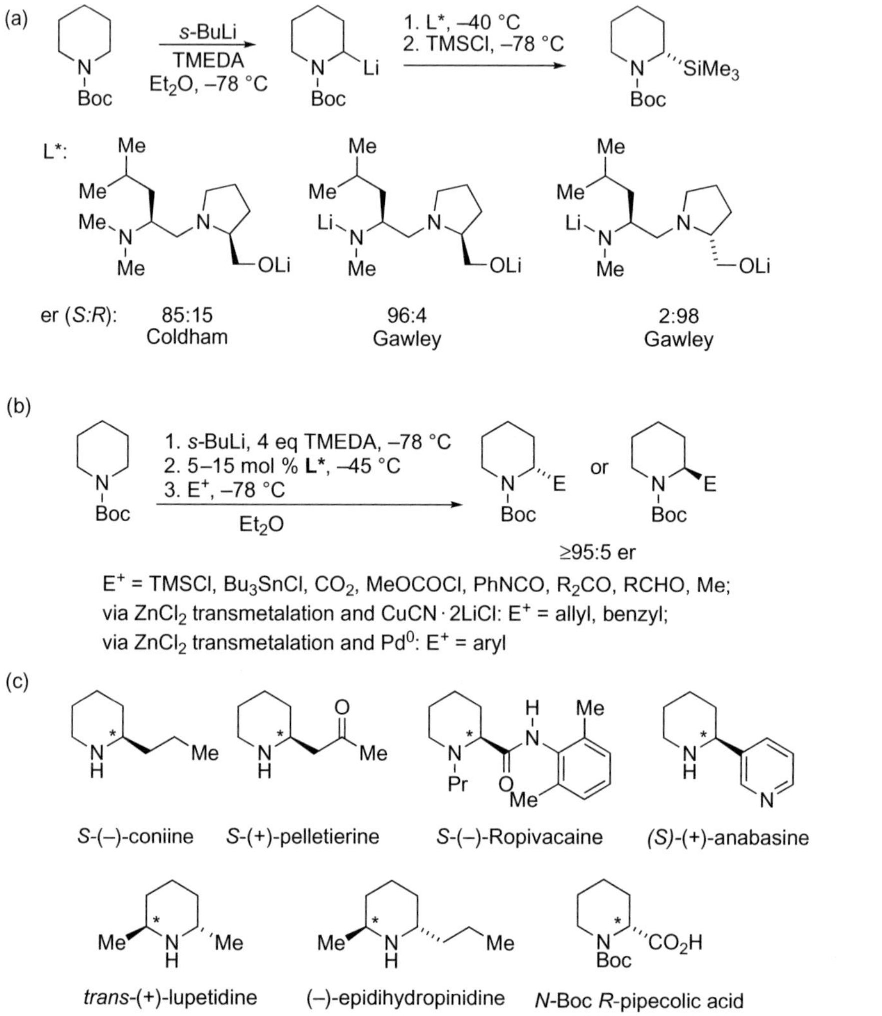

SCHEME 3.59 Dynamic resolution of *N*-Boc-2-lithiopiperidine [390,391,393,395] (a) Resolutions using stoichiometric L*; (b) resolutions using catalytic L*; (c) natural products synthesized using the catalytic resolution procedure. The stereocenter formed in the key step is indicated with *.

α-alkoxyorganolithium of known configuration is predicated on having an α-alkoxyorganostannane of known configuration. These are made by *O*-alkylation of the corresponding α-hydroxystannanes, which are in turn formed by asymmetric reduction (Chapter 7) of an acyl stannane [301,302,397], kinetic resolution using a lipase enzyme [398], or oxidation of α-stannylboronates [309]. Enantiomeric purities of the α-alkoxystannanes thus obtained are often ≥95:5 er.

Reaction with carbonyl electrophiles is possible, so enantiopure stannanes are excellent precursors of enantiopure α-alkoxy tertiary alcohols [299,300], α-alkoxy acids and esters [399], α-alkoxyketones [400], and γ-alkoxyhydrazides (precursors to γ-lactones) [403], as the

TABLE 3.15 Stereospecific Reactions of α-alkoxyorganolithiums with Electrophiles

Entry	Precursor Stannane	Electrophile	Product	% Yield	Ref.
1	Me, Bn, SnBu₃, MOMO	Acetone	Me Me Me, Bn, OH, MOMO	90	[299]
2	OBOM, i-Pr, SnBu₃	CO₂	OBOM, i-Pr, CO₂H	93	[399]
3	OBOM, Et, SnBu₃	ClCO₂Me	OBOM, Et, CO₂Me	71	[399]
4	OMOM, Et, SnBu₃	RCONMe₂	OMOM, Et, (CH₂)₃, Me, O, O, O	76	[400]
5	OMOM, Et, SnBu₃	CH₂=CHCON₂Me₃	OMOM, Et, O, (CH₂)₂, N·NMe₂, Me	50	[401]

examples in Table 3.15 illustrate. Note, however, that addition of these nucleophiles to aldehydes and unsymmetric ketones is not diastereoselective. Unfortunately, the reaction of these α-alkoxyorganolithiums with alkyl halides is usually inefficient and not stereoselective due to the intervention of single electron transfer processes [300]. Methylation can be achieved with dimethyl sulfate, however [299,300], and silylation is stereospecific [299].

The chemistry of lithiated N-methylpiperidines and N-methylpyrrolidines, α-aminoorganolithiums that are not dipole-stabilized, exhibits features that are quite distinct from those found for lithiated dipole-stabilized heterocycles. 2-Lithio-N-methylpiperidine and 2-lithio-N-methylpyrrolidine exhibit remarkable configurational stability: in the presence of TMEDA, they are configurationally stable at temperatures as high as −40 °C, and are more prone to chemical decomposition than racemization [263,402,403]. In 4:1 hexane/ether, the free energy barrier to enantiomerization is ∼22 kcal/mol at −78 °C. α-Lithiated N-alkyl heterocycles react smoothly with alkyl halides (Scheme 3.60), and react more efficiently than either lithiated formamidines [404] or Boc heterocycles [372,405]. Finally, the mechanistic (and stereochemical) course of their electrophilic substitution reactions depends on the electrophile in a unique way [320,406]. These organolithium compounds are obtained by tin−lithium exchange from the corresponding stannane [407,408]; examples of their reactivity are shown in Scheme 3.60 [320,406]. With most carbonyl electrophiles, retention of configuration is observed, whereas with alkyl halides, inversion is observed. When the electrophile is easily reduced, as with benzophenone or tert-butyl bromoacetate, the products are racemic. The reactions affording racemic products proceed by an SET (radical) mechanism [320], whereas the others go by polar $S_E2$ret and $S_E2$inv mechanisms, as shown in the inset in Scheme 3.60 [320,406,409].

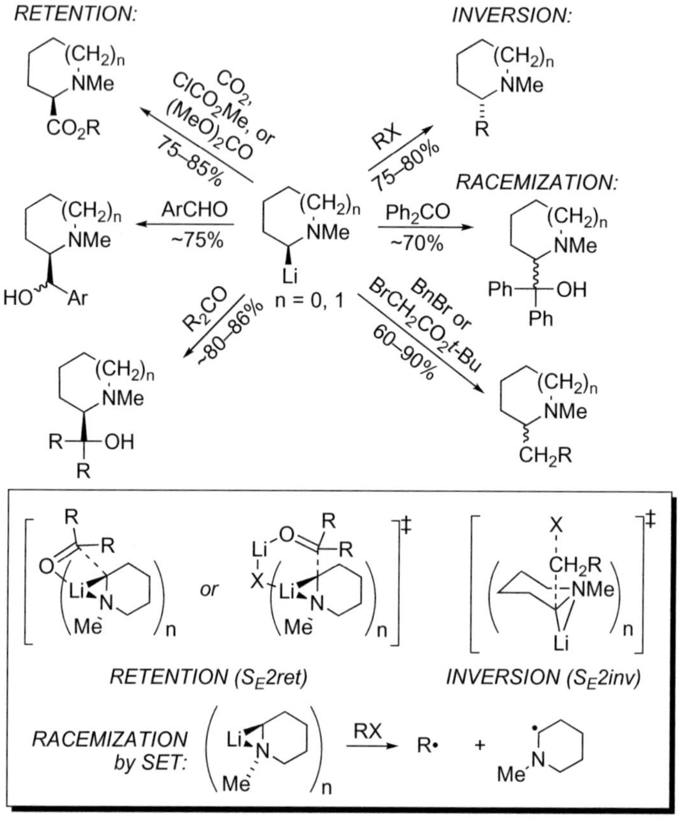

SCHEME 3.60 2-Lithio *N*-methylpiperidines and pyrrolidines are versatile reagents in electrophilic substitutions. The stereochemical course of the reaction depends on the electrophile. *Inset*: proposed transition structures for the invertive and retentive electrophilic substitutions, and SET mechanistic proposal for the electrophiles that afford racemic products [320,406,409].

REFERENCES

[1] Cram, D. J. *Fundamentals of Carbanion Chemistry*; Academic: New York, 1965.
[2] Stowell, J. C. *Carbanions in Organic Sythesis*; Wiley: New York, 1979.
[3] Bates, R. B. *Carbanions*; Springer-Verlag: Berlin, 1983.
[4] *Comprehensive Carbanion Chemistry*; Buncel, E.; Durst, T., Eds.; Elsevier: Amsterdam, 1980.
[5] Seebach, D. *Angew. Chem., Int. Ed. Engl.* **1988**, *27*, 1624–1654.
[6] Boche, G. *Angew. Chem., Int. Ed. Engl.* **1989**, *28*, 277–297.
[7] Buncel, E.; Dust, J. M. *Carbanion Chemistry. Structures and Mechanisms*; American Chemical Society and Oxford University Press: Washington DC and Oxford, 2003.
[8] Clayden, J. *Organolithiums: Selectivity for Synthesis*; Pergamon, 2002.
[9] *The Chemistry of Organolithium Compounds*; Rappoport, Z.; Marek, I., Eds.; Wiley: New York, 2004.
[10] *Organometallics: Compounds of Group 1 (Li ... Cs)*; Majewski, M.; Snieckus, V., Eds.; Thieme: Stuttgart, 2006; Vol. 8a.
[11] *Stereochemical Aspects of Organolithium Compounds*; Gawley, R. E., Ed.; Verlag Helvetica Chimica Acta: Weinheim, 2010; Vol. 26.
[12] Gould, E. S. *Mechanism and Structure in Organic Chemistry*; Holt, Rinehart and Winston: New York, 1959.

[13] Caine, D. In *Carbon-Carbon Bond Formation*; Augustine, R. L., Ed.; Marcel Dekker: New York, 1979, pp. 85–352.
[14] Caine, D. In *Comprehensive Organic Synthesis. Selectivity, Strategy, and Efficiency in Modern Organic Chemistry*; Trost, B. M., Fleming, I., Eds.; Pergamon: Oxford, 1991; Vol. 3, pp. 1–63.
[15] Arya, P.; Qin, H. *Tetrahedron* **2000**, *56*, 917–947.
[16] Lehn, J.-M. *Pure Appl. Chem.* **1978**, *50*, 871–892.
[17] Lehn, J.-M. *Angew. Chem., Int. Ed. Engl.* **1988**, *27*, 89–112.
[18] d'Angelo, J. *Tetrahedron* **1976**, *32*, 2979–2990.
[19] Jackman, L. M.; Lange, B. C. *Tetrahedron* **1977**, *33*, 2737–2769.
[20] Evans, D. A. In *Asymmetric Synthesis*; Morrison, J. D., Ed.; Academic: Orlando, 1984; Vol. 3, pp. 1–110.
[21] Williard, P. G. In *Comprehensive Organic Synthesis. Selectivity, Strategy, and Efficience in Modern Organic Chemistry*; Schreiber, S. L., Ed.; Pergamon: Oxford, 1991; Vol. 1, pp. 1–48.
[22] Collum, D. B.; McNeil, A. J.; Ramirez, A. *Angew. Chem., Int. Ed.* **2007**, *46*, 3002–3017.
[23] Li, D.; Keresztes, I.; Hopson, R.; Williard, P. G. *Acc. Chem. Res.* **2009**, *42*, 270–280.
[24] Hintze, P. G. W. M. J. *J. Am. Chem. Soc.* **1990**, *112*, 8602–8604.
[25] Arnett, E. M.; Fischer, F. G.; Nichols, M. A.; Ribiero, A. A. *J. Am. Chem. Soc.* **1990**, *112*, 801–808.
[26] Sakuma, K.; Gilchrist, J. H.; Romesberg, F. E.; Cajthami, C. E.; Collum, D. B. *Tetrahedron Lett.* **1993**, *34*, 5213–5216.
[27] Edwards, A. J.; Hockey, S.; Mair, F. S.; Raithby, P. R.; Snaith, R.; Simpkins, N. S. *J. Org. Chem.* **1993**, *58*, 6942–6943.
[28] Reetz, M. T.; Hütte, S.; Goddard, R. *J. Am. Chem. Soc.* **1993**, *115*, 9339–9340.
[29] Jackman, L. M.; Dunne, T. S. *J. Am. Chem. Soc.* **1985**, *107*, 2805–2806.
[30] Estermann, H.; Seebach, D. *Helv. Chim. Acta* **1988**, *71*, 1824–1839.
[31] Narasaka, K.; Ukaji, Y.; Watanabe, K. *Chem. Lett.* **1986**, 1755–1758.
[32] Bunn, B. J.; Simpkins, N. S. *J. Org. Chem.* **1993**, *58*, 533–534.
[33] Yasukata, T.; Koga, K. *Tetrahedron: Asymmetry* **1993**, *4*, 35–38.
[34] Hasegawa, Y.; Kawasaki, H.; Koga, K. *Tetrahedron Lett.* **1993**, *34*, 1963–1966.
[35] Creger, P. L. *J. Am. Chem. Soc.* **1970**, *92*, 1396–1397.
[36] Aebi, J. D.; Seebach, D. *Helv. Chim. Acta* **1985**, *68*, 1507–1518.
[37] Laube, T.; Dunitz, J. D.; Seebach, D. *Helv. Chim. Acta* **1985**, *68*, 1373–1393.
[38] Eleveld, M. B.; Hogeveen, H. *Tetrahedron Lett.* **1986**, *27*, 631–634.
[39] Juaristi, E.; Beck, A. K.; Hansen, J.; Matt, T.; Mukhopadhyay, T.; Simson, M.; Seebach, D. *Synthesis* **1993**, 1271–1290.
[40] Vedejs, E.; Lee, N. *J. Am. Chem. Soc.* **1995**, *117*, 891–900.
[41] Seebach, D.; Boes, M.; Naef, R.; Schweizer, W. B. *J. Am. Chem. Soc.* **1983**, *105*, 5390–5398.
[42] Cram, D. J.; Gosser, L. *J. Am. Chem. Soc.* **1964**, *86*, 5457–5465.
[43] Ford, W. T.; Graham, E. W.; Cram, D. J. *J. Am. Chem. Soc.* **1967**, *89*, 689–690.
[44] Ford, W. T.; Graham, E. W.; Cram, D. J. *J. Am. Chem. Soc.* **1967**, *89*, 690–692.
[45] Collum, D. B. *Acc. Chem. Res.* **1992**, *25*, 448–454.
[46] Mekelburger, H. B.; Wilcox, C. S. In *Comprehensive Organic Synthesis. Selectivity, Strategy, and Efficiency in Modern Organic Chemistry*; Trost, B. M., Fleming, I., Eds.; Pergamon: Oxford, 1991; Vol. 2, pp. 99–131.
[47] Kopka, I. E.; Fataftah, Z. A.; Rathke, M. W. *J. Org. Chem.* **1987**, *52*, 448–450.
[48] Ireland, R. E.; Wipf, P.; Armstrong, J. D., III *J. Org. Chem.* **1991**, *56*, 650–657.
[49] Xie, L.; Saunders, W. H., Jr. *J. Am. Chem. Soc.* **1991**, *113*, 3123–3130.
[50] Hall, P. L.; Gilchrist, J. H.; Collum, D. B. *J. Am. Chem. Soc.* **1991**, *113*, 9571–9574.
[51] Romesberg, F. E.; Collum, D. B. *J. Am. Chem. Soc.* **1995**, *117*, 2166–2178.
[52] Sun, X.; Kenkre, S. L.; Remenar, J. F.; Gilchrist, J. H.; Collum, D. B. *J. Am. Chem. Soc.* **1997**, *119*, 4765–4766.
[53] Sun, X.; Collum, D. B. *J. Am. Chem. Soc.* **2000**, *122*, 2452–2458.
[54] Ireland, R. E.; Mueller, R. H.; Willard, A. K. *J. Am. Chem. Soc.* **1976**, *98*, 2868–2877.
[55] Moreland, D. W.; Dauben, W. G. *J. Am. Chem. Soc.* **1985**, *107*, 2264–2273.
[56] Narula, A. S. *Tetrahedron Lett.* **1981**, *22*, 4119–4122.
[57] Aoki, K.; Tomioka, K.; Noguchi, H.; Koga, K. *Tetrahedron* **1997**, *53*, 13641–13656.
[58] Curthbertson, E.; O'Brien, P.; Towers, T. D. *Synthesis* **2001**, 693–695.
[59] Qin, Y.-c.; Stivala, C. E.; Zakarian, A. *Angew. Chem., Int. Ed.* **2007**, *46*, 7466–7469.

[60] Fataftah, Z. A.; Kopka, I. I.; Rathke, M. W. *J. Am. Chem. Soc.* **1980**, *102*, 3959–3960.

[61] Corey, E. J.; Gross, A. W. *Tetrahedron Lett.* **1984**, *25*, 495–498.

[62] Hattori, K.; Yamamoto, H. *J. Org. Chem.* **1993**, *58*, 5301–5303.

[63] Hall, P. L.; Gilchrist, J. H.; Harrison, A. T.; Fuller, D. J.; Collum, D. B. *J. Am. Chem. Soc.* **1991**, *113*, 9575–9585.

[64] Munchhof, M. J.; Heathcock, C. H. *Tetrahedron Lett.* **1992**, *33*, 8005–8006.

[65] Corey, E. J. *J. Am. Chem. Soc.* **1954**, *76*, 175–179.

[66] Corey, E. J.; Sneen, R. A. *J. Am. Chem. Soc.* **1956**, *78*, 6269–6278.

[67] Fleming, I. *Frontier Orbitals and Organic Chemical Reactions*; Wiley-Interscience: New York, 1976.

[68] Fleming, I. *Molecular Orbitals and Organic Chemical Reactions*; Wiley: New York, 2010. Student edition, 376 pages; Reference Edition, 515 pages.

[69] Agami, C. *Tetrahedron Lett.* **1977**, 2801–2804.

[70] Agami, C.; Chauvin, M.; Levisalles, J. *Tetrahedron Lett.* **1979**, 1855–1858.

[71] Agami, C.; Levisalles, J.; Cicero, B. L. *Tetrahedron* **1979**, *35*, 961–967.

[72] Houk, K. N.; Paddon-Row, M. N. *J. Am. Chem. Soc.* **1986**, *108*, 2659–2962.

[73] Rondan, N. G.; Paddon-Row, M. N.; Caramella, P.; Houk, K. N. *J. Am. Chem. Soc.* **1981**, *103*, 2436–2438.

[74] Houk, K. N. *Pure Appl. Chem.* **1983**, *55*, 277–282.

[75] Seebach, D.; Zimmerman, J.; Gysel, U.; Ziegler, R.; Ha, T.-K. *J. Am. Chem. Soc.* **1988**, *110*, 4763–4772.

[76] House, H. O.; Tefertiller, B. A.; Olmstead, H. D. *J. Org. Chem.* **1968**, *33*, 935–942.

[77] Kuwajima, I.; Nakamura, E.; Shimizu, M. *J. Am. Chem. Soc.* **1982**, *104*, 1025–1030.

[78] Hoffmann, R. W. *Chem. Rev.* **1989**, *89*, 1841–1860.

[79] Caramella, P.; Rondan, N. G.; Paddon-Row, M. N.; Houk, K. N. *J. Am. Chem. Soc.* **1981**, *103*, 2438–2440.

[80] McGarvey, G. J.; Williams, J. M. *J. Am. Chem. Soc.* **1985**, *107*, 1435–1437.

[81] Seebach, D.; Naef, R.; Calderari, G. *Tetrahedron* **1984**, *40*, 1313–1324.

[82] Fitzi, R.; Seebach, D. *Tetrahedron* **1988**, *44*, 5277–5292.

[83] Seebach, D.; Fadel, A. *Helv. Chim. Acta* **1985**, *68*, 1243–1250.

[84] Seebach, D.; Aebi, J. D.; Naef, R.; Weber, T. *Helv. Chim. Acta* **1985**, *68*, 144–154.

[85] Seebach, D.; Lamatsch, B.; Amstutz, R.; Beck, A. K.; Dobler, M.; Egli, M.; Fitzi, R.; Gautschi, M.; Herradón, B.; Hidber, P. C.; Irwin, J. J.; Locher, R.; Maestro, M.; Maetzke, T.; Mouriño, A.; Pfammatter, E.; Plattner, D. A.; Schickli, C.; Schweizer, W. B.; Seiler, P.; Stucky, G.; Petter, W.; Escalante, J.; Juaristi, E.; Quintana, D.; Miravitlles, C.; Molins, E. *Helv. Chim. Acta* **1992**, *75*, 913–934.

[86] Seebach, D.; Sting, A. R.; Hoffmann, M. *Angew. Chem., Int. Ed. Engl.* **1996**, *35*, 2708–2748.

[87] Cativiela, C.; Diaz-de-Villegas, M. D. *Tetrahedron: Asymmetry* **1998**, *9*, 3517–3599.

[88] Cativiela, C.; Diaz-De-Villegas, M. D. *Tetrahedron: Asymmetry* **2007**, *18*, 569–623.

[89] Naef, R.; Seebach, D. *Helv. Chim. Acta* **1985**, *68*, 135–143.

[90] Vedejs, E.; Fields, S. C.; Schrimpf, M. R. *J. Am. Chem. Soc.* **1993**, *115*, 11612–11613.

[91] Ley, S. V.; Downham, R.; Edwards, P. J.; Innes, J. E.; Woods, M. *Contemp. Org. Synth.* **1995**, *2*, 365–392.

[92] Ley, S. V.; Baeschlin, D. K.; Dixon, D. J.; Foster, A. C.; Ince, S. J.; Priepke, H. W. M.; Reynolds, D. J. *Chem. Rev.* **2001**, *101*, 53–80.

[93] Ley, S. V.; Polara, A. *J. Org. Chem.* **2007**, *72*, 5943–5959.

[94] Boons, G.-J.; Downham, R.; Kim, K. S.; Ley, S. V.; Woods, M. *Tetrahedron* **1994**, *50*, 7157–7176.

[95] Downham, R.; Kim, K. S.; Ley, S. V.; Woods, M. *Tetrahedron Lett.* **1994**, *35*, 769–772.

[96] Ley, S. V.; Diez, E.; Dixon, D. J.; Guy, R. T.; Michel, P.; Nattrass, G. L.; Sheppard, T. D. *Org. Biomol. Chem.* **2004**, *2*, 3608–3617.

[97] Seebach, D.; Wasmuth, D. *Angew. Chem., Int. Ed. Engl.* **1981**, *20*, 971.

[98] Kawabata, T.; Yahiro, K.; Fuji, K. *J. Am. Chem. Soc.* **1991**, *113*, 9694–9696.

[99] Kawabata, T.; Wirth, T.; Yahiro, K.; Suzuki, H.; Fuji, K. *J. Am. Chem. Soc.* **1994**, *116*, 10809–10810.

[100] Fuji, K.; Kawabata, T. *Chem. Eur. J.* **1998**, *4*, 373–376.

[101] Kawabata, T.; Fuji, K. *Top. Stereochem.* **2003**, *23*, 175–205.

[102] Zhao, H. W.; Hsu, D. C.; Carlier, P. R. *Synthesis* **2005**, 1–16.

[103] Eames, J.; Suggate, M. J. *Angew. Chem., Int. Ed.* **2005**, *44*, 186−189.
[104] Carlier, P. R.; Hsu, D. S.; Bryson, S. A. *Top. Stereochem.* **2010**, *26*, 53−91.
[105] Espenson, J. H. *Chemical Kinetics and Reaction Mechanisms*, 2nd ed.; McGraw-Hill: New York, 2002.
[106] Kawabata, T.; Suzuki, H.; Nagae, Y.; Fuji, K. *Angew. Chem., Int. Ed.* **2000**, *39*, 2155−2157.
[107] Williams, R. M.; Im, M.-N. *J. Am. Chem. Soc.* **1991**, *113*, 9276−9286.
[108] Yamada, S.; Oguri, T.; Shioiri, T. *J. Chem. Soc.* **1976**, 136−137.
[109] Ade, E.; Helmchen, G.; Heiligenmann, G. *Tetrahedron Lett.* **1980**, *21*, 1137−1140.
[110] Schmierer, R.; Grotemeier, G.; Helmchen, G.; Selim, A. *Angew. Chem., Int. Ed. Engl.* **1981**, *20*, 207−208.
[111] Helmchen, G.; Selim, A.; Dorsch, D.; Taufer, I. *Tetrahedron Lett.* **1983**, *24*, 3213−3216.
[112] Helmchen, G.; Schmierer, R. *Tetrahedron Lett.* **1983**, *24*, 1235−1238.
[113] Meyers, A. I.; Mihelich, E. D. *Angew. Chem., Int. Ed. Engl.* **1976**, *15*, 270−281.
[114] Meyers, A. I. *Acc. Chem. Res.* **1978**, *11*, 375−381.
[115] Lutomski, K. A.; Meyers, A. I. In *Asymmetric Synthesis*; Morrison, J. D., Ed.; Academic: Orlando, 1984; Vol. 3, pp. 213−273.
[116] Sonnet, P. E.; Heath, R. R. *J. Org. Chem.* **1980**, *45*, 3137−3139.
[117] Evans, D. A.; Takacs, J. M. *Tetrahedron Lett.* **1980**, *21*, 4233−4236.
[118] Evans, D. A.; Takacs, J. M.; McGee, L. R.; Ennis, M. D.; Mathre, D. J.; Bartroli, J. *Pure Appl. Chem.* **1981**, *53*, 1109−1127.
[119] Evans, D. A.; Ennis, M. D.; Mathre, D. J. *J. Am. Chem. Soc.* **1982**, *104*, 1737−1739.
[120] Gage, J. R.; Evans, D. A. *Org. Synth.* **1993**, *Coll. Vol. 8*, 339−343.
[121] Evans, D. A.; Britton, T. C.; Ellman, J. A. *Tetrahedron Lett.* **1987**, *28*, 6141−6144.
[122] Evans, D. A.; Bartroli, J. *Tetrahedron Lett.* **1982**, *23*, 807−810.
[123] Evans, D. A.; Dow, R. L.; Shih, T. L.; Takacs, J. M.; Zahler, R. *J. Am. Chem. Soc.* **1990**, *112*, 5290−5313.
[124] Hilbert, H. *Tetrahedron* **2001**, *57*, 7675−7683.
[125] Oppolzer, W.; Moretti, R.; Thomi, S. *Tetrahedron Lett.* **1989**, *30*, 5603−5606.
[126] Kim, B. H.; Curran, D. P. *Tetrahedron* **1993**, *49*, 293−318.
[127] Myers, A. G.; Yang, B. H.; Chen, H.; Gleason, J. L. *J. Am. Chem. Soc.* **1994**, *116*, 9361−9362.
[128] Myers, A. G.; Yang, B. H.; Chen, H.; McKinstry, L.; Kopecky, D. J.; Gleason, J. L. *J. Am. Chem. Soc.* **1997**, *119*, 6496−6511.
[129] Myers, A. G.; Yang, B. H.; Chen, H.; Kopecky, D. J. *Synlett* **1997**, 457−459.
[130] Myers, A. G.; Yang, B. H. *Org. Synth.* **2004**, *Coll. Vol. 10*, 12−15.
[131] Myers, A. G.; Yang, B. H.; Chen, H. *Org. Synth.* **2004**, *Coll. Vol. 10*, 509−516.
[132] Seebach, D.; Bossler, H.; Grundler, H.; Shoda, S.; Wenger, R. *Helv. Chim. Acta* **1991**, *74*, 197−224.
[133] Miller, S. A.; Griffiths, S. L.; Seebach, D. *Helv. Chim. Acta* **1993**, *76*, 563−595.
[134] Bossler, H. G.; Seebach, D. *Helv. Chim. Acta* **1994**, *77*, 1124−1165.
[135] Vong, B. G.; Abraham, S.; Xiang, A. X.; Theodorakis, E. A. *Org. Lett.* **2003**, *5*, 1617−1620.
[136] Sandham, D.; Taylor, R.; Carey, J. S.; Fassler, A. *Tetrahedron Lett.* **2000**, *41*, 10091−10094.
[137] Meyers, A. I.; Harre, M.; Garland, R. *J. Am. Chem. Soc.* **1984**, *106*, 1146−1148.
[138] Meyers, A. I.; Lefker, B. A.; Sowin, T. J.; Westrum, L. J. *J. Org. Chem.* **1989**, *54*, 4243−4246.
[139] Meyers, A. I.; Wanner, K. T. *Tetrahedron Lett.* **1985**, *26*, 2047−2050.
[140] Meyers, A. I.; Lefker, B. A.; Wanner, K. T.; Aitken, R. A. *J. Org. Chem.* **1986**, *51*, 1936−1938.
[141] Meyers, A. I.; Lefker, B. A. *Tetrahedron* **1987**, *43*, 5663−5676.
[142] Meyers, A. I.; Berney, D. *Org. Synth.* **1993**, *Coll. Vol. 8*, 241−247.
[143] Romo, D.; Meyers, A. I. *Tetrahedron* **1991**, *47*, 9503−9569.
[144] Meyers, A. I.; Westrum, L. J. *Tetrahedron Lett.* **1993**, *34*, 7701−7704.
[145] Snyder, L.; Meyers, A. I. *J. Org. Chem.* **1993**, *58*, 7507−7515.
[146] Meyers, A. I.; Wallace, R. H. *J. Org. Chem.* **1989**, *54*, 2509−2510.
[147] Meyers, A. I.; Lefker, B. A. *J. Org. Chem.* **1986**, *51*, 1541−1544.
[148] Meyers, A. I.; Hanreich, R.; Wanner, K. T. *J. Am. Chem. Soc.* **1985**, *107*, 7776−7778.
[149] Meyers, A. I.; Stuyers, M. A. *Tetrahedron Lett.* **1989**, *30*, 1741−1744.
[150] Meyers, A. I.; Berney, D. *J. Org. Chem.* **1989**, *54*, 4673−4676.
[151] Enders, D. In *Asymmetric Synthesis*; Morrison, J. D., Ed.; Academic: Orlando, 1984; Vol. 3, pp. 275−339.
[152] Enders, D.; Kipphardt, H.; Fey, P. *Org. Synth.* **1993**, *Coll. Vol. 8*, 403−414.

[153] Bergbreiter, D. E.; Momongan, M. In *Comprehensive Organic Synthesis. Selectivity, Strategy, and Efficiency in Modern Organic Chemistry*; Trost, B. M., Fleming, I., Eds.; Pergamon: Oxford, 1991; Vol. 2, pp. 503−526.

[154] Enders, D. In *Stereoselective Synthesis: Lectures Honouring Prof. Dr. H. C. Rudolf Wiechert*; Ottow, E., Schöllkopf, K., Schulz, B. G., Eds.; Springer: Berlin, 1994, pp. 63−90.

[155] Job, A.; Janek, C. F.; Bettray, W.; Peters, R.; Enders, D. *Tetrahedron* **2002**, *58*, 2253−2329.

[156] Corey, E. J.; Enders, D. *Tetrahedron Lett.* **1976**, 3−6.

[157] Corey, E. J.; Enders, D. *Chem. Ber.* **1978**, *111*, 1337−1361.

[158] Enders, D.; Eichenauer, H. *Chem. Ber.* **1979**, *112*, 2933−2960.

[159] Enders, D.; Eichenauer, H.; Baus, U.; Schubert, H.; Kremer, K. A. M. *Tetrahedron* **1984**, *40*, 1345−1359.

[160] Enders, D.; Bhushan, V. *Z. Naturforsch.* **1987**, *42*, 1595−1596.

[161] Enders, D.; Plant, A. *Synlett* **1990**, 725−726.

[162] Gawley, R. E.; Termine, E. J. *Synth. Commun.* **1982**, *12*, 15−18.

[163] Enders, D.; Dyker, H.; Raabe, G. *Angew. Chem., Int. Ed. Engl.* **1992**, *31*, 618−620.

[164] Davenport, K. G.; Eichenauer, H.; Enders, D.; Newcomb, M.; Bergbreiter, D. E. *J. Am. Chem. Soc.* **1979**, *101*, 5654−5659.

[165] Enders, D. *Chem. Scripta* **1985**, *25*, 139−147.

[166] Enders, D.; Bachstädter, G.; Kremer, K. A. M.; Marsch, M.; Harms, K.; Boche, G. *Angew. Chem., Int. Ed. Engl.* **1988**, *27*, 1522−1524.

[167] Mori, K.; Nomi, H.; Chuman, T.; Kohno, M.; Kato, K.; Noguchi, M. *Tetrahedron* **1982**, *38*, 3705−3711.

[168] Job, A.; Nagelsdiek, R.; Enders, D. *Collect. Czech. Chem. Commun.* **2000**, *65*, 524−538.

[169] Nicolaou, K. C.; Papahatjis, D. P.; Claremon, D. A.; Dolle, R. E., III *J. Am. Chem. Soc.* **1981**, *103*, 6967−6969.

[170] Birkbeck, A. A.; Enders, D. *Tetrahedron Lett.* **1998**, *39*, 7823−7826.

[171] Enders, D.; Ridder, A. *Synthesis* **2000**, 1848−1851.

[172] Lim, D.; Coltart, D. M. *Angew. Chem., Int. Ed.* **2008**, *47*, 5207−5210.

[173] Wengryniuk, S. E.; Lim, D.; Coltart, D. M. *J. Am. Chem. Soc.* **2011**, *133*, 8714−8720.

[174] Beeson, T. D.; Mastracchio, A.; Hong, J.-B.; Ashton, K.; MacMillan, D. W. C. *Science* **2007**, *316*, 582−585.

[175] Jang, H.-Y.; Hong, J.-B.; MacMillan, D. W. C. *J. Am. Chem. Soc.* **2007**, *129*, 7004−7005.

[176] Kim, H.; MacMillan, D. W. C. *J. Am. Chem. Soc.* **2008**, *130*, 398−399.

[177] Seebach, D. *Angew. Chem., Int. Ed.* **1979**, *18*, 239−258.

[178] *Umpoled Synthons: A Survey of Sources and Uses in Synthesis*; Hase, T. A., Ed.; Wiley-Interscience: New York, 1987.

[179] Alemán, J.; Cabrera, S.; Maerten, E.; Overgaard, J.; Jørgensen, K. A. *Angew. Chem., Int. Ed.* **2007**, *46*, 5520−5523.

[180] Simpkins, N. S.; Weller, M. D. *Top. Stereochem.* **2010**, *26*, 1−52.

[181] Hogeveen, H.; Zwart, L. *Tetrahedron Lett.* **1982**, *23*, 105−108.

[182] Cain, C. M.; Simpkins, N. S. *Tetrahedron Lett.* **1987**, *28*, 3723−3724.

[183] Shirai, R.; Tanaka, M.; Koga, K. *J. Am. Chem. Soc.* **1986**, *108*, 543−545.

[184] Sato, D.; Kawasaki, H.; Shimada, I.; Arata, Y.; Okamura, K.; Date, T.; Koga, K. *J. Am. Chem. Soc.* **1992**, *114*, 761−763.

[185] Cousins, R. P. C.; Simpkins, N. S. *Tetrahedron Lett.* **1989**, *30*, 7241−7244.

[186] Aoki, K.; Noguchi, H.; Tomioka, K.; Koga, K. *Tetrahedron Lett.* **1993**, *34*, 5105−5108.

[187] Koga, K. *Pure Appl. Chem.* **1994**, *66*, 1487−1492.

[188] Butler, B.; Schultz, T.; Simpkins, N. S. *Chem. Commun.* **2006**, 3634−3636.

[189] Kim, H.; Kawasaki, H.; Nakajima, M.; Koga, K. *Tetrahedron Lett.* **1989**, *30*, 6537−6540.

[190] Rodeschini, V.; Simpkins, N. S.; Wilson, C. *J. Org. Chem.* **2007**, *72*, 4265−4267.

[191] Sobukawa, M.; Koga, K. *Tetrahedron Lett.* **1993**, *34*, 5101−5104.

[192] Yamashita, T.; Mitsui, H.; Watanabe, H.; Nakamura, N. *Bull. Chem. Soc. Jpn.* **1982**, *55*, 961−962.

[193] Hogeveen, H.; Menge, W. M. P. B. *Tetrahedron Lett.* **1986**, *27*, 2767−2770.

[194] Ando, A.; Shiori, T. *J. Chem. Soc. Chem. Commun.* **1987**, 656−658.

[195] Tomioka, K.; Shindo, M.; Koga, K. *Chem. Pharm. Bull.* **1989**, *37*, 1120−1122.

[196] Murakata, M.; Nakajima, M.; Koga, K. *J. Chem. Soc. Chem. Commun.* **1990**, 1657−1658.

[197] Imai, M.; Hagihara, A.; Kawasaki, H.; Manabe, K.; Koga, K. *J. Am. Chem. Soc.* **1994**, *116*, 8829−8830.
[198] Imai, M.; Hagihara, A.; Kawasaki, H.; Manabe, K.; Koga, K. *Tetrahedron* **2000**, *56*, 179−185.
[199] Matsuo, J.-i.; Kobayashi, S.; Koga, K. *Tetrahedron Lett.* **1998**, *39*, 9723−9726.
[200] Murakata, M.; Yasukata, T.; Aoki, T.; Nakajima, M.; Koga, K. *Tetrahedron* **1998**, *54*, 2449−2458.
[201] Yamashita, Y.; Odashima, K.; Koga, K. *Tetrahedron Lett.* **1999**, *40*, 2803−2806.
[202] Hughes, D. L. In *Comprehensive Asymmetric Catalysis*; Jacobsen, E. N., Pfaltz, A., Yamamoto, H., Eds.; Springer: Berlin, 1999; Vol. 3, pp. 1273−1294.
[203] O'Donnell, M. J. *Aldrichimica Acta* **2001**, *34*, 3−15.
[204] Hughes, D. L. In *Comprehensive Asymmetric Catalysis Supplement 1*; Jacobsen, E. N., Pfaltz, A., Yamamoto, H., Eds.; Springer: Berlin, 2004, pp. 161−169.
[205] O'Donnell, M. J.; Bennett, W. D.; Wu, S. *J. Am. Chem. Soc.* **1989**, *111*, 2353−2355.
[206] Hughes, D. L.; Dolling, U.-H.; Ryan, K. M.; Schoenewaldt, E. F.; Grabowski, E. J. J. *J. Org. Chem.* **1987**, *52*, 4745−4752.
[207] Corey, E. J.; Xu, F.; Noe, M. C. *J. Am. Chem. Soc.* **1997**, *119*, 12414−12415.
[208] Lygo, B.; Wainwright, P. G. *Tetrahedron Lett.* **1997**, *38*, 8595−8598.
[209] O'Donnell, M. J.; Delgado, F.; Pottorf, R. S. *Tetrahedron* **1999**, *55*, 6347−6362.
[210] Chinchilla, R.; Mazón, P.; Nájera, C. *Tetrahedron: Asymmetry* **2000**, *11*, 3277−3281.
[211] Thierry, B.; Plaquevent, J.-C.; Cahard, D. *Tetrahedron: Asymmetry* **2001**, *12*, 983−986.
[212] Okino, T.; Takemoto, Y. *Org. Lett.* **2001**, *3*, 1515−1517.
[213] Mohamadi, F.; Richards, N. G. J.; Guida, W. C.; Liskamp, R.; Lipton, M.; Caufield, C.; Chang, G.; Hendrickson, T.; Still, W. C. *J. Comput. Chem.* **1990**, *11*, 440−467.
[214] Overman, L. E.; Larrow, J. F.; Stearns, B. A.; Vance, J. M. *Angew. Chem., Int. Ed.* **2000**, *39*, 213−215.
[215] Huang, A.; Kodanko, J. J.; Overman, L. E. *J. Am. Chem. Soc.* **2004**, *126*, 14043−14053.
[216] Steven, A.; Overman, L. E. *Angew. Chem., Int. Ed.* **2007**, *46*, 5488−5508.
[217] Takao, K.; Ochiai, H.; Hashizuka, T.; Koshimura, H.; Tadano, K.; Ogawa, S. *Tetrahedron Lett.* **1995**, *36*, 1487−1490.
[218] Hoyt, S. B.; Overman, L. E. *Org. Lett.* **2000**, *2*, 3241−3244.
[219] Vedejs, E.; Daugulis, O. *J. Am. Chem. Soc.* **1999**, *121*, 5813−5814.
[220] Vedejs, E.; Daugulis, O. *J. Am. Chem. Soc.* **2003**, *125*, 4166−4173.
[221] Tsuji, J.; Takahashi, H.; Morikawa, M. *Tetrahedron Lett.* **1965**, *6*, 4387−4388.
[222] Tsuji, J. *Acc. Chem. Res.* **1969**, *2*, 144−152.
[223] Tsuji, J.; Minami, I. *Acc. Chem. Res.* **1987**, *20*, 140−145.
[224] Reiser, O. *Angew. Chem., Int. Ed.* **1993**, *32*, 547−549.
[225] Hayashi, T. In *Catalytic Asymmetric Synthesis*; Ojima, I., Ed.; VCH: New York, 1993, pp. 325−365.
[226] Trost, B. M.; Van Vranken, D. L. *Chem. Rev.* **1996**, *96*, 395−422.
[227] Williams, J. M. J. *Synlett* **1996**, 705−710.
[228] Pfaltz, A.; Lautens, M. In *Comprehensive Asymmetric Catalysis*; Jacobsen, E. N., Pfaltz, A., Yamamoto, H., Eds.; Springer-Verlag: Berlin, 1999; Vol. 2, pp. 833−844.
[229] Helmchen, G. *J. Organomet. Chem.* **1999**, *576*, 203−214.
[230] Helmchen, G.; Pfaltz, A. *Acc. Chem. Res.* **2000**, *33*, 336−345.
[231] Trost, B. M.; Lee, C. In *Catalytic Asymmetric Synthesis*; 2nd ed.; Ojima, I., Ed.; Wiley-VCH: New York, 2000, pp. 593−649.
[232] Trost, B. M.; Crawley, M. L. *Chem. Rev.* **2003**, *103*, 2921−2943.
[233] Kazmaier, U. *Curr. Org. Chem.* **2003**, *7*, 317−328.
[234] Trost, B. M. *J. Org. Chem.* **2004**, *69*, 5813−5837.
[235] Tunge, J. A.; Burger, E. C. *Eur. J. Org. Chem.* **2005**, 1715−1726.
[236] Trost, B. M.; Machacek, M. R.; Aponick, A. *Angew. Chem., Int. Ed.* **2006**, *39*, 747−760.
[237] Braun, M.; Meier, T. *Synlett* **2006**, 661−676.
[238] Trost, B. M.; Fandrick, D. R. *Aldrichimica Acta* **2007**, *40*, 59−72.
[239] Trost, B. M.; Machacek, M. R.; Aponick, A. *Acc. Chem. Res.* **2006**, *39*, 747−760.
[240] Trost, B. M.; Li, L.; Guile, S. D. *J. Am. Chem. Soc.* **1992**, *114*, 8745−8747.
[241] Trost, B. M.; Chupak, L. S.; Lübbers, T. *J. Am. Chem. Soc.* **1998**, *120*, 1732−1740.
[242] Trost, B. M.; Bunt, R. C. *J. Am. Chem. Soc.* **1994**, *116*, 4089−4090.
[243] Trost, B. M.; Krueger, A. C.; Bunt, R. C.; Zambrano, J. *J. Am. Chem. Soc.* **1996**, *118*, 6520−6521.
[244] Trost, B. M.; Radinov, R.; Grenzer, E. M. *J. Am. Chem. Soc.* **1997**, *119*, 7879−7880.

[245] Braun, M.; Laicher, F.; Meier, T. *Angew. Chem., Int. Ed.* **2000**, *39*, 3494−3497.

[246] Braun, M.; Meier, T. *Synlett* **2005**, 2968−2972.

[247] Trost, B. M.; Schroeder, G. M. *Chem. Eur. J.* **2005**, *11*, 174−184.

[248] Burger, E. C.; Tunge, J. A. *Org. Lett.* **2004**, *6*, 4113−4115.

[249] Behenna, D. C.; Stoltz, B. M. *J. Am. Chem. Soc.* **2004**, *126*, 15044−15045.

[250] Trost, B. M.; Xu, J. *J. Am. Chem. Soc.* **2005**, *127*, 2846−2847.

[251] Beak, P.; Basu, A.; Gallagher, D. J.; Park, Y. S.; Thayumanavan, S. *Acc. Chem. Res.* **1996**, *29*, 552−560.

[252] Boche, G.; Walborsky, H. M. *Cyclopropane Derived Reactive Intermediates*; John Wiley and Sons: Chichester, 1990.

[253] Gawley, R. E.; Rein, K. In *Comprehensive Organic Synthesis. Selectivity, Strategy, and Efficiency in Modern Organic Chemistry*; Pattenden, G., Ed.; Pergamon: Oxford, 1991; Vol. 3, pp. 65−83.

[254] Gawley, R. E.; Rein, K. S. In *Comprehensive Organic Synthesis. Selectivity, Strategy, and Efficiency in Modern Organic Chemistry*; Schreiber, S. L., Ed.; Pergamon: Oxford, 1991; Vol. 1, pp. 459−485.

[255] Highsmith, T. K.; Meyers, A. I. In *Advances in Heterocyclic Natural Product Synthesis*; Pearson, W. H., Ed.; JAI: Greenwich, CT, 1991; Vol. 1, pp. 95−135.

[256] Nájera, C.; Yus, M. *Trends Org. Chem.* **1991**, *2*, 155−181.

[257] Meyers, A. I. *Tetrahedron* **1992**, *48*, 2589−2612.

[258] Hoffmann, R. In *Organic Synthesis via Organometallics (OSM 4)*; Enders, D., Gans, H.-J., Keim, W., Eds.; Wieweg: Aachen, 1992; Vol. 4, pp. 79−91.

[259] Gawley, R. E.; Zhang, Q. *Tetrahedron* **1994**, *50*, 6077−6088.

[260] Hoppe, D.; Hintze, F.; Tebben, P.; Paetow, M.; Ahrens, H.; Schwerdtfeger, J.; Sommerfeld, P.; Haller, J.; Guarnieri, W.; Kolczewski, S.; Hense, T.; Hoppe, I. *Pure Appl. Chem.* **1994**, *66*, 1479−1486.

[261] Yus, M. *Chem. Soc. Rev.* **1996**, 155−161.

[262] Yus, M.; Foubelo, F. *Rev. Heteroatom Chem.* **1997**, *17*, 73−107.

[263] Nájera, C.; Yus, M. *Rec. Res. Dev. Org. Chem.* **1997**, *1*, 67−96.

[264] Hoppe, D.; Hense, T. *Angew. Chem., Int. Ed.* **1997**, *36*, 2283−2316.

[265] Gawley, R. E. *Curr. Org. Chem.* **1997**, *1*, 71−94.

[266] Kessar, S. V.; Singh, P. *Chem. Rev.* **1997**, *97*, 721−737.

[267] Foubelo, F.; Yus, M. *Trends Org. Chem.* **1998**, *7*, 1−26.

[268] Guijarro, D.; Yus, M. *Rec. Res. Dev. Org. Chem.* **1998**, *2*, 713−744.

[269] Clayden, J. *Synlett* **1998**, 810−816.

[270] Gawley, R. E. In *Advances in Asymmetric Synthesis*; Hassner, A., Ed.; JAI: Greenwich, CT, 1998; Vol. 3, pp. 77−111.

[271] Ahlbrecht, H.; Beyer, U. *Synthesis* **1999**, 365−390.

[272] Ramón, D. J.; Yus, M. *Eur. J. Org. Chem.* **2000**, 225−237.

[273] Katritzky, A. R.; Piffl, M.; Lang, H.; Anders, E. *Chem. Rev.* **1999**, *99*, 665−722.

[274] Basu, A.; Thayumanavan, S. *Angew. Chem., Int. Ed.* **2002**, *41*, 716−738.

[275] Mealy, M. J.; Bailey, W. F. *J. Organomet. Chem.* **2002**, *646*, 59−67.

[276] *Organolithiums in Enantioselective Synthesis*; Hodgson, D. M., Ed.; Springer: Berlin Heidelberg, 2003.

[277] Nájera, C.; Yus, M. *Curr. Org. Chem.* **2003**, *7*, 867−926.

[278] Tomooka, K.; Ito, M. In *Main Group Metals in Organic Synthesis*; Yamamoto, H., Oshima, K., Eds.; Wiley-VCH Verlag GmbH & Co. KGaA: Weinheim, 2004; Vol. 1, pp. 1−34.

[279] Chinchilla, R.; Nájera, C.; Yus, M. *Tetrahedron* **2005**, *61*, 3139−3176.

[280] Ponthieux, S.; Paulmier, C. *Top. Curr. Chem.* **2000**, *208*, 113−142.

[281] Shimizu, T.; Kamigata, N. *Org. Prep. Proced. Int.* **1997**, *29*, 605−629.

[282] Brown, T. L. *Acc. Chem. Res.* **1968**, *1*, 23−32.

[283] Brown, T. L. In *Advances in Organometallic Chemistry*; Stone, F. G. A., West, R., Eds.; Academic Press: New York, 1965; Vol. 3, pp. 365−395.

[284] Hoppe, D.; Marr, F.; Brüggemann, M. *Top. Organomet. Chem.* **2003**, *5*, 61−138.

[285] Hoppe, D.; Christoph, G. In *The Chemistry of Organolithium Compounds*; Rappoport, Z, Marek, I., Eds.; John Wiley and Sons, Ltd.: Oxford, 2004, pp. 1055−1164.

[286] O'Brien, P. *Chem. Commun.* **2008**, 655−667.

[287] Breuning, M.; Steiner, M. *Synthesis* **2008**, 2841−2867.
[288] Seebach, D.; Beck, A. K. *Chem. Ber.* **1975**, *108*, 314−321.
[289] Cohen, T.; Bhupathy, M. *Acc. Chem. Res.* **1989**, *22*, 152−161.
[290] Cohen, T.; Matz, J. R. *J. Am. Chem. Soc.* **1980**, *102*, 6900−6902.
[291] Reich, H. J.; Bowe, M. D. *J. Am. Chem. Soc.* **1990**, *112*, 8994−8995.
[292] Hiiro, T.; Kambe, N.; Ogawa, A.; Miyoshi, N.; Murai, S.; Sonoda, N. *Angew. Chem., Int. Ed. Engl.* **1987**, *26*, 1187−1188.
[293] Hiiro, T.; Atarashi, Y.; Kambe, N.; Fujiwara, S.-I.; Ogawa, A.; Ryu, I.; Sonoda, N. *Organometallics* **1990**, *9*, 1355−1357.
[294] Reich, H. J.; Green, D. P.; Phillips, N. H. *J. Am. Chem. Soc.* **1991**, *113*, 1414−1416.
[295] Wolckenhauer, S. A.; Rychnovsky, S. D. *Org. Lett.* **2004**, *6*, 2745−2748.
[296] Peterson, D. J. *J. Organomet. Chem.* **1970**, *21*, P63–64.
[297] Peterson, D. J. *J. Am. Chem. Soc.* **1971**, *93*, 4027−4031.
[298] Peterson, D. J.; Ward, J. F. *J. Organomet. Chem.* **1974**, *66*, 209−217.
[299] Still, W. C.; Sreekumar, C. *J. Am. Chem. Soc.* **1980**, *102*, 1201−1202.
[300] Sawyer, J. S.; Kucerovy, A.; Macdonald, T. L.; McGarvey, G. J. *J. Am. Chem. Soc.* **1988**, *110*, 842−853.
[301] Chan, P. C.-M.; Chong, J. M. *J. Org. Chem.* **1988**, *53*, 5584−5586.
[302] Marshall, J. A.; Gung, W. Y. *Tetrahedron Lett.* **1988**, *29*, 1657−1660.
[303] Pearson, W. H.; Lindbeck, A. C. *J. Org. Chem.* **1989**, *54*, 5651−5654.
[304] Reich, H. J.; Borst, J. P.; Coplein, M. B.; Phillips, N. H. *J. Am. Chem. Soc.* **1992**, *114*, 6577−6579.
[305] Tomooka, K.; Igarashi, T.; Watanabe, M.; Nakai, T. *Tetrahedron Lett.* **1992**, *33*, 5795−5798.
[306] Burchat, A. F.; Chong, J. M.; Park, S. B. *Tetrahedron Lett.* **1993**, *34*, 51−54.
[307] Granã, P.; Paleo, M. R.; Sardina, F. J. *J. Am. Chem. Soc.* **2002**, *124*, 12511−12514.
[308] Klein, R.; Gawley, R. E. *J. Am. Chem. Soc.* **2007**, *129*, 4126−4127.
[309] Matteson, D. S.; Tripathy, P. B.; Sarkur, A.; Sadhu, K. N. *J. Am. Chem. Soc.* **1989**, *111*, 4399−4402.
[310] Tomooka, K.; Igarashi, T.; Nakai, T. *Tetrahedron Lett.* **1994**, *35*, 1913−1916.
[311] Kells, K. W.; Chong, J. M. *Org. Lett.* **2003**, *5*, 4215−4218.
[312] Pearson, R. G. *Acc. Chem. Res.* **1971**, *4*, 152−160.
[313] Slack, D. A.; Baird, M. C. *J. Am. Chem. Soc.* **1976**, *98*, 5539−5546.
[314] Jemmis, E. D.; Chandrasekhar, J.; Schleyer, P. v. R. *J. Am. Chem. Soc.* **1979**, *101*, 527−533.
[315] Gawley, R. E. *Tetrahedron Lett.* **1999**, *40*, 4297−4300.
[316] Ingold, C. K. In *Structure and Mechanism in Organic Chemistry*, 2nd ed.; Cornell University: Ithaca, NY, 1969. pp. 563-584.
[317] Curtin, D. Y. *Rec. Chem. Progr.* **1954**, *15*, 111−128.
[318] Seeman, J. I. *Chem. Rev.* **1983**, *83*, 83−134.
[319] Ingold, K. U. *Acc. Chem. Res.* **1980**, *13*, 317−323.
[320] Gawley, R. E.; Low, E.; Zhang, Q.; Harris, R. *J. Am. Chem. Soc.* **2000**, *122*, 3344−3350.
[321] Beak, P.; Anderson, D. R.; Curtis, M. D.; Laumer, J. M.; Pippel, D. J.; Weisenburger, G. A. *Acc. Chem. Res.* **2000**, *33*, 715−727.
[322] Hoffmann, R. W. *Top. Stereochem.* **2010**, *26*, 165−188.
[323] Hirsch, R.; Hoffmann, R. W. *Chem. Ber.* **1992**, *125*, 975−982.
[324] Hoffmann, R. W.; Lanz, J.; Metternich, R.; Tarara, G.; Hoppe, D. *Angew. Chem., Int. Ed. Engl.* **1987**, *26*, 1145−1146.
[325] Basu, A.; Gallagher, D. J.; Beak, P. *J. Org. Chem.* **1996**, *61*, 5718−5719.
[326] Kagan, H. B.; Fiaud, J. C. *Top. Stereochem.* **1988**, *18*, 249−330.
[327] Gawley, R. E. *J. Org. Chem.* **2006**, 71, 2411-2416; corrigendum *J. Org. Chem.* **2008**, *2473*, 6470.
[328] Reich, H. J.; Medina, M. A.; Bowe, M. D. *J. Am. Chem. Soc.* **1992**, *114*, 11003−11004.
[329] Krief, A. *Tetrahedron* **1980**, *36*, 2531−2640.
[330] Rondan, N. G.; Houk, K. N.; Beak, P.; Zajdel, W. J.; Chandrasekhar, J.; Schleyer, P. v. R. *J. Org. Chem.* **1981**, *46*, 4108−4110.
[331] Bach, R. D.; Braden, M. L.; Wolber, G. J. *J. Org. Chem.* **1983**, *48*, 1509−1514.
[332] Bartolotti, L. J.; Gawley, R. E. *J. Org. Chem.* **1989**, *54*, 2980−2982.
[333] Seebach, D.; Hansen, J.; Seiler, P.; Gromek, J. M. *J. Organomet. Chem.* **1985**, *285*, 1−13.

[334] Boche, G.; Marsch, M.; Harbach, J.; Harms, K.; Ledig, B.; Schubert, F.; Lohrenz, J. C. W.; Albrecht, H. *Chem. Ber.* **1993**, *126*, 1887−1894.

[335] Schleyer, P. v. R.; Clark, T.; Kos, A. J.; Spitznagel, G. W.; Rohde, C.; Arad, D.; Houk, K. N.; Rondan, N. G. *J. Am. Chem. Soc.* **1984**, *106*, 6467−6475.

[336] Gawley, R. E.; Klein, R.; Ashweek, N. J.; Coldham, I. *Tetrahedron* **2005**, *61*, 3271−3280.

[337] Low, E.; Gawley, R. E. *J. Am. Chem. Soc.* **2000**, *122*, 9562−9563.

[338] Komine, N.; Wang, L.; Tomooka, K.; Nakai, T. *Tetrahedron Lett.* **1999**, *40*, 6809−6812.

[339] Meyers, A. I.; Dickman, D. A.; Boes, M. *Tetrahedron* **1987**, *43*, 5095−5108.

[340] Seebach, D.; Syfrig, M. A. *Angew. Chem., Int. Ed. Engl.* **1984**, *23*, 248−249.

[341] Huber, I. M. P.; Seebach, D. *Helv. Chim. Acta* **1987**, *70*, 1944−1954.

[342] Seebach, D.; Huber, I. M. P.; Syfrig, M. A. *Helv. Chim. Acta* **1987**, *70*, 1357−1379.

[343] Gawley, R. E.; Smith, G. A. *Tetrahedron Lett.* **1988**, *29*, 301−302.

[344] Rein, K. S.; Gawley, R. E. *J. Org. Chem.* **1991**, *56*, 1564−1569.

[345] Gawley, R. E.; Zhang, P. *J. Org. Chem.* **1996**, *61*, 8103−8112.

[346] Gawley, R. E. *J. Am. Chem. Soc.* **1987**, *109*, 1265−1266.

[347] Meyers, A. I.; Dickman, D. A. *J. Am. Chem. Soc.* **1987**, *109*, 1263−1265.

[348] Rein, K.; Goicoechea-Pappas, M.; Anklekar, T. V.; Hart, G. C.; Smith, G. A.; Gawley, R. E. *J. Am. Chem. Soc.* **1989**, *111*, 2211−2217.

[349] Meyers, A. I.; Dickman, D. A. *J. Am. Chem. Soc.* **1987**, *109*, 1263−1265.

[350] Meyers, A. I.; Warmus, J. S.; Gonzalez, M. A.; Guiles, J.; Akahane, A. *Tetrahedron Lett.* **1991**, *32*, 5509−5512.

[351] Meyers, A. I.; Guiles, J.; Warmus, J. S.; Gonzalez, M. A. *Tetrahedron Lett.* **1991**, *32*, 5505−5508.

[352] Meyers, A. I.; Boes, M.; Dickman, D. A. *Angew. Chem., Int. Ed. Engl.* **1984**, *23*, 458−459.

[353] Loewe, M. F.; Meyers, A. I. *Tetrahedron Lett.* **1985**, *26*, 3291−3294.

[354] Meyers, A. I.; Guiles, J. *Heterocycles* **1989**, *28*, 295−301.

[355] Dickman, D. A.; Meyers, A. I. *Tetrahedron Lett.* **1986**, *27*, 1465−1468.

[356] Meyers, A. I.; Miller, D. B.; White, F. H. *J. Am. Chem. Soc.* **1988**, *110*, 4778−4787.

[357] Nozaki, H.; Aratani, T.; Toraya, T.; Noyori, R. *Tetrahedron* **1971**, *27*, 905−913.

[358] Hoppe, D.; Hintze, F.; Tebben, P. *Angew. Chem., Int. Ed. Engl.* **1990**, *29*, 1422−1423.

[359] Kizirian, J.-C. *Chem. Rev.* **2008**, *108*, 140−205.

[360] Kizirian, J.-C. *Top. Stereochem.* **2010**, *26*, 189−251.

[361] Marsch, M.; Harms, K.; Zschage, O.; Hoppe, D.; Boche, G. *Angew. Chem., Int. Ed. Engl.* **1991**, *30*, 321−323.

[362] Würthwein, E.-U.; Hoppe, D. *J. Org. Chem.* **2005**, *70*, 4443−4451.

[363] Carstens, A.; Hoppe, D. *Tetrahedron* **1994**, *50*, 6097−6108.

[364] Hoppe, D.; Carstens, A.; Krämer, T. *Angew. Chem., Int. Ed. Engl.* **1990**, *29*, 1424−1425.

[365] Gawley, R. E.; Hart, G.; Goicoechea-Pappas, M.; Smith, A. L. *J. Org. Chem.* **1986**, *51*, 3076−3078.

[366] Gawley, R. E.; Hart, G. C.; Bartolotti, L. J. *J. Org. Chem.* **1989**, *54*, 175−181, corrigendum 4726.

[367] Kerrick, S. T.; Beak, P. *J. Am. Chem. Soc.* **1991**, *113*, 9708−9710.

[368] Ashweek, N. J.; Brandt, P.; Coldham, I.; Dufour, S.; Gawley, R. E.; Hæffner, F.; Klein, R.; Sanchez-Jimenez, G. *J. Am. Chem. Soc.* **2005**, *127*, 449−457.

[369] Yousaf, T. I.; Williams, R. L.; Coldham, I.; Gawley, R. E. *Chem. Commun.* **2008**, 97−98.

[370] Coldham, I.; Leonori, D.; Beng, T. K.; Gawley, R. E. *Chem. Commun.* **2009**, 5239−5240, corrigendum **2010**, 9267-9268.

[371] Gawley, R. E. *Top. Stereochem.* **2010**, *26*, 93−133.

[372] Beak, P.; Lee, W.-K. *Tetrahedron Lett.* **1989**, *30*, 1197−1200.

[373] Beak, P.; Kerrick, S. T.; Wu, S.; Chu, J. *J. Am. Chem. Soc.* **1994**, *116*, 3231−3239.

[374] Nikolic, N. A.; Beak, P. *Org. Synth.* **1998**, *Coll. Vol. 9*, 391−396.

[375] Dearden, M. J.; Firkin, C. R.; Hermet, J.-P. R.; O'Brien, P. *J. Am. Chem. Soc.* **2002**, *124*, 11870−11871.

[376] Dixon, A. J.; McGrath, M. J.; O'Brien, P. *Org. Synth.* **2009**, *Coll. 11*, 25−37.

[377] Wu, S.; Lee, S.; Beak, P. *J. Am. Chem. Soc.* **1996**, *118*, 715−721.

[378] Campos, K. R.; Klapars, A.; Waldman, J. H.; Dormer, P. G.; Chen, C. *J. Am. Chem. Soc.* **2006**, *128*, 3538−3539.

[379] Gallagher, D. J.; Kerrick, S. T.; Beak, P. *J. Am. Chem. Soc.* **1992**, *114*, 5872−5873.

[380] Gallagher, D. J.; Beak, P. *J. Org. Chem.* **1995**, *60*, 7092−7093.

[381] Wiberg, K. B.; Bailey, W. F. *Angew. Chem., Int. Ed. Engl.* **2000**, *39*, 2127−2129.
[382] Wiberg, K. B.; Bailey, W. F. *Tetrahedron Lett.* **2000**, *41*, 9365−9368.
[383] Wiberg, K. B.; Bailey, W. F. *J. Am. Chem. Soc.* **2001**, *123*, 8231−8238.
[384] O'Brien, P.; Wiberg, K. B.; Bailey, W. F.; Hermet, J.-P. R.; McGrath, M. J. *J. Am. Chem. Soc.* **2004**, *126*, 15480−15489.
[385] McGrath, M. J.; O'Brien, P. *J. Am. Chem. Soc.* **2005**, *127*, 16378−16379.
[386] Bilke, J. L.; O'Brien, P. *J. Org. Chem.* **2008**, *73*, 6452−6454.
[387] Bailey, W. F.; Beak, P.; Kerrick, S. T.; Ma, S.; Wiberg, K. B. *J. Am. Chem. Soc.* **2002**, *124*, 1889−1896.
[388] Coldham, I.; O'Brien, P.; Patel, J. J.; Raimbault, S.; Sanderson, A. J.; Stead, D.; Whittaker, D. T. E. *Tetrahedron: Asymmetry* **2007**, *18*, 2113−2119.
[389] Coldham, I.; Sheikh, N. S. *Top. Stereochem.* **2010**, *26*, 253−293.
[390] Coldham, I.; Raimbault, S.; Chovatia, P. T.; Patel, J. J.; Leonori, D.; Sheikh, N. S.; Whittaker, D. T. E. *Chem. Commun.* **2008**, 4174−4176.
[391] Beng, T. K.; Gawley, R. E. *J. Am. Chem. Soc.* **2010**, *132*, 12216−12217.
[392] Beng, T. K.; Yousaf, T. I.; Coldham, I.; Gawley, R. E. *J. Am. Chem. Soc.* **2009**, *131*, 6908−6909.
[393] Beng, T. K.; Gawley, R. E. *Org. Lett.* **2011**, *13*, 394−397.
[394] Beak, P.; Lee, W. K. *J. Org. Chem.* **1990**, *55*, 2578−2580.
[395] Beng, T. K.; Gawley, R. E. *Heterocycles* **2012**, *84*, 697−718.
[396] Lehn, J.-M.; Wipff, G. *J. Am. Chem. Soc.* **1976**, *98*, 7498−7505.
[397] Marshall, J. A.; Welmaker, G. S.; Gung, B. W. *J. Am. Chem. Soc.* **1991**, *113*, 647−656.
[398] Chong, J. M.; Mar, E. K. *Tetrahedron Lett.* **1991**, *32*, 5683−5686.
[399] Chan, P. C.-M.; Chong, J. M. *Tetrahedron Lett.* **1990**, *31*, 1985−1988.
[400] Chong, J. M.; Mar, E. K. *Tetrahedron* **1989**, *45*, 7709−7716.
[401] Chong, J. M.; Mar, E. K. *Tetrahedron Lett.* **1990**, *31*, 1981−1984.
[402] Gawley, R. E.; Zhang, Q. *J. Am. Chem. Soc.* **1993**, *115*, 7515−7516.
[403] Gawley, R. E.; Zhang, Q. *Tetrahedron* **1994**, *50*, 6077−6088.
[404] Meyers, A. I.; Edwards, P. D.; Reiker, W. F.; Bailey, T. R. *J. Am. Chem. Soc.* **1984**, *106*, 3270−3276.
[405] Beak, P.; Lee, W. K. *J. Org. Chem.* **1993**, *58*, 1109−1117.
[406] Gawley, R. E.; Zhang, Q. *J. Org. Chem.* **1995**, *60*, 5763−5769.
[407] Gawley, R. E.; Barolli, G.; Madan, S.; Saverin, M.; O'Connor, S. *Tetrahedron Lett.* **2004**, *45*, 1759−1761.
[408] Madan, S.; Milano, P.; Eddings, D. B.; Gawley, R. E. *J. Org. Chem.* **2005**, *70*, 3066−3071.
[409] Gawley, R. E.; Eddings, D. B.; Santiago, M. *Org. Biomol. Chem.* **2006**, *4*, 4285−4291.

Chapter 4

1,2- and 1,4-Additions to C=X Bonds

Some of the earliest attempts to understand stereoselectivity in organic reactions were the rationalizations and predictive models made in the early 1950s by Curtin [1], Cram [2], and Prelog [3] to explain the addition of achiral nucleophiles such as Grignard reagents to the diastereotopic faces of ketones and aldehydes having a proximal stereocenter.[1] In the decades since, there has been a steady stream of additional contributions to the understanding of these phenomena (reviews: [6–8]).

In this book, a distinction is made between additions that involve allylic nucleophiles and those that do not. For the purposes of this discussion, the addition of enolates and allylic nucleophiles will be labeled π-transfers, and nonallylic nucleophiles will be labeled σ-transfers, as illustrated in Figure 4.1. Note that for σ-transfers aggregation is possible, so that the addition may proceed through a transition state featuring either a 4-membered ring or a larger ring. This chapter covers 1,2- and 1,4-additions to carbonyls by σ-transfer; the addition of enolates and allyls (π-transfer) is detailed in Chapter 5.

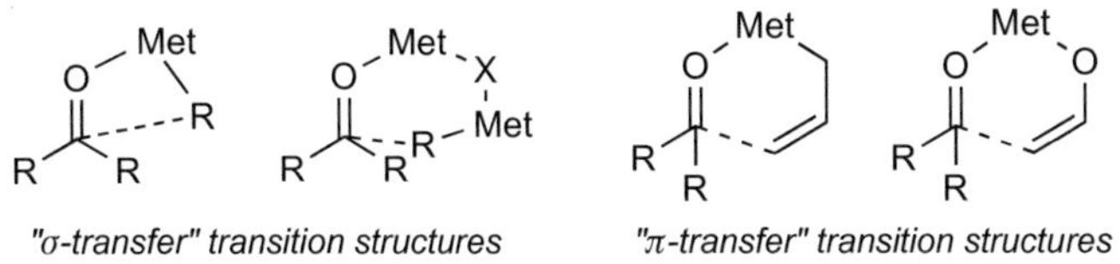

"σ-transfer" transition structures   "π-transfer" transition structures

FIGURE 4.1 Classification of nucleophilic additions to carbonyls.

This chapter begins with a detailed examination of the evolution of the theory of nucleophilic attack on a chiral aldehyde or ketone, from Cram's original "rule of steric control of asymmetric induction" to the current formulation. Then follows a discussion of Cram's simpler "rigid model" (chelate rule), then carbonyl additions using chiral catalysts and chiral (nonenolate) nucleophiles. The chapter concludes with asymmetric 1,4-additions to conjugated carbonyls.

---

1. For a review of the early literature on the stereoselective reactions of chiral aldehydes, ketones, and α-keto esters, and also of the addition of Grignards and organolithiums to achiral ketones and aldehydes in the presence of a chiral complexing agent or chiral solvent, see refs. [4,5].

Principles of Asymmetric Synthesis.
© 2012, Professor Robert E. Gawley and Professor Jeffrey Aubé. Elsevier Ltd. All rights reserved.

4.1 CRAM'S RULE: OPEN-CHAIN MODEL

In 1894, the stereoselective addition of cyanide to chiral carbonyl compounds, the Kiliani–Fischer synthesis of carbohydrates, was proclaimed by Emil Fischer to be "the first definitive evidence that further synthesis with asymmetric systems proceeds in an asymmetric manner" [9]. By the mid-twentieth century, enough experimental data had accumulated that attempts to rationalize the selectivity of such additions could be made. One of the more useful predictive models was formulated by Cram in 1952 (Figure 4.2a, [2]). In this model, Cram proposed that coordination of the metal of (for example) a Grignard reagent to the carbonyl oxygen rendered it the bulkiest group in the molecule. It would tend to orient itself between the two least bulky groups, as shown. In 1959 [10], the model was redrawn as in Figure 4.2b, which also implies a second, less favored conformation, Figure 4.2c.

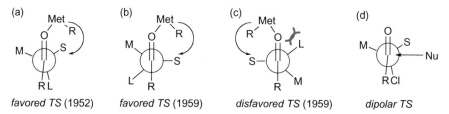

(a)	(b)	(c)	(d)
favored TS (1952)	favored TS (1959)	disfavored TS (1959)	dipolar TS

FIGURE 4.2 (a–c) Cram's models for predicting the major isomer of a nucleophilic addition to a carbonyl having a stereocenter in the α position [2,10]. (d) Cornforth's dipole model for α-chloro ketones [11]. S, M, and L refer to the small, medium, and large groups, respectively.

These models correctly predict the major diastereomer of most asymmetric additions. A notable exception is Grignard addition to α-chloro ketones, which led Cornforth to propose a model where the halogen plays the role of the large substituent so that the C=O and C–Cl dipoles are opposed (Figure 4.2d, [11]). As explained in the following sections, the Cornforth model has not seen much use in addition of nucleophiles such as Grignard reagents and organolithiums, because the more recent Felkin–Anh model (Section 4.1.4) predicts the same result. However, as will be seen in Chapter 5, some cyclic crotyl borations and aldol additions appear to follow this model.

4.1.1 The Karabatsos Model

The predictive value of Cram's rule notwithstanding, the rationale was speculative, and as spectroscopic methods developed, it was called into question. For example, Karabatsos studied the conformations of substituted aldehydes [12] and dimethylhydrazones [13] by NMR, and concluded that one of the ligands at the α position eclipses the carbonyl. It was felt that in the addition reaction, the organometallic probably *did* coordinate to the carbonyl oxygen as Cram had suggested, and Karabatsos used the conformations of the dimethylhydrazone as a model for the metal-coordinated carbonyl. He concluded that since the aldehyde and the hydrazone have similar conformations, so should the metal-complexed carbonyl [14]. He also assumed that the transition state is early, so that there is little bond breaking or bond making in the transition states (Hammond postulate [15]), and that the arrangement of the three ligands on the α-carbon are therefore the same in the transition state as they are in the starting materials: eclipsed.

Thus, Karabatsos concluded that the rationale for Cram's rule was incorrect [14]. In 1967, he published a new model, which took into account the approach of the nucleophile from either side of all three eclipsed conformers [14]. He noted that the enthalpy and entropy of activation for Grignard or hydride additions to carbonyls are $8-15$ kcal/mol and -20 to -40 cal/mol K, respectively. Since the barrier to rotation around the sp^2-sp^3 carbon—carbon bond is much lower [16], the selectivity must arise from Curtin—Hammett kinetics [17,18]. Of the six possible conformers (Figure 4.3), four were considered unlikely due to steric repulsion between the nucleophile and either the medium or large α-substituents (Figure 4.3b, c, e, and f). The two most likely transition states, Figure 4.3a and d, have the nucleophile approaching closest to the smallest group on the α-carbon, and are distinguished by the repulsive interactions between the carbonyl oxygen and the α substituent (either M or L), with Figure 4.3a favored.

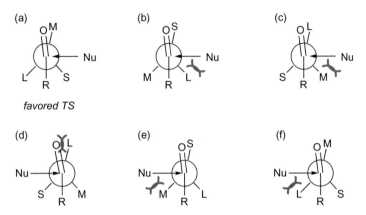

FIGURE 4.3 Karabatsos's transition state models [14].

4.1.2 Felkin's Experiments

In 1968, Felkin noted that neither the Cram nor the Karabatsos models predict the outcome of nucleophilic addition to cyclohexanones [19], and fail to account for the effect of the size of R on the selectivity [20]. The point about cyclohexanones is important since it is unlikely that the mechanisms of Grignard and hydride additions to cyclic and acyclic ketones differ significantly. The data in Table 4.1 indicate that as the size of the other carbonyl substituent increases, so does the selectivity, except for the single example where the "large" substituent on the stereogenic α-carbon is cyclohexyl and the second carbonyl substituent is *tert*-butyl.

TABLE 4.1 Stereoselectivity (% ds) of Reductions of $R_1MeCHC(=O)R_2$ by $LiAlH_4$ [20]

Large Substituents	$R_2 = Me$	$R_2 = Et$	$R_2 = i\text{-}Pr$	$R_2 = t\text{-}Bu$
$R_1 = c\text{-}C_6H_{11}$	62	66	80	62
$R_1 = Ph$	74	76	83	98

To explain these results, Felkin proposed a new model [20], in which the incoming nucleophile attacks the carbonyl from a direction that is *antiperiplanar* to the large substituent (Figure 4.4), while maintaining the notion of an early transition state. Whereas the Cram and Karabatsos models dictate that the nucleophile's approach eclipses (Cram dihedral $0°$) or nearly eclipses (Karabatsos dihedral $30°$) the small substituent on the α-carbon, Felkin proposed that the nucleophile bisects the bond between the medium and small substituents, as in the conformers shown in Figure 4.4a and b (60° dihedral). Felkin suggested that the factor controlling the relative energy of the transition states is the repulsive interaction between R and either the small or medium ligands on the stereocenter, and assumed that there is no energy differential resulting from the interaction between the carbonyl oxygen and either the small or medium substituents on the α-carbon.[2] Thus, the conformer shown in Figure 4.4a is favored. The higher selectivities observed across the board (Table 4.1) when the "large" group is phenyl was explained by the greater electronegativity of phenyl over cyclohexyl (*i.e.*, increased differential between Figure 4.4a and b). Felkin also postulated that when one of the substituents was a chlorine, it would assume the role of the "large" *antiperiplanar* substituent due to polar effects, thus obviating the need for the Cornforth model (Figure 4.2d). To explain the seemingly anomalous result with a *tert*-butyl substituent, Felkin suggested that the normally preferred conformation is destabilized by a severe 1,3-interaction between the large substituent and one of the methyls of the *tert*-butyl, as in Figure 4.4c.[3]

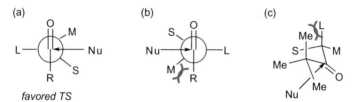

(a)

favored TS

(b)

(c)

FIGURE 4.4 (a) and (b) Felkin's transition state models. (c) Destabilized "favored" transition state with a flanking *tert*-butyl [20].

The factors that influence selectivity in nucleophilic additions to cyclohexanones include those discussed above, plus torsional effects that are not present in acyclic systems, as well as subtle stereoelectronic factors. For a summary of the issues surrounding nucleophilic additions to cyclohexanones, see refs. [19,21−32]).

4.1.3 The Bürgi–Dunitz Trajectory: A Digression

Note that these three models vary in their assumptions about the trajectory of the incoming nucleophile, but *all are entirely speculative.* How might the approach trajectory be determined? The legendary crystallographer Jack Dunitz suggested "turning on the lights."[4] Bürgi, Dunitz, and Schefter took the position that an observed set of static structures, obtained by X-ray crystallography, when arranged in the right sequence might provide a picture of the changes that occur along the reaction pathway [34]. The model system chosen was a nucleophilic approach to a carbonyl by a tertiary amine. Figure 4.5 illustrates the series of compounds whose crystal structures were compared. In the structures of A−E, the nitrogen

---

2. This rationale is a major weakness of Felkin's theory [21]. First, note that the following distances are identical in both transition states: Nu−O, Nu−R, Nu−S, Nu−M. Second, it is hard to accept that R=H is *more* sterically demanding than oxygen, as would be required for aldehydes (H/S and H/M interactions more important than O/S and O/M).
3. This is a 2,3-*P*-3,4-*M gauche* (or *syn*) pentane conformation, which is equivalent to 1,3-diaxial substituents on a cyclohexane. Note that − because the carbonyl substituent is a *tert*-butyl − it cannot be avoided by rotation around the *tert*-butyl−carbonyl bond. For further elaboration of this effect, see Figure 5.5 and the accompanying discussion. For an explanation of the *P,M* terminology, see Glossary, Section 1.11.
4. "The difference between a chemist and a crystallographer can be compared to two people who try to ascertain what furniture is present in a darkened room; one probes around in the dark breaking the china, while the other stays by the door and switches on the light." (J.D. Dunitz, quoted in ref. [33]).

interacts with the carbonyl carbon to varying degrees, while in F it is covalently bonded, making an aminal. It was noted that in all cases the nitrogen, and the carbonyl carbon and oxygen atoms lie in an approximate local mirror plane (the "normal" plane), but that the carbonyl carbon deviates significantly from the plane defined by the oxygen and the two α substituents (see Figure 4.6a). This deviation increased as the N—C distance decreased, but the N—C—O and R—C—R′ angles varied only slightly from their mean values (Figure 4.6b).

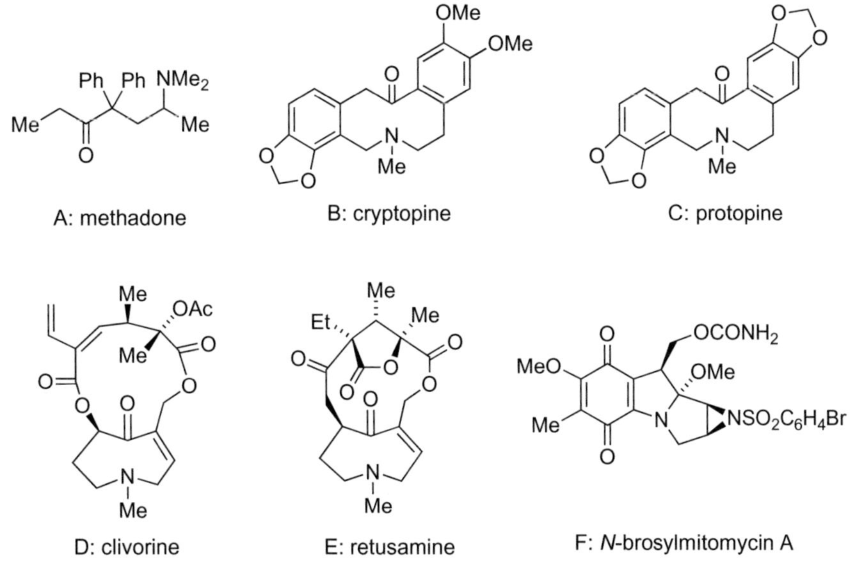

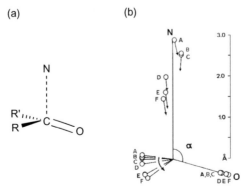

A: methadone B: cryptopine C: protopine

D: clivorine E: retusamine F: *N*-brosylmitomycin A

FIGURE 4.5 Compounds whose X-ray structures provided the basis for the "Bürgi—Dunitz" trajectory.

When the coordinates of the carbonyl carbon atoms and the direction of the C—O bonds are superimposed on a three-dimensional graph, and the position of the nitrogen is plotted on the normal plane, the trajectory of approach is revealed: it "*is not perpendicular to the C—O bond but forms an angle of 107° with it*" (Figure 4.6) [34]. Also revealed is the variation in C—O bond length and the distortion of the RR′C=O plane as the nitrogen approaches bonding distance. The small arrows indicate the presumed direction of the nitrogen lone pair.

(a)   (b)

FIGURE 4.6 (a) Orientation of the superimposed carbonyl and nitrogen atoms. (b) Superimposed plot of the N, C, and O atoms of structures A—F, and the variance of the RR′C plane from the RR′O plane. α is the "Bürgi—Dunitz angle," 107°. Reprinted with permission from ref. [34], ©1973, American Chemical Society.

The crystal structure data are appealing (as far as they go), but the extent to which substituent effects and crystal packing forces influenced the arrangement of the atoms could not be evaluated. Also, the structural data could provide no information about energy variations along (or variant from) the proposed reaction path. In 1974, Bürgi, Dunitz, Lehn, and Wipff studied the approach of hydride to formaldehyde using computational methods [35]. Thus, a hydride was placed at varying distances from formaldehyde and the minimum energy geometry was located. By super-imposing these geometries, the theoretical approach trajectory could be deduced. The results (Figure 4.7), can be summarized as follows. At $H^- -C$ distances of >3.0 Å, the hydride approaches along the X-axis. At an $H^- -C$ distance of 3.0 Å, the H^- and formaldehyde hydrogens are about 2.7 Å apart. At this point, the hydride leaves the HCH plane and glides over the formaldehyde hydrogens until it senses the optimal direction for its attack on the carbonyl, $105 \pm 5°$.

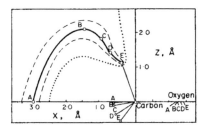 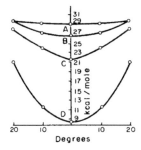

FIGURE 4.7 (a) Minimum energy path for addition of hydride to formaldehyde. Points A, B, C, D, and E correspond to $H^- -C$ distances of 3.0, 2.5, 2.0, 1.5, and 1.12 Å. The dashed and dotted curves show paths that are 0.6 and 6.0 kcal/mol higher than the minimum energy path. (b) Energy profiles for lateral displacement out of the normal (XZ) plane. Reprinted with permission from ref. [35], ©1974, Elsevier Science, Ltd.

A weakness of these computational studies is that they are not particularly instructive in evaluating the addition of organometallics such as organolithiums, organozincs, or Grignard reagents, in that they lack a metal that coordinates to, and activates the carbonyl for subsequent nucleophilic addition. More recent theoretical studies, on the addition of methyllithium and dimethylmagnesium, are helpful in understanding these types of additions [36,37]. Nakamura and Morokuma recognized that the organometallic nucleophile coordinates to an unpaired electron on the carbonyl oxygen, in the nodal plane of the π-bond. This mode of bonding, as opposed to η^2 bonding to the π bond, produces a more reactive complex [38]. Scheme 4.1 shows calculated structures that occur along the reaction coordinate for the addition of $(MeLi)_2$ to CH_3CHO. The ground state is a closed $(MeLi)_2$ dimer. To free the methyl to add to the carbonyl, the bond between the C^2 methyl and Li^1 is broken, affording the highly reactive open dimer,[5] still coordinated to the carbonyl oxygen through Li^1. In this highly reactive intermediate, the C^2 methyl is still in the nodal plane of the carbonyl, so the aldehyde must rotate almost 90° to reach the transition structure. In summary, the Bürgi–Dunitz angle of 107° is achieved in the transition structure, *keeping in mind that the nucleophile only adopts this angle when approaching the carbonyl group*. The direction of rotation of the carbonyl determines which heterotopic face is presented to the nucleophilic carbon. A subtle point that is relevant to asymmetric synthesis is that the transition structure is chiral, even

---

5. Other authors failed to find an open dimer intermediate when the water of solvation is coordinated to Li^2 in the addition of $(MeLi)_2$ to $CH_2C=O$ [39]. For a theoretical study of the addition of MeMgCl to $CH_2C=O$, see ref. [40].

though neither the starting materials nor the product are chiral. Note also the obtuse angle (*) in the side view of the transition structure, revealing that the nucleophilic carbon leans toward the H and away from the methyl. See the discussion below, on the deviation of the nucleophile from the "normal plane" (the "Flippin–Lodge" angle).

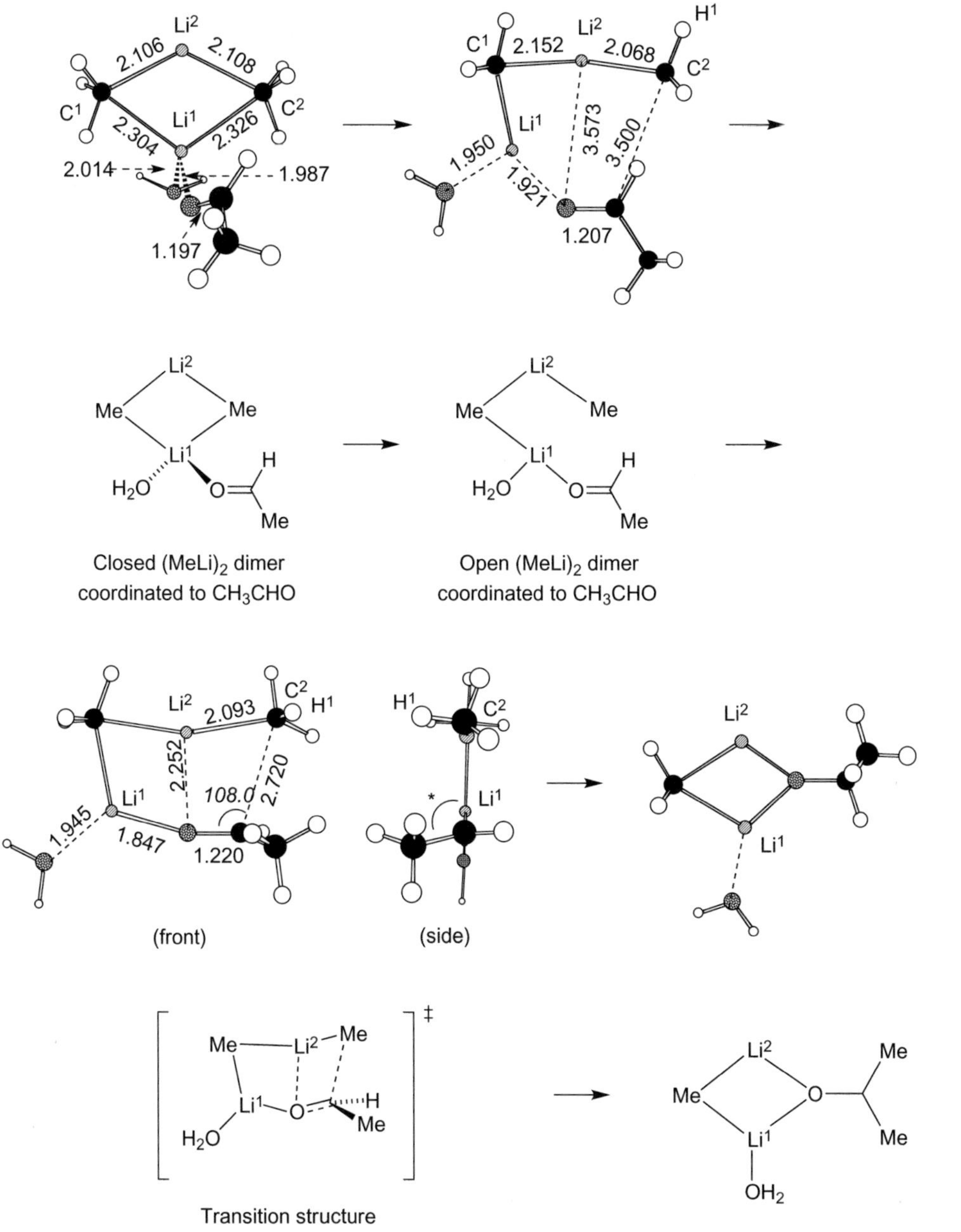

Closed (MeLi)$_2$ dimer coordinated to CH$_3$CHO

Open (MeLi)$_2$ dimer coordinated to CH$_3$CHO

(front)

(side)

Transition structure

SCHEME 4.1 Addition of methyllithium dimer to acetaldehyde. The high-quality computational structures were kindly provided by E. Nakamura, adapted from ref. [36].

The addition of Me_2Mg to acetone has also been modeled by Nakamura and Morokuma [37]. Dimethylmagnesium is monomeric, and when solvated and coordinated to the carbonyl oxygen, it becomes tetrahedral, as shown in Scheme 4.2. The transition structure is early, and features a 4-membered ring, with the position of the nucleophilic methyl at the Dunitz angle of 107°.

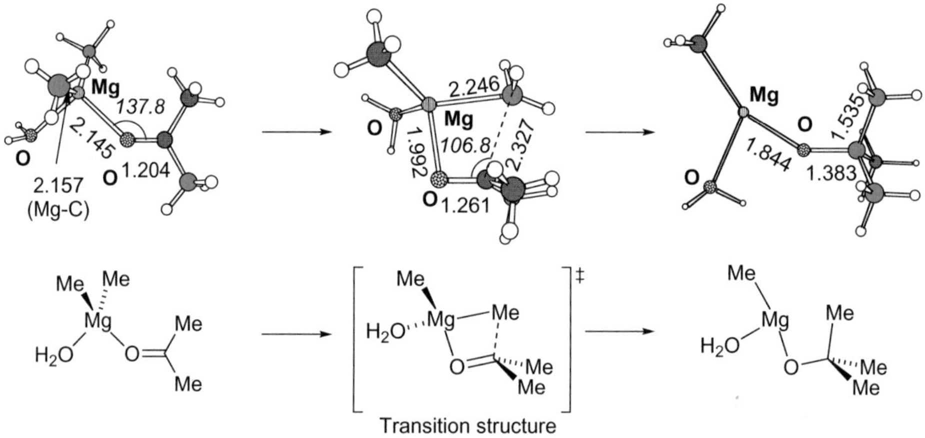

SCHEME 4.2 Computational modeling of the addition of Me_2Mg to acetone. The high-quality computational structures were kindly provided by E. Nakamura, adapted from ref. [37].

4.1.4 Back to the Cram's Rule Problem (Anh's Analysis)

In 1977, Anh [41] used *ab initio* methods to evaluate the energies of all the postulated transition structures (Figures 4.2–4.4) for the reaction of 2-methylbutanal and 2-chloropropanal (the former to test the Cram, Karabatsos, and Felkin models, and the latter to test the Felkin and Cornforth models). The nucleophile was H^-, located 1.5 Å from the carbonyl carbon, at a 90° angle, on each face of the carbonyl. Rotation of the C_1–C_2 carbon–carbon bond then provided an energy trace that included structures close to all of the previously proposed conformational models. The results for both compounds clearly showed the Felkin transition states to be the lowest energy conformers for attack on either face of the carbonyl. Inclusion of a proton or lithium ion, coordinated to the oxygen, produced similar results. It therefore appeared that Felkin's notion of attack *antiperiplanar* to the large substituent was correct.

The Felkin *geometries* have the lowest energy, but that did not necessarily mean that the Felkin *rationale* was correct. Recall that Felkin assumed that a hydrogen is more sterically demanding than an oxygen.[2] In their calculations, Anh and Eisenstein held the geometry of the carbonyl rigidly (in the Felkin conformation) and varied the angle of hydride attack on the two aldehydes coordinated to a cation. They not only found optimum angles of 100°, but also found that the energy difference between the two transition states was *amplified* in this geometry [41]. Thus, the Felkin model was revised to include the Bürgi–Dunitz trajectory. Nonperpendicular attack increases the eclipsing effect with either the small or medium substituents, and also increases the interaction of the nucleophile with R, while decreasing the interaction with the oxygen. With Anh's modifications, the Felkin transition states appear to be on a firm theoretical footing, as illustrated in Figure 4.8.

(a)   (b)

favored TS

FIGURE 4.8   The Felkin−Anh transition state models for asymmetric induction [21,41].

4.1.5 Further Refinements

Heathcock, in 1983 [42], proposed that the increase in selectivity seen as the size of the "other" substituent increased (Table 4.1, [20]), or when the carbonyl is complexed to a Lewis acid [42] might be explained by deviations of the attack trajectory from the normal plane. In 1987 [43], Heathcock reported the results of a semi-empirical study of the angle of approach for the attack of pivaldehyde by hydride. The results, illustrated in Figure 4.9a, illustrate that the approach deviates significantly away from the normal plane, away from the *tert*-butyl group. Although not illustrated, the Bürgi−Dunitz component was variable, but was about the same as found for attack on formaldehyde (108−115°). Although the potential surface near the transition state for nucleophilic additions to unhindered carbonyls is fairly flat [35,44], and has room for some "wobble" in the approach (*cf.* Figure 4.7b), Heathcock showed [43] that constraining the hydride to the normal plane in approach to pivaldehyde is higher in energy, especially at longer bond distances. At 2.5 Å, the energy difference reached its maximum of 0.7 kcal/mol. Figure 4.9b shows Heathcock's rationale for Felkin's observations [20] listed in Table 4.1. When R is small, the "Flippin−Lodge angle", ϕ,[6] is large, and the nonbonded interactions resulting from interaction of the nucleophile with the substituents in R* are diminished. Nakamura's calculated transition structure for the addition of methyllithium dimer to acetaldehyde is consistent with this rationale, showing a small deviation from the normal plane [36]. As the size of R increases, the approach trajectory is pushed back toward the normal plane, increasing the nonbonded interactions with R*, and amplifying the selectivity.

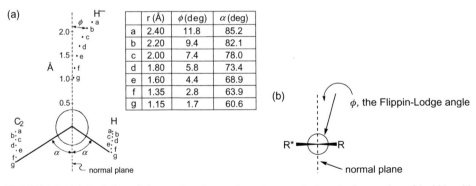

	r (Å)	ϕ (deg)	α (deg)
a	2.40	11.8	85.2
b	2.20	9.4	82.1
c	2.00	7.4	78.0
d	1.80	5.8	73.4
e	1.60	4.4	68.9
f	1.35	2.8	63.9
g	1.15	1.7	60.6

FIGURE 4.9   (a) Deviation of the attack trajectory from the normal plane in the reaction of hydride with pivaldehyde. Reprinted with permission from ref. [43], ©1987, American Chemical Society. (b) Newman projection of a ketone, with an approaching nucleophile, and the Flippin−Lodge angle of deviation from the normal plane, away from the larger substituent, R* (after ref. [45]).

---

6. Heathcock named this angle after his two collaborators, Lee Flippin and Eric Lodge [45].

In his 1977 paper, Anh also addressed the issue of which substituent would assume the role of the "large" substituent *anti* to the incoming nucleophile. A simple rule was offered [41]: the substituents should be ordered according to the energies of the antibonding, σ* orbitals. The preferred *anti* substituent will be the one having the lowest lying σ* orbital, not necessarily the one that is the most demanding sterically. This rule explains the α-chloro ketone anomaly noted by Cornforth, since the σ* orbital of the carbon–chlorine bond is lower in energy than a carbon–carbon bond.

The transition structures in Schemes 4.1 and 4.2 are cyclic, and the orientation of the substituents on the α-carbon, relative to these rings, can influence the relative transition state energies. In 1987, Heathcock evaluated the effect of substituents in the context of aldol additions (6-membered rings) [46], and later Houk [47] and Bienz [48] addressed the issue in the context of ketone reductions (see below).

Specifically, Heathcock examined a series of aldehydes designed to evaluate the relative importance of steric and orbital energy effects (Anh's σ* orbital effect). Aldehydes having a substituent with a low energy σ* orbital (methoxy and phenyl) as well as a sterically variable substituent (methyl, ethyl, isopropyl, *tert*-butyl, phenyl) were synthesized and evaluated. The data are summarized in Table 4.2.

If the *antiperiplanar* substituents in the Felkin–Anh model (L in Figure 4.8) are those with low-lying σ* orbitals (X in Table 4.2), one would expect a gradual increase in selectivity as the steric bulk of the remaining substituent (M in Figure 4.8) increased. The data in Table 4.2 show that this is clearly not the case. In the methoxy series, the expected trend is observed for methyl, ethyl, and isopropyl. But the *tert*-butyl and the phenyl groups are anomalous, if one accepts the standard *A* values[7] as a measure of steric bulk. In the phenyl series, there is no apparent pattern, and when R = *tert*-butyl, the Anh hypothesis predicts the wrong product.

TABLE 4.2 Cram's Rule Stereoselectivities (% ds) for Aldol Additions to Aldehydes (Negative Value Indicates Anti-Cram is Favored), Assuming X is the Large Substituent in the Felkin–Anh Model [46]

X	R = Me	R = Et	R = i-Pr	R = t-Bu	R = Ph
OMe	58	76	92	93	83
Ph	78	86	70	−63	−

These data were interpreted using the four-conformer model shown in Figure 4.10, while also recognizing that the transition structure is likely a closed, 6-membered ring chair (see below and Chapter 5). Simply put, *both steric and electronic effects determine the favored* anti *substituent*. Thus when X = methoxy (Figure 4.10a), conformers A and B are favored when R is methyl, ethyl, or isopropyl, and approach is favored *via* conformer A. Given the differences in *A* values between isopropyl and *tert*-butyl (2.21 kcal/mol *vs.* ~4.8 kcal/mol, respectively), it is perhaps

7. See glossary for definition and partial list. For a complete list, see ref. [49].

surprising that the selectivities are essentially the same. When R is *tert*-butyl, its bulk begins to compensate for the σ* orbital effect, and conformations C and D become important, with D favored. A rationale for the observed (93% ds) selectivity for the *tert*-butyl ligand is that a very high selectivity results from the preference of A over B, but is tempered by an offsetting selectivity of D over C. When R is phenyl, the bulk of the phenyl as well as its low-lying $C_{sp^3}-C_{sp^2}$ σ* orbital play a role. A prediction made on the basis of phenyl's bulk alone (*A* value = 2.8 kcal/mol) would predict a selectivity somewhat greater than when R is isopropyl (still assuming an *anti* methoxy group), but the phenyl σ* orbital is lower in energy than a $C_{sp^3}-C_{sp^3}$ σ* orbital, which increases the importance of conformers C and D (again D is favored).

When X = phenyl (Figure 4.10b), and when R is methyl or ethyl, conformer E is dominant. Note that the selectivity in the phenyl series for methyl and ethyl ligands is greater than in the methoxy series (Table 4.2). This is because the phenyl group is bulky *and* has a low energy σ* orbital, so that the electronic and steric effects act in concert. For the isopropyl and *tert*-butyl ligands, the importance of the G/H conformers increases, and when R is *tert*-butyl they predominate.

Heathcock labeled conformers C, D, G, and H as "non-Anh" conformations, since they have one of the ligands with a *higher* σ* orbital energy *anti* to the nucleophile. The non-Anh conformations are more important in the phenyl series, because there is less difference in the σ* orbital energies between $C_{sp^3}-C_{sp^3}$ and $C_{sp^3}-C_{sp^2}$ bonds than between carbon–carbon and carbon–heteroatom bonds.

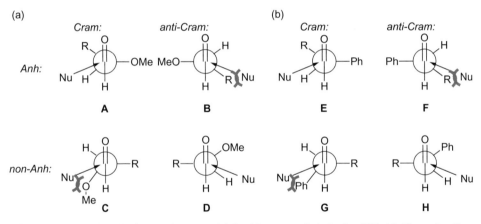

FIGURE 4.10   Heathcock's four-conformer model for 1,2-asymmetric induction [46]. (a) Electronic effects favor methoxy as antiligand (A and B) while steric effects may favor C and D. (b) Electronic effects favor phenyl as antiligand (E and F) while steric effects favor G and H for very large alkyl groups.

Houk [47] and Bienz [48] have employed computational methods to address the issue of stereoselectivity in nucleophilic additions involving 4-membered transition structures (recall Schemes 4.1 and 4.2). In their calculations, they label the three positions in the Felkin–Anh model as *anti* [to the approaching nucleophile], *inside* [the 4-membered ring], and *outside* [the 4-membered ring], as shown in Figure 4.11a. Both groups concluded that the interaction between the nucleophile and the substituent at the *inside* position were more important than interactions between the nucleophile and the substituent at the *outside* position. Thus, the two relevant conformations for most situations, I and J, are as shown in Figure 4.11b, in which the large and medium substituents occupy the *anti* and *inside* positions, while the smallest substituent occupies the *outside* position. The favored transition structure (I) minimizes steric interactions with the *inside* substituent. The *anti*-Cram conformation in J is analogous to conformers D and H in Figure 4.10.

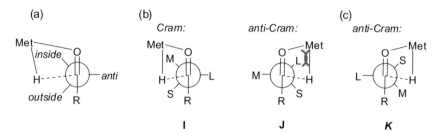

(a)   (b)   (c)
                                 Cram: *anti-Cram:*   *anti-Cram:*

                    I   J   K

FIGURE 4.11 Houk [47] and Bienz [48] postulated transition structures having a 4-membered ring, whereby interactions between the metal and the *inside* substituent are important.

   Bienz tested this analysis in the reduction of the three ketones illustrated in Scheme 4.3 [48], in which the large group was phenyl, cyclohexyl, and *tert*-butyl. For all three ketones the calculated ratios, predicted using *ab initio* methods at several levels of theory with LiH as the reducing agent, predicted the correct diastereomer, but in each case exceeded the experimentally observed ratio. When the large group is *tert*-butyl, reduction by LiAlH$_4$ at $-78\,°C$ favors the *anti*-Cram product [48]. This was rationalized by invoking transition structure K in Figure 4.11c, in which the large *tert*-butyl remains *anti*, and the small hydrogen occupies the *inside* position. Transition structure I was said to be relatively disfavored in this case because the larger "M" substituent occupies the *inside* position. While this theoretical analysis is fine, it is not obvious that reduction by LiAlH$_4$ is the best test, as it is unlikely to proceed through a 4-membered ring transition state.

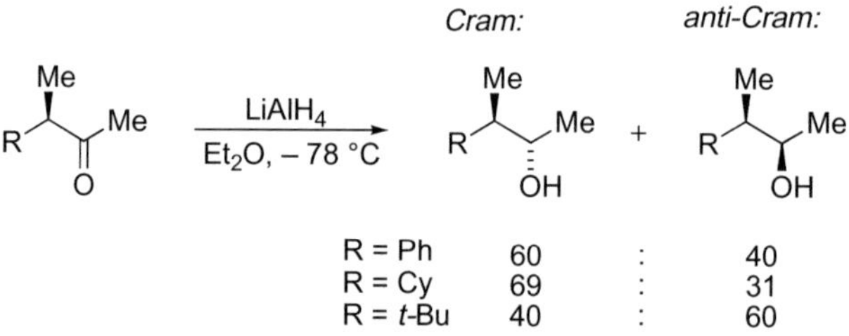

	Cram:		*anti-Cram:*
R = Ph	60	:	40
R = Cy	69	:	31
R = *t*-Bu	40	:	60

SCHEME 4.3   Testing the Felkin–Anh hypothesis in ketone reductions.

   Since the Heathcock work involved aldol additions that proceed through a 6-membered ring transition structure, it is worthwhile to re-examine the Heathcock conformer model (Figure 4.10) in the context of the work by Houk and Bienz, which involves 4-membered ring transition structures. Consider the structures shown in Figure 4.12 for the methoxy series, which correspond to the four structures in Figure 4.10a. Heathcock's analysis favors A over B for the Anh transition structures, and D over C for the non-Anh transition structures. Houk and Bienz conclude that, in most instances, the smallest substituent occupies the *outside* position relative to the 4-membered ring illustrated in Figure 4.11. When extrapolated to a 6-membered ring, the two relevant conformers are, therefore, A and D. The *inside* substituent is "R" in A and "OMe" in D. The analogy breaks down when the chair conformations are examined: it is not likely that "R" in A or "OMe" in D could possibly have any interaction with the metal or its ligands.

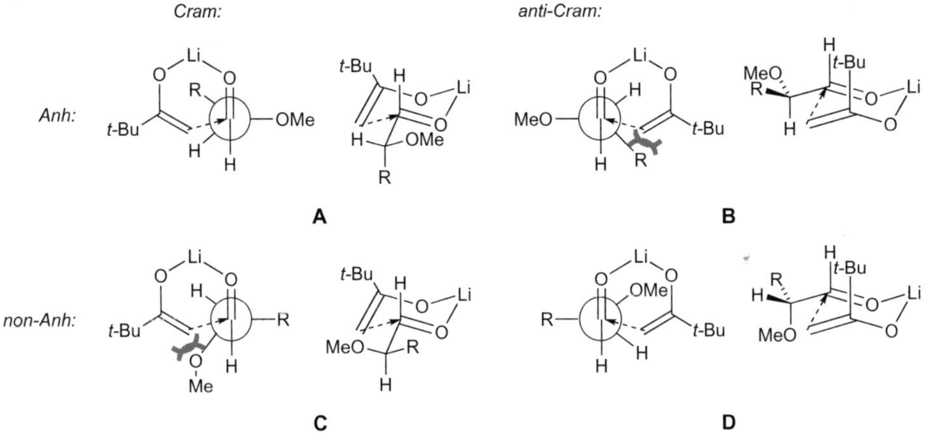

FIGURE 4.12 The Heathcock transition structures reconsidered.

The bottom line. The ratios of *Cram* and *anti-Cram* isomers in the open-chain model are not large, and usually correspond to energy differences of only 1.0−2.0 kcal/mol in competing transition states. That is to say: the factors influencing the stereoselectivity can be very subtle.[8] Nevertheless, the ability to predict a stereochemical outcome is a powerful tool.

4.2 CRAM'S RULE: RIGID, CHELATE, OR CYCLIC MODEL

In his 1952 paper [2], Cram also considered a cyclic model that may be invoked when chelation is possible. In 1959, the model was examined in detail for α-hydroxy and α-amino ketones, since the cyclic and acyclic models predict different outcomes for these systems [10]. The cyclic model (Figure 4.13) has stood the test of time rather well, and has received direct experimental and computational confirmation [37,50]. The cyclic model is applicable to cases where there is a chelating heteroatom on the α-carbon, when that carbon is also a stereocenter (reviews: [6,7,51,52]).

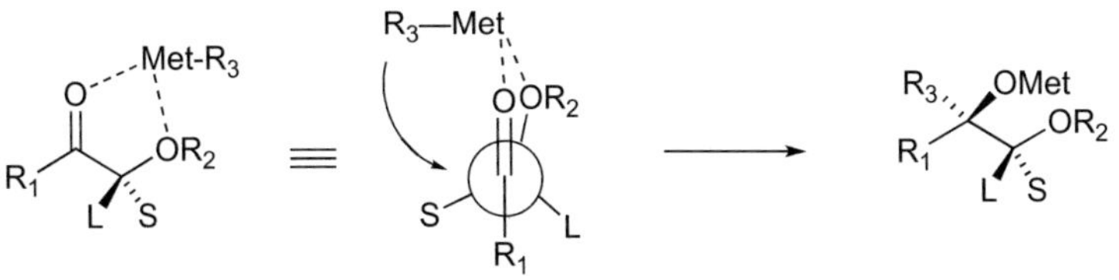

FIGURE 4.13 Cram's cyclic model for asymmetric induction. L and S are large and small substituents, respectively [2,10]).

Table 4.3 lists selected examples where exceptionally high stereoselection has been encountered. Solvent effects play an important role in achieving high selectivity. For example, the >99% diastereoselectivities for the addition of Grignard reagents to α-alkoxy ketones in THF (entry 1) were greatly diminished in ether, pentane, or methylene chloride [53]. Eliel

---

8. Recall that the free energy difference between *synclinal* and *antiperiplanar* conformations of butane is 0.9 kcal/mol.

demonstrated similar selectivites for additions by dimethylmagnesium in THF (entry 2). With aldehydes, there have been conflicting reports. Still reported a 90% diastereoselectivity in the reaction of methylmagnesium bromide with 2-(benzyloxymethoxy)propanal [54], but Eliel [55] and Keck [56] observed poor selectivities in THF. Eliel found good selectivities (90–94% ds) in ether (*e.g.*, entry 3) for the addition of a Grignard to the benzyl or MOM ethers of 2-hydroxyundecanal. For a number of additions of less reactive nucleophiles, Reetz has shown that prior organization of the chelate by complexation with a Lewis acid improves results with aldehydes [57]. Along these lines, Keck has reported [56] that prior coordination of an α-alkoxy aldehyde with magnesium bromide in methylene chloride, followed by addition of a vinyl Grignard affords excellent selectivity (entry 4). To achieve high selectivity, the THF in which the Grignard was formed had to be distilled away and replaced by methylene chloride [56]. Walsh has found that addition of EtZnCl or EtZnBr promotes chelation of even hindered silyloxy groups to achieve high diastereoselectivity in the addition of alkyl and vinyl zinc reagents to aldehydes (entry 7) [58].

TABLE 4.3 Selected Examples of Nucleophilic Addition to α-Alkoxy Carbonyls

Entry	Educt	Conditions	Product (dr)	Ref.
1	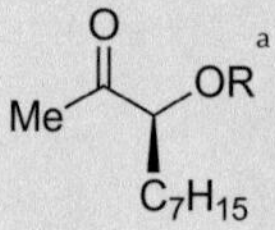	BuMgBr THF,[b] −78 °C	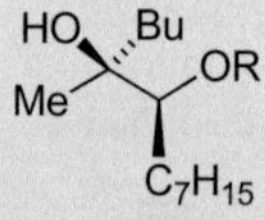 (>99:1)	[53]
2	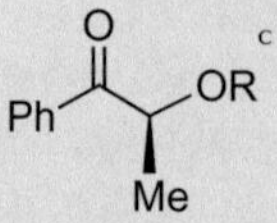	Me$_2$Mg THF, −70 °C	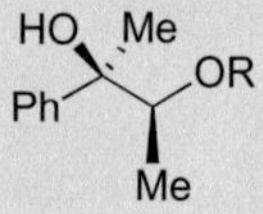 (>99:1)	[50]
3	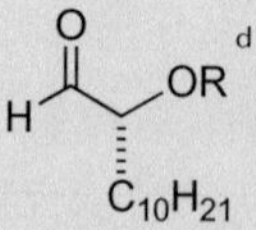	Ph(CH$_2$)$_3$MgBr Et$_2$O,[e] −78 °C	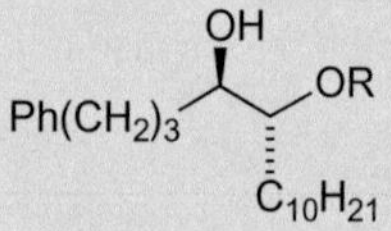 (94:6)	[55]
4	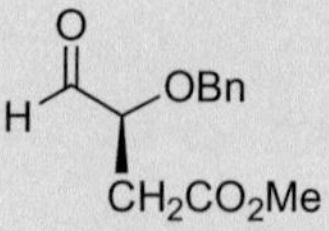	MgBr$_2$·OEt$_2$ CH$_2$=CHMgBr CH$_2$Cl$_2$,[e] −78 °C	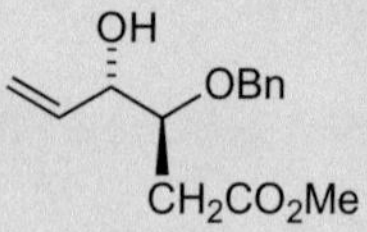 (>99:1)	[56]
5	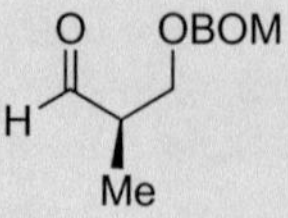	Me$_2$CuLi Et$_2$O, −78 °C	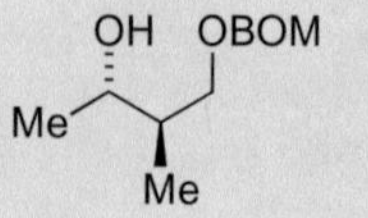 (97:3)	[54]

(*Continued*)

TABLE 4.3 (Continued)

Entry	Educt	Conditions	Product (dr)	Ref.
6	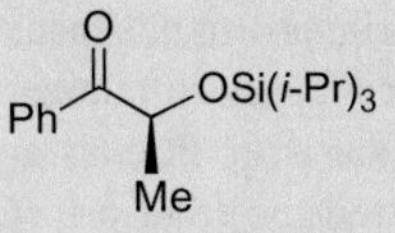 Ph—C(=O)—CH(Me)—OSi(i-Pr)₃	Me₂Mg THF, −70 °C	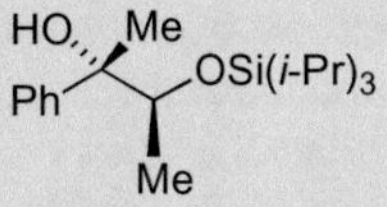 HO, Me / Ph—C—CH(Me)—OSi(i-Pr)₃ / Me (58:42)	[50]
7	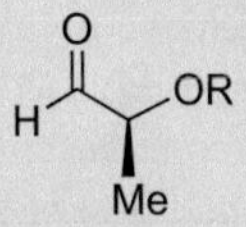 H—C(=O)—CH(Me)—OR R = TBS, TES, TIPS	Et₂Zn 1.5 equiv EtZnCl Toluene, 0 °C to RT	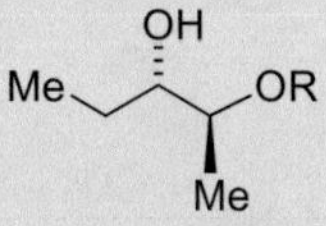 OH / Me—CH—CH(Me)—OR / Me (≥95:5)	[58]

[a]R = MEM (methoxyethoxymethyl-), MOM (methoxymethyl-), MTM (methylthiomethyl-), CH₂—furyl, Bn (benzyl-), BOM (benzyloxymethyl-).
[b]Pentane, ether, and methylene chloride afforded much lower selectivities.
[c]R = Me, SiMe₃.
[d]R = Bn, MOM.
[e]THF affords much lower selectivity.

The cyclic model applies mainly for α-alkoxy carbonyls (*5-membered chelate*), whereas β-alkoxy carbonyls (*6-membered chelate*) are less selective in most cases. An exception is the addition of cuprates to β-alkoxy aldehydes having a stereocenter at the α-position (entry 5).

Two features of the cyclic model are particularly important synthetically. The first is that the selectivities can be significantly higher than for the acyclic category. Compare entries 2 and 6 of Table 4.3: the methoxy and trimethylsilyloxy groups chelate the magnesium (entry 2) whereas the triisopropylsilyloxy group does not (entry 6). This poorly selective example reacts by the acyclic pathway (also compare entries 1−5 and 7 with Tables 4.1 and 4.2). The second noteworthy point is that the product predicted by the cyclic and acyclic models are often different. As shown in Scheme 4.4, the predictions of the acyclic and cyclic models are different for Table 4.3, entry 1 (see also entries 2, 6, and 7).

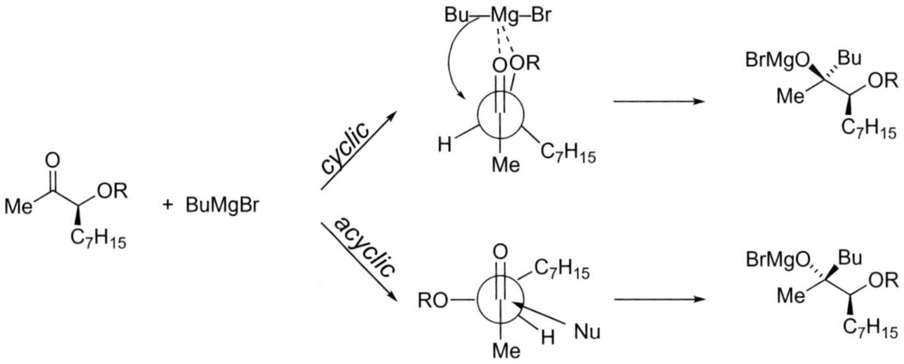

SCHEME 4.4 Cyclic and acyclic models often predict opposite outcomes.

Eliel has examined the mechanism of the addition of dimethylmagnesium (Me_2Mg) to α-alkoxy ketones (*e.g.*, Table 4.3, entries 2 and 6) in detail, since dimethylmagnesium is a well-characterized monomer in THF solution. Scheme 4.5 summarizes the picture of the mechanism that evolved from these studies [50]. Beginning with the educt in the middle of the scheme, there are two competing pathways for the addition reaction. One involves chelated (cyclic) intermediates (to the right of the scheme), while the other involves nonchelated (acyclic) intermediates (shown on the left). One should also recognize that there are two distinct issues that must be considered for these competing pathways: their *relative rates* and their *stereoselectivities*.

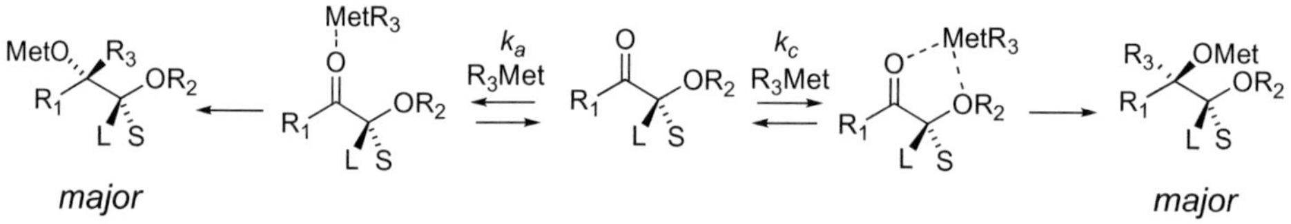

SCHEME 4.5 The acyclic and cyclic mechanisms compete for the consumption of substrate.

The chelate rule will only be applicable if addition via the chelate is faster than addition by the acyclic mechanism (*i.e.*, $k_c > k_a$ in Scheme 4.5). Because the chelate is rigid, it is often considerably more stereoselective as well.[9] *However, the relative rate issue is independent of the stereoselectivities of the two processes.* For example, chelation can be used to control regioselectivity: selective reduction of a diester is achieved by preferential chelation to a 5-membered ring over a 6-membered ring by magnesium bromide (Scheme 4.6, ref. [56]).

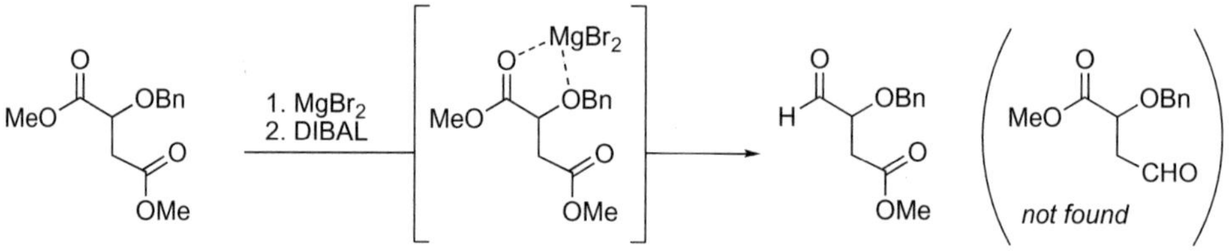

SCHEME 4.6 Independent of stereochemical issues, chelation can determine reactivity [56].

For the chelate path to be faster than the acyclic path, chelation must lower the energy of activation relative to the acyclic path, as shown in Figure 4.14 [50]. The two individual steps illustrated in this diagram deserve comment. First, note that the chelated intermediate is lower in energy and has a smaller energy of activation for its formation than the monodentate intermediate on the acyclic pathway. That the chelate is more stable than the monodentate complex is no surprise. However, the increased organization of the chelated transition state ($\Delta S^{\ddagger}$ less positive) and the increased steric interactions that result ($\Delta H^{\ddagger}$ more positive) would seem to dictate a slower reaction,[10] but these effects are offset by the enthalpy gained by complexation of the alkoxy ligand to the metal and the entropy gained by liberation of an additional solvent from the metal by the bidentate ligand. Regarding the second step, whereby

---

9. This may seem contrary to the reactivity–selectivity principle, wherein one expects a decrease in selectivity to accompany an increase in reactivity, but this principle has a number of limitations. For an extensive discussion of the reactivity–selectivity principle, see ref. [59].
10. Recall (Section 1.5) that $k \propto (e^{-\Delta H^{\ddagger}/RT})(e^{-\Delta S^{\ddagger}/R})$.

the chelate reacts faster than the monodentate complex, it is pertinent that the kinetics of the addition of dimethylmagnesium are first order in organometallic [50]. This requires the intramolecular transfer of an R_3 ligand *via* a 4-membered ring transition state. The distance between the metal ligand (R_3) and the carbonyl carbon is greater in the (linear) acyclic transition state than in the chelated one, so the chelate is farther along the reaction coordinate than the linear complex [50]. The calculated transition structure of the addition of Me_2Mg to methoxyacetone is consistent with this mechanism [37].

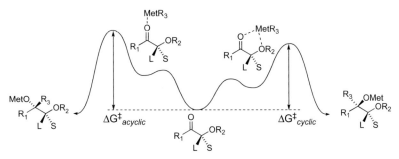

FIGURE 4.14 Energetics of the Cram-chelate (acyclic) model. $\Delta G^{\ddagger}_{acyclic} > \Delta G^{\ddagger}_{cyclic}$ (after ref. [50]; see also ref. [37]).

The relative energies of the intermediates and transition structures along the reaction coordinates are subject to the influence of solvation, which may alter relative stabilities and rates. This may explain the solvent effects discussed earlier (*cf.* Table 4.3, entries 1, 3, and 4). The energetic features outlined above may also explain the lack of selectivity in the nucleophilic additions to β-alkoxy carbonyl compounds. It is possible that even though 6-membered chelates are formed, their rates of formation are slower than addition via the nonchelated path, or that they are less reactive than a 5-membered chelate. Either of these circumstances (or a combination of both) would raise the transition state energy for the chelate path and the primary addition mode could be shifted to the less selective open-chain mechanism.[11]

Nakamura and Morokuma's computational studies on the addition of Me_2Mg to methoxyacetone and methoxyacetaldehyde are consistent with the experimental data, and resemble the addition of Me_2Mg to acetone (Scheme 4.2) [37]. As shown in Scheme 4.7, the tetrahedral magnesium is chelated by the carbonyl (O^1) and ether (O^2) oxygens. In this chelate, note that the nucleophilic methyl (C^2) is already above the plane of the carbonyl and close to the position it occupies in the transition structure — we can again see that it is further along the reaction coordinate than an unchelated complex (*cf.* Scheme 4.1 and Figure 4.14). Note that the 5-membered chelate is puckered in both the ground state and the transition structure, and also note the Dunitz angle (*) in the front view of the calculated transition structure. In an achiral substrate, pseudorotation would allow either methyl (C^2 or $C^{2'}$) to add to the carbonyl (C^1); the transition structures would be enantiomeric and isoenergetic. If a substituent is in the place of H_{Si}, C^2 would encounter steric repulsion as it approaches the transition state (addition to the *Re* face of C^1), whereas if a substituent is in the place of H_{Re}, it would not.

---

11. Another possibility is that the intrinsic selectivity of reaction *via* a 6-membered chelate is lower.

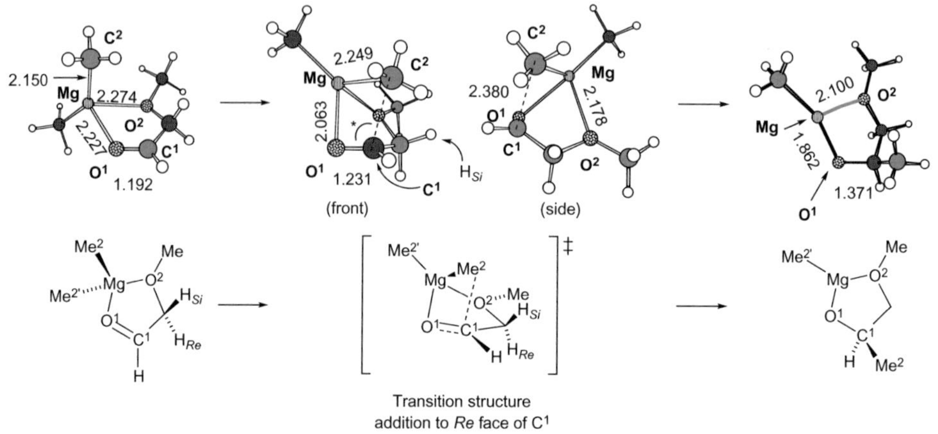

SCHEME 4.7 Computational modeling of the addition of Me₂Mg to methoxyacetaldehyde. The high-quality computational structures were kindly provided by E. Nakamura, adapted from ref. [37].

Because of the high selectivities observed in chelation-controlled additions, they are often used in stereoselective total syntheses. For example, highly selective additions of Grignards were used in the synthesis of monensin [60,61], lasalocid [62,63], formamicinone [64], and soraphen A$_{1\alpha}$ [65], shown in Figure 4.15.

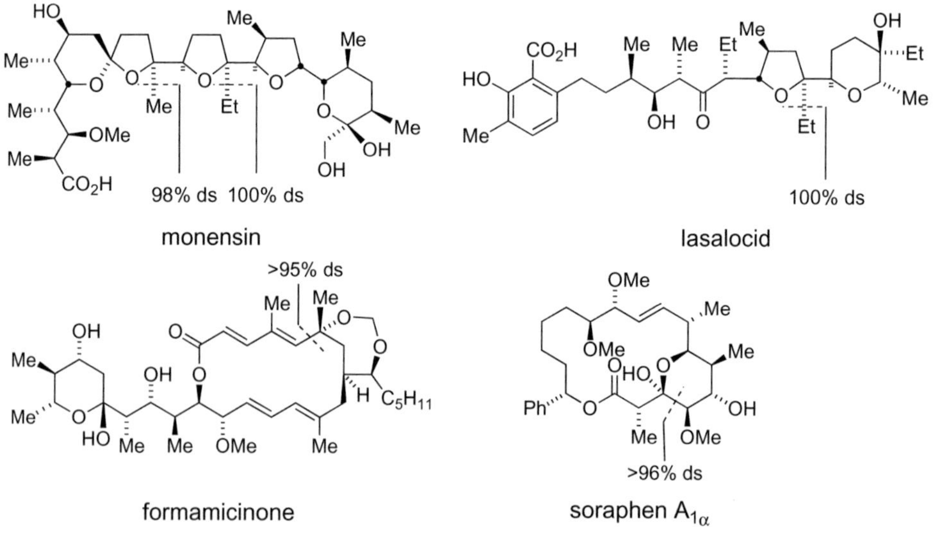

FIGURE 4.15 Chelation-controlled addition of Grignards to ketones and aldehydes figured prominently in syntheses of monensin [60], lasalocid [62,63], formamicinone [64], and soraphen A$_{1\alpha}$ [65]. The disconnections used and the selectivities achieved are indicated for the stereocenters formed by the Grignard addition.

4.3 CHIRAL CATALYSTS AND CHIRAL AUXILIARIES

The preceding discussion summarizes a great deal of work done since the 1950s on the stereoselective additions of achiral carbanionic nucleophiles to carbonyls having a neighboring stereocenter. The knowledge gained during these studies has aided in the development of two different approaches to stereoselective additions to heterotopic carbonyl faces: (i) those using chiral nucleophiles with achiral carbonyl compounds [66−75]; and (ii) a potentially more useful process, one in which neither partner is chiral, but a chiral catalyst is used to induce stereoselectivity (reviews: [76−84]).

All of the reactions discussed in this chapter require coordination of a carbonyl to a metal. This coordination activates the carbonyl toward attack by a nucleophile, and may occur by two intrinsically different bonding schemes: η^1 or η^2 (Figure 4.16). The best evidence indicates that η^1 coordination predominates for Lewis acids such as boron or tin [85,86], and (more importantly) η^1 bonding produces a more reactive species [38]. In the following discussions, it will be assumed that η^1 bonding to the carbonyl oxygen is operative.

FIGURE 4.16   Geometries and relative reactivities of coordinated carbonyls [38].

The potential utility of an asymmetric addition to a prochiral carbonyl can be seen by considering how one might prepare 4-octanol (to take a structurally simple example) by asymmetric synthesis. Figure 4.17 illustrates four possible retrosynthetic disconnections. Note that of these four, two present significant challenges: asymmetric hydride reduction requires discrimination between the enantiotopic faces of a nearly symmetrical ketone (a), and asymmetric hydroboration−oxidation requires a perplexing array of olefin configuration and regiochemical issues (b). In contrast, the addition of a metal alkyl to an aldehyde offers a much more realistic prospect (c) or (d).

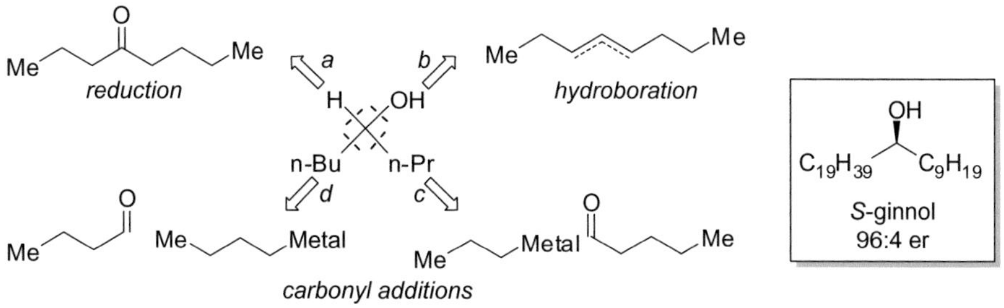

FIGURE 4.17   Simple retrosynthetic strategies for the asymmetric synthesis of 4-octanol. Asymmetric addition to an aldehyde was used to prepare S-ginnol [87].

4.3.1 Catalyzed Additions of Organometallic Compounds to Aldehydes

A number of organometallic compounds have been evaluated for this type of reaction, but the system studied most thoroughly is the addition of organozinc compounds (reviews: [76−78,88−95]).[12,13,14] At or below room temperature the rate of addition of an organozinc

---

12. Organometallics such as organolithium compounds and Grignard reagents are often too reactive to afford selectivity in additions to aldehydes, but some progress has been made [80,96−98].

13. Additions to ketones is more problematic because of the smaller difference in size of the two carbonyl substituents, but again, progress is being made [80,99−101].

14. For a review of enantioselective additions of alkyne nucleophiles, see ref. [79].

compound to an aldehyde is negligible, and organozincs are not basic enough to cause significant deprotonation at an α-carbon. Addition of a Lewis acid, however, causes rapid addition to the carbonyl. Replacement of one of the alkyl substituents on zinc with an alkoxide produces a more reactive species, and amino alcohols have been found to be very useful catalysts for the addition reaction [96,102−105]. At least part of the reason for the increased reactivity is a rehybridization of the zinc from linear to bent upon complexation to an alkoxide, and to tetrahedral upon bidentate coordination. Additionally, donor ligands such as oxygen and nitrogen render the alkyl group more nucleophilic. Figure 4.18 illustrates some of the catalysts that afford good yields and high enantioselectivities in the diethylzinc reaction with benzaldehyde.

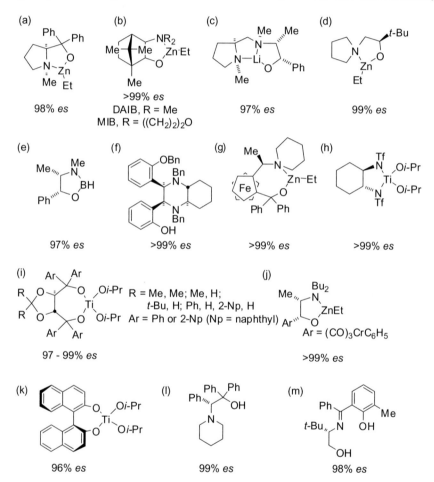

FIGURE 4.18 Catalysts for the diethylzinc reaction with benzaldehyde: (a) [106]; (b) DAIB [107]; MIB [108]; (c) [109]; (d) [110]; (e) [111]; (f) [112]; (g) [113]; (h) [114−116]; (i) [81,117,118]; (j) [119]; (k) [81,120−122]; (l) [123,124]; (m) [125].

The versatility of organozinc additions has been greatly enhanced by new methods of preparing organozinc reagents, some which are summarized in Scheme 4.8 [84,93,95,126−130]. An additional advantage of organozinc reagents is their compatibility with numerous functional groups, such as esters, thioesters, carbamates, amines, sulfides, sulfoxides, sulfones, nitriles, and imides [128]. A disadvantage can be the waste of one of the organic moieties, but solutions to this have been devised in which an unreactive organic ligand ($-CH_2SiMe_3$, $-CH_2t$-Bu, $-CH_2 CMe_2Ph$) is incorporated [131−133].

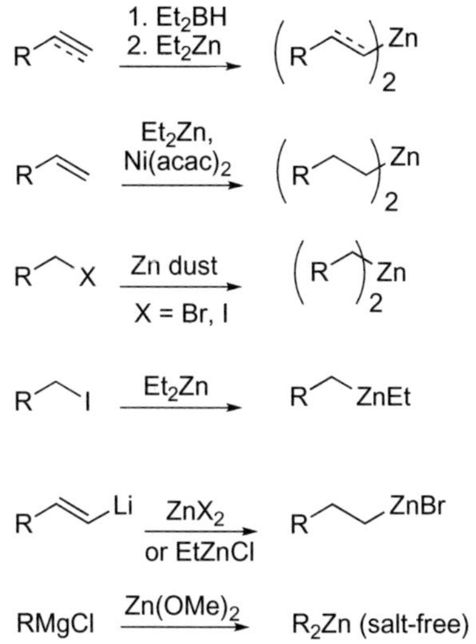

SCHEME 4.8 Selected methods for preparing organozinc compounds [93,95,130].

One of the most thoroughly studied classes of catalysts is the amino alcohols, especially the Noyori catalyst illustrated in Figure 4.18b. The mechanisms that have been proposed for the amino alcohol-catalyzed reaction all involve two zinc atoms, one amino alcohol, and three alkyl groups on the active catalyst [88,107,109,134,135]. A composite mechanism is illustrated in Scheme 4.9 for a "generic" β-amino alcohol.[15] NMR evidence [107] indicates dynamic exchange of the alkyl groups on zinc as shown in the brackets (a bridged species has also been proposed [107]). In experiments done with a polymer-bound amino alcohol catalyst, Frechet noted that the alkoxide product is not bound to the catalyst and that the alkyl transfer must have therefore occured from diethylzinc in solution.

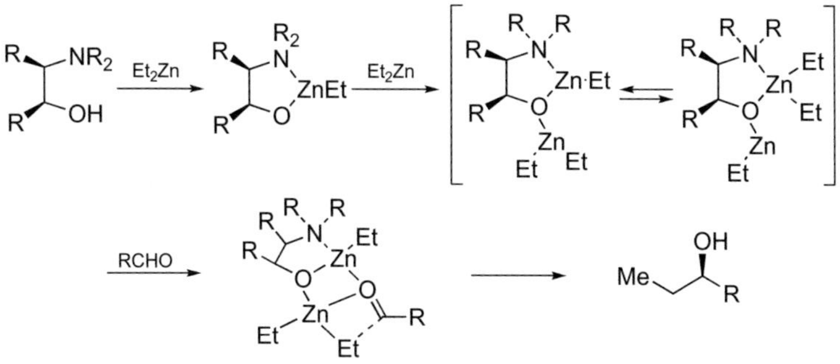

SCHEME 4.9 Proposed mechanistic scheme for amino alcohol-catalyzed diethylzinc reaction [78].

---

15. For discussions of the various mechanistic models and detailed analyses, see refs. [76,136−140].

It might be expected that use of an amino alcohol of less than 100% enantiomeric purity would place an upper limit on the enantiomeric purity of the product. However, Noyori reported that when a catalyst (Figure 4.18b) of 58:42 er was used in the diethylzinc reaction, 1-phenyl-1-propanol of 98:2 er was isolated in 92% yield [107]. In the absence of an aldehyde, the zinc alkoxide, produced from the reaction of the amino alcohol with diethylzinc, is a dimer (Scheme 4.10). When both enantiomers of the amino alcohol are present, both homochiral and heterochiral dimers may be formed. In the case of the Noyori catalyst, the heterochiral dimer is considerably more stable than the homochiral dimer. The latter decomposes to the active, monomeric catalyst immediately upon exposure to a dialkylzinc or an aldehyde, whereas the heterochiral dimer does not [104]. Thus, the minor enantiomer of the catalyst is "tied up" by the major enantiomer.[16] Detailed kinetic analysis also implicates product inhibition in the mechanism [142].

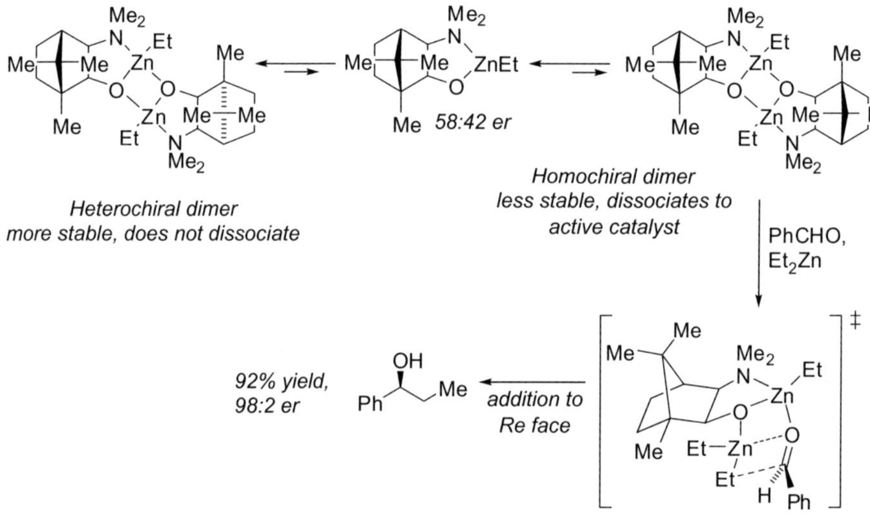

SCHEME 4.10 Amplification of enantiomer ratio by the Noyori catalyst [105,107,143,144].

A particularly striking example of a nonlinear effect is observed when the product of the reaction is also active as the catalyst for its own formation. Such a process is termed autocatalysis. An example from the Soai group is shown in Scheme 4.11 [83,145–147]. In successive reactions in which the product is used to catalyze the next reaction, the enantiomer ratio increases incrementally. After four rounds, a 2% excess of one enantiomer (51:49 er) is amplified to an 88% excess (94:6 er). These spectacular findings have prompted several mechanistic studies, summarized in ref. [140]. Although all of the mechanistic details have not been established, it is clear that a zinc alkoxide dimer (Scheme 4.11, inset) is formed and is inactive as a catalyst.

---

16. Nonlinear effects are sometimes called asymmetric amplification. For detailed analyses, see Section 1.10 and refs. [76,136–141].

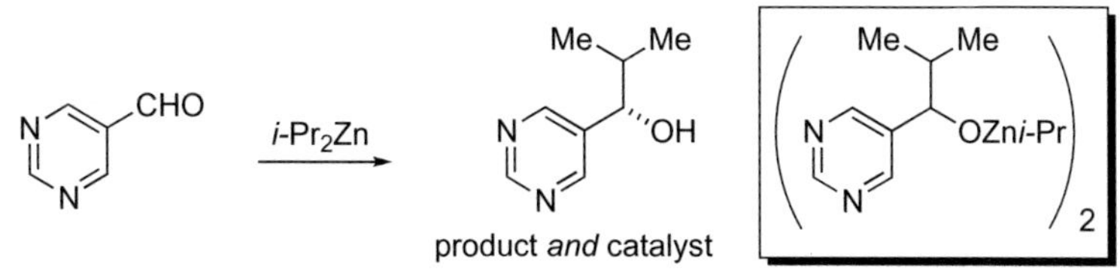

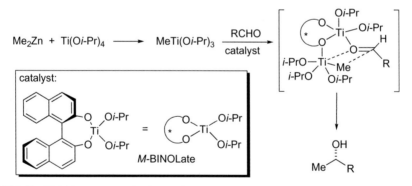

Round 1: catalyst of 51:49 er gave product of 55:45 er
Round 2: catalyst of 55:45 er gave product of 78:22 er
Round 3: catalyst of 78:22 er gave product of 91:9 er
Round 4: catalyst of 91:9 er gave product of 94:6 er

SCHEME 4.11   The Soai autocatalytic system, whereby the product of the reaction is the catalyst for its own formation [83,145−147].

The titanium bis(sulfonamide), TADDOLate, and BINOLate ligands (Figure 4.18h, i, and k) catalyze the addition of organozinc compounds to carbonyls by first transferring the alkyl group from zinc to titanium [118,122,148−150]. Critical to the success of these Ti-catalyzed processes is the presence of Ti(Oi-Pr)$_4$. For both Ti-TADDOLate and Ti-BINOLate ligands, the enantios-electivities are identical, whether R$_2$Zn or RTi(Oi-Pr)$_3$ is used as the source of the alkyl group, suggesting a similar intermediate for both additions [118,122]. Walsh has shown that, for BINOLate titanium-catalyzed reactions, there are two titanium atoms in the transition structure, as illustrated in Scheme 4.12. The evidence points to the RTi(Oi-Pr)$_3$ as the species that ultimately transfers the alkyl group to the aldehyde, probably through a transition structure such as that illustrated. Thus, the alkyl group is transferred to Ti(Oi-Pr)$_4$, and not the titanium BINOLate [122].

SCHEME 4.12   Walsh's mechanism for the Ti-BINOLate catalyzed addition of Me$_2$Zn to aldehydes [122].

Seebach proposed a model to rationalize the face selectivity of Ti-TADDOLate additions of dialkylzinc compounds and alkyltitanium compounds, which is illustrated in Figure 4.19a [118]. In this rationale, it was noted that the most selective TADDOL ligands had clearly defined pseudoaxial and pseudoequatorial aryl substituents. Coordination of the aldehyde to the titanium and orientation away from the proximal Aryl$_{ax}$ group preferentially exposes the Si face of the aldehyde. Implicit in the argument was that the nucleophile would approach from the direction of the viewer. In light of the Walsh studies outlined above, it seems that a slight modification is necessary, as illustrated in Figure 4.19b.

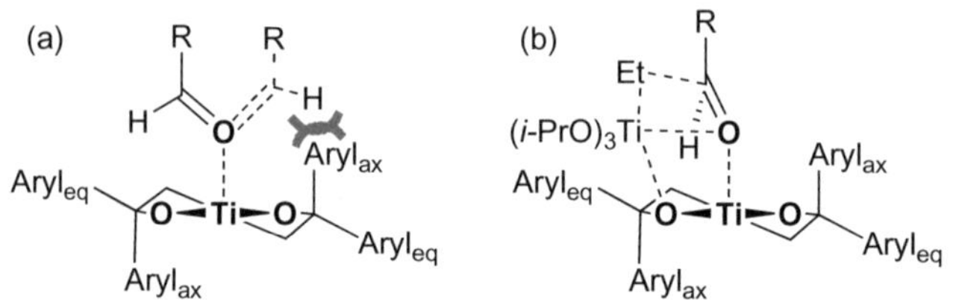

FIGURE 4.19 (a) Seebach's model for predicting the steric course of Ti-TADDOLate-catalyzed addition of R_2Zn or $RTi(Oi\text{-}Pr)_3$ to aldehydes. The structure with the dashed lines represents the less favored orientation [118]. (b) Incorporation of the Walsh transition structure into the Seebach model.

The selectivity of various ligands in additions of diethylzinc to benzaldehyde is indicated in Figure 4.18. To provide a glimpse of the scope of organozinc additions to aldehydes, Table 4.4 lists several examples chosen to illustrate substrate scope for aliphatic, aromatic, alkynyl, and vinyl aldehydes. Note the functionalized organozinc reagents illustrated in entries 8–11, 13, and 14. The trimethylsilylmethyl group in entries 10 and 11 is unreactive, and therefore preserves the more valuable functionalized zinc ligand.

TABLE 4.4 Catalyzed Additions of Organozinc Compounds to Aldehydes[a]

Entry	Carbonyl	RM	Catalyst	% Yield	er	Ref.
1	$n\text{-}C_6H_{13}CHO$	Et_2Zn	4.18a	96	95:5	[106]
2	$n\text{-}C_5H_{11}CHO$	Et_2Zn	4.18b (MIB)	96	95:5	[108]
3	$c\text{-}C_6H_{11}CHO$	Et_2Zn	4.18b (MIB)	94	>99:1	[108]
4	$c\text{-}C_3H_5CHO$	Et_2Zn	4.18b (MIB)	91	99:1	[108]
5	$c\text{-}C_6H_{11}CHO$	Et_2Zn	4.18g	92	99:1	[113]
6	$t\text{-}BuCHO$	Et_2Zn	4.18g	93	99:1	[113]
7	$n\text{-}C_6H_{13}CHO$	Et_2Zn	4.18h	78	>99:1	[114]
8	$BuC{\equiv}CCHO$	$(PivO(CH_2)_3)_2Zn$	4.18h	83	95:5	[151]
9	$TIPSOCH_2CHO$	$(AcO(CH_2)_4)_2Zn$	4.18h	60	92:8	[152]
10	$EtCHO$	$Cl(CH_2)_4Zn\text{-}CH_2SiMe_3$	4.18h	86	>97:3	[132]
11	$PhCHO$	$AcO(CH_2)_5Zn\text{-}CH_2SiMe_3$	4.18h	74	95:5	[132]
12	$c\text{-}C_6H_{11}CHO$	Bu_2Zn	4.18i	35	95:5	[153]
13	$PhCHO$	$(MOMO\text{-}(CH_2)_6)_2Zn$	4.18i	68	92:8	[153]
14	$PhCHO$	$(C_2H_3\text{-}(CH_2)_2)_2\text{-}Zn$	4.18i	83	95:5	[153]

(Continued)

TABLE 4.4 (Continued)

Entry	Carbonyl	RM	Catalyst	% Yield	er	Ref.
15	1- or 2-NpCHO	Et$_2$Zn	4.18j	98	>99:1	[119]
16	2-NpCHO	Et$_2$Zn	4.18k	>90	96:4	[120]
17	n-C$_8$H$_{17}$CHO	Et$_2$Zn	4.18k	94	93:7	[121]
18	c-C$_6$H$_{11}$CHO	Et$_2$Zn	4.18k	75	93:7	[121]
19	n-C$_6$H$_{13}$CHO	Et$_2$Zn	4.18l	>90	96:4	[123]
20	c-C$_6$H$_{11}$CHO	Et$_2$Zn	4.18l	99	99:1	[123]
21	PhCH=CHCHO	Et$_2$Zn	4.18m	69	87:13	[125]
22	c-C$_6$H$_{11}$CHO	Et$_2$Zn	4.18m	67	97:3	[125]

[a]Numbers in the catalyst column refer to Figure 4.18.

4.3.2 Addition of Organometallics to Azomethines

The stereoselective addition of organometallics to azomethines (C=N bond) has not been as fully developed as additions to carbonyls for several reasons (reviews: [154−164]). First, imines are not as electrophilic as carbonyls, and so are less susceptible to nucleophilic attack. Second, azomethines are prone to deprotonation and many organometallic reagents are sufficiently basic that the preferential mode of reaction is abstraction of an α proton. Third, imines are susceptible to E/Z isomerization (often catalyzed by the Lewis acids that are a prerequisite to nucleophilic attack), which complicates the issue of stereochemical predictability. Finally, catalytic approaches have been slow to develop, because the amine product often poisons the catalyst. Nevertheless, the importance of amines in chemistry and medicine has furnished ample motive to pursue this method of synthesis. In fact, since the nitrogen is substituted (C=NR instead of C=O), azomethines provide an opportunity for auxiliary-based stereochemical control that is not available to carbonyls. The following discussion covers some highly utilitarian auxiliary-based approaches and concludes with some of the more effective catalytic methods for addition of organometallic compounds to azomethines.

The addition of organometallics to $SAMP$ and $RAMP$ hydrazones has been studied by the Enders [165−172] and Denmark groups [164,173−176]. The best selectivities result from addition of organolanthanide reagents; Table 4.5 illustrates several highly selective examples. In conjunction with reductive cleavage of the hydrazine by hydrogenolysis [165,166,171], borane reduction [177], or dissolving metal reduction [178], the addition provides a convenient synthesis of α-branched primary amines (compare Figure 4.17 for the synthesis of secondary alcohols). The intermediate hydrazines are somewhat unstable, but N-acylation makes for easier handling [169,178].

Additions to C=N Bonds: Name Reaction Soup

Although additions to C=N bonds are closely related to those described for carbonyls, there are some challenges that are specific to these reactions including (1) the fact that there is often stereochemistry associated with the C=N bond itself, (2) the need to activate imines (which is one reason one often sees additions to variants like oximes, hydrazones, or imines bearing pendant coordinating groups), (3) relative instability of the C=N bond, particularly with simple imines when there's water around. When additions to C=N bonds involve cycloadditions or intramolecular reactions, one gets into heterocyclic territory, which makes these reactions of great interest to heterocyclic chemists—a community to which both of this book's authors proudly belong. Accordingly, we would like to add to the discussions in Sections 4.3.2 and 4.3.3 (additions to C=N bonds), 6.1.2 (hetero-Diels–Alders), and 6.1.3 (1,3-dipolar cycloadditions) by presenting a few more of our favorite reactions of imines and their ilk.

Two classic approaches utilize the chiral pool in powerful ways. On the top, we see the straightforward but eminently practical use of the amino acid tryptophan in the Pictet–Spengler cyclization, in which the iminium ion is formed and subsequently undergoes a Friedel–Crafts-like reaction with the indole ring (for a review of the asymmetric Pictet–Spengler, see ref. [1]). This example is from the laboratory of Cook, who has masterfully used this approach to synthesize numerous alkaloids (e.g., the nifty Dieckmann reaction used in the second step provided here as a mechanism problem [2,3]). On the bottom, a simple condensation reaction provides a versatile scaffold that has been heavily exploited by Husson and Royer in what they term a CN(R,S) approach [4,5].[1] In one such application, the nitrile group can be eliminated in acid to afford an iminium ion that is subject to nucleophilic attack. In that regard, all serious students of heterocyclic stereochemistry are recommended to the powerful guidelines published by Stevens for the outcomes of such reactions [6].

Catalytic means of carrying out Pictet–Spengler or more generally Mannich reactions, began to appear in the 1990s, with important early examples being the organocatalyzed direct

---

1. "CN(R,S)" is a double entendre: it refers both to the use of a nitrile-containing compound to direct the selective formation of either an R or S stereocenter, and to the Centre National de la Recherche Scientifique, where the process was developed.

Mannich between aldehydes and imines reported by Barbas [7] or the Brønsted acid-mediated Pictet–Spengler reaction of List [8]. A version of particular mechanistic interest devised by Jacobsen specifically for the purpose of promoting the cyclization between an indole and an acyliminium species is shown below [9]. The challenge here was to influence the addition of the appended nucleophile to a fully-substituted iminium ion lacking an obvious attachment point for a Lewis acid and that was cationic already, to boot. Success here was attributed to the tight ion pair formed between the thiourea catalyst and the *in situ* generated iminium species.

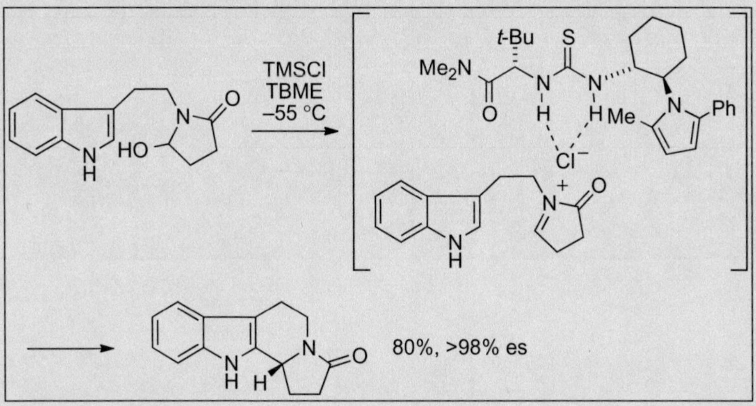

The Staudinger reaction is the [2 + 2] reaction of an imine with a ketene to afford a β-lactam; as such, the reaction is of tremendous practical importance to pharmaceutically minded chemists (for a review, see ref. [10]; also see the related ester enolate–imine cyclocondensation reaction, reviewed in refs. [11,12]). Despite much interest and effort, the asymmetric Staudinger reaction has still only been realized for a limited subset of

useful reaction types. The example shown involves the reaction of an achiral ketene with a chiral imine and demonstrates the general mechanism of the reaction, whereby the initial addition of the imine to the ketene generates a zwitterionic intermediate [13]. The torquoselective collapse of this intermediate is controlled by orbital symmetry (i.e., is conrotatory) and also responds to the presence of the chiral directing group (i.e., undergoes the clockwise rotation shown as opposed to the counterclockwise alternative).

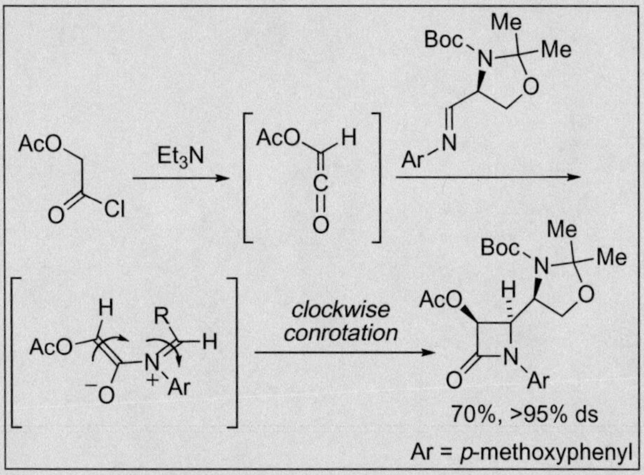

References

[1] Lorenz, M.; Van, L. M. L.; Cook, J. M. *Curr. Org. Synth.* **2010**, *7*, 189–223.
[2] Li, J.; Wang, T.; Yu, P.; Peterson, A.; Weber, R.; Soerens, D.; Grubisha, D.; Bennett, D.; Cook, J. M. *J. Am. Chem. Soc.* **1999**, *121*, 6998–7010.

[3] Cox, E. D.; Cook, J. M. *Chem. Rev.* **1995**, *95*, 1797–1842.

[4] Royer, J.; Bonin, M.; Micouin, L. *Chem. Rev.* **2004**, *104*, 2311–2352.

[5] Husson, H.-P.; Royer, J. *Chem. Soc. Rev.* **1999**, *28*, 383–394.

[6] Stevens, R. V. In *Strategies and Tactics in Organic Synthesis*; Lindberg, T., Ed.; Academic: Orlando, 1984, pp. 275–298.

[7] Córdova, A.; Watanabe, S.-i.; Tanaka, F.; Notz, W.; Barbas, C. F. *J. Am. Chem. Soc.* **2002**, *124*, 1866–1867.

[8] Seayad, J.; Seayad, A. M.; List, B. *J. Am. Chem. Soc.* **2006**, *128*, 1086–1087.

[9] Raheem, I. T.; Thiara, P. S.; Peterson, E. A.; Jacobsen, E. N. *J. Am. Chem. Soc.* **2007**, *129*, 13404–13405.

[10] Palomo, C.; Aizpurua, J. M.; Ganboa, I.; Oiarbide, M. *Curr. Med. Chem.* **2004**, *11*, 1837–1872.

[11] Hart, D. J.; Ha, D. C. *Chem. Rev.* **1989**, *89*, 1447–1465.

[12] Benaglia, M.; Cinquini, M.; Cozzi, F. *Eur. J. Org. Chem.* **2000**, *2000*, 563–572.

[13] Palomo, C.; Cossio, F. P.; Cuevas, C.; Lecea, B.; Mielgo, A.; Roman, P.; Luque, A.; Martinez-Ripoll, M. *J. Am. Chem. Soc.* **1992**, *114*, 9360–9369.

TABLE 4.5 Asymmetric Addition of Organoceriums to Hydrazones

Entry	R_1	R_2	R_3	% Yield	dr	Ref.
1	Me	$(EtO)_2CH$	$EtLi/CeCl_3$	91	96:4	[166]
2	Me	"	$n\text{-}BuLi/CeCl_3$	92	97:3	[166]
	$(CH_2)_2OMe$	$(MeO)_2CH$	$MeLi/CeCl_3$	91	90:10	[176]
3	Me	$Ph(CH_2)_2$	$MeLi/CeCl_3$	81	98:2	[173]
4	Me	"	$PhLi/CeCl_3$	72	96:4	[173]
5	Me	$PhCH_2$	$MeLi/CeCl_3$	66	96:4	[173]
6	Me	$E\text{-}CH_3CH{=}CH$	"	82	96:4	[173]
7	Me	$TBSO(CH_2)_4$	$n\text{-}PrLi/YbCl_3$	83	>99:1	[169]
8	Me	Pr	$TBSO(CH_2)_4Li/YbCl_3$	99	>99:1	[169]
9	$(CH_2)_2OMe$	$Ph(CH_2)_2$	$n\text{-}BuLi/CeCl_3$	72	97:3	[173]
10	"	Me	$Ph(CH_2)_2Li/CeCl_3$	53	97:3	[174]
11	"	$t\text{-}Bu$	"	60	98:2	[174]
12	"	Ph	"	80	97:3	[174]
13	"	$TIPSOCH_2$	$MeLi/CeCl_3$	89	95:5	[176]

In all cases studied, the organometal adds to the *Re* face of the azomethine when the alkoxymethylpyrrolidine stereocenter is *S*. Denmark has studied the steric course for the addition of organocerium compounds lacking a chelating group on the pyrrolidine ring. He found that both the absolute configuration of the product and the diastereoselectivity were the same, whether R_1 was methyl or methoxymethyl (*i.e.*, SAMP) (Scheme 4.13).[17] Based on these results, and an analysis of structural information of hydrazones obtained by X-ray crystallography, two predictive models were proposed (Scheme 4.13). Since a chelating ligand on the pyrrolidine cannot be involved in determining the stereoselectivity, other factors must be responsible. As shown in Scheme 4.13, hypothesis A, coordination of the metal to the lone pair of the azomethine nitrogen and rotation toward the *Re* face (away from R_1) predicts the correct configuration in the product, although intermolecular addition of a second organometal (opposite R_1) is also possible. Hypothesis B posits that the pyrrolidine nitrogen is pyramidalized by the R_1 substituent, and that it acts as a base to coordinate the metal. In this model, the metal is coordinated to the pyrrolidine nitrogen *cis* to R_1, with the hydrazone adopting a configuration in which the azomethine is oriented away from C-5 of the ring to avoid $A^{1,3}$-strain with a pseudoequatorial R_1. Hypothesis A appears to be most consistent with the Nakamura postulates for addition of organometallics to carbonyls (Schemes 4.1 and 4.2), in which metal coordination to a lone pair is the first step, but the aggregation state of the organocerium reagent is not known.

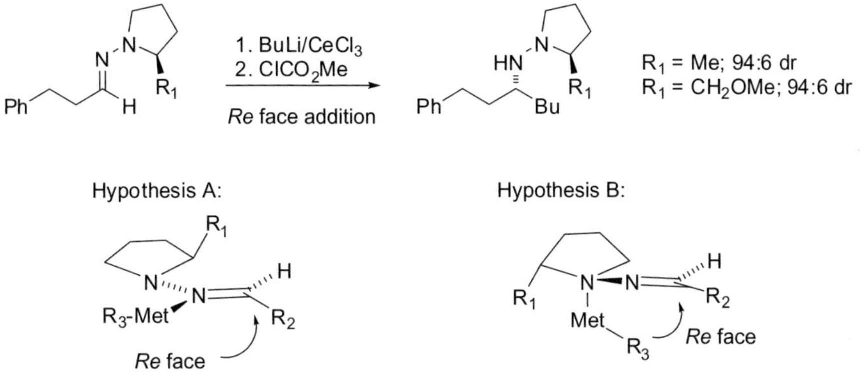

SCHEME 4.13 Revealing results for organocerium additions to hydrazones, and two predictive models [164].

The conjugate addition of organometallics to unsaturated azomethines (in the form of oxazolines, Scheme 4.14a) was one of the first carbon–carbon bond forming reactions that proceeded with >95% selectivity [179–181] (review: ref. [182]). The proposed mechanistic rationale [180,183] has the alkyllithium coordinating to the lone pair of the oxazoline nitrogen, and chelated by the methoxy group at the 4-position.[18] The alkyl group of the

---

17. Note that, if R_1 in Scheme 4.13 is methyl, the CIP priority reverses from *S* to *R*.
18. An alternative transition structure, placing the lithium on the π-cloud of the oxazoline, has also been proposed [183].

organometallic is oriented away from the side of the 4-substituent, and transfer occurs from the *Si* face. This alkyl transfer is reminiscent of a symmetry-allowed [184] suprafacial 1,5-sigmatropic rearrangement [185]. Early on, the Meyers group showed that the 5-phenyl substituent had little effect on the selectivity [180] (see also ref. [185]); more recently [183,186], they have shown that a chelating group at the 4-position is not necessary either (Scheme 4.14b). The conditions necessary to hydrolyze the robust oxazoline nucleus initially limited the usefulness of this method, but subsequent work [185] has shown that the oxazoline may be alkylated with methyl triflate and reduced to an oxazolidine (in one pot), which is then easily hydrolyzed to an aldehyde.

The early (1975) contributions from the Meyers laboratory (Scheme 4.14a) paved the way for a number of related methods in subsequent years. Figure 4.20 illustrates a number of conceptually related conjugate additions. In several of these examples, there is a crucial difference from the examples in Scheme 4.14: in all except Figure 4.20e the α-carbon is prochiral, and two stereocenters are formed in the reaction. Fortunately, it is possible to either alkylate or protonate the azaenolate stereoselectively, such that two new stereocenters are produced in a single operation. Depending on the method, the two alkyl groups may be introduced in either a *cis* or a *trans* fashion. For example, the naphthalene oxazolines (Figure 4.20a−c) alkylate *trans* to the first alkyl group, whereas the cycloalkenyl imines (Figure 4.20f) may alkylate either *cis* or *trans* selectively, depending on the method used. A generalized example (for 1-naphthalenes) is shown in Scheme 4.14c.

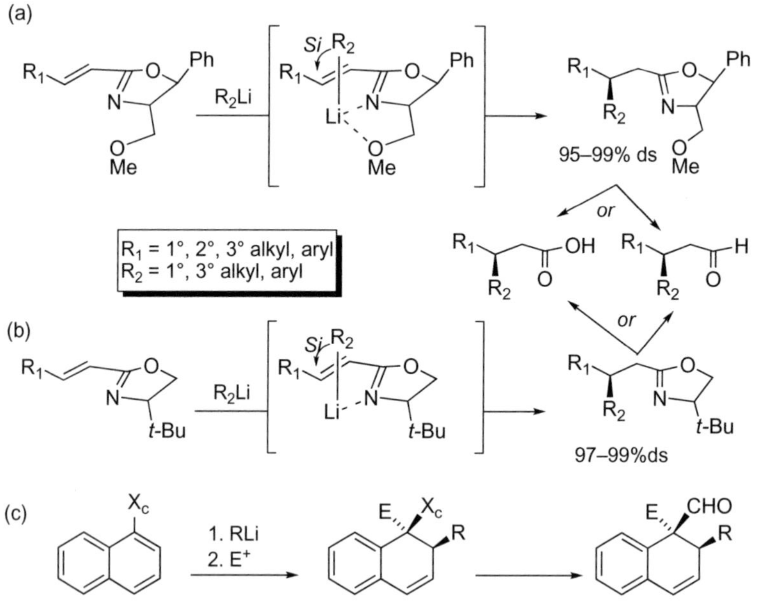

SCHEME 4.14 Asymmetric addition of organolithiums to oxazolines: (a) [179,180]; (b) [183]; (c) tandem asymmetric addition and alkylation of naphthalenes.

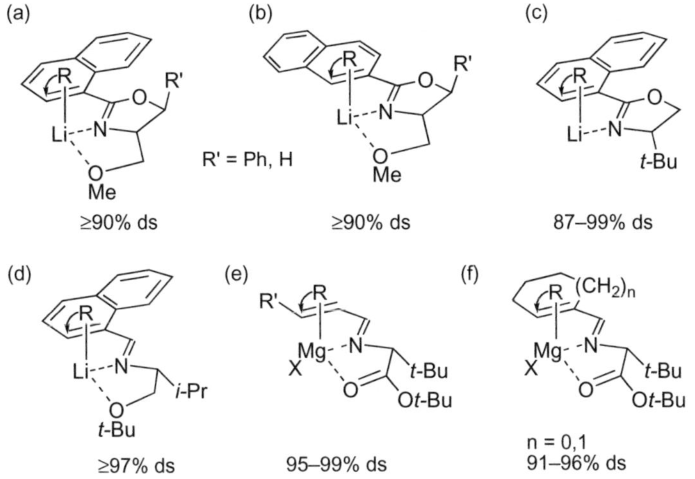

FIGURE 4.20   (a) Addition to 1-naphthyloxazolines [185]; (b) addition to 2-naphthyloxazolines [185]; (c) addition to 1-naphthyloxazolines lacking a chelating group [186]; (d) addition to 1-naphthaldehyde imines [187]; (e) addition to crotyl amino acid imines [188,189]; (f) addition to cyclohexene and cyclopentene aldehyde amino acid imines [190].

A powerful, auxiliary-based method for asymmetric addition to azomethines uses sulfinimines [191]. The Davis group has popularized S-(p-tolyl)-sulfinimines [192−197], while the Ellman group has developed S-(tert-butyl)-sulfinimines [198−200] (Scheme 4.15). The two functional groups have similar steric requirements, and the rationales for stereoinduction are similar. The sulfinyl group activates the imine toward nucleophilic attack, deactivates the imine toward deprotonation, and is easily removable. Moreover, the steric course of nucleophilic addition of an organometallic is readily predictable. The steric course of the addition of organometallics to sulfinimines is consistent with addition *via* a 6-membered chair transition structure, in which the S-substituent is equatorial, and the metal is coordinated to the sulfoximine oxygen.[19] Note also the E configuration at the C=N bond and the equatorial orientation of R_1. Table 4.6 lists several examples.

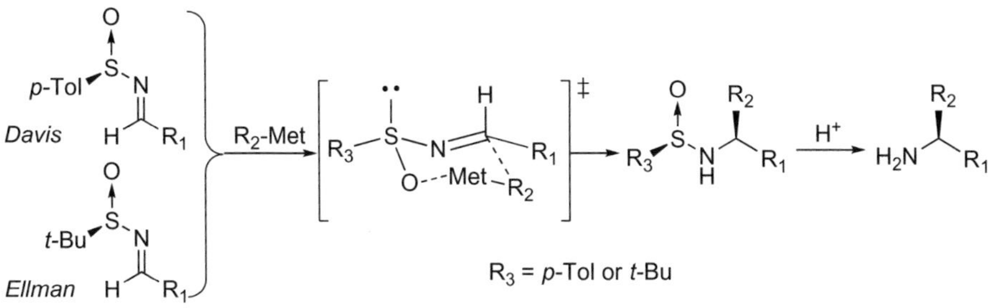

SCHEME 4.15   Davis and Ellman sulfinylimines as chiral auxiliaries for asymmetric addition.

---

19. Lu has shown that the steric course can be affected by choice of solvent [201].

TABLE 4.6 Additions of Organometallic Compounds to Sulfinyl Amines[a]

Entry	R_1	R_2-Met	R_3	% Yield	dr	Ref.
1	Et	MeMgBr	t-Bu	96	93:7	[198]
2	Et	PhMgBr	t-Bu	100	96:4	[198]
3	i-Pr	MeMgBr	t-Bu	99	98:2	[198]
4	Ph	MeMgBr	t-Bu	96	97:3	[198]
5	Ph	(S,S-ring) Me Li	p-Tol	76	96:4	[202]
6	i-Pr	"	p-Tol	84	99:1	[202]
7	i-Pr	(S,S-ring) Li	p-Tol	72	86:14	[202]
8	t-Bu	"	p-Tol	72	98:2	[202]
9	p-ClC$_6$H$_4$ (cyclobutyl)	BrMg～Me～OMgCl	t-Bu	74	97:3	[201]

[a]The R_1 to R_3 groups refer to positions described in Scheme 4.15.

Considerable progress has been made on the catalytic addition of organometallic reagents to azomethines [158,160–163,203]. For example, Charette has found that the BozPHOS ligand, in the presence of copper triflate, catalyzes the addition of organozinc reagents to diphenylphosphinolyl imines in excellent yield and enantioselectivity [160,204]. The BozPHOS ligand is the monoxide of MeDuPHOS (Figure 4.21). Due to the symmetry of MeDuPHOS, the two phosphorous atoms are equivalent, and the phosphorous atoms of MeDuPHOS and BozPHOS are not stereogenic.

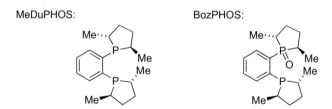

FIGURE 4.21   Methyl DuPHOS and BozPHOS ligands.

The examples in Table 4.7 reveal that conditions have been identified by which organozinc reagents can be added to either aromatic or (via the α-tosyl amine) aliphatic aldehyde imines. Note entries 4–6, in which the organozinc reagent is generated *in situ* from a Grignard reagent, thus expanding the versatility of the method beyond commercially available organozinc compounds. The aliphatic imines are also made *in situ*, first by reaction of the aldehyde with *p*-toluene sulfinic acid and Ph_2PONH_2, followed by elimination of *p*-toluene sulfinic acid during the reaction with the organozinc [160].[20]

TABLE 4.7 Catalytic Enantioselective Addition of Organozinc Reagents to Diphenylphosphinoyl Imines [130,160,204]

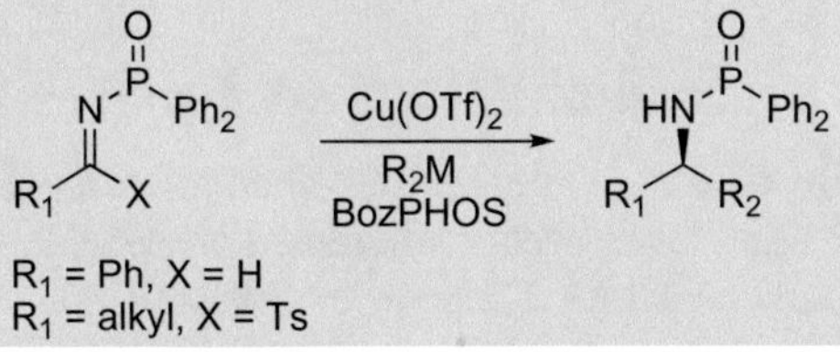

$R_1 = Ph, X = H$
$R_1 = alkyl, X = Ts$

Entry	R_1	R_2M	% Yield	er	Ref.
1	Ph	Me_2Zn	93	98:2	[160]
2	Ph	Bu_2Zn	92	98:2	[160]
3	Ph	$i\text{-}Pr_2Zn$	84	97:3	[160]
4	Ph	$EtMgCl/Zn(OMe)_2$	95	99:1	[130]
5	Ph	$i\text{-}PrMgCl/Zn(OMe)_2$	57	97:3	[130]
6	Ph	$BuMgCl/Zn(OMe)_2$	96	98:2	[130]
7	$Ph(CH_2)_2-$	Et	98	98:2	[160]
8	$Cy\text{-}C_6H_{11}-$	Et	89	98:2	[160]
9	$n\text{-}C_6H_{13}-$	Et	98	97:3	[160]

A mechanism has not been firmly established for these additions, but a predictive model is illustrated in Figure 4.22. It is thought that the alkyl group is first transferred to copper, which is bound to the phosphine of BozPHOS. A zinc atom chelates the P=O of BozPHOS and the P=O of the imine, and R_1–Cu adds across the *Si* face of the azomethine, as illustrated in Figure 4.22 [206]. Note that the azomethine lies in a horizontal plane in an orientation that avoids the two methyl groups of the BozPHOS.

---

20. For a copper catalyzed addition of organozinc compounds to *N*-formyl imines, generated *in situ* by elimination of *p*-toluenesulfinic acid, and using a monodentate phosphoramidite ligand, see ref. [205].

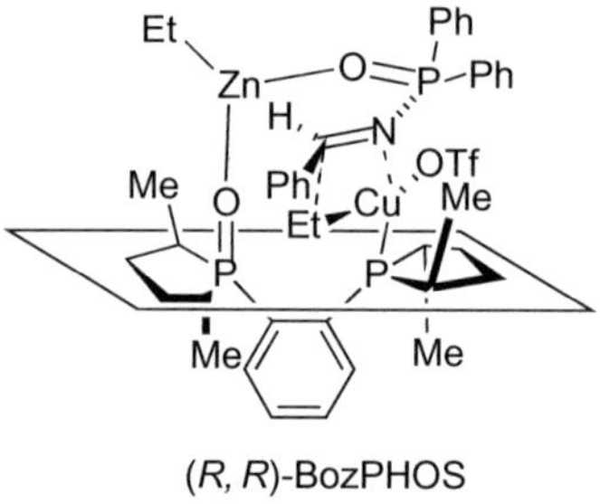

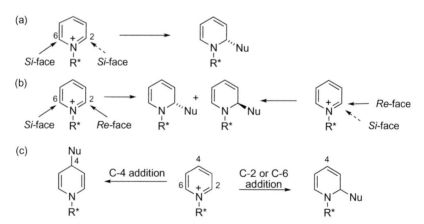

(R, R)-BozPHOS

FIGURE 4.22 Predictive model for the BozPHOS/copper-mediated addition of diethylzinc to diphenylphosphinoyl imines.

4.3.3 Additions of Organometallics to Pyridinium Ions

Stereoselective addition of Grignards to chiral pyridinium ions has been used to gain access to an important class of chiral heterocycles: substituted piperidines. Comins uses an *N*-acylpyridinium obtained by acylation with 2-cumylcyclohexanol [207−212], while Charette has developed an amidinium auxiliary derived from valinol [213−217].

Note that these processes are complicated by the symmetry of the ring system: *Si*-face attack at C-2 and *Si*-face attack at C-6 are equivalent (*i.e.*, the *Si*-faces of C-2 and C-6 are homotopic, Figure 4.23a). As a result of this equivalence, face selectivity at C-2 is topologically equivalent to regioselectivity (C-2 *vs.* C-6) from a single face. Thus, in a transition structure where (for example) attack of a nucleophile comes exclusively from the direction of the viewer, addition to C-2 and C-6 produce the same set of isomers that would result from attack at the front and back of only C-2 (Figure 4.23b). Addition at either C-2 or C-6 constitute a 1,2-addition, but an additional complication arises from the possibility of 1,4-addition at C-4 (Figure 4.23c).

FIGURE 4.23 Complications of pyridinium additions due to ring symmetry. (a) Homotopic faces of C-2 and C-6; (b) equivalence of 100% selective addition to only the front face with no regioselectivity and 100% regioselectivity with no face selectivity; (c) possibility of competing 1,2- *vs.* 1,4-addition.

Two approaches addressing the complications outlined above have been taken. First, the Comins team installed a large (removable) triisopropylsilyl blocking group at C-3, which hinders approach of a nucleophile toward C-2 and C-4 (Figure 4.24a). The phenyl moiety on the alcohol fragment of the carbamate chiral auxiliary then π-stacks with the pyridinium and blocks the front face, as illustrated in Figure 4.24b. For greater synthetic utility, the 4-position of the pyridine is often substituted with a methoxy group.

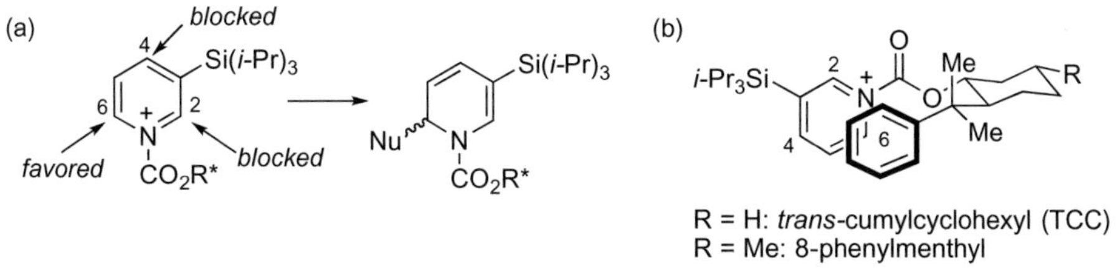

R = H: *trans*-cumylcyclohexyl (TCC)
R = Me: 8-phenylmenthyl

FIGURE 4.24   (a) A triisopropylsilyl group at C-3 of the pyridinium ion blocks approach of a nucleophile at C-2 and C-4; (b) Comins' conformational model favoring *Re*-face (back side) attack at C-6 of an acylpyridinium ion [218].

A complementary approach, using an amidine chiral auxiliary, has been used by the Charette group [213–217,219]. In this approach (Figure 4.25), the chiral auxiliary chelates the metal such that delivery to C-2 is favored over C-4, while intermolecular approach of a nucleophile to C-6 is blocked by the phenyl moiety of the auxiliary. An advantage of this approach is that this auxiliary obviates the need for a large blocking group at C-3. In fact, if there is a methoxy group at C-3, there remains a >95% regioselectivity for delivery of the nucleophile to C-2 [216]. The steric course of the addition can be rationalized by orientation

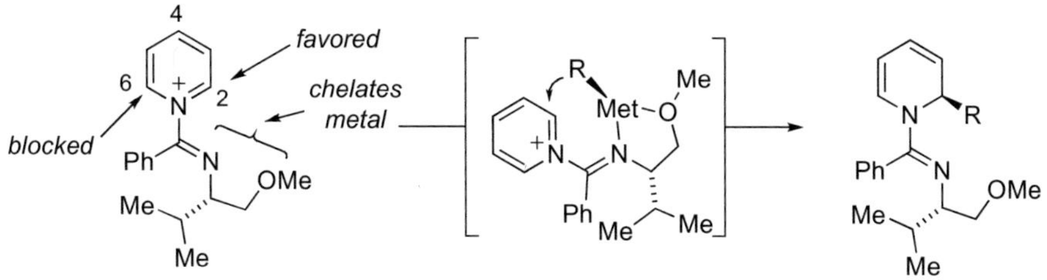

FIGURE 4.25   The amidinium chiral auxiliary developed by the Charette group.

of the nucleophilic R-group *anti* to the isopropyl group of the 5-membered chelate ring, similar to the additions illustrated in Figure 4.20. Note that, due to $A^{1,3}$-strain, it is unlikely that the phenyl and pyridinium rings are in the same plane as the 5-membered chelate ring.

Table 4.8 lists examples that illustrate the levels of diastereoselectivity; in entries 5–8, the regioselectivity for addition to C-2 is >95%. Figure 4.26 illustrates several alkaloids synthesized using these approaches.

TABLE 4.8 Asymmetric Additions to Chiral Pyridinium Salts

Entry	Educt	RM	% Yield	dr	Ref.
1	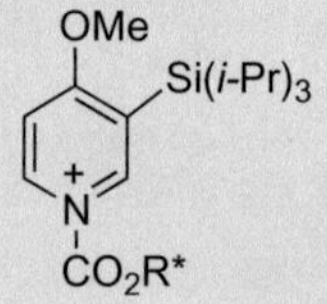	MeMgBr	92	95:5	[218]
2	"	i-BuMgBr	95	96:4	[218]
3	"	c-C₆H₁₁MgBr	90	90:10	[218]
4	"	PhMgBr	88	96:4	[218]
5	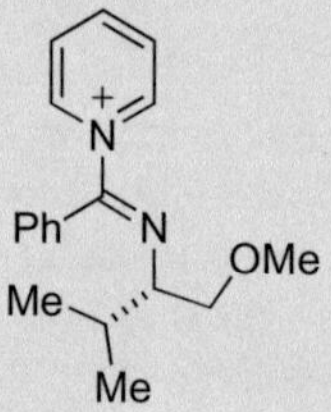	MeMgBr	77	>95:5	[213]
6	"	Et₂Zn	73	>95:5	[213]
7	"	PhMgBr	89	>95:5	[213]
8	"	C₄H₉C≡CMgBr	65	>95:5	[213]

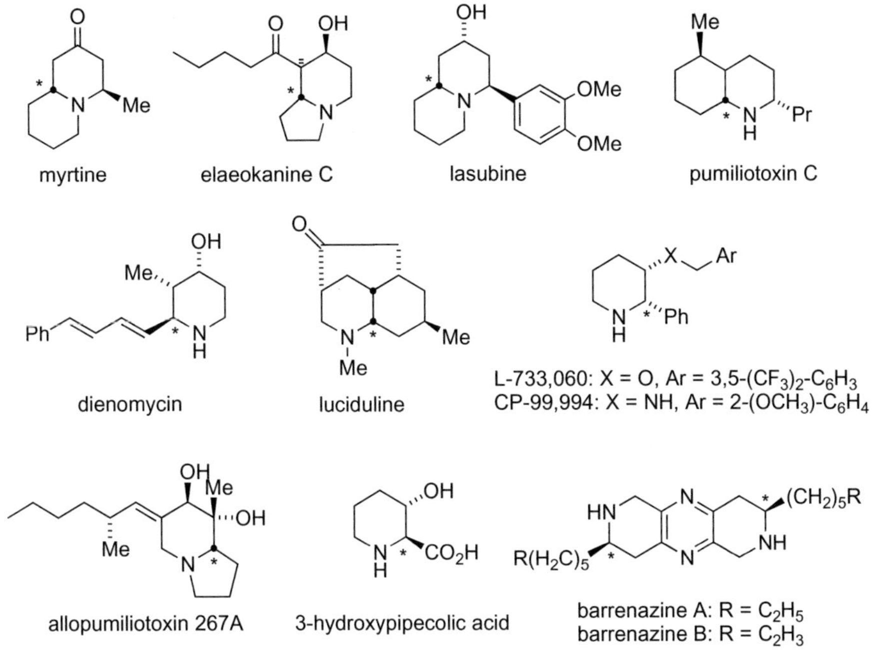

FIGURE 4.26   Alkaloids synthesized by asymmetric addition to chiral pyridiniums: myrtine ([220]; elaeokanine C [221]; lasubine I [220]; pumiliotoxin C [222]; dienomycin [223]; luciduline [224]; substance P antagonists L-733,060 and CP-99,994 [216]; allopumiliotoxin 267A [225]; 3-hydroxypipecolic acid [219]; and the barrenazines [215]. The stereocenter created in the addition reaction is indicated (*).

4.3.4 Hydrocyanations of Carbonyls

The addition of cyanide to an aldehyde or ketone (hydrocyanation) is an old reaction, but it has been the subject of renewed interest since Reetz's discovery that a chiral Lewis acid could be used to catalyze the asymmetric addition of trimethylsilylcyanide to isobutyraldehyde ([226]; reviews: [77,227−236]). The general process, illustrated in Scheme 4.16, usually employs trimethylsilylcyanide (Me_3SiCN), because hydrogen cyanide itself adds to carbonyls as well (nonselectively). An alcohol is slowly added to the reaction mixture, which reacts with TMS-CN to produce HCN or HNC *in situ* [237]. This slow addition can be a drawback to the utility of this reagent. Other cyanide sources include KCN/Ac_2O, and ethyl or methyl cyanoformate ($NCCO_2R$), and can obviate the disadvantages of TMS-CN/ROH reagents. As illustrated in Scheme 4.16, the catalysts often simultaneously bind both the cyanide ion and the aldehyde such that the addition is intramolecular [238]. These bifunctional catalysts have a Lewis acidic site to coordinate the carbonyl oxygen of the electrophile, and a Lewis basic site to bind the HCN or HNC. Successful catalysts are constructed to present only one heterotopic face of the carbonyl to the nucleophile. Catalytic turnover is achieved by reaction of the catalyst-bound cyanohydrin product with TMS-CN or $ROSiMe_3$.

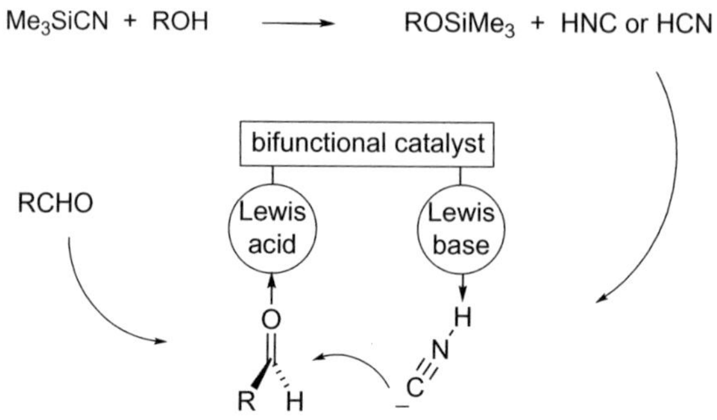

Me₃SiCN + ROH ⟶ ROSiMe₃ + HNC or HCN

RCHO

bifunctional catalyst

Lewis acid   Lewis base

SCHEME 4.16 Addition of trimethylsilylcyanide to an aldehyde on a bifunctional catalyst that has both Lewis acidic and Lewis basic binding sites.

Many catalysts have been screened for enantioselectivity, but few provide high enantios-electivity across a broad range of substrates. For example, many catalysts give good yields and enantioselectivities in the addition of Me₃SiCN to benzaldehyde, but other aryl aldehydes and aliphatic aldehydes fare poorly with the same catalysts. Below are described examples that are highly enantioselective, general in their substrate scope, and whose mechanism and steric course are well understood.

Figure 4.27 illustrates some of the ligands that, when coordinated to titanium, provide high enantioselectivity and are effective catalysts for addition to both aromatic and aliphatic aldehydes. Mechanistic work has shown that, at least for some titanium complexes, the cyanide ion is transferred first to titanium, and then in a bimetallic complex, to an aldehyde that is complexed to a second titanium [239–242].

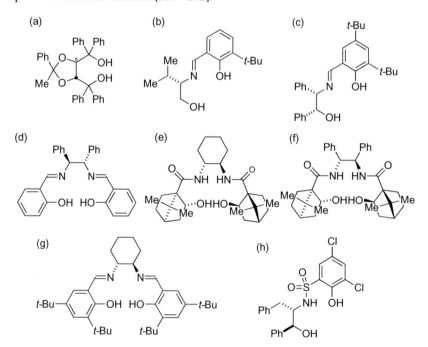

FIGURE 4.27 Ligands for the titanium-mediated hydrocyanation of aldehydes. (a) [243]; (b) [244]; (c) [245]; (d) [246]; (e) [247]; (f) [248]; (g) [241]; (h) [249].

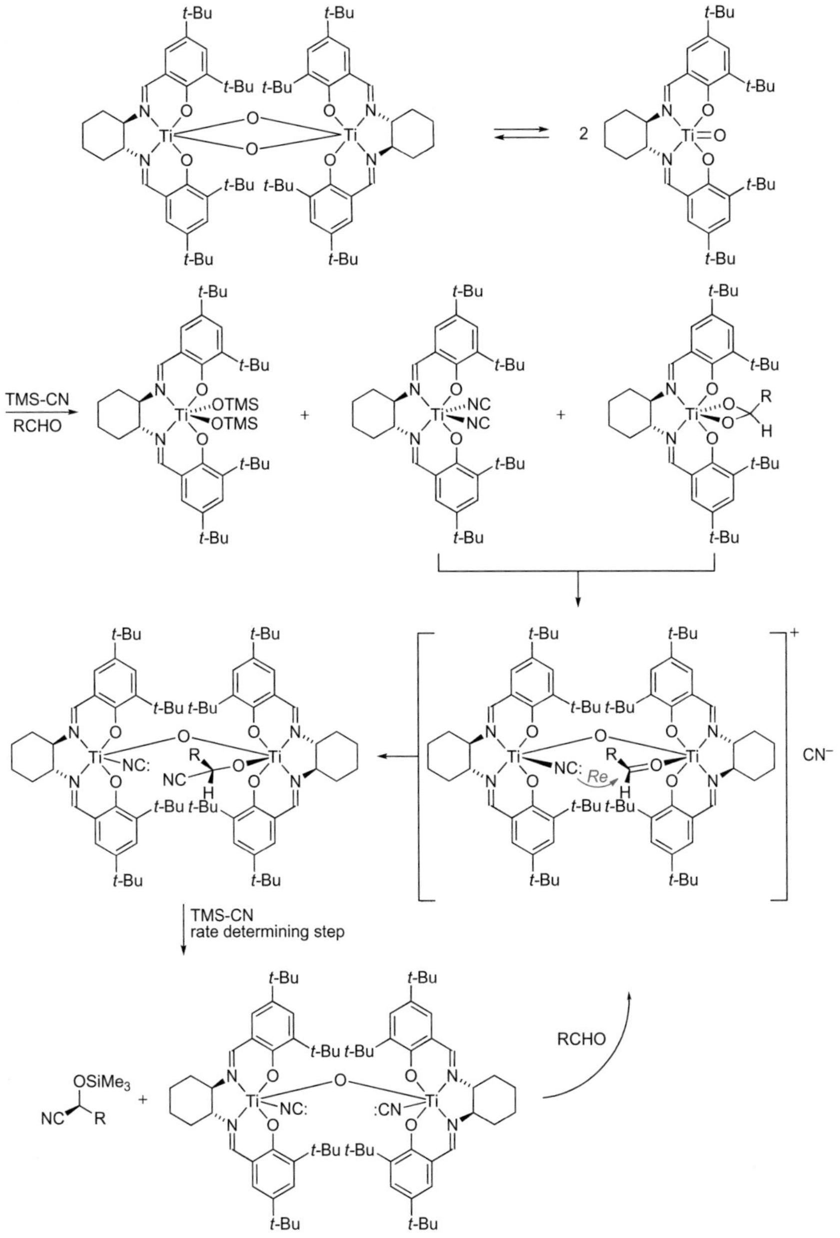

SCHEME 4.17   Mechanism and catalytic cycle for trimethylsilylcyanation of aldehydes with titanium salen catalyst [241,242,252].

Scheme 4.17 outlines the mechanism that has been determined for the salen ligand illustrated in Figure 4.27g, which can use KCN as the cyanide source in additions to aromatic and aliphatic aldehydes, as well as ketones [241,242,250–252]. The titanium complex undergoes a concentration and solvent-dependent dimerization; the kinetic order in the catalyst dimer is 1.3, indicating more than one metal center in the rate-determining step. Furthermore, the reaction is zero order in aldehyde. The dimeric precatalyst dissociates to a titanium oxide complex, which then reacts with TMS-CN to make a mixture of a bis-trimethylsilyloxy titanium complex and a bis-isonitrile complex. Separately, the titanium oxide complex adds to the aldehyde to form a titanium complex coordinated by the hydrated aldehyde. The hydrated aldehyde complex and the bis-isonitrile complex then form the critical, cationic, oxo-bridged bimetallic complex (with a CN^- gegenion) that has the isocyanate coordinated to one titanium, and the aldehyde coordinated to the other. Transfer of the cyanide ion to the *Re* face is accompanied by coordination of the cyanide to the titanium on the left of the figure. The rate-determining step is the reaction of this complex with TMS-CN to form the cyanohydrin and a new oxo-bridged bimetallic complex; addition of the aldehyde to the latter completes the catalytic cycle.

Using combinatorial methods, a bifunctional thiourea catalyst was developed in the Jacobsen group to catalyze the trimethylsilylcyanation of ketones [253,254]. Extensive kinetic studies and molecular modeling work was done to elucidate the mechanism and the steric course of this reaction. As shown in Scheme 4.18, the ketone is hydrogen-bonded by the two NH's of the thiourea (the Lewis acidic site), with the smaller R-group oriented toward the amide NH. The trifluoroethanol solvent converts the TMS-CN to HCN, which is coordinated to the tertiary amine nitrogen on the 6-membered ring (the Lewis basic site), such that the cyanide ion is delivered intramolecularly to the carbonyl carbon.

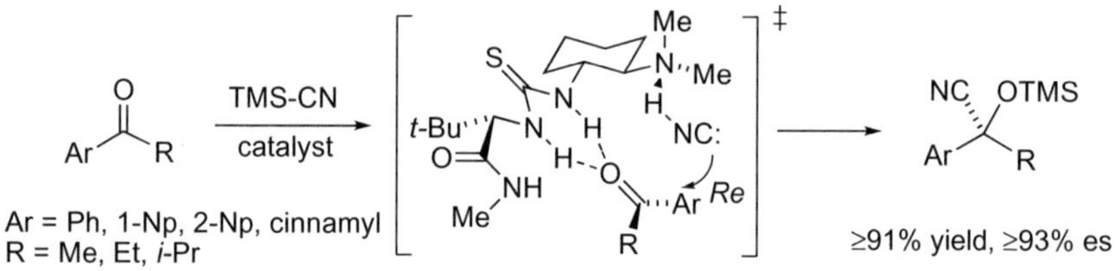

SCHEME 4.18   Thiourea catalyzed trimethylsilylcyanation of ketones [253,254].

A monofunctional oxazaborolidine has been used to catalyze trimethylsilylcyanations (Scheme 4.19), when triphenylphosphine oxide is present as an additive [255].[21] NMR and IR evidence suggest that the phosphine oxide reacts with TMS-CN to make a hypervalent phosphorous isonitrile, $Ph_3P(OTMS)(NC:)$.[22] In the absence of the triphenylphosphine oxide additive, the products were nearly racemic. The steric course of the reaction can be rationalized by coordination of the aldehyde oxygen to the boron atom on the convex face of the 5−5 ring system. A formyl hydrogen bond [259] to the oxygen then exposes the *Si* face of the aldehyde to the nucleophile.

---

21. Phosphine oxide additives are also necessary in some aluminum-BINOL catalyzed trimethylsilylcyanations. See refs. [256,257].
22. See also ref. [258].

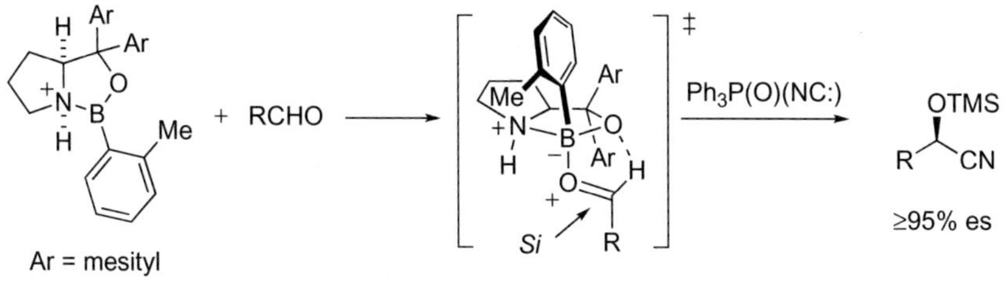

Ar = mesityl

SCHEME 4.19   Oxazaborolidine catalyzed trimethylsilylcyanation of aldehydes [255].

4.3.5 Hydrocyanations of Azomethines (the Strecker Reaction)

The addition of HCN to an azomethine (imine), the Strecker reaction, was discovered in the mid-nineteenth century, and is a common way to prepare α-amino acids by hydrolysis of the α-aminonitrile addition product. Although asymmetric hydrogenation of dehydro amino acids offers an excellent method for synthesis of amino acids in some cases, there are structural limitations (see Section 7.2). The Strecker reaction offers a complementary method for asymmetric synthesis of α-amino acids. Only recently has the Strecker reaction been rendered enantioselective, and some highly effective catalysts have been developed (reviews: [158,229,233,234,238,260,261]).[23] As is true of catalytic cyanide addition to carbonyls, few catalysts give high levels of enantioselectivity for both aliphatic and aromatic imines in the Strecker reaction, and the successful catalysts are often bifunctional (see Scheme 4.16).

Using combinatorial screening, the Jacobsen group has discovered a series of urea and thiourea catalysts for the Strecker reaction [263–268]. In the optimized catalyst (Scheme 4.20), the cyanide can be generated either by *in situ* reaction of TMS-CN with methanol, or by potassium cyanide and acetic acid. In the latter instance, the reaction is scalable to at least multigram scale [268]. The combinatorial approach to catalyst discovery can reveal effective catalysts [269], even without any insight into how they work. In the present case, mechanistic insight came only after catalyst optimization, and extensive kinetic and computational studies [267]. A study of several catalyst structures revealed a high degree of correlation between the rate of addition and the degree of enantioselectivity. This led to the conclusion that the catalyst stabilizes the transition structure leading to the major enantiomer, while simultaneously destabilizing the transition structure leading to the minor enantiomer. Using DCN as the cyanide source, NMR monitoring of the reaction revealed that the HCN is the source of the *N*-proton of the α-aminonitrile, *not* the catalyst. Furthermore, a linear free energy study of substituted benzaldehyde imines revealed negative ρ-values, consistent with a positively charged iminium ion in the transition state.

The proposed mechanism entails complexation of HNC to the thiourea followed by proton transfer to make a series of three iminium/cyanide ion pair structures. In the first, the iminium ion is hydrogen-bonded to the nitrogen of the complexed cyanide, while the carbon is hydrogen-bonded to the thiourea. The cyanide ion then rearranges such that the nitrogen becomes complexed to the thiourea. In the rate-determining step, the iminium ion migrates to the amide carbonyl ion to make the key iminium/cyanide ion pair (boxed structure), which is stabilized by several noncovalent interactions. The cyanide then adds to the iminium ion in the enantiomer-determining step. Catalyst turnover is achieved by release of the α-aminonitrile addition product.

---

23. For an auxiliary-based approach, using *p*-tolylsulfinimines and diethylaluminum cyanide, see ref. [262].

Note the contrast with the bifunctional thiourea catalyst for trimethylsilylcyanation of carbonyls described in Scheme 4.18, in which the Lewis acidic thiourea binds the electrophile, whereas here, the thiourea binds the nucleophile. The difference is that, in the boxed structure of Scheme 4.20, the electrophile is a protonated iminium ion, which requires a Lewis basic site for coordination. The HNC is coordinated to the Lewis acidic site through the Lewis basic (formally negatively charged) carbon.

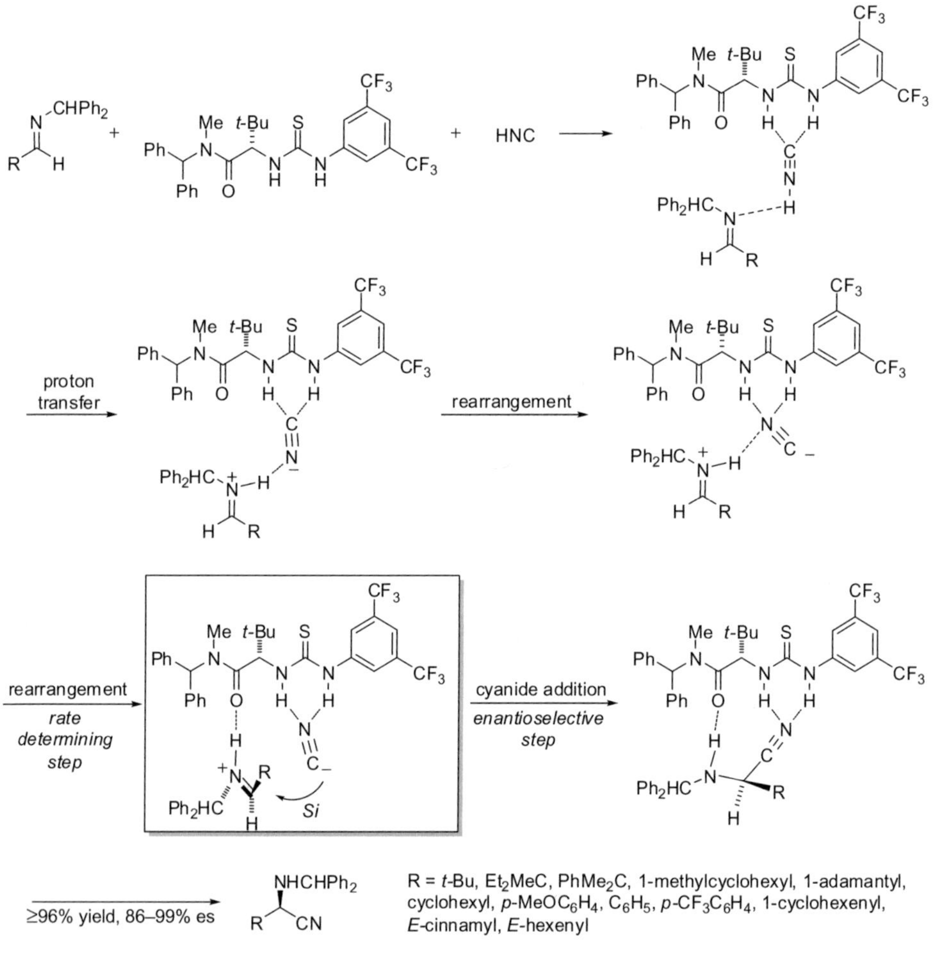

SCHEME 4.20 Mechanism and examples of thiourea catalyzed Strecker reaction [267,268].

Snapper and Hoveyda have used combinatorial methods to identify a novel class of titanium-bound Schiff base tripeptide catalysts for the asymmetric Strecker reaction [270–272]. The optimal catalyst is shown in Scheme 4.21. Mechanistic studies revealed that the reaction is first order in catalyst, but zero order in substrate, TMS-CN, and alcohol [272]. Moreover, the enthalpy of activation is very large and negative, $\Delta S^{\ddagger} = -45.6 \pm 4.1$ cal/mol·K indicating a highly ordered transition state. By varying the amino acid components of the tripeptide, it was determined that the amide carbonyl of the central amino acid (*O-tert*-butyl threonine) plays a critical role as the Lewis basic site, involving a rapid and reversible hydrogen bond formation with the HNC, followed by rate-limiting delivery of cyanide to the *Re* face of the imine that is bound to the Lewis acidic titanium. Molecular modeling suggest that coordination of the imine to the top face of the metal is favored over the bottom face. Note that the imine is bound to titanium with the R-group oriented away from the large *tert*-butyl substituent on the peptide backbone, thus minimizing steric repulsion. In this orientation, the carbonyl-bound HNC is readily delivered to the *Re* face of the C=N bond.

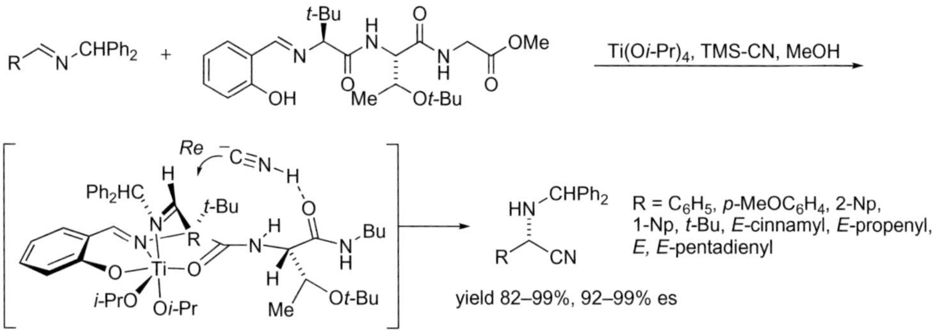

SCHEME 4.21   Snapper/Hoveyda titanium-tripeptide catalyst for the Strecker reaction [270−272].

4.4 CONJUGATE ADDITIONS

Two strategies have been used for asymmetric 1,4-additions: those that are based on a chiral auxiliary that is covalently attached to one of the reactants, and catalytic additions that rely on chiral ligands on a metal (reviews: [203,274−296]).[24] Most of the efforts in recent years have focused on the development of new catalysts, and some of the more selective catalysts are illustrated in Figure 4.28. Reviews of conjugate additions are usually organized according to the ligand or catalyst employed. In contrast, the following discussion is organized by electrophile, and includes both auxiliary-based and catalytic approaches.

---

24. For an early monograph on conjugate additions, see ref. [273].

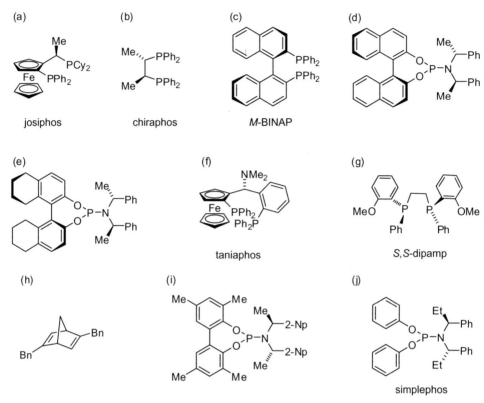

FIGURE 4.28 Ligands for transition metal catalyzed conjugate additions.

4.4.1 Acyclic Esters and Ketones

Since esters exhibit a strong preference for a conformation in which an alkoxy C—H is synperiplanar to the carbonyl, the job of an auxiliary is to then project an appendage back over the enoate π-system, leaving only one face open to preferential approach by a nucleophile. Figure 4.29 illustrates three of the more selective auxiliaries for this purpose. These auxiliaries are illustrated with the esters in their most stable conformations, with the alkoxy C—H and the carbonyl synperiplanar, and the enoate in the s-trans conformation. Presumably the ground state preference for this conformation is also felt in the transition state, which has the rear face shielded. Table 4.9 lists several examples of asymmetric additions to E-enoates. Less success has been realized in asymmetric additions to Z-enoates and to di- and trisubstituted double bonds.

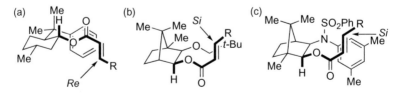

FIGURE 4.29 Chiral auxiliaries for asymmetric 1,4-addition to (the illustrated front face) of esters. Note the C—H/C=O coplanarity and the s-trans enone in the illustrated ground state conformations. (a) [297,298]; (b) [298]; (c) [299].

TABLE 4.9 Asymmetric 1,4-Addition to Unsaturated Esters of Chiral Alcohols

Entry	R	OR*[a]	Nucleophile	% Yield	dr	Ref.
1	Me ·	4.29a	PhCuBF$_3$	76	>99:1	[297]
2	"	4.29c	"	97	>99:1	[299]
3	"	4.29c	VinylCuBF$_3$	94	>99:1	[299]
4	"	4.29c	EtCuBF$_3$	90	>99:1	[299]
5	"	4.29a	BuCuBF$_3$	75	>99:1	[297]
6	"	4.29b	Me$_2$C=C(CH$_2$)$_2$−CuP(Bu)$_3$BF$_3$	81	99:1	[298]
7	"	4.29c	i-PrCuBF$_3$	92	>99:1	[299]
8	Et	4.29c	MeCuBF$_3$	86	99:1	[299]
9	Bu	4.29a	MeCuP(Bu)$_3$BF$_3$	96	93:7	[298]
10	"	4.29b	"	82	97:3	[298]
11	n-C$_8$H$_{17}$	4.29b	"	90	99:1	[298]
12	i-Pr	4.29c	MeCuBF$_3$	92	>99:1	[299]

[a]Entries in the OR* column refer to Figure 4.29.

Several catalysts have been developed for 1,4-addition of organometallics to esters and acyclic ketones. Some of the more popular ligands for esters and ketones are illustrated in Figure 4.28a−e, and selected examples of conjugate additions to esters from the Feringa and Oestrich groups are listed in Table 4.10. Ferrocene-based diphosphine ligands such as josiphos are useful for catalyzing the additions of Grignard/copper(I) reagents; a limitation is that only Grignard reagents derived from primary alkyl halides work well (entries 1−5). BINAP mediates the rhodium-catalyzed addition of diphenylmethyl silane as its pinacol boronate (entry 6), a process that is useful because the silane can be stereoselectively replaced with a hydroxyl group by Fleming oxidation [300]. Organozinc/copper(II) reagents add selectively to α,β-unsaturated malonate esters and Meldrum's acid derivatives (entries 7−9), to provide quarternary stereocenters in excellent yield and enantioselectivity.

TABLE 4.10 Catalyzed 1,4-Additions of Organometallic Nucleophiles to Esters

Entry	Ester	Nucleophile	Ligand[a]	% Yield	er	Ref.
1	Me⌁CO₂Me	EtMgBr·CuBr	4.28a	99	97:3	[301]
2	"	Me₂(CH₂)₂MgBr·CuBr	4.28a	98	98:2	[301]
3	"	CH₂=CH(CH₂)₂MgBr·CuBr	4.28a	85	92:8	[301]
4	Cy⌁CO₂Me	EtMgBr·CuBr	4.28a	86	99:1	[301]
5	Ph⌁CO₂Me	"	4.28a	94	99:1	[301]
6	C₄H₉⌁CO₂Me	Me₂PhSi–BPin·Rh(I)	4.28c	60	>99:1	[302]
7	Pr⌁CO₂Et / CO₂Et	Me₂Zn·Cu(II)	4.28e	98	97:3	[303]
8	(structure with Ph, Me, O)	Et₂Zn·Cu(II)	4.28d	95	92:8	[304]
9	(Meldrum-type structure with indanylidene)	Me₂Zn·Cu(II)	4.28d	98	>99:1	[304]

[a]*Numbers in the ligand column refer to Figure 4.28.*

The catalysts in Figure 4.28 are sufficiently selective such that they can be effectively used to control the absolute configuration in cases of double asymmetric induction, as shown by the examples in Table 4.11.[25] Entries 1 and 2 show the matched and mismatched pair of additions of methyl Grignard/copper(I), catalyzed by the two enantiomers of josiphos. Entries 3 and 4 illustrate *syn* and *anti* addition of dimethylzinc/copper(II) to malonate esters, with identical selectivity, mediated by a phosphoramidite ligand. Entries 5 and 6 illustrate the matched and mismatched addition of methyl Grignard/copper(I) to the homoallylic silane illustrated, using the enantiomers of josiphos. These last two examples illustrate how the relative configuration of 1,3-methyl/hydroxyl diastereomers can be obtained selectively after Fleming oxidation [300].

---

25. Recall the discussion in Chapter 1 on reagent-based stereocontrol (Section 1.6).

TABLE 4.11 Examples of Reagent-Based Stereocontrol in the 1,4-Addition of Organometallics to Esters

Entry	Ester	Nucleophile	Ligand[a]	Product	% Yield	dr	Ref.
1	Me / Bu / O / SEt (enoate)	MeMgBr·CuBr	4.28a	Me Me O / Bu / SEt	83	96:4	[305]
2	"	MeMgBr·CuBr	ent-4.28a	Me Me O / Bu / SEt	77	92:8	[305]
3	Me / Et / CO₂Et / CO₂Et	Me₂Zn·Cu(II)	4.28e	Me Me / Bu / CO₂Et / CO₂Et	100	97:3	[303]
4	"	"	ent-4.28d	Me Me / Bu / CO₂Et / CO₂Et	100	97:3	[303]
5	SiMe₂Ph O / Bu / SEt	MeMgBr·CuBr	4.28a	PhMe₂Si Me O / Bu / SEt	80	>95:5	[302]
6	"	"	ent-4.28a	PhMe₂Si Me O / Bu / SEt	80	85:15	[302]

[a]Numbers in the ligand column refer to Figure 4.28.

The Miyaura group has developed methods for the enantioselective 1,4-addition of aryl groups to acyclic ketones, starting with either aryltrifluoroborate salts or triaryl bismuth compounds mediated by Pd(II) or Cu(II), respectively, with the chiraphos ligand [306,307]; several examples are listed in Table 4.12, entries 1–4 [307–310]. The mechanism of the reaction is discussed later, in Section 4.4.3 on cyclic ketones (Scheme 4.26). The Feringa and Minnaard groups have developed methodology for the enantioselective copper-mediated conjugate addition of Grignard reagents to acyclic enones using the josiphos ligand, as shown in Scheme 4.22 [289,311–313]. The regioselectivity for 1,4- over 1,2-addition is good, as are the enantioselectivities. Several examples are listed in Table 4.12, entries 5–10. The 1,2-addition products were found to be racemic in all cases, indicating a nonselective reaction of the Grignard reagent that is not mediated by the copper.

The fact that esters react more slowly than ketones allowed study of the mechanism of the Grignard/copper(I) additions. The Feringa group elucidated the mechanism illustrated in Scheme 4.22 for the josiphos-mediated addition to enones [313]. The resting state of the josiphos·CuBr complex is the dimer shown at the top of the scheme. Addition of Grignard reagent generates the active catalyst in which the copper adopts a distorted tetrahedral geometry. The enone approaches the copper from the least hindered side allowing the copper to

TABLE 4.12 Conjugate Additions to Acyclic Ketones

Entry	Enone	Nucleophile	Ligand[a]	% Yield	er (R:S)	Ref.
1	C_5H_{11} ⟶ O, Me	$PhBF_3K \cdot Pd(II)$	4.28b	93	9:91	[307]
2	i-Pr ⟶ O, Me	$PhBF_3K \cdot Pd(II)$	4.28b	65	9:91	[307]
3	"	$Ph_3Bi + Cu(BF_4)_2$	4.28b	63	8:92	[307]
4	Ph ⟶ O, Me	$3\text{-MeOC}_6H_4BF_3K \cdot Pd(II)$	4.28b	90	3:97	[307]
5	C_5H_{11} ⟶ O, Me	$PrMgBr \cdot CuBr$	4.28a	84	95:5	[312]
6	C_4H_9 ⟶ O, Me	$EtMgBr \cdot CuBr$	4.28a	91	95:5	[312]
7	"	$MeMgBr \cdot CuBr$	4.28a	86	99:1	[312]
8	Me ⟶ O, Me	$BuMgBr \cdot CuBr$	4.28a	78	96:4	[312]
9	i-Pr ⟶ O, Me	$MeMgBr \cdot CuBr$	4.28a	52	93:7	[312]
10	Ph ⟶ O, Me	$MeMgBr \cdot CuBr$	4.28a	73	98:2	[312]

[a]The numbers in the ligand column refer to Figure 4.28.

reversibly coordinate to the π-bond of the enone while the magnesium coordinates the carbonyl oxygen in a π-complex. At the bottom of the scheme is a perspective drawing illustrating the postulated square-pyramidal geometry of the π-complex. This geometry is stabilized by the coordination of the double bond to the copper and by the coordination of the magnesium to the carbonyl oxygen, which adopts a skewed conformation. Rearrangement then gives the unstable Cu(III) σ-complex shown in the lower left corner of the scheme. It is likely that the σ- and π-complexes are in a rapid equilibrium. In the rate-determining step, reductive elimination from the σ-complex produces the magnesium bromide enolate product as a second Grignard adds to the CuBr · josiphos complex to regenerate the active catalyst.

Chong has shown that diiodobinaphthol catalyzes the enantioselective addition of vinyl boronate esters to unsaturated ketones, as shown in Table 4.13 [314]. The reaction has some notable features. First of all, as little as 3 mol% of the binaphthol is required to effect addition, which is conducted at 40 °C in dichloromethane. Moreover, aliphatic diols such as diisopropyl tartrate failed to catalyze the reaction. Water and methanol inhibit the reaction, but this effect

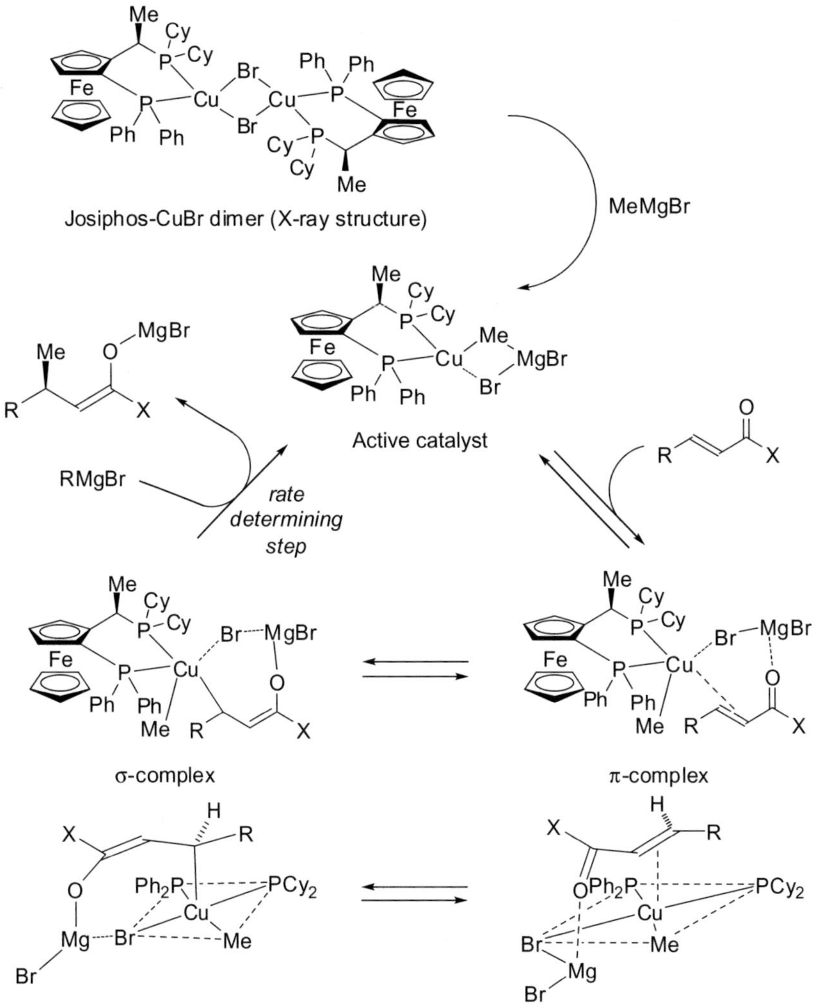

SCHEME 4.22 Mechanism of the josiphos · CuBr-catalyzed conjugate addition of Grignard reagents to enones [313].

can be suppressed by the addition of molecular sieves to the reaction mixture. Unsubstituted binaphthols and binaphthols with electron donating substituents such as a methyl group catalyze the reaction, but only in very low yield. Electron withdrawing substituents on the binaphthol work much better. The optimal catalyst was found to be the diiodobinaphthol illustrated in Scheme 4.23. The mechanism of the reaction involves transesterification of the dimethyl boronate ester with *M*-diiodobinaptol to make an electron deficient binaphthol boronate ester, which is more Lewis acidic than an aliphatic ester, and therefore more reactive. Calculations done by Paton, Goodman, and Pellegrinet reveal that the transition structure is a sofa-like 6-membered ring that, although there is no stereochemical consequence, shields the *Si* face of the vinyl group, which then adds to the *Re* face (R$_2$ = Ph) of the enone [315]. A second transesterification with the dimethyl boronate ester generates the dimethoxy boron enolate of the ketone and releases the binaphthol to complete the catalytic cycle.

TABLE 4.13 Asymmetric Conjugate Addition Vinyl Boronate Esters to Enones [314][a]

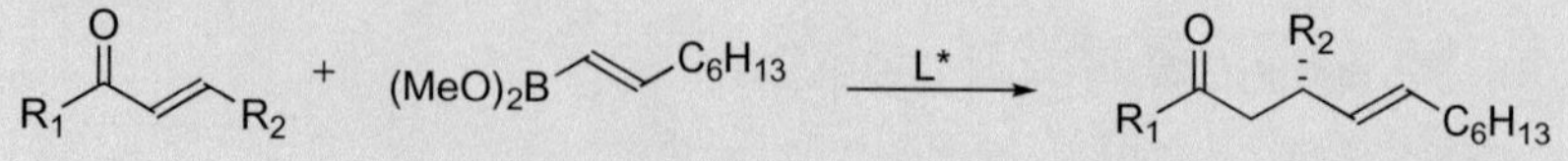

Entry	R_1	R_2	% Yield	er
1	Ph	Ph	93	99:1
2	Ph	Cinnamyl	84	99:1
3	Ph	Me	95	98:2
4	Ph	i-Pr	92	97:3
5	Me	Ph	81	99:1

[a]For the structure of L*, see Scheme 4.23.

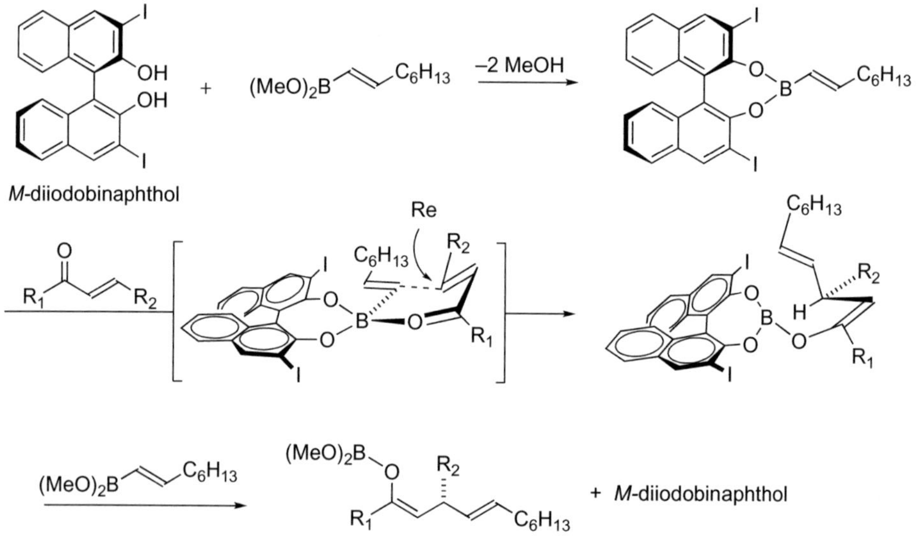

SCHEME 4.23 Mechanism of the diiodobinaphthol-catalyzed alkenylation of enones [314,315].

Figure 4.30 illustrates several natural products synthesized by asymmetric addition of organometallic compounds to esters. Particularly noteworthy is the ability of the method to produce the correct relative and absolute configuration of the alkyl branches on these acyclic frameworks. The illustrated structure for norpectinatone is the one originally postulated [316], but was proven incorrect by asymmetric synthesis [317]. Iterative addition allows for the formation of the all-*syn* deoxypropionate lipids found in (−)-lardolure, (−)-mycocerosic acid, and phthioceranic acid.

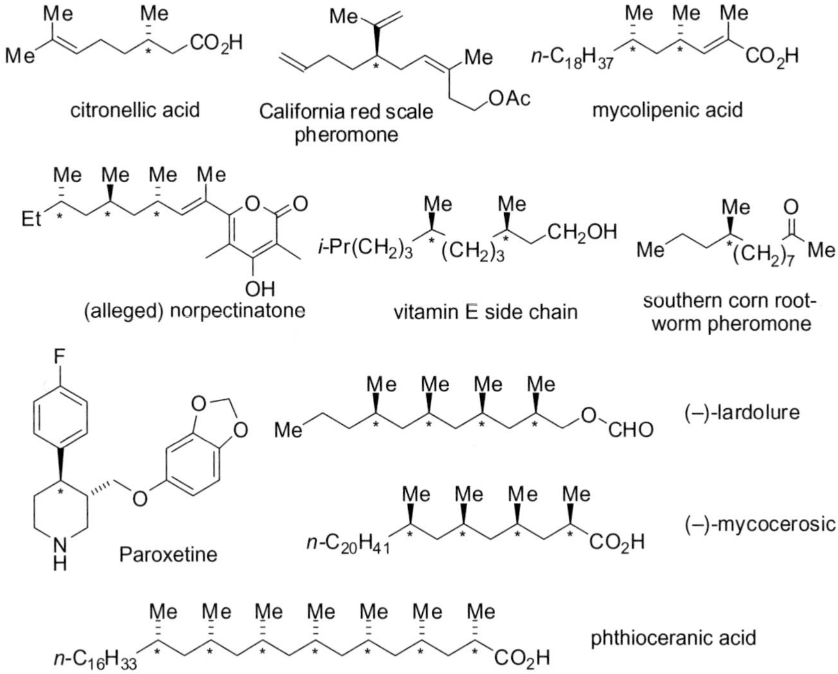

FIGURE 4.30 Natural products synthesized by asymmetric 1,4-addition of cuprates to esters: citronellic acid [298]; California red scale pheromone [318]; mycolipenic acid (W. Oppolzer; T. Godel, unpublished, quoted in [319]); the alleged norpectinatone [317]; vitamin E side chain (W. Oppolzer; R. Moretti, unpublished, quoted in [319]); southern corn rootworm pheromone [320]; paroxetine [321]; (−)-lardolure [322]; (−)-mycocerosic acid [323]; phthioceranic acid [324]. Stereocenters created in the asymmetric conjugate addition are marked (*).

4.4.2 Acyclic Amides and Imides

A number of chiral auxiliaries have been developed in conjugate additions of organometallics to acyclic enamides [275]. Some are listed in Scheme 4.24, along with the rationale for the steric course of the addition. Examples using these auxiliaries are listed in Table 4.14. Both ephedrine and pseudoephedrine amides of butenoic acid have been used. Mukaiyama's ephedrine amides (Scheme 4.24a) require excess Grignard, and work best with organomagnesium *bromides* [325]. The transition structure shown was proposed by the authors to account for the configuration of the product and involves chelation by a magnesium aggregate to the enone carbonyl and a heteroatom on the auxiliary. This chelation reduces conformational motion in the ground state as well as the transition state, and reduces the possible number of competing nucleophile approach trajectories. The stereocenters on the auxiliary are quite remote from the site of attack. Although attack on the face opposite the methyl and phenyl groups in this chelate (as drawn) accounts for the configuration of the product, it is not clear how this steric effect is transmitted across the metallocycle chelate to the external double bond. It may be that the methyl and phenyl substituents induce a curved shape to the chelate ring that favors approach from the convex face, or perhaps the substrate is an aggregate of unknown structure.

Addition of organolithium compounds to amides of pseudoephedrine (Scheme 4.24b) effect addition to the same face of the β-carbon as the ephedrine, but the authors propose a different rationale to explain the steric course of the addition [326]. In this case, a

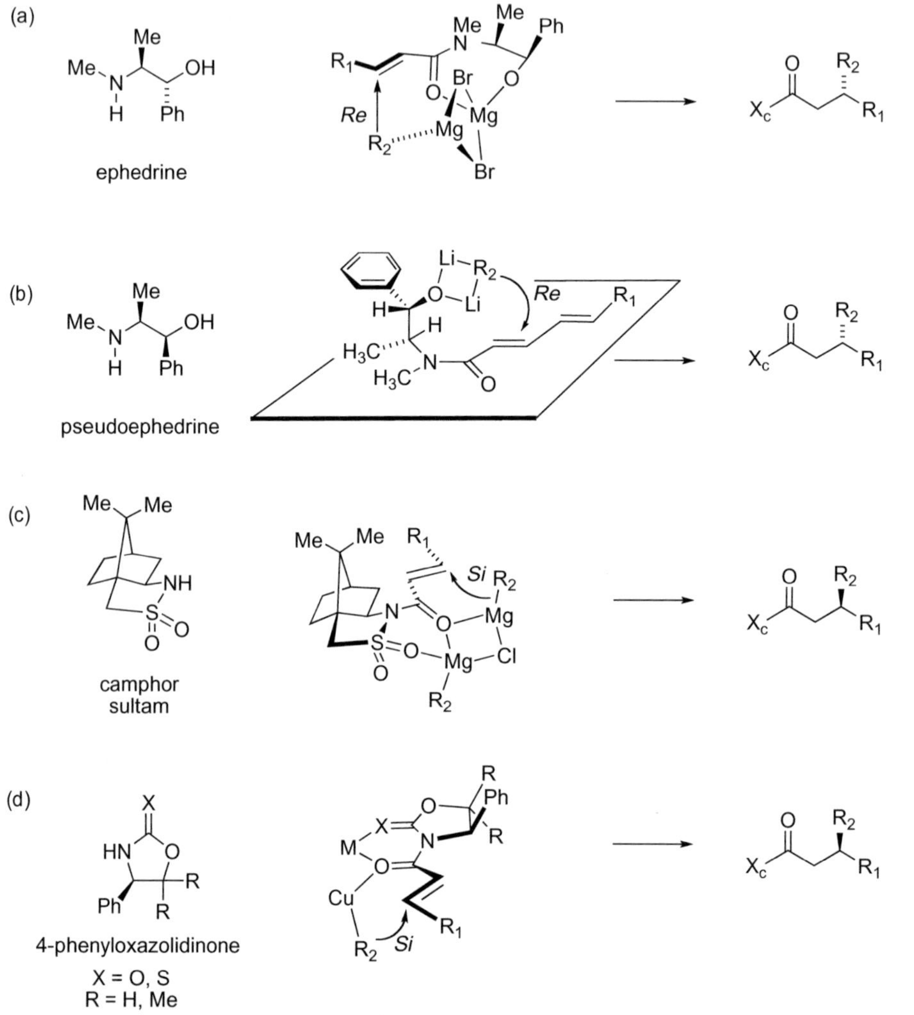

SCHEME 4.24 Auxiliaries for the asymmetric 1,4-addition of Grignards, organolithiums, and cuprates to acyclic amides, with a rationale for the observed steric course of the reaction. (a) Ephedrine [325]; (b) pseudoephedrine [326]; (c) sultam [330,331]; (d) 4-phenyloxazolidinones (X = O, R = H) [332,334]; 4-phenyloxazolidinethiones (X = S, R = H, Me) [333].

conformation of the pseudoephedrine amide is thought to be similar to the enolate structures proposed by Myers (Scheme 3.23 [327−329]). In this instance, the organolithium could aggregate with the lithium alkoxide, and be directed to the upper *Re* face of the enone. Complexation of the organolithium with the lithium alkoxide could also account for the preponderance of 1,4-addition products. The results of the ephedrine reaction with a Grignard could also be rationalized using this model while exchanging the positions of the phenyl and hydrogen of the pseudoephedrine.

For the sultam (Scheme 4.24c), the situation is more clear: the bridge methyls of the camphor hinder approach from the upper *Re* face [330,331], similar to the situation with enolate alkylation of the same auxiliary (Scheme 3.22).

Additions to the oxazolidinone and oxazolidinethione auxiliaries (Scheme 4.24d) are rationalized by positing chelation of the amide oxygen with the oxygen or sulfur of the heterocycle, then adding the cuprate to the face opposite the phenyl substituent [332,333]. In all four of these rationales, note that the conformation of the enone is *s-cis*, which avoids $A^{1,3}$-strain that would destabilize an *s-trans* conformation.

TABLE 4.14 Asymmetric 1,4-Additions to Enamides

Entry	R_1	Nucleophile	Auxiliary[a]	% Yield	dr	Ref.
1	Me	PhMgBr	4.24a	63	95:5	[325]
2	Me	EtMgBr	4.24a	79	98:2	[325]
3	Ph	"	4.24a	48	98:2	[325]
4	Et	PhMgBr	4.24a	76	93:7	[325]
5	Bu	EtMgBr	4.24a	69	99	[325]
6	Et	BuMgBr	4.24a	59	79:21	[325]
7	Me (1,4-addition)	PhLi	4.24b	80	97:3	[326]
8	H (1,4-addition)	PhLi	4.24b	65	97:3	[335]
9	Me	EtMgCl	4.24c	80	94	[330]
10	Me	i-Pr	4.24c	92	86	[330]
11	Et	BuMgCl	4.24c	89	95	[330]
12	Ph	EtMgBr	4.24c	83	95:5	[331]
13	Ph	PrMgBr	4.24c	87	97:3	[331]
14	Ph	i-PrMgBr	4.24c	59	80:20	[331]
15	Me	EtMgBr CuBr + BF₃	4.24d X = O, R = H	89	83:17	[332]
16	Me	EtMgBr CuBr + Me₂AlCl	4.24d X = O, R = H	88	94:6	[332]
17	Me	i-PrMgBr CuBr + BF₃	4.24d X = O, R = H	99	90:10	[332]
18	Me	PhMgBr CuBr + BF₃	4.24d X = O, R = H	65	>98:2	[332]
19	Me	PhMgBr CuBr + TMSI	4.24d X = S, R = H	70	90:10	[333]
20	Me	PhMgBr CuI + TMSI	4.24d X = S, R = Me	70	90:10	[333]
21	Me	EtMgBr CuI + TMSI	4.24d X = S, R = Me	85	98:2	[333]

[a]Numbers in the auxiliary column are illustrated in Scheme 4.24 along with a rationale for the steric course.

The enantiomer of the phosphoramidite ligand illustrated in Figure 4.28d was evaluated by Pineschi in the copper-mediated conjugate addition of dialkylzinc compounds to a number of acyclic amides and imides. Of several amides and imides tested, the imide derived from crotonic acid and 2-pyrrolidinone proved optimal. Surprisingly, pyrrolidine amides and N,N-diethyl amides failed to react. Moreover, imides derived from 6- and 7-membered lactams gave poor enantioselectivity. The scope of the conjugate addition was then evaluated; selected results are listed in Table 4.15.

TABLE 4.15 Catalytic Enantioselective Additions of Dialkylzinc Compounds to Acyclic Imides [336][a]

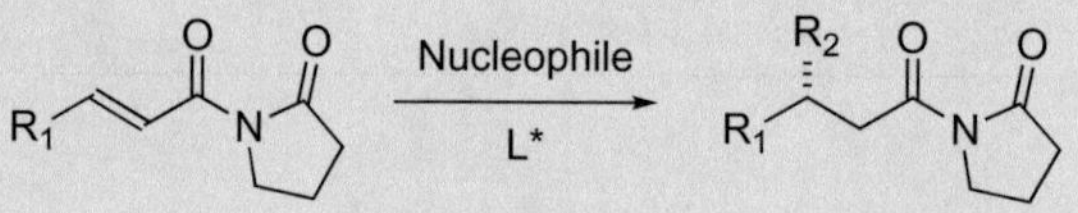

Entry	R_1	Nucleophile	% Yield	er
1	Pr	$Et_2Zn \cdot Cu(II)$	80	92:8
2	Me	$i\text{-}Pr_2Zn \cdot Cu(II)$	78	80:20
3	i-Pr	$Et_2Zn \cdot Cu(II)$	75	97:3
4	Ph	"	78	99:1
5	Ph	$i\text{-}Pr_2Zn \cdot Cu(II)$	75	90:10

[a]The chiral ligand, L*, is the enantiomer of that illustrated in Figure 4.28d.

Natural products that involve conjugate additions to amides and imides as the key step are shown in Figure 4.31. The bakuchiol example is particularly noteworthy because of the highly diastereoselective (95:5) formation of a quaternary stereocenter.

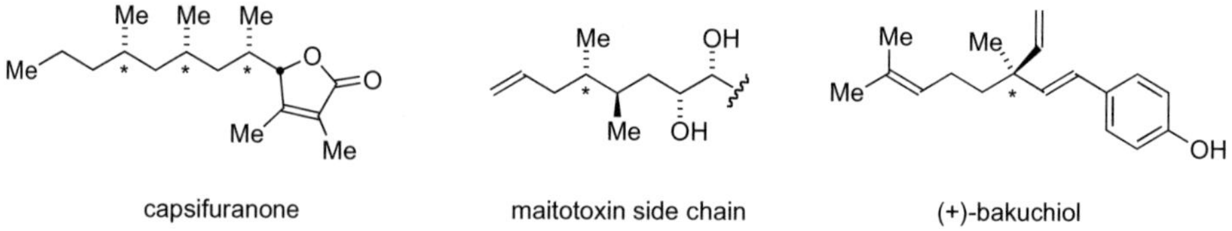

capsifuranone   maitotoxin side chain   (+)-bakuchiol

FIGURE 4.31 Natural products formed by conjugate addition of organometallics to acyclic imides: capsifuranone [337]; maitotoxin side chain [338]; (+)-bakuchiol [334].

4.4.3 Cyclic Ketones and Lactones

Incorporation of an auxiliary into a cyclic system has been used for the diastereoselective addition of cuprates to unsaturated 6-membered ring dioxinones, which are perhaps less important for their synthetic potential than for the mechanistic insight they provide. The dioxinones shown in Scheme 4.25a were obtained from R-3-hydroxybutanoic acid using the

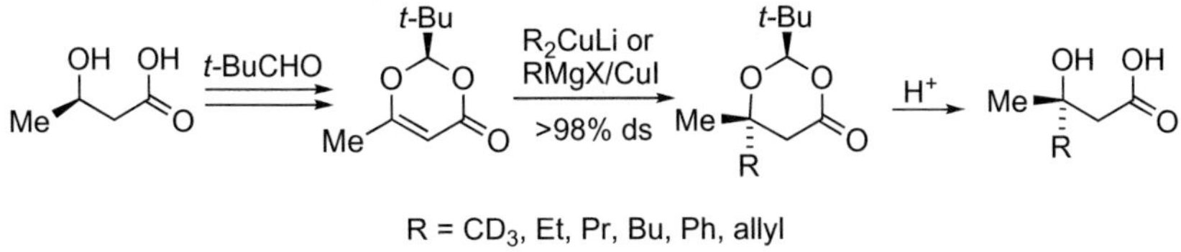

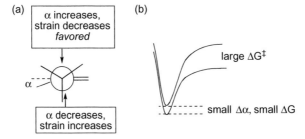

R = CD$_3$, Et, Pr, Bu, Ph, allyl

SCHEME 4.25 Asymmetric conjugate addition of cuprates to dioxinones [339].

"self-regeneration of chirality centers" concept discussed in Chapter 3 (*cf.*, Scheme 3.11 and 3.12). After the addition, hydrolytic removal of the "achiral auxiliary" (pivaldehyde) liberates a 3-alkyl-3-hydroxybutyrate that is essentially enantiomerically pure [339].

The additions are all >98% diastereoselective (the limit of detection), which is surprising since the dioxinone ring is in a sofa conformation, with only the acetal carbon significantly out of plane, leaving approach from either face essentially unhindered (recall the low selectivities for alkylation of *tert*-butylcyclohexanone enolates, Scheme 3.9). Interestingly, examination of a number of X-ray crystal structures revealed that dioxinone acetals such as these have the common feature of pyramidalized carbonyl and β-carbon atoms [339]. Empirically, additions occur from the direction of the β-carbon's pyramidalization (see also ref. [340]). The reason for the pyramidalization in the substrate is the relief of torsional strain (however, calculations indicated that the energy required to flatten the pyramidal atoms is very small, ∼0.1 kcal/mol). Seebach suggests [339] that approach of the nucleophile from the direction of pyramidalization should minimize the strain even more (see also ref. [19]). Since the reaction is kinetically controlled, and the selectivity is therefore determined in the transition state ($\Delta\Delta G^{\ddagger}$), this hypothesis (which is based on ground state arguments) may seem a risky infringement of the Curtin–Hammet principle [17,18]. Nevertheless, the strain that produces the pyramidalization ($\Delta G°$ for the flat and pyramidal geometries) in the ground state and the energy differences in the transition state ($\Delta\Delta G^{\ddagger}$) have the same origin, and approach from the direction of pyramidalization relieves the strain while approach from the opposite direction increases it (Figure 4.32a). Thus, the energy difference between the two pyramidal ground states is *amplified* in the transition state (see also the two Morse curves in Figure 4.32b). Seebach also noted that the pyramidalization was evident in a computed model structure, which (since X-ray structural information is not always available) makes the following hypothesis all the more valuable: "*The steric course of attack on a trigonal center can be predicted from the direction of its pyramidalization*" [339].

(a)

α increases, strain decreases favored

α

α decreases, strain increases

(b)

large $\Delta G^{\ddagger}$

small $\Delta\alpha$, small ΔG

FIGURE 4.32 (a) Schematic showing how torsional strain is affected by the direction of attack on a pyramidalized trigonal center. (b) Linear perturbation of a Morse function that produces small distortions in the ground state can lead to large energy differences in the transition state [339].

Catalytic methods for the 1,4-addition to cyclic ketones of aryl boron and bismuth compounds, as well as Grignard reagents and organozinc compounds have been developed, and are discussed in that order below.

Boron and bismuth. As indicated in a previous section, the Miyaura group has developed methods for the enantioselective 1,4-addition of [ArBF$_3$]K and Ar$_3$Bi groups to enones using a Pd(II) complexes of the chiraphos (Figure 4.28b) or dipamp (Figure 4.28g) ligands [306,307]; several examples of additions to cyclic ketones are listed in Table 4.16 [307–310]. Hayashi has introduced a chiral diene ligand (Figure 4.28h) that provides excellent enantioselectivities in the Rh(I)-mediated conjugate addition of aryl boronic acids to enones [341].

TABLE 4.16 Pd(II)-Diphosphine Catalyzed Addition of Aryl Groups to Enones

Entry	Enone	Nucleophile	Ligand[a]	% Yield	er (*R:S*)	Ref.
1	O (6-membered enone)	PhBF$_3$K + Pd(II)	4.28g	99	96:4	[307]
2	"	Ph$_3$Bi + Cu(BF$_4$)$_2$	4.28g	92	96:4	[307]
3	"	PhB(OH)$_2$ + Rh(I)	4.28h	94	98:2	[341]
4	"	*p*-F$_3$C–C$_6$H$_4$B(OH)$_2$ + Rh(I)	4.28h	90	>99:1	[341]
5	"	C$_5$H$_{11}$CH=CHB(OH)$_2$ + Rh(I)	4.28h	73	94:6	[341]
6	O (5-membered enone)	PhBF$_3$K + Rh(I)	4.28b	60	3:97	[307]
7	"	Ph$_3$Bi + Cu(BF$_4$)$_2$	4.28b	85	3:97	[306]
8	"	PhB(OH)$_2$ + Rh(I)	4.28h	78	98:2	[341]
9	O (7-membered enone)	PhBF$_3$K + Rh(I)	4.28g	91	94:6	[307]
10	"	Ph$_3$Bi + Cu(BF$_4$)$_2$	4.28g	89	88:12	[306]
11	"	PhB(OH)$_2$ + Rh(I)	4.28h	81	95:5	[341]

[a]*Numbers in the ligand column refer to Figure 4.28.*

The mechanism of the Pd(II)-mediated reaction using the dipamp and chiraphos ligands is shown in Scheme 4.26a, starting clockwise from the upper left. It involves oxidative addition of the aryltrifluoroborate to the palladium, coordination of the enone, insertion, and hydrolysis to free the addition product and regenerate the catalyst. The steric course of the reaction is rationalized on the basis of X-ray crystal structures of the dicationic palladium catalysts (the full structures are shown in Figure 4.28b and g, p. 222). A 9–10° dihedral angle was observed between the planes defined by the P–Pd–P and S–Pd–S atoms ("S" = solvent). The twist in the square–planar geometry is induced by steric repulsion between the pseudoequatorial *P*-phenyl groups and the solvent molecules. In the reaction, the ligands replace the solvent in the catalytic cycle, and it is assumed that the twist is maintained throughout. Partial structures, illustrating the structure in the lower right of the catalytic cycle, are shown in the inset of Scheme 4.26, in which the aromatic ring (Ar) and the enone are shown in place of the two solvent molecules. Thus, for chiraphos, the lower right and upper left quadrants can best accommodate the nucleophilic aryl group and the enone. Molecular modeling of a cyclic enone coordinated to the Ar—Pd species (lower right in catalytic cycle) revealed several low energy conformations, but the one in which the Ar—Pd and C=C bonds are *antiperiplanar* correctly predicts the observed configuration in the product, and in this conformation, the bulk of the enone occupies the vacant quadrant of the catalyst.

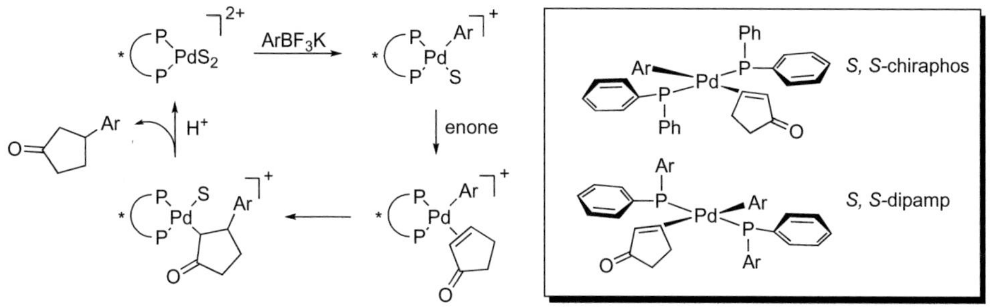

SCHEME 4.26 Catalytic cycle and stereochemical rationale for the Pd(II)-diphosphine catalyzed addition of aryl groups to enones [307]. "S" indicates solvent. See Figure 4.28 for the complete ligand structures.

The steric course of the rhodium · diene-mediated addition of aryl boronic acids is rationalized by the structures of the intermediates shown in Scheme 4.27 [341]. The C-2-symmetric *R,R*-diene ligand places bulky benzyl groups in the upper left and lower right quadrant of the square—planar complex, when viewed with the rhodium ligands in a horizontal plane. To avoid steric interference with the benzyl group, the enone preferentially complexes the rhodium so that the *Re* face of the α-carbon is coordinated to the metal. Reductive elimination then gives the *R*-configured addition product.

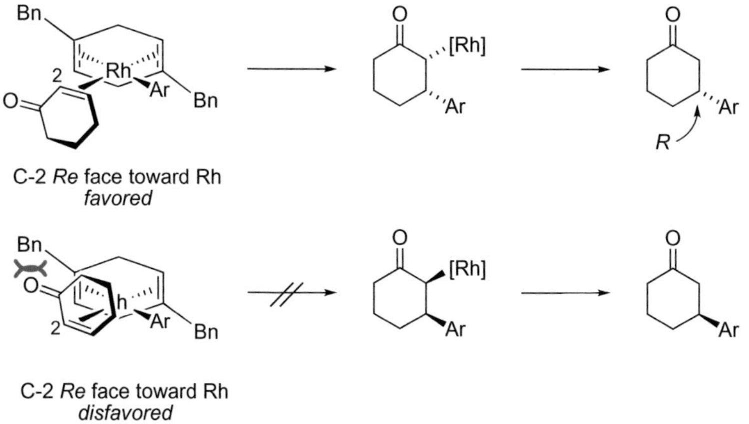

C-2 *Re* face toward Rh
favored

C-2 *Re* face toward Rh
disfavored

SCHEME 4.27 Rationale of the steric course of the rhodium-catalyzed addition of boronic acids to enones [341]. For the complete structure of the diene ligand, see Figure 4.28h.

Magnesium. The ligands developed by the Feringa and Minnaard groups for the enantiose-lective copper-mediated conjugate addition of Grignard reagents to cyclic enones are bispho-sphines [289,311−313]. The examples listed in Table 4.17 illustrate the excellent regio- and enantioselectivity observed. The 1,2-addition products were found to be racemic in all cases, indicating a nonselective reaction of the Grignard reagent that is not mediated by the copper. The mechanism of the reaction was explained earlier in Scheme 4.22 [313].

TABLE 4.17 Copper(I) Catalyzed 1,4-Addition of Grignard Reagents to Enones Using Taniaphos (See Figure 4.28f) as the Chiral Ligand [311]

Entry	Enone	Grignard[a]	Regioselectivity (% 1,4-addition)	% Yield	er
1		EtMgBr · Cu(I)	95	69	98:2
2	"	MeMgBr · Cu(I)	83	—	95:5
3	"	PrMgBr · Cu(I)	81	—	97:3
4		EtMgBr · Cu(I)	80	60	93:7

[a]*The copper loading is 5 mol%.*

Zinc and aluminum. The Charette group has shown that organozinc reagents generated from Grignard reagents add in a 1,4-fashion to cyclohexenone as illustrated by the examples in Table 4.18, entries 1−4. These additions are mediated by Cu(OTf)$_2$, and a phosphoramidite ligand based on α-methylbenzyl amine and BINOL [130]. Issues of functional group compat-ibility and versatility (note the addition of secondary organometallics in entries 2 and 3) may render the conversion of a Grignard reagent (compare Table 4.17) to an organozinc reagent favorable. The examples in Table 4.18, entries 5−8, reported by the Alexakis group, illustrate the addition of a methyl or ethyl group in a 1,4-fashion to 3-substituted cyclohexenones in excellent yield and enantioselectivity to make quaternary stereocenters [342,343].

TABLE 4.18 Copper-Mediated Addition of Organozinc Compounds to 3-Substituted (R) Cyclohexenones

Entry	R	Nucleophile	Ligand[a]	% Yield	er	Ref.
1	H	Me$_2$Zn Cu(II)	ent-4.28d	88	>99:1	[130]
2	H	i-Pr$_2$Zn Cu(II)	"	93	97:3	[130]

(Continued)

TABLE 4.18 (Continued)

Entry	R	Nucleophile	Ligand[a]	% Yield	er	Ref.
3	H	Cy$_2$Zn	"	94	97:3	[130]
		Cu(II)				
4	H	Ph(CH$_2$)Zn	"	97	97:3	[130]
		Cu(II)				
5	Et	Me$_3$Al	4.28i	80	97:3	[342]
		Cu(I)				
6	4-pentenyl	"	"	76	97:3	[342]
7	–(CH$_2$)$_3$ ⟨O⟩	"	"	81	97:3	[342]
8	Me	Et$_3$Al	4.28j	71	96:4	[343]
		Cu(I)				

[a]Entries in the ligand column refer to Figure 4.28.

REFERENCES

[1] Curtin, D. Y.; Harris, E. E.; Meislich, E. K. *J. Am. Chem. Soc.* **1952**, *74*, 2901−2904.
[2] Cram, D. J.; Elhafez, F. A. A. *J. Am. Chem. Soc.* **1952**, *74*, 5828−5835.
[3] Prelog, V. *Helv. Chim. Acta.* **1953**, *36*, 308−319.
[4] Morrison, J. D.; Mosher, H. S. *Asymmetric Organic Reactions*; Prentice-Hall: Englewood Cliffs, NJ, 1971.
[5] *Asymmetric Synthesis*; Morrison, J. D., Ed.; Academic: Orlando, 1983−1985.
[6] Reetz, M. T. *Angew. Chem., Int. Ed. Engl.* **1984**, *23*, 556−569.
[7] Mengel, A.; Reiser, O. *Chem. Rev.* **1999**, *99*, 1191−1223.
[8] Reetz, M. T. *Chem. Rev.* **1999**, *99*, 1121−1162.
[9] Fischer, E. *Chem. Ber.* **1894**, *27*, 3189−3232. see p. 3210.
[10] Cram, D. J.; Kopecky, K. R. *J. Am. Chem. Soc.* **1959**, *81*, 2748−2755.
[11] Cornforth, J. W.; Cornforth, R. H.; Matthew, K. K. *J. Chem. Soc.* **1959**, 112−127.
[12] Karabatsos, G. J.; Hsi, N. *J. Am. Chem. Soc.* **1965**, *87*, 2864−2870.
[13] Karabatsos, G. J.; Taller, R. A. *Tetrahedron* **1968**, *24*, 3923−3937.
[14] Karabatsos, G. J. *J. Am. Chem. Soc.* **1967**, *89*, 1367−1371.
[15] Hammond, G. S. *J. Am. Chem. Soc.* **1955**, *77*, 334−338.
[16] Karabatsos, G. J.; Fenoglio, D. J. *Top. Stereochem.* **1970**, *5*, 167−203.
[17] Curtin, D. Y. *Rec. Chem. Progr.* **1954**, *15*, 111−128.
[18] Seeman, J. I. *Chem. Rev.* **1983**, *83*, 83−134.
[19] Chérest, M.; Felkin, H. *Tetrahedron Lett.* **1968**, 2205−2208.
[20] Chérest, M.; Felkin, H.; Prudent, N. *Tetrahedron Lett.* **1968**, 2199−2204.
[21] Anh, N. T. *Top. Curr. Chem.* **1980**, *88*, 145−162.
[22] Frenking, G.; Köhler, K. F.; Reetz, M. T. *Angew. Chem., Int. Ed. Engl.* **1991**, *30*, 1146−1149.
[23] Huang, X. L.; Dannenberg, J. J. *J. Am. Chem. Soc.* **1993**, *115*, 6017−6024.
[24] Gung, B. W. *Tetrahedron* **1996**, *52*, 5263−5301.
[25] Gung, B. W. *Chem. Rev.* **1999**, *99*, 1377−1386.
[26] Adcock, W.; Trout, N. A. *Chem. Rev.* **1999**, *99*, 1415−1435.

[27] Mehta, G.; Chandrasekhar, J. *Chem. Rev.* **1999**, *99*, 1437−1467.

[28] Cieplak, A. S. *Chem. Rev.* **1999**, *99*, 1265−1336.

[29] Kaselj, M.; Chung, W.-S.; le Noble, W. J. *Chem. Rev.* **1999**, *99*, 1387−1413.

[30] Tomoda, S. *Chem. Rev.* **1999**, *99*, 1243−1263.

[31] Senda, Y. *Chirality* **2002**, *14*, 110−120.

[32] Cheung, C. K.; Tseng, L. T.; Lin, M.-H.; Srivastava, S.; le Noble, W. J. *J. Am. Chem. Soc.* **1986**, *108*, 1598−1605.

[33] Seebach, D. *Angew. Chem., Int. Ed. Engl.* **1990**, *29*, 1320−1367.

[34] Bürgi, H. B.; Dunitz, J. D.; Schefter, E. *J. Am. Chem. Soc.* **1973**, *95*, 5065−5067.

[35] Bürgi, H. B.; Dunitz, D.; Lehn, J. M.; Wipff, G. *Tetrahedron* **1974**, *30*, 1563−1572.

[36] Nakamura, M.; Nakamura, E.; Koga, N.; Morokuma, K. *J. Am. Chem. Soc.* **1993**, *115*, 11016−11017.

[37] Mori, S.; Nakamura, M.; Nakamura, E.; Koga, N.; Morokuma, K. *J. Am. Chem. Soc.* **1995**, *117*, 5055−5065.

[38] Klein, D. P.; Gladysz, J. A. *J. Am. Chem. Soc.* **1992**, *114*, 8710−8711.

[39] Haeffner, F.; Sun, C.; Williard, P. G. *J. Am. Chem. Soc.* **2000**, *122*, 12542−12546.

[40] Yamazaki, S.; Yamabe, S. *J. Org. Chem.* **2002**, *67*, 9346−9353.

[41] Anh, N. T.; Eisenstein, O. *Nouv. J. Chimie* **1977**, *1*, 61−70.

[42] Heathcock, C. H.; Flippin, L. A. *J. Am. Chem. Soc.* **1983**, *105*, 1667−1668.

[43] Lodge, E. P.; Heathcock, C. H. *J. Am. Chem. Soc.* **1987**, *109*, 2819−2820.

[44] Scheiner, S.; Lipscomb, W. N.; Kleier, D. A. *J. Am. Chem. Soc.* **1976**, *98*, 4770−4777.

[45] Heathcock, C. H. *Aldrichimica Acta* **1990**, *23*, 99−111.

[46] Lodge, E. P.; Heathcock, C. H. *J. Am. Chem. Soc.* **1987**, *109*, 3353−3361.

[47] Wu, Y.-D.; Houk, K. N. *J. Am. Chem. Soc.* **1987**, *109*, 908−910.

[48] Smith, R. J.; Trzoss, M.; Bühl, M.; Bienz, S. *Eur. J. Org. Chem.* **2002**, 2770−2775.

[49] Eliel, E. L.; Wilen, S. H.; Mander, L. N. *Stereochemistry of Organic Compounds*; Wiley-Interscience: New York, 1994.

[50] Chen, X.; Hortelano, E. R.; Eliel, E. L.; Frye, S. V. *J. Am. Chem. Soc.* **1992**, *114*, 1778−1784.

[51] Eliel, E. L. In *Asymmetric Synthesis*; Morrison, J. D., Ed.; Academic: Orlando, 1983; Vol. 2, pp. 125−155.

[52] Guillarme, S.; Plé, K.; Banchet, A.; Liard, A.; Haudrechy, A. *Chem. Rev.* **2006**, *106*, 2355−2403.

[53] Still, W. C.; McDonald, J. H., III. *Tetrahedron Lett.* **1980**, *21*, 1031−1034.

[54] Still, W. C.; Schneider, J. A. *Tetrahedron Lett.* **1980**, *21*, 1035−1038.

[55] Ko, K.-Y.; Eliel, E. L. *J. Org. Chem.* **1986**, *51*, 5353−5362.

[56] Keck, G. E.; Andrus, M. B.; Romer, D. R. *J. Org. Chem.* **1991**, *56*, 417−420.

[57] Reetz, M. T. *Acc. Chem. Res.* **1993**, *26*, 462−468.

[58] Stanton, G. R.; Johnson, C. N.; Walsh, P. J. *J. Am. Chem. Soc.* **2010**, *132*, 4399−4408.

[59] Giese, B. *Angew. Chem., Int. Ed. Engl.* **1977**, *16*, 125−136.

[60] Collum, D. B.; McDonald, J. H.; Still, W. C. *J. Am. Chem. Soc.* **1980**, *102*, 2118−2120.

[61] Collum, D. B.; McDonald, J. H.; Still, W. C. *J. Am. Chem. Soc.* **1980**, *102*, 2120−2121.

[62] Nakata, T.; Kishi, Y. *Tetrahedron Lett.* **1978**, 2745−2748.

[63] Nakata, T.; Schmid, G.; Vranesic, B.; Okigawa, M.; Smith-Palmer, T. *J. Am. Chem. Soc.* **1978**, *100*, 2933−2935.

[64] Durham, T. B.; Blanchard, N.; Savall, B. M.; Powell, N. A.; Roush, W. R. *J. Am. Chem. Soc.* **2004**, *126*, 9307−9317.

[65] Abel, S.; Faber, D.; Hüter, O.; Giese, B. *Synthesis* **1999**, 188−197.

[66] Solladié, G. In *Asymmetric Synthesis*; Morrison, J. D., Ed.; Academic: Orlando, 1983; Vol. 2, pp. 157−199.

[67] Gawley, R. E. In *Advances in Asymmetric Synthesis*; Hassner, A., Ed.; JAI: Greenwich, CT, 1998; Vol. 3, pp. 77−111.

[68] Gawley, R. E. In *Grignard Reagents: New Developments*; Richey, H. G., Ed.; John Wiley and Sons: New York, 2000, pp. 139−164.

[69] Gawley, R. E.; Zhang, Q.; McPhail, A. T. *Tetrahedron: Asymmetry* **2000**, *11*, 2093−2106.

[70] Gaul, C.; Arvidsson, P. I.; Bauer, W.; Gawley, R. E.; Seebach, D. *Chem. Eur. J.* **2001**, *7*, 4117−4125.

[71] Gawley, R. E.; Campagna, S. A.; Santiago, M.; Ren, T. *Tetrahedron: Asymmetry* **2002**, *13*, 29−36.

[72] Papillon, J. P. N.; Taylor, R. J. K. *Org. Lett.* **2002**, *4*, 119−122.

[73] Boudier, A.; Knochel, P. *Tetrahedron Lett.* **1999**, *40*, 687−690.
[74] Boudier, A.; Darcel, C.; Flachsmann, F.; Micouin, L.; Oestreich, M.; Knochel, P. *Chem. Eur. J.* **2000**, *6*, 2748−2761.
[75] Boudier, A.; Hupe, E.; Knochel, P. *Angew. Chem., Int. Ed. Engl.* **2000**, *39*, 2294−2297.
[76] Noyori, R.; Kitamura, M. *Angew. Chem., Int. Ed. Engl.* **1991**, *30*, 49−69.
[77] Duthaler, R. O.; Hafner, A. *Chem. Rev.* **1992**, *92*, 807−832.
[78] Soai, K.; Niwa, S. *Chem. Rev.* **1991**, *92*, 833−856.
[79] Trost, B. M.; Weiss, A. H. *Adv. Synth. Catal.* **2009**, *351*, 963−983.
[80] Luderer, M. R.; Bailey, W. F.; Luderer, M. R.; Fair, J. D.; Dancer, R. J.; Sommer, M. B. *Tetrahedron: Asymmetry* **2009**, *20*, 981−998.
[81] Walsh, P. J. *Acc. Chem. Res.* **2003**, *36*, 739−749.
[82] Ramón, D.; Yus, M. *Chem. Rev.* **2006**, *106*, 2126−2208.
[83] Soai, K.; Kawasaki, T.; Sato, I. In *New Frontiers in Asymmetric Catalysis*; Mikami, K., Lautens, M., Eds.; Wiley: Hoboken, 2007, pp. 259−274.
[84] Knochel, P. *Synlett* **1995**, 393−403.
[85] Reetz, M. T.; Hüllman, M.; Massa, W.; Berger, S.; Rademacher, P.; Heymanns, P. *J. Am. Chem. Soc.* **1986**, *108*, 2405−2408.
[86] Denmark, S. E.; Almstead, N. G. *J. Am. Chem. Soc.* **1993**, *115*, 3133−3139.
[87] Langer, F.; Schwink, L.; Devasagayaraj, A.; Chavant, P.-Y.; Knochel, P. *J. Org. Chem.* **1996**, *61*, 8229−8243.
[88] Evans, D. A. *Science* **1988**, *240*, 420−426.
[89] Blystone, S. L. *Chem. Rev.* **1989**, *89*, 1663−1679.
[90] Tomioka, K. *Synthesis* **1990**, 541−549.
[91] Erdik, E. *Tetrahedron* **1992**, *48*, 9577−9648.
[92] Erdik, E. *Organozinc Reagents in Organic Synthesis*; CRC Press: Boca Raton, FL, 1996.
[93] Knochel, P.; Perea, J. J. A.; Jones, P. *Tetrahedron* **1998**, *54*, 8275−8319.
[94] *Organozinc Reagents*; Knochel, P.; Jones, P., Eds.; Oxford University Press: New York, 1999.
[95] Lemire, A.; Coté, C.; Janes, M. K.; Charette, A. B. *Aldrichimica Acta* **2009**, *42*, 71−83.
[96] Mukaiyama, T.; Soai, K.; Sato, T.; Shimizu, H.; Suzuki, K. *J. Am. Chem. Soc.* **1979**, *101*, 1455−1460.
[97] Duguet, N.; Petit, S. M.; Marchand, P.; Harrison-Marchand, A.; Maddaluno, J. *J. Org. Chem.* **2008**, *73*, 5397−5409.
[98] Da, C.-S.; Wang, J.-R.; Yin, X.-G.; Fan, X.-Y.; Liu, Y.; Yu, S.-L. *Org. Lett.* **2009**, *11*, 5578−5581.
[99] Dosa, P. I.; Fu, G. C. *J. Am. Chem. Soc.* **1998**, *120*, 445−446.
[100] Weber, B.; Seebach, D. *Angew. Chem., Int. Ed. Engl.* **1992**, *31*, 84−86.
[101] García, C.; LaRochelle, L. K.; Walsh, P. J. *J. Am. Chem. Soc.* **2002**, *124*, 10970−10971.
[102] Sato, T.; Soai, K.; Suzuki, K.; Mukaiyama, T. *Chem. Lett.* **1978**, 601−604.
[103] Yamakawa, M.; Noyori, R. *J. Am. Chem. Soc.* **1995**, *117*, 6327−6335.
[104] Kitamura, M.; Oka, H.; Noyori, R. *Tetrahedron* **1999**, *55*, 3605−3614.
[105] Yamakawa, M.; Noyori, R. *Organometallics* **1999**, *18*, 128−133.
[106] Soai, K.; Ookawa, A.; Kaba, T.; Ogawa, K. *J. Am. Chem. Soc.* **1987**, *109*, 7111−7115.
[107] Kitamura, M.; Okada, S.; Suga, S.; Noyori, R. *J. Am. Chem. Soc.* **1989**, *111*, 4028−4036.
[108] Nugent, W. A. *Chem. Commun.* **1999**, 1369−1370.
[109] Itsuno, S.; Fréchet, J. M. J. *J. Org. Chem.* **1987**, *52*, 4140−4142.
[110] Noyori, R.; Suga, S.; Kawai, K.; Okada, S.; Kitamura, M.; Oguni, N.; Hayashi, M.; Kaneko, T.; Matsuda, Y. *J. Organomet. Chem.* **1990**, *382*, 19−37.
[111] Joshi, N. N.; Srebnik, M.; Brown, H. C. *Tetrahedron Lett.* **1989**, *30*, 5551−5554.
[112] Shono, T.; Kise, N.; Shirakawa, E.; Matsumoto, H.; Okazaki, E. *J. Org. Chem.* **1991**, *56*, 3063−3067.
[113] Watanabe, M.; Araki, S.; Butsugan, Y.; Uemura, M. *J. Org. Chem.* **1991**, *56*, 2218−2224.
[114] Yoshioka, M.; Kawakita, T.; Ohno, M. *Tetrahedron Lett.* **1989**, *30*, 1657−1660.
[115] Knochel, P.; Brieden, W.; Rozema, M. J.; Eisenberg, C. *Tetrahedron Lett.* **1993**, *34*, 5881−5884.
[116] Balsells, J.; Walsh, P. J. *J. Am. Chem. Soc.* **2000**, *122*, 3250−3251.
[117] Seebach, D.; Plattner, D. A.; Beck, A. K.; Wang, Y. M.; Hunziker, D.; Petter, W. *Helv. Chim. Acta* **1992**, *75*, 2171−2209.
[118] Ito, Y. N.; Beck, A. K.; Bohác, A.; Ganter, C.; Gawley, R. E.; Kühnle, F. N. M.; Piquer, J. A.; Tuleja, J.; Wang, Y. M.; Seebach, D. *Helv. Chim. Acta* **1994**, *77*, 2071−2110.

[119] Heaton, S. B.; Jones, G. B. *Tetrahedron Lett.* **1992**, *33*, 1693−1696.
[120] Zhang, F.-J.; Yip, C.-W.; Cao, R.; Chan, A. S. C. *Tetrahedron: Asymmetry* **1997**, *8*, 585−589.
[121] Mori, M.; Nakai, T. *Tetrahedron Lett.* **1997**, *38*, 6233−6236.
[122] Balsells, J.; Davis, T. J.; Carroll, P.; Walsh, P. J. *J. Am. Chem. Soc.* **2002**, *124*, 10336−10348.
[123] Sola, L.; Reddy, K. S.; Vidal-Ferran, A.; Moyano, A.; Pericas, M. A.; Riera, A.; Alvarez-Larena, A.; Piniella, J.-F. *J. Org. Chem.* **1998**, *63*, 7078−7082.
[124] Vázquez, J.; Pericas, M. A.; Maseras, F.; Lledós, A. *J. Org. Chem.* **2000**, *65*, 7303−7309.
[125] Tanaka, T.; Yasuda, Y.; Hayashi, M. *J. Org. Chem.* **2006**, *71*, 7091−7093.
[126] Chou, T.-S.; Knochel, P. *J. Org. Chem.* **1990**, *55*, 4791−4793.
[127] Knochel, P.; Rozema, M. J.; Tucker, C. E.; Retherford, C.; Furlong, M.; Achyutharao, S. *Pure Appl. Chem.* **1992**, *64*, 361−369.
[128] Knochel, P.; Leuser, H.; Gong, L.-Z.; Perrone, S.; Kneisel, F.f. In *Handbook of Functionalized Organometallics; Applications in Synthesis*; Knochel, P., Ed.; Wiley-VCH: Weinheim, 2005, pp. 251−346.
[129] Krasovskiy, A.; Malakhov, V.; Gavryushin, A.; Knochel, P. *Angew. Chem., Int. Ed.* **2006**, *45*, 6040−6044.
[130] Cote, A.; Charette, A. *J. Am. Chem. Soc.* **2008**, *130*, 2771−2773.
[131] Berger, S.; Langer, F.; Lutz, C.; Knochel, P.; Mobley, T. A.; Reddy, C. K. *Angew. Chem., Int. Ed. Engl.* **1997**, *36*, 1496−1498.
[132] Lutz, C.; Knochel, P. *J. Org. Chem.* **1997**, *62*, 7895−7898.
[133] Lutz, C.; Jones, P.; Knochel, P. *Synthesis* **1999**, 312−316.
[134] Soai, K.; Yokoyama, S.; Hayasaka, T. *J. Org. Chem.* **1991**, *56*, 4254−4268.
[135] Corey, E. J.; Hannon, F. *Tetrahedron Lett.* **1987**, *28*, 5233−5236.
[136] Guillaneaux, D.; Zhao, S.-H.; Samuel, O.; Rainford, D.; Kagan, H. B. *J. Am. Chem. Soc.* **1994**, *116*, 9430−9439.
[137] Kitamura, M.; Suga, S.; Niwa, M.; Noyori, R. *J. Am. Chem. Soc.* **1995**, *117*, 4832−4842.
[138] Blackmond, D. G. *Acc. Chem. Res.* **2000**, *33*, 402−411.
[139] Klussmann, M.; Mathew, S. P.; Iwamura, H.; Wells, D. H., Jr.; Armstrong, A.; Blackmond, D. G. *Angew. Chem., Int. Ed.* **2006**, *45*, 7989−7992.
[140] Satyanarayana, T.; Abraham, S.; Kagan, H. B. *Angew. Chem., Int. Ed.* **2009**, *48*, 456−494.
[141] Kagan, H. B. *Synlett* **2001**, 888−899.
[142] Rosner, T.; Sears, P. J.; Nugent, W. A.; Blackmond, D. G. *Org. Lett.* **2000**, *2*, 2511−2513.
[143] Rasmussen, T.; Norrby, P.-O. *J. Am. Chem. Soc.* **2001**, *123*, 2464−2465.
[144] Kozlowski, M. C.; Dixon, S. L.; Panda, M.; Lauri, G. *J. Am. Chem. Soc.* **2003**, *125*, 6614−6615.
[145] Soai, K.; Shibata, T.; Morioka, H.; Choji, K. *Nature* **1995**, *378*, 767−768.
[146] Soai, K.; Shibata, t.; Sato, I. *Acc. Chem. Res.* **2000**, *33*, 382−390.
[147] Singleton, D. A.; Vo, L. K. *Org. Lett.* **2003**, *5*, 4337−4339.
[148] Weber, B.; Seebach, D. *Tetrahedron* **1994**, *50*, 7473−7484.
[149] Seebach, D.; Beck, A. K.; Heckel, A. *Angew. Chem., Int. Ed.* **2001**, *40*, 92−138.
[150] Pellissier, H. *Tetrahedron* **2008**, *64*, 10279−10317.
[151] Lütjens, H.; Nowotny, S.; Knochel, P. *Tetrahedron: Asymmetry* **1995**, *6*, 2675−2678.
[152] Eisenberg, C.; Knochel, P. *J. Org. Chem.* **1994**, *59*, 3760−3761.
[153] Bussche-Hunnefeld, J. L. v. d.; Seebach, D. *Tetrahedron* **1992**, *48*, 5719−5730.
[154] Volkmann, R. A. In *Comprehensive Organic Synthesis. Selectivity, Strategy, and Efficiency in Modern Organic Chemistry*; Trost, B. M., Fleming, I., Eds.; Pergamon: Oxford, 1991; Vol. 5, pp. 355−396.
[155] Denmark, S. E.; Nicaise, O. J.-C. *Chem. Commun.* **1996**, 999−1004.
[156] Enders, D.; Reinhold, U. *Tetrahedron: Asymmetry* **1997**, *8*, 1895−1946.
[157] Bloch, R. *Chem. Rev.* **1998**, *98*, 1407−1438.
[158] Kobayashi, S.; Ishitani, H. *Chem. Rev.* **1999**, *99*, 1069−1094.
[159] Alvaro, G.; Savoia, D. *Synlett* **2002**, 651−673.
[160] Charette, A. B.; Boezio, A. A.; Coté, A.; Moreau, E.; Pytkowicz, J.; Desrosiers, J.-N.; Legault, C. *Pure Appl. Chem.* **2005**, *77*, 1259−1267.
[161] Friestad, G. K.; Mathies, A. K. *Tetrahedron* **2007**, *63*, 2541−2569.
[162] Petrini, M.; Torregiani, E. *Synthesis* **2007**, 159−186.
[163] Vilaivan, T.; Bhanthumnavin, W.; Sritana-Anant, Y. *Curr. Org. Chem.* **2005**, *9*, 1315−1392.

[164] Denmark, S. E.; Edwards, J. P.; Weber, T.; Piotrowski, D. W. *Tetrahedron: Asymmetry* **2010**, *21*, 1278−1302.
[165] Enders, D.; Schubert, H.; Nübling, C. *Angew. Chem., Int. Ed. Engl.* **1986**, *25*, 1109−1110.
[166] Enders, D.; Funk, R.; Klatt, M.; Raabe, G.; Hovestreydt, E. R. *Angew. Chem., Int. Ed. Engl.* **1993**, *32*, 418−420.
[167] Enders, D.; Bartzen, D. *Liebigs Ann. Chem.* **1993**, 569−574.
[168] Enders, D.; Schankat, J. *Helv. Chim. Acta.* **1993**, *76*, 402−406.
[169] Enders, D.; Tiebes, J. *Liebigs Ann. Chem.* **1993**, 173−177.
[170] Enders, D.; Klatt, M.; Funk, R. *Synlett* **1993**, 226−228.
[171] Enders, D.; Nübling, C.; Schubert, H. *Liebigs Ann./Recl.* **1997**, 1089−1100.
[172] Job, A.; Janek, C. F.; Bettray, W.; Peters, R.; Enders, D. *Tetrahedron* **2002**, *58*, 2253−2329.
[173] Denmark, S. E.; Weber, T.; Piotrowski, D. W. *J. Am. Chem. Soc.* **1987**, *109*, 2224−2225.
[174] Weber, T.; Edwards, J. P.; Denmark, S. E. *Synlett* **1989**, *1*, 20−22.
[175] Denmark, S. E.; Edwards, J. P.; Nicaise, O. *J. Org. Chem.* **1993**, *58*, 569−578.
[176] Nicaise, O.; Denmark, S. E. *Bull. Soc. Chim. Fr.* **1997**, *134*, 395−398.
[177] Enders, D.; Lochtman, R.; Meiers, M.; Müller, S. F.; Lazny, R. *Synlett* **1998**, 1182−1184.
[178] Denmark, S. E.; Nicaise, O.; Edwards, J. P. *J. Org. Chem.* **1990**, *55*, 6219−6223.
[179] Meyers, A. I.; Whitten, C. E. *J. Am. Chem. Soc.* **1975**, *97*, 6266−6267.
[180] Meyers, A. I.; Smith, R. K.; Whitten, C. E. *J. Org. Chem.* **1979**, *44*, 2250−2256.
[181] Ziegler, F. E.; Gilligan, P. J. *J. Org. Chem.* **1981**, *46*, 3874−3880.
[182] Lutomski, K. A.; Meyers, A. I. In *Asymmetric Synthesis*; Morrison, J. D., Ed.; Academic: Orlando, 1984; Vol. 3, pp. 213−273.
[183] Meyers, A. I.; Shipman, M. *J. Org. Chem.* **1991**, *56*, 7098−7102.
[184] Woodward, R. B.; Hoffmann, R. *The Conservation of Orbital Symmetry*; Academic: New York, 1970.
[185] Meyers, A. I.; Roth, G. P.; Hoyer, D.; Barner, B. A.; Laucher, D. *J. Am. Chem. Soc.* **1988**, *110*, 4611−4624.
[186] Rawson, D. J.; Meyers, A. I. *J. Org. Chem.* **1991**, *56*, 2292−2294.
[187] Meyers, A. I.; Brown, J. D.; Laucher, D. *Tetrahedron Lett.* **1987**, *28*, 5279−5282.
[188] Hashimoto, S.; Yamada, S.; Koga, K. *J. Am. Chem. Soc.* **1976**, *98*, 7450−7452.
[189] Hashimoto, S.; Yamada, S.; Koga, K. *Chem. Pharm. Bull.* **1979**, *27*, 771−782.
[190] Kogen, H.; Tomioka, K.; Hashimoto, S.; Koga, K. *Tetrahedron* **1981**, *37*, 3951−3956.
[191] Senanayake, C. H.; Krishnamurthy, D.; Lu, Z.-H.; Han, Z.; Gallou, I. *Aldrichimica Acta* **2005**, *38*, 93−104.
[192] Davis, F. A.; Friedman, A. J.; Kluger, E. W. *J. Am. Chem. Soc.* **1974**, *96*, 5000−5001.
[193] Annunziata, R.; Cinquini, M.; Cozzi, F. *J. Chem. Soc. Perkin 1* **1982**, 339−343.
[194] Davis, F. A.; Reddy, R. E.; Szewczyk, J. M.; Reddy, G. V.; Portonovo, P. S.; Zhang, H.; Fanelli, D.; Reddy, R. T.; Zhou, P.; Carroll, P. J. *J. Org. Chem.* **1997**, *62*, 2555−2563.
[195] Davis, F. A.; Zhou, P.; Chen, B.-C. *Chem. Soc. Rev.* **1998**, *27*, 13−18.
[196] Davis, F. A.; Zhang, Y.; Andemichael, Y.; Fang, T.; Fanelli, D. L.; Zhang, H. *J. Org. Chem.* **1999**, *64*, 1403−1406.
[197] Fanelli, D. L.; Szewczyk, J. M.; Zhang, Y.; Reddy, G. V.; Burns, D. M.; Davis, F. A. *Org. Synth.* **2004**, *Coll. Vol. 10*, 47−53.
[198] Liu, G.; Cogan, D. A.; Ellman, J. A. *J. Am. Chem. Soc.* **1997**, *119*, 9913−9914.
[199] Ellman, J. A.; Owens, T. D.; Tang, T. P. *Acc. Chem. Res.* **2002**, *35*, 984−995.
[200] Robak, M. T.; Merbage, M. A.; Ellman, J. A. *Chem. Rev.* **2010**, *110*, 3600−3740.
[201] Lu, B. Z.; Senanayake, C.; Li, N.; Han, Z.; Bakale, R. P.; Wald, S. A. *Org. Lett.* **2005**, *7*, 2599−2602.
[202] Davis, F. A.; Ramachandan, T.; Liu, H. *Org. Lett.* **2004**, *6*, 3393−3395.
[203] Yamada, K.-i.; Tomioka, K. *Chem. Rev.* **2008**, *108*, 2874−2886.
[204] Coté, A.; Boezio, A. A.; Charette, A. B. *Angew. Chem., Int. Ed.* **2004**, *43*, 6525−6528.
[205] Pizzuti, M. G.; Minnaard, A. J.; Feringa, B. L. *J. Org. Chem.* **2008**, *73*, 940−947.
[206] A. Charette, private communication.
[207] Comins, D. L.; Joseph, S. P.; Goehring, R. R. *J. Am. Chem. Soc.* **1994**, *116*, 4719−4728.
[208] Comins, D. L.; Joseph, S. P. In *Advances in Nitrogen Heterocycles*; Moody, C. J., Ed.; JAI: Greenwich, 1996; Vol. 2, pp. 251−294.
[209] Comins, D. L.; Chen, X.; Morgan, L. A. *J. Org. Chem.* **1997**, *62*, 7435−7438.

[210] Joseph, S.; Comins, D. L. *Curr. Opin. Drug Disc. Dev.* **1998**, *5*, 870−880.
[211] Comins, D. L. *J. Heterocycl. Chem.* **1999**, *36*, 1491−1500.
[212] Comins, D. L.; Higuchi, K. *Beilstein J. Org. Chem.* **2007**, *3*, article 42; doi 10.1186/1860-5397-1183-1142.
[213] Charette, A. B.; Grenon, M.; Lemire, A.; Pourashraf, M.; Martel, J. *J. Am. Chem. Soc.* **2001**, *123*, 11829−11830.
[214] Larivée, A.; Charette, A. B. *Org. Lett.* **2006**, *8*, 3955−3957.
[215] Focken, T.; Charette, A. B. *Org. Lett.* **2006**, *8*, 2985−2988.
[216] Lemire, A.; Grenon, M.; Pourashraf, M.; Charette, A. B. *Org. Lett.* **2004**, *6*, 3517−3520.
[217] Lemire, A.; Charette, A. B. *Org. Lett.* **2005**, *7*, 2747−2750.
[218] Comins, D. L.; Goehring, R. R.; Joseph, S. P. *J. Org. Chem.* **1990**, *55*, 2574−2576.
[219] Lemire, A.; Charette, A. B. *J. Org. Chem.* **2010**, *75*, 2077−2080.
[220] Comins, D. L.; La Munyon, D. H. *J. Org. Chem.* **1992**, *57*, 5807−5809.
[221] Comins, D. L.; Hong, H. *J. Am. Chem. Soc.* **1991**, *113*, 6672−6673.
[222] Comins, D. L.; Dehghani, A. *Tetrahedron Lett.* **1991**, *32*, 5697−5700.
[223] Comins, D. L.; Green, G. M. *Tetrahedron Lett.* **1999**, *40*, 217−218.
[224] Comins, D. L.; Brooks, C. A.; Al-awar, R. S.; Goehring, R. R. *Org. Lett.* **1999**, *1*, 229−232.
[225] Comins, D. L.; Huang, S.; McArdle, C. L.; Ingalls, C. L. *Org. Lett.* **2001**, *3*, 469−471.
[226] Reetz, M. T.; Kunisch, F.; Heitmann, P. *Tetrahedron Lett.* **1986**, *27*, 4721−4724.
[227] Narasaka, K. *Synthesis* **1991**, 1−11.
[228] Gregory, R. J. H. *Chem. Rev.* **1999**, *99*, 3649−3682.
[229] Gröger, H. *Chem. Eur. J.* **2001**, *7*, 5247−5251.
[230] North, M. *Tetrahedron: Asymmetry* **2003**, *14*, 147−176.
[231] Brunel, J.-M.; Holmes, I. P. *Angew. Chem., Int. Ed.* **2004**, *43*, 2752−2778.
[232] North, M.; Usanov, D. L.; Young, C. *Chem. Rev.* **2008**, *108*, 5146−5226.
[233] Wang, J.; Wang, W. T.; Li, W.; Hu, X. L.; Shen, K.; Tan, C.; Liu, X. H.; Feng, X. M. *Chem. Eur. J.* **2009**, *15*, 11642−11659.
[234] Liu, X. H.; Lin, L. L.; Feng, X. M. *Chem. Commun.* **2009**, 6145−6158.
[235] Belokon, Y. N.; Clegg, W.; Harrington, R. W.; Maleev, V. I.; North, M.; Pujol, M. O.; Usanov, D. L.; Young, C. *Chem. Eur. J.* **2009**, *15*, 2148−2165.
[236] Moberg, C.; Wingstrand, E. *Synlett* **2010**, 355−367.
[237] Mai, K.; Patil, G. *J. Org. Chem.* **1986**, *51*, 3545−3548.
[238] Shibasaki, M.; Kanai, M.; Mita, T. *Org. React.* **2008**, *70*, 1−119.
[239] Hayashi, M.; Matsuda, T.; Oguni, N. *J. Chem. Soc., Chem. Commun.* **1990**, 1364−1365.
[240] Hayashi, M.; Matsuda, T.; Oguni, N. *J. Chem. Soc., Perkin Trans. 1* **1992**, 3135−3140.
[241] Belokon, Y. N.; Caveda-Cepas, S.; Green, B.; Ikonnikov, N. S.; Khrustalev, V. N.; Larichev, V. S.; Moscalenko, M. A.; North, M.; Orizu, C.; Tararov, V. I.; Tasinazzo, M.; Timofeeva, G. I.; Yashkina, L. V. *J. Am. Chem. Soc.* **1999**, *121*, 3968−3973.
[242] Belokon, Y. N.; Green, B.; Ikonnikov, N. S.; Larichev, V. S.; Lokshin, B. V.; Moscalenko, M. A.; North, M.; Orizu, C.; Peregudov, A. S.; Timofeeva, G. I. *Eur. J. Org. Chem.* **2000**, 2655−2661.
[243] Narasaka, K.; Yamada, T.; Minamikawa, H. *Chem. Lett.* **1987**, *16*, 2073−2076.
[244] Hayashi, M.; Miyamoto, Y.; Inoue, t.; Oguni, N. *J. Chem. Soc., Chem. Commun.* **1991**, 1752−1753.
[245] Jiang, Y.; Zhou, X.; Hu, W.; Li, Z.; Mi, A. *Tetrahedron: Asymmetry* **1995**, *6*, 2915−2916.
[246] Pan, W.; Feng, X.; Gong, L.; Hu, W.; Li, Z.; Mi, A.; Jiang, Y. *Synlett* **1996**, 337−338.
[247] Hwang, C.-D.; Hwang, D.-R.; Uang, B.-J. *J. Org. Chem.* **1998**, *63*, 6762−6763.
[248] Chang, C.-W.; TzuYang, C.; Hwang, C.-D.; Uang, B.-J. *J. Chem. Commun.* **2002**, 54−55.
[249] You, J.-S.; Gau, H.-M.; Choi, M. C. K. *Chem. Commun.* **2000**, 1963−1964.
[250] Belokon, Y. N.; Gutnov, A. V.; Moskalenko, M. A.; Yashkina, L. V.; Lesovoy, D. E.; Ikonnikov, N. S.; Larichev, V. S.; North, M. *Chem. Commun.* **2002**, 244−245.
[251] Belokon, Y.; Carta, P.; Gutnov, A.; Maleev, V.; Moskalenko, M.; Yashkina, L.; Ikonnikov, N.; Voskoboev, N.; Khrustalev, V.; North, M. *Helv. Chim. Acta* **2002**, *85*, 3301−3312.
[252] Belokon, Y. N.; Blacker, A. J.; Carta, P.; Clutterbuck, L. A.; North, M. *Tetrahedron* **2004**, *60*, 10433−10447.
[253] Fuerst, D. E.; Jacobsen, E. N. *J. Am. Chem. Soc.* **2005**, *127*, 8964−8965.
[254] Zuend, S. J.; Jacobsen, E. N. *J. Am. Chem. Soc.* **2007**, *129*, 15872−15883.
[255] Ryu, D. H.; Corey, E. J. *J. Am. Chem. Soc.* **2004**, *126*, 8106−8107.

[256] Hamashima, Y.; Sawada, D.; Kanai, M.; Shibasaki, M. *J. Am. Chem. Soc.* **1999**, *121*, 2641−2642.
[257] Casas, J.; Nájera, C.; Sansano, J. M.; Sansano, J. M. *Org. Lett.* **2002**, *4*, 2589−2592.
[258] Shibasaki, M.; Kanai, M.; Funabashi, K. *Chem. Commun.* **2002**, 1989−1999.
[259] Corey, E. J.; Lee, T. W. *Chem. Commun.* **2001**, 1321−1329.
[260] Ohfune, Y.; Shinada, T. *Eur. J. Org. Chem.* **2005**, 5127−5143.
[261] Connon, S. J. *Angew. Chem., Int. Ed.* **2008**, *47*, 1176−1178.
[262] Davis, F. A.; Portonovo, P. S.; Reddy, R. E.; Chiu, Y.-h *J. Org. Chem.* **1996**, *61*, 440−441.
[263] Sigman, M. S.; Jacobsen, E. N. *J. Am. Chem. Soc* **1998**, *120*, 4901−4902.
[264] Vachal, P.; Jacobsen, E. N. *J. Am. Chem. Soc.* **2002**, *124*, 10012−10014.
[265] Vachal, P.; Jacobsen, E. N. *Org. Lett.* **2000**, *2*, 867−870.
[266] Sigman, M. S.; Vachal, P.; Jacobsen, E. N. *Angew. Chem., Int. Ed.* **2000**, *39*, 1279−1281.
[267] Zuend, S. J.; Jacobsen, E. N. *J. Am. Chem. Soc.* **2009**, *131*, 15358−15374.
[268] Zuend, S. J.; Coughlin, M. P.; Lalonde, M. P.; Jacobsen, E. N. *Nature* **2009**, *461*, 968−970.
[269] Fonseca, M. H.; List, B. *Curr. Opin. Chem. Biol.* **2004**, *8*, 319−326.
[270] Krueger, C. A.; Kuntz, K. W.; Dzierba, C. D.; Wirschun, W. G.; Gleason, J. D.; Snapper, M. L.; Hoveyda, A. H. *J. Am. Chem. Soc.* **1999**, *121*, 4284−4285.
[271] Porter, J. R.; Wirschun, W. G.; Kuntz, K. W.; Snapper, M. L.; Hoveyda, A. H. *J. Am. Chem. Soc.* **2000**, *122*, 2657−2658.
[272] Josephsohn, N. S.; Kuntz, K. W.; Snapper, M. L.; Hoveyda, A. H. *J. Am. Chem. Soc.* **2001**, *123*, 11594−11599.
[273] Perlmutter, P. *Conjugate Addition Reactions in Organic Synthesis*; Pergamon: Oxford, 1992.
[274] Tomioka, K.; Koga, K. In *Asymmetric Synthesis*; Morrison, J. D., Ed.; Academic: Orlando, 1983; Vol. 2, pp. 201−224.
[275] Rossiter, B. E.; Swingle, N. M. *Chem. Rev.* **1992**, *92*, 771−806.
[276] Schmalz, H.-G. In *Comprehensive Organic Synthesis. Selectivity, Strategy, and Efficiency in Modern Organic Chemistry*; Trost, B. M., Fleming, I., Eds.; Pergamon: Oxford, 1991; Vol. 4, pp. 199−236.
[277] Kanai, M.; Nakagawa, Y.; Tomioka, K. *J. Syn. Org. Chem. Jpn.* **1996**, *54*, 474−480.
[278] Krause, N. *Angew. Chem., Int. Ed.* **1998**, *37*, 283−285.
[279] Feringa, B. L. *Acc. Chem. Res.* **2000**, *33*, 346−353.
[280] Sibi, M. P.; Manyem, S. *Tetrahedron* **2000**, *56*, 8033−8061.
[281] Krause, N.; Hoffmann-Röder, A. *Synthesis* **2001**, 171−196.
[282] Berner, O. M.; Tedeschi, L.; Enders, D. *Eur. J. Org. Chem.* **2002**, 1877−1894.
[283] Chelucci, G.; Orru, G.; Pinna, G. A. *Tetrahedron* **2003**, *59*, 9471−9515.
[284] Hayashi, T.; Yamasaki, K. *Chem. Rev.* **2003**, *103*, 2829−2844.
[285] Bolm, C.; Legros, J.; Le Paih, J.; Zani, L. *Chem. Rev.* **2004**, *104*, 6217−6254.
[286] Vicario, J. L.; Badia, D.; Carrillo, L.; Etxebarria, J.; Reyes, E.; Ruiz, N. *Org. Prep. Proced. Int.* **2005**, *37*, 513−538.
[287] Ballini, R.; Bosica, G.; Fiorini, D.; Palmieri, A.; Petrini, M. *Chem. Rev.* **2005**, *105*, 933−971.
[288] Christoffers, J.; Koripelly, G.; Rosiak, A.; Rössle, M. *Synthesis* **2007**, 1279−1300.
[289] Lopez, F.; Minnaard, A. J.; Feringa, B. L. *Acc. Chem. Res.* **2007**, *40*, 179−188.
[290] Harutyunyan, S. R.; den Hartog, T.; Geurts, K.; Minnaard, A. J.; Feringa, B. L. *Chem. Rev.* **2008**, *108*, 2824−2852.
[291] Breit, B.; Schmidt, Y. *Chem. Rev.* **2008**, *108*, 2928−2951.
[292] Ay, S.; Nieger, M.; Brase, S. *Chem. Eur. J.* **2008**, *14*, 11539−11556.
[293] Alexakis, A.; Vuagnoux-d'Augustin, M.; Martin, D.; Kehrli, S.; Palais, L.; Henon, H.; Hawner, C. *Chimia* **2008**, *62*, 461−464.
[294] Alexakis, A.; Backvall, J. E.; Krause, N.; Pamies, O.; Dieguez, M. *Chem. Rev.* **2008**, *108*, 2796−2823.
[295] Jerphagnon, T.; Pizzuti, M. G.; Minnaard, A. J.; Feringa, B. L. *Chem. Soc. Rev.* **2009**, *38*, 1039−1075.
[296] Teichert, J. F.; Feringa, B. L. *Angew. Chem., Int. Ed.* **2010**, *49*, 2486−2528.
[297] Oppolzer, W.; Löher, H. J. *Helv. Chim. Acta* **1981**, *64*, 2808−2811.
[298] Oppolzer, W.; Moretti, R.; Godel, T.; Meunier, A.; Löher, H. *Tetrahedron Lett.* **1983**, *24*, 4971−4974.
[299] Helmchen, G.; Wegner, G. *Tetrahedron Lett.* **1985**, *26*, 6051−6054.
[300] Fleming, I.; Henning, R.; Parker, D. C.; Plaut, H. E.; Sanderson, P. E. J. *J. Chem. Soc. Perkin 1* **1995**, 317−337.
[301] Lopez, F.; Harutyunyan, S. R.; Meetsma, A.; Minnaard, A. J.; Feringa, B. L. *Angew. Chem., Int. Ed.* **2005**, *44*, 2752−2756.
[302] Hartmann, E.; Oestreich, M. *Angew. Chem., Int. Ed.* **2010**, *49*, 6195−6198.

[303] Schuppan, J.; Minnaard, A. J.; Feringa, B. L. *Chem. Commun.* **2004**, 792−793.

[304] Fillion, E.; Wilsily, A. *J. Am. Chem. Soc.* **2006**, *128*, 2774−2775.

[305] den Hartog, T.; Jan van Dijken, D.; Minnaard, A. J.; Feringa, B. L. *Tetrahedron: Asymmetry* **2010**, *21*, 1574−1584.

[306] Nishikata, T.; Yamamoto, Y.; Miyaura, N. *Chem. Commun.* **2004**, 1822−1823.

[307] Nishikata, T.; Yamamoto, Y.; Gridnev, I. D.; Miyaura, N. *Organometallics* **2005**, *24*, 5025−5032.

[308] Nishikata, T.; Yamamoto, Y.; Miyaura, N. *Angew. Chem., Int. Ed.* **2003**, *42*, 2768−2770.

[309] Nishikata, T.; Yamamoto, Y.; Miyaura, N. *Organometallics* **2004**, *23*, 4317−4324.

[310] Nishikata, T.; Yamamoto, Y.; Miyaura, N. *Chem. Lett.* **2005**, *34*, 720−721.

[311] Feringa, B. L.; Badorrey, R.; Peña, D.; Harutyunyan, S. R.; Minnaard, A. J. *Proc. Nat. Acad. Sci.* **2004**, *101*, 5834−5838.

[312] López, F.; Harutyunyan, S. R.; Minnaard, A. J.; Feringa, B. L. *J. Am. Chem. Soc.* **2004**, *126*, 12784−12785.

[313] Harutyunyan, S. R.; López, F.; Browne, W. R.; Correa, A.; Peña, D.; Badorrey, R.; Meetsma, A.; Minnaard, A. J.; Feringa, B. L. *J. Am. Chem. Soc.* **2006**, *128*, 9103−9118.

[314] Wu, T. R.; Chong, J. M. *J. Am. Chem. Soc.* **2007**, *129*, 4908−4909.

[315] Paton, R. S.; Goodman, J. M.; Pellegrinet, S. C. *J. Org. Chem.* **2008**, *73*, 5078−5089.

[316] Capon, R. J.; Faulkner, D. J. *J. Org. Chem.* **1984**, *49*, 2506−2508.

[317] Oppolzer, W.; Moretti, R.; Bernardinelli, G. *Tetrahedron Lett.* **1986**, *27*, 4713−4716.

[318] Oppolzer, W.; Stevenson, T. *Tetrahedron Lett.* **1986**, *27*, 1139−1140.

[319] Oppolzer, W. *Tetrahedron* **1987**, *43*, 1969−2004. correctly printed in erratum, p. 4057.

[320] Oppolzer, W.; Dudfield, P.; Stevenson, T.; Godel, T. *Helv. Chim. Acta* **1985**, *68*, 212−215.

[321] Murthy, K. S. K.; Rey, A. W.; Tjepkema, M. *Tetrahedron Lett.* **2003**, *44*, 5355−5358.

[322] Des Mazery, R.; Pullez, M.; López, F.; Harutyunyan, S. R.; Minnaard, A. J.; Feringa, B. L. *J. Am. Chem. Soc* **2005**, *127*, 9966−9967.

[323] ter Horst, B.; Feringa, B. L.; Minnaard, A. J. *Chem. Commun.* **2007**, 489−491.

[324] ter Horst, B.; Feringa, B. L.; Minnaard, A. J. *Org. Lett.* **2007**, *9*, 3013−3015.

[325] Mukaiyama, T.; Iwasawa, N. *Chem. Lett.* **1981**, 913−916.

[326] Chea, H.; Sim, H.-S.; Yun, J. *Adv. Synth. Catal.* **2009**, *351*, 855−858.

[327] Myers, A. G.; Yang, B. H. *Org. Synth.* **2004**, *Coll Vol. 10*, 12−15.

[328] Myers, A. G.; Yang, B. H.; Chen, H.; Gleason, J. L. *J. Am. Chem. Soc.* **1994**, *116*, 9361−9362.

[329] Myers, A. G.; Yang, B. H.; Chen, H.; McKinstry, L.; Kopecky, D. J.; Gleason, J. L. *J. Am. Chem. Soc.* **1997**, *119*, 6496−6511.

[330] Oppolzer, W.; Poli, G.; Kingma, A. J.; Starkemann, C.; Bernardinelli, G. *Helv. Chim. Acta* **1987**, *70*, 2201−2214.

[331] Cao, X.; Liu, F.; Lu, W.; Chen, G.; Yu, G.-A.; Liu, S. H. *Tetrahedron* **2008**, *64*, 5629−5636.

[332] Williams, D. R.; Kissel, W. S.; Li, J. J. *Tetrahedron Lett.* **1998**, *39*, 8593−8596.

[333] Sabala, R.; Hernández-García, L.; Ortiz, A.; Romero, M.; Olivo, H. F. *Org. Lett.* **2010**, *12*, 4268−4270.

[334] Esumi, T.; Shimizu, H.; Kashiyama, A.; Sasaki, C.; Toyota, M.; Fukuyama, Y. *Tetrahedron Lett.* **2008**, *49*, 6846−6849.

[335] Ocejo, M.; Carrillo, L.; Badía, D.; Vicario, J. L.; Fernández, N.; Reyes, E. *J. Org. Chem.* **2009**, *74*, 4404−4407.

[336] Pineschi, M.; Del Moro, F.; Di Bussolo, V.; Macchia, F. *Adv. Synth. Catal.* **2006**, *348*, 301−304.

[337] Williams, D. R.; Nold, A. L.; Mullins, R. J. *J. Org. Chem.* **2004**, *69*, 5374−5382.

[338] Morita, M.; Ishiyama, S.; Koshino, H.; Nakata, T. *Org. Lett.* **2008**, *10*, 1675−1678.

[339] Seebach, D.; Zimmerman, J.; Gysel, U.; Ziegler, R.; Ha, T.-K. *J. Am. Chem. Soc.* **1988**, *110*, 4763−4772.

[340] Meyers, A. I.; Leonard, W. R., Jr; Romine, J. L. *Tetrahedron Lett.* **1991**, *32*, 597−600.

[341] Hayashi, T.; Ueyama, K.; Tokunaga, N.; Yoshida, K. *J. Am. Chem. Soc.* **2003**, *125*, 11508−11509.

[342] d'Augustin, M.; Palais, L.; Alexakis, A. *Angew. Chem., Int. Ed.* **2005**, *44*, 1376−1378.

[343] Palais, L.; Mikhel, I. S.; Bournaud, C.; Micouin, L.; Falciola, C. A.; Vuagnoux-d'Augustin, M.; Rosset, S.; Bernardinelli, G.; Alexakis, A. *Angew. Chem., Int. Ed.* **2007**, *46*, 7462−7465.

Aldol and Michael Additions of Allyls, Enolates, and Enolate Equivalents

In this chapter, the discussion of additions to carbonyls continues with the aldol addition reaction and the mechanistically similar allyl addition reactions, both examples of "π-transfer" additions illustrated in Figure 4.1. Also discussed are asymmetric Michael addition reactions and conjugate additions of nitrogen nucleophiles.

The aldol condensation is one of the oldest reactions in organic chemistry, dating back to the first half of the nineteenth century, but about 1980 it underwent a renaissance after irreversible and highly stereoselective methods were developed. Much of the excitement and interest in asymmetric synthesis in the following decades was due to the development of aldol addition reactions and the mechanistically similar allyl addition reactions and their deployment in natural product total synthesis, particularly in the areas of polypropionate or macrolide chemistry. More recently, the area was renovated yet again by the advent of catalytic means of accomplishing this chemistry, including those utilizing "organocatalysts." We begin our discussion with allylation chemistry because allyl addition reactions are irreversible and the transition state assemblies are somewhat less complex than those of aldol additions.

Scheme 5.1 illustrates the transition structure most often invoked to explain the selectivities observed in π-transfer 1,2-carbonyl additions (*cf.* Figure 4.1): the so-called Zimmerman–Traxler transition structure [1]. This model, which was originally proposed to rationalize the selectivity of the Ivanov reaction, has its shortcomings (as will be seen) and suffers from an oversimplification when applied to enolates, in that it illustrates a monomeric enolate (*cf.* Section 3.1 and refs. [2–4]). Nevertheless, it serves the very useful purpose of providing a simple means to rationalize relative and absolute configurations in almost all of the asymmetric 1,2-additions we will see. The favored transition structure has *lk* topicity (*Si/Si* illustrated) because the alternative has a pseudo 1,3-diaxial interaction between the aldehyde phenyl and the magnesium alkoxide. Because the magnesium alkoxide is on a trigonal carbon in the 6-membered ring, this repulsive interaction is not large, and the selectivity for the *anti* product is only 76% [1].[1]

---

1. We will use the *syn/anti* nomenclature [5] to describe the relative configuration of aldol stereoisomers, and the *lk/ul* nomenclature [6] to describe the topicity of the reaction. For definitions, see glossary, Section 1.11.

Principles of Asymmetric Synthesis.
© 2012, Professor Robert E. Gawley and Professor Jeffrey Aubé. Elsevier Ltd. All rights reserved.

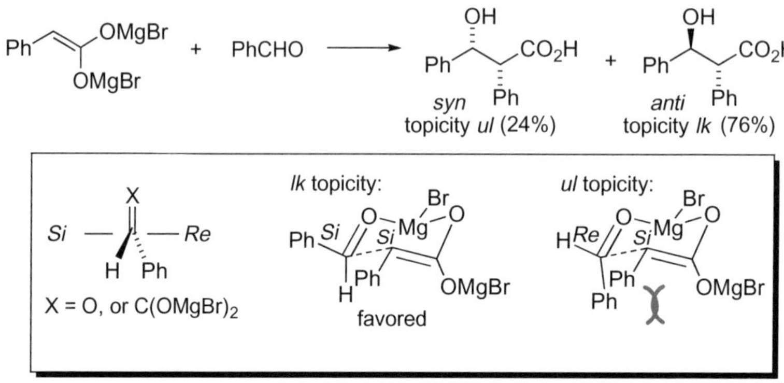

SCHEME 5.1 The Zimmerman−Traxler transition state model for the Ivanov reaction [1].

5.1 1,2-ALLYLATIONS AND RELATED REACTIONS

Most allylic organometallic or organometalloid systems are reactive enough to add to aldehyde carbonyls without the aid of additional Lewis acids, the notable exception being allylsilanes (reviews: refs. [7−14]). Often, the allylic metal or metalloid atom itself activates the carbonyl and a highly organized six-membered ring transition structure similar to the Zimmerman−Traxler model results. We mostly consider cases where chiral ligands on the metal or on an acid catalyst induce selectivity by interligand asymmetric induction. Reactions of allyl metal compounds in which the metal-bearing carbon is stereogenic are not covered.

To explain the chemistry of allylic metals, the reactions of allylic boron compounds [8,11−13] are covered in detail. The boron chemistry is divided into four parts: simple enantioselectivity (addition of CH_2=$CHCH_2$−, creating one new stereocenter), simple diastereoselectivity of crotyl additions (relative configuration after CH_3CH=$CHCH_2$− addition, where neither reagent is chiral), single asymmetric induction with chiral allyl boron compounds (one and two new stereocenters), and double asymmetric induction (both reactants chiral, one and two new stereocenters). The related classes of reactions, allenylations and propargylations, are also briefly covered, followed by a brief discussion of other allyl metal systems.

5.1.1 Simple Enantioselectivity

Scheme 5.2 illustrates the enantiomeric chair transition structures and products for the addition of an allyl borane to acetaldehyde. Note that in assembly a, the Re face of the aldehyde is attacked, producing the S-alcohol. Conversely, attack on the Si face of the aldehyde produces the R-alcohol (assembly b). In the inset are shown two alternative chair transition structures, which originate by reversing the position of the aldehyde methyl and hydrogen substituents of assemblies a and b (or equivalently, by flipping the chair). These are destabilized by severe 1,3-diaxial interactions between the aldehyde methyl and one of the ligands on boron. Note that the boron ligand is fully axial (unlike the pseudoaxial magnesium alkoxide in Scheme 5.1), and the boron−oxygen bond is fairly short.[2] These two differences mean that the repulsive interaction is quite strong, and the aldehyde is preferentially oriented with

---

2. A B−O bond is 1.36−1.47 Å, whereas a Mg−O bond is 2.0−2.1 Å [15].

its nonhydrogen substituent equatorially. Thus the simple concepts of conformational analysis of substituted cyclohexanes, applied to the Zimmerman−Traxler model, provide a basis for a "first approximation analysis" of these closed (cyclic) transition structures.

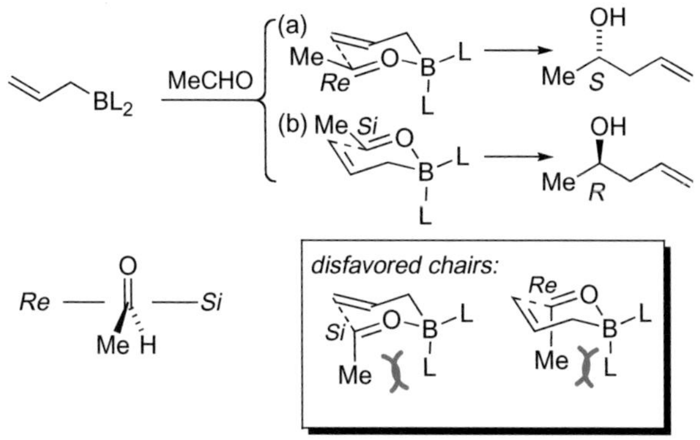

SCHEME 5.2   Cyclic transition states for allyl boron additions.

Unless there is a chiral ligand on boron, assemblies *a* and *b* of Scheme 5.2 are enantiomeric and the product will be racemic. If the ligand is chiral, then the transition structures are diastereomeric and the products will be formed in unequal amounts under conditions of kinetic control (Section 1.4). Figure 5.1 illustrates several chiral boron reagents that have been tested in the allyl boration reaction, with typical enantioselectivities for each.

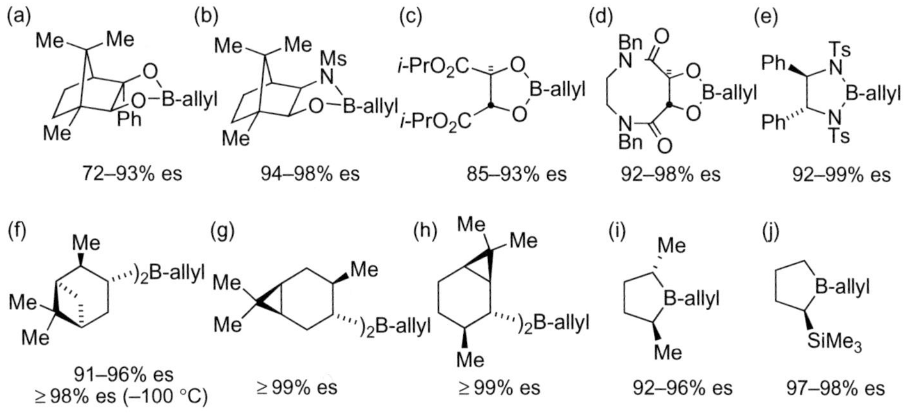

FIGURE 5.1   Chiral boron compounds for asymmetric allyl addition to achiral aldehydes, and representative enantioselectivities (a−e at 78 °C, g−j at −100 °C). (a) [16]; (b) [17]; (c) [18]; (d) [18]; (e) [19]; (f−h) [20−23]; (i−j) [24].

A particularly useful class of borane-based allylation reagents has been introduced by Soderquist. Prepared by ring expansion of B-methoxy-9-borabicyclononane followed by resolution *via* complexation of an ephedrine agent, these reagents undergo highly selective allylation and crotylation reactions with aldehydes (Scheme 5.3 [25]). They have the additional benefits of stability in air, of being generally crystalline, and of recyclability.

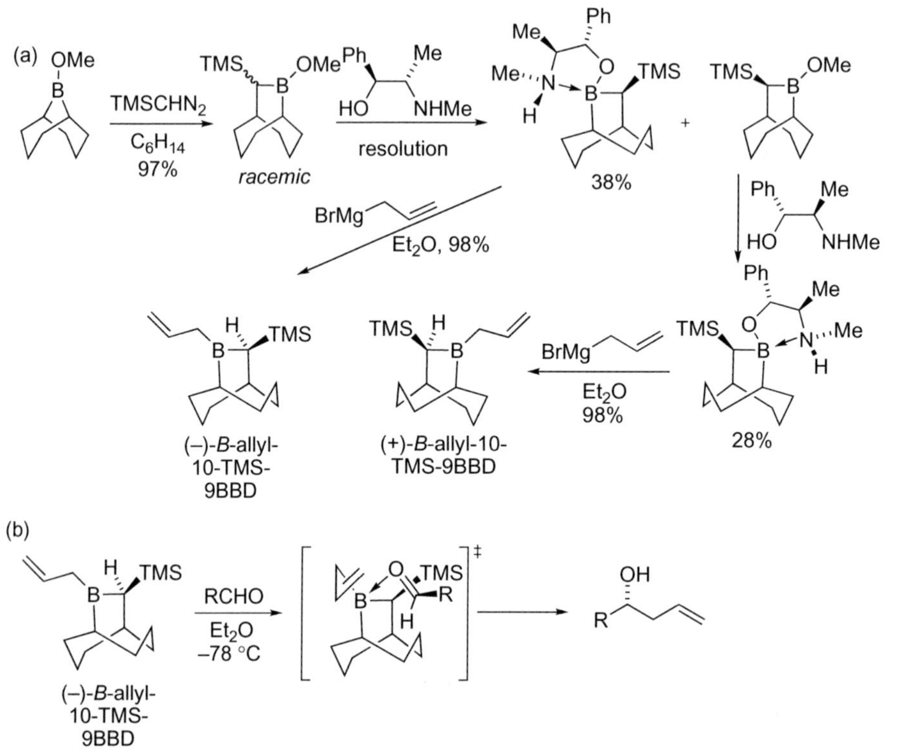

SCHEME 5.3 (a) Preparation, reactions of 10-(trimethylsilyl)-9-borabicyclo[3.3.2]decanes (10-TMS-9-BBDs) [25]. (b) A proposed transition state assembly.

5.1.2 Simple Diastereoselectivity

When there is a substituent on the allyl double bond, geometric isomers are possible and two new stereocenters are formed. The transition structures in Scheme 5.4 illustrate how the E-crotyl boron compound affords racemic *anti* addition product and the Z-crotyl compound affords the *syn* product.[3] For the E-isomer, the most stable chair presents the *Re* face of the aldehyde to the *Re* face of the double bond, or *vice versa* (*lk* topicity). These two transition structures are enantiomeric (and therefore isoenergetic in the absence of a chiral influence), as are the *anti* products. Likewise, the Z-crotyl species assembles with *ul* topicity, presenting the *Re* face of the aldehyde to the *Si* face of the double bond, or *vice versa*, which produces the *syn* addition product.

Note that reversing the face of only one component of the assembly reverses the topicity and the relative configuration of the stereocenters in the product. For example, exchanging the positions of the methyl and hydrogen in either the aldehyde or the crotyl moiety of the *lk* transition structure changes the topicity to *ul*, and the *syn* product would be produced. As before (Scheme 5.2), exchanging the aldehyde substituents causes severe 1,3-interactions

---

3. This assumes that there are no isomerizations that precede the addition. For discussions of such phenomena for boranes and boronates, see ref. [26].

with the axial boron ligand. Therefore, the tendency is for *lk* topicity for *E*-crotyl species, giving *anti* products and *ul* topicity for *Z*-crotyl compounds, giving *syn* products.

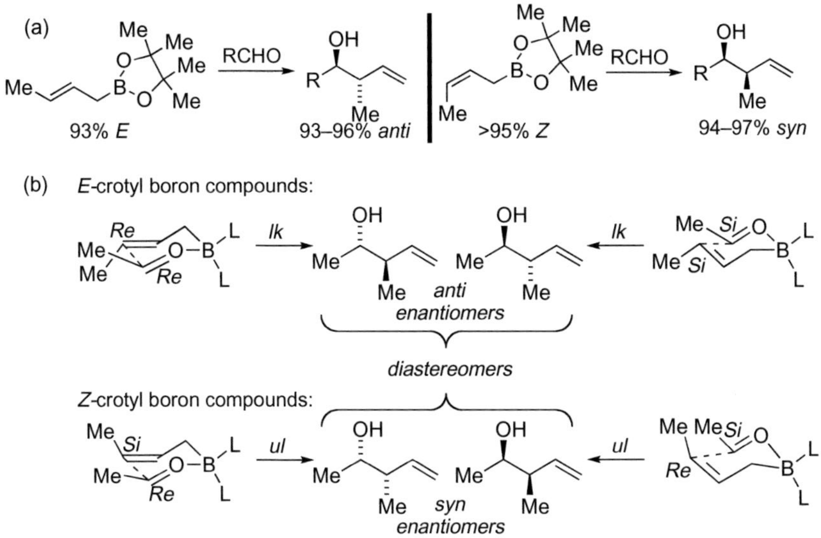

SCHEME 5.4 (a) Stereospecificity (within experimental error) of crotyl borane additions to aldehydes, R=Me, Et, *i*-Pr, Ph [26]. (b) Transition structures for stereospecific addition of crotyl boron compounds to aldehydes.

5.1.3 Single Asymmetric Induction

Figure 5.1 lists a number of auxiliaries for asymmetric allyl addition to aldehydes. Substituted allyl boron compounds have also been used in reactions with achiral aldehydes. Table 5.1 lists several examples of 2- and 3-substituted allyl boron compounds, and the products derived from their addition. Note that for the *E*- and *Z*-crotyl compounds, the enantioselectivity indicated is for the isomer illustrated. In some cases, there were more than one of the other three possible isomers found as well.

TABLE 5.1 Asymmetric Addition of Substituted Allyl Boron Compounds to Aldehydes[a]

Entry	RCHO	L₂BR	Product	% Yield	% es	Ref.
1	*E*-cinnamyl	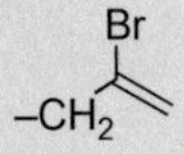 Br, −CH₂, L₂ = ent-5.1e	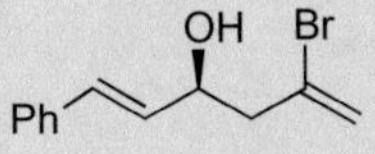 Ph ⤳ OH Br	79	94	[19]
2	*n*-C₅H₁₁−	"	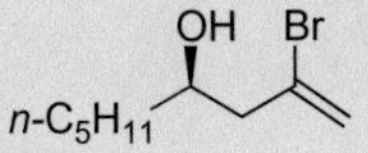 *n*-C₅H₁₁ OH Br	77	>99	[19]

(*Continued*)

TABLE 5.1 (Continued)

Entry	RCHO	L₂BR	Product	% Yield	% es	Ref.
3	Me—	Z-crotyl L₂ = 5.1f	(OH, Me, Me)	75	95	[27]
4	Me—	E-crotyl L₂ = 5.1f	(OH, Me, Me)	78	95	[27]
5	Me—	—CH₂ OMe L₂ = 5.1f	(OH, R, OMe) (R = Me)	59	95	[28]
6	Ph—	"	" (R = Ph)	75	95	[28]
7	n-C₆H₁₃—	—CH₂ SiMe₂N(i-Pr)₂ L₂ = 5.1f	(OH, R, SiMe₂N(i-Pr)₂) (R = n-C₆H₁₃)	52	>95	[29]
8	c-C₆H₁₁—	"	" (R = c-C₆H₁₁)	63	>95	[29]
9	Ph—	"	" (R = Ph)	50	>95	[29]
10	Me—	(B, H, TMS)	(OH, Me)	71	98	[25]
11	Me—	Me (B, H, TMS)	(OH, Me, Me)	68	97	[25]
12	Me—	Me (B, H, TMS)	(OH, Me, Me)	73	98	[25]

[a]Ligands are illustrated in Figure 5.1.

5.1.4 Double Asymmetric Induction

When the boron ligands and the aldehyde are both chiral, the inherent stereoselectivities of each partner may be matched or mismatched (Section 1.6). In principle, a chiral aldehyde could derive facial selectivity from either the Felkin−Anh−Heathcock model (Figures 4.8 and 4.10) or the Cram−chelate model (Figure 4.13). However, because the boron of these reagents can accept only one additional ligand, chelation is not possible. Therefore only the Felkin−Anh−Heathcock effects are operative in these reactions, and they are usually relatively weak, with diastereoselectivities of $\leq 70\%$. The high diastereoselectivities of many of the auxiliaries illustrated in Figure 5.1 can therefore be used to control the relative and absolute configuration of both of the new stereocenters in the addition product. Table 5.2 lists selected examples of double asymmetric induction with two α-alkoxyaldehydes and several auxiliaries (the 4,5-*anti* isomer is favored by Cram's rule).

TABLE 5.2 Double Asymmetric Induction in Addition of Allyl Boron Compounds to Aldehydes[a]

Entry	RCHO	L$_2$BR	Product	% Yield	% ds	Ref.
1	Ph⌄CHO, ŌMOM	allyl L$_2$ = 5.1e	Ph⌄5⌄4⌄, OH, ŌMOM	80	96	[19]
2	Ph⌄CHO, OMOM	allyl L$_2$ = 5.1e	Ph⌄5⌄4, OH, ŌMOM	–	98	[19]
3	Me⌄Me, O O, CHO	allyl L$_2$ = 5.1a	Me⌄Me, O O, 5 4, ŌH	87	96	[30]
4	"	allyl L$_2$ = 5.1c	"	85	93	[18,31]
5	"	allyl L$_2$ = 5.1d	"	84	98	[18]
6	"	allyl L$_2$ = *ent*-5.1c	Me⌄Me, O O, 5 4, ŌH	90	98	[31]
7	"	allyl L$_2$ = *ent*-5.1d	"	81	>99	[18,31]
8	"	*E*-crotyl L$_2$ = 5.1c	Me⌄Me, O O, 5 4 3 Me, ŌH	85	96	[8,32,33]

(Continued)

TABLE 5.2 (Continued)

Entry	RCHO	L₂BR	Product	% Yield	% ds	Ref.
9	"	E-crotyl L₂ = 5.1i	"	74	86	[24]
10	(aldehyde structure)	E-crotyl L₂ = 5.1a	(product structure)	85	72	[30]
11	"	E-crotyl L₂ = ent-5.1c	"	87	87	[8,32,33]
12	"	E-crotyl L₂ = ent-5.1i	"	71	96	[24]
13	"	Z-crotyl L₂ = 5.1a	(product structure)	86	>98	[30]
14	"	Z-crotyl L₂ = 5.1c	"	90	76	[8,33]
15	"	Z-crotyl L₂ = ent-5.1c	"	84	>99	[8,33]
16	"	Z-crotyl L₂ = ent-5.1i	"	66	92	[34]
17	"	Z-crotyl L₂ = 5.1i	(product structure)	65	82	[34]

[a]Ligands are illustrated in Figure 5.1.

Noteworthy among these examples is the ability to achieve high diastereoselectivity for both the 3,4-*syn* and 3,4-*anti* isomers, almost independent of the chirality sense of the aldehyde. Comparison of several examples show the expected trends for matched and mismatched pairs (*cf.* entry pairs 1/2, 4/6, 5/7, 9/12, 16/17). Note that either 3,4-*anti* diastereomer can be obtained with 96% ds (entries 8 and 12); the two 3,4-*syn* isomers are also available selectively (entries 13–16 and 17), although only one ligand (Figure 5.1i) is selective for the 3,4-*syn*-4,5-*syn* product (entry 17) that is a mismatched pair (*cf.* entry 16). Note that with Roush's tartrate ligand (Figure 5.1c), the E-crotyl mismatched pair is more selective than the matched pair (entries 8/11; for a rationale, see ref. [33]), and the matched and mismatched pair give the same major product isomer with the Z-crotyl compound (entries 14/15).

Several substituted allyl and crotyl derivatives have been designed to increase the usefulness of the boron-mediated allyl addition of aldehydes. For example, silanes such as those shown in Table 5.1, entries 7–9, can be stereospecifically converted to hydroxyls [29,35] or transformed into alcohols with a formal 1,3-hydroxy migration [36]. Additionally, vinyl bromides such as those shown in Table 5.1, entries 1 and 2 can be converted into a number of functional groups by standard chemical means [19]. Examples of these transformations are shown in Scheme 5.5. Note also that ozonolysis of any of these adducts give "aldol" adducts (Section 5.2).

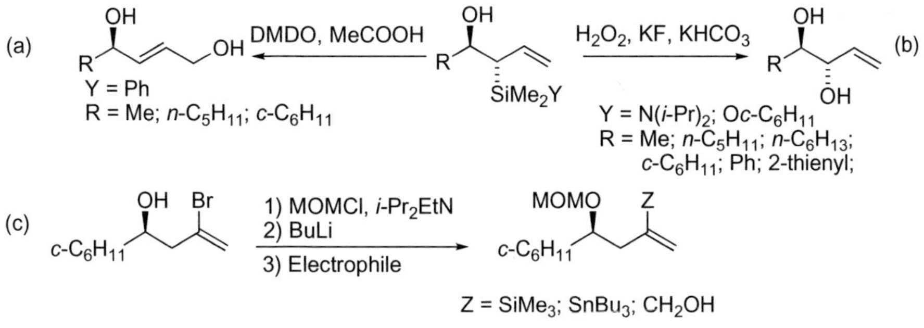

SCHEME 5.5 Transformations of functionalized addition products. (a) [36]; (b) [29,35]; (c) [19].

The concept of allylation can be readily extended to the related reactions of allenylboranes, which add to carbonyl compounds to afford homopropargylic alcohols, and propargylboranes, which give allenes as products. The concept is illustrated in the context of the Soderquist reagents shown in Scheme 5.6, which are prepared analogously to the method depicted in Scheme 5.3 [37]. These examples, which are particularly impressive because they involve ketones as electrophiles (as expected, aldehydes also work well [38]), are complementary to allylation methods because the products can be converted to a variety of other materials. For example, the propargylic reagent affords allenes that, after ozonolysis, provide products that are formally equivalent to the additions of chiral acyl anions.

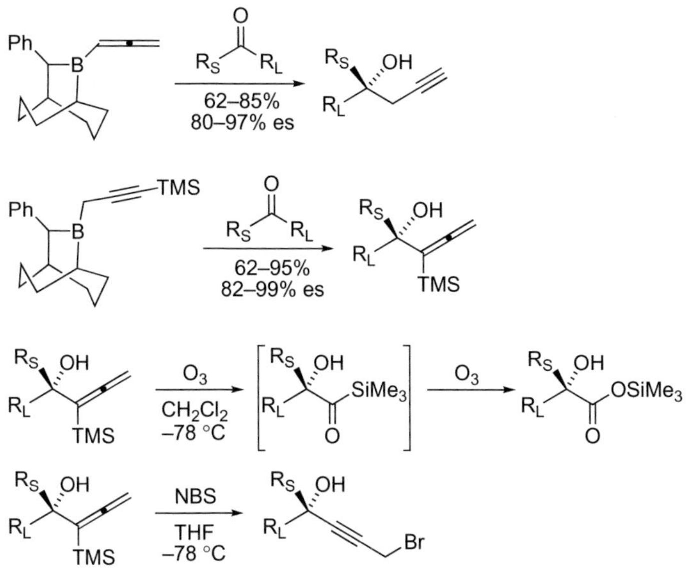

SCHEME 5.6 Additions of allenyl and propargyl boranes to ketones, along with representative transformations of the products [37].

5.1.5 Other Allyl Metals

Numerous other metals have been used in π-transfer addition reactions (reviews: refs. [7–10, 13,14,39]). Based on stereochemical tendencies and mechanistic considerations, these reagents have been classified into three groups, as illustrated in Scheme 5.7 [8,40]:

> *Type 1.* Reagents that are stereospecific in the sense that an *E*-crotyl isomer affords the *anti* addition product (*lk* topicity) and a *Z*-crotyl isomer affords the *syn* product (*ul* topicity). The transition structure is thought to be a closed chair, analogous to the Zimmerman– Traxler transition structure (Scheme 5.1).
>
> *Type 2.* Reactions that are catalyzed by Lewis acids and are stereoconvergent to *syn* adducts for either the *E*- or the *Z*-crotyl organometallics (*ul* topicity). The transition structure is usually considered to be open (acyclic), but the exact nature of the transition state is still a matter of discussion [8,9].
>
> *Type 3.* Allyl organometallics that are (usually) generated *in situ* and which equilibrate to the more stable *E*-crotyl species, then add *via* a closed, Zimmerman–Traxler transition structure, producing *anti* adducts preferentially [8].

The boron-containing compounds discussed in the previous sections are typical of Type 1 reagents. Also included in this group are reactions of allyl aluminums and uncatalyzed reactions of allyl tin reagents [8,40].

Reactions that fit Type 2 are catalyzed by Lewis acids that coordinate to the carbonyl oxygen of the aldehyde, thereby precluding coordination by the allyl metal. Such reactions proceed *via* an open transition state. As indicated previously, allylsilanes are not reactive enough to add to aldehydes without acid catalysis, so they fall into this category [40]. Allylstannane additions may be catalyzed by Lewis acids, so stannanes sometimes fall into

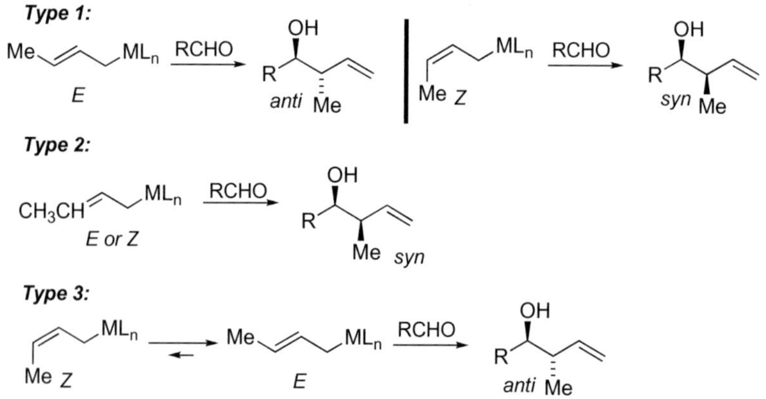

SCHEME 5.7 Mechanistic types for allyl addition to carbonyls. Types 1 and 3 proceed through transition structures similar to those in Scheme 5.4 [8,40].

this group as well [41,42], as do allyl titanium reagents [8,9,40]. Scheme 5.8 shows some enantioselective examples of allylsilane [43] and allylstannane [44,45] additions; many enantioselective additions of allylstannanes involve chirality transfer from the stannane where the allylic carbon bearing the tin is stereogenic [9,14], and are not discussed herein.

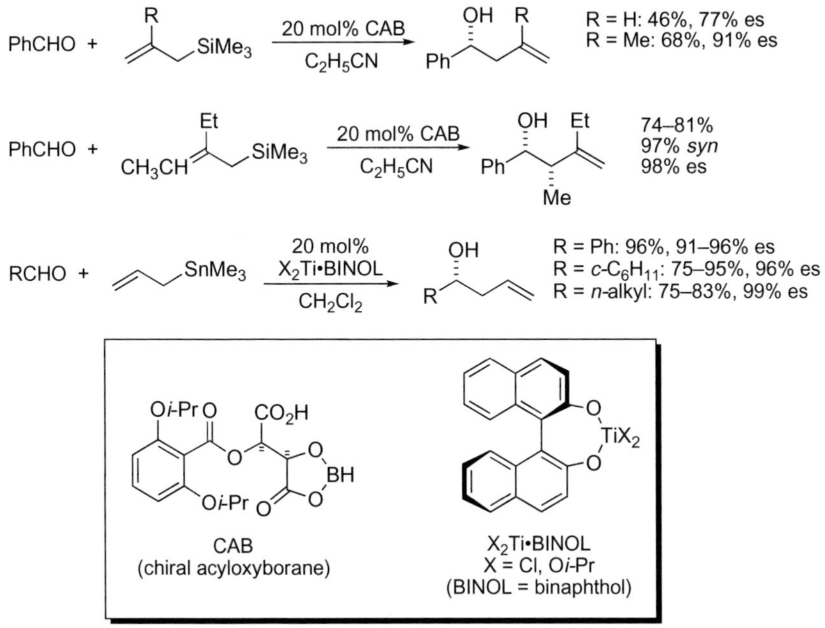

SCHEME 5.8 Enantioselective additions of allylsilanes [43] and allylstannanes [44,45], mediated by chiral catalysts.

Figure 5.2 illustrates the six possible open transition structures for the Lewis acid mediated addition of allyl metals to an aldehyde. Note that for each topicity, there are two *syn-clinal* arrangements and one *antiperiplanar*. Several factors must be considered in explaining the observed *ul* topicity of these reactions (giving *syn* relative configuration in the products), and a number of rationales have been offered. If one assumes that the conformation is *antiperiplanar* in the transition state, then structure *a* would be favored over *d*, since this arrangement minimizes the interaction between in the aldehyde substituent, R, and the methyl of the crotyl group.

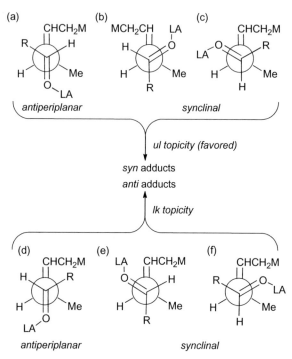

FIGURE 5.2 Newman projections of possible open transition structures for Lewis acid (LA) catalyzed additions to aldehydes.

On the other hand, Seebach suggested in 1981 that the topicity of a number of reactions (including these) may be explained by having the double bonds oriented in a *synclinal* arrangement.[4] He reasoned that steric repulsion between the R and CHCH$_2$M moieties would favor *b* over *c* and *e* over *f*. Then, assuming that the nucleophile approaches the carbonyl along the Bürgi–Dunitz trajectory (Section 4.1.3), either the hydrogen (in *b*) or the methyl group (in *e*) must be squeezed in between the alkyl group and hydrogen of the aldehyde. The former would be favored. This hypothesis was offered as a "topological rule" (not a mechanism).

Later, studies of intramolecular silane [40] and stannane [42] additions offered a direct comparison between *synclinal* arrangement *c* and *antiperiplanar* arrangement *d*, the former being favored. Because of the intramolecular nature of the addition, conformations analogous to the other possibilities were not possible.

The reactions of some electron-deficient allylsilanes can be promoted by Lewis base catalysts (for a general review of this subject see ref. [46]). Following up on early work by Kobayashi [47], who showed that the reactions of allyltrichlorosilane were accelerated by the additives such as HMPA or dimethylformamide, Denmark developed a series of chiral phosphoramides that promoted stereospecific (*i.e.*, Type 1) allylations with good enantioselectivities (Scheme 5.9a) [48]. A recent example is given in Scheme 5.9b [49]. The transition structure features the simultaneous

---

4. For a discussion of the Seebach rule as applied to the Michael reaction, see Figure 5.12 and the accompanying discussion.

complexation and activation of both aldehyde and allylsilane in a familiar chair-like assembly. Notably, this is an example of "organocatalysis" (because there is no metal Lewis acid involved), a topic of which more will be said later in this chapter.

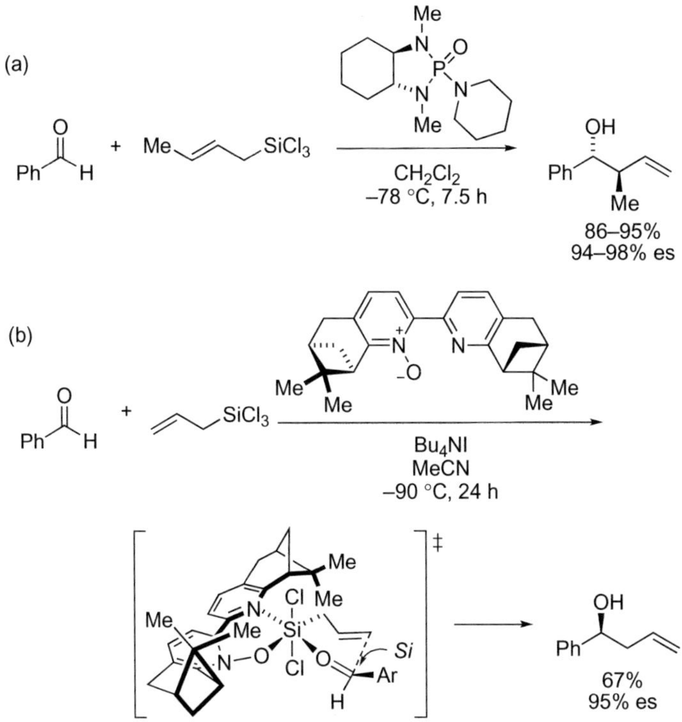

SCHEME 5.9   Examples of Lewis base catalysis of silane additions by (a) Denmark [48], and (b) Malkov and Kočovsky [49].

Allyl chromium, titanium, and zirconium reagents fall into the Type 3 category. Enantioselective reactions in this class are relatively rare, although the diastereoselectivities can be quite high (reviews: [7,8,14]).

Allyl- and crotyl-silanes remain popular tools in organic synthesis because they are versatile and easy to use. An attractive alternative to carrying out asymmetric synthesis with an achiral silane reaction partner has been extensively explored by Panek (review: ref. [50]). Scheme 5.10 shows a sophisticated example of this chemistry wherein a chiral, enantiomerically enriched allylsilane reagent reacts with a chiral aldehyde [51]. In both reactions, strict Felkin control leads to exclusive addition from the *Si* face of the aldehyde. With respect to the allylsilane, open transition states in which *anti* approach of the electrophile relative to the C−Si bond, which is perpendicular to the plane of the alkene group to maximize stabilization of the incipient β-cationic center in the transition state, are observed [52]. When combined, the two cases shown switch between *synclinal* and *antiperiplanar* transition states to afford *anti* vs. *syn* products with high diastereoselectivity.

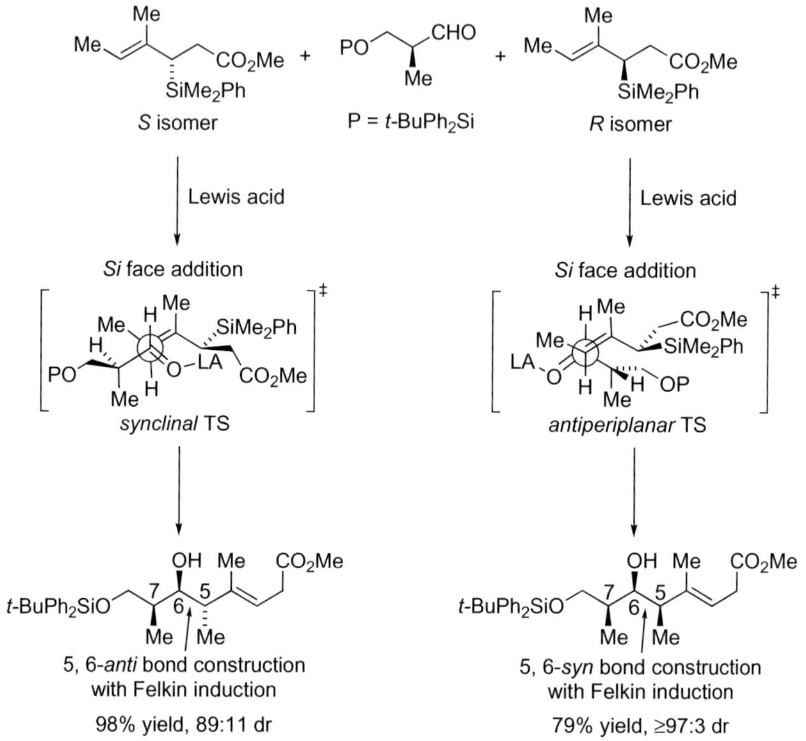

SCHEME 5.10 Examples of the addition of a chiral allylsilane to an aldehyde. Note the proposed dependence of transition states upon Lewis acid [51].

5.2 ALDOL ADDITIONS[5]

The Ivanov reaction (Scheme 5.1) is an early example of an aldol addition reaction that proceeded selectively. There has been an enormous amount of work done in this area, and only a small amount of the developmental work will be covered here. A large number of chiral auxiliaries and catalysts have been developed, but we will concentrate on only a few, which suffice to provide an understanding of the structural factors that influence selectivity. The transition structures presented in the following discussion are oversimplifications in that the enolate and its metal are represented as monomers, when in fact they are not [2–4]. On the other hand, much of the available data may be rationalized on the basis of these structures, so the simplification is justified in the absence of detailed structural and configurational information about mixed aggregate transition structures. We will discuss aldol reactions in two contexts. Here, we concentrate on metal-mediated aldol reactions, with an emphasis on important methods that use chiral auxiliaries and Lewis acid catalysts. Later, we address the exciting new dimensions that organocatalysis has brought to this area.

---

5. Note the distinction between the aldol *condensation*, in which α,β-unsaturated carbonyls are formed, and the aldol *addition,* which is stopped at the β-hydroxy carbonyl stage. For reviews of the early literature, mostly focusing on the aldol condensation, see refs. [53,54]. Numerous reviews of the aldol addition have appeared [15,54–57] (Li and Mg enolates), [58,59] (B and Al enolates), and [60,61] (transition metal enolates). For a two-volume treatise on the subject see refs. [62,63].

5.2.1 Simple Diastereoselectivity

Kinetic control. The Zimmerman–Traxler model, as applied to propionate and ethyl ketone aldol additions, is perhaps the most commonly invoked model of stereochemical induction in organic chemistry (Scheme 5.11; note the similarity to the boron-mediated allyl additions in Scheme 5.2). One main feature of this model is that it predicts a significant dependence of stereoselectivity on the enolate geometry, which is in turn dependent on the nature of X and the deprotonating agent (see section 3.1.1). In addition, the configuration and selectivity of the kinetically controlled aldol addition is dependent on the size of the substituents on the two reactants.

Figure 5.3 illustrates both enantiomers of most of the possible transition structures that have been postulated for aldol additions of R_1CH_2COX enolates. In the closed transition structures (Figure 5.3a and b), the chair conformations would normally be expected to pre-dominate, but in certain instances a boat may be preferable.[6] For the open transition structures, study of the intramolecular addition of silyl ethers, catalyzed by Lewis acids, showed a moderate preference for an *anti* conformation [69]. In intermolecular cases, the choice between open structures of *ul* or *lk* topicity will be governed by the relative magnitude of the gauche interactions between R_1 and either R_2 or ML_n on the aldehyde.

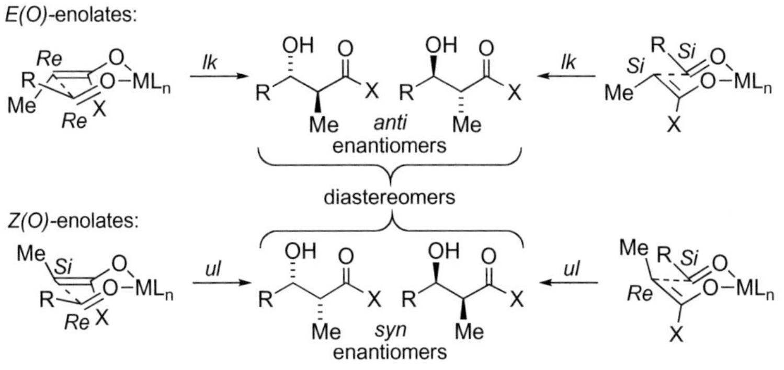

SCHEME 5.11 Transition structures for stereoselective propionate additions to aldehydes.

---

6. Computational studies predict that the geometry (chair, half chair, boat, etc.) depends on the nature of R_1, R_2, and M. Theory also predicts that Z-enolates prefer a closed chair, but that E-enolates may prefer a boat [64–66]. For an empirical rule for predicting aldol topicity, see ref. [67]. For an investigation into the effect of metal and solvent on transition structures, see ref. [68].

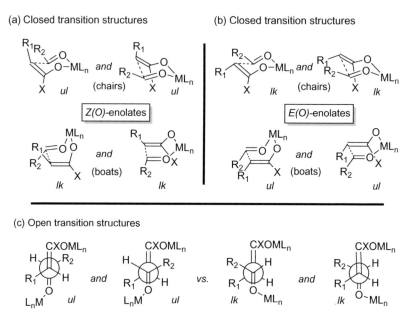

FIGURE 5.3 (a) Chair and boat transition structures for $Z(O)$-enolates. (b) Chair and boat transition structures for $E(O)$-enolates. (c) *ul* and *lk* open transition structures. Note that in all cases, the topicity is such that *ul* → *syn*; *lk* → *anti*.

The stereoselectivity of the aldol addition often depends on the selectivity of enolate formation. Ireland's rationale for the selective formation of lithium $Z(O)$-enolates of ketones, amides, and imides, and the selective formation of ester $E(O)$-enolates was discussed in Section 3.1.1 [70]. The rationale for the selective formation of $E(O)$- and $Z(O)$-boron enolates by reaction with dialkylboron triflate and a tertiary amine [71] is shown in Scheme 5.12 [58,72]. The boron triflate coordinates to the carbonyl oxygen, thereby increasing the acidity of the α-proton so that it can be removed by amine bases, as shown in Scheme 5.12a. In most cases, the stereochemical situation is as shown in Scheme 5.12b. The boron is *trans* to the CH_2R_1, R_1 is *antiperiplanar* to X, and removal of the H_{Re} proton gives the $Z(O)$-enolate. Note that for amides (X = NR_2), $A^{1,3}$ strain between R_1 and NR_2 particularly destabilizes the $E(O)$-enolate. In certain instances, a repulsive van der Waals interaction between the X and BR_2OTf moieties may be particularly severe (*e.g.*, *t*-BuS— and Bu_2BOTf), such that the boron is oriented *trans* to X, which forces R_1 *synperiplanar* to X to avoid the boron ligands, as illustrated in Scheme 5.12c. Removal of the H_{Si} proton then gives the $E(O)$-enolate.

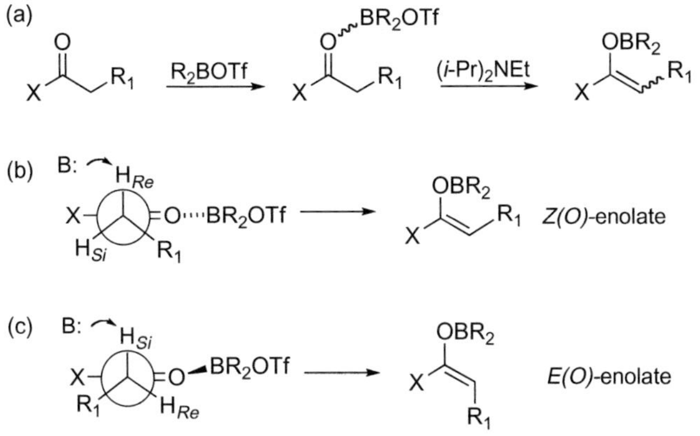

SCHEME 5.12   (a) Rationale for the stereoselective formation of boron enolates [72]. (b) When the boron is *cis* to X, R$_1$ may orient *anti* to X, and the Z(O)-enolate ensues. (c) If the boron is *trans* to X, A^{1,3} strain considerations force R$_1$ *syn* to X, and removal of the proton from a conformation in which the C—H bond is perpendicular to the carbonyl affords the E(O)-enolate.

Not all aldol additions exhibit a dependence of product configuration on enolate geometry. Acid-catalyzed aldols [55], some base-catalyzed aldols [68], and aldols of some transition metal enolates [73,74] show no such dependency. For example, zirconium enolates afford *syn* adducts (*ul* topicity) independent of enolate geometry for a number of propionates [73,74]. As shown in Scheme 5.13, two explanations have been proposed to explain the behavior of zirconium enolates. One explanation (Scheme 5.13a) is that the closed transition structure changes from a chair for the Z(O)-enolate to a boat for the E(O)-enolate [15,73,75]. Another hypothesis is that these additions occur *via* an open transition structure. Although the original authors [74] suggested an open transition structure, they did not provide an illustration. Heathcock has proposed an open transition structure similar to the one illustrated in Scheme 5.13b for an acid-catalyzed aldol addition where the Lewis acid on the oxygen is small [76]. According to this rationale, the topicity is determined by the relative energies of the van der Waals interactions between the methyl group and either the Lewis acid or R group [76]. Heathcock postulates that when the Lewis acid is small, *ul* topicity is preferred, since it minimizes the gauche interactions between the methyl and R in the forming bond. In the case of the zirconium enolates, there is an equivalent of lithium chloride present from transmetalation of the lithium enolate with Cp$_2$ZrCl$_2$, which can act as a (small) Lewis acid. The transition structure illustrated in Scheme 5.13b is then favored because it relieves the gauche interaction between the methyl and R in the forming bond.

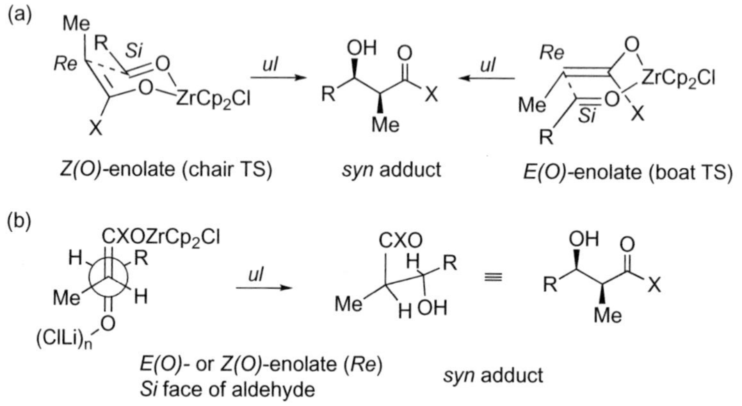

SCHEME 5.13 Explanations for the *ul* selectivity of *E(O)*- and *Z(O)*-zirconium enolates: (a) *Z(O)*-enolate chair and *E(O)*-enolate boat [15,73]; (b) open structure [74] (see also ref. [76]).

Thus, two explanations rationalize the same result. There is a valuable lesson here: although transition state models may serve a useful predictive value, they may or may not depict reality. The scientific method allows you to test a hypothesis, but consistency with a hypothesis does not constitute a proof: it constitutes a failure to disprove the hypothesis.

Thermodynamic control. It is also possible for the aldolate adduct to revert to aldehyde and enolate, and equilibration to the thermodynamic product may afford a different diastereomer (the *anti* aldolate is often the more stable). The tendency for aldolates to undergo the retro aldol addition increases with the acidity of the enolate: amides < esters < ketones (the more stable enolates are more likely to fragment), and with the steric bulk of the substituents (bulky substituents tend to destabilize the aldolate and promote fragmentation). On the other hand, a highly chelating metal stabilizes the aldolate and retards fragmentation. The slowest equilibration is with boron aldolates, and increases in the series lithium < sodium < potassium, and (with alkali metal enolates) also increases in the presence of crown ethers.[7]

To achieve kinetic control in the aldol addition reaction, it does not matter if the rate for the retro aldol is fast, as long as the difference between the relative rates for *syn* vs. *anti* addition is large. As an example, consider the following "case study." Let us assume that the rate of *ul* addition for a *Z(O)*-enolate (to give *syn* adduct) is significantly faster than the rate of *lk* addition (giving *anti* adduct), such that $k_{syn}/k_{anti} = 100$. Under these conditions, a retro aldol must occur 100 times before one *syn* → *anti* isomerization can occur. The rates of these individual processes can be measured with experiments such as those illustrated in Scheme 5.14 [78]. In Scheme 5.14a, aldehyde exchange clearly involves a retro aldol, and has a half-life of 15 min. In Scheme 5.14b, isomerization to the more stable *anti* isomer has a half-life of 8 h at a higher temperature. Because the retro aldol and the *ul* addition are both much faster than the unfavored *lk* addition, even the crossover is *syn*-selective.

---

7. For a thorough discussion of the factors affecting the equilibration of aldolates, see ref. [15]. For a procedure for thermodynamic equilibration, see ref. [77].

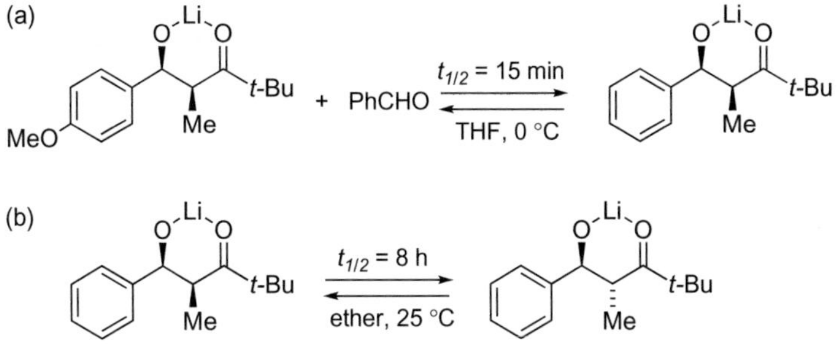

SCHEME 5.14 (a) Aldehyde exchange and (b) *syn-anti* isomerization of aldolates [78].

In summary, the following generalizations have emerged for aldol additions under kinetic control:

- $Z(O)$-enolates are highly *syn*-selective (*ul* topicity) when X is fairly large [79].
- $Z(O)$-enolates with a large R_1 (such as an isopropyl or *tert*-butyl) give *anti* products (*lk* topicity) selectively [79].
- $E(O)$-enolates are highly *anti*-selective (*lk* topicity) only with a very large X group (such as 2,6-di-*tert*-butylphenol) [79].
- For a closed transition structure, shorter M–O bond lengths amplify the van der Waals interactions between R_1, R_2, and X relative to enolates with longer bond lengths, resulting in higher stereoselectivities [15]. With boron enolates for example, $Z(O)$-enolates are highly *syn*-selective [58].

A general solution to the problem of generating *anti* products (*lk* topicity) in high selectivity across a broad range of substrate types has not yet been found. A few special conditions that accomplish this in a limited setting are known (*e.g.*, the Paterson aldol of lactate-derived ketones [59]) but this remains an area of active research.

5.2.2 Single Asymmetric Induction

For the addition of acetate and methyl ketone enolates (one new stereocenter), a number of approaches have been taken to induce enantioselectivity (review: ref. [80]); one of these methods will be mentioned in the succeeding section, along with the propionate and ethyl ketone additions. In the open transition structures of Figure 5.3, each illustrated *lk* or *ul* pair is enantiomeric in the absence of any stereocenters in the two reactants. Introduction of a chirality element converts the paired transition structures (*i.e.*, transition structures of the same topicity) and products from enantiomers to diastereomers, and allows diastereoselection under either kinetic or thermodynamic control. There are three opportunities for introduction of chirality: a chiral auxiliary (X*), and the two sites of Lewis acid ($ML*_n$) coordination: the enolate and the aldehyde oxygens. In principle, either one, two, or three could be chiral, allowing for the possibility of single, double, and even triple asymmetric induction.

The following discussion is organized by the "location" of the introduced chirality: X (intraligand asymmetric induction) or ML_n (interligand asymmetric induction). Additionally, there is the possibility of a stereogenic center in the aldehyde, which will normally have an observable influence only in cases where the stereocenter is close to the carbonyl (*i.e.*, Cram's rule situations – see Section 4.1). Most of the examples that have been published to date include stereogenic centers in either X or ML_n, but not both.

Intraligand asymmetric induction. The first example of an auxiliary-based asymmetric aldol addition was reported by the Enders group, who used the enolate of a *SAMP* hydrazone in a crossed aldol (see Section 3.13 for the use of these auxiliaries in alkylation chemistry, ref. [81]). This method afforded good yields, but only modest selectivities. Introduction of chirality in X (Figure 5.3) produces an enolate that affords much higher selectivities. Some of the more popular and effective auxiliaries are shown in Figure 5.4. The first of these (Figure 5.4a, R = methyl) was evaluated in racemic form by Heathcock in 1979, as its lithium Z(O)-enolate [82,83]. Later, a synthesis of the *S*-enantiomer from *S-tert*-leucine (*S-tert*-butylglycine) was reported [84]. A similar auxiliary was reported by the Masamune group in 1980 (Figure 5.4b, R = methyl), which afforded outstanding selectivities as its boron Z(O)-enolate. Initially [5] the racemate was resolved, but subsequently a synthesis from mandelic acid was reported [85]. Both the Heathcock and the Masamune auxiliaries are self-immolative (*cf.* Section 1.2, p. 2): "removal" of the auxiliary by oxidative cleavage of the α-alkoxyketone to a carboxylic acid destroys the stereocenter. Figure 5.4c illustrates one of the most frequently used auxiliaries, the oxazolidinone imides introduced by the Evans laboratory in 1981 [86]. These auxiliaries, which are made from amino alcohols such as valinol and phenylalaninol, can be cleaved to acid, aldehyde, or an alcohol [87,88] (*cf.* Scheme 3.16 and 3.17), and the auxiliary can be recovered in good yield. Another particularly useful reaction is the reduction of the auxiliary-containing amide by Cp$_2$Zr(H)Cl, which affords an aldehyde directly [89].

Reaction of either the boron Z(O)-enolates [86] or the zirconium E(O)- or Z(O)-enolates [90] are highly selective. Most of the auxiliaries shown in Figure 5.4 are poorly selective when R is hydrogen (*i.e.*, "acetate" enolates). The exceptions are the Crimmins thiazolidinethione (Figure 5.4c, X = S) and the Helmchen acetate shown in Figure 5.4d. Another regularly used class of aldol auxiliaries (Figure 5.4e) are the camphor sultams developed by Oppolzer [91−94] (*cf.* Scheme 3.18). Most of these auxiliaries (Figure 5.4a being the exception) are available as either enantiomer, making either enantiomer of any aldol adduct available. In the following discussions, only one enantiomer is illustrated, and it should be recognized that the other is also generally available.

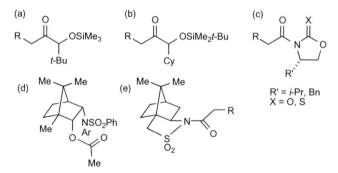

FIGURE 5.4 Chiral auxiliaries for asymmetric aldol additions. (a) As racemate [82,83], enantiopure from *tert*-leucine (*tert*-butyl glycine) [84]; (b) from mandelic acid [5,85]; (c) from valine or phenylalanine [86,95]; (d) from camphor [96]; *(e)* from camphor [92,93].

The Heathcock and Masamune auxiliaries (Figure 5.4a,b) are structurally and conceptually similar, and will be discussed together. Scheme 5.15 illustrates two possibilities that can arise in these systems, depending on the metal and the substituents on silicon: a chelated or nonchelated orientation in the transition structure. Note that, for the *S*-enantiomer illustrated, the chelated enolate has the R group (*tert*-butyl or cyclohexyl) oriented to the rear, and the *front* face of the enolate is most accessible to the electrophile. Conversely, the nonchelated structure has the dipoles of the C=O and C−O bonds aligned in opposition, with the R group now projecting to the front of the structure leaving the *rear* face more accessible.

If both the Z(O)- and the E(O)-enolates can be made, and if both follow the Zimmerman−Traxler models (*i.e.*, chair transition structures), then both *syn* and *anti* adducts should be available (Scheme 5.15, path a *vs.* b or c *vs.* d). Since both enantiomers of the auxiliary are available, any desired combination of relative and absolute configurations in the products would be available.

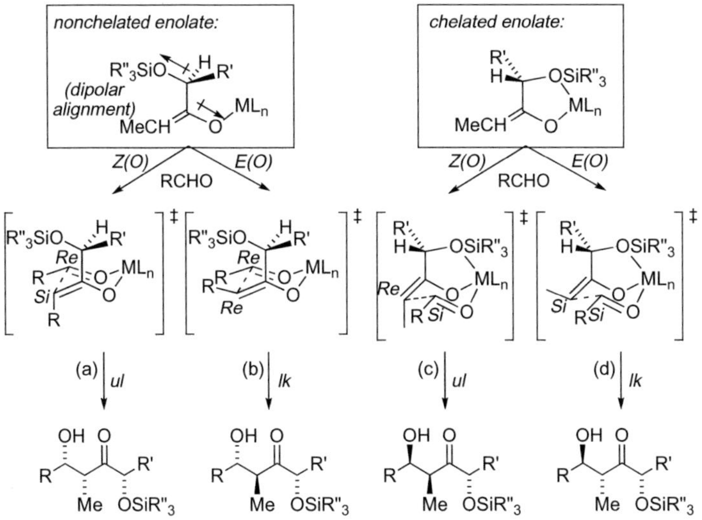

SCHEME 5.15 Chelated and nonchelated pathways to aldol adduct diastereomers for the Heathcock and Masamune auxiliaries.

Note that each E(O)- or Z(O)-enolate will have a choice of two Zimmerman−Traxler transition structures. Thus (see Scheme 5.15), a Z(O)-enolate may add through nonchelated path a or chelated path c, both of which afford *syn* adducts, but of opposite absolute configuration at the two new stereocenters. Likewise, an E(O)-enolate may add *via* path b or d, affording diasteomeric *anti* adducts.

Highly selective additions of these auxiliaries have been achieved *via* all four of the postulated pathways. Table 5.3 lists several examples. For example, Z(O)-dibutylboron enolates (entries 1, 2) often have selectivities of >99%, and are postulated to proceed through nonchelated path a [84,85]. The reason path c cannot compete is that the boron cannot accommodate more than four ligands, and two of the ligands are nonexchangeable alkyl groups. Additionally, boron enolates are not reactive enough to add to aldehydes unless the latter are coordinated to a Lewis acid. In the absence of external acids then, the boron of the enolate must activate the aldehyde by coordination and its two available ligand sites are occupied by the enolate and the aldehyde oxygens.

When the α-hydroxyl is silylated with a *tert*-butyldimethylsilyl group, chelation is difficult no matter what the metal. Lithium enolates of the TBS ethers are not particularly selective in their additions to aldehydes, but transmetalation to titanium affords enolates that are highly selective in their addition reactions (Table 5.3, entry 3). Acylation of the oxygen with a benzoyl group and deprotonation with LDA affords an enolate that gives the relative configuration shown in path a, although chelation by the benzoyl carbonyl oxygen is postulated (entry 4). With a smaller trimethylsilyl group, a lithium cation can simultaneously coordinate the enolate oxygen, the silyloxyl oxygen, and the aldehyde oxygen. Thus, the $Z(O)$-lithium enolate affords *syn* adducts according to path c (entry 5).

TABLE 5.3 Asymmetric Additions of the Heathcock-Masamune Enolates

Entry	Enolate	Path[a]	RCHO	% Yield	% ds	Ref.
1	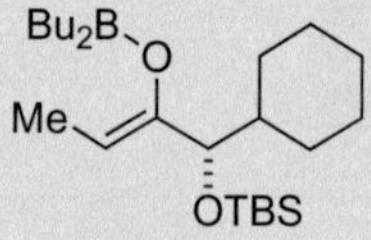 Bu$_2$B, Me, OTBS (cyclohexyl)	a	Et, *i*-Pr, Ph, BnO(CH$_2$)$_2$	70–85	≥97	[85]
2	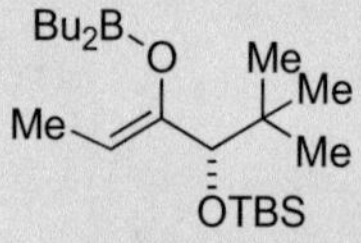 Bu$_2$B, Me, OTBS, Me/Me/Me	a	*i*-Pr, *t*-Bu, Ph, BnO(CH$_2$)$_2$	75–80	>95	[84]
3	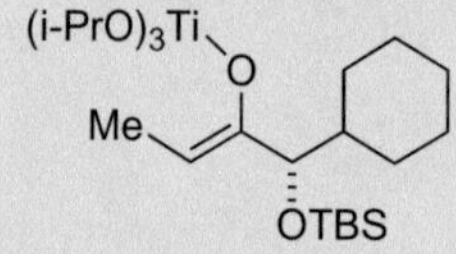 (*i*-PrO)$_3$Ti, Me, OTBS (cyclohexyl)	a	Et, *i*-Pr, *t*-Bu, Ph	75–88	≥98	[97]
4	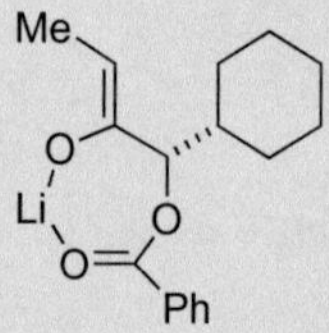 Me, Li, O, O=C–Ph (cyclohexyl)	a	Et, *i*-Pr, Ph	67–96	86–97	[98]
5	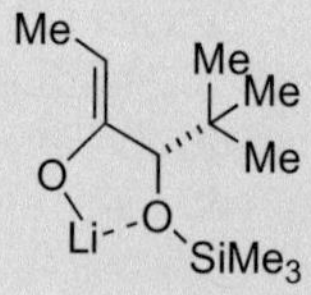 Me, Li···O–SiMe$_3$, Me/Me/Me	c	*i*-Pr, *t*-Bu, Ph	75–80	>95	[84]
6	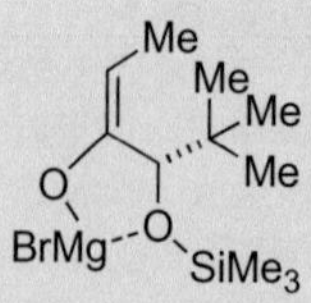 Me, BrMg···O–SiMe$_3$, Me/Me/Me	d	*i*-Pr, *t*-Bu, Ph	75–85	92–95	[84]
7	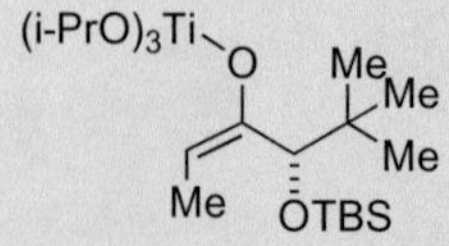 (*i*-PrO)$_3$Ti, Me, OTBS, Me/Me	b	Me, *i*-Pr, *t*-Bu, Ph	85–88	80–>95	[84]

[a]*The "Path" column indicates the product configuration and the proposed transition structure from Scheme 5.15.*

Deprotonation of the ketone educt with *N*-(bromomagnesio)-2,2,6,6-tetramethylpiperi-dide affords the *E(O)*-enolate selectively. Addition of the magnesium *E(O)*-enolate having a trimethylsilyloxy group affords *anti* adducts (entry 6) and is postulated to occur *via* chelated path d (Scheme 5.15). Transmetalation of the *tert*-butyldimethylsilyloxy-protected magnesium *E(O)*-enolate affords a titanium enolate that cannot chelate, and adds to aldehydes *via* path b (entry 7). In this case, only benzaldehyde afforded selectivity lower than 95%.

Probably the single most used class of chiral auxiliaries in the aldol addition is the oxazolidinone imide introduced by Evans (Figure 5.4c, see also Scheme 3.21). These auxiliaries are prepared by condensing amino alcohols [88,99] with diethyl carbonate [88]. Deprotonation by either LDA or dibutylboron triflate and a tertiary amine affords only the *Z(O)*-enolate. Scheme 5.16 illustrates open and closed transition structures that have been postulated for these *Z(O)*-enolates under various conditions, and Table 5.4 lists typical selectivities for the various protocols. The first to be reported (and by far the most selective) was the dibutylboron enolate (Table 5.4, entry 1), which cannot activate the aldehyde and simultaneously chelate the oxazolidinone oxygen [86]. Dipolar alignment of the auxiliary and approach of the aldehyde from the *Re* face of the enolate affords *syn* adduct with outstanding diastereoselection, presumably *via* the closed transition structure illustrated in Scheme 5.16a [86]. The other *syn* isomer can be formed under two different types of conditions. In one, a

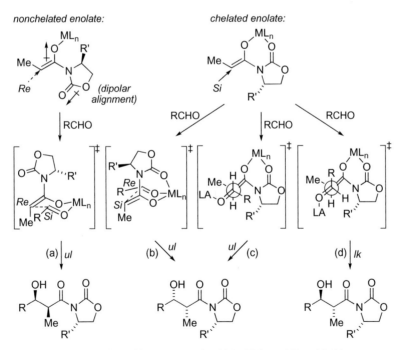

SCHEME 5.16   Open and closed transition structures for aldol additions of Evans' imides.

titanium enolate is postulated to chelate the oxazolidinone oxygen [100][8] or sulfur of an oxa-zolidinethione [101] exposing the *Si* face of the enolate (Scheme 5.16b). Additional coordination of the aldehyde and addition *via* the closed transition structure shown in Scheme 5.16b affords excellent selectivity (Table 5.4, entries 2 and 3).

If the boron enolate is allowed to react with an aldehyde in the presence of another Lewis acid (LA), the addition is thought to occur *via* the open transition structures shown in Scheme 5.16c and d [76]. If the Lewis acid is small, the preferred orientation is as shown in Scheme 5.16c, which minimizes the gauche interaction between the methyl and R groups on the forming bond (Table 5.4, entry 4). Both $SnCl_4$ and $TiCl_4$ are relatively "small" because of the long metal–oxygen bond. If the Lewis acid is large, the interaction between the Lewis acid and the methyl may outweigh the methyl/R gauche interaction. When the aldehyde is complexed to diethylaluminum chloride, the Lewis acid is effectively larger than either the tin or titanium complexes because of the shorter Al–O bond compared to either Sn–O or Ti–O, and because the ligands on the aluminum are relatively bulky. In this instance, the other face of the aldehyde will present itself to the enolate affording the *anti* adduct, as shown in Scheme 5.16d (Table 5.4, entry 5).

TABLE 5.4 Asymmetric Additions of the Evans Imide Enolates

Entry	Enolate	Path[a]	RCHO	% Yield	% ds	Ref
1	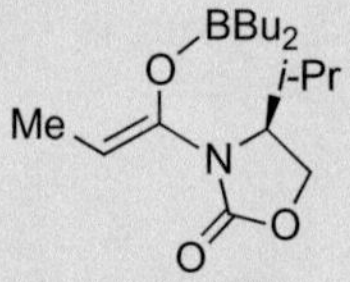	a	Bu, *i*-Pr, Ph	75–88	>99	[86]
2	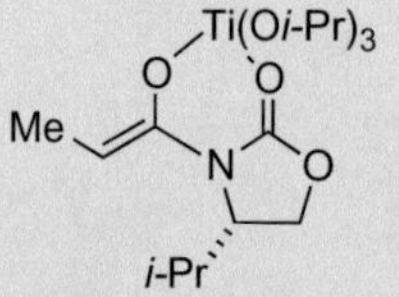	b	Bu, *i*-Pr, Ph	56–75	85–92	[100]
3	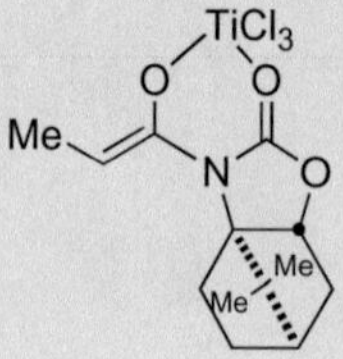	b	*n*-Pr, *i*-Pr, 1-propenyl, Ph	84–88	97–99	[101]

(Continued)

8. Recall that the titanium enolates of the Heathcock and Masamune auxiliaries (Table 5.3, entries 3 and 7) were postulated to occur by a *nonchelating* pathway. However, in those cases, the potential chelating atom was a severely crowded TBS ether oxygen, as opposed to the more basic and less crowded urethane carbonyl oxygen in the Evans auxiliary.

TABLE 5.4 (Continued)

Entry	Enolate	Path[a]	RCHO	% Yield	% ds	Ref
4		c	Et, i-Pr, i-Bu, t-Bu, 2-propenyl, Ph (·SnCl$_4$ or TiCl$_4$)	50−68	87−93	[76]
5	"	d	Et, i-Pr, i-Bu, t-Bu, 2-propenyl, Ph (·Et$_2$AlCl)	62−86	74−95	[76]

[a]The "Path" column indicates the product configuration and the proposed transition structure from Scheme 5.16

For syntheses requiring the *syn* adducts, it is practical to use the boron enolate without additional Lewis acids [86], since the auxiliary is available as either enantiomer and is recoverable.[9] On the other hand, the *anti* adducts are (so far) only available by the diethyl-aluminum chloride/boron enolate protocol [76]. Similar principles may be used to prepare *syn* and *anti* halohydrins by aldol addition of α-halo acetate enolates of Evans imides [102,103].

Crimmins has devised a series of sulfur-substituted versions of the classic Evans auxiliary (Scheme 5.17a) [95,104]. These auxiliaries function in a similar manner to their forebears, but with the valuable new property that it is possible to prepare either relative configuration of the *syn* aldol product using the same auxiliary but by changing reaction conditions. Specifically, the formation of the titanium enolate from the thiazolidinethione precursor shown afforded a *syn* product analogous to that observed using the Evans auxiliary, presumably by the intermediacy of a Zimmerman−Traxler like transition state shown, in which the Ti center engages in coordination to the enolate the aldehdye and the thiocarbonyl group of the auxiliary (*cf.* path b in Scheme 5.16). In contrast, carrying out the reaction in the presence of a good coordinating ligand like TMEDA or (−)-sparteine disrupts the coordination with the thiocarbonyl group and instead the reaction proceeds through a transition state like that in path a in Scheme 5.16, affording the opposite diastereomer (or enantiomer, once the chiral auxiliary is removed). Both "Evans" and "non-Evans" products are *syn* as usual.

9. Originally, Evans used the illustrated auxiliary (R′ = i-Pr) for one product configuration and a similar norephedrine-derived auxiliary (R′ = Me, plus a Ph at C-5) for the other [86]. Since that time, experience has shown [88] that a phenylalanine-derived auxiliary (R′ = Bn) is usually better for practical reasons, and is available as either enantiomer.

A weakness of the many aldol auxiliaries is their inability to selectively add acetate enolates. A Crimmins auxiliary equipped with a bulky mesityl substituent is able to overcome this limitation as shown in Scheme 5.17b [105]. Two rationales to account for this have been

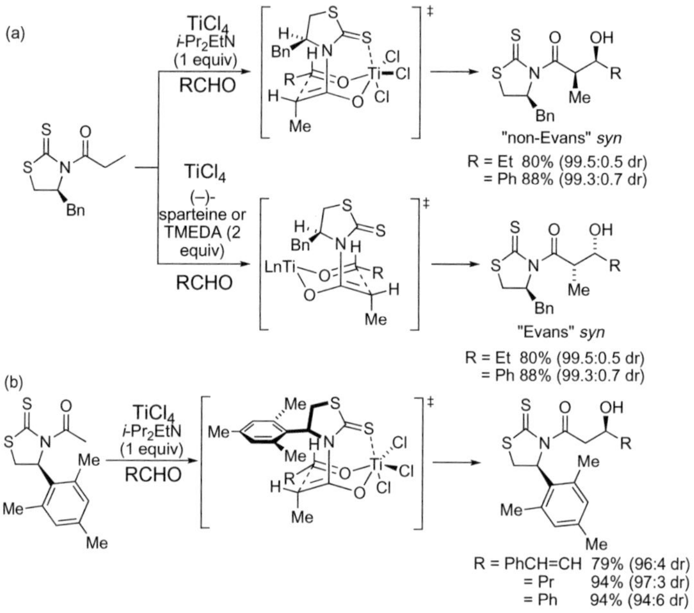

SCHEME 5.17 Crimmins's thiazolidinethione auxiliaries are useful for (a) providing either diastereomer of a given *syn* aldol product from the same auxiliary and (b) for obtaining high selectivity in some aldol reactions arising from acetate enolates. Note that the two adducts arising from the Evans and non-Evans pathways would afford enantiomers following the removal of the chiral auxiliary.

advanced, one involving a chair-like transition state bearing additional coordination between the thiocarbonyl of the auxiliary; the second features a nonchelated transition state, with a boat-like conformation.

Another excellent auxiliary for this purpose is the ester developed by Helmchen, shown in Scheme 5.18 [96], which gave yields in the 50–70% range. The authors proposed a closed Zimmerman–Traxler transition structure, as shown on the lower left [96], however the open structure shown in the lower right, which does not require coordination of the bulky silyloxy group to titanium, should also be considered. The aldehyde may be oriented to avoid the large *tert*-butyldimethylsilyl (TBS) group as shown, with the R group away from the TBS. Both of these models have the *Re* face of the aldehyde approaching the enol ether from the front face of the enolate, opposite the side that is shielded by the sulfonamide. Note also that the silyloxy group is oriented downward, to avoid the sulfonamide. An *anti*-selective addition (92% ds) was also reported for the reaction of the *E(O)*-enol ether of this auxiliary with isobutyraldehyde [96]. Another excellent auxiliary for the asymmetric aldol addition is the camphor sultam developed in the Oppolzer laboratories (Figure 5.4e). A significant feature of this auxiliary is the crystallinity of the aldol adducts, which often simplifies purification and diastereomer enrichment. As the trimethylsilyl enol ethers (ketene acetals), acetate aldol additions afford good selectivities at −78 °C and purification by recrystallization affords adducts that are >99% pure in most cases [94].

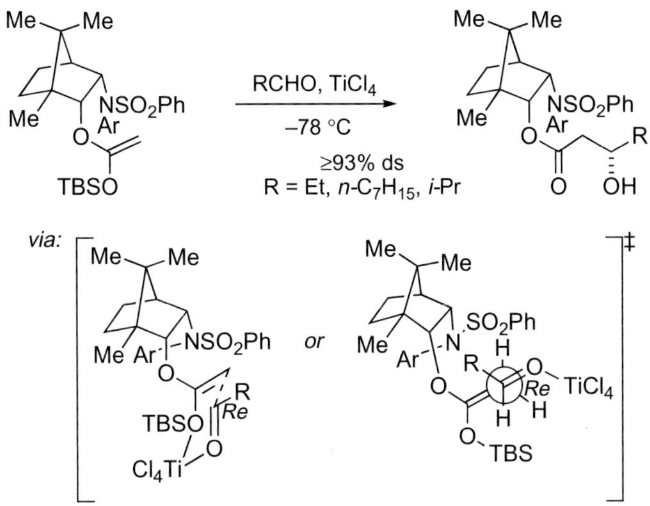

SCHEME 5.18   Asymmetric addition of acetate enol silyl ethers to aldehydes [96].

Interligand asymmetric induction with stoichiometric inducers. Asymmetric induction by chiral ligands on the enolate metal has the advantage that the chiral moiety does not have to first be attached to one of the reactants and later removed (or destroyed). It is present only after enolate formation, and can be recovered for reuse. The introduction of chirality in the enolate metal (or metalloid) and its ligands was the intellectual stepping stone toward developing asymmetric catalysis for the aldol addition reaction, in that the stereogenic unit responsible for the asymmetric induction is not covalently bonded to either reactant. Additionally, chiral ligands on the metal allow double asymmetric induction when one of the reactants is chiral, and triple asymmetric induction when both are. Most of the work that has been done in this area uses the same metals discussed in the previous section: boron, lithium, titanium, and tin.

In 1986, the groups of Masamune [106] and Paterson [59,107] reported (virtually simultaneously) that boron enolates containing C_2-symmetric "BR_2" moieties are effective mediators in asymmetric aldol additions. The Masamune group [106] studied the aldol addition of boron esters of *tert*-heptyl thiol acetate and propionate $E(O)$-enolates. As shown in Scheme 5.19, both types of reagents were highly selective. When R_1 is hydrogen, the selectivities are somewhat lower, because the *tert*-heptyl group can rotate away from the C_2-symmetric boracycle. When R_1 is an alkyl group, $A^{1,3}$ strain forces the *tert*-heptyl group toward the boracycle, crowding the transition structure and increasing the free energy difference ($\Delta\Delta G^{\ddagger}$) between the two illustrated transition structures. The product esters could be reduced to the corresponding primary alcohols. In spite of the high selectivities, the method has the disadvantage that the chiral boron compound is difficult to make.

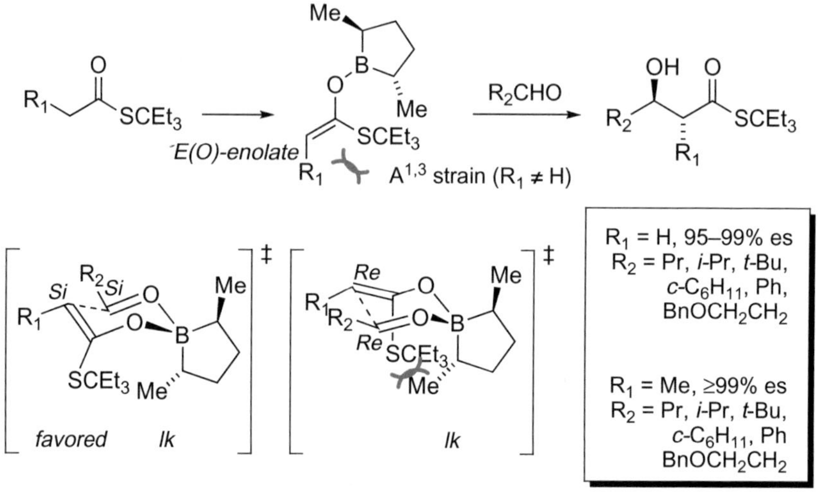

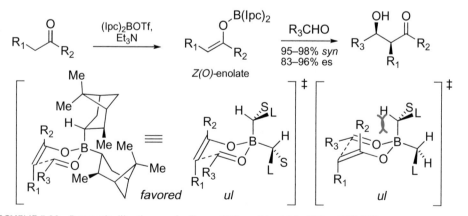

SCHEME 5.19 Masamune's chiral boron enolate aldol additions [106].

Following an early lead from the Meyers group [108], Paterson used the readily available diisopinocampheyl (Ipc) boron triflate to make $Z(O)$-boron enolates of 3-pentanone [107] and other ketones [109], which add to aldehydes to produce *syn* adducts in 83–96% es (Scheme 5.20 and Table 5.5). Based on molecular mechanics calculations [65,66], the transition structure analysis shown in Scheme 5.20 was suggested to rationalize the enantioselectivity. The axial boron ligand rotates so that the C—H bond is over the top of the Zimmerman—Traxler six-membered ring, and the equatorial ligand orients with its C—H bond toward the axial ligand. It is interesting to note that, because of severe van der Waals interactions, the two boron—carbon bonds are conformationally locked. Note that the two methyls of the isopinocampheyl moieties are both similarly oriented, with the equatorial Ipc-methyls pointed toward the viewer. A simpler representation is to depict the carbon attached to boron as shown in the middle, with "L" and "S" representing the CHMe and CH_2 ligands respectively. The favored transition structure has the enolate oriented away from the "L" ligand to avoid van der Waals repulsion between "L" and the pseudoaxial R_2 moiety [65,66,109].

SCHEME 5.20 Paterson's diisopinocampheylboron $Z(O)$-enolate aldol addition [107,109].

Use of diisopinocampheyl boron chloride in place of the triflate affords $E(O)$-enolates, but the isopinocampheyl ligands were ineffective for *anti* aldol reactions [110]. Encouraged by the molecular mechanics analysis of the $Z(O)$-enolate additions, Gennari and Paterson used computational methods to design a new boron ligand for use with $E(O)$-enolates [111]. The design was cued by Still's comment [112] that *cis*-2-ethyl-1-isopropylcyclohexane has only one conformation that avoids 2,3-P-3,4-M gauche pentane interactions (Figure 5.5).[10] This conformation is analogous to a diaxial *cis*-1,3-dimethylcyclohexane interaction. They realized that replacement of the ethyl group with a CH_2B moiety would afford a molecule that is similarly conformationally constrained, and that such a molecule was available from menthone (Figure 5.5). Molecular mechanics calculations suggested that aldol additions of E (O)-enolates using this ligand on boron would be enantioselective [111].

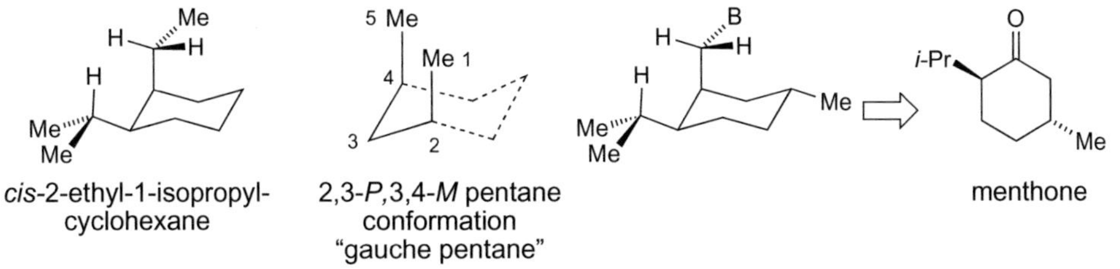

cis-2-ethyl-1-isopropyl-cyclohexane

2,3-P,3,4-M pentane conformation "gauche pentane"

menthone

FIGURE 5.5 The illustrated conformation of *cis*-2-ethyl-1-isopropylcyclohexane is the only one that has no destabilizing "gauche pentane" interactions [112]; similar interactions restrict the conformational motion of a boron ligand available from menthone [111].

When the "methylmenthyl" (MeMn) ligand was evaluated for selectivity in the addition of $E(O)$-enolates [111], it was found that the adducts were 86–100% *anti*, and the enantioselectivities were 78–94% (Scheme 5.21 and Table 5.5). The transition structure suggested to explain the chirality sense of the products again features the pseudoaxial R_2 avoiding interaction with the larger of the ligands (*i.e.*, menthyl) on the axial carbon bonded to boron. In the isopinocampheyl ligand (Scheme 5.20), the "large" and "small" ligands were rather similar (CH_2 vs. $CHMe$); in the present instance, the difference is huge (H vs. menthyl). Note that the indicated (*) bond is the one that is restricted by the "gauche pentane" interactions in the menthyl moiety.

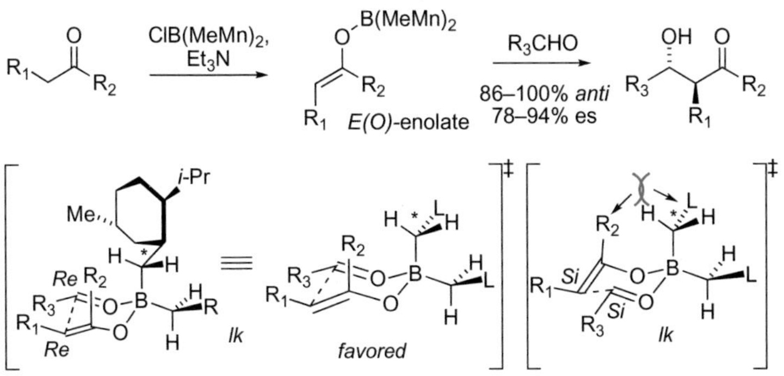

SCHEME 5.21 Paterson–Gennari di(methylmenthyl) (MeMn) boron enolate aldol additions [111].

---

10. See the glossary (Chapter 1) for an explanation of the P, M terminology. The energy of a gauche pentane interaction is *ca.* 3.7 kcal/mol [113].

TABLE 5.5 Asymmetric Aldol Additions of Ketone Enolates Using Chiral Ligands on Boron (Ipc = isopinocampheyl; MeMn = methylmenthyl). See Schemes 5.20 and 5.21

$$R_1 \overset{O}{\underset{}{\wedge}} R_2 \xrightarrow[\text{2. R}_3\text{CHO}]{\text{1. R}_2\text{BOTf, Et}_3\text{N}} R_3 \overset{OH}{\underset{R_1}{\wedge}} \overset{O}{\underset{}{\wedge}} R_2$$

Entry	BR₂	R₁	R₂	R₃	syn:anti	% yield	% es	Ref.
1	Ipc	Me	Et	Me	97:3	91	91	[109]
2	Ipc	Me	Et	2-propenyl	98:2	78	95	[109]
3	Ipc	Me	Et	n-Pr	97:3	92	90	[109]
4	Ipc	Me	Et	E-C₃H₅	98:2	75	93	[109]
5	Ipc	Me	Ph	2-propenyl	98:2	97	95	[109]
6	Ipc	Me	i-Pr	2-propenyl	95:5	99	94	[109]
7	MeMn	Me	Et	2-propenyl	3:97	62	88	[111]
8	MeMn	Me	i-Pr	2-propenyl	0:100	51	94	[111]
9	MeMn	-(CH₂)₃-		2-propenyl	0:100	60	87	[111]
10	MeMn	-(CH₂)₄-		2-propenyl	0:100	59	78	[111]
11	MeMn	Me	Ph	2-propenyl	14:86	60	93	[111]
12	MeMn	H	i-Pr	2-propenyl	–	66	88	[111]
13	MeMn	H	i-Bu	2-propenyl	–	80	77	[111]
14	MeMn	H	t-Bu	2-propenyl	–	62	88	[111]
15	MeMn	H	Me	2-propenyl	–	65	80	[111]
16	MeMn	H	Ph	2-propenyl	–	81	85	[111]

In addition to their usefulness for the asymmetric addition of achiral aldehydes, it will be seen in Section 5.2.4 that the Paterson strategy is particularly useful for the aldol addition of chiral fragments such as the large, polyfunctional ketone and aldehyde fragments needed for convergent macrolide synthesis.

In 1989, Corey reported that diazaborolidines are efficient reagents for asymmetric aldol additions of acetate and propionate thioesters [114]. Thiophenyl esters add to aldehydes giving *syn* adducts, whereas *tert*-butyl esters give *anti* adducts [115]. Both react *via* closed, Zimmerman–Traxler transition states; the difference in the topicity is due to different enolate geometries for the two ester types. Corey's rationale for the divergent enolate geometries involves competing mechanisms for deprotonation of the zwitterion shown in Scheme 5.22. Complexation of the boron reagent with the ester produces the zwitterionic complex (boxed), which may undergo either E_1 or E_2 elimination of HBr. Corey suggests that both E_1 and E_2 reactions occur from the illustrated conformation of the zwitterion [115]. Ionization (E_1) is favored when RX is thiophenyl and disfavored when RX is *tert*-butoxy. Note that E_1 ionization can only occur when X can easily stabilize the positive charge by resonance, which is only possible when the substituent on X (R) becomes coplanar with the rest of the molecule. E_1 is also favored when the base is bulky, while smaller bases facilitate E_2 reaction. Deprotonation by an E_2 mechanism is faster with the less bulky triethyl amine than with diisopropylethyl amine.

For *tert*-butyl esters, deprotonation is effected with triethyl amine (which favors E_2), while E_1 ionization is disfavored because it requires moving the bulky *tert*-butyl group into planarity. For thiophenylesters, E_1 reaction is facilitated by the thiophenyl group, while E_2 reaction is slowed by use of the bulky diisopropyl ethyl amine.[11]

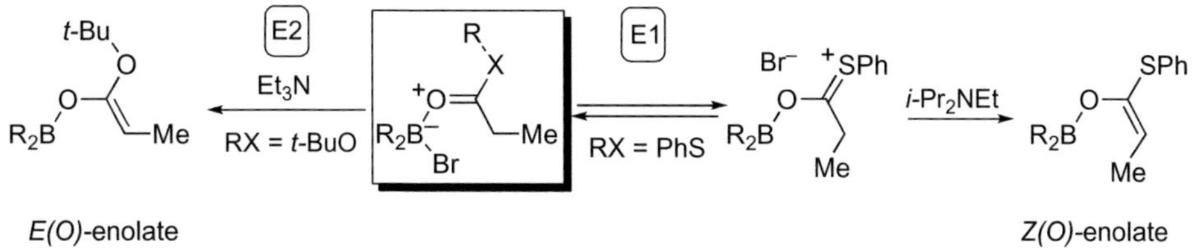

SCHEME 5.22   Rationale for boron enolate configuration [115].

The diazaborolidines mediate the diastereoselective and enantioselective formation of *syn* [114,115] and *anti* aldols [115], as summarized in Scheme 5.23. The aryl group of the sulfonamide must be electron withdrawing, or else the boron is not a strong enough Lewis acid to mediate the process. The process has also been used for the formation of *anti* halohydrins [117,118], and in aldol additions to azomethines [116]. The illustrated transition structure (Scheme 5.23a) has been postulated to account for the observed enantioselectivity in the Z(O)-enolate addition [114]. Corey suggests that the *trans* phenyl substituents force the sulfonamide aryl groups into a conformation that places each aryl ring in a *trans* orientation to its neighbor. This *conformation* is reminiscent of the *configuration* engineered by Masamune earlier (Scheme 5.19), and may have similar control features. It is interesting to note, however, that a similar chair transition structure employing the E(O)-enolate predicts the wrong enantiomer (Scheme 5.23b) of the *anti* addition product [119]!

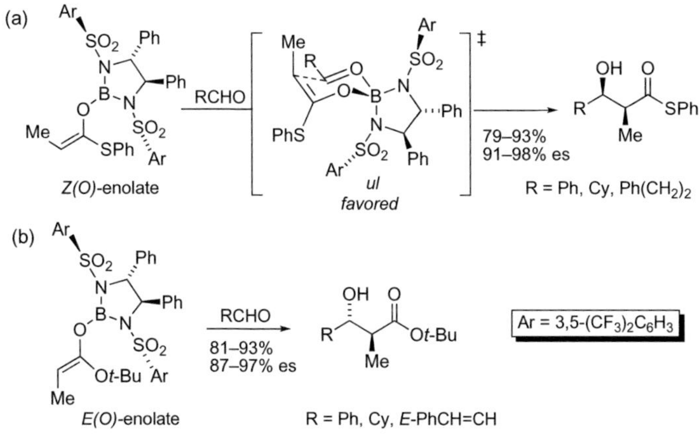

SCHEME 5.23   Corey's diazaborolidine-mediated aldol additions [114,115].

Asymmetric catalysis: metal-based catalysts. The 1990s saw an explosion in the development of effective catalysis of the aldol reaction, with an emphasis on variations such as the

---

11. A rationale similar to this may be used to explain the selective formation of E(O)-enolates of *tert*-heptylthio- and *tert*-butylthio-propionates (Scheme 5.19 and ref. [116]): the Et$_3$CS— replaces the Me$_3$CO— in the Scheme 5.22 rationale) and the selective formation of ketone Z(O)-enolates with dialkyl boron triflates and E(O)-enolates with dialkylboron halides (Schemes 5.20 and 5.21: the triflates are more likely to ionize than the halides, thus favoring ionization over direct deprotonation of the zwitterion).

Mukaiyama aldol reaction [120]. In this reaction, silyl enol ethers react with aldehydes with the aid of a chiral Lewis acid. These reactions proceed *via* open transition structures such as those shown in Figure 5.3c.[12] In this context, the primary role of the Lewis acid is to differentiate the enantiotopic faces of the electrophilic aldehyde, although more sophisticated designs also provide additional organization of the aldol transition state.

In 1991 and 1992, several groups reported boron-based Lewis acids for catalytic Mukaiyama aldol additions (Figure 5.6). Three of these are oxazaborolidines derived from the reaction of borane with amino acid derivatives (Figure 5.6a−c), while the fourth (Figure 5.6d) is derived from tartrate. Examples of aldol additions using these catalysts are listed in Table 5.7. The turnover numbers are not large (20−100 mol% of catalyst being required), and the enol ether variability is somewhat limited. The Kiyooka catalyst (Figure 5.6a; Table 5.6, entries 1−3) and the Masamune catalyst (Figure 5.6b; Table 5.6, entry 4) are similar, and have been evaluated for the asymmetric addition of ketene acetals. The Kiyooka catalyst only becomes catalytic (*cf.* entries 1 and 2) in nitromethane solvent. The Corey (Table 5.6, entry 5) and Yamamoto (Table 5.6, entry 6) catalysts are effective with enol ethers of ketones, but not ketene acetals.

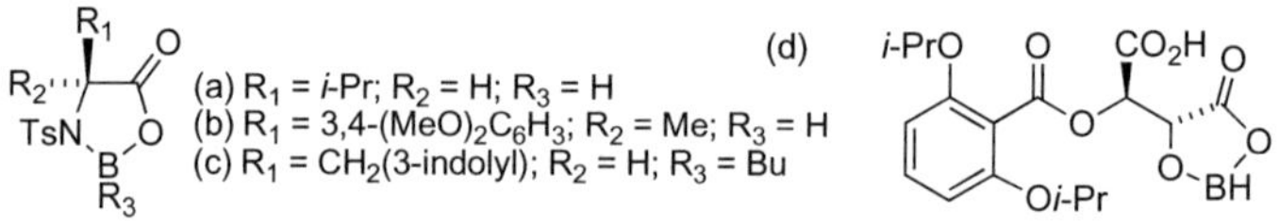

(a) R_1 = *i*-Pr; R_2 = H; R_3 = H
(b) R_1 = 3,4-(MeO)$_2$C$_6$H$_3$; R_2 = Me; R_3 = H
(c) R_1 = CH$_2$(3-indolyl); R_2 = H; R_3 = Bu

FIGURE 5.6 Boron-centric catalysts for the Mukaiyama aldol reaction: (a) Kiyooka catalyst [122,123]; (b) Masamune catalyst [124]; (c) Corey catalyst [125]; (d) Yamamoto catalyst [126,127].

TABLE 5.6 Catalytic Mukaiyama Aldol Additions

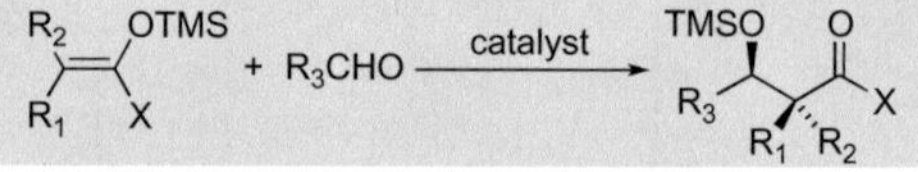

Entry	Catalyst (mol%)[a]	R_1, R_2	X	R_3	% Yield	% es	Ref.
1	a (100)	Me, Me	OEt	Ph, E-PhCH=CH−, Ph(CH$_2$)$_2$−	80−87	91−96	[122]
2	a (20)	Me, Me	OEt	*i*-Pr, Ph, E-PhCH=CH−, Ph(CH$_2$)$_2$−	60−97	91−98	[123]
3	a (20)	H, H	OPh	Ph	66	90	[123]
4	b (17)	Me, Me	OEt	Pr, *i*-Bu, Ph, c-C$_6$H$_{11}$, Ph(CH$_2$)$_2$−, BnO(CH$_2$)$_2$−	68−86	92−99	[124]

(Continued)

12. For an asymmetric Mukaiyama aldol that proceeds by an "ene" mechanism, see ref. [121].

TABLE 5.6 (Continued)

Entry	Catalyst (mol%)[a]	R_1, R_2	X	R_3	% Yield	% es	Ref.
5	c (20)	H, H	Bu, Ph	Pr, c-C_6H_{11}, 2-furyl, Ph	56–100	93–96	[125]
6	d (20)	H, Me (E, Z mix)	Et, Bu, Ph	Pr, Bu, E-$CH_3CH{=}CH-$, E-$PhCH{=}CH-$, Ph	55–99 (80– >95% syn)	77–96	[126]

[a]*The catalyst column refers to the structures in Figure 5.6.*

An interesting point is the difference between the Masamune (Figure 5.6b) and Kiyooka (Figure 5.6a) catalysts. One works with a sub-stoichiometric amount of catalyst and the other does not (Table 5.6, entries 1 and 4). Masamune screened a number of catalysts, including ones similar to Kiyooka's (Figure 5.6a), and suggested the catalytic cycle illustrated in Scheme 5.24, in which step two is thought to be the slow step. The critical difference is the quaternary stereocenter in the Masamune catalyst ($R_2 \neq H$). When the stereocenter in the catalyst is quaternary, the N–C–CO bond angle is compressed (Thorpe–Ingold effect – see glossary, Section 1.11), thereby accelerating the silicon transfer (note the cyclic transition state) so that the catalyst can turn over more efficiently.

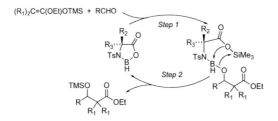

SCHEME 5.24 Catalytic cycle for oxazaborolidine catalyzed Mukaiyama aldol addition [125].

Throughout the 1990s, a number of approaches to the aldol reaction using transition metal or lanthanide catalysis appeared (Figure 5.7); these are now the mainstay of activity on the metal side of the aldol ledger (reviews: refs. [128–131]). In an early example, tin triflates ligated by chiral diamines (Figure 5.7a and b) were found to activate aldehydes toward addition by silyl enol ethers of acetate and $E(O)$-propionate thioesters (Table 5.7, entries 1–3). The catalytic version is thought to go by the two-step process shown in Scheme 5.25, with the slow step again being release of the alkoxide adduct by silylation [132,133]. In related work, highly selective aldol reactions starting with a silyl ketene acetal were reported by Carreira in 1994 (entry 4).

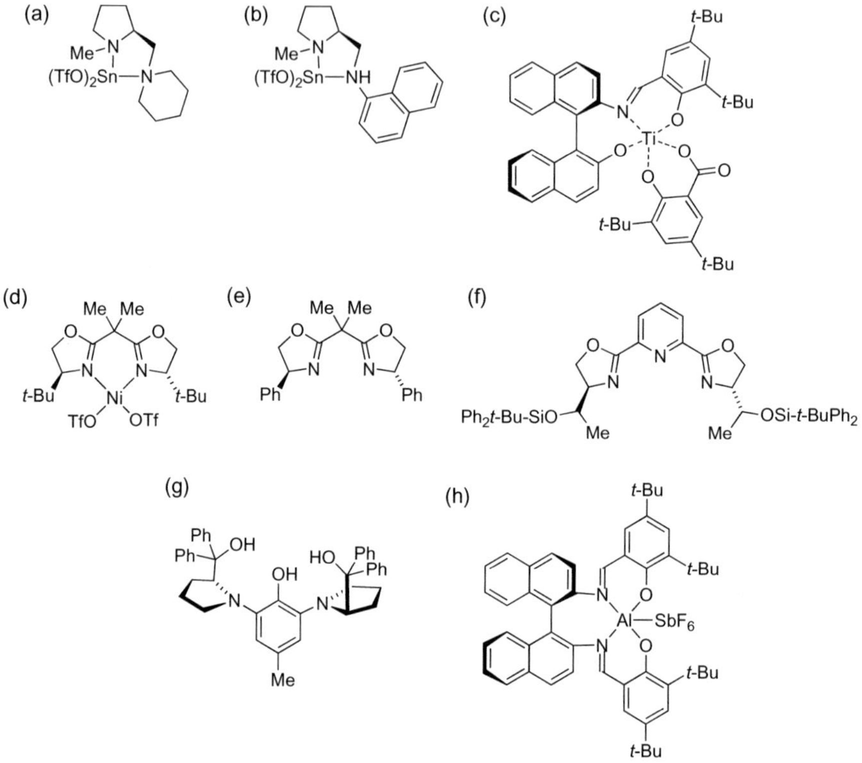

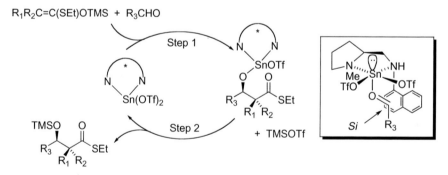

FIGURE 5.7 Examples of chiral Lewis acids and ligands used as aldol addition catalysts: (a) and (b) Kobayashi catalysts [135]; (c) Carriera catalyst [134]; (d) a nickel (bis)-oxazoline (Box) complex [136]; (e) Box catalyst (review: [137]); (f) a pyridyl (bis)-oxazoline (PyBox) catalyst [138]; (g) Trost's ProPhenol ligand [139]; (h) an Evans Salen catalyst [140].

SCHEME 5.25 Proposed catalytic cycle for the Kobayashi–Mukaiyama aldol addition. Inset: proposed model for the aldehyde *Si*-face selectivity due to the catalyst [132,133].

These Are a Few of Our Favorite Things

Writing this book has given us the chance to reflect upon both trends in the field as well as its Greatest Hits (for lack of a better phrase). While we have endeavored to point out the most useful methods in organic stereochemistry and asymmetric synthesis, we thought it would also be fun to identify a few of our personal favorite molecules. Here follows a list of our highly subjective selections for your consideration. You will, no doubt, have others of your own.

Carbohydrates and derivatives. We begin with tartaric acid because it has been involved in so many firsts [1]. As its sodium—ammonium salt, its enantiomers were separated with tweezers by Pasteur in his historic resolution experiment; as its sodium rubidium salt, it was the first molecule to have its absolute configuration determined by anomalous dispersion X-ray analysis. Chemically, it plays a starring role in the Sharpless asymmetric epoxidation (Chapter 8), and it crystallizes out in bottles of nicely aging wines. What's not to like? We also include glucose as an exemplar of how carbohydrates can serve as starting materials for complex synthesis. This field exploded in the 1980s and played a major role in the "chiron approach" to the synthesis of enantiopure compounds, espoused by Hanessian (a chiron is a chiral starting material, *i.e.*, a member of the chiral pool, see ref. [2]). Another chiron of interest is the acetal of glyceraldehydes, prepared as drawn from mannitol and a useful building block for chemical synthesis [3]. Perhaps one might think that chiral pool approaches using carbohydrates are inelegant and old hat. If so, perhaps it would be worth thinking about what our descendants will be using as starting materials when all of the petroleum products run out .

Amines and amino acids. The other major branch of the family of chirons, a case can be made for the inclusion of many amino acids here but we'll limit ourselves to a few that have had impressive runs. In terms of simple, straightforward amines, it's hard to beat α-methylbenzylamine. It is a useful resolving agent (and still used as such), a reliably decent and occasionally terrific chiral auxiliary, and as far as chiral amines go, a solid value pricewise. Valine is one of the most commonly used amino acids in organic synthesis, in large part due to its appearance in multiple chiral auxiliaries (e.g., Evans, Schöllkopf, Meyers...), often as the reduced form valinol. Serine can be converted to the Garner aldehyde, which is a nitrogenous cousin to glyceraldehyde as a starting material for asymmetric synthesis [4]. And proline, which has been dubbed the "simplest enzyme" [5], played a major role in getting the organocatalysis ball rolling.

The ligands.[1] Here, our personal choices include sparteine, in part because it is one of best overall ligands for coordinating lithium

1. One of these three is not like the other ones. How?

atoms associated with anions and for asymmetric deprotonations. It has also challenged chemists to (1) try to explain why this particular structure, and not close analogs, is such an effective chiral promoter [6], and (2) deal with the problem of easy availability of only one enantiomer from natural sources [7]. Each of the last two selections narrowly edged out competitors for their appearances here. BINOL was chosen over TADDOL (and appears on our cover!) as an example of an outstanding diol ligand that has found a use in too many asymmetric catalysts to even mention (one could start with the index of this book). It also won out because of its importance in Noyori's work (it is hard to argue with a Nobel Prize). Finally, we chose the chiral salen complex popularized by Jacobsen as an example of a privileged catalyst [8]. One could also make a good case for the bisoxazoline (Box) family of catalysts, which have enjoyed comparable utility across the universe of asymmetric catalysis, but the modularity of the salen design swung the pendulum in its favor.

Biologically active molecules. Although it may look like a CIP priority assignment problem for a sophomore organic chemistry student, isoflur-

isoflurane   L-DOPA   dextrorphan

ane represents a compelling issue for both organic chemists and biomedical scientists alike. It was so hard to obtain in enantiomerically pure form that its resolution was published in *Science* [9]. Moreover, the mechanism of action of general anesthetics continues to be one of the significant unsolved problems of modern pharmacology [10]. L-DOPA has not only ameliorated the suffering of millions afflicted with Parkinson's disease, but it has served as the poster child for the importance of enantioselective synthesis of pharmaceutical agents and has even become a Hollywood star through its appearance in the book and subsequent film *Awakenings* [11]. Finally, of the hundreds of possible choices of examples where pharmacological activity depends on which enantiomer is used, consider dextrorphan. Although the enantiomer of the synthetic opioid narcotic levorphanol, dextrorphan is a non-opioid cough suppressant (it binds instead to receptors for *N*-methyl-D-aspartate [12]).

In the end, for many of us our favorite chiral molecule may be the one we are currently working on or with.

References

[1] Nagendrappa, G. *Resonance* **2007**, *12*, 38–48.
[2] Hanessian, S.; Giroux, S.; Merner, B. L. *Design and Strategy in Organic Synthesis: From the Chiron Approach to Catalysis*; Wiley-VCH: Weinheim, 2012.
[3] Jurczak, J.; Pikul, S.; Bauer, T. *Tetrahedron* **1986**, *42*, 447–488.
[4] Liang, X.; Andersch, J.; Bols, M. *J. Chem. Soc., Perkin Trans.* **2001**, *1*, 2136–2157.
[5] Movassaghi, M.; Jacobsen, E. N. *Science* **2002**, *298*, 1904–1905.
[6] Hoppe, D.; Hense, T. *Angew. Chem., Int. Ed. Engl.* **1997**, *36*, 2283–2316.
[7] O'Brien, P. *Chem. Commun.* **2008**, 655–667.
[8] Yoon, T. P.; Jacobsen, E. N. *Science* **2003**, *299*, 1691–1693.
[9] Meinwald, J.; Thompson, W. R.; Pearson, D. L.; Konig, W. A.; Runge, T.; Francke, W. *Science* **1991**, *251*, 560–561.
[10] Campagna, J. A.; Miller, K. W.; Forman, S. A. *New Engl. J. Med.* **2003**, *348*, 2110–2124.
[11] Sacks, O. *Awakenings*; Vintage Books: New York, 1999.
[12] Choi, D. W.; Viseskul, V. *Eur. J. Pharmacol.* **1988**, *155*, 27–35.

TABLE 5.7 Catalytic Mukaiyama Aldol Additions

$$\text{Donor} + R_3\text{CHO} \xrightarrow{\text{catalyst}} \text{Product}$$

Entry	Donor	Product	Catalyst (mol%) + condns[a]	% Yield	% es	Ref.
1	R₂, OTMS / R₁, X (silyl enol ether)	O / X, R₃, R₁ R₂, OTMS	a (100)	77–90	91–99	[141]
2	"	"	b (100), Bu₂Sn(OAc)₂	70–96 (100% syn)	>99	[141]
3	"	"	b (20)	71–80 (80–100% syn)	94–99	[133]
4	OSiMe₃ / OMe	O OH / MeO, R₃	c	72–98	>98	[134]
5	S, O / S, N, Me (thiazolidinethione)	S, O, OH / S, N, R₃, Me	d (10) +2,6 lutidine (2.3 equiv) + TMSOTf (0.4 M in CH₂Cl₂)	46–84	95–99 (syn:anti 88:12–98:2)	[136]
6	O, O / PhS, OH, Me	O, OH / PhS, R₃, Me	e (13) + Cu(OTf)₂ (10 mol%)	48–83	94–98 (syn:anti 69:31–97:3)	[142]
7[b]	OTMS / Ph	O Me OH / Ph, COOEt	f (10) +10 mol% Sc(OTf)₃	92	96	[138]
8	O / Me, Ar	O, OH / Ar, R₃	g (5) + Et₂N (10 mol%) + Ph₃P=S (15 mol %)	33–79	56– >99	[139]

[a]The catalyst column refers to the structures in Figure 5.7.
[b]Here, the electrophile is methyl pyruvate and not an aldehyde.

A major advance occurred in 2003 with the publication by Evans [136] of a highly selective *direct* aldol reaction, that is, one in which the prior generation of a nucleophilic species was not required (entry 5, Scheme 5.26a, for a review of direct aldol reactions see ref. [128]). The reaction was best accomplished using a preformed Ni (bis)-oxazoline (or Box) complex [137], which was proposed to coordinate to a thiazolidinethione amide. The ensuing complex is then deprotonated by exogenous base to afford the $Z(O)$-enolate, which selectively approaches the aldehyde partner from the face opposite to that of the large *tert*-butyl groups (*ul* topicity). Added trimethylsilyl triflate then silylates the aldolate, thus triggering turnover.

Shair published a reaction in which the reactive thiopropionate was generated by the *in situ* decarboxylation of a malonic half thioester precursor, also promoted by a Lewis acid (here, Cu(OTf)$_2$) coordinated to a Box ligand (Table 5.7, entry 6). Numerous oxazoline-based ligands have been used in all manner of chiral catalysis research, including one additional example shown here (Table 5.7, entry 7), in which the ketene acetal reacted not with an aldehyde, but rather a ketone (albeit a particularly reactive pyruvate) in the presence of a pyridine (bis)-oxazoline (PyBox). Trost developed a set of proline-derived ligands (eventually to be known as ProPhenol ligands) that proved highly effective in direct aldol reactions of aryl methyl ketones (Table 5.7, entry 8). The proposed pre-aldol complex, depicted in Scheme 5.26b, features a pair of zinc atoms (derived from added Et$_2$Zn) able to coordinate both the ketone and aldehyde reactants. As in the Ni-Box example, the ketone is rendered acidic to generate a metalloenolate *in situ* that attacks the aldehyde.

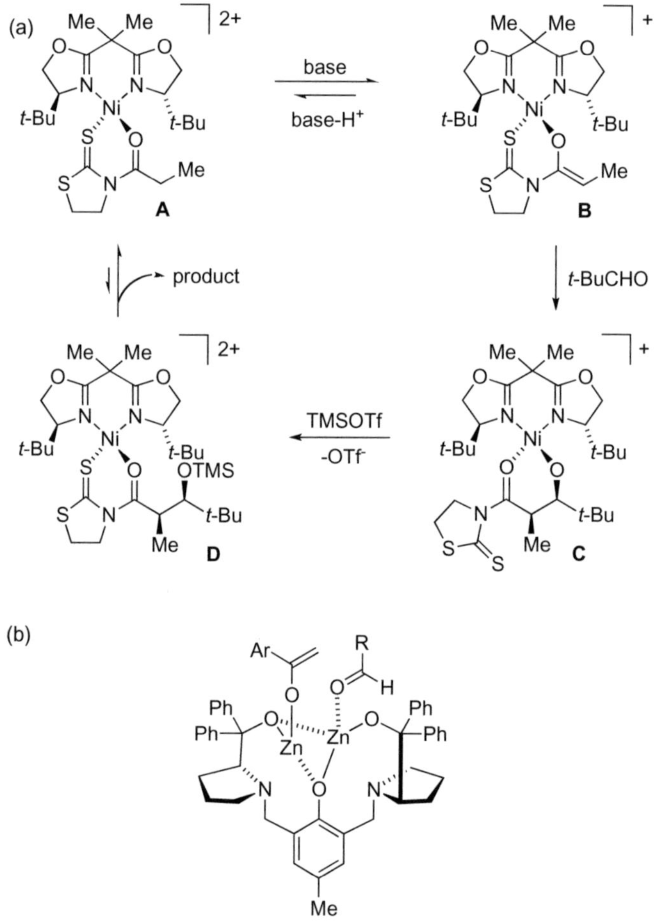

SCHEME 5.26 (a) Direct aldol as promoted by a preformed Ni-Box complex [136]. (b) Proposed reactive complex in Trost's ProPhenol-promoted direct aldol reaction [139].

The appeal of the direct aldol is obvious, as is the main limitation that they are gener-ally limited to carbonyl compounds able to enolize only on one side of the carbonyl or to symmetrical substrates. To appreciate this, consider the use of 2-butanone in an aldol reac-tion as shown in Scheme 5.27a. The modest pK_a and steric differences between protons at C-1 and C-3 in this precursor are such that use of the ketone in a direct aldol reaction under thermodynamic conditions would almost certainly result in a complex mixture of products, but the aldol product arising from the C-1 enolate is achievable with an appropriate cata-lytic method. However, the alternative C-3 enolate cannot be formed in this way. One approach to this problem is by generating a specific enol or enol equivalent by some means other than deprotonation. In particular, catalytic reduction of an α,β-unsaturated carbonyl species to the corresponding enolate (known for years to be accomplished by reduction with Na in liquid ammonia) has been examined in the context of reductive aldol reactions (Table 5.8).

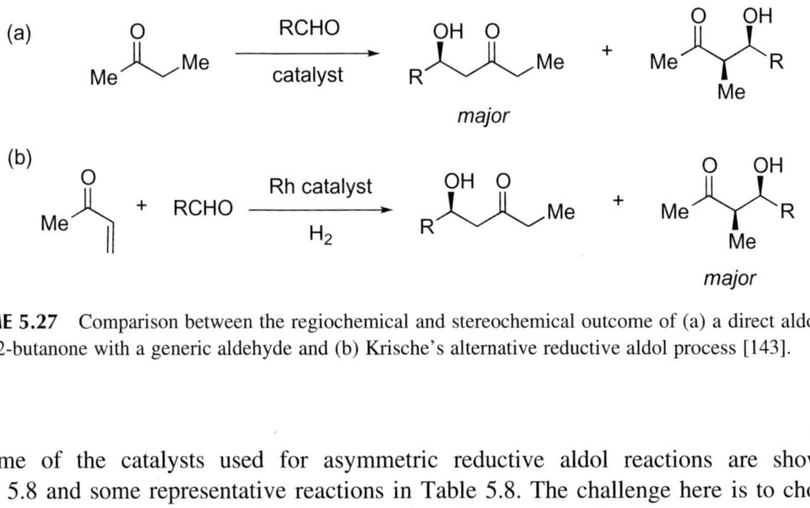

SCHEME 5.27 Comparison between the regiochemical and stereochemical outcome of (a) a direct aldol reac-tion of 2-butanone with a generic aldehyde and (b) Krische's alternative reductive aldol process [143].

Some of the catalysts used for asymmetric reductive aldol reactions are shown in Figure 5.8 and some representative reactions in Table 5.8. The challenge here is to choose a catalyst able to carry out the reductive generation of an enolate that is also compatible with the ensuing aldol reaction. Most of the early versions of this chemistry used a silane as the terminal reducing agent and a Rh-, Ir-, or Cu-based catalyst (either preformed as in Figure 5.8c or generated *in situ*). Particularly interesting is the ability to switch from γ to α substitu-tion by a change of ligand (*cf.* entries 5 and 6). Finally, Krische has reported a reductive aldol reaction of methyl vinyl ketone under near-classic hydrogenation conditions (entry 7), afford-ing the long-sought regioisomer switch noted in Scheme 5.27b [143].

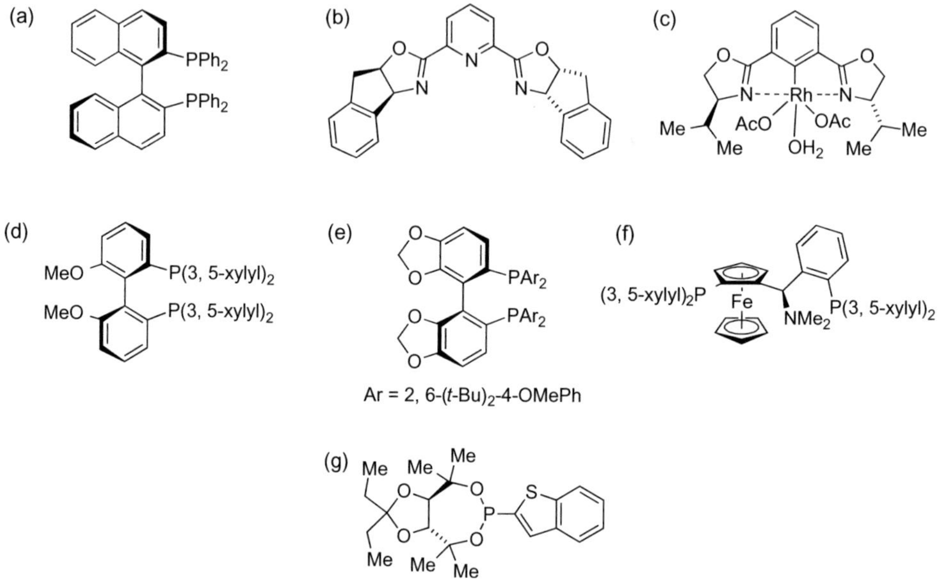

FIGURE 5.8 Asymmetric reductive aldol ligands and catalysts. (a) [144,145]; (b) [144,145]; (c) [146,147]; (d) [148]; (e) [149]; (f) [149]; (g) [143].

TABLE 5.8 Asymmetric Examples of Reductive Aldol Reactions

Entry	Substrate	Product	Catalyst, Ligand (mol%), and Conditions	% Yield (dr)	% es	Ref.
1	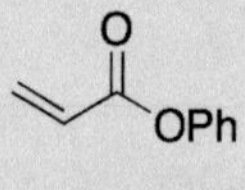	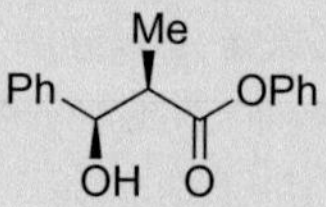	a (6.5), [Rh(cod) Cl]₂ (2.5), Et₂MeSiH	72 (77:23)	94	[144,145]
2	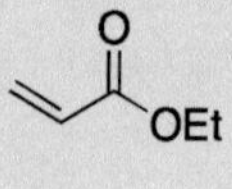	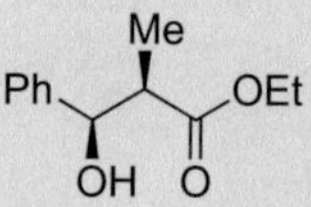	b (7.5), [Ir (coe)₂Cl]₂ (5.0), Et₂MeSiH	68 (87:13)	99	[144,145]
3	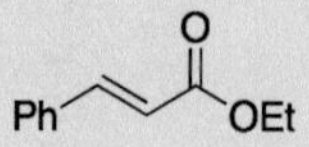	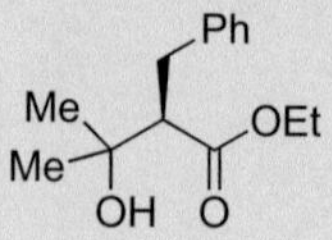	c (1.0), MePh₂SiH	83	98	[146,147]

(Continued)

TABLE 5.8 (Continued)

Entry	Substrate	Product	Catalyst, Ligand (mol%), and Conditions	% Yield (dr)	% es	Ref.
4	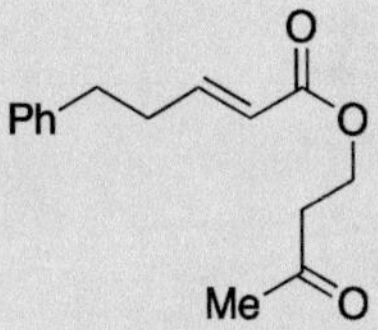	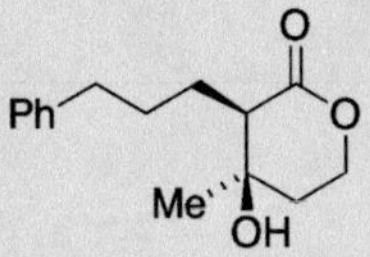	e (5), Cu (OAc)$_2$·H$_2$O (5), (Me$_2$SiH)$_2$O	61	90	[148]
5	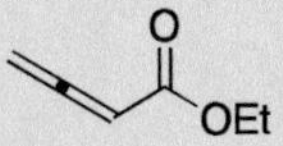	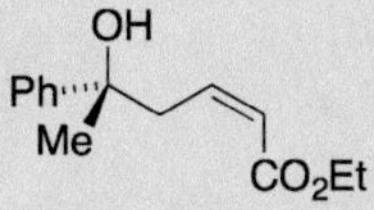	f (5), CuOAc (2.5), PCy$_3$ (5), pinacolborane	96	99	[149]
6	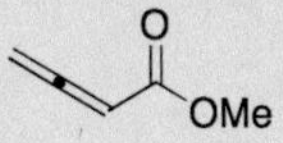	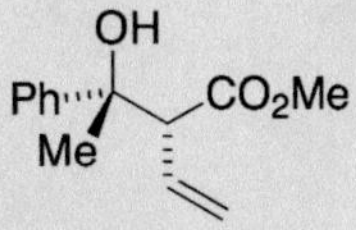	g (5), CuF (Ph$_3$P)$_3$·2EtOH (2.5), pinacolborane	90	92	[149]
7	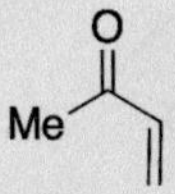	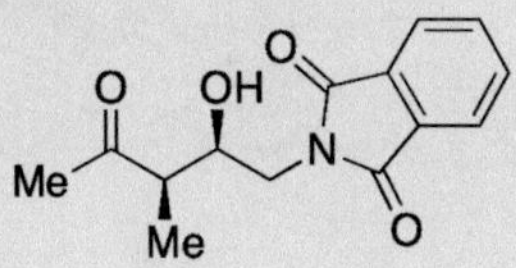	d (12), Rh (cod)$_2$OTf (5), H$_2$ (1 atm)	88	98	[143]

5.2.3 Organocatalysis of the Aldol Reaction and its Variants

Although the term "organocatalysis" has only been popular since the beginning of the twenty-first century (see Seebach's foreword for the citation to the first occurrence), organocatalysis has operated continuously since before life began. For the aldol reaction, organocatalysis usually entails the *in situ* formation of enamine and/or iminium ion intermediates from aldehyde and ketone precursors to afford either aldol addition products (aldols) or α,β-unsaturated ketones resulting from the aldol condensation (reviews: refs. [62,128,150]). These reactions may be considered biomimetic insofar as numerous biosynthetic pathways and aldolase enzymes use similar intermediates. However, the concept can also include other non-metal-centric methods, including enzyme, catalytic antibody, and organic Brønsted acid promoters.

The dawn of the organocatalysis renaissance is generally traced to the independent discoveries by Eder, Sauer, and Wiechert [151] and Hajos and Parrish [152] that symmetrical ketones of the type shown in Scheme 5.28 undergo stereoselective intramolecular aldol reactions when treated with proline or other amino acids. These remarkable papers established, right out of the gate, several features that have drawn numerous researchers to the field of organocatalysis: use of an incredibly simple and readily available catalyst (Jacobsen has aptly referred to proline as "the simplest enzyme" [153]) and an experimentally simple — although not necessarily mechanistically simple — way to obtain a valuable chiral product. The

particular examples initially reported were of interest not only because they were synthetically important (the utility of these building blocks, including the [4.3.0]bicyclic
Wieland–Miescher ketone for steroid synthesis is clearly evident) but they also featured a
rare early example of an asymmetric synthesis based on enantiotopic group selectivity using
a nonenzymatic catalyst.

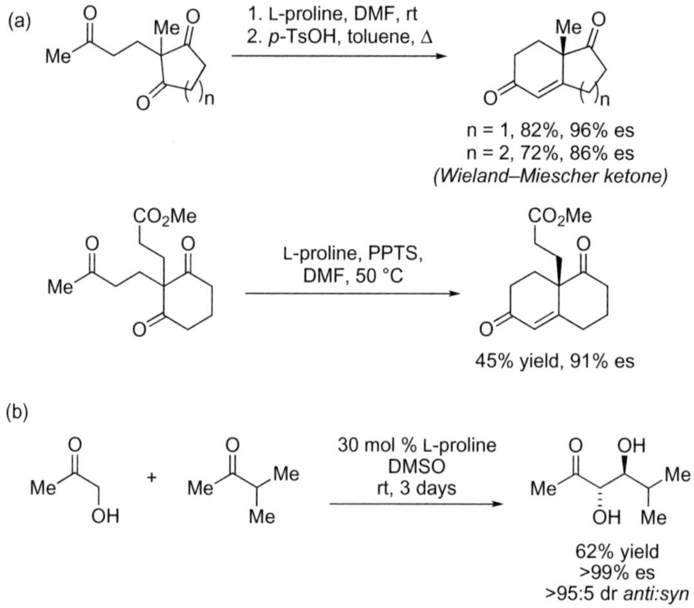

SCHEME 5.28 (a) The original Hajos–Parrish–Eder–Sauer–Weichert reactions [151,152]. (b) An early
intermolecular aldol addition reaction published by List [154].

Nearly 30 years later, List, Lerner, and Barbas published a proline-catalyzed aldol carried
out in an intermolecular fashion. The example shown in Scheme 5.28b is remarkable in terms
of exquisite control of both relative and absolute configuration and also for the direct use of
an α-hydroxyketone as the enolate precursor. This disclosure (which coincided closely with
the introduction of the organocatalytic Diels–Alder reaction, Section 6.1) led to a virtual
explosion of activity in the organocatalysis of the aldol reaction.

The mechanism of the proline-promoted aldol reaction has been much discussed, with
most authors preferring a mechanism introduced by List and Houk (left side of Scheme 5.29
[155]). First, the catalyst generates a nucleophilic enamine that adds to the carbonyl reaction
partner. In the addition step leading to the transition state shown, the carboxylic acid moiety
of the proline hydrogen bonds to the carbonyl oxygen, directing the aldehyde to the *Re* face
of the enamine, while simultaneously providing internal acid catalysis and providing a measure of conformational rigidity (participation by the basic nitrogen of the proline is also
invoked). In the model shown, the *lk* topicity arises from the preference of R_3 to remain in a
pseudoequatorial orientation in the transition state. Water promotes turnover by facilitating
the hydrolysis of the iminium ion intermediate. The Hajos–Parrish–Eder–Sauer–Weichert
reaction was rationalized *via* a similar model (inset).

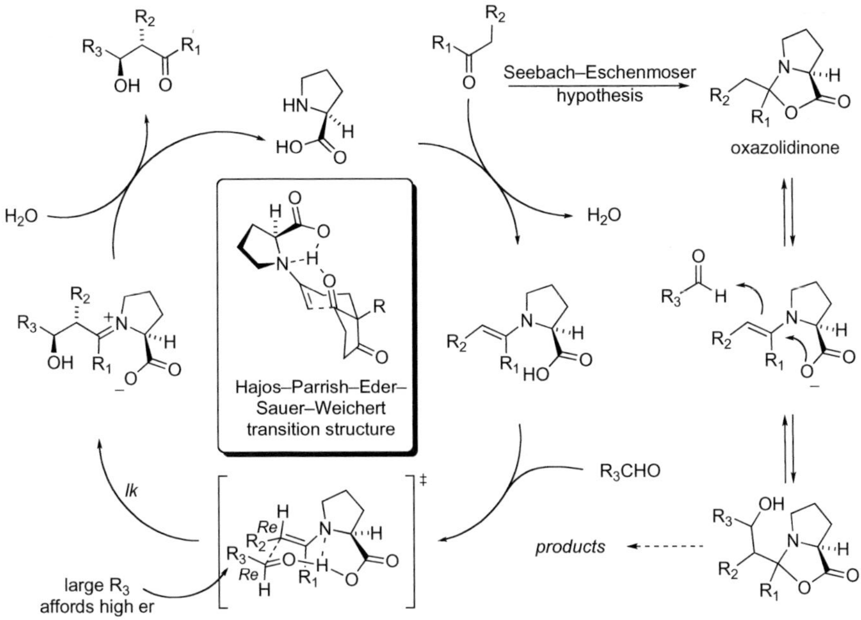

SCHEME 5.29 Catalytic cycle of the proline-promoted aldol reaction as proposed by List and Houk [155], with the transition state for the special case of the group-selective HPESW reaction also depicted (inset). The basic elements of an alternative proposal by Seebach and Eschenmoser are shown to the right [156].

The importance of this reaction is such that many authors have weighed in with alternative mechanistic rationales. In particular, Seebach and Eschenmoser suggested an alternative whereby an initially formed oxazolidinone intermediate previously proposed as an off-pathway (or more colorfully, "parasitic") intermediate instead was intrinsically involved in the mechanism. They suggested that opening of the oxazolidinone ring would afford a carboxylate intermediate that would then attack the aldehyde partner to ultimately afford product [156]. The interested reader is directed to the literature for continued discussion of the actual pathway of this now-iconic example of organocatalysis (*e.g.*, see refs. [157−160]).

Since its introduction, numerous investigators have sought to improve upon proline as a mediator of the aldol addition reaction. This has resulted in catalysts with additional acidic or hydrogen bonding isosteres to replace the carboxylic acid of proline (see, *e.g.*, (*b*)−(*f*) in Figure 5.9). Interestingly, most retain a five-membered ring basic amine core, presumably because it has an optimal combination of properties such as nucleophilicity and the ability to carry an adjacent substituent in an orientation able to participate in the transition state of the aldol event. The axially dissymmetric catalyst in Figure 5.9g and the *trans*-diamine of Figure 5.9h provide some exceptions, whereas the imidazolone catalyst in Figure 5.9i returns to a five-membered ring motif [161].[13] As one might expect, transition structures

---

13. For the use of this chemotype in this context of the Diels−Alder reaction, see Section 6.1.

analogous to those shown in Scheme 5.29 are generally proposed for the structurally similar catalysts. One additional example involving the highly selective axially chiral example is illustrated in Figure 5.9j. The catalyst in Figure 5.9i is incapable of delivering the carbonyl electrophile *via* hydrogen bonding and so the model shown in Figure 5.9k has been offered for this example.

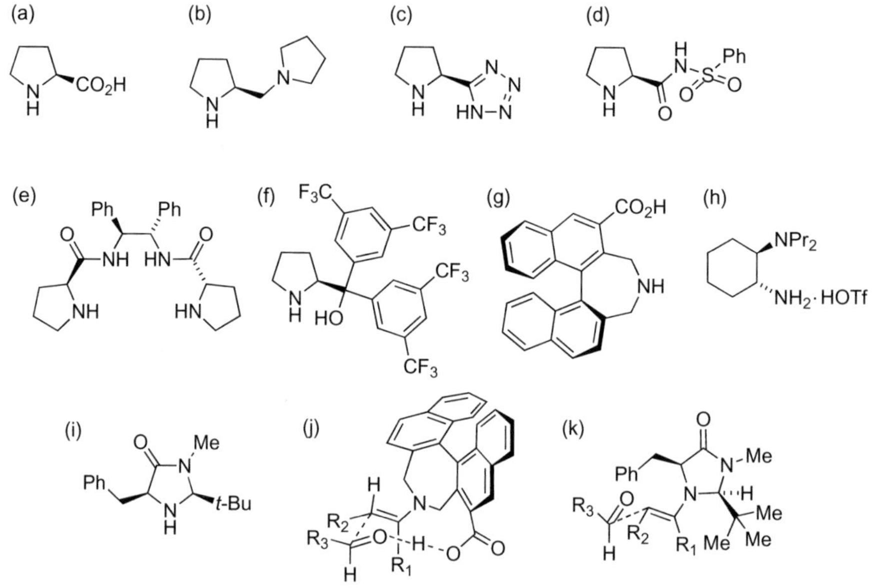

FIGURE 5.9 Organocatalysts for the asymmetric aldol reaction, beginning with (a) proline that inspired them all. (b) Yamamoto [162]; (c) Arvidsson [163]; (d) Ley [164]; (e) Zhao [165]; (f) Hayashi [166]; (g) Maruoka [167]; (h) Cheng [168]; (i) MacMillan [161]. Finally, proposed transition structures for the catalysts depicted in (g) and (i) are shown in (j) and (k), respectively.

Organocatalytic aldol reactions would not receive much interest if the only innovation was replacing a metal-based catalyst by a wholly organic one (although avoiding metals might be viewed as one approach to "green" methodology). However, a perusal of Table 5.9 does reveal some attractive features of this methodology. First of all, the ability to use straightforward reaction conditions that do not require the *in situ* generation of a hard enolate species is attractive. Secondly, organocatalytic methods often use low-molecular-weight ketones and aldehydes that, while not strictly excluded from other catalytic systems, fit in very nicely with reactions promoted by the catalysts depicted in Figure 5.9. A notable example of this is *unprotected* 2-hydroxyacetone reported by List (Scheme 5.28b; [169]). In addition, the easy formation of an imine/enamine equilibrium mixture from an enolizable aldehyde is especially convenient when one wishes to carry out an aldol with an aldehyde enolate equivalent. A primary inconvenience often cited with organocatalysis in general and apparent in these aldol examples is that it is often necessary to use relatively high catalyst loadings; note that all but one of the examples in Table 5.9 require 10–30% catalyst loadings for sufficiently zippy reactions.

TABLE 5.9 Representative Organocatalytic Aldol Reactions

Entry	Reactant(s)	Major Product	Catalyst + conditions	% Yield	er (dr)	Ref.
1			a (10 mol%), DCM	95	99:1 (91:9)	[170]
2			a (30 mol%), DMSO, 8–10 h	62	86:14	[169]
3			a (30 mol%), DMSO, 8–10 h	97	98:2	[169]
4			a (30 mol%), DMSO	68	99:1	[171]
5			a (50 mol%), DCM	94	94:6	[172]
6			a (10 mol%), DMF, 4 d	51	99:1 (80:20)	[173]

(Continued)

TABLE 5.9 (Continued)

Entry	Reactant(s)	Major Product	Catalyst + conditions	% Yield	er (dr)	Ref.
7	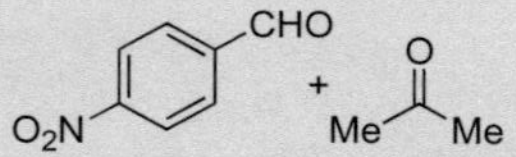 O_2N-C$_6$H$_4$-CHO + Me-CO-Me	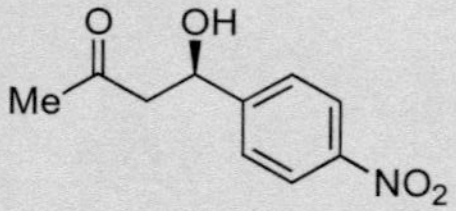 4-NO_2-C$_6$H$_4$-CH(OH)-CH$_2$-CO-Me	b (20 mol%), Acetone	60	94:6	[162]
8	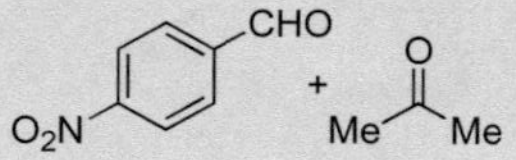 O_2N-C$_6$H$_4$-CHO + Me-CO-Me	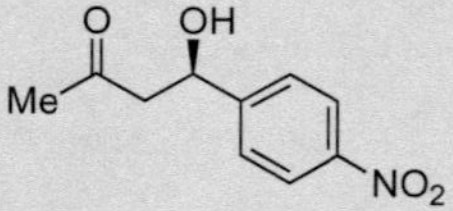 4-NO_2-C$_6$H$_4$-CH(OH)-CH$_2$-CO-Me	c (20 mol%), DMSO	82	90:10	[163]
9	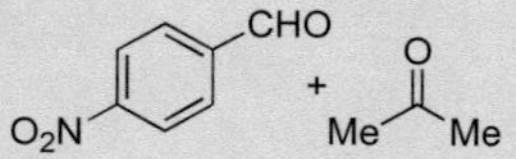 O_2N-C$_6$H$_4$-CHO + Me-CO-Me	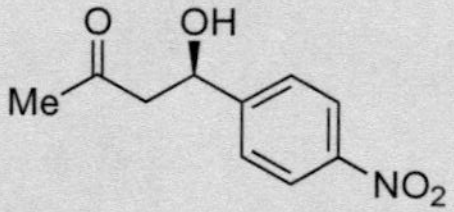 4-NO_2-C$_6$H$_4$-CH(OH)-CH$_2$-CO-Me	d (10 mol%), Acetone	100	96:4	[164]
10	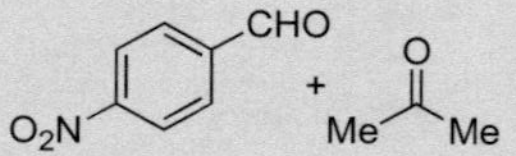 O_2N-C$_6$H$_4$-CHO + Me-CO-Me	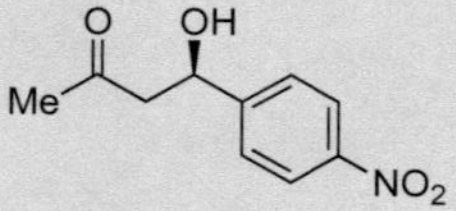 4-NO_2-C$_6$H$_4$-CH(OH)-CH$_2$-CO-Me	e (10 mol%), Acetone	88	99:1	[165]
11	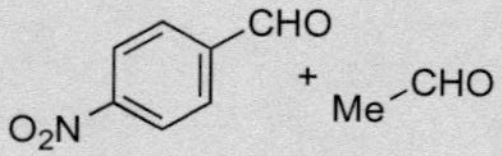 O_2N-C$_6$H$_4$-CHO + Me-CHO	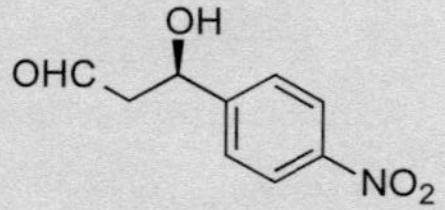 4-NO_2-C$_6$H$_4$-CH(OH)-CH$_2$-CHO	f (10 mol%), DMF, 1 d	56	91:9	[166]

TABLE 5.9 (Continued)

Entry	Reactant(s)	Major Product	Catalyst + conditions	% Yield	er (dr)	Ref.
12	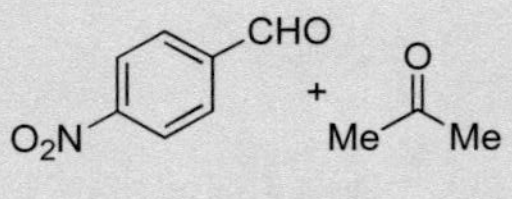	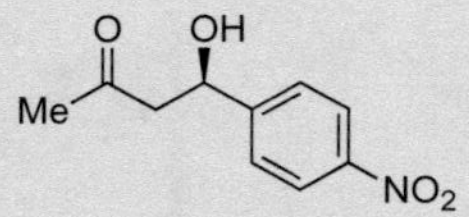	g (5 mol%), DMF	82	98:2	[167]
13	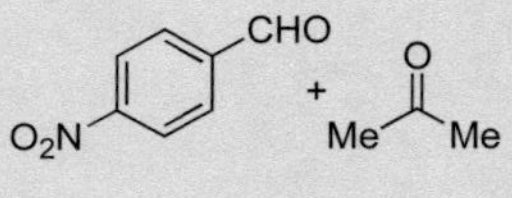	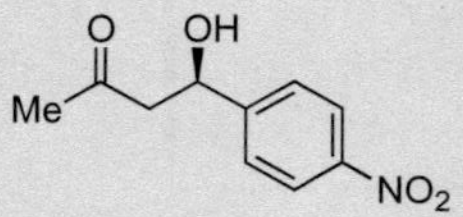	h (10 mol%), Acetone	94	98:2	[168]
14	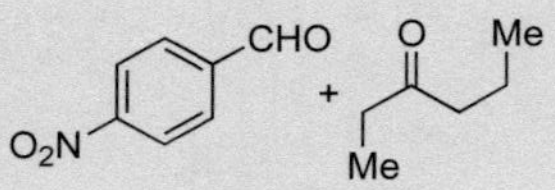	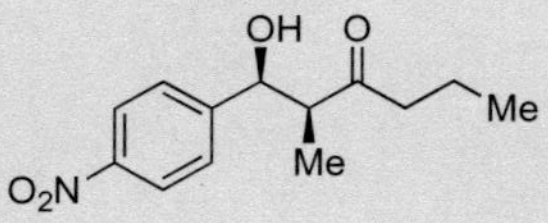	h (10 mol%), Acetone	75	98:2 (80:20)	[168]
15	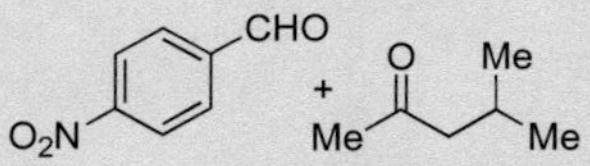	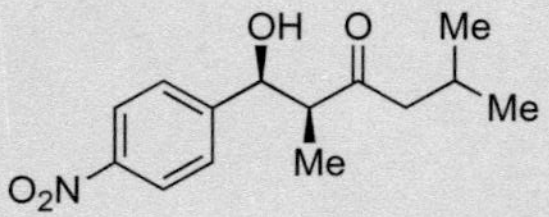	h (10 mol%), Acetone	56	93:7	[168]
16	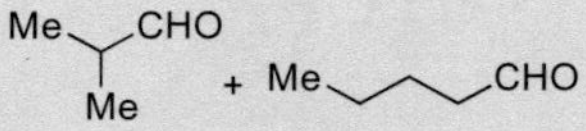	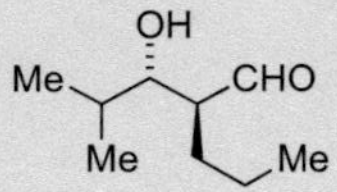	i (10–20 mol%), Et₂O, 4 °C, Amberlyst-15, MeOH	72	88:12 (86:14)	[161]

Organocatalysis is beginning to find its way into total synthesis (review: [174]). Two examples where the aldol reaction has been particularly well deployed are shown in Scheme 5.30. Part (a) summarizes some highlights of MacMillan's route to callipeltoside C [175]. Scheme 5.30b illustrates Nakamura's straightforward aldol reaction between acetone and a reactive indole-derived ketone to generate the sole stereocenter in the entertainingly named R-convolutamydine A [176].

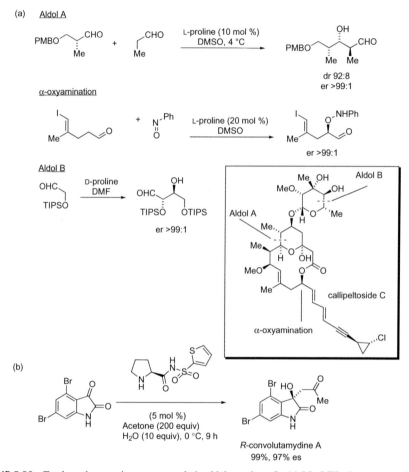

SCHEME 5.30   Total syntheses using organocatalytic aldol reactions. In (a) MacMillan's route to callipelto-side C [175], several aldol disconnections and an organocatalytic α-oxyamination were key steps (see Section 8.32). (b) Nakamura's route to convolutamydine used a catalyst analogous to that shown in Figure 5.9d, except that the phenyl group was replaced by a 2-thiophene [176].

An important organocatalytic process that is closely related to the aldol is the Morita−Baylis−Hillman (MBH) reaction whereby (1) an α,β-unsaturated carbonyl compound or related Michael acceptor is converted into an enolate by the conjugate addition reaction of a catalyst, (2) the enolate undergoes an aldol event with an appropriate electrophile, and (3) elimination then regenerates a double bond analogous to that in the original starting material (Scheme 5.31a; reviews: [177,178]). A number of asymmetric variants of this process have been reviewed, beginning with the chiral auxiliary approach given in Table 5.10, entries 1 and 2. Here, note the interesting but so far unexplained dependence of selectivity upon reaction pressure. High reaction pressures are often used to overcome the generally sluggish nature of the Morita−Baylis−Hillman reaction, which typically entails a large, negative volume of activation. The chiral amines in entries 3 and 4 differentiate the two faces of the generated enolate through the temporary attachment of these units onto the β position (*i.e.*, adjacent to the enolate) in the first step of the reaction. Scheme 5.31b illustrates the rationale for entry 4. Induction of chirality at the emerging secondary alcohol stereocenter requires high diastereofacial selectivity with respect to the incoming aldehyde (note that the equilibrating intermediates have the same relative topicities between the short-lived α-stereocenter and the new secondary alcohol), which is favored by the hydrogen bonding motif proposed in Scheme 5.31b [179]. Two examples of an approach introduced by Schaus and involving catalysis by an achiral phosphine with a chiral BINOL-derived additive are given in entries 5 and 6; these presumably operate by forming a Brønsted acid complex in the reaction intermediate but the details of this are currently unknown [182]. Finally, an example of an aza-MBH reaction, entailing a Mannich-type of addition onto an imine reaction partner, is provided to suggest the ever-expanding scope of this reaction.

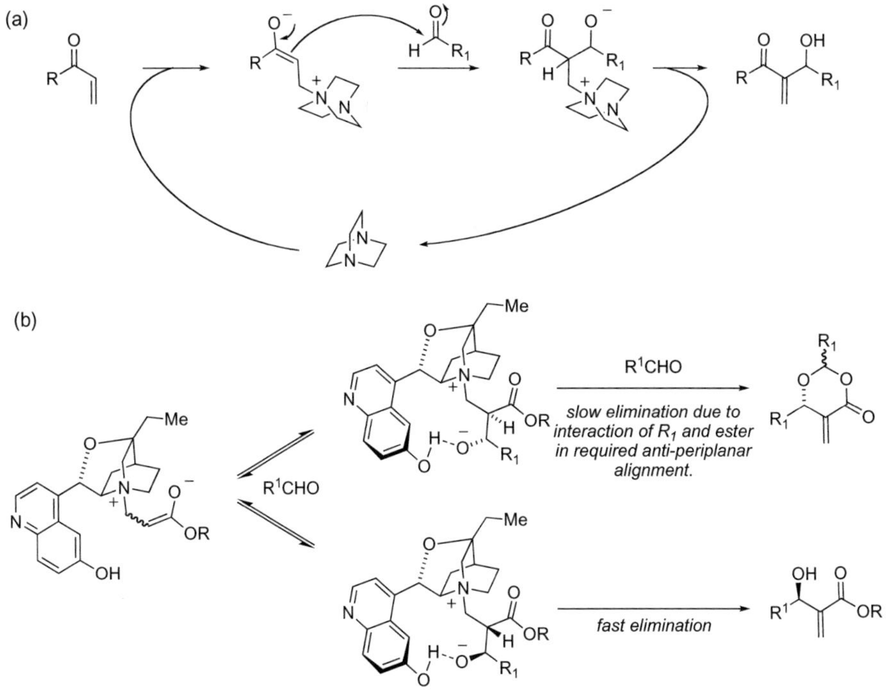

SCHEME 5.31 (a) General mechanism of the Morita−Baylis−Hillman reaction. (b) Proposed transition structures for the reaction given in entry 4, Table 5.10 [179]. Note the formation of the acetal-containing byproduct, which assists in the separation of the unwanted isomer.

TABLE 5.10 Asymmetric Morita–Baylis–Hillman Reactions

Entry	Substrate	Product	Catalyst (mol%) and Conditions	% Yield	% es	Ref.
1			(100) 1 bar	93 (61:39 dr)		[180]
2			(100) 7.5 kbar	42 (99:1 dr)		[180]
3			(10), 3 kbar	50	73	[181]
4			(10)	50	96	[179]
5			(10), PEt₃	88	95	[182]

(*Continued*)

TABLE 5.10 (Continued)

Entry	Substrate	Product	Catalyst (mol%) and Conditions	% Yield	% es	Ref.
6	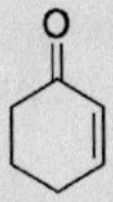	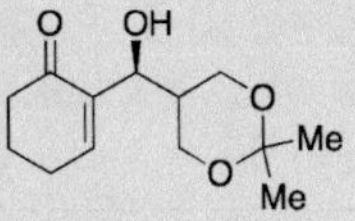	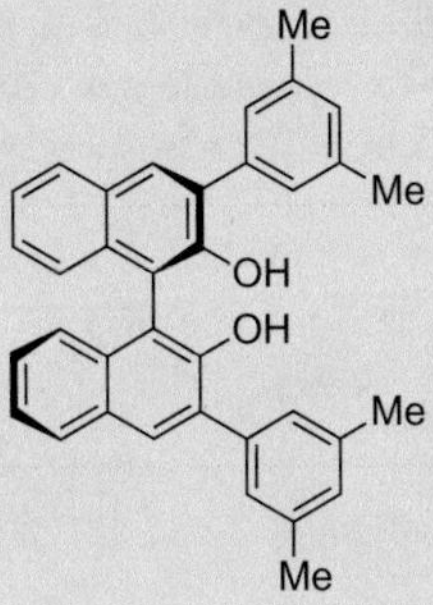 (20), PEt₃	70	96	[182]
7	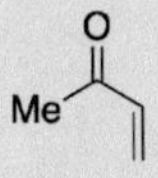	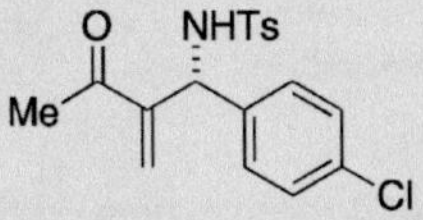	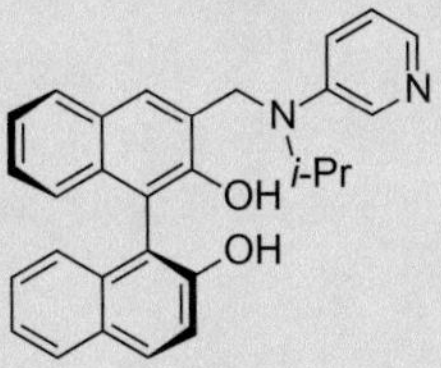 (10)	96	98	[183]

5.2.4 Double Asymmetric Induction and Beyond: Synthetic Applications of the Aldol Reaction

The asymmetric aldol reaction has been very extensively used in the total synthesis of natural products, particularly those that are biogenetically derived from propionate and acetate building blocks. In those settings, a critical question is how to choose aldol disconnections that are consistent both with synthetic efficiency (often tied to convergence) and with maximum stereoselectivity. In considering the latter, one needs to consider the combined effects of the face selectivities of both of the aldol components and in many cases, second-order influences of other nearby substituents.

As seen in previous sections, there are numerous chiral enolates or equivalents thereof that have very high diastereofacial preferences in their reactions with electrophiles. In contrast, most chiral aldehydes do not have a very high facial bias (see Cram's rule, Section 4.1),[14] meaning that the overall stereoselectivity of a given aldol reaction will very often be determined by the face selectivity of the enolate component. When this happens, one

---

14. For a thorough analysis of stereoselective aldol additions of achiral lithium and boron enolates to chiral aldehydes, see ref. [184].

says that the reaction is subject to "reagent-based stereocontrol" and that it may be subject to double asymmetric induction (see Section 1.6). For example, as part of a synthesis of 6-deoxyerythronolide-B and the Prelog–Djerassi lactone [185], Masamune examined the selectivity of each enantiomer of his chiral enolate with a chiral aldehyde, as shown in Scheme 5.32. The aldehyde, which itself has a low inherent bias, and shows only 60% diastereoselectivity when allowed to react with an achiral boron enolate, may be converted selectively into either of the two possible *syn* adducts with 94% and 98% diastereoselectivity for the mismatched and matched cases, respectively [185].

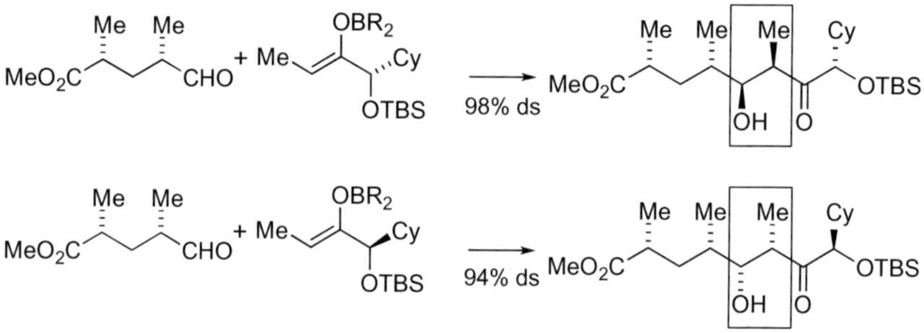

SCHEME 5.32 Matched and mismatched asymmetric aldol additions of the Masamune enolate [185].

In his synthesis of the Prelog–Djerassi lactone, Evans tested two auxiliaries that give products with opposite absolute configurations at the two new stereocenters [186]. Here again, the facial bias inherent in the aldehyde is low. Since the two chiral enolates are not enantiomers, we can only guess which is the matched and which is the mismatched case, but it hardly matters: the selectivity is $\geq 99.8\%$ for both (Scheme 5.33).

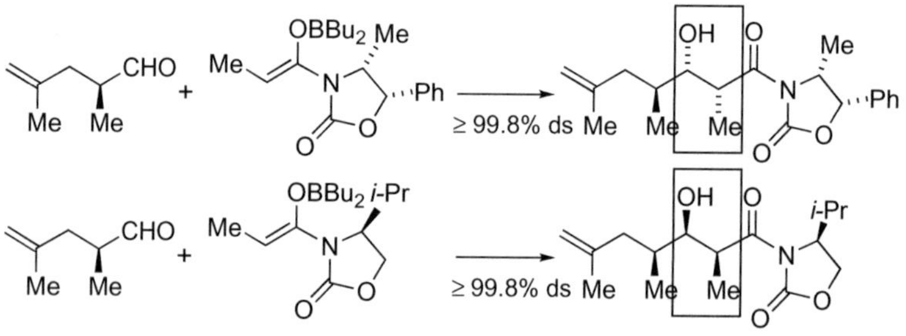

SCHEME 5.33 Reagent-based stereocontrol in aldol additions using Evans imide enolates [186]. Note that the boxed areas of the molecule respond only to the configuration of the auxiliary-containing enolates.

One cannot always bank on reagent-based stereocontrol, even with reagents as selective as the Evans imide enolates. For example, during the course of a synthesis of cytovaricin [187], the enolate shown in Scheme 5.34 was expected to afford the *syn* adduct when added to the aldehyde illustrated. Instead, an *anti* aldol adduct was formed as a single diastereomer. Note that the *Re* face of the enolate is preferred according to the transition state analysis presented in Scheme 5.13a, and the *Si* face of the aldehyde is preferred according to the Felkin−Anh theory (Section 4.1 and Figure 4.8 or see glossary, Section 1.11). Analysis of the product configuration, as shown in the inset, indicates that the preferred faces of both the enolate and aldehyde were coupled. Apparently, the *Si* facial preference of the aldehyde was sufficiently strong to disrupt the *lk* topicity preferred by the enolate.

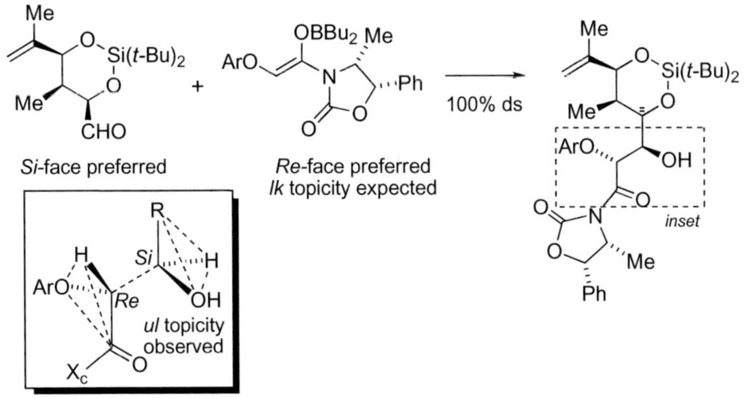

SCHEME 5.34   A rare case of mismatched double asymmetric induction that is (essentially) 100% diastereoselective [187].

Aldol additions may also be used to couple two large fragments in a convergent synthesis, but such reactions are not amenable to auxiliary-based approaches. Instead, the chemist must choose disconnections for which combining the face selectivities of the enolate and aldol components would lead naturally to the desired aldol outcome. A successful example of this is shown in Scheme 5.35 [188,189]. Note the dependence of stereoselectivity on whether the enolate is bound to a dialkyl boron atom or a lithium ion.

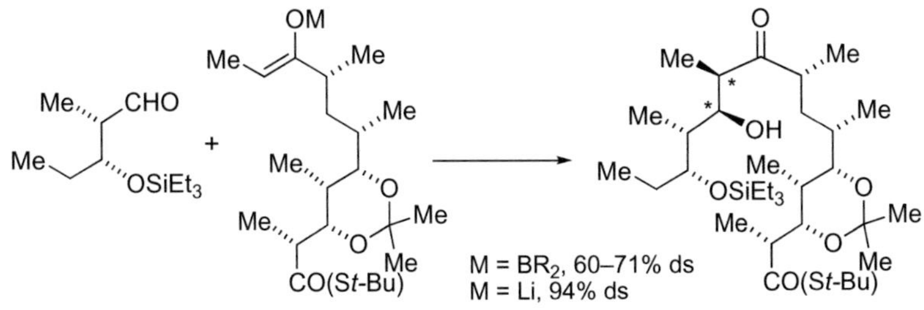

SCHEME 5.35   Selective coupling of two chiral fragments (double asymmetric induction) in the asymmetric synthesis of 6-deoxyerythronolide B [185,188]. For a similar reaction in the synthesis of erythronolide B, see ref. [190].

Analysis of the major addition product of Scheme 5.35 (Figure 5.10a) indicates that the *Si* face of the enolate adds to the *Re* face of the aldehyde; the latter corresponds to *anti*-Cram selectivity (Section 4.1). Two explanations were offered to explain the selectivity of this aldehyde. Masamune originally suggested that the enolate adds to the aldehyde through a boat transition state that is also chelated by the silyloxy group (Cram cyclic model, Section 4.2), as illustrated in Figure 5.10b. Weaknesses of this postulate are that Cram's cyclic model is more often effective when the chelate is a five-membered ring, and that the triethylsilyloxy group probably is not a good chelator [191]. Ten years after Masamune's original hypothesis was offered, Roush analyzed a considerable amount of data accumulated in the interim, and pointed out that a Zimmerman—Traxler chair, adding to the *Si* face of the aldehyde (as expected by the Felkin—Anh model), is destabilized by a 2,3-*P*,3,4-*M* gauche pentane interaction (*cf.* Figure 5.5), as indicated in Figure 5.10c. Roush suggested that an *anti* Felkin—Anh (anti-Cram) chair transition structure more adequately explains the facts, as shown in Figure 5.10d [184]. Whatever the mechanism, achiral lithium enolates add to the aldehyde of Scheme 5.35 with selectivities in the 80—90% ds range; the higher selectivity observed with the chiral enolate may therefore be attributed to matched pair double asymmetric induction [188]. However, note that if the target had had the opposite absolute configuration at the indicated stereocenters, it would have been the minor isomer under any of the conditions examined.

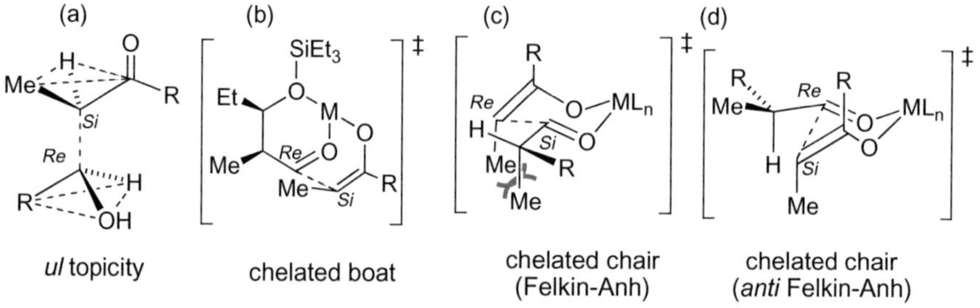

FIGURE 5.10 Analysis of possible transition structures for the aldol addition in Scheme 5.35: (a) The observed topicity; (b) boat transition structure postulated by Masamune [188]; (c) gauche pentane interaction that destabilizes the Cram (or Felkin—Anh) selectivity of the aldehyde; (d) *anti*-Cram (*anti* Felkin—Anh) addition *via* a chelated chair [184].

The stereoselectivity of the aldol additions shown in Schemes 5.34 and 5.35 is obviously the result of a complex series of factors, foremost among which are the Felkin—Anh preference dictated by the α-substituent on the aldehyde and the proximal stereocenters on the enolate. It has also been noted that more remote stereocenters, such as at the β-position of the aldehyde, may also influence the selectivity of these types of reactions. Evans has studied some of the more subtle effects on crossed aldol selectivity, such as protecting groups at a remote site on the enolate [192], β-substituents on the aldehyde component [193], and also of matched and mismatched stereocenters at the α and β positions of an aldehyde (double asymmetric induction) [194]. Further, the effect of chiral enolates adding to α,β-disubstituted aldehydes has been evaluated [195]. Scheme 5.36 depicts a particularly detailed analysis wherein the overall reaction responds to (1) the usual Felkin—Ahn considerations leading to aldehyde face selectivity, (2) the face selectivity of a powerful boron enolate, and (3) two sets of 1,3-diasteomeric preferences arising from stereocenters marked α and γ in the figure [196].

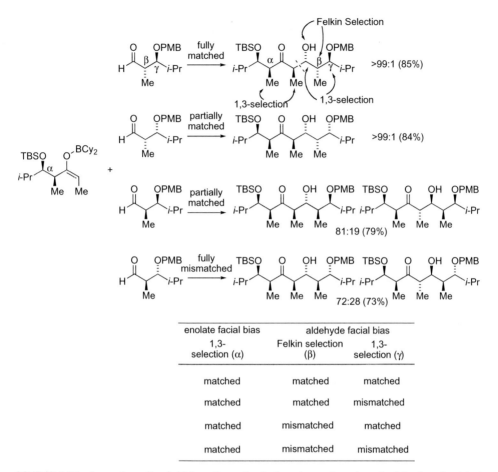

enolate facial bias	aldehyde facial bias	
1,3-selection (α)	Felkin selection (β)	1,3-selection (γ)
matched	matched	matched
matched	matched	mismatched
matched	mismatched	matched
matched	mismatched	mismatched

SCHEME 5.36   A complex suite of aldol reactions wherein the outcome depends on the interplay of matched, partially matched, and mismatched facial biases [195].

The aldol addition reaction and the related crotyl metal additions (Section 5.1), have figured prominently in the total synthesis of complex natural products (reviews: [197−201]). The examples shown in Figure 5.11 clearly demonstrate the power of this technique in modern asymmetric synthesis. In two of the cases shown ((a) vs. (b) and (d) vs. (e)), the alternative disconnection patterns emphasize the versatility of aldol technology even when applied to a specific target. Finally, the Novartis synthesis of discodermolide on 60 g scale [202−206] using a combination of approaches pioneered by Smith [207,208] and Paterson [209] provides an impressive demonstration of the utility of this time-honored but still evolving reaction in organic synthesis.

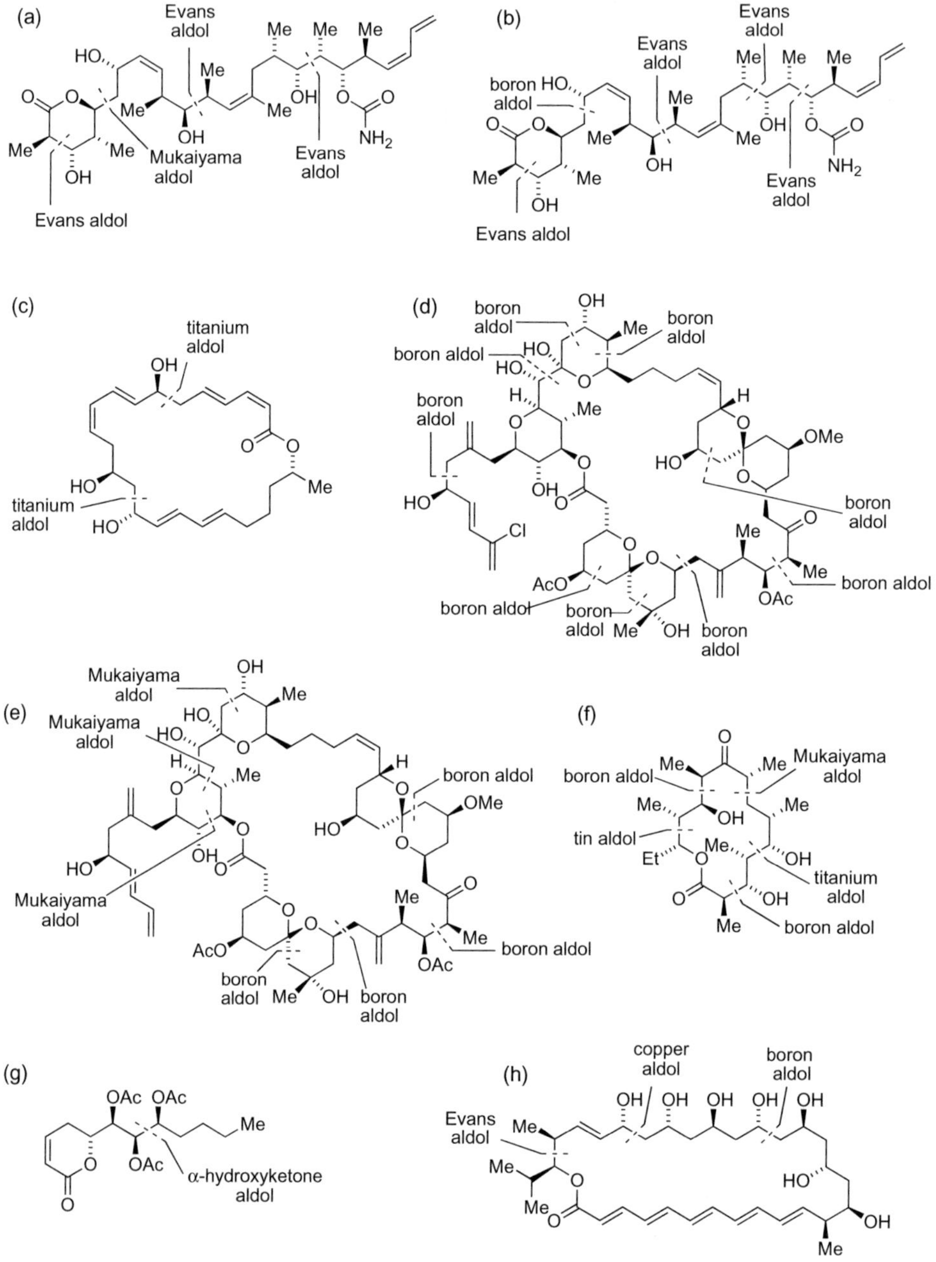

FIGURE 5.11 Natural products synthesized using aldol methodology along with key retrosynthetic disconnections: discodermolide by (a) Smith [210] and (b) the Novartis research team [202–206]; (c) macrolactin A by Carriera [211], spongistatin by (d) Paterson [212] and (e) Evans [213–215]; (f) oleandolide by Evans [216]; (g) boronolide by Trost [217]; (h) roxaticin by Evans [218].

5.3 MICHAEL ADDITIONS[15]

The term "Michael addition" has been used to describe 1,4 (conjugate) additions of a variety of nucleophiles including organometallics, heteroatom nucleophiles such as sulfides and amines, enolates, and allylic organometals to the so-called "Michael acceptors" such as α,β-unsaturated aldehydes, ketones, esters, nitriles, sulfoxides, and nitro compounds. Here, we will mainly discuss the classical Michael reaction, which employs resonance-stabilized anions such as enolates and azaenolates, but will also include a few examples of enamines and heteroatom-based nucleophiles either because of close similarity or synthetic utility. 1,4-Addition reactions of organometallic agents such as cuprates are discussed in Section 4.4.

5.3.1 Simple Diastereoselectivity: Basic Transition State Analysis

When a prochiral acceptor ($R_1CH=A$) and a prochiral donor ($R_2CH=D$) react, the stereoisomers are labeled as either *syn* or *anti* based on the relative configurations of R_1 and R_2 when the Michael adduct is drawn in a zig-zag projection, as shown in Scheme 5.37. Using the *Re/Si* nomenclature and assuming that the CIP rank is $A > R_1 > H$ and $D > R_2 > H$, the *syn* adducts arise from *lk* topicity and *anti* adducts arise from *ul* topicity.

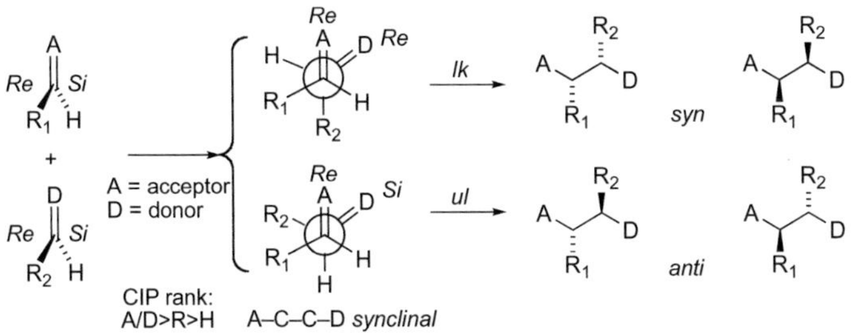

SCHEME 5.37 Topicity [6,67] and adduct [220] nomenclature for Michael additions.

Seebach suggested in 1981 [67] that enolate donors and carbonyl acceptors are probably *synclinal* in the transition state. Steric repulsion between R_1 and the donor D, is proposed to orient R_1 *antiperiplanar* to D; pyramidalization (*cf.* Figure 3.5 and refs. [222−225]) and tilting of the donor to accomodate the Bürgi−Dunitz angle of 107° (*cf.* Figures 4.6 and 4.7, and refs. [226−228]) are proposed to disfavor the *lk* topicity, since R_2 is more sterically demanding than hydrogen (Figure 5.12).

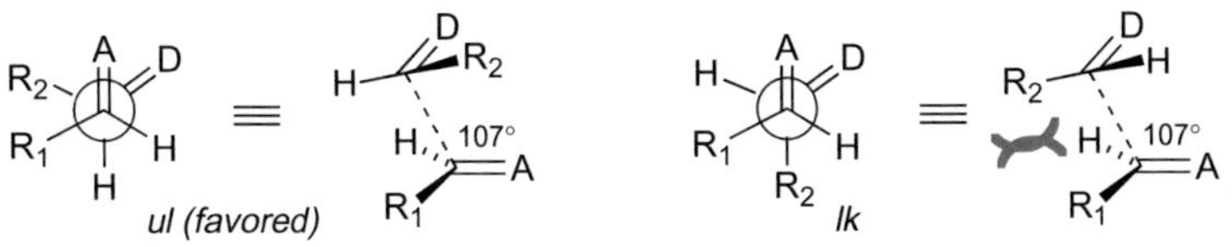

FIGURE 5.12 Pyramidalization of the donor and the Bürgi−Dunitz trajectory contribute to destabilization of the *lk* topicity combination according to Seebach [67].

---

15. For a comprehensive coverage of conjugate addition reactions, see ref. [219]. For a comprehensive review of the stereochemical aspects of base-promoted Michael reaction, see ref. [220]; for a similarly comprehensive review of acid-catalyzed Michael reactions and conjugate additions of enamines, see ref. [221].

Analysis of numerous examples [220,221] and mechanistic studies [229,230] led Heathcock to refine these hypotheses and, in consideration of the actual substituents (R_1, R_2, A, and D), place them on firmer mechanistic grounds.[16] The four transition structures in Scheme 5.38 are direct extensions of those in Scheme 5.37 and Figure 5.12. For ketone and ester enolates, there is a strong correlation between the relative configuration of the product and the enolate geometry: $Z(O)$-enolates give *anti* products and $E(O)$-enolates give *syn* adducts [229]. The rationale for this is that transition structures for paths a and c (Scheme 5.38) are favored due to repulsive interactions between Y and R_3 in paths b and d. The selectivity of $Z(O)$-enolates appears to be higher than that of $E(O)$-enolates, probably due to the destabilization of path c by the pyramidalization and trajectory considerations illustrated in Figure 5.12, which intrinsically favor paths a and d, in which a hydrogen is *antiperiplanar* to the enone double bond.

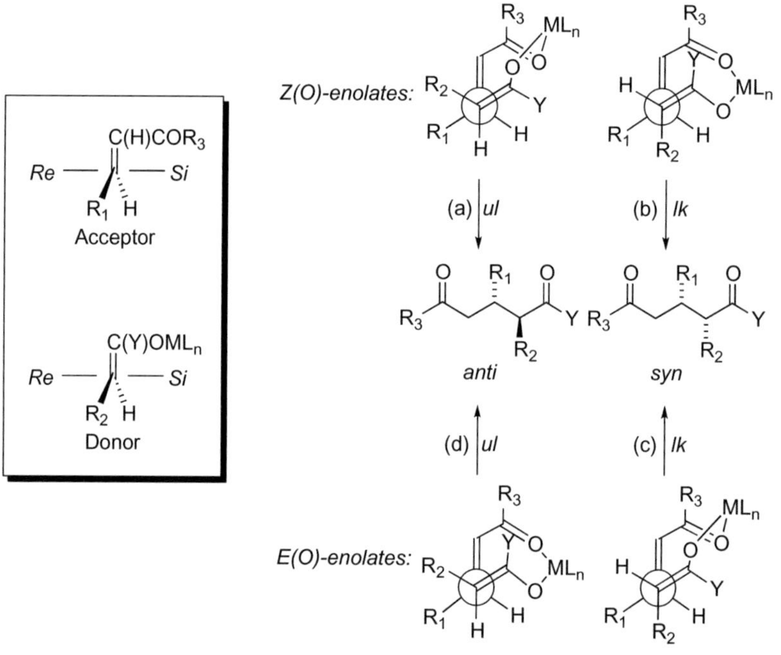

SCHEME 5.38 Proposed chelated transition structures (and topicities) for Michael additions of lithium enolates of ketones, esters, and amides to enones [229,230]. Only one enantiomeric transition structure and product is shown for each topicity (*Si* face of the acceptor).

---

16. On the other hand, a computational study [231] of the Michael addition of propionaldehyde lithium enolate adding to *E*-crotonaldehyde indicates an *anticlinal* conformation around the forming bond (*i.e.*, A eclipsing R_2 in the *ul* topicity and A eclipsing H in the *lk* topicity of Figure 5.12).

For amide enolates, the situation is similar in that, when R_3 and Y are large, the transition structures of paths a and c are favored [230]. However, recall that acyclic amides invariably form $Z(O)$-enolates, so amide $E(O)$-enolates are only possible when R_2 and Y are joined, that is, in a lactam. In contrast to ketone and ester enolates, however, the transition structures of paths b and d appear to be intrinsically favored when Y and R_3 are small. This latter trend is (at least partly) contrary to what would be expected based on the simple analysis of Figure 5.12, but can be rationalized as follows. For the lactams, the R_2 and Y substituents present a rather flat profile, so that interaction with R_3 in path d is minimal. Additionally, the R_2–Y ring "eclipses" the β-hydrogen of the enone in c, destabilizing this structure. For amide $Z(O)$-enolates and acceptors with an R_3 substituent such as a phenyl, there may actually be an attractive interaction between Y and R_3, favoring path b. Clearly each case must be analyzed separately, but these transition structures serve as a starting point for such analyses. Note also that the structures of Scheme 5.38 all have enones in an *s-cis* conformation, which is not available to cyclic acceptors such as cyclohexenone, cyclopentenone, and unsaturated lactones.

For the purpose of asymmetric synthesis, we are interested in expanding on simple diastereoselectivity and differentiating between the two *lk* transition structures (*Re*–*Re* and *Si*–*Si*) and the two *ul* transition structures (*Re*–*Si* and *Si*–*Re*) for each enolate geometry. This is done by rendering the *Re* and *Si* faces of either component diastereotopic by the introduction of a stereogenic element. For asymmetric Michael reactions, a stereocenter in a removable substituent on the acceptor or in Y or the metal (ML_n) of the donor have been used to this end. The following discussion is organized in historical order: auxiliary-based approaches, interligand asymmetric induction, and catalysis.

5.3.2 Chiral Donors

Ester enolates. Oppolzer showed in 1983 that the $Z(O)$-dienolate shown in Scheme 5.39a adds to cyclopentenone with 63% diastereoselectivity [232]. Additionally, the enolate adduct can be allylated selectively, thereby affording (after purification) a single stereoisomer having three contiguous stereocenters in 48% yield. The transition structure illustrated is not analogous to any of those illustrated in Scheme 5.38 because cyclopentenone is an *s-trans-Z*-enone, whereas the enones in Scheme 5.38 are *s-cis-E*. In 1985, Corey reported the asymmetric Michael addition of the $E(O)$-enolate of phenylmenthone propionate to *E*-methyl crotonate as shown in Scheme 5.39b [233]. The product mixture was 90% *syn*, and the *syn* adducts were produced in a 95:5 ratio, for an overall selectivity of 86% for the illustrated isomer. The transition structure proposed by the authors to account for the observed selectivity is similar to that shown in Scheme 5.38c, but with the enone illustrated in an *s-trans* conformation. Intramolecular variations of these reactions were reported by Stork in 1986, as illustrated in Scheme 5.39c and 5.39d [234]. Two features of these reactions deserve comment. First, the carbonyl of the acceptor is not chelated to the enolate metal, and second, the selectivity of the camphor-derived ester is significantly higher than the similar reaction in Scheme 5.39a. The latter effect seems to be due to the position of the bridgehead methyl, which helps restrict conformational motion when it neighbors the ester enolate [234]. Note that hydrolysis of the adducts from the products in Scheme 5.39c and d affords enantiomeric cyclopentanones.

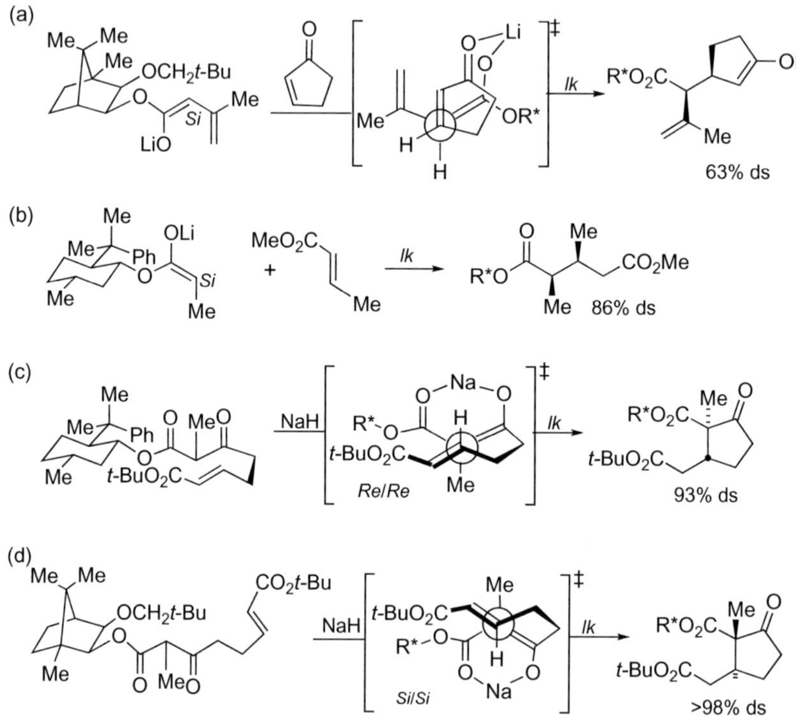

SCHEME 5.39 Asymmetric Michael additions of ester enolates. (a) [232]; (b) [233]; (c) and (d) [234].

Amide and imide enolates. Scheme 5.40 illustrates several examples of asymmetric Michael additions of chiral amide and imide enolates. Evans' imides, as their titanium enolates, afforded the results shown in Schemes 5.40a and b [235,236]. The yields and selectivities for the reaction with acrylates and vinyl ketones are excellent, but the reaction is limited to β-unsubstituted Michael acceptors: β-substituted esters and nitriles did not react, and β-substituted enones added with no selectivity [236]. A pseudoephedrine approach reported by Murry and coworkers led to an interesting result whereby it proved possible to change the major diastereoisomer from *anti* to *syn* through the addition of LiCl, as shown in Scheme 5.40c [237]. These workers carried out NMR studies to show that while a Z(O)-enolate was formed under both reaction conditions, a chelated delivery of the Michael acceptor to the top face of the enolate was favored under standard conditions (Scheme 5.40d) while added salt was suggested to lead to a change in face selectivity of the enolate coupled with a change in relative topicity (Scheme 5.40e).

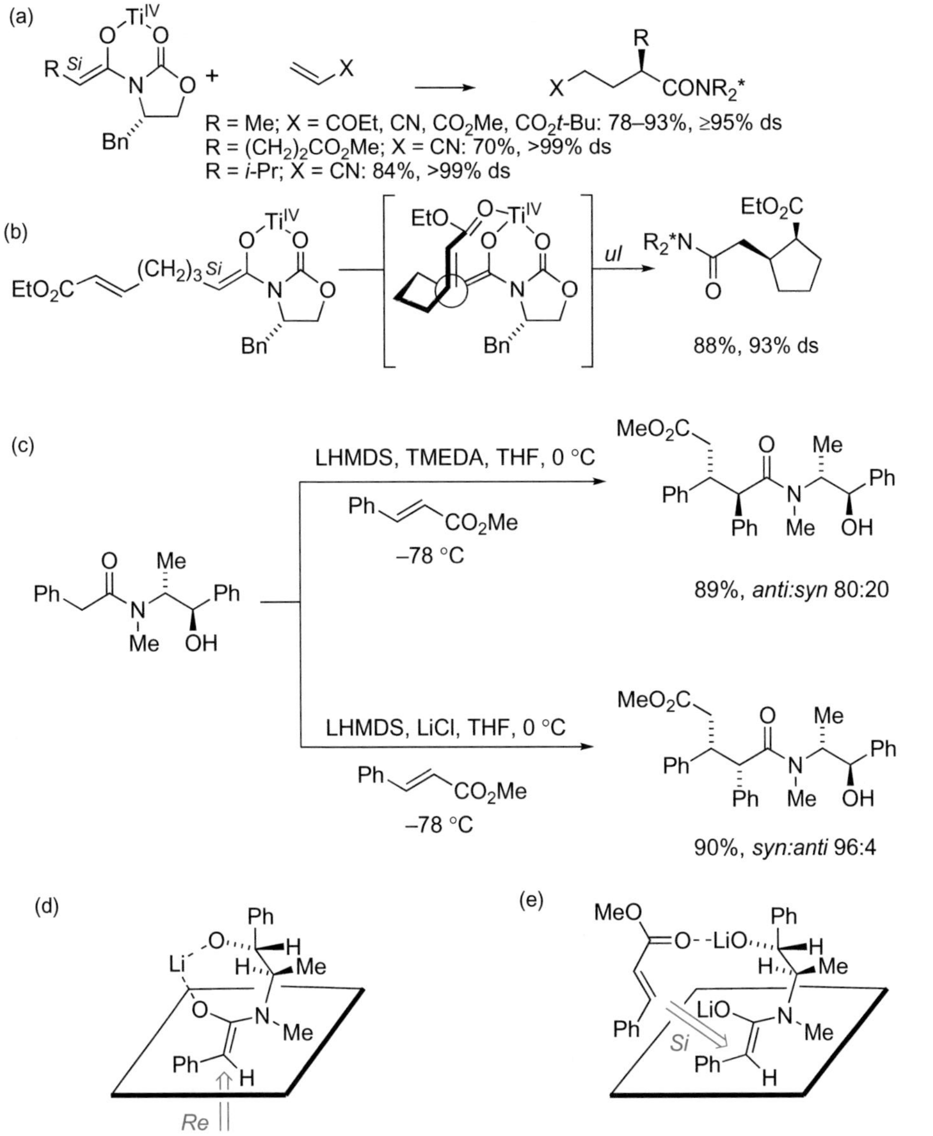

(a)

R = Me; X = COEt, CN, CO$_2$Me, CO$_2$t-Bu: 78–93%, ≥95% ds
R = (CH$_2$)$_2$CO$_2$Me; X = CN: 70%, >99% ds
R = i-Pr; X = CN: 84%, >99% ds

(b)

88%, 93% ds

(c)

LHMDS, TMEDA, THF, 0 °C

Ph‿CO$_2$Me

−78 °C

89%, *anti:syn* 80:20

LHMDS, LiCl, THF, 0 °C

Ph‿CO$_2$Me

−78 °C

90%, *syn:anti* 96:4

(d)

Re ‖

(e)

SCHEME 5.40 Asymmetric Michael addition of amide and imide enolates mediated by chiral auxiliaries: Evans in (a) [235,236] and (b) [236] and pseudoephedrine in (c)−(e) ([237]); *cf.* with Myers' rationale for alkylation reactions in Scheme 3.23.

Ketone and aldehyde azaenolates. Perhaps the most versatile of the auxiliaries for the asymmetric alkylation of ketones and aldehydes are the SAMP/RAMP hydrazones developed by Enders (*cf.* Schemes 3.26, 3.27, and Table 3.10). These hydrazones, as their lithium $E(O)$-enolates, also undergo highly selective Michael additions [238−241]. Several examples are illustrated in Scheme 5.41a. The rationale for the formation of the $E(O)$-enolate and for the *Re* facial selectivity of SAMP hydrazones is illustrated in Scheme 3.27. The Michael acceptors also react on the *Re* face of SAMP hydrazones, and the *ul* topicity at the new bond can be rationalized by Seebach's postulate (Figure 5.12 and Scheme 5.38d), that places the β-substituent of the Michael acceptor (R_3 in Scheme 5.41a) *antiperiplanar* to the double bond of the donor (= D in Figure 5.12), and has the acceptor double bond (= A in Figure 5.12) bisecting the angle between the donor double bond (= D) and the donor substituent (R_2 in Scheme 5.41a). Scheme 5.41b illustrates an extension for the synthesis of substituted cycloalkanes. The indicated (*) stereocenters are formed in the Michael addition; the other is formed by internal 1,2-asymmetric induction.

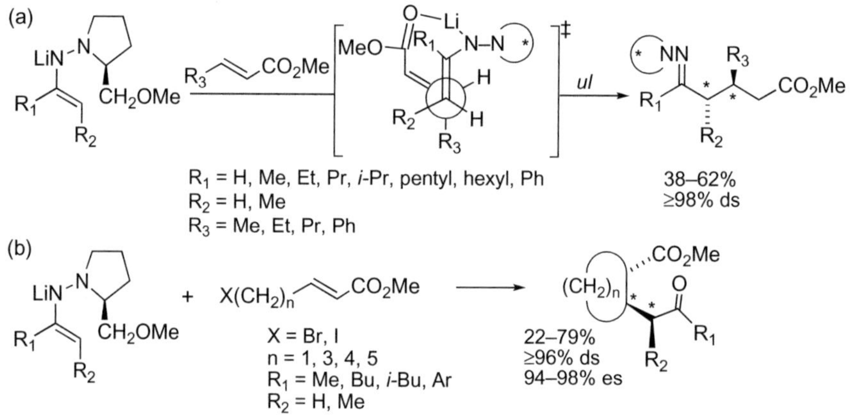

SCHEME 5.41   Michael additions of SAMP/RAMP hydrazones. (a) [238−240]; (b) [241]. Stereocenters formed in the Michael reaction are indicated (*).

The Koga group has investigated the asymmetric Michael addition of β-keto esters, as their valine lithium enamides, as shown in Scheme 5.42 [242,243]. The lithium derivative adds directly to methylene malonic esters without further activation [242], but is not reactive enough to add to methyl vinyl ketone or ethyl acrylate unless trimethylsilyl chloride is also added [243]. Interestingly, the absolute configuration of the product changes when HMPA is added to the reaction mixture. The rationale for this observation is that in the absence of HMPA, the electrophile coordinates to the lithium, taking the position of L in the chelated structure shown in the inset, thus delivering the electrophile to the *Re* (rear) face. When the strongly coordinating HMPA is present, it occupies the "L" position and blocks the *Re* face, thereby directing the electrophile to the *Si* face.

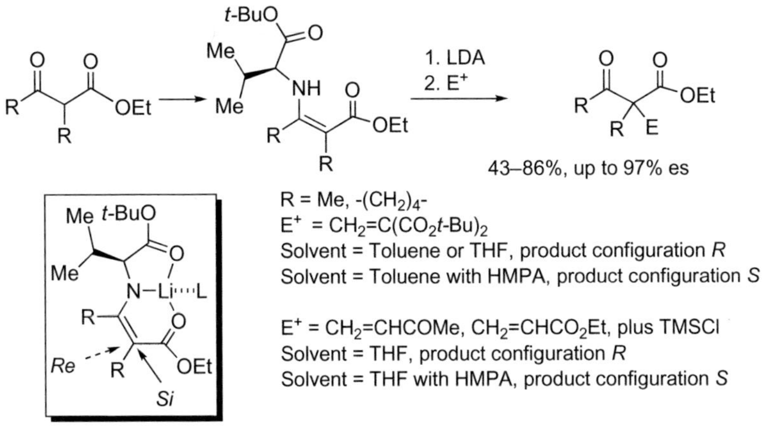

SCHEME 5.42   Koga's asymmetric Michael additions of valine enamines of β-keto esters [242,243].

Enamine intermediates.[17] The condensation of a secondary amine and a ketone to make an enamine is a well-known reaction which has seen wide use in organic synthesis [245−247]. Imines of a primary amine and a ketone exist in a tautomeric equilibrium between the imine and secondary enamine forms, although in the absence of additional stabilization factors (*cf.* Scheme 5.42), the imine is usually the only detectable tautomer. Nevertheless, the enamine tautomer is very reactive toward electrophiles and Michael additions occur readily [248]. Moreover, since the enamine is so much more reactive than the corresponding enol equivalent, the formation of enamines *in situ* is a common and successful tactic in organocatalytic approaches to the Michael reaction.

The mechanism of the Michael additions of preformed tertiary and secondary enamines are shown in Scheme 5.43. For tertiary enamines, the Michael addition is accompanied by proton transfer from the α'-position to either the α-carbon or a heteroatom in the acceptor, affording the regioisomeric enamine as the initial adduct [249]. The proton transfer and the carbon−carbon bond forming operations may not be strictly concerted, but they are nearly so, since conducting the addition in deuterated methanol led to no deuterium incorporation [249]. With secondary enamines, there is also transfer of a proton, but this time from the nitrogen. Again, isotope labeling studies [250] suggest that the two steps are "more or less concerted" [248], in a reaction that resembles the ene reaction (Scheme 5.43b).

Theoretical studies indicate that these transition structures are probably influenced by frontier molecular orbitals (in addition to steric effects), as indicated in Scheme 5.43c [251]. For the reaction of aminoethylene (a primary enamine) and acrolein, the enamine HOMO and the enone LUMO have the most attractive interactions when aligned in the chair configuration shown, which has the enone in an *s-cis* conformation. Note that this orientation places the NH and the electrophile α-carbon in close proximity for proton transfer *via* the "ene" transition structure.

---

17. For reviews of enamine Michael additions, see refs. [221,244].

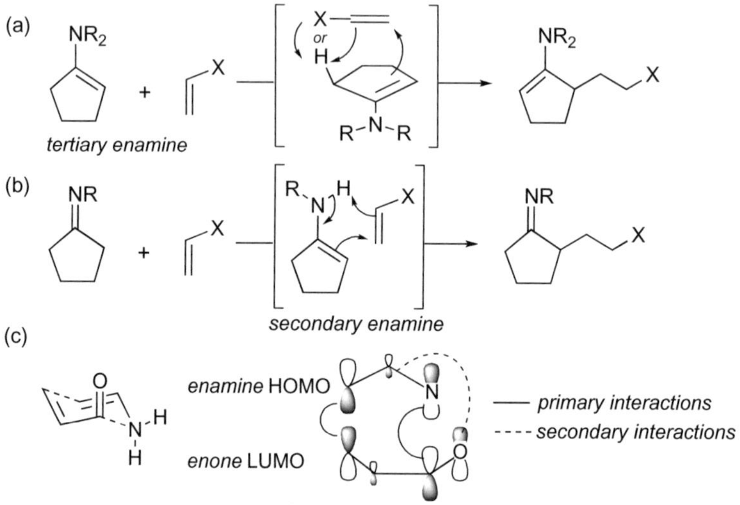

SCHEME 5.43 (a) Suprafacial Michael addition-proton transfer of a tertiary enamine [249]. (b) Aza-ene-like transition structure for secondary enamine Michael additions [248]. (c) Molecular orbital analysis of enamine and enone interactions [251].

Because the amines are removed in the subsequent hydrolytic workup, enamines are obviously amenable to an auxiliary-based asymmetric synthesis using a chiral amine. It is especially significant from a preparative standpoint that unsymmetrical ketones alkylate at the less substituted position *via* tertiary enamines (*e.g.*, C_6 of 2-methylcyclohexanone) whereas the *more* hindered position is alkylated preferentially with secondary enamines (*e.g.*, C_2 of 2-methylcyclohexanone). Since both enantiomeric ketones are converted into a single reactive enamine in the latter cases, they have been dubbed "deracemizations" by d'Angelo and coworkers [252].

In 1969, Yamada demonstrated that the cyclohexanone enamine derived from proline methyl ester would add to acrylonitrile or methyl acrylate with 70−80% enantioselectivity (Scheme 5.44a [253]), but Ito later showed that the selectivity was much better if a prolinol ether was used instead (Scheme 5.44b [254]). Seebach investigated the asymmetric Michael addition of enamines of prolinol methyl ether, as shown in the examples of Scheme 5.44c [255,256], and likewise found high selectivities. These examples share a common sense of asymmetric induction at C_2 of the cyclohexanone, and the origin of the asymmetric induction is of interest. The example in Scheme 5.44d shows that the effect is not steric in origin, since the propyl group is isosteric and isoelectronic with the methoxymethyl group, but the selectivity is essentially lost without the oxygen (*cf.* Scheme 5.44c and d [256]). Two explanations are consistent with these results. First, we assemble the two reactants in a *synclinal* orientation with the aryl group *antiperiplanar* to the donor double bond (Scheme 5.45, upper left; *cf.* Figure 5.12). One possibility is that the dipole of the methoxymethyl then stabilizes a zwitterionic intermediate in the nonpolar solvent, as shown in Scheme 5.45a.[18] Another is that the oxygen serves as a relay atom for a hydrogen transfer (an "internal solvation" effect) such as illustrated in Scheme 5.45b (*cf.* Scheme 5.43a).

---

18. *Ab initio* studies suggest that a zwitterionic intermediate may normally be too high in energy to be kinetically accessible [251], but "internal solvation" by the methoxymethyl may lower the barrier.

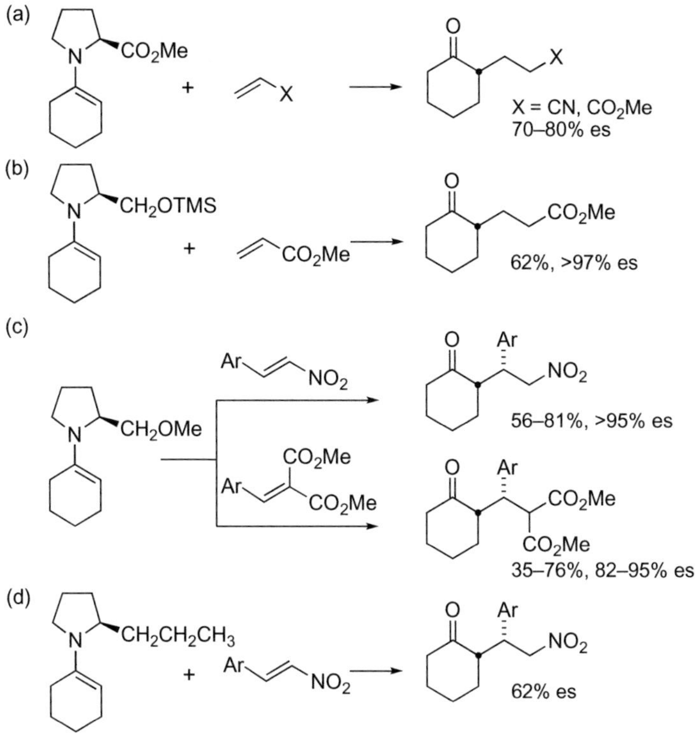

SCHEME 5.44 Asymmetric Michael additions of preformed tertiary enamines. (a) [253]; (b) [254]; (c) [255,256]; (d) [256].

Other examples shed some light on the importance of the proton transfer in these enamine Michael additions. For example, the $\Delta^{1,2}$ enamine of β-tetralone (Scheme 5.45c) afforded high yields of *3-substituted* $\Delta^{1,2}$ enamine products, even though the $\Delta^{2,3}$ enamine isomer was not present in the reaction mixture [257] (see also ref. [258]). Under the reaction conditions (toluene or ether, stirring for 3–4 days), the $\Delta^{1,2}$ isomer must isomerize to the $\Delta^{2,3}$ isomer that reacts much faster, probably due to the greater acidity of the benzylic proton of the $\Delta^{2,3}$ isomer compared to the C_3-proton of the $\Delta^{1,2}$ isomer.

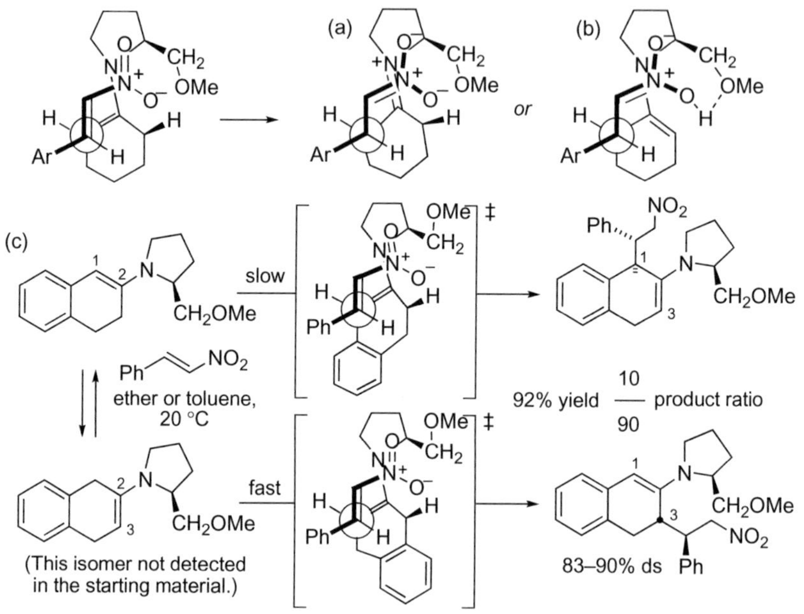

SCHEME 5.45 Involvement of the methoxymethyl in the asymmetric Michael addition: (a) by dipolar stabilization of a zwitterion intermediate, or (b) by assisting in the proton transfer to the Michael acceptor. (c) Isolation of an addition product *via* enamine rearrangement [257].

The asymmetric Michael addition of secondary enamines has been reviewed by d'Angelo [248]. Some of the more selective examples of this type of reaction are listed in Table 5.11. It is significant that these Michael additions are highly regioselective, reacting virtually exclusively at the more highly substituted carbon, which affords α,α-disubstituted (quaternary) cyclopentanones, cyclohexanones, furans, and pyrans in excellent yields and selectivities. An important advantage of this process is that it is stereoconvergent: racemic 2-substituted ketones are converted into nearly enantiopure products. The mild conditions of these reactions (nonpolar solvents, room temperature) and the high overall yields make this an attractive process for large-scale applications. The products of these reactions will catch the eye of anyone familiar with the Robinson annelation and related reactions [259,260], as these types of compounds are often used as key building blocks in synthesis.

TABLE 5.11 Michael Additions to Cyclic Ketones and Lactones *via* Their Secondary Enamines

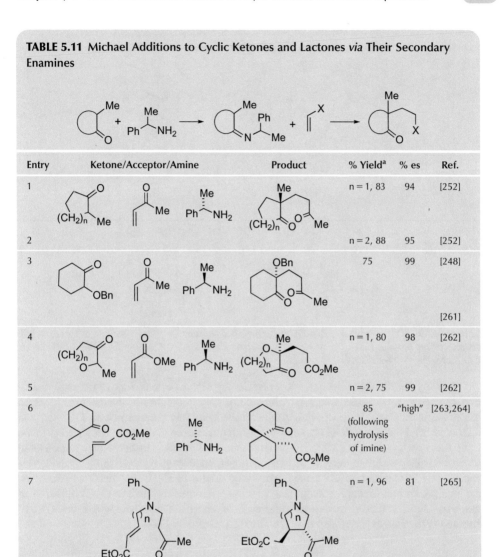

Entry	Ketone/Acceptor/Amine	Product	% Yield[a]	% es	Ref.
1			n = 1, 83	94	[252]
2			n = 2, 88	95	[252]
3			75	99	[248]
					[261]
4			n = 1, 80	98	[262]
5			n = 2, 75	99	[262]
6			85 (following hydrolysis of imine)	"high"	[263,264]
7			n = 1, 96	81	[265]
8			n = 2, 82	95	

[a]The yields listed are for the overall conversion of the ketone educt into the diketone or keto ester product.

What is the origin of the regioselectivity and what determines the face selectivity of the Michael addition? The regioselectivity results from the aza-ene-like mechanism of this reaction. As shown in Scheme 5.46, although both enamines may form, reaction of the less substituted isomer is retarded by $A^{1,3}$ strain effects. Note that in the aza-ene transition structure (Scheme 5.43b), the NH must be *syn* to the enamine double bond. Thus, the more highly substituted enamine isomer, in its most stable conformation, is in the proper conformation for Michael addition. In contrast, the reactive conformer of the less substituted isomer is destabilized by severe $A^{1,3}$ strain between the ring substituent and the nitrogen substituent, which increases as the carbon—nitrogen bond gains double bond character in the transition state.

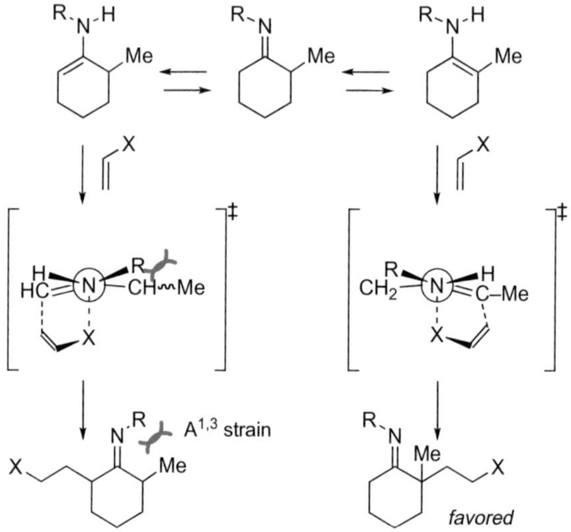

SCHEME 5.46 $A^{1,3}$ strain raises the energy of the transition structure for Michael addition of the less substituted secondary enamine [248]. The dashed lines in the transition structures indicate the primary MO interactions, according to Scheme 5.43c.

The origin of the face selectivity was revealed by MNDO calculations of the chair transition structures shown in Scheme 5.47, which differ only in the face of the enamine to which the enone is attached. By constraining the two reactants into parallel planes 3 Å apart and rotating around the indicated (*) bond of each structure, the conformations shown were found to be the lowest in energy [248]. The ground state conformation probably approximates the center structure, with the benzylic carbon—hydrogen bond *synperiplanar* to C_1 of the cyclopentene due to repulsion of the methyl and phenyl groups by C_5. In the transition structures, the benzylic carbon—hydrogen bond rotates 60° and becomes *synclinal* to C_1. Comparison of these structures indicated an energy difference of about 1.1 kcal/mol, which corresponds closely to the value expected based on the observed selectivity [248].

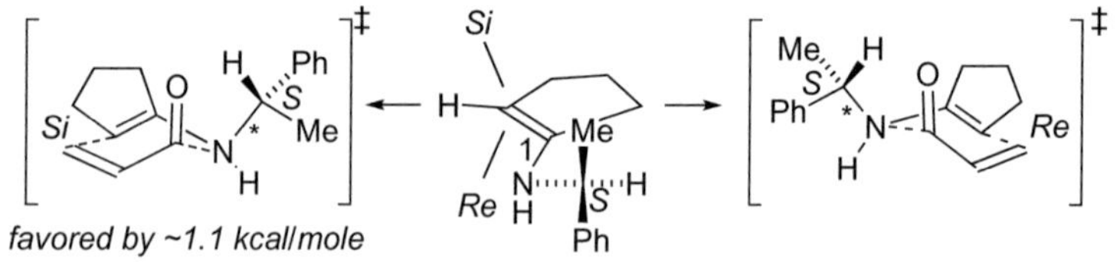

SCHEME 5.47 Calculated low energy conformers for *Re* and *Si* attack of acrolein on cyclopentanone *S*-phenethyl enamine [248].

Allyl anions. The sulfur and phosphorous-stabilized allyl anions shown in Figure 5.13 have been examined by the Hua and Hanessian groups in asymmetric Michael additions to several enones. In these auxiliaries, the sulfur and the phosphorous are stereogenic, and the phosphorous additionally has chiral ligands. Some of the more selective examples of Michael additions using these ligands are listed in Table 5.12.

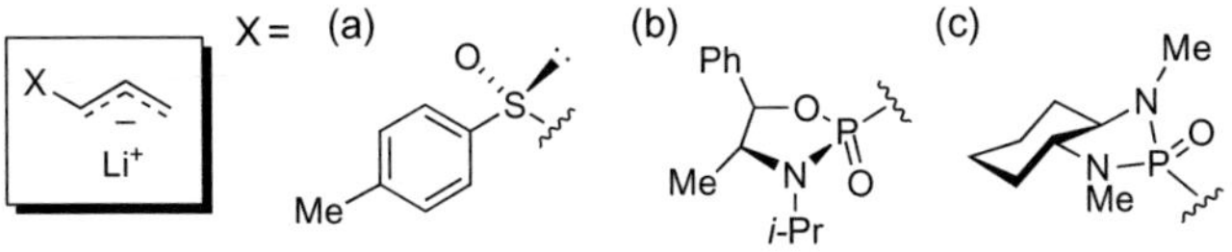

FIGURE 5.13 Auxiliaries for asymmetric Michael addition of allyl anions: (a) [266]; (b) [267]; (c) [268].

TABLE 5.12 Asymmetric Michael Additions of Sulfur and Phosphorous-stabilized Allyllithiums

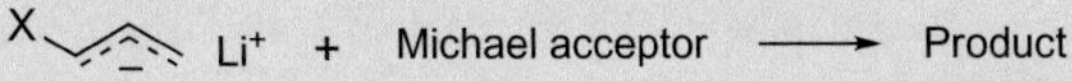

Entry	X[a]	Acceptor	Product	% Yield	% ds	Ref.
1	a	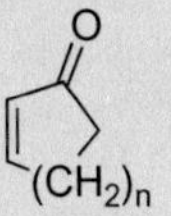	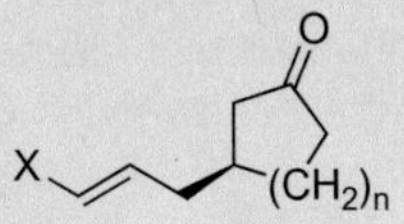	n = 1, 91	98	[266]
2	b			n = 1, 79	99	[267]
3	c			n = 1, 88	96	[268]
4	b			n = 2, 70	94	[267]
5	b			n = 3, 71	97	[267]
6	a	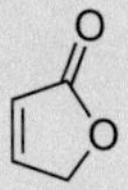	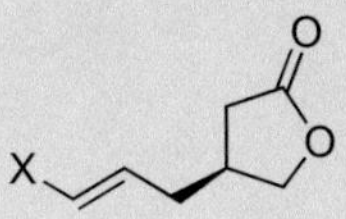	82	85	[266]
7	c			93	>99	[268]
8	a	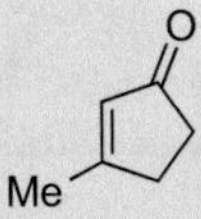	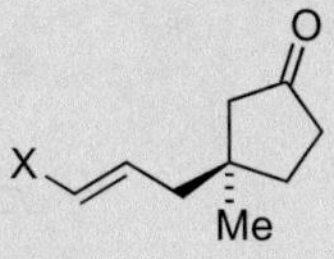	80	97	[266]
9	c			75	97	[266]
10	c	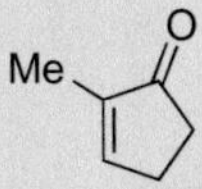	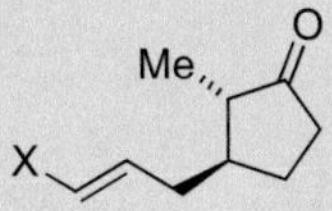	80	94	[268]
11	c	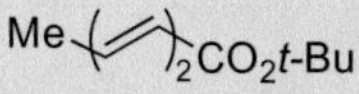	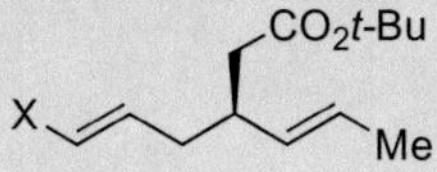	76	>99	[268]

[a] The X column refers to the auxiliaries in Figure 5.13.

The mechanism of allylic sulfoxide addition is proposed to occur through a chelated 10-membered ring transition structure [269], as shown in Scheme 5.48a. The illustrated conformation features the favorable alignment of the molecular orbitals illustrated in the inset (*cf.* Scheme 5.43). However, it also has been suggested [220] that the reaction may proceed by sequential 1,2-addition followed by an alkoxide-accelerated Cope rearrangement,[19] as shown in Scheme 5.48b. Note that the same conformation and orbital alignment are operative in this mechanism. For the addition of the phosphorous-stabilized allyllithium of Figure 5.13b and c, 10-membered rings are postulated [267,268]. The 10-membered rings shown in Schemes 5.48c and d have conformations similar to that shown in Scheme 5.48a; conceivably the tandem 1,2-carbonyl addition/3,3-Cope rearrangement suggested for the sulfoxides could intervene in these cases as well. For these auxiliaries, the site of lithium coordination to the phosphoryl group determines the chirality sense of the products. For the phosphaoxazolidine (Scheme 5.48c), the lithium coordinates *anti* to the bulky *N*-isopropyl substituent, but for the phosphaimidazolidine (Scheme 5.48d) the reason for the similar placement is not as obvious. The inset illustrates the five-membered heterocycle in a half-chair conformation, with the *N*-methyls in pseudoequatorial configurations (the half chair is held rigid by the *trans* fused cyclohexane, which is deleted for clarity). The *N*-methyls are labeled according to their relative configurations on the stereogenic phosphorous. Note that coordination of the lithium *syn* to the N_{Re}-methyl generates a lithium/methyl interaction reminiscent of 2,3-*P*,3,4-*M* gauche pentane (*cf.* Figure 5.5), but coordination to the N_{Si}-methyl does not. Thus the latter site is preferred.

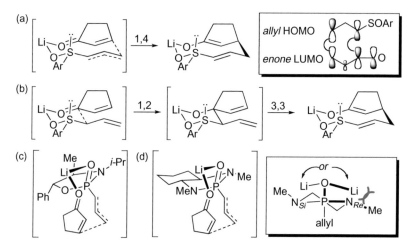

SCHEME 5.48 Allyl sulfoxide additions: (a) 1,4-mechanism [269]; (b) tandem 1,2-addition/3,3-rearrangement mechanism [220, 270]. (c) and (d) Transition structures for allyl phosphine oxides [267,268]. *Inset:* Gauche pentane interaction between lithium and the N_{Re}-methyl.

---

19. Such a mechanism has been demonstrated in the addition of dithianyl allyl lithiums [270].

5.3.3 Chiral Michael Acceptors

Posner has shown that enones having a chiral sulfoxide in the α-position are excellent receptors for conjugate addition of organometallics (Scheme 4.14), [271], and may also be used as Michael acceptors in enolate additions [272–274]. The face selectivity can be rationalized based on either chelation of the metal by the enone and sulfoxide oxygens (Figure 5.14a) or by dipole alignment (Figure 5.14b) (*cf.* Scheme 4.16). In the following examples, which are chosen from others that are not as selective, the following trend emerges: enolates that are monosubstituted at the α-position follow the nonchelate (dipole) model, while α,α-disubstituted enolates follow the chelate model [275].

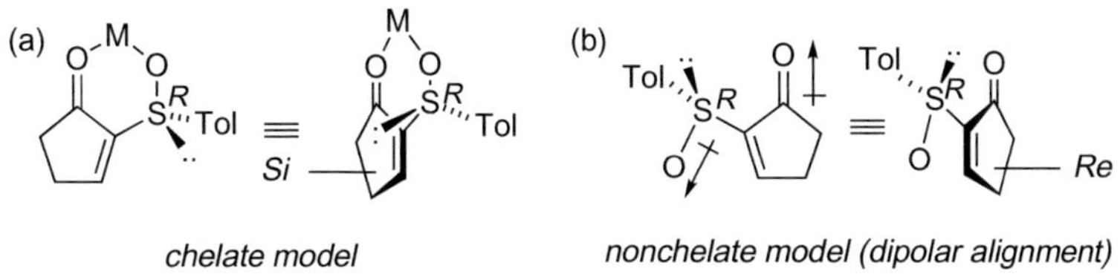

FIGURE 5.14 Models for face-selective addition of enolates to *R*-sulfoxides. (a) Chelate model predicts nucleophilic attack on *Si* face. (b) Nonchelate model, which has the C=O and S—O bonds *antiperiplanar*, predicts *Re* face attack.

The lithium enolate of methyl trimethylsilyl acetate adds to cyclopentenone and cyclohexenone sulfoxides by the nonchelate model with good to excellent selectivity, as shown in Scheme 5.49a [274]. After the Michael addition, the sulfoxide and trimethylsilyl groups are removed, and the selectivity is assessed by determining the enantiomeric purity of the β-substituted ketone. Similarly, lithium enolates of phenylthioacetate esters add to five and six-membered lactones as shown in Scheme 5.49b [274]. The chirality sense of these products is consistent with a nonchelate model: the nucleophile adds to the *Si* face of the *S*-sulfoxide (*lk* topicity).

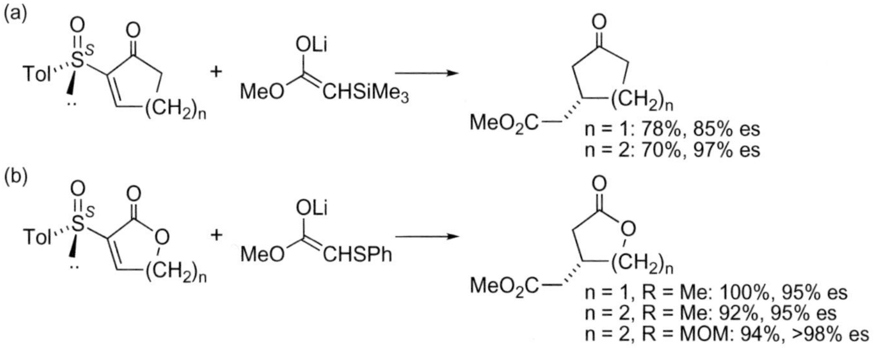

SCHEME 5.49 Sulfoxide mediated asymmetric Michael additions to (a) cycloalkenones and (b) lactones. Both are postulated to proceed *via* the nonchelate model (Figure 5.14b [274]).

Scheme 5.50 illustrates three applications of this methodology to total synthesis. The first example is taken from Posner's synthesis of estrone and estradiol [275], the second from Posner's synthesis of methyl jasmonate [276], and the third from Holton's synthesis of aphidicolin [277]. The latter is particularly noteworthy in that two contiguous quaternary centers are created in the asymmetric addition with excellent selectivity. In the estrone synthesis, the chirality sense of the product is consistent with the nonchelate model, but the other two examples adhere to a chelate model. Note that the difference is the degree of substitution at the α-position of the enolate.

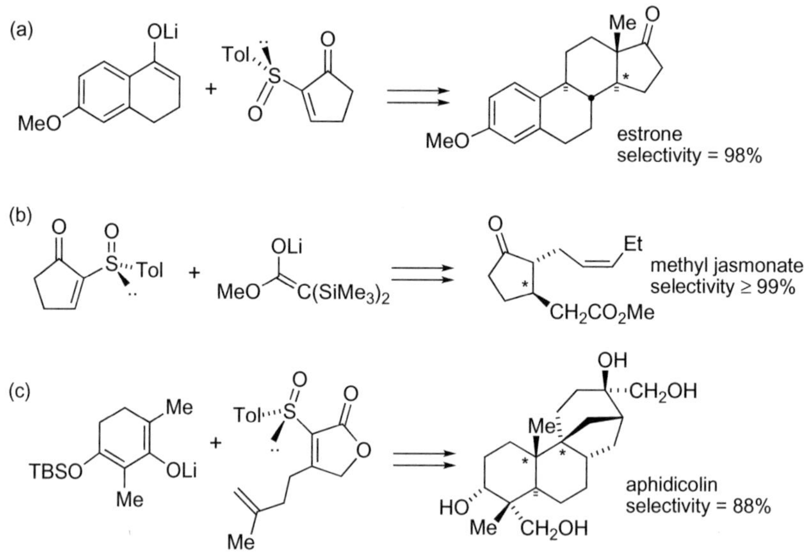

SCHEME 5.50 Applications of sulfoxide Michael additions in natural product synthesis: (a) estrone [275]; (b) methyl jasmonate [276]; (c) aphidicolin [277]. Stereocenters formed in the Michael addition are indicated (*).

5.3.4 Interligand Asymmetric Induction and Catalysis

Asymmetric Michael additions can also be carried out under the influence of a chiral ligand or catalyst, although development of these methods has lagged behind that of the aldol reaction (also see Section 4.4 for the synthetically related but mechanistically distinct problem of carrying out organometal-mediated conjugate addition reactions onto enones).

In early work, several groups demonstrated that the complexation of potassium enolates with chiral crown ethers led to good selectivity in Michael reactions. For example, Cram used C_2-symmetric crowns based on binaphthol to catalyze the addition of 2-carbomethoxyindanone to methyl vinyl ketone (Scheme 5.51a [278]). A second example, the addition of methyl 2-phenylpropionate to methyl acrylate is shown in Scheme 5.51b [278]. These examples share the common feature of an acidic carbon stabilized by two functional groups, which permits employment of catalytic amounts of base and crown. The catalytic cycle is probably as follows [278]:

1. $\text{Crown} \cdot K^+ t\text{-BuO}^- + H\text{-R} \rightarrow \text{Crown} \cdot K^+ R^- + t\text{-BuOH}$
2. $\text{Crown} \cdot K^+ R^- + C{=}C{-}C{=}O \rightarrow R\text{-}C\text{-}C{=}C\text{-}O^- K^+ \cdot \text{Crown}$
3. $R\text{-}C\text{-}C{=}C\text{-}O^- K^+ \cdot \text{Crown} + H\text{-}R \rightarrow R\text{-}C\text{-}CH\text{-}C{=}O + \text{Crown} \cdot K^+ R^-$

The key step for catalyst turnover is the last one, whereby the enolate adduct deprotonates the next molecule of starting carbonyl. Clearly the initial carbon acid must be more acidic than the Michael product for this step to proceed. The stabilized enolates used in

Schemes 5.51a and b ensure this fact, but epimerization of the product can also be a problem if there are any α protons left [279,280].

The Seebach group showed that cyclohexanone lithium enolates show good selectivities when complexed to chiral diamines, chiral lithium amides (Scheme 5.51c [3]), or chiral diethers (Scheme 5.51d [281]). The Mukaiyama group explored the use of tin enolates complexed to chiral diamines as shown in Schemes 5.51e and f. The propionate imide enolate shown in Scheme 5.51e (when used in excess) adds to benzal acetone with excellent selectivity [282]. The topicity of this addition is consistent with a mechanism similar to that shown in Scheme 5.38a, but note that titanium enolates of chiral imides added with low selectivity to β-substituted enones (Schemes 5.40d and e: ref. [236]). If the dithioketene acetal shown in Scheme 5.51f is added slowly to a mixture of an enone, tin triflate, and chiral diamine, good enantioselectivities are achieved with catalytic amounts of tin and diamine [283]. The slow addition is necessary to keep a low concentration of the dithioketene acetal so as to minimize a competitive nonselective addition.

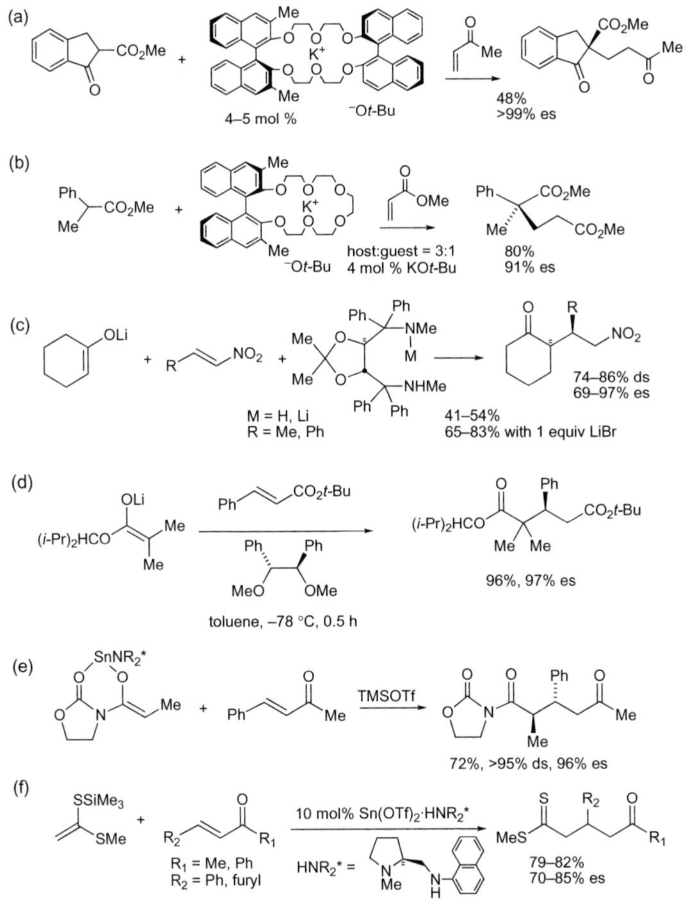

SCHEME 5.51   (a and b) Cram's C_2-symmetric chiral crowns for asymmetric Michael addition [278]. (c) Seebach's investigation of cyclohexanone lithium enolate complexed to chiral diamines with extra lithium [3]. (d) Tomioka's Michael addition of an ester enolate [281]. (e) Mukaiyama's imide tin enolate and chiral diamine [282]. (f) Mukaiyama's catalytic tin dithioenolate Michael addition (absolute configuration unassigned) [283].

The success with stoichiometric formation of enamines as nucleophiles described in Section 5.3.2 has led in the twenty-first century to catalytic versions wherein the nucleophilic species is generated *in situ*. In fact, an early example of organocatalysis can be found in the example shown in Scheme 5.52a. Following a 1973 lead by Långström and Bergson, who used a partially resolved amino alcohol as an asymmetric Michael catalyst [284], Wynberg used quinine as a catalyst for the asymmetric addition of 2-carbomethoxy-1-indanone to methyl vinyl ketone, obtaining 88% enantioselectivity in an optimized case (Scheme 5.52a), but the absolute configuration of the product was not determined and the scope of this reaction was not great [285].

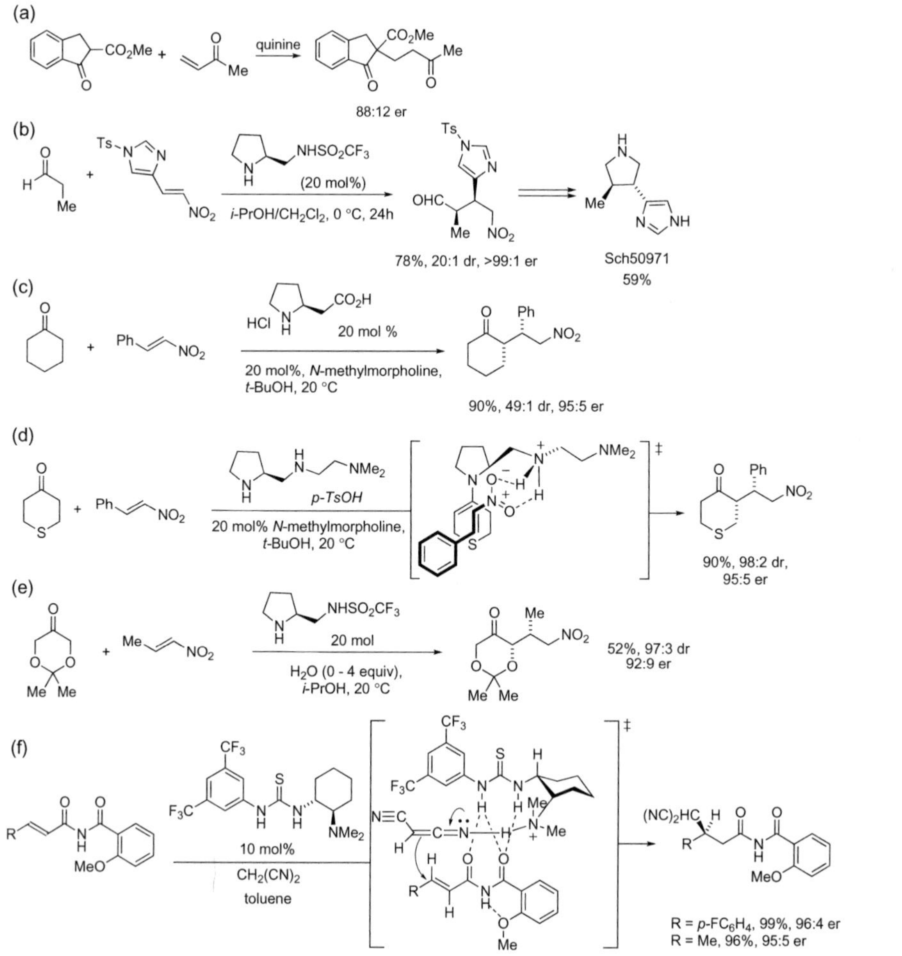

SCHEME 5.52 Organocatalysis approaches to Michael addition chemistry using (a) quinine [285] or (b)–(e) enamine catalysis (HOMO raising) [286–289]. In contrast, in (f) the reaction is accelerated by hydrogen bonding activation (LUMO lowering) [290].

Scheme 5.52b−e provides more recent examples of organocatalytic Michael reactions. Enamines, whether premade or formed *in situ*, accelerate reactions by raising the LUMO of the nucleophilic species[20]; three examples of this are demonstrated in Scheme 5.52(b)−(d). In general, the catalysts employed are similar or identical to those used in the aldol chemistry discussed in Section 5.2.3. One often sees nitroalkanes as the Michael acceptor in these reactions (*e.g.*, the conversion of the adduct in Scheme 5.52b to the histamine-3 agonist Sch50971 [286]), perhaps because the nitro group provides a convenient two-point binding site to ensure high selectivity in the conjugate addition reaction. The *synclinal* transition structure proposed by Pansare for the example shown in Scheme 5.52d demonstrates this nicely [289]. In Scheme 5.52f a single example of a mechanistically distinct approach to asymmetric Michael reaction chemistry is shown, wherein the Michael acceptor is activated for attack by a stabilized anion through LUMO lowering combined with reagent delivery through an additional hydrogen bonding interaction engaging the tertiary amine group.

5.3.5 Conjugate Addition of Nitrogen Nucleophiles

As important as carbon−carbon bond formation is to the organic chemist, it is occasionally necessary to form carbon−heteroatom bonds as well. For example, the conjugate addition of nitrogen nucleophiles to esters or their equivalents is important partially because the resulting β-amino acid derivatives can be converted into β-lactams and are monomers used for the construction of β-peptides [292−294]. Although there has been less work on analogous oxygen or sulfur nucleophiles, some instructive examples will be included at the end of the chapter. We begin with a short discussion of nitrogen nucleophile conjugate additions (no longer strictly "Michael additions" because the nucleophile is not an enolate; review: [295]). This will follow familiar lines to readers who have read this far: older approaches using chiral reagents and chiral auxiliaries, with a burst of innovation in the 1990s and beyond to promote the reaction using asymmetric catalysis.

Early approaches utilized one or more covalently bound chiral groups to direct the addition event, followed by removal of the auxiliary or chiral inducing group (Scheme 5.53). Good results have been achieved for the direct conjugate addition of an amine using the conjugate base or high pressure. Scheme 5.53a shows Davies' straightforward addition of a chiral amide derived from the workhorse α-methylbenzylamine[21] to *tert*-butyl cinnamic ester. Simple protic quench of the resulting enolate affords the conjugate addition product whereas methylation leads to the 2,3-*syn* dialkylated product; both reactions occur with high diastereoselectivity. Alternatively, conjugate addition to the 2-methylcinnamic ester shown in the third example of Scheme 5.53a affords the *anti* isomer following quench with a bulky proton source (not shown). Removal of the α-methylbenzyl group is accomplished under reductive conditions. Since the nitrogen atom of the chiral reagent ends up in the product, this is not a chiral auxiliary approach but rather an application of a chiral reagent. Differentially protected versions of this reagent [296,297] as well as other chiral *N*-nucleophiles have been reported [298]. The bottom part of Scheme 5.53 provides examples of straightforward chiral auxiliary approaches, with a model for face selectivity of one shown in Scheme 5.53b.

---

20. For a general organizational chart of the various kinds of organocatalysis, see ref. [291].
21. The low cost and ready availability of this amine as either enantiomer (or racemic) form made it a favorite of early practitioners of asymmetric synthesis, despite its usually modest stereoinduction. The high selectivities observed here and in Table 5.11 are among the exceptions.

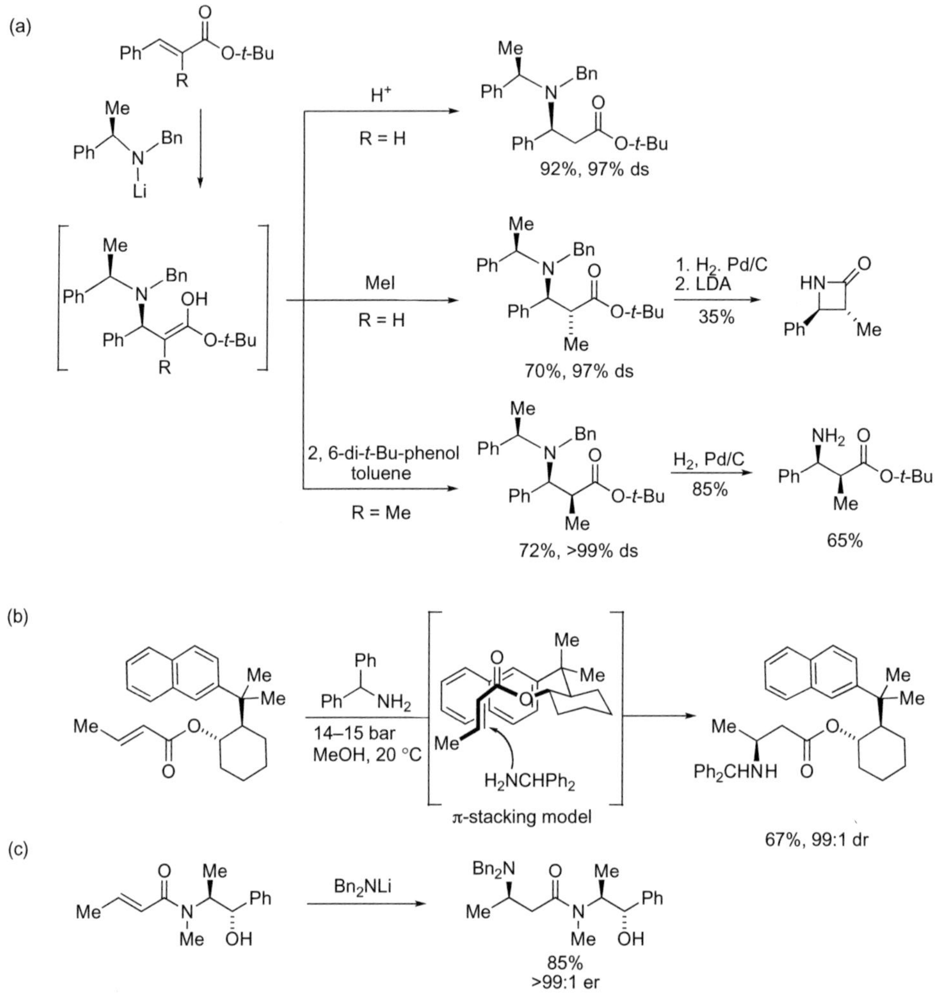

(a)

(b)

π-stacking model

67%, 99:1 dr

(c)

85%
>99:1 er

SCHEME 5.53 Conjugate addition reactions of amines using (a) a chiral nitrogen species [299,300] or (b) and (c) chiral auxiliaries [301,302].

The approach using chiral nitrogen nucleophiles is hampered by the limitations of commercially available nitrogen sources or in preparing made-to-order reagents, providing strong motivation to develop catalytic methods. Since conjugate addition reactions are generally promoted by Lewis acid activation of the electrophilic α,β-unsaturated system, chiral Lewis acid catalysis was an obvious approach to the problem. However, a complicating factor is that Lewis acids are rendered less active through coordination to basic nitrogen nucleophiles such as amines. A way around this is to use either N- or O-alkylated hydroxylamines, which are certainly less basic than simple amines and possibly better nucleophiles as well due to the α-effect.[22]

---

22. The "α-effect" occurs when a nucleophile attached to another atom having a lone pair is more nucleophilic than expected by its basicity; for a review discussing the origin and solvent dependence of this phenomenon, see ref. [303].

Baldwin published the uncatalyzed, thermal addition of *a*-substituted hydroxylamines to α,β-unsaturated esters (Scheme 5.54: ref. [304]). This poorly selective process entails a conjugate addition reaction followed by oxygen attack onto the ester to afford an isoxazolidinone. The *O*-acyl group assists in the reductive cleavage of the N—O bond. This reaction was further developed by Saito, who added chiral esters and Lewis acids to the mix [305]. All of these reactions were proposed to proceed with hydrogen bonding between the hydroxylamine and the carbonyl of the unsaturated ester to help organize the transition state. This method has been employed in the synthesis of monomers needed for the synthesis of β-peptides [306]. Sibi has reported catalytic versions using achiral hydroxylamines [307,308] as well as hydrazines [309] using bisoxazoline-derived ligands (Scheme 5.54b and c). A key part of developing both reactions was the recognition that an appropriate amide derivative would permit two-point binding on the α,β-unsaturated substrate, which is incorporated into the nicely differentiated octahedral transition state organized around magnesium. This basic organization has been retained in many conjugate additions of heteroatoms promoted by chiral catalysts.

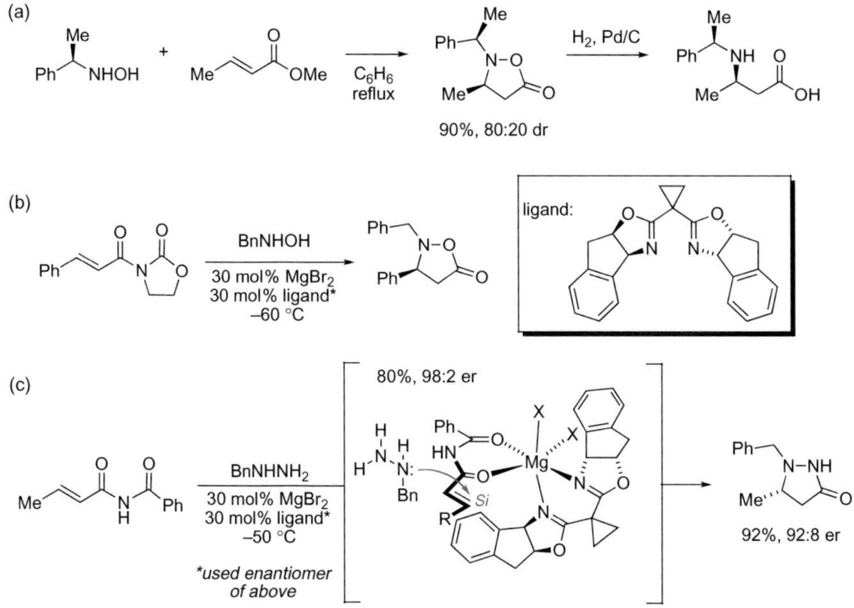

SCHEME 5.54 Asymmetric cyclocondensation reactions using (a) a chiral hydroxylamine [304], or a chiral Lewis acid and (b) an achiral N-alkylhydroxylamine [307] or (c) an alkylhydrazone [309]. The proposed transition structure for the hydrazone reaction is shown in (c); note that enantiomeric catalysts were used in (b) vs. (c).

Catalysts for the conjugate addition of nitrogen nucleophiles are shown in Figure 5.15 and representative reactions in Table 5.13. The first example of a conjugate addition of an achiral N-nucleophile was reported by Jørgensen (entry 1). This was followed quickly by Sibi, both in the context of cyclocondensation chemistry as shown in Scheme 5.54b, but also by using the mono-nucleophilic BnONH$_2$ in a simple conjugate addition reaction (entry 2). The reactions that use oxazoline-based catalysts are carried out using acceptors that have an additional binding moiety on the acyl group to help bind a metal ion and assemble the transition structure prior to addition (*e.g.*, Scheme 5.54c). A common challenge is the need to avoid catalyst deactivation by the basic amine nucleophile, thus leading to the popularity of relatively nonbasic arylamines or hydroxylamine derivatives (entries 3 and 4). This same issue also contributes to high catalyst loadings used in many amine nucleophile additions. Shibasaki was able to avoid this using the ytterbium/BINOL system shown in entry 4, which made it possible to carry out the reactions at low loadings and using ordinary reaction concentrations [310]. A clever approach later published by Sodeoka involved the controlled release of reactive amine from its triflate salt by the catalyst (Figure 5.15e and Table 5.13, entries 5 and 6: ref. [311]). Jacobsen used salen catalysts to promote the insertion of numerous nucleophiles, including hydrazoic acid (entry 7) to α,β-unsaturated imide [312] and ketone [313] acceptors; this system also led to highly selective additions of other nucleophiles including HCN, HSAr, oximes, and carbon acids to these acceptors. More recent work has seen conjugate additions to α,β-unsaturated ketones [314] or aldehydes [315] promoted by organocatalysts (entries 8 and 9). The intramolecular example in entry 9 is particularly impressive because it entails the reaction of a poorly nucleophilic carbamate nitrogen, succeeding in part because of the highly electrophilic nature of the iminium ion intermediate.

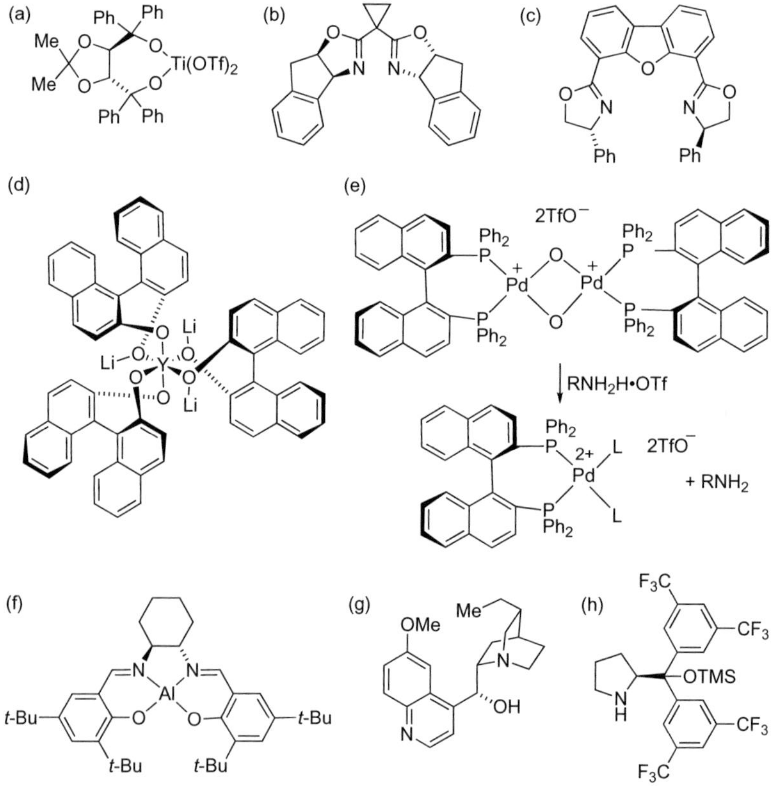

FIGURE 5.15 Catalysts used for conjugate addition reactions of achiral nitrogen nucleophiles. See Table 5.13 for reactions and references.

TABLE 5.13 Catalytic Asymmetric Conjugate Additions of Nitrogen Nucleophiles

$$R_1 \diagdown XR \ (O) \diagdown R_2 \ + \ Nu^{\ominus} \longrightarrow R_1 \diagdown \overset{Nu}{\diagup} \diagdown \overset{O}{\diagup} XR \ R_2$$

Entry	Nucleophile	Acceptor	Product	Catalyst/Ligand, Conditions[a]	% Yield	% es	Ref.
1	BnONH₂	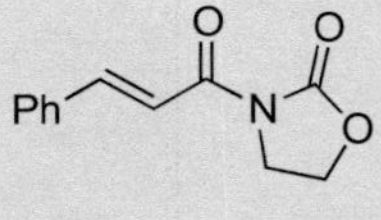	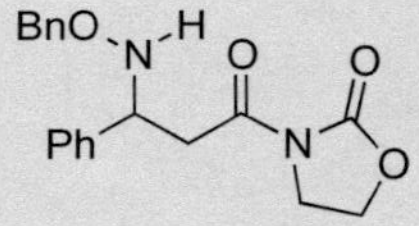	a (10 mol%), RT	69	71	[316]
2	BnONH₂	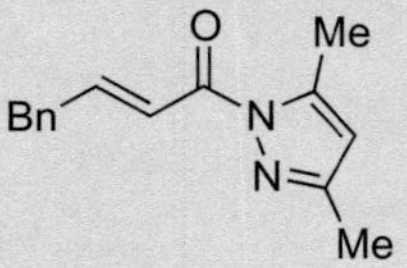	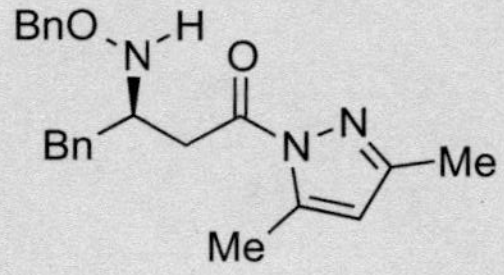	a (30 mol%), MgBr₂, −60 °C	80	97	[308]
3	PhNHMe	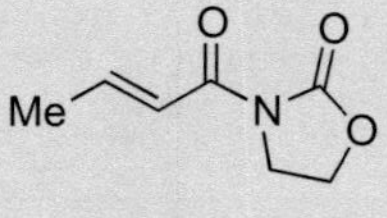	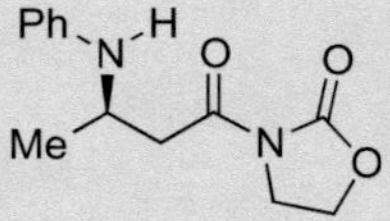	c (5 mol%) Ni(ClO₄)₂ · 6H₂O (5 mol%)	62	95	[317]
4	MeONH₂	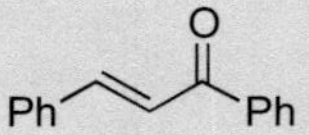	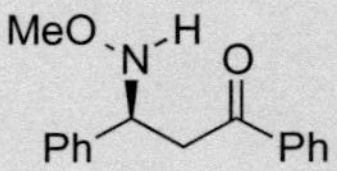	d (3 mol%), drierite, −20 °C, 42 h	97	97	[310]
5	PhNH₂[b]	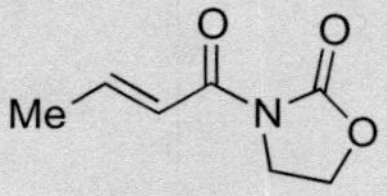	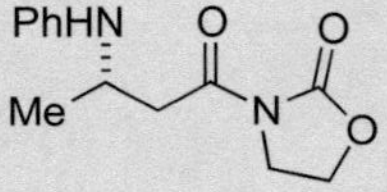	e (0.2 mol%), THF, RT	83	98	[311]

(Continued)

TABLE 5.13 (Continued)

Entry	Nucleophile	Acceptor	Product	Catalyst/Ligand, Conditions[a]	% Yield	% es	Ref.
6	BnNH$_2$	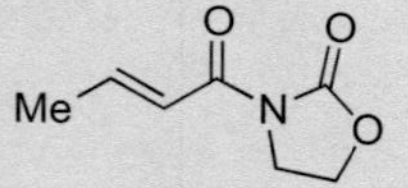 (oxazolidinone crotonyl)	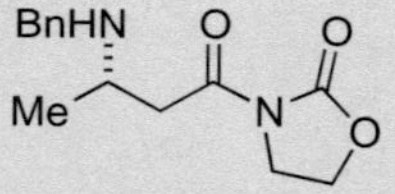 (BnHN, Me, oxazolidinone)	e (0.2 mol%), THF, 40 °C	75	93	[311]
7	NaN$_3$/HCl	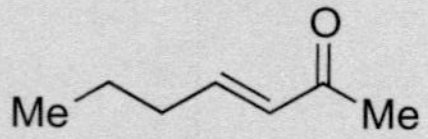 (ketone, Me)	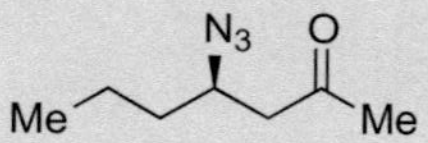 (N$_3$, Me)	f (2.5 mol%), methylcyclohexane, −78 to −40 °C	88	97	[313]
8	ArCH=N–NHMe	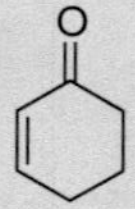 (cyclohexenone)	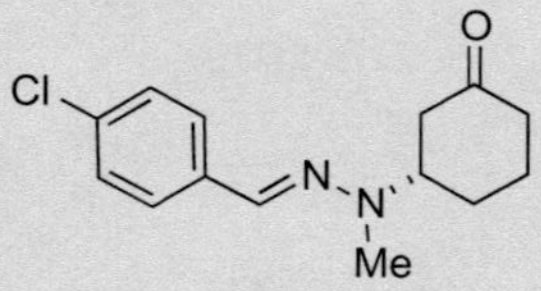 (hydrazone, Me, Cl-aryl)	g (20 mol%), (Et$_2$O), 5 days	75	87	[314]
9	Cbz–NH, CHO (structure)	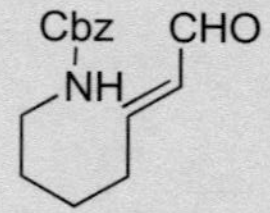	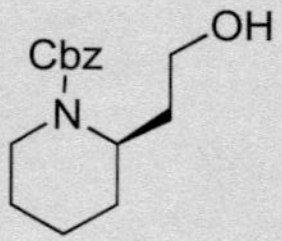 (OH, Cbz–N structure)	h (20 mol%), 1:1 MeOH/DCE, −25 °C, 3 days, then NaBH$_4$	70	97	[315]

[a]The catalyst column refers to the structures in Figure 5.15.
[b]Added as TfOH salts.

Much less work has been reported on the addition of oxygen nucleophiles to α,β-unsaturated carbonyl compounds, partly due to the availability of a highly attractive alternative approach of carrying out an asymmetric reduction of a corresponding 1,3-dicarbonyl compound (Scheme 5.55a). Nonetheless, activation of appropriate acceptors using chiral Lewis acids has been accomplished; two highly selective examples using oxime nucleophiles are depicted in Scheme 5.55b and c [318]. As above, the N—O bond in the product can be reduced to reveal the formal addition product of a "chiral hydroxide" moiety.

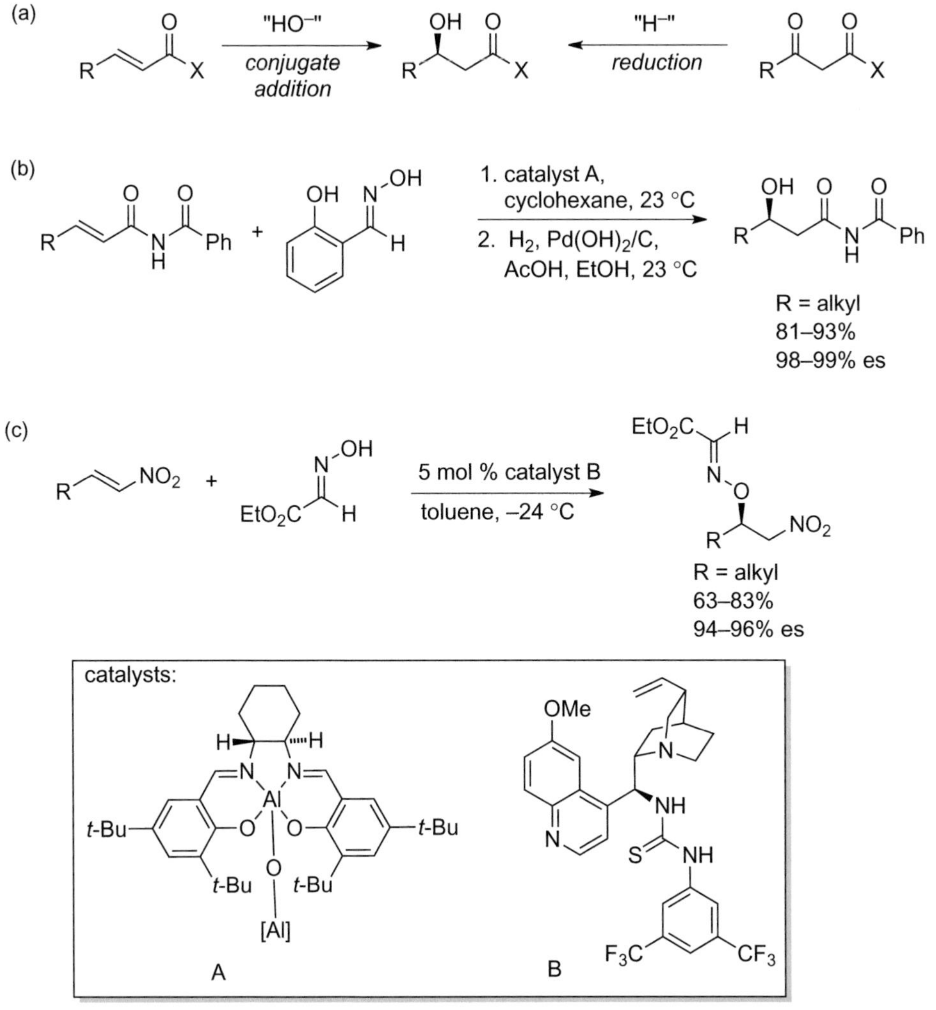

SCHEME 5.55   (a) Comparison between conjugate addition and asymmetric reduction routes to β-hydroxycarbonyl compounds. Asymmetric O-conjugate addition reactions employing oxime nucleophiles by (b) Jacobsen [319] and (c) Jørgensen [318]. Catalyst A exists as a dimer.

Conjugate addition of sulfur is perhaps a specialty area but one with potential impact in medicinal chemistry due to the existence of stereogenic sulfur-containing centers in various drug molecules (Scheme 5.56, review: ref. [320]). The example shown in Scheme 5.56a is a conjugate addition of sulfur to afford a product having some resemblance to the important blood-pressure regulating drug captopril [321]. Note, however, that the stereochemistry-defining step in this reaction is an asymmetric protonation as opposed to an asymmetric conjugate addition reaction. Another auxiliary-based method permits the coordinated introduction of both sulfur and oxygen substituents (Scheme 5.56b), useful for a synthesis of diltiazem [322]. Finally, an organocatalytic approach reported by Xiao has resulted in the highly selective formation of compounds containing fully substituted, stereogenic carbon atoms [323].

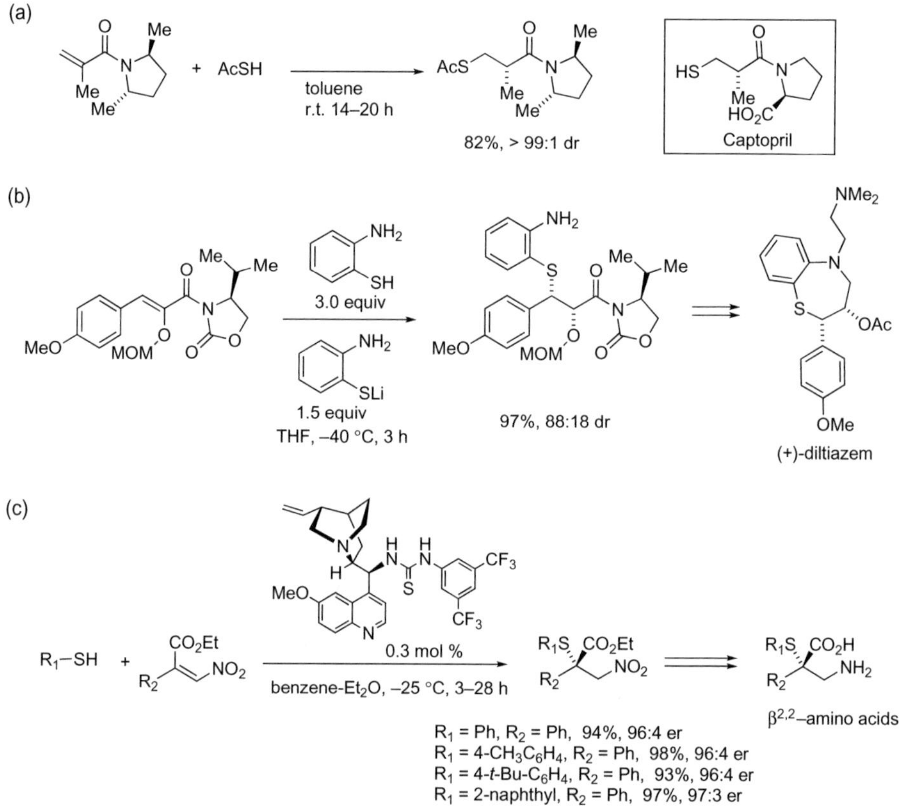

SCHEME 5.56 (a) and (b) Auxiliary- [321,322] and (c) catalyst-based [323] examples of asymmetric sulfur conjugate addition reactions.

REFERENCES

[1] Zimmerman, H. E.; Traxler, M. D. *J. Am. Chem. Soc.* **1957**, *79*, 1920–1923.

[2] Seebach, D. *Angew. Chem., Int. Ed. Engl.* **1988**, *27*, 1624–1654.

[3] Juaristi, E.; Beck, A. K.; Hansen, J.; Matt, T.; Mukhopadhyay, T.; Simson, M.; Seebach, D. *Synthesis* **1993**, *12*, 1271–1290.

[4] Williard, P. G.; Liu, Q.-Y. *J. Am. Chem. Soc.* **1993**, *115*, 3380–3381.

[5] Masamune, S.; Ali, S. A.; Snitman, D. L.; Garvey, D. S. *Angew. Chem., Int. Ed. Engl.* **1980**, *19*, 557–558.

[6] Seebach, D.; Prelog, V. *Angew. Chem., Int. Ed. Engl.* **1982**, *21*, 654−660.

[7] Hoffmann, R. *Angew. Chem., Int. Ed. Engl.* **1982**, *21*, 555−566.

[8] Roush, W. R. In *Comprehensive Organic Synthesis. Selectivity, Strategy, and Efficiency in Modern Organic Chemistry*; Trost, B. M., Fleming, I., Eds.; Pergamon: Oxford, 1991; Vol. 2, pp. 1−53.

[9] Fleming, I. In *Comprehensive Organic Synthesis. Selectivity, Strategy, and Efficiency in Modern Organic Chemistry*; Trost, B. M., Fleming, I., Eds.; Pergamon: Oxford, 1991; Vol. 2, pp. 563−593.

[10] Yamamoto, Y.; Maruyama, K. *Heterocycles* **1982**, *18*, 357−386.

[11] Hoffmann, R. W.; Niel, G.; Schlapbach, A. *Pure Appl. Chem.* **1990**, *62*, 1993−1998.

[12] Brown, H. C.; Ramachandran, P. V. *Pure Appl. Chem.* **1991**, *63*, 307−316.

[13] Yamamoto, Y. *Acc. Chem. Res.* **1987**, *20*, 243−249.

[14] Yamamoto, Y.; Asao, N. *Chem. Rev.* **1993**, *93*, 2207−2293.

[15] Evans, D. A. *Top. Stereochem.* **1982**, *13*, 1−115.

[16] Hoffmann, R. W.; Herold, T. *Chem. Ber.* **1981**, *114*, 375−383.

[17] Reetz, M. T.; Zierke, T. *Chemistry & Industry (London)* **1988**, *20*, 663−664.

[18] Roush, W. R.; Walts, A. E.; Hoong, L. K. *J. Am. Chem. Soc.* **1985**, *107*, 8186−8190.

[19] Corey, E. J.; Yu, C.-M.; Kim, S. S. *J. Am. Chem. Soc.* **1989**, *111*, 5495−5496.

[20] Racherla, U. S.; Brown, H. C. *J. Org. Chem.* **1991**, *56*, 401−404.

[21] Brown, H. C.; Racherla, U. S.; Liao, Y.; Khanna, V. V. *J. Org. Chem.* **1992**, *57*, 6608−6614.

[22] Racherla, U. S.; Liao, Y.; Brown, H. C. *J. Org. Chem.* **1992**, *57*, 6614−6617.

[23] Brown, H. C.; Kulkarni, S. V.; Racherla, U. S. *J. Org. Chem.* **1994**, *59*, 365−369.

[24] Short, R. P.; Masamune, S. *J. Am. Chem. Soc.* **1989**, *111*, 1892−1894.

[25] Burgos, C. H.; Canales, E.; Matos, K.; Soderquist, J. A. *J. Am. Chem. Soc.* **2005**, *127*, 8044−8049.

[26] Hoffmann, R. W.; Zeiss, H.-J. *J. Org. Chem.* **1981**, *46*, 1309−1314.

[27] Brown, H. C.; Bhat, K. S. *J. Am. Chem. Soc.* **1986**, *108*, 293−294.

[28] Brown, H. C.; Jadhav, P. K.; Bhat, K. S. *J. Am. Chem. Soc.* **1988**, *110*, 1535−1538.

[29] Barrett, A. G. M.; Malecha, J. W. *J. Org. Chem.* **1991**, *56*, 5243−5245.

[30] Hoffmann, R. W.; Endesfelder, A.; Zeiss, H.-J. *Carbohydr. Res.* **1983**, *123*, 320−325.

[31] Roush, W. R.; Hoong, L. K.; Palmer, M. A. J.; Park, J. C. *J. Org. Chem.* **1990**, *55*, 4109−4117.

[32] Roush, W. R.; Halterman, R. L. *J. Am. Chem. Soc.* **1986**, *108*, 294−296.

[33] Roush, W. R.; Hoong, L. K.; Palmer, M. A. J.; Straub, J. A.; Palkowitz, A. D. *J. Org. Chem.* **1990**, *55*, 4117−4126.

[34] Garcia, J.; Kim, B. M.; Masamune, S. *J. Org. Chem.* **1987**, *52*, 4831−4832.

[35] Roush, W. R.; Grover, P. T.; Lin, X. *Tetrahedron Lett.* **1990**, *31*, 7563−7566.

[36] Roush, W. R.; Grover, P. T. *Tetrahedron Lett.* **1990**, *31*, 7567−7570.

[37] Hernandez, E.; Burgos, C. H.; Alicea, E.; Soderquist, J. A. *Org. Lett.* **2006**, *8*, 4089−4091.

[38] Lai, C.; Soderquist, J. A. *Org. Lett.* **2005**, *7*, 799−802.

[39] Lu, Z.; Ma, S. *Angew. Chem., Int. Ed.* **2008**, *47*, 258−297.

[40] Denmark, S. E.; Weber, E. J. *Helv. Chim. Acta* **1983**, *66*, 1655−1660.

[41] Yamamoto, Y.; Yatagai, H.; Naruta, Y.; Maruyama, K. *J. Am. Chem. Soc.* **1980**, *102*, 7107−7109.

[42] Denmark, S. E.; Weber, E. J. *J. Am. Chem. Soc.* **1984**, *106*, 7970−7971.

[43] Ishihara, K.; Mouri, M.; Gao, Q.; Maruyama, T.; Furuta, K.; Yamamoto, H. *J. Am. Chem. Soc.* **1993**, *115*, 11490−11495.

[44] Costa, A. L.; Piazza, M. G.; Tagliavini, E.; Trombini, C.; Umani-Ronchi, A. *J. Am. Chem. Soc.* **1993**, *115*, 7001−7002.

[45] Keck, G. E.; Tarbet, K. H.; Geraci, L. S. *J. Am. Chem. Soc.* **1993**, *115*, 8467−8468.

[46] Denmark, S. E.; Beutner, G. L. *Angew. Chem., Int. Ed.* **2008**, *47*, 1560−1638.

[47] Kobayashi, S.; Nishio, K. *J. Org. Chem.* **1994**, *59*, 6620−6628.

[48] Denmark, S. E.; Coe, D. M.; Pratt, N. E.; Griedel, B. D. *J. Org. Chem.* **1994**, *59*, 6161−6163.

[49] Malkov, A. V.; Orsini, M.; Pernazza, D.; Muir, K. W.; Langer, V.; Meghani, P.; Kočovský, P. *Org. Lett.* **2002**, *4*, 1047−1049.

[50] Masse, C. E.; Panek, J. S. *Chem. Rev.* **1995**, *95*, 1293−1316.

[51] Jain, N. F.; Takenaka, N.; Panek, J. S. *J. Am. Chem. Soc.* **1996**, *118*, 12475−12476.

[52] Panek, J. S.; Cirillo, P. F. *J. Org. Chem.* **1993**, *58*, 999−1002.

[53] Nielsen, A. T.; Houlihan, W. J. *Org. React.* **1968**, *16*, 1−438.

[54] Heathcock, C. H. In *Comprehensive Organic Synthesis. Selectivity, Strategy, and Efficiency in Modern Organic Chemistry*; Trost, B. M., Fleming, I., Eds.; Pergamon: Oxford, 1991; Vol. 2, pp. 133−179.

[55] Mukaiyama, T. *Org. React.* **1982**, *28*, 203−331.

[56] Heathcock, C. H. In *Asymmetric Synthesis*; Morrison, J. D., Ed.; Academic: Orlando, 1984; Vol. 3, pp. 111−212.

[57] Masamune, S.; Choy, W.; Petersen, J. S.; Sita, L. R. *Angew. Chem., Int. Ed. Engl.* **1985**, *24*, 1−76.

[58] Kim, B. S.; Williams, S. F.; Masamune, S. In *Comprehensive Organic Synthesis. Selectivity, Strategy, and Efficiency in Modern Organic Chemistry*; Trost, B. M., Fleming, I., Eds.; Pergamon: Oxford, 1991; Vol. 2, pp. 239−275.

[59] Cowden, C. J.; Paterson, I. *Asymmetric Aldol Reactions Using Boron Enolates*; John Wiley & Sons, Inc., 2004.

[60] Paterson, I. In *Comprehensive Organic Synthesis. Selectivity, Strategy, and Efficiency in Modern Organic Chemistry*; Trost, B. M., Fleming, I., Eds.; Pergamon: Oxford, 1991; Vol. 2, pp. 301−319.

[61] Arya, P.; Qin, H. *Tetrahedron* **2000**, *56*, 917−947.

[62] Mahrwald, R.; Ed. *Modern Aldol Reactions, Vol. 1: Enolates, Organocatalysis, Biocatalysis and Natural Product Synthesis*; Wiley-VCH: Weinheim, 2004.

[63] Mahrwald, R.; Ed. *Modern Aldol Reactions, Volume 2: Metal Catalysis*; Wiely-VCH: Weinheim, 2004.

[64] Li, Y.; Paddon-Row, M. N.; Houk, K. N. *J. Org. Chem.* **1990**, *55*, 481−493.

[65] Bernardi, A.; Capelli, A. M.; Gennari, C.; Goodman, J. M.; Paterson, I. *J. Org. Chem.* **1990**, *55*, 3576−3581.

[66] Bernardi, A.; Capelli, A. M.; Comotti, A.; Gennari, C.; Gardner, M.; Goodman, J. M.; Paterson, I. *Tetrahedron* **1991**, *47*, 3471−3484.

[67] Seebach, D.; Golinski, J. *Helv. Chim. Acta* **1981**, *64*, 1413−1423.

[68] Denmark, S. E.; Henke, B. R. *J. Am. Chem. Soc.* **1989**, *111*, 8032−8034.

[69] Denmark, S. E.; Lee, W. *J. Org. Chem.* **1994**, *59*, 707−709.

[70] Ireland, R. E.; Mueller, R. H.; Willard, A. K. *J. Am. Chem. Soc.* **1976**, *98*, 2868−2877.

[71] Mukaiyama, T.; Inoue, T. *Chem. Lett.* **1976**, 559−562.

[72] Evans, D. A.; Nelson, J. V.; Vogel, E.; Taber, T. R. *J. Am. Chem. Soc.* **1981**, *103*, 3099−3111.

[73] Evans, D. A.; McGee, L. R. *Tetrahedron Lett.* **1980**, *21*, 3975−3978.

[74] Yamamoto, Y.; Maruyama, K. *Tetrahedron Lett.* **1980**, *21*, 4607−4610.

[75] Reetz, M. T.; Peter, R. *Tetrahedron Lett.* **1981**, *22*, 4691−4694.

[76] Walker, M. A.; Heathcock, C. H. *J. Org. Chem.* **1991**, *56*, 5747−5750.

[77] Swiss, K. A.; Choi, W.-B.; Liotta, D. C.; Abdel-Magid, A. F.; Maryanoff, C. A. *J. Org. Chem.* **1991**, *56*, 5978−5980.

[78] Heathcock, C. H.; Buse, C. T.; Kleschick, W. A.; Pirrung, M. C.; Sohn, J. E.; Lampe, J. *J. Org. Chem.* **1980**, *45*, 1066−1081.

[79] Heathcock, C. H. In *Comprehensive Organic Synthesis. Selectivity, Strategy, and Efficiency in Modern Organic Chemistry*; Trost, B. M., Fleming, I., Eds.; Pergamon: Oxford, 1991; Vol. 2, pp. 181−238.

[80] Braun, M. *Angew. Chem., Int. Ed. Engl.* **1987**, *26*, 24−37.

[81] Eichenauer, H.; Friedrich, E.; Lutz, W.; Enders, D. *Angew. Chem., Int. Ed. Engl.* **1978**, *17*, 206−208.

[82] Heathcock, C. H.; Pirrung, M. C.; Buse, C. T.; Hagen, J. P.; Young, S. D.; Sohn, J. E. *J. Am. Chem. Soc.* **1979**, *101*, 7077−7078.

[83] Heathcock, C. H.; Pirrung, M. C.; Lampe, J.; Buse, C. T.; Young, S. D. *J. Org. Chem.* **1981**, *46*, 2290−2300.

[84] Draanen, N. A. V.; Arseniyades, S.; Crimmins, M. T.; Heathcock, C. H. *J. Org. Chem.* **1991**, *56*, 2499−2506.

[85] Masamune, S.; Choy, W.; Kerdesky, F. A. J.; Imperiali, B. *J. Am. Chem. Soc.* **1981**, *103*, 1566−1568.

[86] Evans, D. A.; Bartroli, J.; Shih, T. L. *J. Am. Chem. Soc.* **1981**, *103*, 2127−2129.

[87] Evans, D. A.; Britton, T. C.; Ellman, J. A. *Tetrahedron Lett.* **1987**, *28*, 6141−6144.

[88] Gage, J. R.; Evans, D. A. *Org. Synth.* **1993**, *Coll. Vol. 8*, 339−343.

[89] White, J. M.; Tunoori, A. R.; Georg, G. I. *J. Am. Chem. Soc.* **2000**, *122*, 11995−11996.

[90] Evans, D. A.; McGee, L. R. *J. Am. Chem. Soc.* **1981**, *103*, 2876−2878.

[91] Oppolzer, W. *Pure Appl. Chem.* **1988**, *60*, 39−48.

[92] Oppolzer, W.; Blagg, J.; Rodriguez, I.; Walther, E. *J. Am. Chem. Soc.* **1990**, *112*, 2767−2772.

[93] Oppolzer, W.; Starkemann, C.; Rodriguez, I.; Bernardinelli, G. *Tetrahedron Lett.* **1991**, *32*, 61−64.

[94] Oppolzer, W.; Starkemann, C. *Tetrahedron Lett.* **1992**, *33*, 2439−2442.

[95] Crimmins, M. T.; King, B. W.; Tabet, E. A. *J. Am. Chem. Soc.* **1997**, *119*, 7883−7884.
[96] Helmchen, G.; Leikauf, U.; Taufer-Knöpfel, I. *Angew. Chem., Int. Ed. Engl.* **1985**, *24*, 874−875.
[97] Siegel, C.; Thornton, E. R. *J. Am. Chem. Soc.* **1989**, *111*, 5722−5728.
[98] Choudhury, A.; Thornton, E. R. *Tetrahedron Lett.* **1993**, *34*, 2221−2224.
[99] Dickman. D. A.; Meyers, A. I.; Smith, G. A.; Gawley, R. E. *Org. Synth.* **1990**, *Coll. Vol. 7*, 530−533.
[100] Nerz-Stormes, M.; Thornton, E. R. *J. Org. Chem.* **1991**, *56*, 2489−2498.
[101] Yan, T.-H.; Tan, C.-W.; Lee, H.-C.; Lo, H.-C.; Huang, T.-Y. *J. Am. Chem. Soc.* **1993**, *115*, 2613−2621.
[102] Evans, D. A.; Sjogren, E. B.; Weber, A. E.; Conn, R. E. *Tetrahedron Lett.* **1987**, *28*, 39−42.
[103] Pridgen, L. N.; Abdel-Magid, A. F.; Lantos, I.; Shilcrat, S.; Eggleston, D. S. *J. Org. Chem.* **1993**, *58*, 5107−5117.
[104] Crimmins, M. T.; Chaudhary, K. *Org. Lett.* **2000**, *2*, 775−777.
[105] Crimmins, M. T.; Shamszad, M. *Org. Lett.* **2007**, *9*, 149−152.
[106] Masamune, S.; Sato, T.; Kim, B. M.; Wollmann, T. A. *J. Am. Chem. Soc.* **1986**, *108*, 8279−8281.
[107] Paterson, I.; Lister, M. A.; McClure, C. K. *Tetrahedron Lett.* **1986**, *27*, 4787−4790.
[108] Meyers, A. I.; Yamamoto, Y. *J. Am. Chem. Soc.* **1981**, *103*, 4278−4279.
[109] Paterson, I.; Goodman, J. M.; Anne Lister, M.; Schumann, R. C.; McClure, C. K.; Norcross, R. D. *Tetrahedron* **1990**, *46*, 4663−4684.
[110] Paterson, I. *Pure Appl. Chem.* **1992**, *64*, 1821−1830.
[111] Gennari, C.; Hewkin, C. T.; Molinari, F.; Bernardi, A.; Comotti, A.; Goodman, J. M.; Paterson, I. *J. Org. Chem.* **1992**, *57*, 5173−5177.
[112] Still, W. C.; Cai, D.; Lee, D.; Hauck, P.; Bernardi, A.; Romero, A. *Lect. Heterocycl. Chem.* **1987**, *9*, 33−42.
[113] Allinger, N. L.; Miller, M. A. *J. Am. Chem. Soc.* **1961**, *83*, 2145−2146.
[114] Corey, E. J.; Imwinkelried, R.; Pikul, S.; Xiang, Y. B. *J. Am. Chem. Soc.* **1989**, *111*, 5493−5495.
[115] Corey, E. J.; Kim, S. S. *J. Am. Chem. Soc.* **1990**, *112*, 4976−4977.
[116] Corey, E. J.; Decicco, C. P.; Newbold, R. C. *Tetrahedron Lett.* **1991**, *32*, 5287−5290.
[117] Corey, E. J.; Choi, S. *Tetrahedron Lett.* **1991**, *32*, 2857−2860.
[118] Corey, E. J.; Lee, D.-H.; Choi, S. *Tetrahedron Lett.* **1992**, *33*, 6735−6738.
[119] Corey, E. J.; Lee, D.-H. *Tetrahedron Lett.* **1993**, *34*, 1737−1740.
[120] Mukaiyama, T.; Banno, K.; Narasaka, K. *J. Am. Chem. Soc.* **1974**, *96*, 7503−7509.
[121] Mikami, K.; Matsukawa, S. *J. Am. Chem. Soc.* **1993**, *115*, 7039−7044.
[122] Kiyooka, S.-i.; Kaneko, Y.; Komura, M.; Matsuo, H.; Nakano, M. *J. Org. Chem.* **1991**, *56*, 2276−2278.
[123] Kiyooka, S.-i.; Kaneko, Y.; Kume, K.-i. *Tetrahedron Lett.* **1992**, *33*, 4927−4930.
[124] Parmee, E. R.; Tempkin, O.; Masamune, S.; Abiko, A. *J. Am. Chem. Soc.* **1991**, *113*, 9365−9366.
[125] Corey, E. J.; Cywin, C. L.; Roper, T. D. *Tetrahedron Lett.* **1992**, *33*, 6907−6910.
[126] Furuta, K.; Maruyama, T.; Yamamoto, H. *J. Am. Chem. Soc.* **1991**, *113*, 1041−1042.
[127] Maruoka, K.; Yamamoto, H. In *Catalytic Asymmetric Synthesis*; Ojima, I., Ed.; VCH: New York, 1993, pp. 413−440.
[128] Trost, B. M.; Brindle, C. S. *Chem. Soc. Rev.* **2010**, *39*, 1600−1632.
[129] Geary, L. M.; Hultin, P. G. *Tetrahedron: Asymmetry* **2009**, *20*, 131−173.
[130] Mlynarski, J.; Paradowska, J. *Chem. Soc. Rev.* **2008**, *37*, 1502−1511.
[131] Machajewski, T. D.; Wong, C.-H. *Angew. Chem., Int. Ed.* **2000**, *39*, 1352−1375.
[132] Mukaiyama, T.; Kobayashi, S.; Uchiro, H.; Shiina, I. *Chem. Lett.* **1990**, 129−132.
[133] Kobayashi, S.; Fujishita, Y.; Mukaiyama, T. *Chem. Lett.* **1990**, *8*, 1455−1458.
[134] Carreira, E. M.; Singer, R. A.; Lee, W. *J. Am. Chem. Soc.* **1994**, *116*, 8837−8838.
[135] Kobayashi, S.; Uchiro, H.; Shiina, I.; Mukaiyama, T. *Tetrahedron* **1993**, *49*, 1761−1772.
[136] Evans, D. A.; Downey, C. W.; Hubbs, J. L. *J. Am. Chem. Soc.* **2003**, *125*, 8706−8707.
[137] Desimoni, G.; Faita, G.; Jørgensen, K. A. *Chem. Rev.* **2006**, *106*, 3561−3651.
[138] Desimoni, G.; Faita, G.; Piccinini, F.; Toscanini, M. *Eur. J. Org. Chem.* **2006**, *23*, 5228−5230.
[139] Trost, B. M.; Ito, H. *J. Am. Chem. Soc.* **2000**, *122*, 12003−12004.
[140] Evans, D. A.; Janey, J. M.; Magomedov, N.; Tedrow, J. S. *Angew. Chem., Int. Ed.* **2001**, *40*, 1884−1888.
[141] Kobayashi, S.; Uchiro, H.; Fujishita, Y.; Shiina, I.; Mukaiyama, T. *J. Am. Chem. Soc.* **1991**, *113*, 4247−4252.

[142] Magdziak, D.; Lalic, G.; Lee, H. M.; Fortner, K. C.; Aloise, A. D.; Shair, M. D. *J. Am. Chem. Soc.* **2005**, *127*, 7284–7285.

[143] Bee, C.; Han, S. B.; Hassan, A.; Iida, H.; Krische, M. J. *J. Am. Chem. Soc.* **2008**, *130*, 2746–2747.

[144] Zhao, C. X.; Duffey, M. O.; Taylor, S. J.; Morken, J. P. *Org. Lett.* **2001**, *3*, 1829–1831.

[145] Taylor, S. J.; Duffey, M. O.; Morken, J. P. *J. Am. Chem. Soc.* **2000**, *122*, 4528–4529.

[146] Shiomi, T.; Nishiyama, H. *Org. Lett.* **2007**, *9*, 1651–1654.

[147] Nishiyama, H,; Shiomi, T. In *Metal Catalyzed Reductive C-C Bond Formation: A Departure from Preformed Organometallic Reagents*, Krische, M. J., Ed.; Springer: Berlin, **2007**; pp. 105–137.

[148] Lam, H. W.; Joensuu, P. M. *Org. Lett.* **2005**, *7*, 4225–4228.

[149] Zhao, D. B.; Oisaki, K.; Kanai, M.; Shibasaki, M. *J. Am. Chem. Soc.* **2006**, *128*, 14440–14441.

[150] Zlotin, S. G.; Kucherenko, A. S.; Beletskaya, I. P. *Russ. Chem. Rev.* **2009**, *78*, 737–784.

[151] Eder, U.; Sauer, G.; Wiechert, R. *Angew. Chem., Int. Ed. Engl.* **1971**, *10*, 496–497.

[152] Hajos, Z. G.; Parrish, D. R. *J. Org. Chem.* **1974**, *39*, 1615–1621.

[153] Movassaghi, M.; Jacobsen, E. N. *Science* **2002**, *298*, 1904–1905.

[154] Notz, W.; List, B. *J. Am. Chem. Soc.* **2000**, *122*, 7386–7387.

[155] Hoang, L.; Bahmanyar, S.; Houk, K. N.; List, B. *J. Am. Chem. Soc.* **2002**, *125*, 16–17.

[156] Seebach, D.; Beck, A. K.; Badine, D. M.; Limbach, M.; Eschenmoser, A.; Treasurywala, A. M.; Hobi, R.; Prikoszovich, W.; Linder, B. *Helv. Chim. Acta* **2007**, *90*, 425–471.

[157] Bahmanyar, S.; Houk, K. N.; Martin, H. J.; List, B. *J. Am. Chem. Soc.* **2003**, *125*, 2475–2479.

[158] Schmid, M. B.; Zeitler, K.; Gschwind, R. M. *Angew. Chem., Int. Ed.* **2010**, *49*, 4997–5003.

[159] Zotova, N.; Broadbelt, L. J.; Armstrong, A.; Blackmond, D. G. *Bioorg. Med. Chem. Lett.* **2009**, *19*, 3934–3937.

[160] Zotova, N.; Moran, A.; Armstrong, A.; Blackmond, D. G. *Adv. Synth. Catal.* **2009**, *351*, 2765–2769.

[161] Mangion, I. K.; Northrup, A. B.; MacMillan, D. W. C. *Angew. Chem., Int. Ed.* **2004**, *43*, 6722–6724.

[162] Nakadai, M.; Saito, S.; Yamamoto, H. *Tetrahedron* **2002**, *58*, 8167–8177.

[163] Hartikka, A.; Arvidsson, P. I. *Eur. J. Org. Chem.* **2005**, *20*, 4287–4295.

[164] Cobb, A. J. A.; Shaw, D. M.; Longbottom, D. A.; Gold, J. B.; Ley, S. V. *Org. Biomol. Chem.* **2005**, *3*, 84–96.

[165] Samanta, S.; Liu, J.; Dodda, R.; Zhao, C.-G. *Org. Lett.* **2005**, *7*, 5321–5323.

[166] Hayashi, Y.; Itoh, T.; Aratake, S.; Ishikawa, H. *Angew. Chem., Int. Ed.* **2008**, *47*, 2082–2084.

[167] Kano, T.; Takai, J.; Tokuda, O.; Maruoka, K. *Angew. Chem., Int. Ed.* **2005**, *44*, 3055–3057.

[168] Luo, S.; Xu, H.; Li, J.; Zhang, L.; Cheng, J.-P. *J. Am. Chem. Soc.* **2007**, *129*, 3074–3075.

[169] List, B.; Lerner, R. A.; Barbas, C. F. *J. Am. Chem. Soc.* **2000**, *122*, 2395–2396.

[170] Pidathala, C.; Hoang, L.; Vignola, N.; List, B. *Angew. Chem., Int. Ed.* **2003**, *42*, 2785–2788.

[171] List, B. *Tetrahedron* **2002**, *58*, 5573–5590.

[172] Bogevig, A.; Kumaragurubaran, N.; Anker Jorgensen, K. *Chem. Commun.* **2002**, *6*, 620–621.

[173] Cordova, A.; Engqvist, M.; Ibrahem, I.; Casas, J.; Sunden, H. *Chem. Commun.* **2005**, *15*, 2047–2049.

[174] Marques-Lopez, E.; Herrera, R. P.; Christmann, M. *Nat. Prod. Rep.* **2010**, *27*, 1138–1167.

[175] Carpenter, J.; Northrup, A. B.; Chung, d.; Wiener, J. J. M.; Kim, S.-G.; MacMillan, D. W. C. *Angew. Chem., Int. Ed.* **2008**, *47*, 3568–3572.

[176] Nakamura, S.; Hara, N.; Nakashima, H.; Kubo, K.; Shibata, N.; Toru, T. *Chem.–Eur. J* **2008**, *14*, 8079–8081.

[177] Basavaiah, D.; Rao, A. J.; Satyanarayana, T. *Chem. Rev.* **2003**, *103*, 811–891.

[178] Basavaiah, D.; Reddy, B. S.; Badsara, S. S. *Chem. Rev.* **2010**, *110*, 5447–5674.

[179] Iwabuchi, Y.; Nakatani, M.; Yokoyama, N.; Hatakeyama, S. *J. Am. Chem. Soc.* **1999**, *121*, 10219–10220.

[180] Gilbert, A.; Heritage, T. W.; Isaacs, N. S. *Tetrahedron: Asymmetry* **1991**, *2*, 969–972.

[181] Marko, I. E.; Giles, P. R.; Hindley, N. J. *Tetrahedron* **1997**, *53*, 1015–1024.

[182] McDougal, N. T.; Schaus, S. E. *J. Am. Chem. Soc.* **2003**, *125*, 12094–12095.

[183] Matsui, K.; Takizawa, S.; Sasai, H. *J. Am. Chem. Soc.* **2005**, *127*, 3680–3681.

[184] Roush, W. R. *J. Org. Chem.* **1991**, *56*, 4151–4157.

[185] Masamune, S.; Hirama, M.; Mori, S.; Ali, S. A.; Garvey, D. S. *J. Am. Chem. Soc.* **1981**, *103*, 1568–1571.

[186] Evans, D. A.; Bartroli, J. *Tetrahedron Lett.* **1982**, *23*, 807–810.

[187] Paterson, I.; Anne Lister, M. *Tetrahedron Lett.* **1988**, *29*, 585–588.

[188] Masamune, S. In *Organic Synthesis: Today and Tomorrow*; Trost, B. M., Hutchinson, C. R., Eds.; Pergamon: Oxford, 1981, pp. 197–215.

[189] Martin, S. F.; Lee, W.-C. *Tetrahedron Lett.* **1993**, *34*, 2711–2714.

[190] Martin, S. F.; Pacofsky, G. J.; Gist, R. P.; Lee, W.-C. *J. Am. Chem. Soc.* **1989**, *111*, 7634–7636.

[191] Chen, X.; Hortelano, E. R.; Eliel, E. L.; Frye, S. V. *J. Am. Chem. Soc.* **1992**, *114*, 1778–1784.

[192] Evans, D. A.; Calter, M. A. *Tetrahedron Lett.* **1993**, *34*, 6871–6874.

[193] Evans, D. A.; Duffy, J. L.; Dart, M. J. *Tetrahedron Lett.* **1994**, *46*, 8537–8540.

[194] Evans, D. A.; Dart, M. J.; Duffy, J. L.; Yang, M. G.; Livingston, A. B. *J. Am. Chem. Soc.* **1995**, *117*, 6619–6620.

[195] Evans, D. A.; Dart, M. J.; Duffy, J. L.; Rieger, D. L. *J. Am. Chem. Soc.* **1995**, *117*, 9073–9074.

[196] Evans, D. A.; Yang, M. G.; Dart, M. J.; Duffy, J. L.; Kim, A. S. *J. Am. Chem. Soc.* **1995**, *117*, 9598–9599.

[197] Masamune, S.; McCarthy, P. A. In *Macrolide Antibiotics. Chemistry, Biology, and Practice*; Academic: Orlando, 1984, pp. 127–198.

[198] Paterson, I.; Mansuri, M. M. *Tetrahedron* **1985**, *41*, 3569–3624.

[199] Hoffmann, R. W. *Angew. Chem., Int. Ed. Engl.* **1987**, *26*, 489–594.

[200] Schetter, B.; Mahrwald, R. *Angew. Chem., Int. Ed.* **2006**, *45*, 7506–7525.

[201] Li, J.; Menche, D. *Synthesis* **2009**, *14*, 2293–2315.

[202] Mickel, S. J.; Sedelmeier, G. H.; Niederer, D.; Daeffler, R.; Osmani, A.; Schreiner, K.; Seeger-Weibel, M.; Berod, B.; Schaer, K.; Gamboni, R. *Org. Process Res. Dev.* **2004**, *8*, 92–100.

[203] Mickel, S. J.; Sedelmeier, G. H.; Niederer, D.; Schuerch, F.; Grimler, D.; Koch, G.; Daeffler, R.; Osmani, A.; Hirni, A.; Schaer, K.; Gamboni, R. *Org. Process Res. Dev.* **2004**, *8*, 101–106.

[204] Mickel, S. J.; Sedelmeier, G. H.; Niederer, D.; Schuerch, F.; Koch, G.; Kuesters, E.; Daeffler, R.; Osmani, A.; Seeger-Weibel, M.; Schmid, E.; Hirni, A.; Schaer, K.; Gamboni, R. *Org. Process Res. Dev.* **2004**, *8*, 107–112.

[205] Mickel, S. J.; Sedelmeier, G. H.; Niederer, D.; Schuerch, F.; Seger, M.; Schreiner, K.; Daeffler, R.; Osmani, A.; Bixel, D.; Loiseleur, O.; Cercus, J.; Stettler, H.; Schaer, K.; Gamboni, R.; Bach, A.; Chen, G. P.; Chen, W. C.; Geng, P.; Lee, G. T.; Loeser, E.; McKenna, J.; Kinder, F. R.; Konigsberger, K.; Prasad, K.; Ramsey, T. M.; Reel, N.; Repic, O.; Rogers, L.; Shieh, W. C.; Wang, R. M.; Waykole, L.; Xue, S.; Florence, G.; Paterson, I. *Org. Process Res. Dev.* **2004**, *8*, 113–121.

[206] Mickel, S. J.; Niederer, D.; Daeffler, R.; Osmani, A.; Kuesters, E.; Schmid, E.; Schaer, K.; Gamboni, R.; Chen, W. C.; Loeser, E.; Kinder, F. R.; Konigsberger, K.; Prasad, K.; Ramsey, T. M.; Repic, J.; Wang, R. M.; Florence, G.; Lyothier, I.; Paterson, I. *Org. Process Res. Dev.* **2004**, *8*, 122–130.

[207] Smith, A. B.; Kaufman, M. D.; Beauchamp, T. J.; LaMarche, M. J.; Arimoto, H. *Org. Lett.* **1999**, *1*, 1823–1826.

[208] Smith, A. B.; Beauchamp, T. J.; LaMarche, M. J.; Kaufman, M. D.; Qiu, Y.; Arimoto, H.; Jones, D. R.; Kobayashi, K. *J. Am. Chem. Soc.* **2000**, *122*, 8654–8664.

[209] Paterson, I.; Florence, G. J.; Gerlach, K.; Scott, J. P. *Angew. Chem., Int. Ed.* **2000**, *39*, 377–380.

[210] Smith, A. B.; Freeze, B. S.; Xian, M.; Hirose, T. *Org. Lett.* **2005**, *7*, 1825–1828.

[211] Kim, Y.; Singer, R. A.; Carreira, E. M. *Angew. Chem., Int. Ed.* **1998**, *37*, 1261–1263.

[212] Paterson, I.; Chen, D. Y. K.; Coster, M. J.; Acena, J. L.; Bach, J.; Gibson, K. R.; Keown, L. E.; Oballa, R. M.; Trieselmann, T.; Wallace, D. J.; Hodgson, A. P.; Norcross, R. D. *Angew. Chem., Int. Ed.* **2001**, *40*, 4055–4060.

[213] Evans, D. A.; Trotter, B. W.; Coleman, P. J.; Cote, B.; Dias, L. C.; Rajapakse, H. A.; Tyler, A. N. *Tetrahedron* **1999**, *55*, 8671–8726.

[214] Evans, D. A.; Trotter, B. W.; Cote, B.; Coleman, P. J.; Dias, L. C.; Tyler, A. N. *Angew. Chem., Int. Ed.* **1997**, *36*, 2744–2747.

[215] Evans, D. A.; Trotter, B. W.; Cote, B.; Coleman, P. J. *Angew. Chem., Int. Ed.* **1997**, *36*, 2741–2744.

[216] Evans, D. A.; Kim, A. S.; Metternich, R.; Novack, V. J. *J. Am. Chem. Soc.* **1998**, *120*, 5921–5942.

[217] Trost, B. M.; Yeh, V. S. C. *Org. Lett.* **2002**, *4*, 3513–3516.

[218] Evans, D. A.; Connell, B. T. *J. Am. Chem. Soc.* **2003**, *125*, 10899–10905.

[219] Perlmutter, P. *Conjugate Addition Reactions in Organic Synthesis*; Pergamon: Oxford, 1992.

[220] Oare, D. A.; Heathcock, C. H. *Top. Stereochem.* **1989**, *19*, 87–170.

[221] Oare, D. A.; Heathcock, C. H. *Top. Stereochem.* **1991**, *20*, 227–407.

[222] Rondan, N. G.; Paddon-Row, M. N.; Caramella, P.; Houk, K. N. *J. Am. Chem. Soc.* **1981**, *103*, 2436–2438.

[223] Houk, K. N. *Pure Appl. Chem.* **1983**, *55*, 277–282.

[224] Seebach, D.; Zimmerman, J.; Gysel, U.; Ziegler, R.; Ha, T.-K. *J. Am. Chem. Soc.* **1988**, *110*, 4763–4772.

[225] Laube, T.; Dunitz, J. D.; Seebach, D. *Helv. Chim. Acta* **1985**, *68*, 1373–1393.

[226] Bürgi, H. B.; Dunitz, J. D.; Schefter, E. *J. Am. Chem. Soc.* **1973**, *95*, 5065–5067.

[227] Anh, N. T.; Eisenstein, O. *Nouv. J. Chimie* **1977**, *1*, 61–70.

[228] Bürgi, H. B.; Dunitz, D.; Lehn, J. M.; Wipff, G. *Tetrahedron* **1974**, *30*, 1563–1572.

[229] Oare, D. A.; Heathcock, C. H. *J. Org. Chem.* **1990**, *55*, 157–172.

[230] Oare, D. A.; Henderson, M. A.; Sanner, M. A.; Heathcock, C. H. *J. Org. Chem.* **1990**, *55*, 132–157.

[231] Bernardi, A.; Capelli, A. M.; Cassinari, A.; Comotti, A.; Gennari, C.; Scolastico, C. *J. Org. Chem.* **1992**, *57*, 7029–7034.

[232] Oppolzer, W.; Pitteloud, R.; Bernardinelli, G.; Baettig, K. *Tetrahedron Lett.* **1983**, *24*, 4975–4978.

[233] Corey, E. J.; Peterson, R. T. *Tetrahedron Lett.* **1985**, *26*, 5025–5028.

[234] Stork, G.; Saccomano, N. A. *Nouv. J. Chimie* **1986**, *10*, 677–679.

[235] Evans, D. A.; Urpí, F.; Somers, T. C.; Clark, J. S.; Bilodeau, M. T. *J. Am. Chem. Soc.* **1990**, *112*, 8215–8216.

[236] Evans, D. A.; Bioldeau, M. T.; Somers, T. C.; Clardy, J.; Cherry, D.; Kato, Y. *J. Org. Chem.* **1991**, *56*, 5750–5752.

[237] Smitrovich, J. H.; DiMichele, L.; Qu, C. X.; Boice, G. N.; Nelson, T. D.; Huffman, M. A.; Murry, J. *J. Org. Chem.* **2004**, *69*, 1903–1908.

[238] Enders, D.; Papadopoulos, K. *Tetrahedron Lett.* **1983**, *24*, 4967–4970.

[239] Enders, D.; Papadopoulos, K.; Rendenbach, B. E. M. *Tetrahedron Lett.* **1986**, *27*, 3491–3494.

[240] Enders, D.; Rendenbach, B. E. M. *Chem. Ber.* **1987**, *120*, 1223–1227.

[241] Enders, D.; Scherer, H. J.; Runsink, J. *Chem. Ber.* **1993**, *126*, 1929–1944.

[242] Tomioka, K.; Ando, K.; Yasuda, K.; Koga, K. *Tetrahedron Lett.* **1986**, *27*, 715–716.

[243] Tomioka, K.; Seo, W.; Ando, K.; Koga, K. *Tetrahedron Lett.* **1987**, *28*, 6637–6640.

[244] Mukherjee, S.; Yang, J. W.; Hoffmann, S.; List, B. *Chem. Rev.* **2007**, *107*, 5471–5569.

[245] *Enamines: Synthesis, Structure, and Reactions*; 2nd ed; Cook, A. G., Ed.; Marcel Dekker: New York, 1988.

[246] *Enamines: Synthesis, Structure, and Reactions*; Cook, A. G., Ed.; Marcel Dekker: New York, 1969.

[247] Dyke, S. F. *The Chemistry of Enamines*; Cambridge: Cambridge, 1973.

[248] d'Angelo, J.; Desmaële, D.; Dumas, F.; Guingant, A. *Tetrahedron: Asymmetry* **1992**, *3*, 459–505.

[249] Pandit, U. K.; Huisman, H. O. *Tetrahedron Lett.* **1967**, *8*, 3901–3905.

[250] de Jeso, B.; Pommier, J.-C. *J. Chem. Soc., Chem. Commun.* **1977**, 565–566.

[251] Sevin, A.; Tortajada, J.; Pfau, M. *J. Org. Chem.* **1986**, *51*, 2671–2675.

[252] Pfau, M.; Revial, G.; Guingant, A.; d'Angelo, J. *J. Am. Chem. Soc.* **1985**, *107*, 273–274.

[253] Yamada, S.; Hiroi, K.; Achiwa, K. *Tetrahedron Lett.* **1969**, *48*, 4233–4236.

[254] Ito, Y.; Sawamura, M.; Kominami, K.; Saegusa, T. *Tetrahedron Lett.* **1985**, *26*, 5303–5306.

[255] Blarer, S. J.; Schweizer, W. B.; Seebach, D. *Helv. Chim. Acta* **1982**, *65*, 1637–1654.

[256] Blarer, S. J.; Seebach, D. *Chem. Ber.* **1983**, *116*, 2250–2260.

[257] Blarer, S.; Seebach, D. *Chem. Ber.* **1983**, *116*, 3086–3096.

[258] Pitacco, G.; Colonna, F. P.; Valentin, E.; Risalti, A. *J. Chem. Soc., Perkin Trans. 1* **1974**, 1625–1627.

[259] Gawley, R. E. *Synthesis* **1976**, *12*, 777–794.

[260] Jung, M. E. *Tetrahedron* **1976**, *32*, 3–31.

[261] Desmaele, D.; d'Angelo, J. *Tetrahedron Lett.* **1989**, *30*, 345–348.

[262] Desmaele, D.; d'Angelo, J.; Bois, C. *Tetrahedron: Asymmetry* **1990**, *1*, 759–762.

[263] D'Angelo, J.; Ferroud, C.; Riche, C.; Chiaroni, A. *Tetrahedron Lett.* **1989**, *30*, 6511–6514.

[264] Le Dreau, M. A.; Desmaele, D.; Dumas, F.; d'Angelo, J. *J. Org. Chem.* **1993**, *58*, 2933–2935.

[265] Hirai, Y.; Terada, T.; Yamazaki, T.; Momose, T. *J. Chem. Soc., Perkin Trans. 1* **1992**, 509–516.

[266] Hua, D. H.; Venkataraman, S.; Coulter, M. J.; Sinai-Zingde, G. *J. Org. Chem.* **1987**, *52*, 719–728.

[267] Hua, D. H.; Chan-Yu-King, R.; McKie, J. A.; Myer, L. *J. Am. Chem. Soc.* **1987**, *109*, 5026–5029.

[268] Hanessian, S.; Gomtsyan, A.; Payne, A.; Hervé, Y.; Beaudoin, S. *J. Org. Chem.* **1993**, *58*, 5032–5034.

[269] Binns, M. R.; Chai, O. L.; Haynes, R. K.; Katsifis, A. A.; Shober, P. A.; Vonwiller, S. C. *Tetrahedron Lett.* **1985**, *26*, 1569–1572.

[270] Ziegler, F. E.; Chakraborty, U. R.; Webster, R. T. *Tetrahedron Lett.* **1982**, *23*, 3237−3240.
[271] Posner, G. In *Asymmetric Synthesis*; Morrison, J. D., Ed.; Academic: Orlando, 1983; Vol. 2, pp. 225−241.
[272] Posner, G. *Acc. Chem. Res.* **1987**, *20*, 72−78.
[273] Posner, G. H. In *The Chemistry of Sulphones and Sulphoxides*; Patai, S., Rapaport, Z., Stirling, C., Eds.; Wiley: New York, 1988, pp. 823−849.
[274] Posner, G. H.; Weitzberg, M.; Hamill, T. G.; Asirvatham, E.; Cun-heng, H.; Clardy, J. *Tetrahedron* **1986**, *42*, 2919−2929.
[275] Posner, G. H.; Switzer, C. *J. Am. Chem. Soc.* **1986**, *108*, 1239−1244.
[276] Posner, G. H.; Asirvatham, E. *J. Org. Chem.* **1985**, *50*, 2589−2591.
[277] Holton, R. A.; Kennedy, R. M.; Kim, H.-B.; Krafft, M. E. *J. Am. Chem. Soc.* **1987**, *109*, 1597−1600.
[278] Cram, D. J.; Sogah, G. D. Y. *J. Chem. Soc., Chem. Commun.* **1981**, 625−628.
[279] Alonso-Lopez, M.; Jimenez-Barbero, J.; Martin-Lomas, M.; Penades, S. *Tetrahedron* **1988**, *44*, 1535−1543.
[280] Takasu, M.; Wakabayashi, H.; Furuta, K.; Yamamoto, H. *Tetrahedron Lett.* **1988**, *29*, 6943−6946.
[281] Yamamoto, Y.; Suzuki, H.; Yasuda, Y.; Iida, A.; Tomioka, K. *Tetrahedron Lett.* **2008**, *49*, 4582−4584.
[282] Yura, T.; Iwasawa, N.; Mukaiyama, T. *Chem. Lett.* **1988**, 1021−1024.
[283] Yura, T.; Iwasawa, N.; Narasaka, K.; Mukaiyama, T. *Chem. Lett.* **1988**, *63*, 1025−1026.
[284] Långström, B.; Berson, G. *Acta Chem. Scand.* **1973**, *27*, 3118−3119.
[285] Hermann, K.; Wynberg, H. *J. Org. Chem.* **1979**, *44*, 2238−2244.
[286] Wang, J.; Li, H.; Lou, B.; Zu, L.; Guo, H.; Wang, W. *Chem.−Eur. J.* **2006**, *12*, 4321−4332.
[287] Enders, D.; Chow, S. *Eur. J. Org. Chem.* **2006**, *71*, 4578−4584.
[288] Terakado, D.; Takano, M.; Oriyama, T. *Chem. Lett.* **2005**, *34*, 962−963.
[289] Pansare, S. V.; Pandya, K. *J. Am. Chem. Soc.* **2006**, *128*, 9624−9625.
[290] Inokuma, T.; Hoashi, Y.; Takemoto, Y. *J. Am. Chem. Soc.* **2006**, *128*, 9413−9419.
[291] Grondal, C.; Jeanty, M.; Enders, D. *Nature Chem.* **2010**, *2*, 167−178.
[292] Seebach, D.; Matthews, J. L. *Chem. Commun.* **1997**, *21*, 2015−2022.
[293] Seebach, D.; Gardiner, J. *Acc. Chem. Res.* **2008**, *41*, 1366−1375.
[294] Gelman, M. A.; Gellman, S. H., Eds.; *Beta-peptides*; John Wiley & Sons, Inc., 2005.
[295] Krishna, P. R.; Sreeshailam, A.; Srinivas, R. *Tetrahedron* **2009**, *65*, 9657−9672.
[296] Davies, S. G.; Fenwick, D. R. *J. Chem. Soc., Chem. Commun.* **1995**, *11*, 1109−1110.
[297] Sewald, N.; Hiller, K. D.; Krner, M.; Findeisen, M. *J. Org. Chem.* **1998**, *63*, 7263−7274.
[298] Enders, D.; Wahl, H.; Bettray, W. *Angew. Chem., Int. Ed. Engl.* **1995**, *34*, 455−457.
[299] Davies, S. G.; Ichihara, O.; Walters, I. A. S. *J. Chem. Soc., Perkin Trans. 1* **1994**, 1141−1147.
[300] Davies, S. G.; Garrido, N. M.; Ichihara, O.; Walters, L. A. S. *J. Chem. Soc., Chem. Commun.* **1993**, 1153−1155.
[301] Dumas, F.; Mezrhab, B.; d'Angelo, J.; Riche, C.; Chiaroni, A. *J. Org. Chem.* **1996**, *61*, 2293−2304.
[302] Etxebarria, J.; Vicario, J. L.; Badia, D.; Carrillo, L.; Ruiz, N. *J. Org. Chem.* **2005**, *70*, 8790−8800.
[303] Buncel, E.; Um, I.-H. *Tetrahedron* **2004**, *60*, 7801−7825.
[304] Baldwin, S. W.; Aubé, J. *Tetrahedron Lett.* **1987**, *28*, 179−182.
[305] Ishikawa, T.; Nagai, K.; Kudoh, T.; Saito, S. *Synlett* **1995**, *11*, 1171−1173.
[306] Lee, H.-S.; Park, J.-S.; Kim, B. M.; Gellman, S. H. *J. Org. Chem.* **2003**, *68*, 1575−1578.
[307] Sibi, M. P.; Liu, M. *Org. Lett.* **2000**, *2*, 3393−3396.
[308] Sibi, M. P.; Shay, J. J.; Liu, M.; Jasperse, C. P. *J. Am. Chem. Soc.* **1998**, *120*, 6615−6616.
[309] Sibi, M. P.; Soeta, T. *J. Am. Chem. Soc.* **2007**, *129*, 4522−4523.
[310] Yamagiwa, N.; Matsunaga, S.; Shibasaki, M. *J. Am. Chem. Soc.* **2003**, *125*, 16178−16179.
[311] Hamashima, Y.; Somei, H.; Shimura, Y.; Tamura, T.; Sodeoka, M. *Org. Lett.* **2004**, *6*, 1861−1864.
[312] Myers, J. K.; Jacobsen, E. N. *J. Am. Chem. Soc.* **1999**, *121*, 8959−8960.
[313] Taylor, M. S.; Zalatan, D. N.; Lerchner, A. M.; Jacobsen, E. N. *J. Am. Chem. Soc.* **2005**, *127*, 1313−1317.
[314] Perdicchia, D.; Jrgensen, K. A. *J. Org. Chem.* **2007**, *72*, 3565−3568.
[315] Carlson, E. C.; Rathbone, L. K.; Yang, H.; Collett, N. D.; Carter, R. G. *J. Org. Chem.* **2008**, *73*, 5155−5158.
[316] Falborg, L.; Jorgensen, K. A. *J. Chem. Soc., Perkin Trans. 1* **1996**, 2823−2826.
[317] Zhuang, W.; Hazell, R. G.; Jorgensen, K. A. *Chem. Commun.* **2001**, *14*, 1240−1241.

[318] Diner, P.; Nielsen, M.; Bertelsen, S.; Niess, B.; Jorgensen, K. A. *Chem. Commun.* **2007**, 3646−3648.

[319] Vanderwal, C. D.; Jacobsen, E. N. *J. Am. Chem. Soc.* **2004**, *126*, 14724−14725.

[320] Enders, D.; Lüttgen, K.; Narine, A. A. *Synthesis* **2007**, *2007*, 959−980.

[321] Kim, B. H.; Lee, H. B.; Hwang, J. K.; Kim, Y. G. *Tetrahedron: Asymmetry* **2005**, *16*, 1215−1220.

[322] Miyata, O.; Shinada, T.; Ninomiya, I.; Naito, T. *Tetrahedron* **1997**, *53*, 2421−2438.

[323] Lu, H.-H.; Zhang, F.-G.; Meng, X.-G.; Duan, S.-W.; Xiao, W.-J. *Org. Lett.* **2009**, *11*, 3946−3949.

Chapter 6

Cycloadditions and Rearrangements

This chapter examines reactions that involve cycloadditions and a variety of molecular rearrangements. These reactions are popular among synthetic chemists as they permit the rapid escalation of chemical complexity, often with high levels of diastereoselectivity and enantioselectivity. The first part of this chapter will begin with the Diels−Alder reaction, for reasons obvious to any practicing organic chemist, and will briefly consider a selection of other cycloaddition reactions. In the second part, rearrangements, many of which are pericyclic processes, will be discussed. Our discussion will not be restricted to concerted, pericyclic reactions, however. Often, stepwise processes that involve a net transformation equivalent to a pericyclic reaction are catalyzed by transition metals. Although we will refer to the standard classifications for pericyclic reactions (*e.g.*, [1,3], [4 +2]), this is for convenience only and does not imply mechanism or adherence to the pericyclic selection rules.

6.1 CYCLOADDITIONS

Cycloaddition reactions have considerable value in organic synthesis for a number of reasons, not the least of which are that two bonds are formed in one operation and that the reactions often exhibit high stereoselectivities. Even if this huge field were limited only to examples that fall into the category of asymmetric synthesis, it would take several volumes to do it justice. Luckily, the basic principles that underlie many synthetically important cycloadditions can be exemplified by careful consideration of several iconic examples. We will accordingly concentrate on discussions of the Diels−Alder reaction and heteroatom-incorporating variants of it, 1,3-dipole cycloadditions routes to heterocycles, and cyclopropanations.

6.1.1 The Diels−Alder Reaction

Many reactions may compete for the title of "the most important process in organic chemistry," but none can challenge the Diels−Alder reaction when it comes to synthetic utility in the formation of 6-membered rings. The enormous body of work that includes synthetic applications and mechanistic investigations of this venerable reaction cannot be adequately summarized in anything less than a monograph. Even the literature limited to the asymmetric Diels−Alder reaction is formidable,[1] and the following discussion is therefore selective. Our

---

1. For reviews of the asymmetric Diels−Alder reaction, see refs. [1−6].

Principles of Asymmetric Synthesis.
© 2012, Professor Robert E. Gawley and Professor Jeffrey Aubé. Elsevier Ltd. All rights reserved.

goal is to illustrate some of the methods that have been developed for the synthesis of enantiopure cyclohexenes and for which transition state models have been proposed. It is hoped that this sampling will afford the reader a taste for the breadth of the process, as well as a basic knowledge of the types of transition state assemblies that favor stereoselective cycloadditions.

The historical development of the asymmetric Diels–Alder reaction begins with auxiliary-based methods for (covalently) modifying the cycloaddition reactants, and has now progressed through chiral (stoichiometric) catalysts, to true catalysts that are efficient in both enantioselectivity and turnover. Thus, the development of the asymmetric Diels–Alder reaction is a microcosm of the entire field of asymmetric synthesis. The following discussion is organized according to the strategy employed: auxiliaries for dienophile modification, diene auxiliaries, and chiral catalysts.

6.1.1.1 Chiral Auxiliaries

In general, cycloadditions catalyzed by Lewis acids proceed at significantly lower temperatures and with higher selectivities than their uncatalyzed counterparts. Factors that contribute to the increased selectivity of the catalyzed reactions include lower temperatures and more organized transition states. For enthalpy-controlled reactions, lowering temperatures increases selectivity (recall Sections 1.4 and 1.5 and Equation 1.9). Coordination of a Lewis acid to the enone carbonyl not only activates the enone by electron withdrawal, it also restricts conformational motion and thereby reduces the number of competing transition states. Figure 6.1 illustrates several chiral auxiliaries for dienophile modification that have been used in the Diels–Alder reaction.

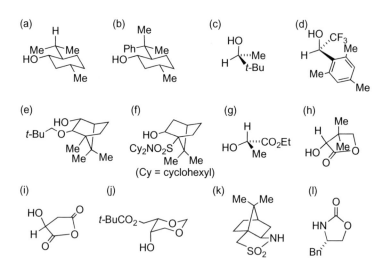

FIGURE 6.1 Dienophile chiral auxiliaries for the asymmetric Diels–Alder reaction. (a) [7], (b) [8–10], (c) [7,11], (d) [12], (e) [13], (f) [14], (g) [15], (h) [16,17], (i) [17], (j) [18], (k) [19], (l) [20].

It is convenient to separately consider chiral auxiliaries that contain a second carbonyl and those that do not. The former may function by chelating the metal of the Lewis acid catalyst, while the latter can only act as monodentate ligands to the metal of the Lewis acid. Figure 6.2a illustrates the probable transition state conformations of ester dienophiles when bound as monodentate ligands to the Lewis acid catalyst, M (auxiliaries 6.1a–f). The $C(=O)-O$ bond prefers the Z (or *cis*) conformation for a variety of reasons [21] and the preference is large: probably >4 kcal/mol. Because of this constraint, the $C-O$ bond may be considered to be similar to a double bond (hence the *E/Z* or *cis/trans* designation). A subtle consequence of this constraint is the effect it has on the conformation of the $O-C$ bond connected to the stereocenter of the chiral auxiliary. To avoid $A^{1,3}$ strain, the carbinol carbon adopts a conformation in which the $C-H$ bond eclipses the carbonyl in the lowest energy conformation, which places the other two substituents (*Large* and *Small* in Figure 6.2a) above and below the plane of the enone. When bound to a Lewis acid, the most stable conformation about the C_1-C_2 bond of the acrylate is *s-trans*, as shown, again to avoid $A^{1,3}$ strain [22].[2] Approach of the diene from the direction of the C_2 *Re* face is favored since this is the face having the least steric interactions with the auxiliary (S *vs.* L). Note also that for an endo transition state, the diene should be oriented *toward* the ester auxiliary. The specific example shown is Corey's mesityl trifluoroethanol auxiliary (Figure 6.1d, [12]).

   Figure 6.1g–l illustrates auxiliaries that may chelate the metal of the Lewis acid catalyst. In these cases, the metal is coordinated *anti* to the olefin and the preferred conformation of the C_1-C_2 bond is *s-cis* to avoid $A^{1,3}$ strain, as shown in Figure 6.2b. Again, the preferred approach of the diene is from the direction of the viewer, but because of the different conformation of the enone, it is now the C_2 *Si* face. The example is Evans' oxazolidinone [20]. In this example, the Lewis acid is Et_2AlCl, but more than one molar equivalent is required for optimum results [20]. Castellino has shown by NMR that Et_2AlCl initially binds in a monodentate fashion, but excess acid creates a bidentate dione $\cdot$ $AlEt_2^+$ complex having a $Cl_2AlEt_2^-$ gegenion [23], providing a reminder that care should always be taken in assessing which complex is actually on the reaction pathway for a particular reaction.

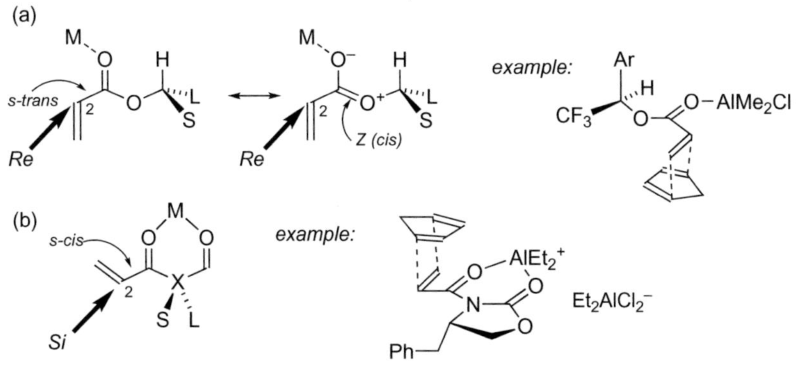

FIGURE 6.2 Probable transition state conformations of (a) a monodentate dienophile complex such as Corey's mesityl trifluoroethanol auxiliary (Figure 6.1d [12]), and (b) a bidentate dienophile such as Evans' oxazolidinone (Figure 6.1l, [20]). S and L refer to the small and large substituents of the auxiliary.

Acrylates. Cyclopentadiene is *very* often used to evaluate asymmetric Diels–Alder reactions. Table 6.1 lists the selectivities found for acrylate cycloadditions using the auxiliaries shown in Figure 6.1 under conditions that are optimized for each auxiliary. Note that there are

---

2. In the absence of a Lewis acid catalyst, both *s-cis* and *s-trans* conformers are present.

four possible norbornene stereoisomers, two endo and two exo. In accord with Alder's endo rule, the endo is heavily favored in all these examples. Although several authors report selectivities in these reactions in terms of selectivity for one endo adduct over the other, the selectivities indicated in Table 6.1 reflect the *total* diastereoselectivity of the major adduct over the other three, if this information could be deduced from the information provided in the paper.

TABLE 6.1 Asymmetric Diels–Alder Reactions of Cyclopentadiene and Acrylates

Entry	$X_c{}^a$	Lewis Acid	Probable TS	Temperature (°C)	% Yield	% ds	Ref.
1	a	$BF_3 \cdot OEt_2$	Non-chelated	−70	80	91	[7]
2	b	$SnCl_4$	Non-chelated	0	95	75	[8,9]
3	c	$BF_3 \cdot OEt_2$	Non-chelated	−70	75	87	[7,11]
4	d	Me_2AlCl	Non-chelated	−78	96	97	[12]
5	e	$TiCl_2(Oi\text{-}Pr)_2$	Non-chelated	−20	96	96	[13]
6	f	$TiCl_2(Oi\text{-}Pr)_2$	Non-chelated	−20	97	89	[14]
7	g	$TiCl_4$	Chelated[b]	−63	88	91	[15]
8	h	$TiCl_4$	Chelated[b]	−64	81	95	[16,17]
9	i	$TiCl_4$	Chelated[b]	−78	86	97	[17]
10	j	Et_2AlCl	Chelated[c]	−70	88	99	[18]
11	k	Et_2AlCl	Chelated	−130	96	95	[19]
12	l	Et_2AlCl	Chelated	−100	94	78	[20]

[a] *The X_c column refers to the auxiliaries in Figure 6.1; the probable transition state (TS) conformations of the dienophile are illustrated in Figure 6.2; the % ds refers to the formation of one of the four possible products (two endo and two exo isomers).*
[b] *In this case, the $TiCl_4$ is thought to shield one face of the enone [24].*
[c] *Chelation is postulated to occur at a ring oxygen.*

An analysis of the auxiliary in Figure 6.1e and two close relatives illustrate how structural changes can affect the selectivity of the cycloaddition and how conformational principles can explain the effects. Scheme 6.1 shows three acrylate/cyclopentadiene cycloadditions with three very similar auxiliaries, run with the same catalyst at similar temperatures, but which exhibit markedly different stereoselectivities. All of these auxiliaries were designed to place the acrylate and a shielding neopentyl group on a rigid scaffolding (camphor skeleton) such that the enone and a *tert*-butyl group lie (more or less) parallel, and they are thought to react *via* a nonchelated conformation analogous to Figure 6.2a. Scheme 6.1a duplicates the data listed in Table 6.1, entry 5 [13]. This auxiliary, developed by Oppolzer, shows outstanding selectivity at −20 °C, but its close counterpart, shown in Scheme 6.1b, exhibits significantly lower (although still useful) selectivity [13]. The only difference is the relationship of the bridgehead methyl to the neopentyl. In the absence of the bridgehead methyl, the *tert*-butyl can rotate away from the acrylate, leaving the *Si* face more accessible. The auxiliary in Scheme 6.1c was prepared to further probe the effects of conformation on selectivity [25]. In this case, an oxygen has been replaced by a methylene. The most likely rationale for the further lowering of selectivity (compare Scheme 6.1b and c) is that the protons of the methylene experience unfavorable van der Waals repulsion with the indicated methyl in the conformation that best shields the acrylate *Si* face. Population of other (unspecified) conformations results in both lower endo selectivity, and lower *Re* facial selectivity within the endo manifold. It is interesting to recall the discussion in Section 1.4 on selectivity (*cf.* Figure 1.3 and the accompanying discussion), which emphasized the small energetic differences that can result in large changes in selectivity. In Scheme 6.1, the selectivities for the three examples correspond to differences in energies of activation ($\Delta\Delta G^{\ddagger}$) of 2.9, 1.7, and 0.8 kcal/mol for the examples in Schemes 6.1a–c, respectively, for the two *endo* isomers. In each case, an increment of approximately 1 kcal/mol (about the same as the energy difference between *gauche* and *anti* butane) has a profound effect on the observed selectivity.

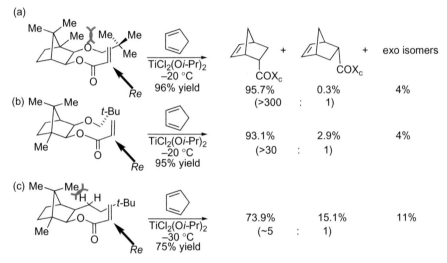

SCHEME 6.1 Camphor-derived auxiliaries for asymmetric Diels–Alder cycloadditions. (a) and (b) [13]. (c) [25]. The auxiliary illustrated in (a) is the enantiomer of that reported in ref. [13].

Intramolecular cycloadditions. Diels–Alder reactions[3] having diene and dienophile connected by three or four atom carbon chains are selective for trans-fused bicyclic adducts only when the dienophile is *trans* and when Lewis acid catalysis is employed. The competing transition states are illustrated in Scheme 6.2a [27]. The auxiliaries illustrated in Figure 6.1a, k, and 1 were used to modify the dienophile fragment for asymmetric intramolecular Diels–Alder reactions for trienes having these attributes. The examples shown in Scheme 6.2b–d reveal that the facial selectivity, dictated by chelating and non-chelating auxiliaries as rationalized in Figure 6.2, determine the chirality sense of the *trans*-fused product.[4] Thus, the absolute configuration of the product obtained using the menthyl auxiliary (Scheme 6.2b) is consistent with an *s-trans* C_1–C_2 conformation (*cf.* Figure 6.2a) and an *anti* transition state. The camphor sultam (Scheme 6.2c) and oxazolidinone imide (Scheme 6.2d) appear to react through an *s-cis* C_1–C_2 conformation (*cf.* Figure 6.2b) in the *anti* transition state.

---

3. For reviews of the intramolecular Diels–Alder reaction, see refs. [26–29].
4. For ease of comparison, the chirality sense of the camphor sultam is inverted from that reported in the literature [30] so that the favored approach at C_2 is toward the *Si* face for all three examples.

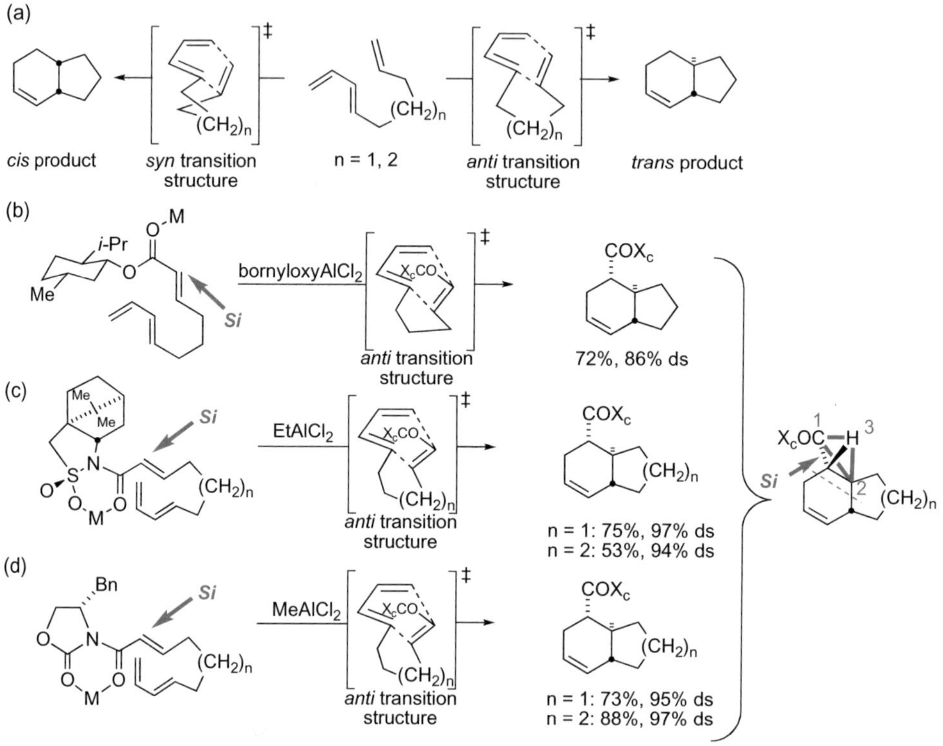

SCHEME 6.2 (a) *syn* and *anti* transition states for the intramolecular Diels–Alder reaction [27]. (b) The contribution of the chiral Lewis acid to the stereoselectivity was negligible [31]. (c) ref. [30]; the illustrated examples are enantiomeric to those reported. (d) ref. [20]. The structure on the right shows the two new bonds (red dashed line), while the solid red lines and numbers indicating CIP priority confirm that the new bond was formed on the *Si* face of the indicated carbons.

Fumarates and maleates are particularly useful dienophiles due to their high reactivity and because they carry carboxylic esters that can be subsequently functionalized. Asymmetric cycloaddition to fumarates has been accomplished by modification of either one or both ends of the diacid. In fact, addition of butadiene to dimenthyl fumarate, reported by Walborsky in 1961, was the first highly selective (89% ds) asymmetric Diels–Alder reaction ever recorded (Scheme 6.3a, refs. [32,33]). Selective reactions of cyclic dienes are also known (Scheme 6.3b, refs. [34,35]). The reactive conformation features an *s-trans* conformation at C_1–C_2, *cis* orientation of the ester ligand relative to the carbonyl oxygen, and orientation of the menthyl moiety to relieve $A^{1,3}$ strain. In this conformation, preferred approach of the diene is from the (rear) C_2 *Si* face.

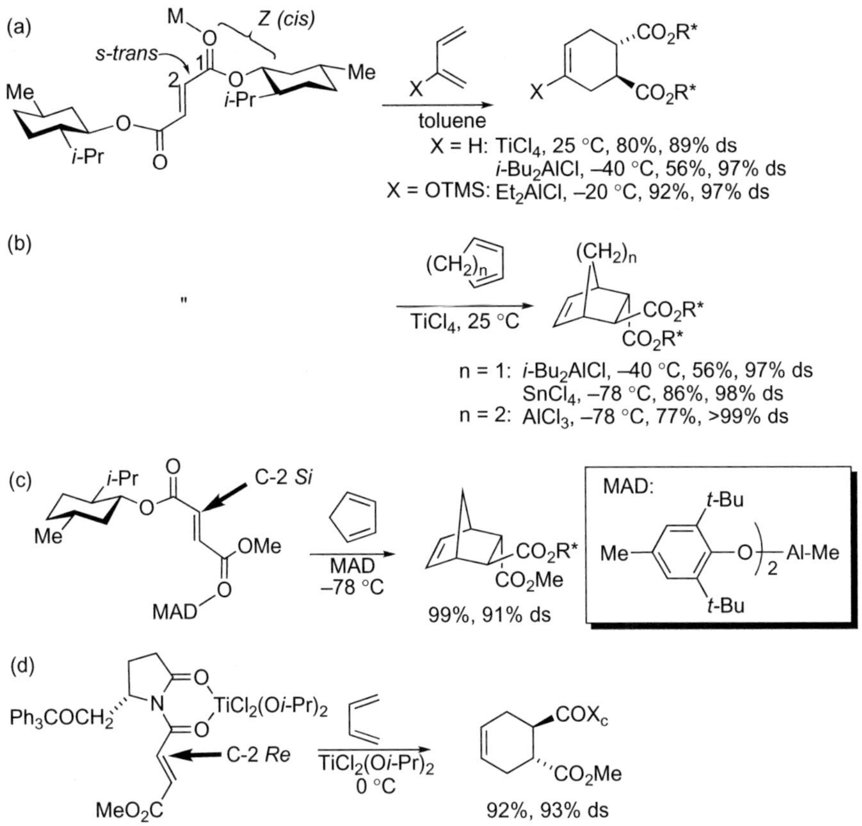

(a)

toluene
X = H: TiCl$_4$, 25 °C, 80%, 89% ds
i-Bu$_2$AlCl, –40 °C, 56%, 97% ds
X = OTMS: Et$_2$AlCl, –20 °C, 92%, 97% ds

(b)

TiCl$_4$, 25 °C

n = 1: i-Bu$_2$AlCl, –40 °C, 56%, 97% ds
SnCl$_4$, –78 °C, 86%, 98% ds
n = 2: AlCl$_3$, –78 °C, 77%, >99% ds

(c)

MAD
–78 °C

99%, 91% ds

MAD:

(d)

TiCl$_2$(Oi-Pr)$_2$
0 °C

92%, 93% ds

SCHEME 6.3 Asymmetric cycloadditions to doubly modified fumarates. (a) X = H [32−34]; X = OTMS [34], (b) n = 1 [34,35]; n = 2 [35], (c) [36], and (d) [37]

Later, Yamamoto demonstrated that high selectivity could also be obtained using a fumarate with only a single chiral ester provided the "exceptionally bulky" methylaluminum bis (2,6-di-*tert*-butyl-4-methylphenoxide) (MAD) was used (Scheme 6.3, [36]). In contrast to the excellent results shown in Scheme 6.3c, use of Et$_2$AlCl or simply heating a fumarate with a single chiral ester and a diene afforded comparatively random product sets. These authors proposed that the MAD binds to the methyl ester in favor of the menthyl ester, but that the face selectivity is determined by the menthyl auxiliary (*cf.* Figure 6.2a). In addition to the face selectivity, the reaction is also selective for the isomer having the menthyl in the exo position, due to the diene orienting away from the MAD and menthyl moieties, and toward the methoxy. Similarly, Koga studied the pyrrolidinone auxiliary shown in Scheme 6.3d [37], which showed excellent selectivity with a titanium catalyst in cycloadditions with butadiene. Other chiral esters used in this context were reported by Corey [12] and Helmchen [38].

An interesting situation arises from cycloaddition of a symmetrical diene such as butadiene or cyclopentadiene to maleic acid or a symmetrical derivative: the reaction affords achiral (meso) adducts (Scheme 6.4a). To break the symmetry, either the diene or the dienophile must be unsymmetrical. For example, cycloaddition of an unsymmetric diene would give a chiral adduct, and Scheme 6.4b shows one such approach. Maleimide having an α-methylbenzyl auxiliary on nitrogen is selective when there is a large substituent at the diene 2-position [39]. A second tactic is the same as the fumarate approach in Scheme 6.4c: attach an auxiliary to only one carboxyl group. After considerable experimentation, Yamamoto showed that 2-phenylcyclohexanol is an excellent auxiliary for *tert*-butyl maleate, as shown in Scheme 6.4c [40]. In this case, the catalyst is thought to chelate the two carbonyls with the phenyl group interacting with the double bond in a π-stacking arrangement.

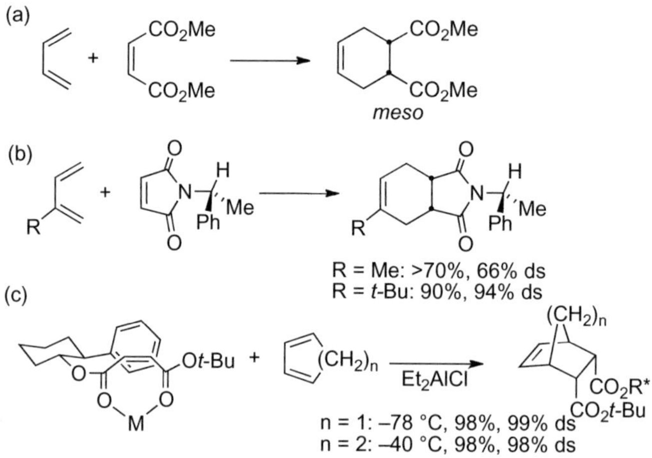
(a)
CO2Me
CO2Me
CO2Me
CO2Me
meso
(b)
H
Me
Ph
R
R = Me: >70%, 66% ds
R = t-Bu: 90%, 94% ds
(c)
Ot-Bu
(CH2)n
Et2AlCl
CO2R*
CO2t-Bu
M
n = 1: –78 °C, 98%, 99% ds
n = 2: –40 °C, 98%, 98% ds

SCHEME 6.4 (a) Cycloadditions of symmetrical dienes to maleates gives achiral products. (b) Asymmetric cycloadditions to a chiral maleimide [39]. (c) Asymmetric cycloadditions to chiral maleic ester [40].

There have been relatively few reports of chiral auxiliaries for the diene component (Scheme 6.5). Successful examples typically connect the chiral directing group to the diene through a heteroatom, both for synthetic simplicity and because the electron-donating heteroatom group raises the HOMO of the diene and therefore leads to greater reactivity. For example, Scheme 6.5a illustrates an enamino diene developed in the Enders laboratory [41,42]. Excellent selectivities were achieved with β-nitrostyrenes as dienophiles, although the yields were modest. Hydrolysis of the enamine on workup afforded a mixture of 2-methyl diastereomers with 75–95% ds. The proposed transition state for the cycloaddition is shown in the inset, although an alternative two-step mechanism (two conjugate additions) was not ruled out [41]. A version where the amino group now occupies C-1 and the diene is additionally activated by a silyloxy group was later reported by Rawal [43]. In 1980, Trost introduced a useful chiral auxiliary derived from mandelic acid, which is available as either enantiomer [44]. The original diastereoselectivity reported for addition to acrolein was 82% at −20 °C (Scheme 6.5c), but Thornton later reported 94% ds at −78 °C [45]. The other examples in Scheme 6.5 illustrate similarly high selectivities and yields, although the Thornton paper does not report specific yields for each example [45]. For the *S* auxiliary, addition to the C_1 *Si* face of the diene is preferred (relative topicity *lk*). Figure 6.3 illustrates two conformational models that were proposed to rationalize this preference. Trost suggested that π-stacking of the diene over the face of the phenyl group shields the C_1 *Re* face, as shown in Figure 6.3a [44]. However, noting that reduction of the phenyl to a cyclohexyl group afforded an auxiliary that is equally selective [44,45], Thornton proposed that the cycloaddition takes place in a diene conformation in which the bond from the α-carbon to the phenyl (or cyclohexyl) is perpendicular to the plane of the ester, as shown in Figure 6.3b [45]. Thornton asserted that the conformation in which the methoxy is nearest the carbonyl is preferred, but no explanation for this preference was offered. Nevertheless, crystal structures of three cycloadducts exhibit this conformation.

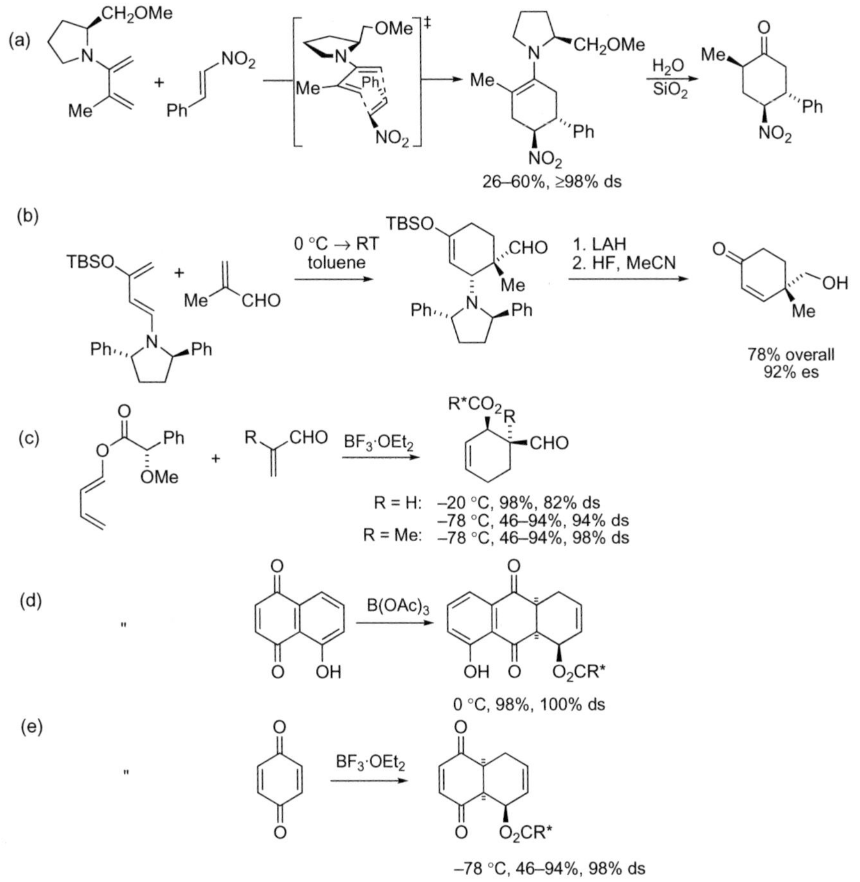

SCHEME 6.5 Asymmetric Diels—Alder reaction of chiral dienes. (a) Enders [41], (b) Rawal [43], (c) Trost, where R = H [44,45] or R = Me [45], (d) Trost [44], and (e) Thornton [45].

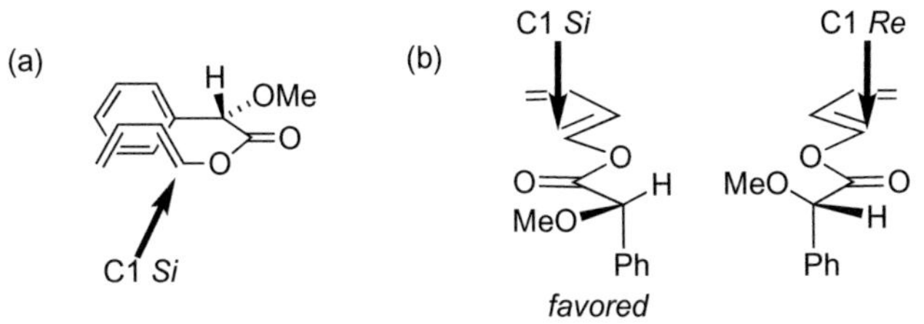

FIGURE 6.3 Two conformational models to explain the relative topicity of the Trost auxiliary for asymmetric Diels—Alder reactions. (a) Trost model [44]. (b) Thornton model [45].

6.1.1.2 Chiral Catalysts: Lewis Acids

Most Diels−Alder reactions proceed faster when the dienophile is electron poor.[5] One obvious way of increasing the speed of a given Diels−Alder reaction, then, is to coordinate a carbonyl-containing dienophile with a Lewis acid (Figure 6.4). This has the effect of lowering the LUMO of the 2-carbon reactant, which results in accelerated reactions over the thermal reaction background.[6] More recent approaches involving organocatalysis often take the related tack of converting an α,β-unsaturated aldehyde or ketone into a reactive, electron poor iminium ion (Figure 6.4). In each case, the activated dienophile now contains a handle for asymmetric induction through interligand chirality transfer or intraligand chirality transfer, respectively.

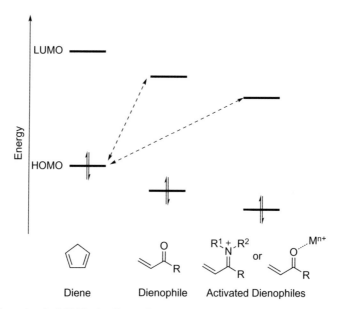

FIGURE 6.4 Lowering the LUMO of a dienophile by either coordination with a Lewis acid or conversion to an iminium ion accelerates the Diels−Alder reaction while simultaneously providing a convenient location for a chiral auxiliary. (*Note*: the relative energies of activated dienophiles vary greatly; only a single energy is shown as an arbitrary qualitative example.)

Quite a number of ligand/metal combinations have been evaluated as chiral catalysts for the Diels−Alder reaction. Much of the effort has been occupied in ligand synthesis and design, but the effort has largely been empirically driven (*i.e.*, trial and error). Figure 6.5 shows some

---

5. Of course, the major exception to this occurs in *inverse demand Diels−Alder reactions*, where the electronic roles of the diene and dienophile are switched. Most all-carbon versions of these reactions, of which there are many, are used for the synthesis of aromatic rings (*e.g.*, ref. [46]). We will limit our discussion of inverse demand Diels−Alders in this book to *hetero Diels−Alder reactions*, a synthetically important class of inverse demand cycloaddition reactions toward heterocycles; see Section 6.1.2.

6. For our money, the best single source for learning about molecular orbitals in organic chemistry is Ian Fleming's *Molecular Orbitals and Organic Chemical Reactions* (Wiley, 2010; Student Edition 376 pages; Reference Edition 515 pages).

of the many complexes that have been tested as catalysts in the Diels−Alder reaction and which show both high diastereoselectivity and high enantioselectivity.[7] Among the metals, the most commonly used are boron and titanium, but copper [47,48], magnesium [49], and lanthanides [50,51] have also found some use. For the purpose of providing an overview and putting each of these catalyst systems on a level playing field, Table 6.2 collects a number of cycloadditions of cyclopentadiene, which is by far the most common diene used to study new asymmetric Diels−Alder reactions (for a specialized review, see ref. [52]). Additional references can be found in the broader review literature [3,5,53−56]. In the following section, we present more detailed analyses of a few of these catalysts, chosen to illustrate current levels of understanding.

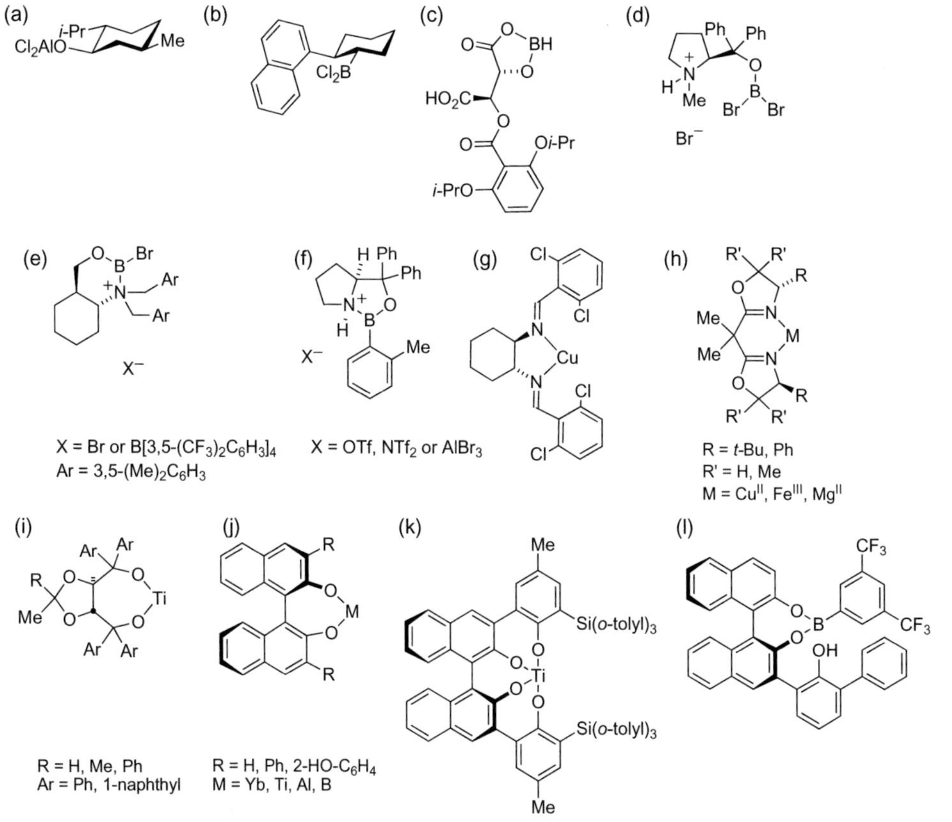

(a) Cl_2AlO—(i-Pr)...Me

(b) Cl_2B (naphthyl/cyclohexyl)

(c) O=...O—BH ; HO_2C ; O=...Oi-Pr ; i-PrO

(d) Ph Ph ; H-N$^+$-Me ; O-B ; Br Br ; Br^-

(e) O-B-Br ; $^+$N ; Ar ; Ar ; X^-
X = Br or B[3,5-$(CF_3)_2C_6H_3]_4$
Ar = 3,5-$(Me)_2C_6H_3$

(f) H Ph ; Ph ; N$^+$-O ; H B ; Me ; X^-
X = OTf, NTf_2 or $AlBr_3$

(g) Cl ; N ; Cl ; Cu ; N ; Cl ; Cl

(h) R' R' ; O ; R ; Me ; N ; M ; Me ; N ; O ; R ; R' R'
R = t-Bu, Ph
R' = H, Me
M = Cu^{II}, Fe^{III}, Mg^{II}

(i) Ar Ar ; R O ; O ; Ti ; Me O ; O ; Ar Ar
R = H, Me, Ph
Ar = Ph, 1-naphthyl

(j) R ; O ; M ; O ; R
R = H, Ph, 2-HO-C_6H_4
M = Yb, Ti, Al, B

(k) Me ; Si(o-tolyl)$_3$; O ; O ; Ti ; O ; O ; Si(o-tolyl)$_3$; Me

(l) CF_3 ; O-B ; CF_3 ; OH

FIGURE 6.5 Selected catalysts for the asymmetric Diels−Alder reaction. (a) [57], (b) [58,59], (c) [60−62], (d) [63], (e) [64], (f) [65−67], (g) [47], (h) R = Ph [48,68], R = t-Bu [48], (i) [69−72], (j) R = H [50,51,73], R = Ph [74,75], R = 2-HO-C_6H_4 [76], (k) [67], (l) [77].

---

7. Cyclopentadiene addition to an acrylate gives two diastereomers (*endo* and *exo*), each of which has two enantiomers. In the presence of a chiral auxiliary, these four stereoisomers are all diastereomers, so that selectivity for one of the four can be expressed as percent diastereoselectivity, % ds. But when using chiral catalysts, it is necessary to express selectivity in terms of both diastereoselectivity (*endo/exo* or *cis/trans*) *and* enantioselectivity (% es for the major diastereomer).

TABLE 6.2 Comparison of Different Catalysts in the Diels–Alder Reactions of Cyclopentadiene

Entry	Cycloadduct	Catalyst	Conditions	Exo/ Endo	er	% Yield	Ref.
1	HO₂C ... (cyclohexyl ester cyclopentadiene adduct)	a	1.25 mol% catalyst, −10 °C, toluene	−	73:27	86	[52]
2	Et–C(O)–C(Me) adduct	b	10 mol% catalyst, −78 °C, CH₂Cl₂	≥ 99:1	91:9	83	[78]
3	OHC, Me adduct	c	20 mol% catalyst, −78 °C, CH₂Cl₂	−	98:2	−	[62]
4	OHC, Me adduct	d	20 mol% catalyst, −78 °C, CH₂Cl₂	−	98:2	84	[63]
5	OHC, Br adduct	e	10 mol% catalyst (X = Br), −94 °C, CH₂Cl₂	94:6	97:3	99	[64]
6	MeO, Me, OTIPS adduct	f	20 mol% catalyst (X = OTf), −50 °C, toluene	−	96:4	98	[67]
7	H, S Me thiazolidinethione adduct	g	10 mol% catalyst, CH₂Cl₂, −30 °C, 16 h	7:93	95:5	86	[47]
8	H, S Me thiazolidinethione adduct	h, R = t-Bu, R′ = H, M = Cu(II)	10 mol% catalyst, CH₂Cl₂, −45 °C, 36 h	4:96	97:3	82	[48]
9	MeO, Me, Me adduct	i	−78 °C, CH₂Cl₂	−	90:10	54	[72]
10	Ph, oxazolidinone adduct	j	20 mol% catalyst, CH₂Cl₂, 0 °C, 20 h	97:3	98:2	77	[51]

(Continued)

TABLE 6.2 (Continued)

Entry	Cycloadduct	Catalyst	Conditions	Exo/ Endo	er	% Yield	Ref.
11	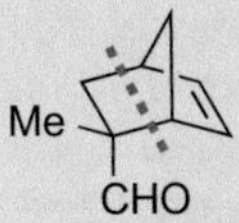 Me, CHO	k	10 mol% catalyst, CH$_2$Cl$_2$, −78 °C, 2 h	99:1	97:3	75	[79]
12	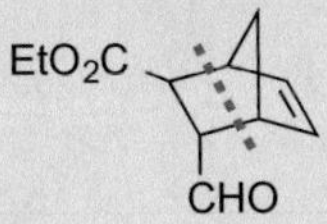 EtO$_2$C, CHO	l	5 mol% catalyst, CH$_2$Cl$_2$, −78 °C, 24 h	2:98	97:3	91	[77]

Monodentate dienophiles. The first chiral Lewis acid catalyst to show high selectivity (86% es in the cycloaddition of cyclopentadiene to 2-methylacrolein) was a dichloroaluminum alkoxide derived from menthol (Figure 6.5a, [57]). This catalyst has C_1-symmetry, but many of the catalysts shown in Figure 6.5 are C_2-symmetric. This feature reduces the number of competing transition states, which is especially important when the ligand sphere of the metal is greater than 4-coordinate. Because of fewer possible coordination sites, the binding and face-selectivities of catalysts containing boron, aluminum, or other tetravalent metals are more easily understood than those of octahedral complexes, so these are examined first.

Figure 6.6 shows Hawkins' (2-aryl)cyclohexylboron dichloride catalyst (Figure 6.5b [58,59]) coordinated to methyl acrylate (*cf.* Figure 6.2). Note that monodentate coordination of the ester is thought to occur *trans* to the alkoxy group, which forces the enone into an *s-trans* conformation to avoid A^{1,3} strain. The geometry shown in Figure 6.6 was observed in the crystal of five related catalyst·crotonate complexes [59] and in solution by NMR [58]. Comparison of these five structures indicates that, as the polarizability of the aryl group increases, a dipole-induced dipole attraction draws the polar ester group of the boron-bound crotonate toward the arene (the five complexes are, in increasing order of polarizability: Ar = phenyl, 3,5-dimethylphenyl, 3,5-dichlorophenyl, 3,5-dibromophenyl, and 1-naphthyl). Since this effect correlates with enantioselectivity, Hawkins concluded that the effect is operative in the transition state [59]. In this conformation, the rear (C_2 *Re*) face of the dienophile is blocked by the aryl group, and approach of the diene toward the face of the crotonate that is not blocked by the aryl moiety is favored. The 2-(1-naphthyl)-cyclohexyl boron catalyst produces ≥ 95% enantioselectivities in the addition of cyclopentadiene to methyl acrylate, methyl crotonate, and dimethyl fumarate [58]. Hawkins later followed up this work with a related approach to the enantioselective Diels−Alder reaction of α,β-unsaturated acid chlorides [78].

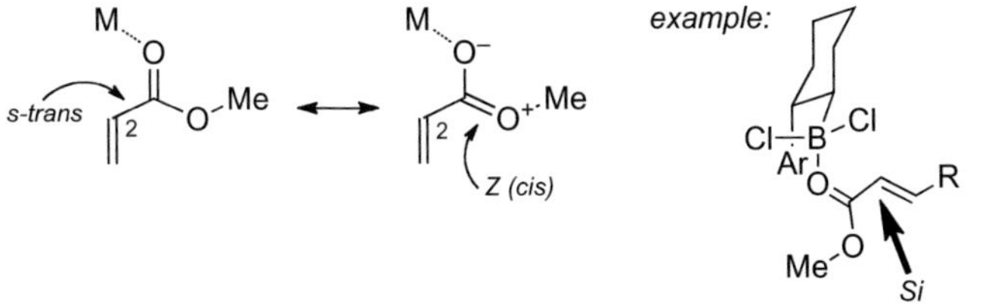

FIGURE 6.6 Methyl acrylate coordinated to Hawkins' 2-arylcyclohexyl boron catalyst [58,59].

Some important Lewis acids used to promote the Diels–Alder reaction are boron-containing catalysts introduced by Corey in the early 1990s (Scheme 6.6). Scheme 6.6 shows the reaction of 2-bromoacrolein and cyclopentadiene catalyzed by the indolyl oxazaborolidine shown [80,81]. This reaction, which is both highly diastereoselective and enantioselective, is thought to react *via* the *s-cis* conformation shown in the inset of Scheme 6.6. This catalyst conformation is suggested by nuclear Overhauser effects in the NMR spectrum of the catalyst–dienophile complex, and by chemical shift changes upon complexation to boron trifluoride [81]. Also, a 1:1 complex of the catalyst and 2-bromoacrolein is orange-red at 210 K, a color that is attributed to charge–transfer complexation between the indole ring and the boron-bound aldehyde [81]. Similar catalysts with different substituents on the nitrogen [81] or the carbon of the oxazaborolidine [81,82] show significantly lower selectivities. For example, the oxaza-borolidine having a 2-naphthyl group (comparable in size, but not as good a π-donor) in place of the indole exhibits only 88% es. Moreover, phenyl, cyclohexyl, or isopropyl groups give only about 65% es with *opposite* topicity [81].

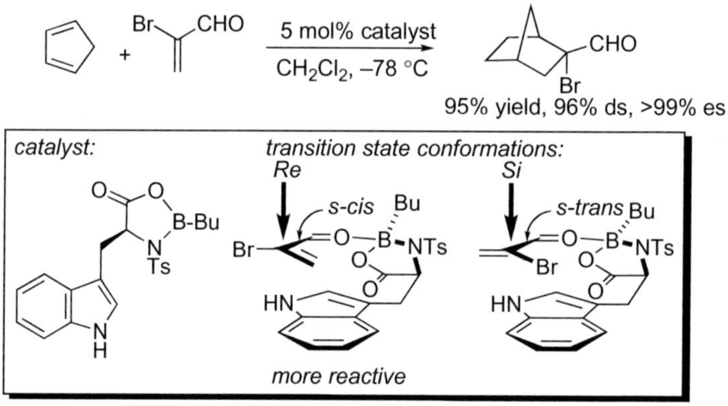

SCHEME 6.6 Asymmetric cycloaddition of 2-bromoacrolein and cyclopentadiene using Corey's indolyl oxazaborolidine catalyst [80,81]

Note that the illustrated conformation has the acrolein oriented in an *s-cis* conformation. This is in contrast to the usual *s-trans* conformation of acroleins coordinated to a Lewis acid (Figure 6.2a), but it is supported by the observation that cyclopentadiene adds to the opposite face of acrolein itself with the same catalyst [81]. It is likely that both *s-cis* and *s-trans* dieno-phile conformers are present, and that the *s-cis* conformer is more reactive. In other words, Curtin–Hammett kinetics are operative (see Section 1.8). The rationale for this increased reactivity is as follows: the *s-trans* conformation of 2-bromoacrolein would place the bromine above the indole ring. Cycloaddition to the top (*Si*) face of the *s-trans* conformer would force the bromine into closer proximity to the indole as C_2 rehybridizes from sp^2 to sp^3, a situation that is avoided in cycloaddition to the top (*Re*) face of the *s-cis* conformer.

Bidentate dienophiles. When a dienophile such as *N*-acryloyloxazolidinone coordinates the metal in a bidentate fashion (*cf.* Figure 6.2b), $A^{1,3}$ strain between the enone β-carbon and the oxazolidinone C-4 methylene forces the enone into an *s-cis* conformation, as shown in Scheme 6.7a. Interactions between the other ligands on the metal, the coordinated dienophile, and the approaching diene then determine the topicity of the cycloaddition. The exact nature of the interactions will depend on the coordination sphere of the metal.

For example, Scheme 6.7b and c shows examples of similar C_2-symmetric ligands (Figure 6.5h) coordinated to metals having different tetravalent geometries and which result in enantiomeric cycloadducts, but with excellent selectivity in both cases. The explanation for the topicity of the two catalysts is revealed by the examination of the proposed arrangements of the catalyst/dienophile complexes, as shown in Scheme 6.7d. The tetrahedral magnesium [49] complex facilitates addition to the C_2 *Si* face because the rear phenyl is blocking the *Re* face [49]. In contrast, the square–planar copper complex facilitates C_2 *Re* addition because the *Si* face is blocked by the *tert*-butyl group [48]. It should be noted, however, that in these two examples, the geometry of the coordination complex appears to be inferred (at least partly) from the topicity of the cycloaddition (note the absence of any anionic ligands in these models).

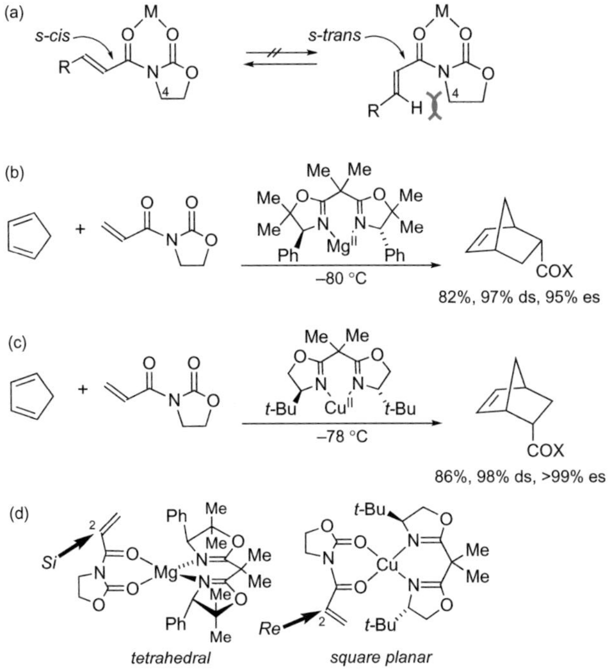

SCHEME 6.7 (a) Acryloyloxazolidinone in bidentate coordination. $A^{1,3}$ strain favors the *s-cis* conformation. (b) Cycloaddition of C_2-symmetric bisoxazoline–magnesium complex [49]. (c) Cycloaddition of C_2-symmetric bisoxazoline–copper complex [48]. (d) Rationale for the different topicities of the bisoxazoline complexes, even though both ligands have the same absolute configuration. The dienophile is drawn in the plane of the paper, and the favored approach is from the direction of the viewer.

Ligands having C_2-symmetry have also been used with metals that are undoubtedly octahedral; however, the analysis of facial selectivity in octahedral complexes is complicated by several possible competing coordination modes of the dienophile. One class of ligands that has been well studied are the TADDOLs (TADDOL is an acronym for $\alpha,\alpha,\alpha',\alpha'$-tetraaryl-1,3-dioxolane-4,5-dimethanol). Both the Narasaka [83] and the Seebach [71] groups have evaluated a number of TADDOLs as ligands for titanium in the asymmetric Diels–Alder reaction. Table 6.3 lists selected data from two extensive reports, which illustrates not only the utility of the titanium TADDOLate complex as an asymmetric catalyst, but also some subtle differences that are not readily explained. For example, Narasaka found that the tetraphenyl dimethyldioxolane ligand (R = R′ = Me; Ar = Ph) promoted the reaction (88% yield) when used in stoichiometric quantities (entry 1), whereas Seebach found that a catalytic amount was not as effective (25% yield) under similar conditions (entry 2). Note the difference in diastereo- and enantioselectivity for these two entries, as well. In contrast, replacing the phenyl group with a 2-naphthyl group affords an outstanding catalyst (entry 3), that gives excellent yields and selectivities on a multigram scale [71]. Entries 4 and 5 illustrate the tetraphenyl methyl−phenyl dioxolane catalyst (R = Me, R′ = Ph; Ar = Ph), which affords outstanding yields and selectivities in either stoichiometric or catalytic modes [83]. Comparison of entry 2 with entry 5 is particularly puzzling: replacement of one of the dioxolane substituents (a position remote from the catalytic site) results in an amazing improvement in catalyst efficiency and selectivity.[8]

TABLE 6.3 Asymmetric Cycloadditions of Crotyloxazolidinones and Cyclopentadiene Catalyzed by Titanium TADDOLate Complexes

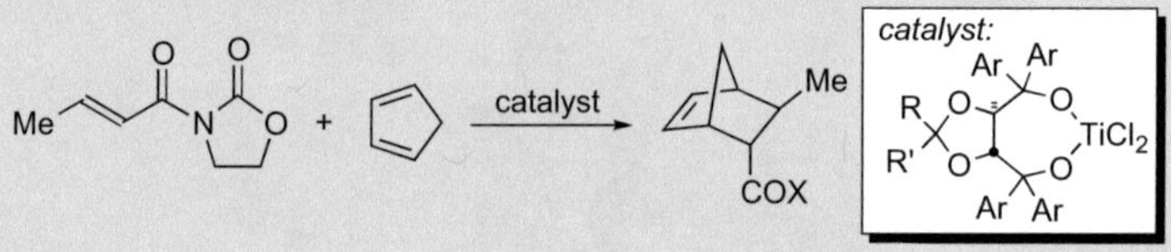

Entry	R/R′	Ar	Temperature (°C)	Equiv. cat.	% Yield	% ds	% es	Ref.
1	Me/Me	Ph	−15	1	88	93	77	[83]
2	Me/Me	Ph	−15	0.15	25	83	72	[71]
3	Me/Me	2-Naphthyl	−15	0.15	96[a]	90	94	[71]
4	Me/Ph	Ph	−15	2	93	90	96	[83]
5	Me/Ph	Ph	0	0.10	87	92	95	[83]

[a]This experiment was done on a >4 g scale.

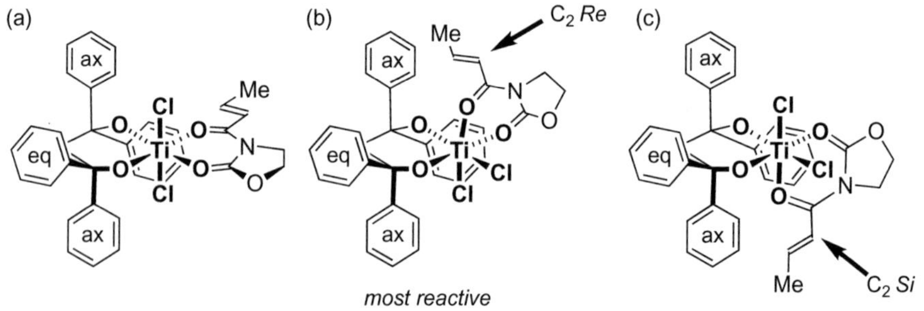

FIGURE 6.7 Titanium TADDOLate−crotyloxazolidinone complexes. The dioxolane ring of the chiral ligand (Figure 6.5i) is deleted for clarity, and the phenyl groups are labeled as axial (ax) or equatorial (eq). (a) Symmetrical complex found by NMR to be the predominant species in solution [84], and also characterized crystallographically [85]. (b) Complex judged to be most likely to be responsible for the asymmetic cycloaddition [71,84]. (c) This complex is probably less reactive, since approach of the dienophile is hindered by the axial phenyl [71].

8. Although entries 2 and 5 are from different laboratories, Seebach's group has reported results similar to those of entry 5: 99% conversion, 88% ds, and 94% es using 15 mol% catalyst at −5 °C [71].

NMR studies have shown that at least three hexacoordinate catalyst oxazolidinone complexes exist in solution [84]. The most abundant has been assigned a structure that has the oxazolidinone oxygens *trans* to the TADDOL oxygens and the chlorines *trans* to each other, as shown in Figure 6.7a. This species has also been characterized crystallographically [85]. There are four other possible complexes, two of which are illustrated in Figure 6.7b and c.[9] It is not known whether these two complexes are the ones that are observed in the NMR [84], but these two are judged to be more reactive, since in these structures, the enone oxygen is *trans* to the weaker π-donor ligand (chlorine) and may therefore experience a higher degree of Lewis acid activation. NMR studies show that one of the axial phenyls undergoes restricted rotation when bidentate ligands are bound to the titanium TADDOLate [84]. When the oxazolidinone ligand is oriented as shown in Figure 6.7b, the dienophile and the axial phenyl are in close proximity and approach of the diene from the direction of the viewer (toward the C_2 *Re* face) is unhindered, and would result in a cycloadduct with the observed absolute configuration [71]. The alternative geometry, shown in Figure 6.7c, is judged to be less reactive, since the diene must approach either from the direction of the viewer (toward the C_2 *Si* face), where it may encounter the nearby axial phenyl, or from the rear, where it is blocked by the equatorial phenyl [71].[10]

As with all speculations regarding transition state structure, it is important to recall that reality is often more complex than the relatively simple explanations typically put forth for understanding the outcomes of particular reactions. In this case, it is very likely that facile ligand exchange of alcohols and phenols on titanium results in numerous possible reactive species in reactions containing TADDOL or BINOL-type ligands [86]. Accordingly, caution should always be applied when considering explanations for the stereochemical outcome of a given reaction.

6.1.1.3 Chiral Catalysts: Organocatalysis

One of the most wide-reaching advances in the asymmetric Diels−Alder reaction in the twenty-first century has been the development of organocatalysts by MacMillan [87].[11] Like the metal-based Lewis acids in the previous section, organocatalysts accelerate the Diels−Alder reaction by lowering the LUMO of the dienophiles but they do so not by simple coordination to the carbonyl group but rather through a catalytic cycle that involves conversion of the dienophile's carbonyl group to an iminium ion that then undergoes cycloaddition reaction: hence "iminium ion catalysis" (see Scheme 6.8). Calculations suggest that the Diels−Alder transition state of an iminium-containing dienophile with cyclopentadiene is a whopping 11−13 kcal/mol lower in energy than the corresponding uncatalyzed process [92]. Once the cycloadduct is formed, *in situ* hydrolysis occurs to yield the aldehyde-containing product and to regenerate the catalyst. Iminium ion activation is an extremely portable concept that has been applied to numerous chemical reactions beyond cycloadditions, including conjugate addition chemistry.[11]

---

9. The other two have the oxazolidinone transposed such that the enone oxygen is *trans* to a TADDOL oxygen.
10. Jørgensen has proposed another rationale based on the geometry observed in the crystal structure [85].
11. Readers interested in organocatalysis are advised to begin with a dedicated issue of *Chemical Reviews* devoted to the topic, beginning with an overview by List [88], or one of several monographs that have appeared [89−91].

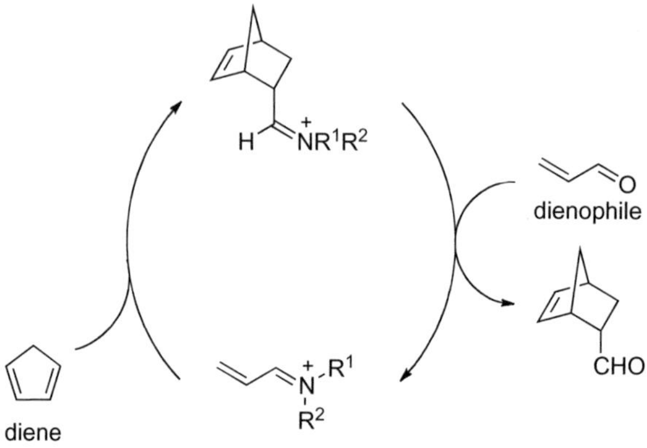

SCHEME 6.8 Catalytic cycle of an organocatalytic Diels—Alder reaction. The turnover of catalyst is facilitated by carrying out the reaction in the presence of water.

The development of iminium ion-forming catalysts for the Diels—Alder reaction is summarized in Table 6.4 for reactions of cyclopentadiene. Having initially examined proline[12] and several C_2-symmetrical catalysts (entries 1–3), MacMillan found that a chiral imidazolidinone gave excellent overall yields and enantioselectivities, albeit as a mixture of comparable amounts of exo and endo cycloadducts (entry 4). The tendency to afford exo isomers as major products is a general tendency of organocatalytic Diels—Alder reactions and one way in which they contrast with the thermal or Lewis acid-promoted versions discussed in Section 6.1.1.2. Another immediately evident contrast is the use of aldehyde dienophiles, which are sufficiently reactive to facilitate iminium ion formation whereas Lewis acid activation benefits from the presence of the more highly basic amides or imides typically used in those settings. The ready conversion of the aldehyde products to numerous downstream functional groups may be considered added value. Replacement of the C-2 dimethyl groups in the catalyst with a furan moiety resulted in highly enantioselective Diels—Alder reactions of simple ketones, with a concomitant switch to quite high endo selectivity as well (entries 7 and 8). Numerous researchers have investigated other secondary amines or close derivatives in similar reactions. For example, Hayashi and Maruoka and their respective co-workers developed exo-selective Diels—Alder of α,β-aldehydes using catalysts shown in entries 9 and 10.

---

12. MacMillan's paper on the organocatalyzed Diels—Alder reaction came out nearly simultaneously with the disclosure by List et al. that proline promoted efficient asymmetric aldol reactions [93].

TABLE 6.4 Comparison of Organocatalytic Diels−Alder Reactions

Entry	R_1	R_2	Catalyst (conditions)	% Yield	Ratio exo (% es)/ endo (% es)	Ref.
1	Ph	H	S-Pro-OMe · HCl (10 mol%) MeOH · H₂O, 23 °C	81	73 (74[a]):27 (ND)	[87]
2	Ph	H	MeO₂C‴⟨N,H⟩CO₂Me ·HCl (10 mol%) MeOH · H₂O, 23 °C	92	72 (78[a]):28 (ND)	[87]
3	Ph	H	MeO₂C CO₂Me / Bn N Bn / H ·HCl (10 mol%) MeOH · H₂O, 23 °C	82	78 (87[a]):22 (ND)	[87]
4	Ph	H	imidazolidinone (N-Me, 2,2-diMe, 5-Bn) ·HCl (10 mol%) MeOH · H₂O, 23 °C	99	57 (97):43 (ND)	[87]
5	i-Pr	H	" (5 mol%) MeOH · H₂O, 23 °C	81	50 (92):50 (97)	[87]
6	Me	Et	imidazolidinone (N-Me, 2,2-diMe, 5-Bn) ·HClO₄ (20 mol%) H₂O, 0 °C	20	13 (ND):88 (racemic)	[94]
7	Me	Me	imidazolidinone (N-Me, 2-(5-Me-furyl), 5-Bn) ·HClO₄ (20 mol%), H₂O, 0 °C	85	7 (ND):93 (81)	[94]

(Continued)

TABLE 6.4 (Continued)

Entry	R$_1$	R$_2$	Catalyst (conditions)	% Yield	Ratio exo (% es)/ endo (% es)	Ref.
8	Me	Et	''	89	4 (ND):96 (95)	[94]
9	Ph	H	Ar = 3,5-(F$_3$C)$_2$C$_6$H$_3$ (10 mol%) CF$_3$CO$_2$H (20 mol%), toluene, rt	100	85 (99):15 (94)	[95]
10	Ph	H	Ar = 4-t-BuC$_6$H$_4$ (12 mol%) pTsOH·H$_2$O (10 mol%) PhCF$_3$, −20 °C	80	93 (96):7 (96)[a]	[96]

[a]Major product formed in the enantiomeric series to that shown. ND = not determined.

These Diels−Alder reactions are of significant scope, leading to a variety of products as shown in Scheme 6.9 (the bonds formed are indicated in each product by red hashes). Structures accessible through this chemistry include fairly simple cyclohexenes, bicyclic structures, and in some cases products containing fully-substituted carbon atoms [97,98]. Appealingly complex structures can also be obtained through intramolecular variants of the process [99] or using a vinylindole diene [100] as shown in Scheme 6.10.

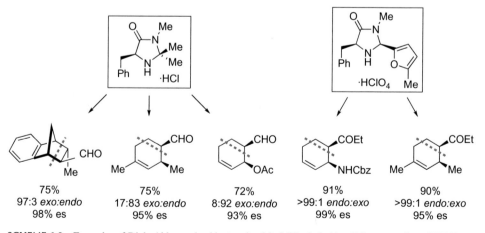

75%
97:3 exo:endo
98% es

75%
17:83 exo:endo
95% es

72%
8:92 exo:endo
93% es

91%
>99:1 endo:exo
99% es

90%
>99:1 endo:exo
95% es

SCHEME 6.9 Examples of Diels-Alder cycloadducts using MacMillan's imidazolidinone catalysts [97,98].

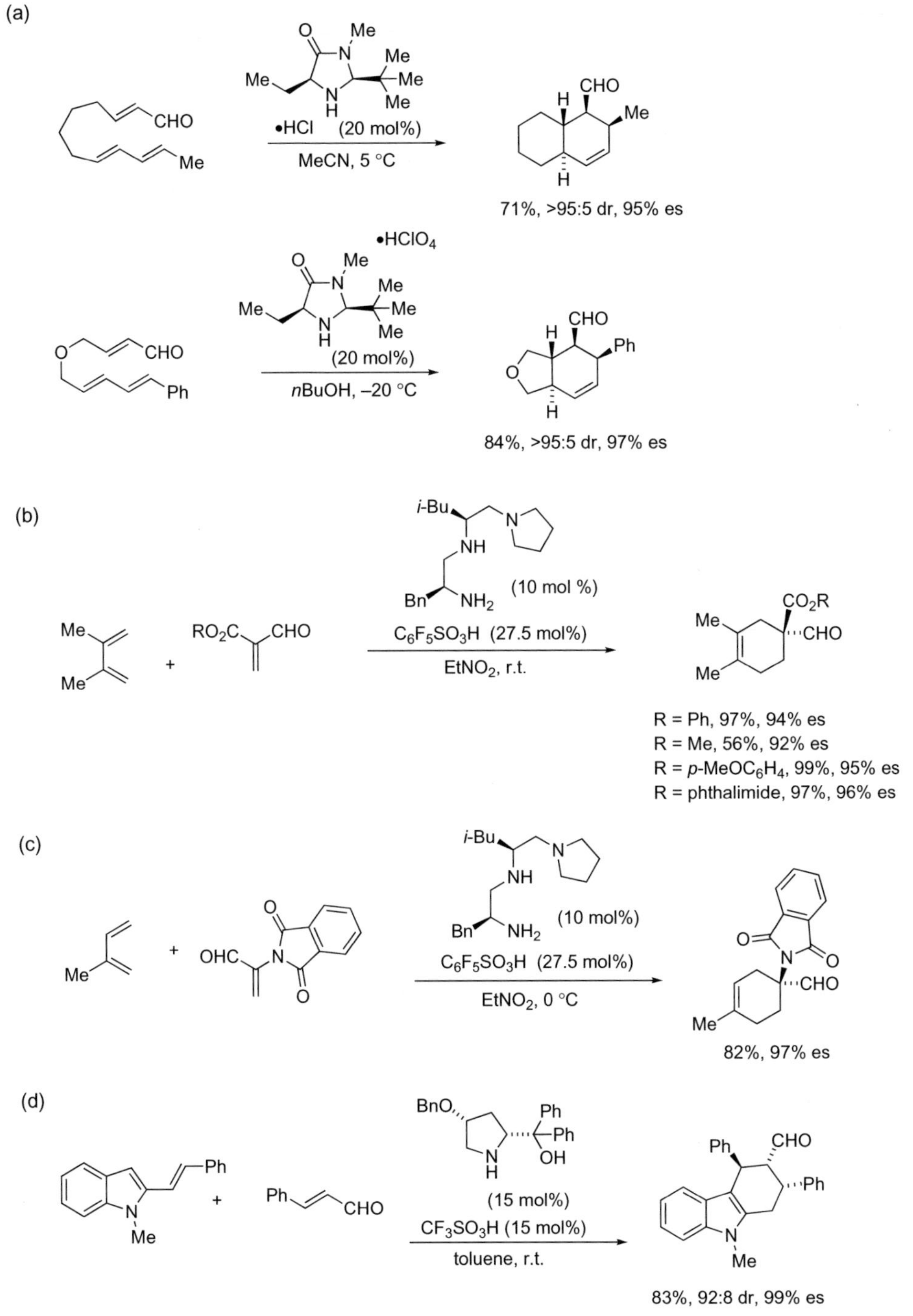
(a)
•HCl (20 mol%)
MeCN, 5 °C
71%, >95:5 dr, 95% es
•HClO4
(20 mol%)
nBuOH, –20 °C
84%, >95:5 dr, 97% es
(b)
(10 mol %)
C6F5SO3H (27.5 mol%)
EtNO2, r.t.
R = Ph, 97%, 94% es
R = Me, 56%, 92% es
R = p-MeOC6H4, 99%, 95% es
R = phthalimide, 97%, 96% es
(c)
(10 mol%)
C6F5SO3H (27.5 mol%)
EtNO2, 0 °C
82%, 97% es
(d)
(15 mol%)
CF3SO3H (15 mol%)
toluene, r.t.
83%, 92:8 dr, 99% es

SCHEME 6.10 Additional examples of iminium ion catalyzed Diels–Alder reaction.

MacMillan has proposed that the reactions occur through the conformations depicted in Scheme 6.11. For adducts formed from the condensation of an aldehyde with the first-generation (2,2-dimethylimidazolidinone) catalyst, the cycloadducts were proposed to result from approach to the *Si* face (rear face, as drawn in Scheme 6.11a), away from the bulky benzyl group at C-5. Here, the geminal dimethyl substitution presumably leads to the exclusive formation of the iminium ion having *E* geometry as shown. The situation is more complex for the case of ketones since both substituents on the erstwhile carbonyl carbon have significant steric bulk. This was evidenced by the observation that carrying out the Diels–Alder reaction of pent-3-ene-2-one in the presence of the original 2,2-dimethyl catalyst, so useful in the aldehyde cases, gave racemic product (Table 6.4, entry 6 [94]). Houk later ascribed this result to a failure to form the iminium ion and only back-ground, uncatalyzed reaction [92]. To accommodate this, the catalyst was re-designed to open up more room; screening of various candidates led to the effective furan-containing version shown in Table 6.4, entries 7 and 8. In addition, it became necessary to consider transition states arising from each iminium ion geometry. For all of the ketone reactions, the major product arises from the favored Z-iminium ion, which reacts preferentially from the *Si* face, as observed for aldehydes. In the special case of iminium species arising from methyl ketones (i.e., R = H in Scheme 6.11b), however, poorer enantioselectivities were proposed to arise from competitive reaction *via* the Re face of the *E*-iminium ion. For ketones having a larger alkyl substituent, i.e., those in which R = Me or greater, the *Re* face of the *E*-iminium ether is now blocked and superior selectivities result (cf. entries 7 and 8 in Table 6.4). Interestingly, the results with α,β-unsaturated ketones led to product

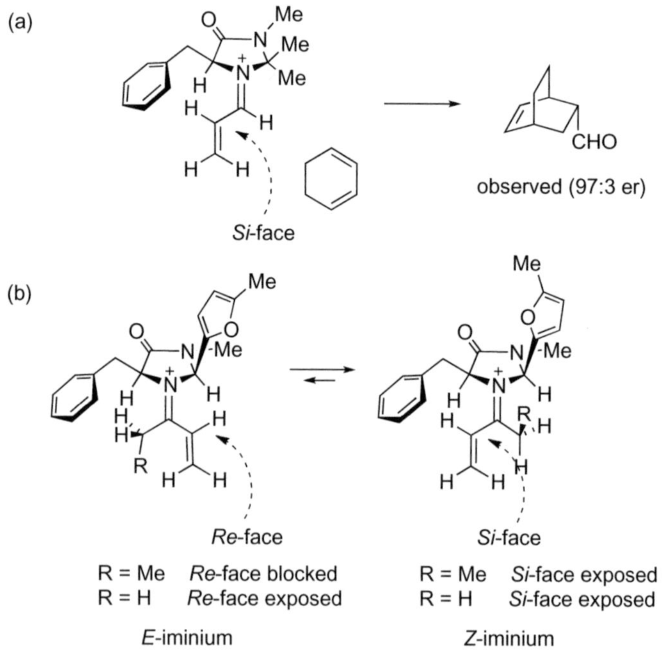

SCHEME 6.11 MacMillan's models for the observed stereoselectivity of iminium-catalyzed Diels–Alder reactions of (a) aldehydes and (b) ketones. These conformations were the results of calculations, which coincidentally placed the furan rings in the different conformations shown [94].

distributions heavily tilted toward the *endo* isomers, again in contrast to the aldehyde reactions. Overall, these initial hypotheses were largely supported by a detailed theoretical study published by Houk in 2006, including the proposed intermediate shown in Scheme 6.11a and the combination of high enantioselectivities and poor *exo/endo* ratios obtained in this case (both verified to arise from concerted, asynchronous pathways [92]). A Z-iminium conformation very much like the one depicted in Scheme 6.11b was in fact found as the lowest energy conformation, but perhaps surprisingly, the corresponding *E* isomer was only slightly destabilized in comparison (0.2 kcal/mol); a number of related conformers were also found. The high *endo* selectivity and enantioselectivity was ascribed to preferential reaction through the Z-iminium species.

6.1.1.4 Chiral Catalysts: Hydrogen-Bonding Activation and Brønsted Acid Catalysis

A more recently developed class of organocatalysts promote the Diels−Alder reaction in much the same way as do Lewis acids: by binding to a carbonyl on the dienophile and lowering its LUMO. Since the early 2000s, these concepts have taken an important place in the chemist's toolbox (for general overviews of hydrogen-bonding and Brønsted acid catalysts, see refs. [101,102]). They differ from simple proton activation by the degree of organization and therefore the selectivity imparted to the activated reactant complexes.

Two important early clues that proton-based catalysis would be useful in the Diels−Alder reaction are found in work by Wassermann [103] and Kelly [104], who showed that cycloadditions could be accelerated by the addition of strong carboxylic acids or phenols, respectively. In 2004, Rawal showed that TADDOL derivatives effectively promoted the reaction between methacrolein and the electron-rich diene shown in Scheme 6.12a [105]. Scheme 6.12b shows the proposed activated complex between the reactive aldehyde and the chiral diol, in which one-point binding places acrolein in an environment where one of its enantiotopic faces is shielded by an aryl group on the catalyst (*cf.* the TADDOL/Ti system in Figure 6.5i). Soon thereafter, Yamamoto published a highly selective reaction of a silyloxydiene with an α,β-unsaturated ketone using a strong organic acid [106]. In this case, the NH proton on the phosphoramide shown is acidified further by the attachment of an N-triflate group. As above, protonation of the dienophile accelerates the reaction and the binaphthyl moiety provides a chiral pocket to engender high selectivity in the reaction.

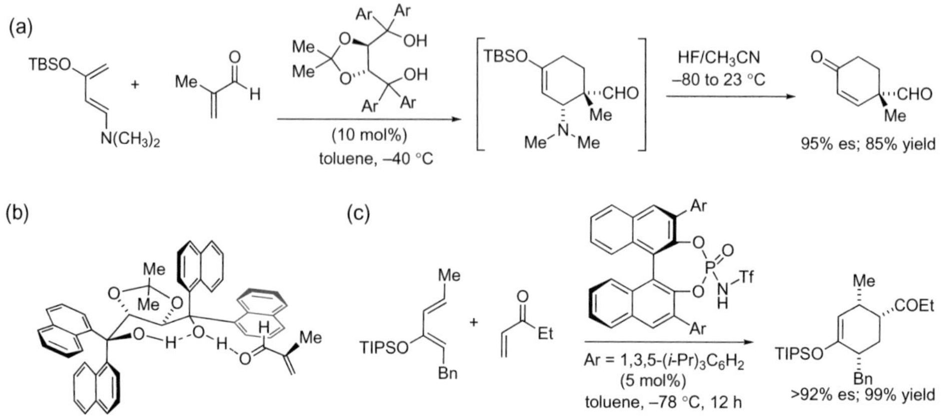

SCHEME 6.12 Diels—Alder reactions promoted by hydrogen bond catalysis reported by (a) Rawal (pre-transition state complex for (a) shown in (b), ref. [105]), and (c) Yamamoto [106].

6.1.1.5 Total Synthesis Using the Asymmetric Diels—Alder Reaction

The Diels—Alder reaction has played a starring role in stereocontrolled total synthesis since the field has existed.[13] We will cite only a few cases out of the hundreds of published examples to exemplify the variety of ways natural products chemists have used this versatile reaction. Scheme 6.13 shows examples where chiral auxiliaries (examples (a) and (b)) and metal-based catalysts ((c)–(g)) were used to control relative and absolute configuration in their work. In each case, the new stereocenters resulting from the Diels—Alder reaction are indicated with an asterisk. In comparing the asterisks in the cycloadducts with where they eventually end up in the final natural product target, the reader will note that not all are self-evident. In fact, three of these examples involve the conversion of a carbon—carbon bond to a carbon—heteroatom bond *via* a rearrangement or insertion reaction (which typically go with retention of configuration, so the Diels—Alder step still gets credit for installing the center in question). Overall, the power of the Diels—Alder is such that it is pressed into service even in non-obvious cases such as these even if doing so requires additional steps to get from adduct to the target.

---

13. Some chemists consider the first stereocontrolled synthesis of a natural product to be Stork's 1953 route to cantharidin, which involved a Diels—Alder reaction of dimethyl acetylenedicarboxylate with furan [107].

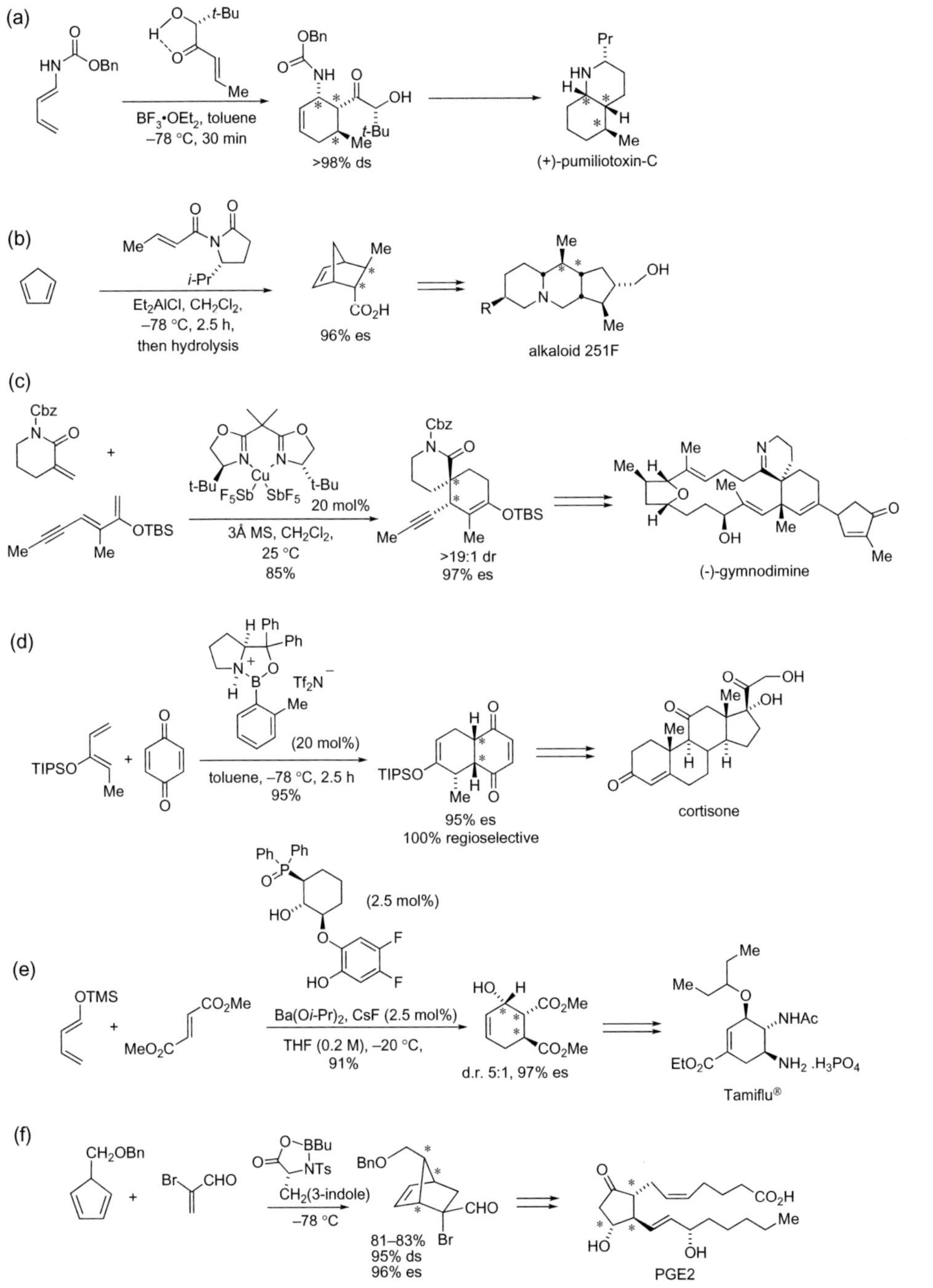

(a)

(+)-pumiliotoxin-C

(b)

alkaloid 251F

(c)

(-)-gymnodimine

(d)

cortisone

(e)

Tamiflu®

(f)

PGE2

SCHEME 6.13 Use of asymmetric Diels−Alder reactions in total synthesis efforts, part 1. Chiral auxiliary approaches by (a) Masamune [108] and (b) Aubé [109]. Metallocatalytic approaches by (c) Romo [110], (d) Corey [111], (e) Yamamoto [112], and (f) Corey [80].

Additional examples, this time employing organocatalysts, are shown in Scheme 6.14. Of these, the most straightforward is the intramolecular approach to solanopyrone D, shown in Scheme 6.14a [99]. The Diels–Alder in Scheme 6.14b is followed by an acid-promoted cyclization sequence that forms an additional ring and a separate step to reduce the aldehyde to the terminal alcohol in the product [113]. The synthesis of tubifolidine entailed a highly successful asymmetric Diels–Alder of maleimide made possible by the strong electronic bias of the diene and strong face selectivity imparted by the thiourea-based catalyst (Scheme 6.14c).

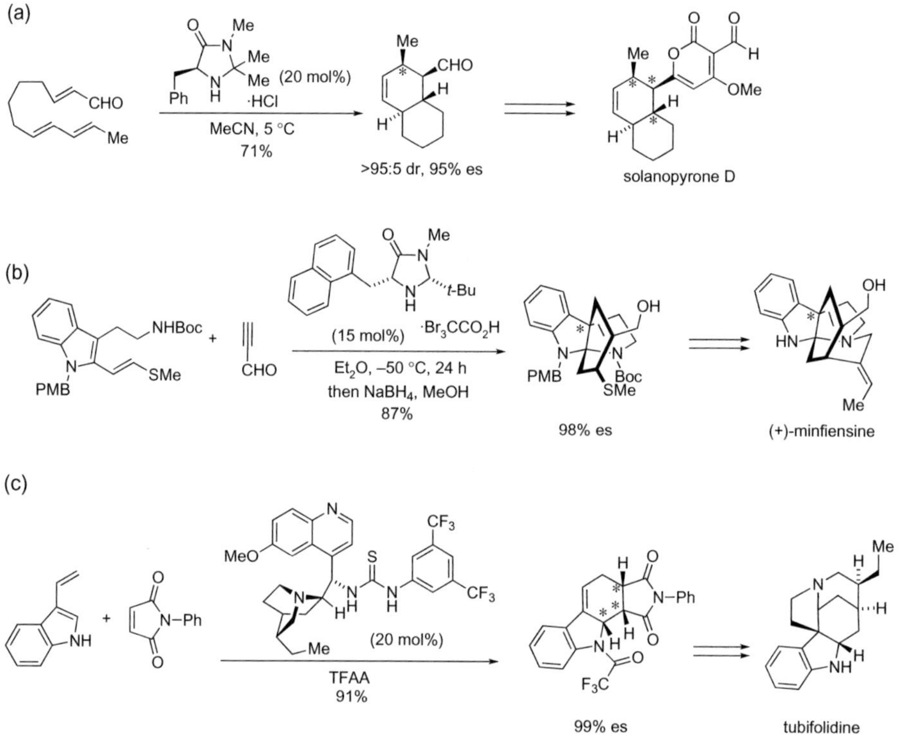

SCHEME 6.14 Asymmetric Diels–Alder reactions in total synthesis, part 2. Organocatalysis routes by (a) and (b) MacMillan [99,113] and (c) Bernardi [114].

6.1.2 Hetero Diels–Alder Reaction

The term "hetero Diels–Alder" refers to any reaction in which one or more of the carbon atoms on a diene or dienophile is replaced by a heteroatom. This definition accommodates an enormous number of possible combinations, of which many have been realized due to the importance of heterocycles as reaction products and synthetic intermediates.[14]

---

14. For a monograph covering [4 +2] cycloadditions that form heterocycles, see ref. [115]. For reviews, see refs. [116–118].

Fortunately, the field can be managed once one appreciates that many of the ways of inducing asymmetry into the all-carbon Diels−Alder reaction have been applied to the heteroatom variants as well. The main differences that one sees between the two broad classes generally result from the more polarized nature of heteroatom-containing dienes or dienophiles, meaning that hetero Diels−Alder reactions tend toward more asynchronous, concerted or even stepwise mechanisms (recall the liberal definition of a cycloaddition put forth at the beginning of this chapter). We will first focus on the reactions of heteroatom-containing dienophiles (aldehydes and imine derivatives) and then consider heterodienes. In both cases, we will present only a few instructive or historically important cases and direct the reader to consult the excellent review literature on the subject for more information and examples [118−120].

It is straightforward to replace a typical Diels−Alder dienophile by an aldehyde or imine because both are intrinsically electron-deficient due to the presence of an electronegative atom. Moreover, as we have seen in previous chapters, both are easily activated by the addition of a Lewis acid, which also provides an obvious opportunity for asymmetric variants based on chiral catalysts. In general, the hetero Diels−Alder reaction of aldehydes requires acid promotion of some type and highly electron-rich dienes such as the silyloxy diene shown in Scheme 6.15a. This particular diene is commonly known as Danishefsky's diene and much of the early work in this area used it and related compounds [121]. Silyloxydienes remain popular because of their high reactivity, the combination of good stereochemical control and flexibility, and the overall utility of the products. Mechanistically, these hetero Diels−Alder reactions depend strongly on reaction conditions that range from a strict pericyclic pathway (*i.e.*, when promoted by milder lanthanide catalysts, which in some cases allow for the isolation of the initial cycloadducts) to a sequential Mukaiyama aldol/cyclization pathway [122]. Since the product of this hetero Diels−Alder reaction is a β-hydroxyketone (albeit one that is masked and cyclic), it has been applied in many of the same synthetic contexts as the aldol reaction. In this context, it is highly responsive to many of the same stereochemical determinants that inform traditional aldol approaches to these targets such as Cram-controlled *vs.* chelation-controlled additions (Sections 4.1 and 4.2) to chiral aldehydes (*e.g.*, see ref. [123]). However, early attempts to adapt the reaction for asymmetric induction using chiral auxiliaries met with limited success (Scheme 6.15b), likely due to rotational flexibility around the two C−O bonds separating auxiliary from the diene [124]. The best diastereomer ratios were obtained when the chiral lanthanide complex Eu(hfc)$_3$ was combined with the auxiliary-containing diene. An interesting feature of this paper was that the good (for its time) selectivity obtained through this combination was opposite to that observed using the auxiliary without a catalyst.[15]

---

15. This work is also noteworthy for being a very early (1983) example of using a lanthanide as a Lewis acid catalyst in organic synthesis.

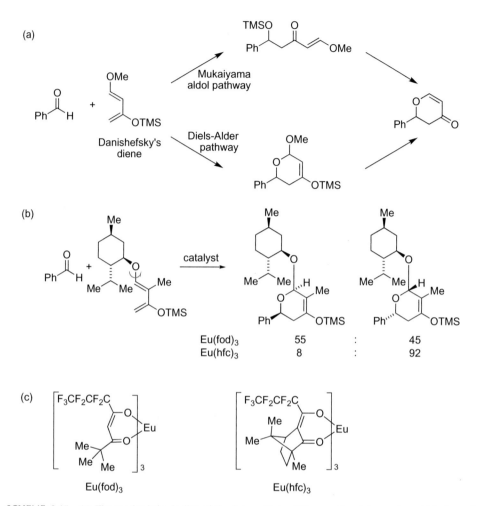

SCHEME 6.15 (a) The mechanistic duality of the hetero Diels–Alder reaction between benzaldehyde and Danishefsky's diene. (b) Use of menthol as a chiral auxiliary combined with (c) two lanthanide catalysts (Eu(fod)$_3$ is achiral and Eu(hfc)$_3$ is chiral, ref. [124]).

The first highly selective example of an asymmetric hetero Diels–Alder route using a chiral catalyst was published by Yamamoto in 1988 using the BINOL-derived aluminum Lewis acids shown in Figure 6.8a [125]. The excellent results shown in Table 6.5 indicate that this reagent was able to nicely differentiate the enantiotopic faces of benzaldehyde; additional bulk on the catalyst effectively improved the endo/exo ratio of the products (entries 1 and 2).[16] Jacobsen developed chromium-based catalysts like the salen-derived cationic species shown in Figure 6.8b (entry 3 [126]). Note that the Jacobsen catalysts appear to operate through a true cycloaddition reaction (surely highly asynchronous) and provide cycloadducts as the primary products. In the cases of silyl enol ethers, these initial adducts are generally converted to ketones *in situ* using trifluoroacetic acid. These important developments have been abetted by numerous other catalyst types able to promote highly selective additions to aldehydes; nearly every catalyst type useful in the all-carbon Diels−Alder reaction (Figure 6.5) has appeared in the heterocyclic variation as well [118−120]. Just a hint of the scope of this reaction can be found in the last entry of Table 6.5, where a coordinated copper catalyst is shown to facilitate the hetero Diels−Alder reaction of a nitroso dienophile [127].

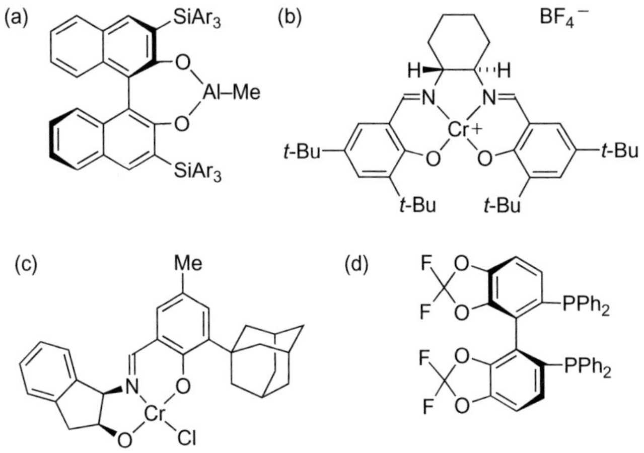

FIGURE 6.8 Representative Lewis acids used to promote hetero Diels−Alder reactions by (a) Yamamoto [125], (b) and (c) Jacobsen [126,128], and (d) Yamamoto again [127].

---

16. It is not obvious that hetero Diels−Alder reactions ought to give products corresponding to an endo transition state, considering that many aldehydes that participate in these reactions do not have groups capable of secondary orbital interactions. For those reactions viewed as true cycloadditions, "endo" products may arise from placement of the bulky Lewis acid *trans* to the alkyl or aryl aldehyde substituent and placement of the typically large coordinated acid in an orientation *anti* to the diene, as illustrated below.

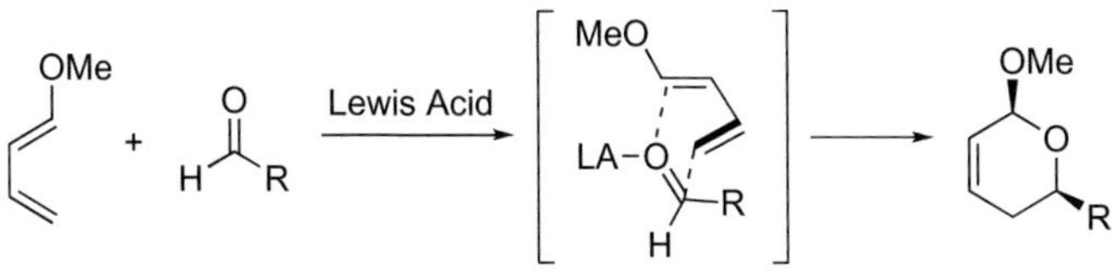

TABLE 6.5 Hetero Diels−Alder Reactions of Aldehydes

Entry	Diene	Aldehyde	Product	Catalyst,[a] Conditions	% Yield	% es of major isomer	Ref.
1	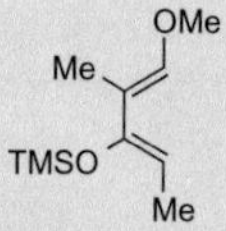 Me, OMe, TMSO, Me	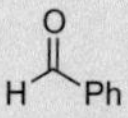 O H Ph	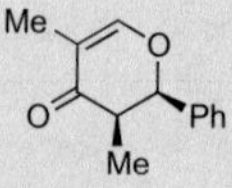 Me O O Ph Me	a (R = Ph, 10 mol%), toluene, −20 °C	77 (+7% trans isomer)	97	[125]
2	"	"	"	a (R = 3,5-xylyl, 10 mol%), toluene, −20 °C	93 (+3% trans isomer)	98	[125]
3	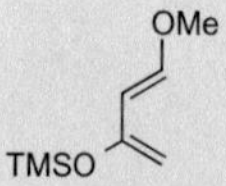 OMe, TMSO	"	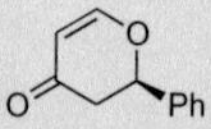 O O Ph	b (3 mol%), −30 °C, t-BuOMe, 4Å sieves	98	82	[126]
4	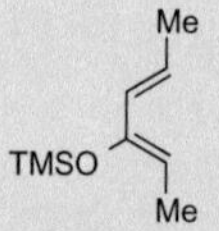 Me, TMSO, Me	"	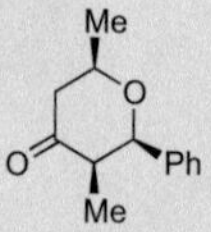 Me O O Ph Me	c (3 mol%), 4Å sieves, rt, 16−40 h, then TBAF, AcOH, MeOH	50	90	[128]
5	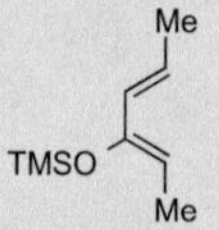 Me, TMSO, Me	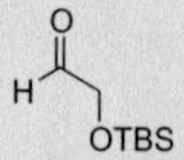 O H OTBS	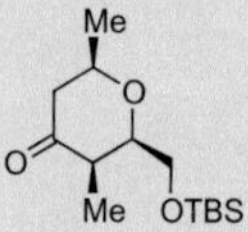 Me O Me OTBS	"	88	99	[128]
6	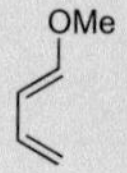 OMe	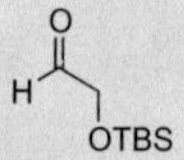 O H OTBS	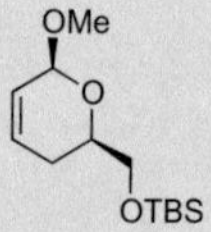 OMe O OTBS	"	91	>99	[128]
7	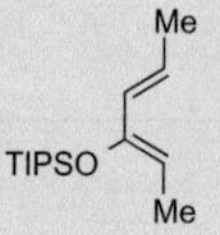 Me, TIPSO, Me	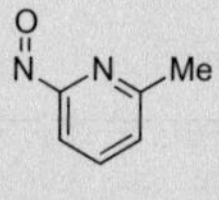 O=N N Me (N-aryl)	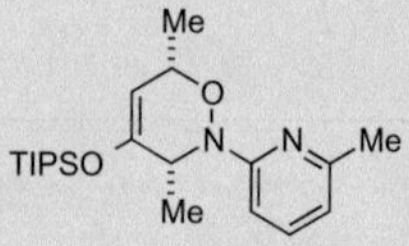 Me O N N Me TIPSO Me	d (10.5 mol%), Cu[(MeCN)$_4$]PF$_6$ (10 mol%), CH$_2$Cl$_2$, −85 to −20 °C	99	>99	[127]

[a]The catalyst structures are shown in Figure 6.8.

A major advance was the discovery by Rawal that some hetero Diels−Alder reactions could be efficiently catalyzed by TADDOL derivatives [129,130]. This contribution was significant in part because it introduced a valuable and reactive diene in the form of the dimethylamino-substituted compound shown in Scheme 6.16a [131]; these compounds even undergo smooth hetero Diels−Alder reactions with unactivated ketones [132]. More important was the discovery that TADDOLs are effective hetero Diels−Alder catalysts even without added metals, thus

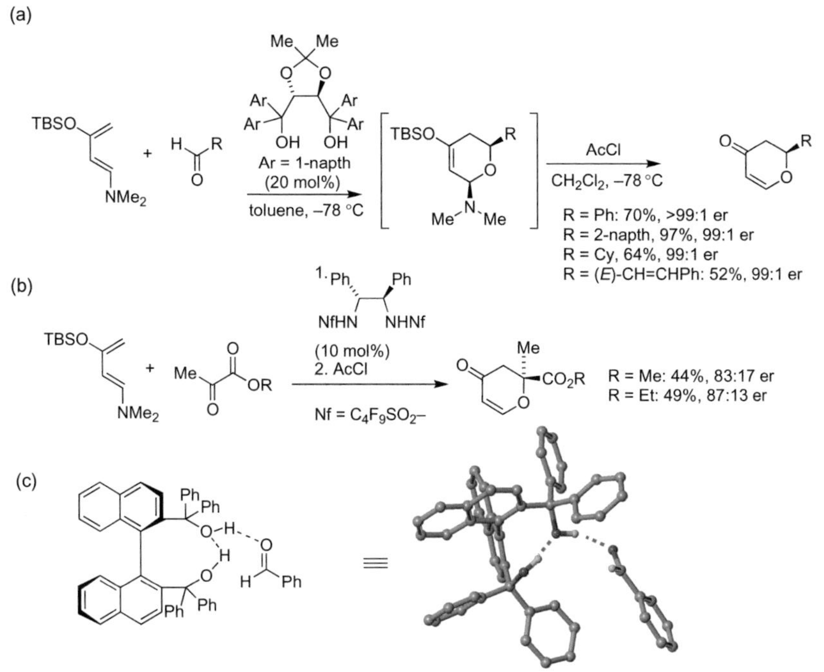

SCHEME 6.16 Hydrogen bond catalysis of the hetero Diels–Alder reaction as reported by (a) Rawal [129,130] and (b) Jørgensen [134]. (c) Ball and stick depiction of the X-ray structure of a complex containing a TADDOLesque diol and benzaldehyde [130].

opening the door for the field of hydrogen bond catalysis, which has subsequently undergone explosive development [129,130]. Interestingly, this discovery was based in part on the observation that Rawal's hetero Diels–Alder reactions were accelerated in chloroform (which is mildly acidic) [133]. Some examples demonstrating the concept are provided in Scheme 6.16; the concept is also portable to the all-carbon Diels–Alder reaction (cf. Scheme 6.12). Indeed, the concepts of hydrogen-bonding catalysis and its close cousin Brønsted acid catalysis are now mainstays of asymmetric synthesis in many areas. Jørgensen applied this concept to a reaction using an α-keto ester as the dienophile (Scheme 6.16b [134]); while α-keto aldehydes have had a long history in the hetero Diels–Alder reaction, the use of ketones of any ilk in this chemistry is still rare. Mechanistically, hydrogen bond catalysis is explained by the complexation and diastereofacial events invoked for Lewis acid-mediated reactions, except that the activating group is a proton rather than a metal. Scheme 6.16c depicts the X-ray structure of an aldehyde–diol complex obtained by Rawal. This somewhat rare snapshot of a commonly proposed activation mode demonstrates both the single-point intermolecular hydrogen bonding of the aldehyde as well as partial immobilization of the catalyst through intramolecular hydrogen bonding [130].

The use of imines as dienophiles follows much the same pattern as described for aldehydes but with a few complicating wrinkles (reviewed in ref. [135]). For example, the increased basicity of imines in general and, especially, amine-containing cycloadducts may deactivate catalysts and accordingly require greater catalyst loadings. In addition, imines are not particularly reactive and are subject to side reactions such as tautomerization to an enamine or, in many cases, hydrolysis to aldehydes and amines. Moreover, transition state analysis can be complicated by the existence of *cis* and *trans* imine isomers (which are often interconvertible under common reaction conditions). Nonetheless, numerous advances have been registered in this area, with a few examples shown in Scheme 6.17. Oh reported a chiral auxiliary approach using the Danishefsky diene (Scheme 6.17a). One obvious opportunity afforded by using imines rather than aldehydes for the hetero Diels–Alder reaction is that the "extra" valency of nitrogen provides a convenient spot for attaching a chiral auxiliary sufficiently close to the bond-forming centers to result in effective diastereofacial selectivity. In this case, the authors invoked a chelation model when an appropriate coordinating group was present and a Felkin–Anh-inspired explanation when one was not (Scheme 6.17b [136]).

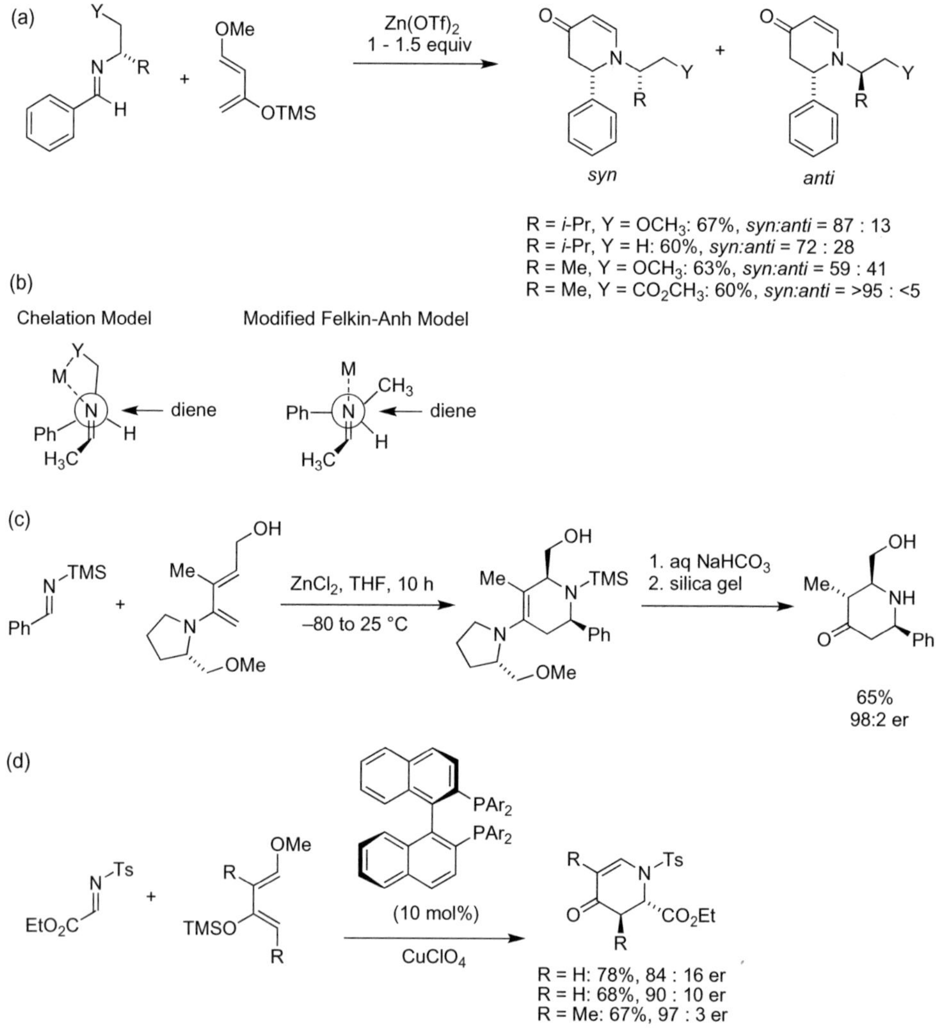
(a)
Zn(OTf)2
1 - 1.5 equiv
syn
anti
R = i-Pr, Y = OCH3: 67%, syn:anti = 87 : 13
R = i-Pr, Y = H: 60%, syn:anti = 72 : 28
R = Me, Y = OCH3: 63%, syn:anti = 59 : 41
R = Me, Y = CO2CH3: 60%, syn:anti = >95 : <5
(b)
Chelation Model
Modified Felkin-Anh Model
diene
diene
(c)
ZnCl2, THF, 10 h
–80 to 25 °C
1. aq NaHCO3
2. silica gel
65%
98:2 er
(d)
(10 mol%)
CuClO4
R = H: 78%, 84 : 16 er
R = H: 68%, 90 : 10 er
R = Me: 67%, 97 : 3 er

SCHEME 6.17 Representative imino Diels–Alder reactions using chiral auxiliaries on the (a) imine [136] or (c) diene [137]. (b) Transition states proposed for the reaction in (a). (d) An example of a chiral catalyst in this reaction [120].

High selectivity was possible but only when appropriate groups were present. Higher selectivity and more generality was obtained by Barluenga, who activated the diene component with an electron-rich, chiral enamine feature that was nearly contemporary with Rawal's initial work in the amino diene area described above (Scheme 6.17c; compare Scheme 6.16a [137]). This brief survey concludes with one example of the many applications of organocatalysis to nitrogenous hetero Diels–Alder chemistry, this time by Jørgensen, who has contributed much to this particular reaction and to organocatalysis in general (Scheme 6.17d [120]).

The versatility of the hetero Diels–Alder concept is enhanced when one considers reactions in which the diene itself incorporates heteroatoms. Because doing so automatically renders the diene electron-deficient, [4 +2] reactions using such substrates are considered to be *inverse electron demand Diels–Alder reactions*. As such, they typically require electron-rich dienophiles for smooth reaction; enol ethers and enamines have been particularly popular for this purpose and provide useful locations for chiral auxiliaries. The example shown in Scheme 6.18a uses a chiral enol ether derivative of menthol for the cycloaddition reaction,

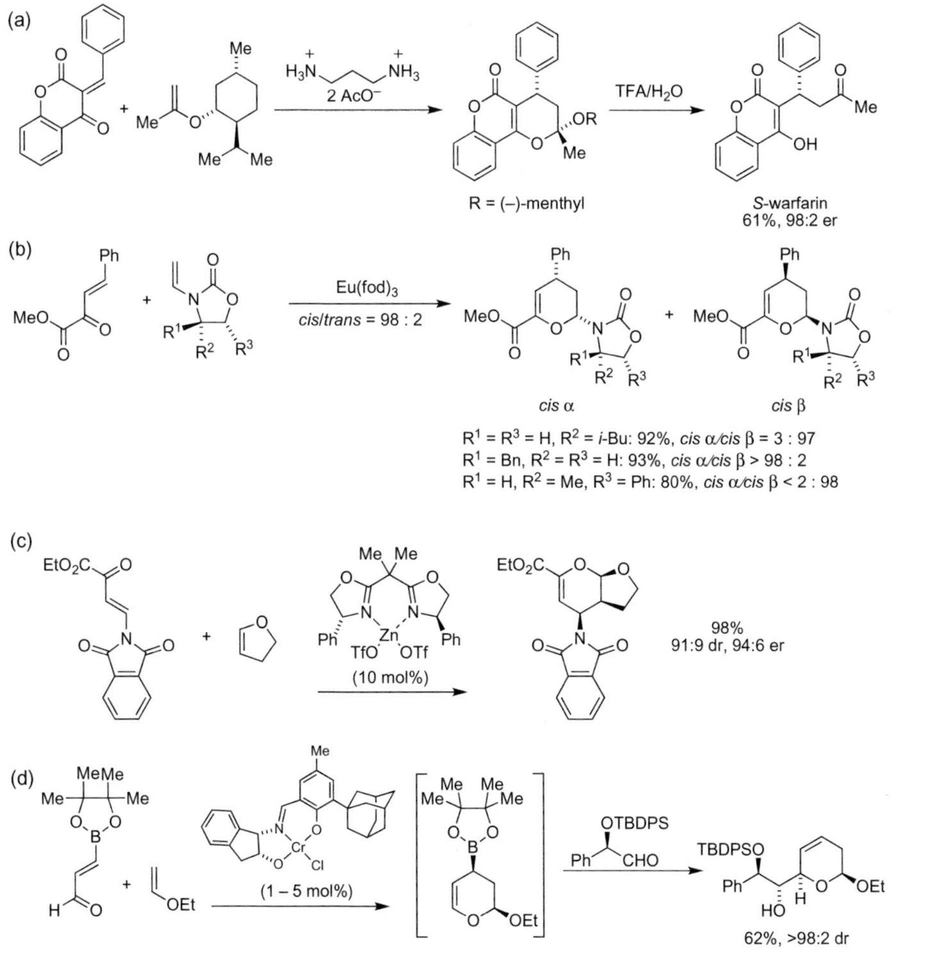

SCHEME 6.18 Examples of inverse demand hetero Diels–Alder reactions utilizing (a) and (b) chiral auxiliaries [138,139] or (c) and (d) metal catalysts [140,143].

which affords as the primary product a mixed ketal that can easily be hydrolyzed to afford, in this case, the *S* enantiomer of the anticoagulant agent warfarin [138]. A set of enamines containing the familiar oxazolidinone auxiliary demonstrate the versatility of this template in that substituents could be conveniently placed on either carbon of the auxiliary to afford high selectivities of the pyran products following auxiliary removal (Scheme 6.18b [139]). Similar substrates can also be subjected to metal catalysts like the Box system shown in Scheme 6.18c [140]. In each of the above cases, note that diene reactivity has been enhanced by additional electron-withdrawing groups (and in Scheme 6.18c, the nitrogen is deactivated by placing it in a phthalimide). Also, pyrans prepared using these reactions have more than passing resemblance to the pyranose moiety found in many sugars and in fact this type of hetero Diels−Alder reaction has found many applications in the synthesis of carbohydrate congeners [141,142]. An interesting diversion from these types of substrates arises from the use of a diene containing a boronic ester at one of the terminal carbons (Scheme 6.18d [143]). Cycloaddition of this compound effected by the Jacobsen catalyst (Figure 6.8c) affords an allylic species that can undergo *in situ* addition to the chiral aldehyde, rendering a nice multicomponent approach to several natural products of interest [143].

Organocatalysis of the inverse demand hetero Diels−Alder reaction is also known (Scheme 6.19a). In an early example, Jørgensen used an enamine generated *in situ* to carry out chemistry very much in analogy with the traditional approaches discussed above [144]. The reaction proceeds as shown in Scheme 6.19b. Scheme 6.19c provides an impressive demonstration of hydrogen bond catalysis for the synthesis of nitrogen heterocycles [145]. In this example of the *Povarov reaction* [146], the "diene" is partly embedded in an aromatic ring. Activation of the imine dienophile takes place to afford the reactive ensemble shown in brackets; the available evidence (including kinetics and isotope effect studies) suggests that the ensuing reaction is a highly asynchronous cycloaddition process followed by re-aromatization. The utility of these urea-derived catalysts derives from their ability to provide a highly organized transition state that effectively lowers the LUMO of the imine and hence its susceptibility to react with an electron-rich partner. In addition, the modularity of these catalysts facilitates tuning for a particular application.

6.1.3 1,3-Dipolar Cycloadditions

Along with the hetero Diels−Alder reaction, the major class of cycloadditions that leads to a heterocyclic product utilize 1,3-dipoles as the 4π component. This diverse family of reactions affords 5-membered rings containing anywhere from zero to five heteroatoms, many of which not only are important in their own right but also serve as useful synthetic intermediates en

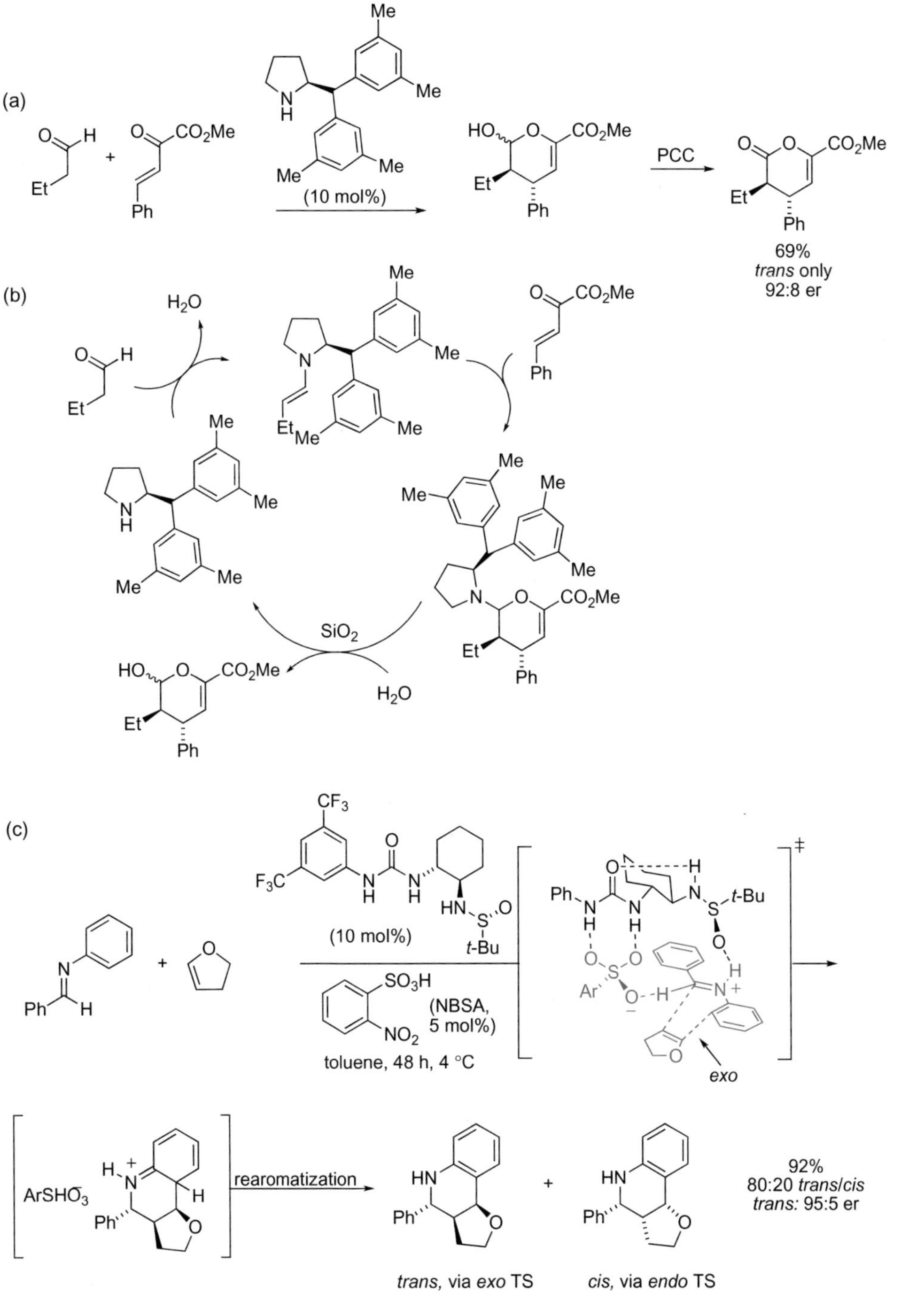

SCHEME 6.19 (a) An inverse demand hetero Diels–Alder reaction promoted by enamine catalysis and (b) its proposed mechanism [144]. (c) A Povarov reaction promoted by hydrogen-bonding catalysts along with a proposed transition state ensemble [145].

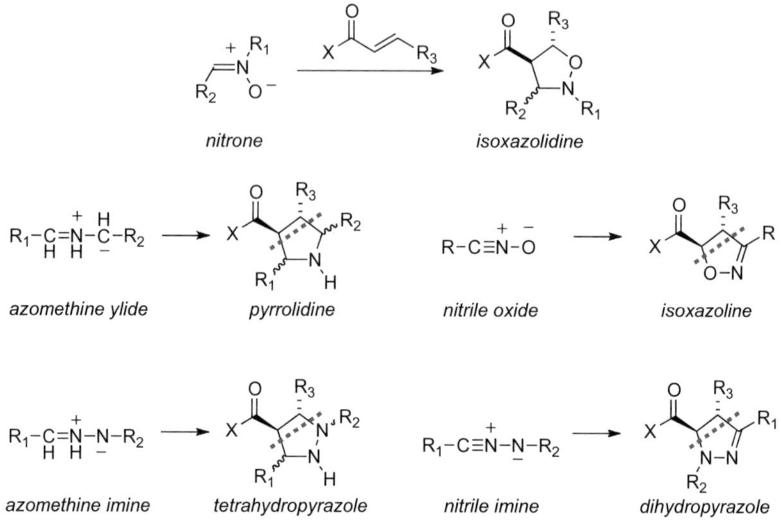

SCHEME 6.20   Representative 1,3-dipoles and their cycloaddition products with a generic dipolarophile.

route to other kinds of ring systems. Scheme 6.20 gives a selection of useful examples. Huisgen is credited with pioneering the development of this field in the 1960s [147], and it continues to evolve to the present day [148,149]. The development of asymmetric 1,3-dipolar cycloaddition chemistry has largely paralleled in form but lagged temporally to that of the asymmetric Diels−Alder reaction, beginning first with the use of chiral substrates and auxiliary-containing reactants, continuing through metal- and organocatalytic techniques. We focus here on nitrones as a commonly used representative class of 1,3-dipoles and provide a few more examples to give the reader a taste of the diversity of the field. A more expansive treatment of the topic can be found in several excellent review articles [150−153].

Despite the many similarities between the 1,3-dipolar cycloaddition reactions and the Diels−Alder, important differences also exist, beginning with fundamental issues of regiochemistry and relative configuration (*e.g.*, *exo vs. endo* transition structures). Although students of organic chemistry are taught that these parameters are best explained through molecular orbital theory, more often than not one can predict the regiochemical outcome of a Diels−Alder reaction using classical arrow-pushing reasoning. In contrast, the HOMO/ LUMO energies of many 1,3-dipoles relative to those of the alkene reactants are such that predicting regiochemistry in this way is much less reliable (the breakthrough papers on this matter were published by Houk in 1973 [154,155]). Similarly, energy differences leading to a preference for *endo vs. exo* transition structures may exist in many 1,3-dipoles but are relatively slight so that the stereochemical outcome is affected strongly by steric or other considerations, particularly in intramolecular cycloaddition reactions. Practically, this means that those wishing to use these reactions should seek out the closest possible literature precedents prior to applying them in a particular situation.

The basic concepts and a sense of the development of the field can be gleaned by examining the development of a single reaction, namely the 1,3-dipolar cycloaddition of nitrones with olefins. Along with the analogous chemistry of nitrile oxides, this has been one of the leading reactions in the field because the isoxazoline products that result are synthetic equivalents of 1,3-amino alcohols, which appear in many different contexts. Scheme 6.21a lays out some of the challenges facing early workers who utilized chiral auxiliary approaches: the high temperatures necessary for conversion combined with modest exo/endo transition

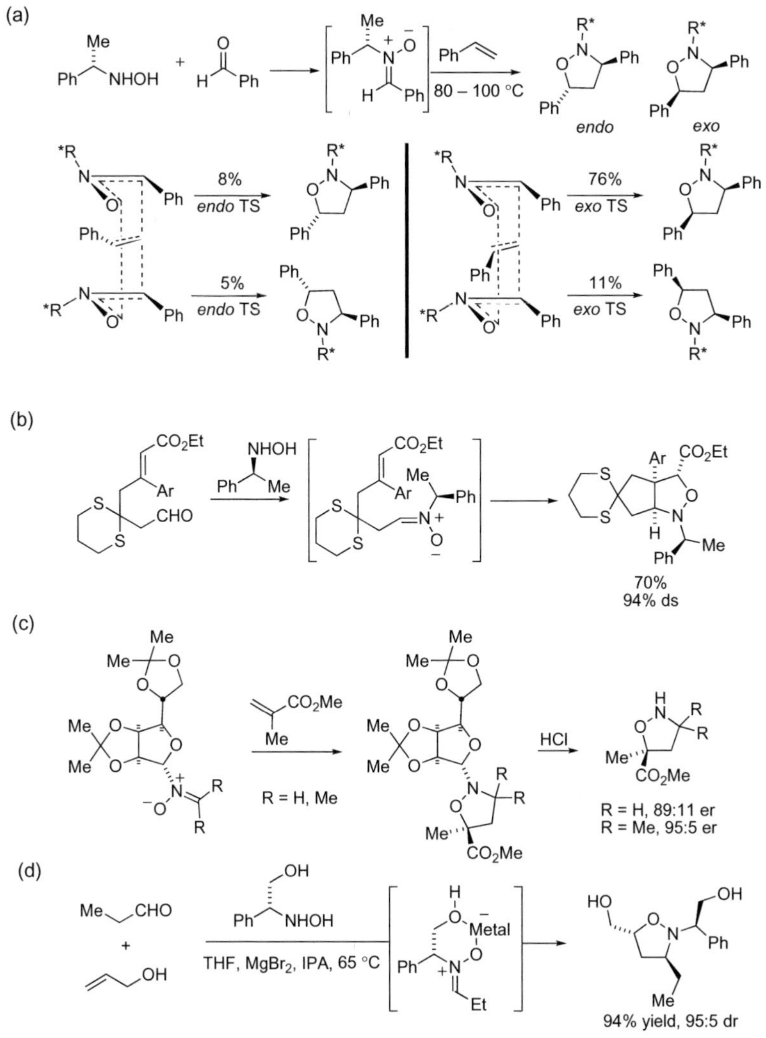

SCHEME 6.21   The development of the asymmetric cycloaddition of nitrones with alkenes: the chiral dipole/ auxiliary years. Approaches carried out (a)–(c) thermally [156,157,160] or (d) with a metal additive [162].

structure and diastereofacial preferences. In the reaction shown, which used an α-methylbenzyl group as the auxiliary, a mixture of all four possible products was observed [156]. However, better selectivity could be achieved using this auxiliary in an intramolecular version (Scheme 6.21b) [157]. The higher selectivity in this case is likely due to its intramolecular nature, which allowed it to occur at room temperature. Numerous investigators have used carbohydrate-derived nitrones, which often afford useful levels of stereoselectivity [158,159]. A single example is shown in Scheme 6.21c [160]. The virtue of using intramolecular chelation to limit the conformational mobility of the chiral-inducing group [161] can be seen in Scheme 6.21d, where the addition of a hydroxyl group onto the N-substituent resulted in very good overall selectivity in the cycloaddition event [162].

More recent approaches utilize the ability to render the dipolarophile faces diastereotopic by coordination with a catalyst. The bidentate coordination and activation of a chiral

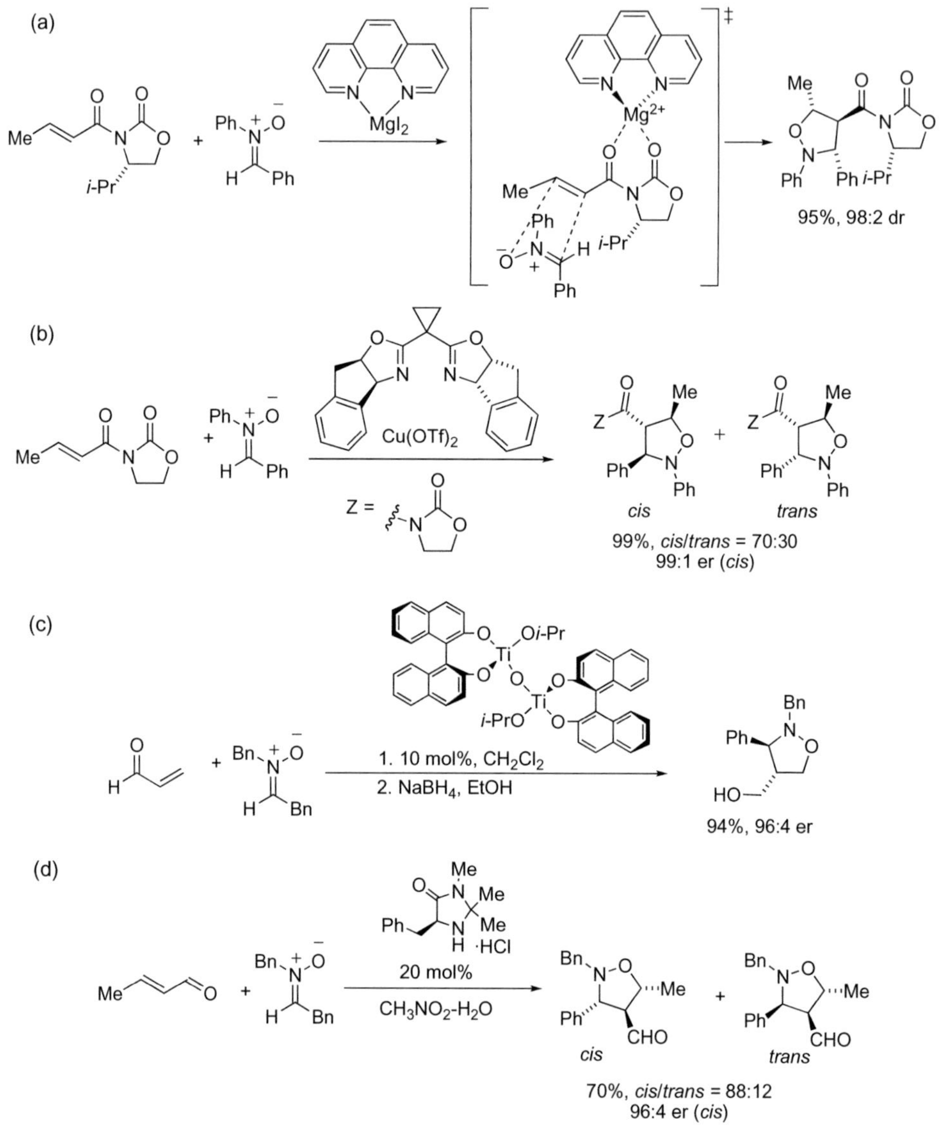

SCHEME 6.22 The development of the asymmetric cycloaddition of nitrones with alkenes: the chiral cataly-sis years. (a) [163,164]; (b) [165]; (c) [167]; (d) [168].

dipolarophile with magnesium maximized the potential inherent in a chiral auxiliary (Scheme 6.22a [163,164]) and directly led to the copper-based catalytic version shown in Scheme 6.22b [165]. Of course, this approach has also been spectacularly successful in the context of the Diels–Alder and numerous conjugate addition reactions (Sections 6.1.1 and 4.4.2), such that it would be reasonable to consider it a *privileged*[17] mode of asymmetric synthesis. Stronger Lewis

---

17. Although the phrase is subject to overuse, the concept of a "privileged catalyst" — a class of catalyst that has utility in the context of numerous different reactions — has undeniable taxonomical value [166].

acid activation of dipolarophiles is less common, possibly due to the existence of competing Lewis basic sites in the nitrone, but asymmetric examples are known (Scheme 6.22c [167]). Finally, orga-nocatalytic activation of α,β-unsaturated aldehydes also leads to effective asymmetric induction for 1,3-dipolar cycloaddition reactions in a manner analogous to its utility in other cycloadditions as shown in the early example reported by MacMillan, shown in Scheme 6.22d [168].

The reactions in Scheme 6.23 demonstrate two non-standard applications of nitrones: an example of the Kinugasa reaction with alkynes (Scheme 6.23a, asymmetric variations are reviewed in ref. [153]) and an unusual reaction with an activated cyclopropane [169].

SCHEME 6.23   (a) Kinugasa reaction of a nitrone and alkyne [153]; (b) Sibi's addition of a nitrone to an activated cyclopropane in a formal [3 +3] process (the absolute configuration of the product was not determined) [169].

Scheme 6.24 hints at the diversity of products and approaches possible from the larger class of 1,3-dipoles. Substituted pyrrolidines are readily produced through reactions of azo-methine ylides, represented here by an example from Williams, who used a chiral auxiliary-decorated version in an asymmetric route to spirotryprostatin A (Scheme 6.24a [170]). In this example, note that $A^{1,3}$ strain places the proximal phenyl group in an axial position, directing the dipolarophile to the opposite face. An attractive feature of azomethine ylides containing an electron-withdrawing group is that the dipole reactant can be generated in situ. A catalytic version is provided in Scheme 6.24b [171]. Adding another nitrogen into the dipole brings us to azomethine imines, which may be generated in situ from hydrazones or, as shown in Schemes 6.24c and d, acylhydrazones. Kobayashi utilized a zirconium-mediated intramolecular reaction, proposed to proceed in a concerted fashion [172]. In contrast, Leighton reported a chiral silicon-based Lewis acid to engender high selectivity in the reactions of the dipole with an enol ether (Scheme 6.24d [173]). In the latter case, a stepwise pathway was proposed. Finally, all-carbon 1,3-dipole equivalents are also known (Scheme 6.24e). An intriguing

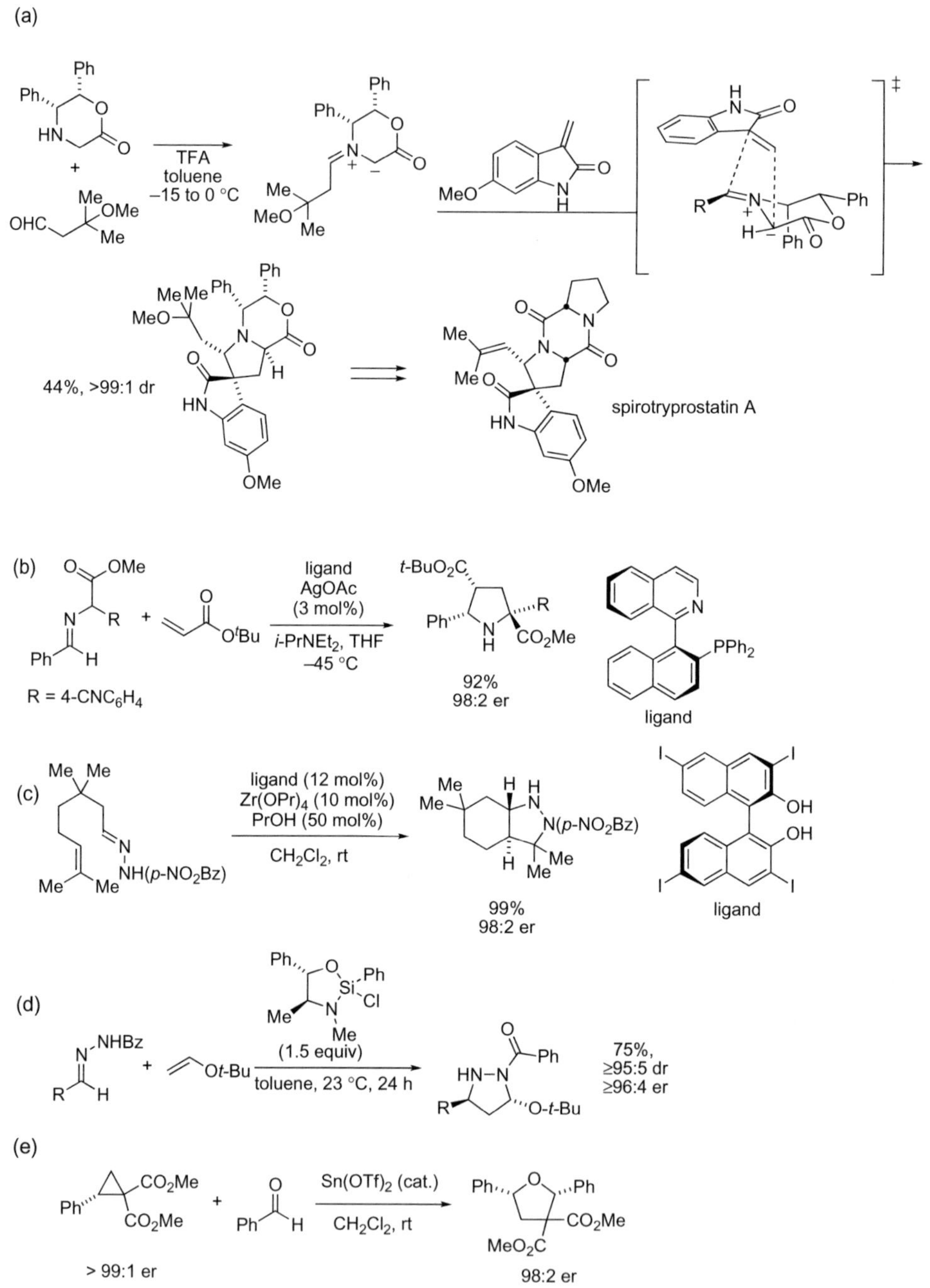
(a)
TFA
toluene
–15 to 0 °C
44%, >99:1 dr
spirotryprostatin A
(b)
R = 4-CNC6H4
ligand
AgOAc
(3 mol%)
i-PrNEt2, THF
–45 °C
92%
98:2 er
ligand
(c)
ligand (12 mol%)
Zr(OPr)4 (10 mol%)
PrOH (50 mol%)
CH2Cl2, rt
99%
98:2 er
ligand
(d)
(1.5 equiv)
toluene, 23 °C, 24 h
75%,
≥95:5 dr
≥96:4 er
(e)
Sn(OTf)2 (cat.)
CH2Cl2, rt
> 99:1 er
98:2 er

SCHEME 6.24 A taste of other chemistry involving 1,3-dipoles. Routes to pyrrolidines using (a) a chiral azomethine ylide [170], and (b) asymmetric catalysis [171]. Azomethine imine cycloadditions promoted by (c) Zr [172] or (d) Si-based [173] Lewis acids. (e) A formal reaction of an all-carbon 1,3-dipole of mechanistic interest [174].

example set forth by Johnson involves the reaction of a chiral cyclopropane with an aldehyde; the high degree of stereoselectivity observed here suggests that ring opening of the activated 3-membered ring and formation of the new bond are correlated events [174].

6.1.4 Summary

The concepts described in this chapter have been extended to nearly every imaginable type of cycloaddition process. Space precludes mentioning any more of them, but the reader should find the principles described herein to be portable to many contexts:

- The best cycloadditions are electronically matched between electron-rich and electron-deficient partners.
- Chiral auxiliaries can be used on either reaction component; the closer to the bond-forming site, the better. The best auxiliaries combine relatively little freedom of movement (often aided by metal chelation) with substantial steric differences between the diastereotopic faces.
- Lewis acid and hydrogen bond coordination to the more electron-deficient reaction partner can both accelerate reactions above nonselective background and provide the opportunity for the incorporation of chiral ligands.
- Carbonyl-containing compounds may be usefully subjected to either iminium ion or enamine catalysis.

6.1.5 [2 +1]-Cyclopropanations and Related Processes[18]

Although the addition of a carbene to a double bond to make a cyclopropane is well known, it is not particularly useful synthetically because of the tendency for extensive side reactions and lack of selectivity for thermally or photochemically generated carbenes. Similar processes involving carbenoids (species that are not free carbenes) are much more useful from the preparative standpoint [175,176]. For example, metal-catalyzed decomposition of diazoalkanes usually results in addition to double bonds without the interference of side reactions such as C−H insertions. Consider the possible retrosynthetic approaches to a 1,2-disubstituted cyclopropane shown in Figure 6.9. Disconnection *a* entails the addition of methylene across a double bond, a conversion that is often stereospecific (*e.g.*, the Simmons−Smith reaction [177]). Disconnections *b* and *c* are more problematic, since the issue of *cis/trans* product isomers (simple diastereoselection) arises.

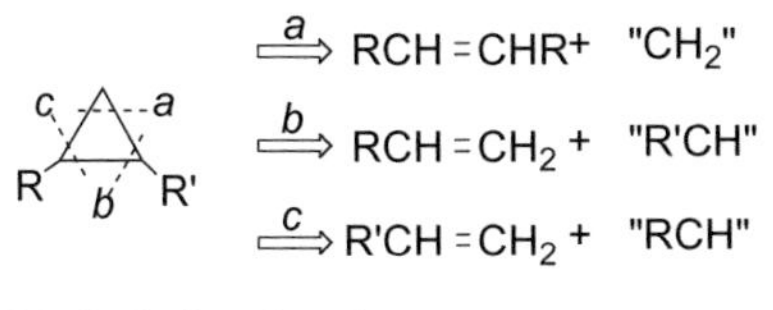

$$\overset{a}{\Longrightarrow} RCH = CHR^+ \quad "CH_2"$$

$$\overset{b}{\Longrightarrow} RCH = CH_2 + \quad "R'CH"$$

$$\overset{c}{\Longrightarrow} R'CH = CH_2 + \quad "RCH"$$

FIGURE 6.9 Retrosynthesis of 1,2-disubstituted cyclopropanes.

18. Not covered in this section are cyclopropanations that involve initial 1,3-dipolar cycloadditions of diazoalkanes to give pyrazolines, followed ring contraction and nitrogen extrusion.

Three main strategies have been taken to apply cyclopropanations to asymmetric synthesis: auxiliary-based methods whereby a covalently attached adjuvant renders either the olefin or the cyclopropanating reagent chiral, and processes that utilize a chiral ligand on a metal catalyst. Scheme 6.25 illustrates these approaches as applied to the more complex case of disconnections *b* and *c* of Figure 6.9. Scheme 6.25a and b show chiral auxiliaries (R*) in the olefin and carbenoid moieties, respectively, while Scheme 6.25c shows a chiral ligand on the metal. Since the transition states of both processes still involve the metal, asymmetric syntheses using these reactions may occur by intraligand or interligand asymmetric induction or a combination thereof. Finally, tandem 1,4-addition−intramolecular alkylation reactions do not involve carbenoids at all but accomplish a similar transformation, also by intraligand asymmetric induction (Scheme 6.25d).

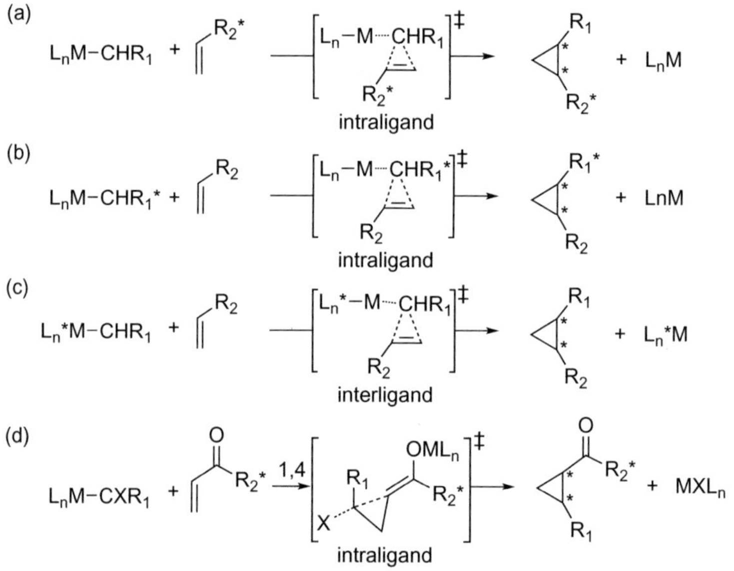

SCHEME 6.25   General strategies for asymmetric induction in cyclopropanations.

The issue of simple stereoselectivity in cyclopropanations of the types shown in Figure 6.9, disconnections *b* and *c*, is not a trivial one, and relatively few additions of ketocarbenoids (by far the most common type of carbenoid studied) show high selectivity. The difficulty can be seen by inspection of the transition states of Scheme 6.26. The transition state leading to the *trans* isomer (*lk* topicity) is usually favored, but unless the COZ group is quite large, the *trans*-selectivities are not great. For example, Doyle reported that if Z = OEt (*i.e.*, ethyl diazoacetate), the $Rh_2(OAc)_4$-catalyzed cyclopropanation of alkenes having R = *n*-alkyl, Ph, and *i*-Pr is only about 60−70% *trans*-selective. With R = *tert*-butyl, the selectivity is 81%. If the olefin is in a ring, the selectivity is not much better [178]. If hindered esters (Z = OCMe (*i*-Pr)$_2$) or amides (Z = N(*i*-Pr)$_2$) are used, the *trans*-selectivity for the $Rh_2(OAc)_4$-catalyzed cyclopropanation of styrene can be improved to 71% and 98%, respectively [179]. BHT esters (Z = *O*-2,6-di-*t*-Bu-4-Me-C$_6$H$_2$) also give good *trans*-selectivity (71−97%) with a variety of alkenes [180]. With $Rh_2(NHCOMe)_4$ as catalyst, these selectivities can be increased further

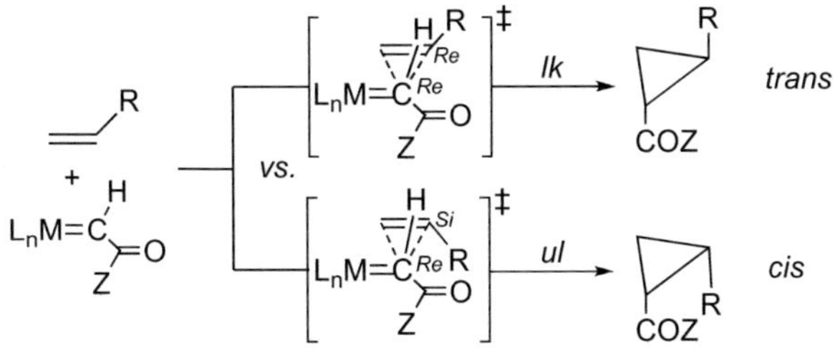

SCHEME 6.26 Transition states and relative topicities for cycloaddition of ketocarbenoids and monosubstituted alkenes.

due to the decreased reactivity of the rhodium carbenoid, which results in a more selective reaction [179,180].

Davies has found that vinyl carbenoids tend toward high selectivities in $Rh_2(OAc)_4$-catalyzed cycloadditions, as shown by the examples in Scheme 6.27 [181]. It is also important to note that the stereoselectivities of the cyclopropanations shown in Schemes 6.26 and 6.27 are not due only to steric effects. For example, changing R_1 in Scheme 6.27 from butyl to *tert*-butyl lowers the selectivity from 85% to 78% ($R_2 = CO_2Et$), while changing R_2 from phenyl to CO_2Et ($R_1 = Ph$), lowers the selectivity from >95% to 89% [181]. Presumably there is a contribution to the relative stabilities of the transition states by both electronic and steric effects, but these have not been quantified.

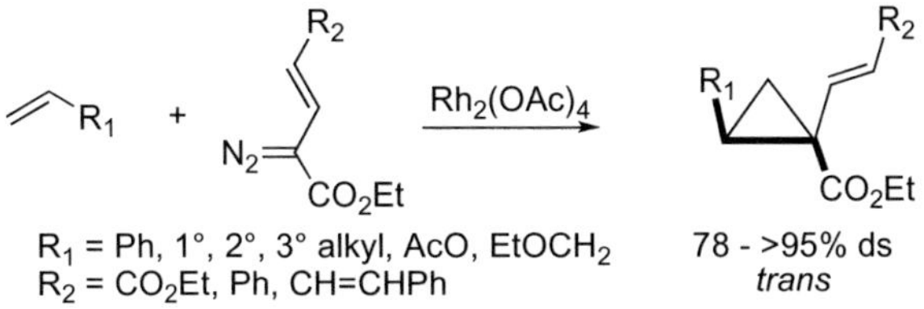

$R_1 = Ph, 1°, 2°, 3°$ alkyl, AcO, $EtOCH_2$ 78 - >95% ds
$R_2 = CO_2Et$, Ph, CH=CHPh *trans*

SCHEME 6.27 Diastereoselective cyclopropanations of vinyl carbenoids [181]. For disubstituted carbenes, *cis/trans* nomenclature is used to describe relative configuration, referring to R_1 relative to the carbonyl moiety, as shown in bold.

The following discussion is organized along the lines of the examples in Scheme 6.25.[19] First, auxiliary-based methods are discussed, followed by methods using chiral catalysts, including examples of double asymmetric induction employing chiral catalysts on chiral substrates and substrates having chiral auxiliaries attached, and finally stepwise cyclopropanation sequences. Within each section, the addition of "CH$_2$" is covered first (*i.e.*, disconnection *a* in Figure 6.9), followed by examples of the addition of "RCH" (*i.e.*, disconnections *b* and *c* of Figure 6.9).

6.1.5.1 Chiral Auxiliaries for Carbenoid Cyclopropanations

Cyclopropanations of functionalized alkenes using the Simmons–Smith reaction [177], or a similar cyclopropanation, have been developed by modifying carbonyl and hydroxyl groups with chiral auxiliaries. A single example was reported by Carrié in 1982 (Scheme 6.28a, [184]),

---

19. For reviews of asymmetric cyclopropanations, see refs. [182,183].

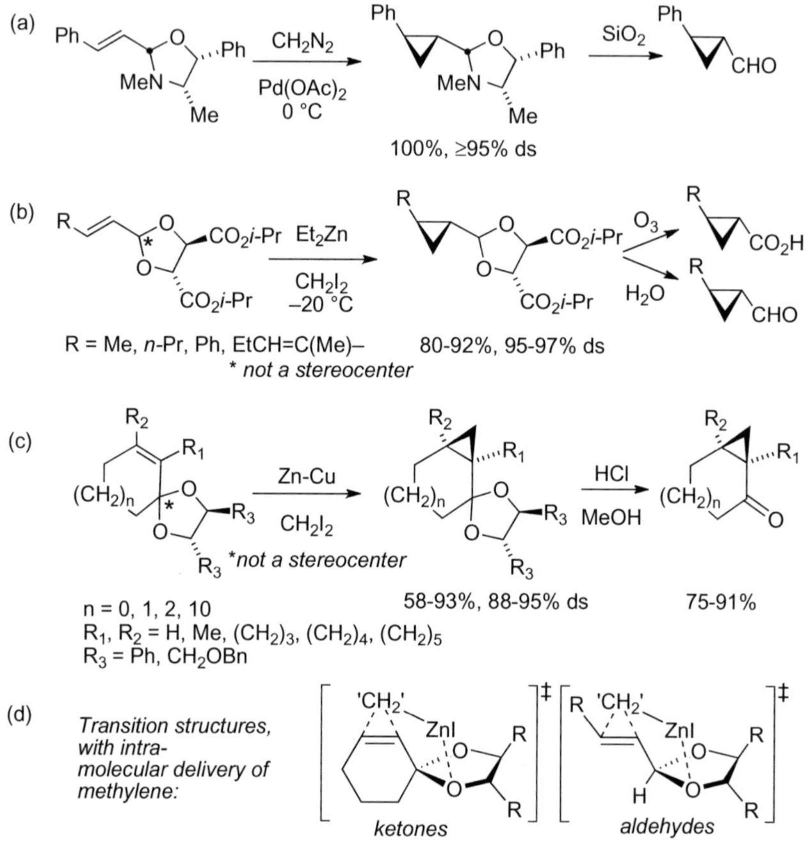

SCHEME 6.28 Auxiliary-based asymmetric cyclopropanations (addition of "CH₂") of α,β-unsaturated aldehydes and ketones. (a) [184]; (b) [185,186]; (c) [187–190]; (d) proposed transition structures [190]. Only one zinc and the transfer methylene are shown; other atoms associated with the Simmons–Smith reagent are deleted for clarity.

whereby the oxazolidine derived from condensation of (−)-ephedrine and cinnamaldehyde was cyclopropanated with diazomethane using palladium acetate as catalyst. The yield was high and the selectivity was ≥95%, but no further examples were provided. More systematic studies were undertaken by the groups of Yamamoto [185,186] and Mash [187–190]. Both of these groups used C_2-symmetric acetals as auxiliaries, as shown in Scheme 6.28b and c and showed that the acetal could be hydrolyzed in the normal manner to the corresponding carbonyl compound. In addition, Yamamoto also converted the acetal to a carboxylic acid using ozone. Note that in both the aldehyde and ketone acetals, the acetal carbon is not stereogenic due to the C_2-symmetry of the starting diol. For the ketone acetals, there is no conformational ambiguity, and the mechanistic rationale shown in Scheme 6.28d was proposed to account for the selectivity of the reaction [190]. Thus, coordination of the zinc to one of the diastereotopic oxygens and oriented *anti* to the adjacent dioxolane substituent placed the "transfer methylene" on the face of the olefin toward the viewer, consistent with the observed absolute configuration. Note that coordination to the other oxygen and orienting *anti* to the adjacent substituent would place the "transfer methylene" distal to the double bond. A related process for the asymmetric cyclopropanation of the enol ethers of cyclic and acyclic ketones has been developed by Tai [191,192].

Charette showed that allylic alcohols can be effectively cyclopropanated by attaching a chiral directing group in the form of a glucose derivative [193] or *trans*-1,2-cyclohexane diol [194], as shown in Scheme 6.29a. The topicity was rationalized by chelation of one of the zinc atoms of the Simmons–Smith reagent by the hydroxyl and the ether oxygen and intra-molecular delivery of the methylene to the olefin in the conformation shown. A drawback of this early method was that removal of these chiral-inducing groups required multistep, destructive procedures [193–195]. Aggarwal later extended this concept to chiral derivates of allylic amines (Scheme 6.29c [196]).

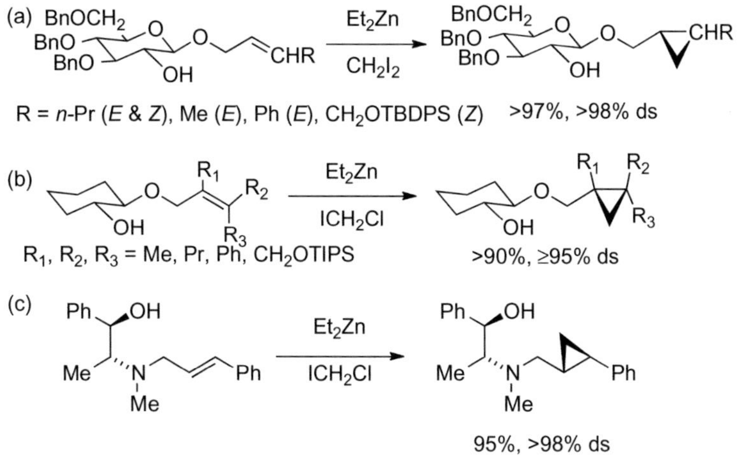

SCHEME 6.29 Asymmetric cyclopropanation of allylic alcohols using (a) a glucose-derived directing group [193] or (b) a cyclohexanediol auxiliary [194]. (c) Analogous reaction of an allylic amine [196].

Cyclopropanation reactions of diazoalkanes catalyzed by transition metals involve metal carbenes as intermediates. Scheme 6.30 illustrates the proposed catalytic cycle for such pro-cesses [197]. The catalyst, L_nM, is coordinatively unsaturated and therefore electrophilic. Loss of nitrogen from the zwitterion at the top affords the metal carbene shown at the right.

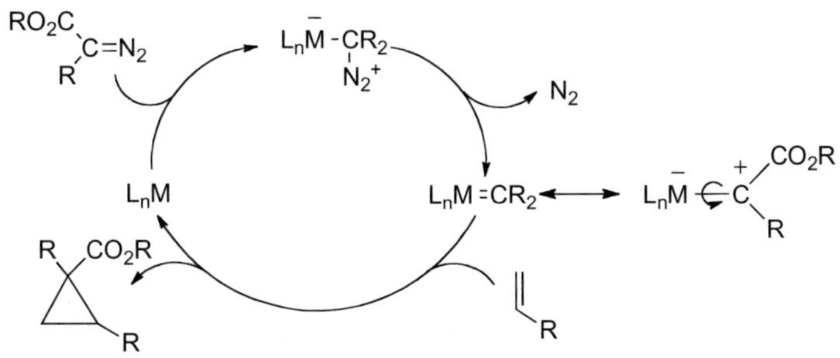

SCHEME 6.30 Catalytic cycle for the transition metal-catalyzed cyclopropanation of olefins by diazoalkanes (after [197] and [198]).

Two canonical forms for the metal carbenoid are shown. For rhodium carbenes, it is thought that they tend to resemble metal-stabilized carbocations, with a low barrier to rotation [197,198]. Also, recall (Scheme 6.26) that the simple diastereoselectivity (relative configuration) in these processes is not high unless very bulky esters are used. In light of this, it is not surprising that asymmetric cyclopropanations of styrene using bornyl, menthyl, and 2-phenyl-cyclohexyl esters of diazoacetic acid afforded both poor *cis/trans* selectivity and low enantioselectivity [199,200]. In contrast, vinyl carbenoids (Scheme 6.26) show good simple diastereoselection [181], and Davies has shown that pantolactone is an excellent chiral auxiliary, as shown in Scheme 6.31 [201–203]. The mechanistic hypothesis involves intramolecular interaction of the pantolactone carbonyl with the electrophilic carbenoid carbon, which shields the *Re* face of the carbene. Note that the conformer in which the carbene's *Si* face is shielded suffers severe steric interactions between the catalyst "wall" and the pantolactone moiety. Approach of the alkene toward the *Si* face of the carbene, coupled with diastereoselectivity favoring *lk* relative topicity, afforded a mixture containing only the two *trans* diastereomers. A weakness of this transition state model is that, although the absolute configuration is rationalized, it is not obvious why the *cis/trans* selectivity (*lk* topicity) should be 100%.

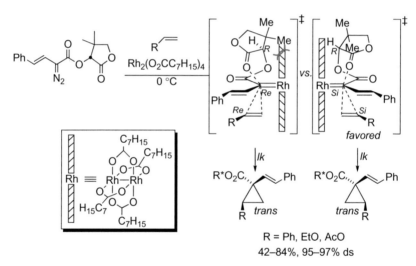

SCHEME 6.31   Diastereoselective cyclopropanation of olefins with vinyl carbenes [201]. Note that only two of the four possible stereoisomers were found in the product mixture. The *trans* nomenclature refers to the relative configuration of R and CO$_2$R*, consistent with that of Scheme 6.25.

The *cis* relationship between the vinyl group and the R group of the olefin presents an interesting opportunity: if the R group is also a vinyl substituent, the product of the cyclopropanation is a *cis*-divinylcyclopropane, precursor to a Cope rearrangement [204]. Although the Cope rearrangement destroys the stereocenters created in the cyclopropanation, it creates others, as shown by the examples in Scheme 6.32.

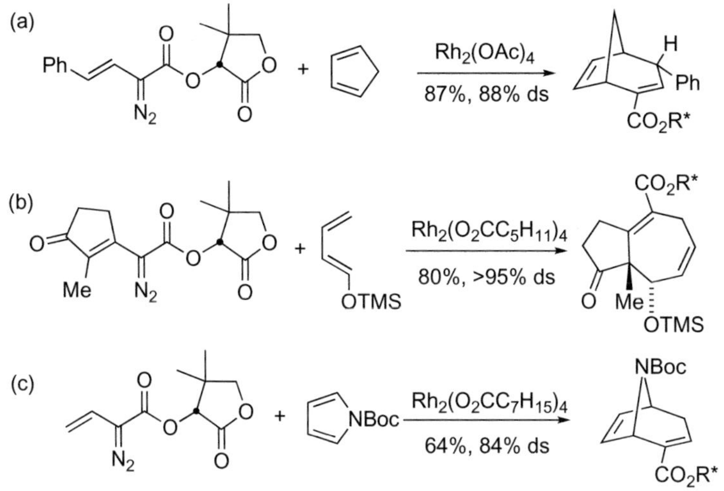

(a)

(b)

(c)

SCHEME 6.32 Synthetic applications of vinylcarbene cyclopropanations coupled with a Cope rearrangement. (a) and (b) [201]; (c) [203].

6.1.5.2 Chiral Catalysts for Carbenoid Cyclopropanations

Scheme 6.33 shows three catalysts for the enantioselective Simmons–Smith cyclopropanation of *trans*-cinnamyl alcohol. Charette's dioxaborolane (Scheme 6.33d, [205–207]) is both highly selective and high yielding, although this procedure is only suitable for small scale.[20] With other olefins, such as *cis* and *trans* disubstituted alkenes and β,β-trisubstituted alkenes, the yields are nearly as good and the enantioselectivities are 96–97%. An important finding in this study [205] was that, in addition to the Lewis acid (boron) that binds the alcohol, a second atom to chelate the zinc was also necessary. In the Charette catalyst (Scheme 6.33d), this atom was determined to be the amide carbonyl oxygen because when this group was eliminated, the reaction gave racemic product. In 2005, Charette reported a catalytic method that utilized a BINOL-equipped phosphoric acid to transfer the zinc-activated methylene unit to the substrate (Scheme 6.33e [208]). An analogous phosphoric acid utilized a TADDOL platform, but had generally lower selectivities [209]. Other chiral ligands used in Simmon–Smith approaches to cyclopropanes include dipeptides [210], disulfonamides [211], and complex chiral alcohols [212].

---

20. Charette has noted an explosion hazard on scale-up of the original procedure [206], and has published an alternative procedure [207].

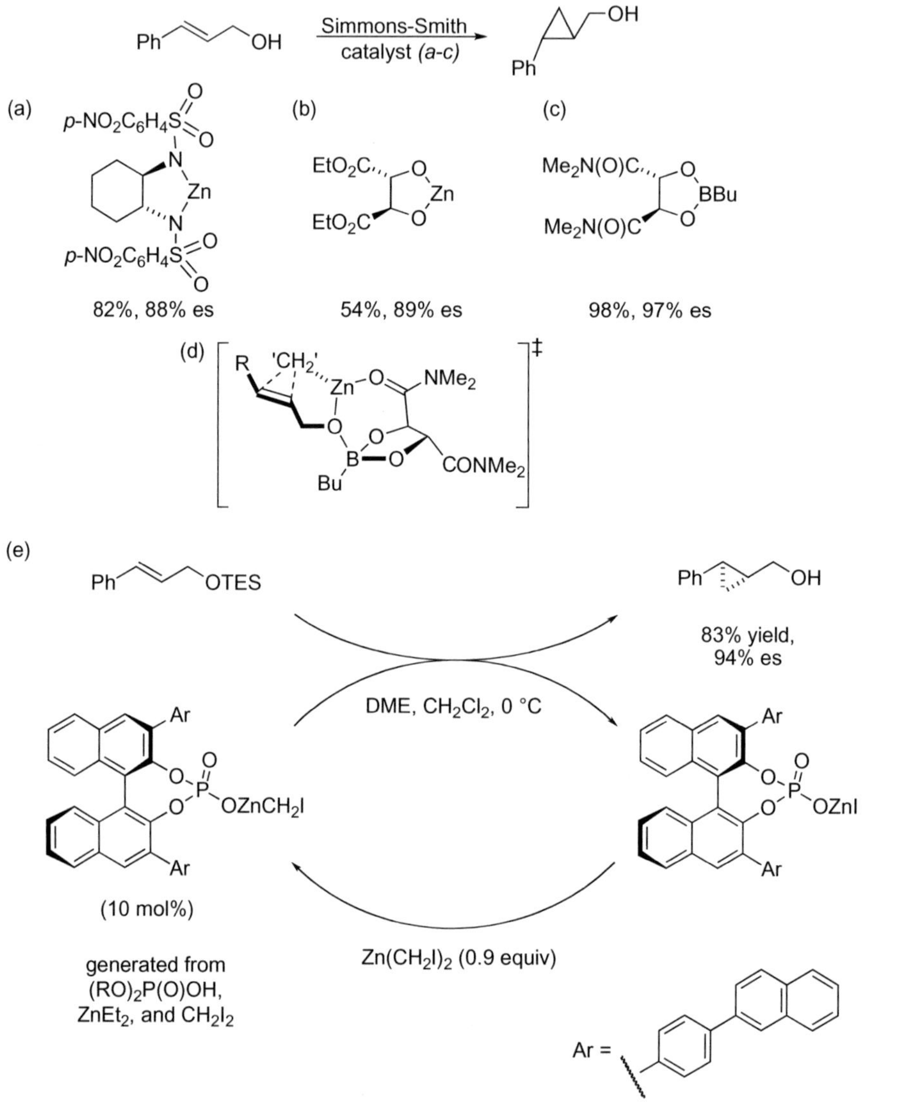

SCHEME 6.33 Asymmetric catalysts for the Simmons–Smith cyclopropanation of *trans*-cinnamyl alcohol: (a) [213]. (b) [214]. (c) [205,207]. (d) Transition state model for catalyst *c* [205]. (e) A catalytic means utilizing a chiral phosphoric acid [208]. Only one zinc and the transfer methylene are shown; other atoms associated with the Simmons–Smith reagent are deleted for clarity.

In 1966, Nozaki et al., reported the first example of an asymmetric cyclopropanation using a chiral copper (II) catalyst [215]. Although the enantioselectivities were low (<20% es), this was the first example of an asymmetric synthesis using a chiral, homogeneous transition metal catalyst. Subsequently, Aratani optimized the ligand design and reported a number of asymmetric cyclopropanations, as shown in Scheme 6.34 [216–218]. For symmetrical *trans*-olefins, relative configuration is not an issue, and better selectivity is achieved with *l*-menthyl (from (−)-menthol) diazoacetate than with the ethyl ester (matched pair double asymmetric induction, Section 1.6 [217]). Cyclopropanation of isobutene is used on a factory scale for the commercial manufacture of the drug cilistatin (Scheme 6.34b) [218]. With monosubstituted olefins, relative as well as absolute configuration is an issue, but *trans* is favored and double asymmetric induction again increases the stereoselectivity (Scheme 6.34c, [217]). Trisubstituted, unconjugated alkenes favor the *cis* relative configuration, as shown by the example in Scheme 6.34d, used in the synthesis of the *cis* isomer of the insecticide permethric acid [217]. Dienes, on the other hand, favor the *trans*-isomer, as shown by the synthesis of chrysanthemic acid shown in Scheme 6.34e [216,218].

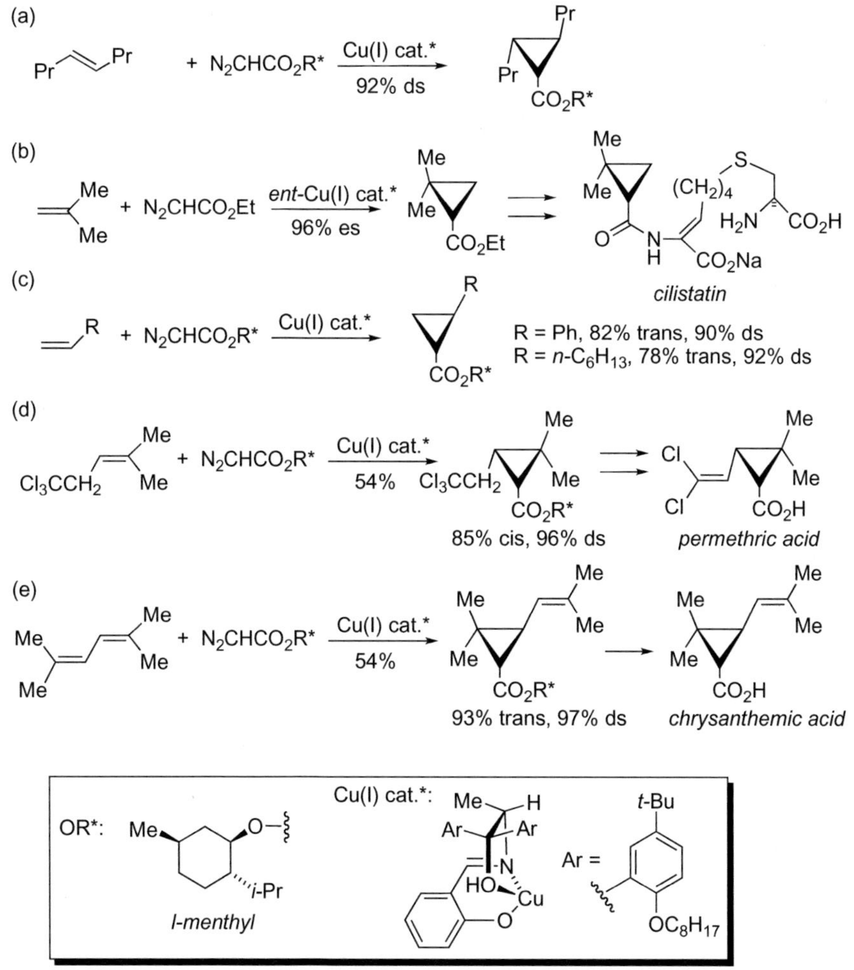

SCHEME 6.34 Aratani's copper-catalyzed asymmetric cyclopropanation of olefins. (*a*) *Trans*-1,2-disubstituted [217]; (*b*) 1,1,-disubstituted [218]; (*c*) monosubstituted, *trans* favored [217]; (*d*) trisubstituted, *cis* favored [217], and (*e*) dienes, *trans* favored [216,218]. *Inset:* chiral auxiliary and coordinatively unsaturated chiral catalyst.

The mechanism that has been proposed to explain the relative and absolute configurations of these examples is illustrated in Scheme 6.35 [218]. The catalyst, shown on the left of the scheme, is coordinatively unsaturated. Reaction with the diazoalkane affords the copper carbenoid shown at the top. The olefin approaches from the less hindered rear, *Re* face (note that the absolute configuration of the carbene carbon is set at this point), such that the indicated carbon (*, which is the one most able to stabilize a cationic charge) is oriented toward the carbene carbon. This is consistent with the metal atom acting as a Lewis acid. A metallacyclobutane is thought to be a discrete intermediate (bottom), and as it is formed, the hydroxyl is released from the copper. Steric repulsion by the large aryl substituents of the chiral ligand tends to force R_1 downward, *cis* to the ester function. Similarly, steric repulsion tends to favor R_2 in a position *trans* to the ester. Collapse of the metallacyclobutane releases the cyclopropane and regenerates the catalyst.

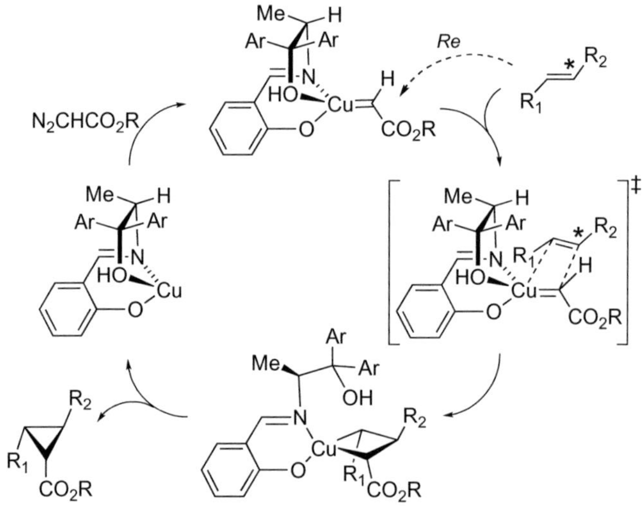

SCHEME 6.35   Proposed catalytic cycle for asymmetric cyclopropanation using Aratani's copper catalyst [218].

This rationale may be used to explain the apparent reversal of both relative and absolute configuration preferences exhibited by the examples in Scheme 6.34d and e. In Scheme 6.34d, R_1 is Cl_3CCH_2-; attack of the copper occurs at the secondary carbon and the carbene carbon attaches to the tertiary site (*), as shown in Scheme 6.36a. The controlling elements are the tertiary carbon of the olefin attaching to the carbene carbon, while the bulky Cl_3CCH_2- is oriented away from the nitrogen ligand. In the example in Scheme 6.34e, the more stable carbocation is allylic, so the trisubstituted olefin "turns around," as shown in Scheme 6.36b. Here, the controlling element is the *trans* orientation of the ester with respect to the isobutenyl group [218].

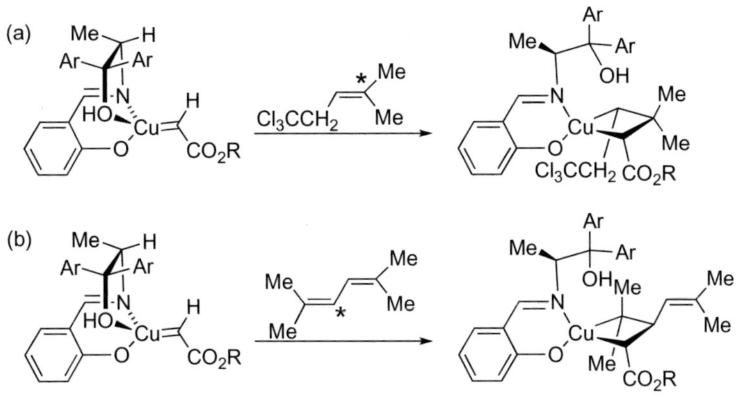

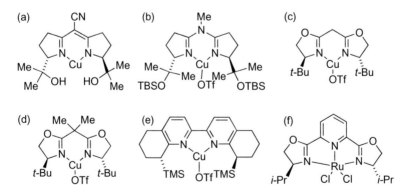

SCHEME 6.36 Rationale for the relative and absolute configuration of the examples from (a) Scheme 6.34d, and (b) Scheme 6.34e [218].

Several other groups have used C_2-symmetric ligands with copper and ruthenium as cyclopropanation catalysts. These ligands, shown in Figure 6.10, are generally more selective than the Aratani ligands. The first to be introduced was the semicorrin of Pfaltz (Figure 6.10a), and most of the others bear a close structural resemblance in that they all have pyrroline, oxazoline or bipyridine ligands chelating the metal. Copper(I) is the oxidation state of the active catalyst for all complexes containing copper, and the mechanism of the cyclopropanation using these catalysts is probably similar to that illustrated above (Schemes 6.35 and 6.36): electrophilic attack by copper, metallacyclobutane formation, and so on. Table 6.6 lists selected examples for each ligand. It was generally found that bulky esters (*e.g.*, *tert*-butyl, BHT, menthyl) are more selective than less bulky ethyl esters (not listed).

FIGURE 6.10 C_2-symmetric catalysts for cyclopropanation chemistry: (a) Pfaltz [219]; (b) Pfaltz [220]; (c) Masamune [221] (see also [222]); (d) Evans [223]; (e) Katsuki [224]; (f) Nishiyama [225].

TABLE 6.6 Asymmetric Cyclopropanations

Entry	Cat.[a]	N_2CO_2R	Alkene	% Yield	% trans	% es	Ref.
1	a	t-Bu	$PhCH{=}CH_2$	60	81	96	[219]
2	b	t-Bu	$PhCH{=}CH_2$	75	81	97	[220]
3	c	menthyl	$PhCH{=}CH_2$	72	86	99[b]	[221]
4	d	BHT	$Me_2C{=}CH_2$	91	–	>99	[223]
5	e	t-Bu	$PhCH{=}CH_2$	75	86	96	[224]
6	f	t-Bu	$PhCH{=}CH_2$	81	97	97	[225]

[a]The "Cat." column refers to the catalysts in Figure 6.10. For the structure of menthyl, see Scheme 6.34.
[b]The absolute configuration reported in this paper is correct (1R, 2R), but it is drawn incorrectly.

In all cases, the *MM* ligand[21] affords the 1*S*,2*S*-*trans* product and the *PP* ligand affords the 1*R*,2*R*-*trans* product. The sense of diastereoselectivity and enantioselectivity are consistent with the models (not mechanisms) shown in Figure 6.11. Because of the C_2-symmetry of the ligands, the configuration of the carbene is the same whether the ester moiety is drawn up or down. Note that the vertical orientation of the carbene and the horizontal orientation of the ligand divide the reagent into four quadrants. Only in the *S*,*S*-*trans* product (from the *MM* complex) are steric interactions between the olefinic substituent, the carbene ester, *and* the ligand substituent avoided (*i.e.*, the olefin substituent is in the lower right quadrant). All other orientations produce repulsive interactions between the olefin and either the ester moiety or the ligand substituent. For ligands having the *PP* configuration, the preferred product is the *R*,*R*-*trans*-cyclopropane.

Weaknesses of the model in Figure 6.11 include the fact that there may be other ligands on the metal that are not taken into consideration here, and that it assumes a similar geometry of the carbene relative to the chelating ligand for all the complexes. On the other hand, the formation of metallacyclobutanes in copper-catalyzed cyclopropanations appears to be an accepted hypothesis [197,223], and the consistency of these representations with an accumulating body of fact makes them useful predictive models, and a good starting point for developing more detailed mechanistic hypotheses.

---

21. Because of fluctuations in atom priority using the CIP sequencing rules (*i.e.*, in spite of their obvious differences, the CIP descriptor for the stereocenters in all ligands except e is *S*), it is useful to define the chirality sense of these ligands using the *P/M* nomenclature [226], applied to the R−C−N−M bond (see the inset in Figure 6.11). Thus, the ligands in Figure 6.10a and b have the *MM* configuration, while those in Figure 6.10c, d, and f have the *PP* configuration. Ligand 6.6e has an extra carbon and is not strictly definable by this system, but its symmetry features are similar to ligands 6.10a and b, so it is considered along with them.

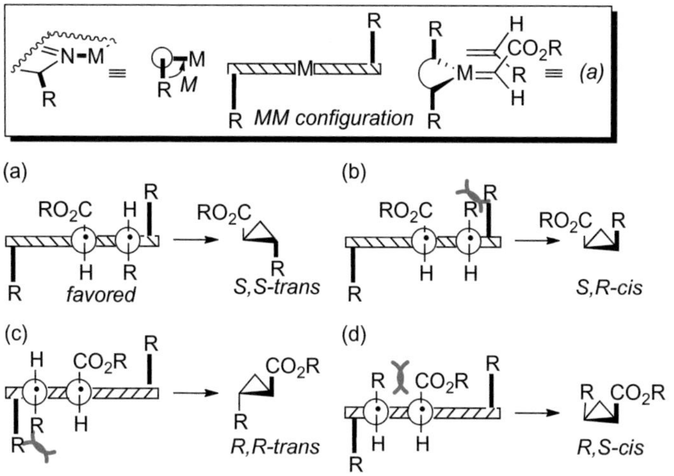

FIGURE 6.11 *Inset:* Definition of *M* configuration of metal complexes, and generalized side view of an *MM*—metal carbene complex with the olefin approaching from the rear (equivalent to the Newman projection shown in *a*). (a) Favored approach, leading to the *S,S-trans* product. (b) Disfavored approach, leading to *S,R-cis* product. (c) Disfavored approach leading to the *R,R-trans* product. (d) Disfavored approach leading to the *R,S-cis* product [197].

It was noted in the previous section that rhodium acetate-catalyzed cyclopropanations of chiral diazo acetates afforded poor diastereoselectivity. Using achiral diazo acetates and methyl 2-oxopyrrolidinone-carboxylate (MEPY) as a chiral ligand on rhodium, reasonable *trans* selectivity and moderate enantioselectivity can be achieved, as shown by the example in Scheme 6.37a [227]. Davies reported that the rhodium prolinate-catalyzed addition of vinyl carbenes to alkenes is 100% selective for the *E*-diastereomers, which are formed in a ≥95:5 ratio for several alkenes, as shown in Scheme 6.37b [228]. Additional fine-tuning in the Davies laboratory led to Rh$_2$[S-biTISP]$_2$, an extraordinarily efficient catalyst in terms of turnover number (92,000) and frequency (4000 h^{-1}); the reaction shown in Scheme 6.37c [229] was carried out on 46 g scale (for a review that includes an excellent discussion of the logic behind the development of this ligand, see ref. [230]). A simpler ligand, *S-N*-1,8-naphthanoyl-*tert*-leucine (*S*-NTTL) was introduced by Müller in 2004 [231] and permitted high diastereo- and enantioselectivities in a variety of settings, such as in the dihydrofuran substrate shown in Scheme 6.37d [232].

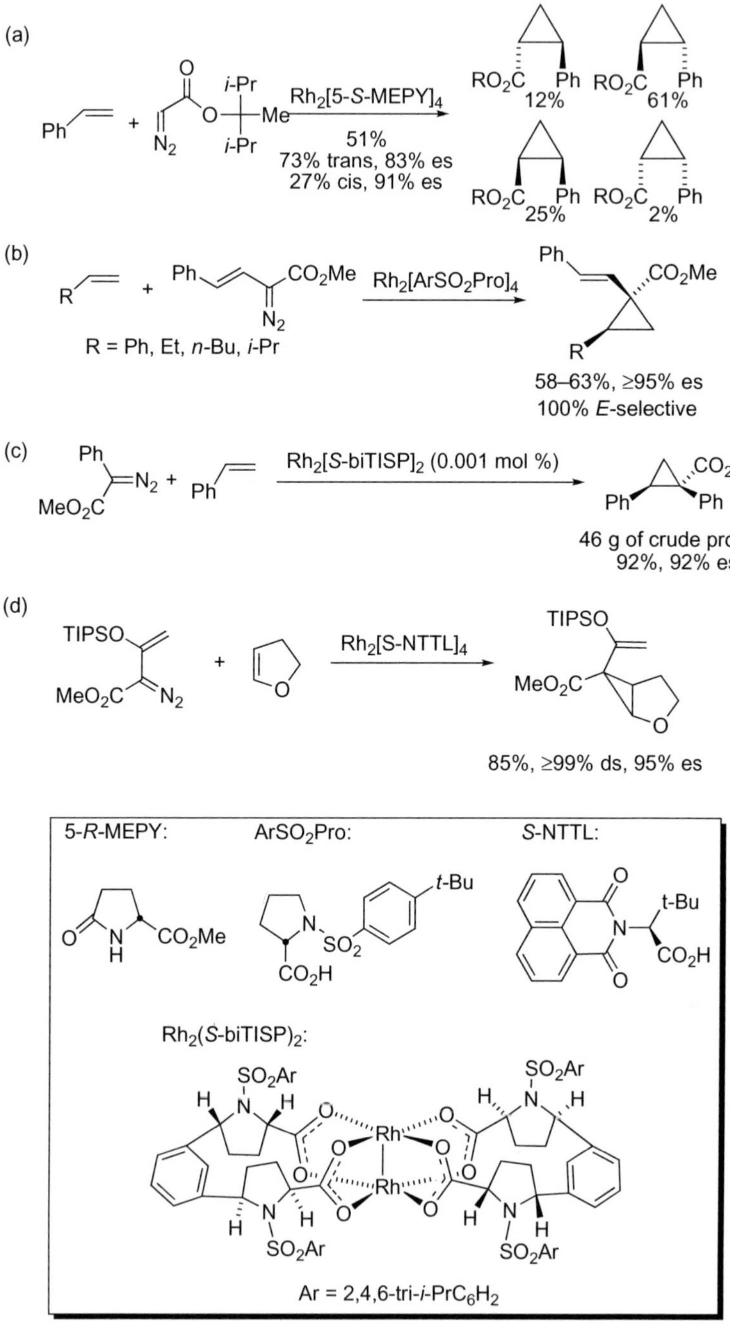

SCHEME 6.37 (a) Asymmetric cyclopropanation of styrene [227]; (b) asymmetric cyclopropanation of alkenes with vinyl carbenes [228]; (c) a highly efficient cyclopropanation [229]; (d) cyclopropanation of dihydrofuran [232]. *Inset:* Ligand structures.

The number of possible transition state geometries is restricted in intramolecular cyclopropanations. These can be quite selective, as shown by the examples illustrated in Scheme 6.38a [233,234] and subject to double asymmetric induction, as shown by the series of examples in Scheme 6.38b. For all of the latter substrates, the *trans* product is slightly preferred when cyclopropanation is mediated by an achiral catalyst [235], but this selectivity is reversed dramatically when the *S* ester is allowed to react with the 5-*S*-MEPY catalyst. Note also that when the chirality sense of the substrate and the catalyst are mismatched (*S* substrate and *R* catalyst), the *cis* selectivities are low, unless R_1/R_2 are trimethylsilyl. For the matched case of double asymmetric induction, the same features that cause the *cis* selectivity lead to the enantioselective (group-selective) cyclopropanation of the divinyl alcohol illustrated in Scheme 6.38c. The group selectivity is significantly diminished, however, if there are substituents on the double bonds [235].

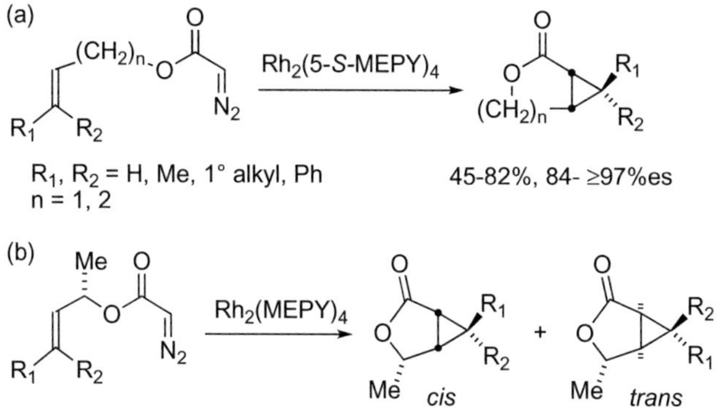

	R$_1$	R$_2$	MEPY	cis:trans	% Yield
	H	n-Bu	S	>95:5	80
	H	n-Bu	R	40:60	39
	n-Bu	H	S	86:14	77
	n-Bu	H	R	50:50	42
	H	TMS	S	>95:5	74
	H	TMS	R	37:63	31
	TMS	H	S	95:5	76
	TMS	H	R	92:8	47

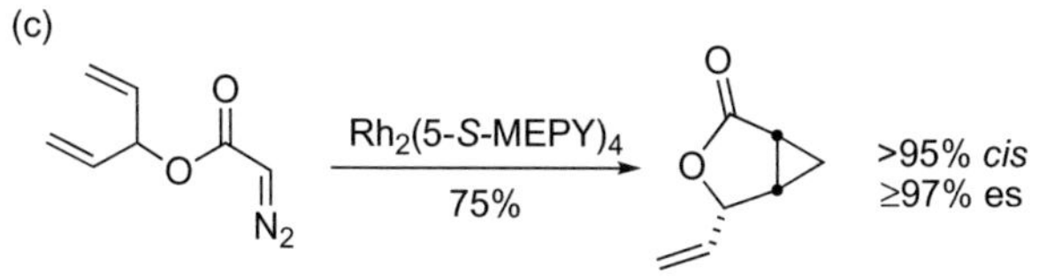

SCHEME 6.38 (a) Enantioselectivity in intramolecular cyclopropanations [233,234]. (b) Double asymmetric induction in intramolecular cyclopropanations [235]. (c) Group-selective asymmetric cyclopropanation [235].

If imitation is indeed the sincerest form of flattery, perhaps it is not surprising that the success of Rh-promoted asymmetric cyclopropanations led scientists to investigate other metals that would catalyze similar processes. The most important of these is ruthenium, which has grown in popularity despite the fact that carbenoids based on this metal are less electrophilic and reactive. However, Ru is inexpensive relative to Rh and numerous cyclopropanation catalysts using it have been reported, beginning with the PyBox-based system reported by Nishiyama (Scheme 6.39a [225]; for a brief review see ref. [236]). Note that some improvement in *trans* selectivity was observed by pairing the chiral catalyst with a chiral substrate. A turnaround to very high *cis* selectivity was realized, along with very high enantioselectivity, using a catalyst containing a SALEN-esque PNNP ligand as shown in Scheme 6.39b [237]. Numerous other catalyst types have been investigated in this context, including the heavily fortified porphyrin reported by Berkessel [238].

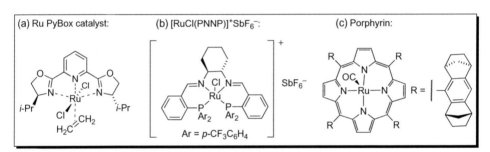

				Ratio (es)		
With RuPyBox:	R = Me	74%	92 (94% es)	:	8 (85% es)	
	R = menthol	82%	97 (93% es)	:	3 (98% es)	
With [RuCl(PNNP)]+ SbF6−:	R = Et	69%	1 (79% es)	:	99 (98% es)	
With porphyrin:	R = Et	80%	96 (93% es)	:	4 (57% es)	

(a) Ru PyBox catalyst: **(b) [RuCl(PNNP)]+SbF6−:** **(c) Porphyrin:**

Ar = p-CF3C6H4

R =

SCHEME 6.39 Asymmetric cyclopropanations promoted by Ru complexes: (a) ref. [225], (b) ref. [237], (c) ref. [238]. *Inset*: catalyst structures.

6.1.5.3 Stepwise Cyclopropanations

Cyclopropanes can also be made *via* ionic reactions since 3-membered rings undergo kinetically favorable ring closure reactions. For an early example, chiral malonate esters were used successfully in asymmetric cyclopropanations, such as *via* the intramolecular S_N2' alkylation shown by the example in Scheme 6.40 [239,240]. The rationale for the diastereoselectivity is shown in the illustrated transition structure. Note that the enolate has C_2-symmetry, so it doesn't matter which face of the enolate is considered. The illustrated conformation has the ester residues *syn* to the enolate oxygens to relieve $A^{1,3}$ strain, with the enolate oxygens and the carbinol methines eclipsed. The allyl halide moiety is oriented away from the dimethylphenyl substituent, exposing the alkene *Re* face to the enolate. The crude selectivity is about 90%, but a single diastereomer was isolated in 67% yield by preparative HPLC.

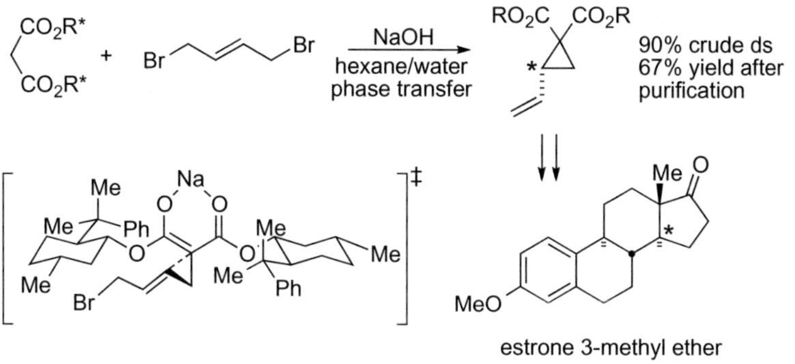

SCHEME 6.40 Asymmetric cyclopropanation of malonate enolates and an application to steroid synthesis [239,240].

Another strategy for stepwise asymmetric cyclopropanation entails the use of chiral electrophiles. Two examples of doing this are shown in Scheme 6.41. In one, Meyers has used bicyclic lactams (*cf.* Schemes 3.24 and 3.25) [241,242] as electrophilic auxiliaries in sulfur ylide cyclopropanations [243]. These auxiliaries, for reasons that are not entirely clear, are preferentially attacked from the α-face. After separation of the diastereomers, the amino alcohol auxiliary may be removed by refluxing in acidic methanol or reductively [241]. For the synthesis of cyclopropyl amino acids, Williams has used an oxazinone auxiliary (*cf.* Scheme 3.17) as an electrophilic component in a sulfur ylide cyclopropanation using Johnson's sulfoximines, as illustrated in Scheme 6.41 [244]. Surprisingly, the sulfur ylide approached from the β face; the authors speculated that this may be due to π-stacking between the phenyls on the oxazinone ring and the phenyl in the sulfoximine [245]. With Corey's [243] dimethylsulfonium methylide, the diastereoselectivity was only about 75%, but with Johnson's sulfoximines (used in racemic form), only one diastereomer could be detected for most substrates studied (with the exception of R = H) [245]. Dissolving metal reduction afforded moderate yields of the cyclopropyl amino acids.

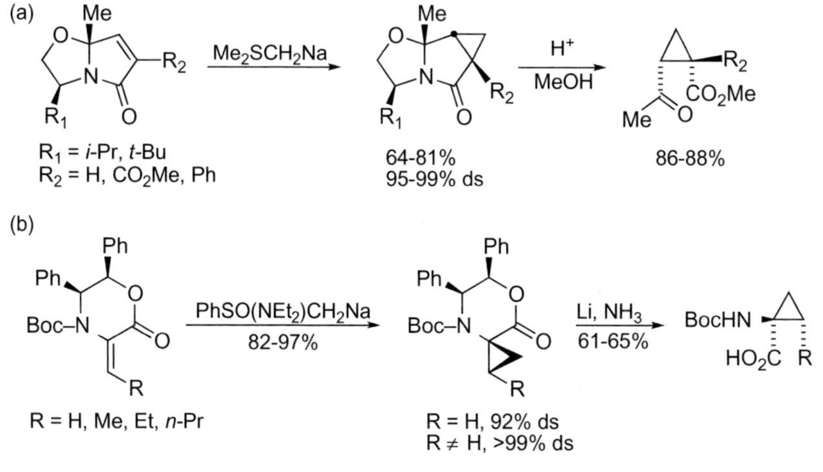

SCHEME 6.41 (a) Meyers' asymmetric cyclopropanations using the bicyclic lactam auxiliary [241]. (b) Williams' asymmetric synthesis of cyclopropyl amino acids [245].

Sulfur ylide-based cyclopropanations have also been carried out using other modern asymmetric synthesis strategies; a few representative examples are collected in Scheme 6.42. Aggarwal has developed a number of chiral sulfur ylides for various applications [246], including the stabilized ylide shown in Scheme 6.42a [247]. We close this topic with three catalytic approaches shown in Scheme 6.42b−d [248−250]. In each case, the initial addition step creates a new stereocenter that subsequently undergoes internal nucleophilic displacement to afford the cyclopropane product in a second step; the exclusive formation of *trans* cyclopropanes results from the minimization of steric interactions in the course of the second bond-forming event.

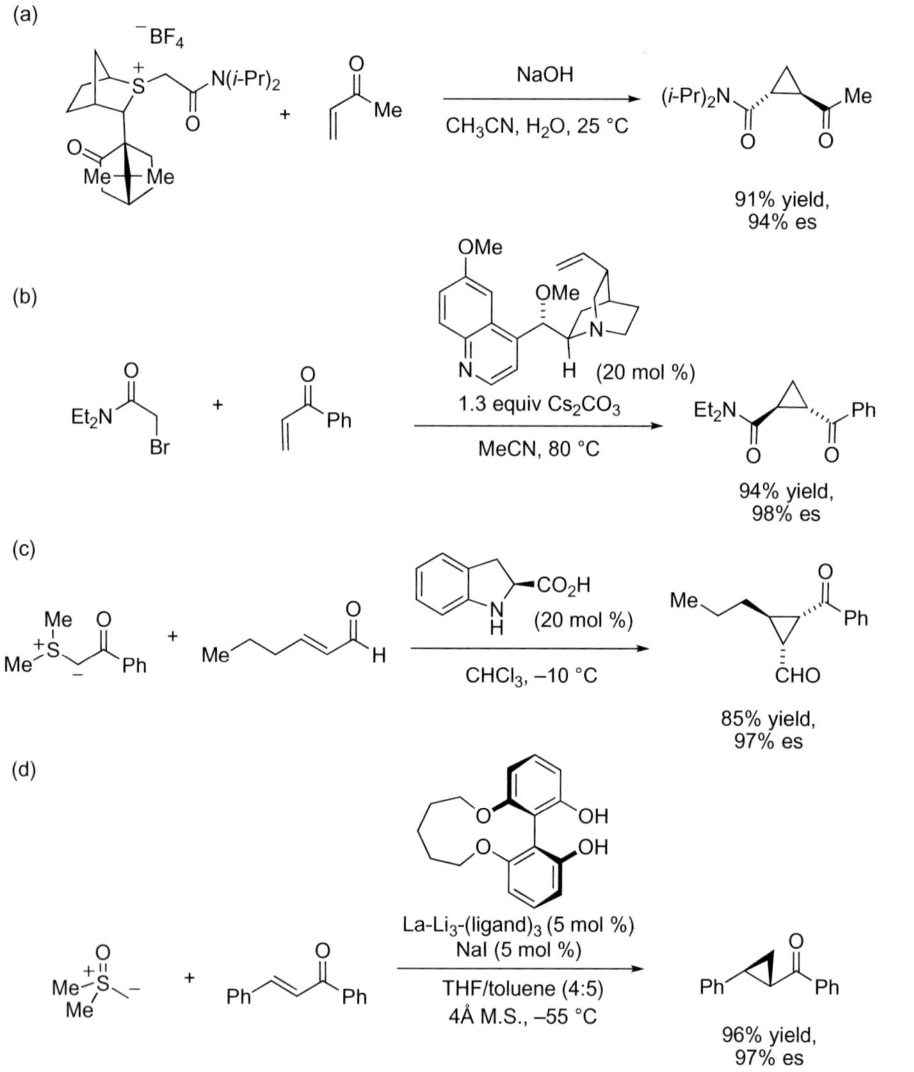

SCHEME 6.42 Approaches to chiral cyclopropanes using (a) a chiral sulfur ylide [247]; and (b)−(d) chiral catalysts [248−250].

Asymmetric Reactions Involving Alkenes

On a cold and rainy evening in late 1984. A student is taking a train for a weekend of vaguely unauthorized fun in New York City, and to salve his guilt about being away from the lab for the weekend he decides to pass the time with the latest issue of the *Journal of the American Chemical Society*. Upon opening the journal, he sees an article title that contains the words "asymmetric olefination." Although well aware that the double bond isomers are diastereomers of each other (a fact unnoticed by more than a few chemistry students over the years), the idea that one could carry out a truly asymmetric reaction by generating a double bond was a new concept to at least one reader back then.

That paper described the use of a chiral phosphonamide reagent that was able to convert achiral 4-*tert*-butylcyclohexanone to a corresponding olefin [1]. In the course of the reaction, an element of "axial chirality" was generated, which is how in this case

major isomer
82%, 95% es

"asymmetric olefination" was even possible. The reaction is mechanistically complex, requiring the diastereofacially selective attack onto the carbonyl relative to the resident *tert*-butyl group, stereoselective formation of an oxaphosphetane, and the stereoselective breakdown of that intermediate to give the olefin (as is usually observed in olefination chemistry). Other researchers have examined this concept (a nice mechanistic discussion and leading references can be found in ref. [2]) and extended it to allenes as well [3].[1] Although tempting to dismiss such reactions as curios, it's worth keeping in mind the potential of axially chiral compounds for their chiroptical properties or as components of molecular rotors [4].

Hoveyda has also reported innovative examples of chiral olefinations as part of his historic collaboration with Richard Schrock on olefin metathesis. Consider the example shown, in which a meso norbornene is transformed into a fused dihydropyran derivative via asymmetric ring opening/ring closing metathesis [5]. Several mechanisms can be drawn, but however one looks at it the asymmetry in the final product arises from the selective differentiation of the two ends of the endocyclic ole-

54%, 96% es

fin (the diallyl ether additive only helps convert the catalyst *in situ* to a more efficient version).

One can barely have a conversation about olefins and stereochemistry without at least mentioning the squalene to lanosterol conversion. Following the enzymatic formation of squalene oxide, enzyme-induced epoxide opening is followed by an olefin to cation cyclization, then another, and another, and another finally affording lanosterol after deprotonation. Besides its

---

1. Try sketching this out before looking up the reference. Also: How many stereogenic elements are in the major product and what is/are the absolute configuration(s)?

importance in biochemistry, this reaction led to the famous Stork–Eschenmoser hypothesis, which was one of the first serious attempts to correlate conformation and stereochemical outcome in a reaction and which also explicitly stated the role of *anti*-addition to an olefin [6]. Mimicking this reaction in the laboratory was perhaps *the* preeminent biomimetic problem of the 1960s and a tremendous amount of work has been done on this problem—too many to even reference here (we'll cite two, including the first article ever published in *Accounts of Chemical Research* [7,8]). For just a flavor of how this work continues to inspire, consider the example of asymmetric iodination published by Ishihara [9], which demonstrates once again how much one can learn by studying nature.

References

[1] Hanessian, S.; Delorme, D.; Beaudoin, S.; Leblanc, Y. *J. Am. Chem. Soc.* **1984**, *106*, 5754–5756.
[2] Gramigna, L.; Duce, S.; Filippini, G.; Fochi, M.; Franchini, M. C.; Bernardi, L. *Synlett* **2011**, 2745–2749.
[3] Pinho e Melo, Teresa M. V. D.; Cardoso, Ana L.; d'A. Rocha Gonsalves, António M.; Costa Pessoa, J.; Paixão, José A.; Beja, Ana M. *Eur. J. Org. Chem.* **2004**, *2004*, 4830–4839.
[4] Koumura, N.; Zijlstra, R. W. J.; van Delden, R. A.; Harada, N.; Feringa, B. L. *Nature* **1999**, *401*, 152–155.
[5] Weatherhead, G. S.; Cortez, G. A.; Schrock, R. R.; Hoveyda, A. H. *Proc. Natl. Acad. Sci. U.S.A.* **2004**, *101*, 5805–5809.
[6] Yoder, R. A.; Johnston, J. N. *Chem. Rev.* **2005**, *105*, 4730–4756.
[7] Wendt, K. U.; Schulz, G. E.; Corey, E. J.; Liu, D. R. *Angew. Chem., Int. Ed.* **2000**, *39*, 2812–2833.
[8] Johnson, W. S. *Acc. Chem. Res.* **1968**, *1*, 1–8.
[9] Ishihara, K.; Sakakura, A.; Ukai, A. *Nature* **2007**, *445*, 900.

6.2 REARRANGEMENTS

Many rearrangements are highly stereoselective reactions that have found considerable application in organic synthesis. Here, we focus on two types of rearrangements: [1,3]-hydrogen shifts and [2,3]-Wittig rearrangements. The former is a transition metal-catalyzed reaction that has found tremendous importance in industry. The latter involves anionic intermediates and provides an opportunity to discuss, in detail, the interplay between simple stereoselection and enantiomeric control in one particular reaction type. Of course, the most common sigmatropic rearrangements are [3,3]-rearrangements such as the Cope and the many variants of the Claisen rearrangement. The latter, in particular, has been expertly reviewed in great detail [15,251−257] so we will limit our discussion here to a brief overview of the Claisen and a few other rearrangement types that have been adapted for asymmetric synthesis.[22]

6.2.1 [1,3]-Hydrogen Shifts

It has long been recognized that certain transition metal complexes can catalyze the migration of carbon−carbon double bonds. When the catalyst is a transition metal hydride, the mechanism involves initial reversible addition of the metal hydride across the double bond to produce a metal σ-alkyl. A double bond is regenerated by elimination of the metal hydride, and if a different hydrogen is eliminated, the net result is olefin migration (Scheme 6.43a) [258]. This mechanism is, therefore, not a direct sigmatropic shift but rather its formal equivalent. If the catalyst is not a metal hydride, the first step is π-complexation of the metal to the double bond, followed by oxidative addition of the metal, producing a π-allyl metal hydride, then reductive elimination at the other end of the allyl system (Scheme 6.43b) [258]. If the olefin has an allylic heteroatom, a third mechanism may intervene. With allylic amines for example (Scheme 6.43c) [259], initial coordination occurs at nitrogen, and β-elimination yields a π-complexed iminium metal hydride. Insertion then yields a bidentate enamine−metal complex, and dissociation liberates the enamine.

All of these processes are under thermodynamic control, so the migration is only useful when there is an isomer that is in a thermodynamic well, that is, when the rearrangement affords a more highly substituted alkene or when the double bond moves into conjugation with a functional group such as a carbonyl (or, alternatively, leads to an enamine). The net rearrangement can involve several individual "[1,3]-rearrangement" steps, such as migration around a ring. Such sequential shifts are blocked by a quaternary carbon. For the purposes of asymmetric synthesis, the initial alkene must be either 1,1-disubstituted or trisubstituted so that the rearrangement produces a new stereogenic center. As shown in Figure 6.12, this is often contrathermodynamic, *but not in the case of compounds with allylic heteroatoms.*

---

22. Some reactions that are not discussed here, but which mechanistically involve rearrangement steps include ring-expansion reactions such as the Baeyer-Villiger and Schmidt reactions. These are covered in Chapter 8 because they replace a C-C bond with a C-heteroatom bond and so are formally oxidations.

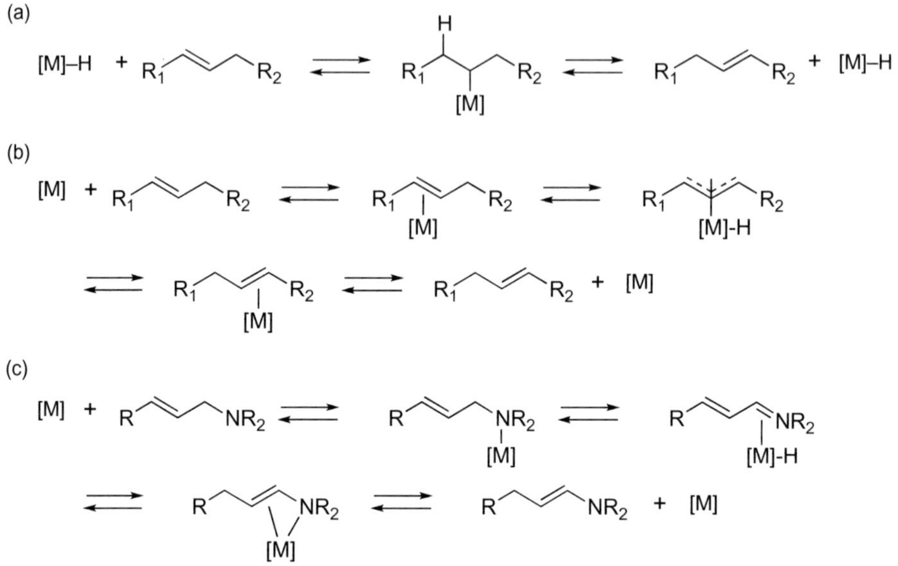

(a)

(b)

(c)

SCHEME 6.43   Transition metal-catalyzed 1,3-hydrogen shifts. (a) Metal hydride catalyst. (b) Metal catalyst. (c) Metal-catalyzed rearrangement of allylic amines to enamines.

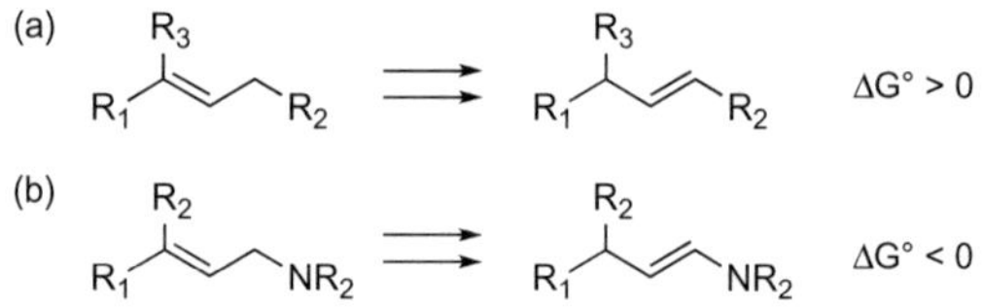

FIGURE 6.12   (a) Contrathermodynamic isomerization of a trisubstituted alkene to a disubstituted one. (b) Thermodynamically favored isomerization of an allylic amine to an enamine.

Following years of modestly successful attempts by other groups (<53% es, reviewed in refs. [260,261] and pp. 266–303 of ref. [258]), Otsuka reported in 1978 that allylic amines could be rearranged to enamines with a chiral Co(II) catalyst with modest 66% enantioselectivity [262]. Further studies [259,263,264] revealed that a cationic Rh(I) catalyst having arylphosphine ligands (the best being BINAP, 2,2'-bis(diphenylphosphino)-1, 1'-binaphthyl) afforded excellent selectivity (97–99% es) with very high catalyst turnover (300,000). This reaction has been scaled up and is now known as the "Takasago process" (Scheme 6.44). It is used for the commercial manufacture of ~1500 tons/year (nearly 40% of the world market) of citronellal and menthol [259,264], and has been described as "the most impressive achievement to date in the area of asymmetric catalysis" [265]. Note that, although citronellal is available from natural sources, the enantiomer ratio of the natural product is only 90:10.

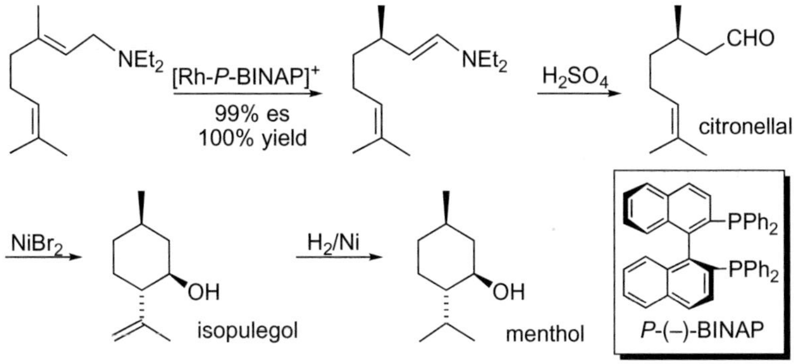

SCHEME 6.44 The Takasago process for the commercial manufacture of citronellal, isopulegol, and menthol [264].[23]

Two aspects of the reaction are stereospecific. The first is that geometric isomers of the allylic amines afford enantiomeric enamines, as shown in Scheme 6.45a [267]. Note that the geometry of the enamine double bond is *not* dependent on the configuration of the double bond of the allylic amine, however. The second stereospecific feature is revealed by the deuterium labeling studies shown in Scheme 6.45b: the R-C_1-d allylic amine, when subjected to enantiomeric rhodium catalysts, undergoes deuterium migration with M-BINAP, and hydrogen migration with R-BINAP [259]. An isotope effect was not observed, indicating that the carbon−hydrogen (or deuterium) bond-breaking step is not rate determining. Furthermore, experiments (not shown) using a mixture of $-CD_2NEt_2$ and $-CH_2NEt_2$ amines revealed no crossover, indicating that the migration is intramolecular [259].

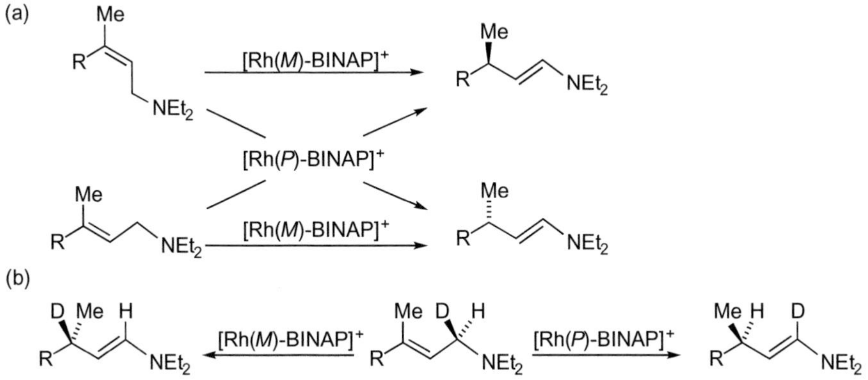

SCHEME 6.45 Stereospecificity of rhodium-catalyzed asymmetric [1,3]-hydrogen shifts.

There are two limiting possibilities that could explain the enantioselectivity: a group-selective metal insertion distinguishing the enantiotopic allylic protons, or a face-selective

---

23. In accord with the recommendation of Prelog and Helmchen, the P,M nomenclature system is used to describe the configuration of molecules containing chirality axes and planes [266]. Note that, for axial chirality, $R = M$ and $S = P$. See the glossary, Section 1.11, for an explanation of these terms.

addition that distinguishes the enantiotopic double bond faces. Figure 6.13 illustrates the conformational analysis of the intermediates involved in the sequence (Scheme 6.43c). First of all, the lowest energy conformation around the $N-C_1-C_2-C_3$ bond (*) of the allylic amine is *antiperiplanar* due to $A^{1,3}$ strain in the *synclinal* conformation (Figure 6.13a). Coordination of the catalyst to the nitrogen (step 1 in Scheme 6.43c) will only increase the energetic bias in favor of the *antiperiplanar* conformation.[24] The second step of the reaction sequence is the oxidative addition of the metal into the C_1-H bond to give a π-bonded α,β-unsaturated iminium ion. Figure 6.13b shows that only the *s-trans* conformer of this species is accessible, because of severe $A^{1,3}$ interactions between the diethylamino substituents and the C_3 substituent in the *s-cis* conformation.

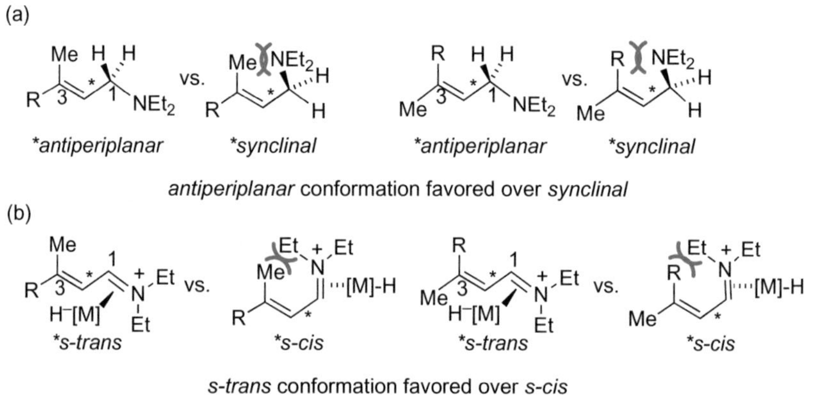

(a)

antiperiplanar   vs.   *synclinal*   *antiperiplanar*   vs.   *synclinal*

antiperiplanar conformation favored over *synclinal*

(b)

s-trans   vs.   *s-cis*   *s-trans*   vs.   *s-cis*

s-trans conformation favored over *s-cis*

FIGURE 6.13 Severe conformational restrictions due to $A^{1,3}$ strain are placed on the intermediates in the asymmetric [1,3]-rearrangement of allylic amines. (a) The starting material (as well as the nitrogen-coordinated rhodium complex) favors the *antiperiplanar* conformation. (b) The π-bonded metal hydride intermediate is restricted to the *s-trans* conformation.

Examination of the conformers illustrated in Figure 6.13b reveals the origin of the *Re/Si* face selectivity in the transfer of hydrogen to C_3. The illustrated conformers have the metal hydride bound to the *Re* face of the iminium ion. Since the *s-cis* conformation is not accessible, and since the rearrangement corresponding to the third step of Scheme 6.43c is suprafacial, the step that determines the configuration of the π-bonded iminium metal hydride also determines the absolute configuration at C_3 in the product. This step is the oxidative addition of the metal into the C_1-H bond (*i.e.*, step 2 of Scheme 6.43c). Thus, the Takasago process is an example of a group-selective insertion of a metal into one of two enantiotopic carbon–hydrogen bonds. The *M*-BINAP rhodium inserts into the $C-H_{Re}$ bond and the *P*-BINAP rhodium inserts into the $C-H_{Si}$ bond.

The interligand asymmetric induction from the binaphthyl moiety to C_3 of the allylic amine covers a considerable distance and deserves comment. As noted above, the enantioselectivity of the overall process is determined in the step where the metal inserts into one of the enantiotopic C_1-protons. The solid-state conformation of the *P*-BINAP ligand has been established by two X-ray crystal structures of ruthenium complexes [268,269], and is

---

24. Low temperature ^1H and ^{31}P NMR studies indicated that only the nitrogen of allylic amines is bound to the metal. No evidence could be found for an N-π-chelate that might stabilize the *synclinal* conformation [259].

illustrated in Figure 6.14a, with the other ligands removed for clarity. Note that the chirality sense of the binaphthyl moiety places the four *P*-phenyl substituents in *quasi*-axial and *quasi*-equatorial orientations. It is apparent that the "upper right" and "lower left" quadrants (which are equivalent due to symmetry) have the most free space for accommodating additional bound ligands. Attachment of the allylic amine in the *antiperiplanar* C_1–C_2 conformation to the square–planar rhodium complex is illustrated in Figure 6.14b. (The two possible binding sites are equivalent due to symmetry.) The β-elimination step must occur through a 4-membered ring transition structure, and the two possibilities are illustrated in Figures 6.14c and d. Note that insertion into the H_{Re}–C bond forces the double bond moiety into close proximity with the *quasi*-equatorial phenyl on the left (Figure 6.14c), whereas metal insertion into the H_{Si}–C bond moves the double bond into the vacant lower left quadrant (Figure 6.14d). The latter is favored.

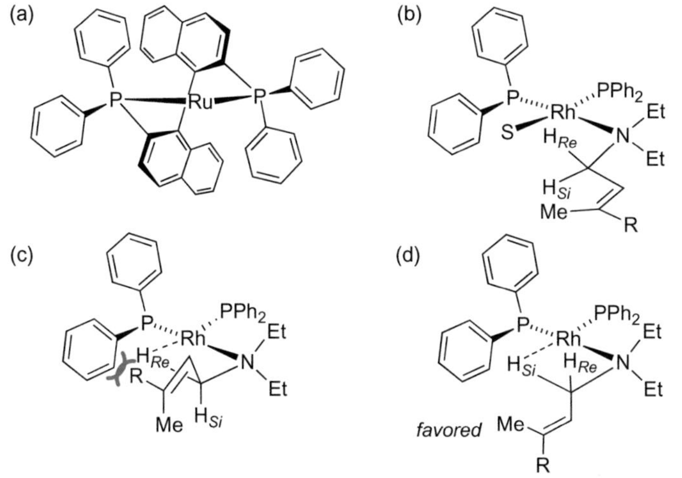

FIGURE 6.14 (a) Conformation of *P*-BINAP in two crystal structures [268,269]. (b) Partial structure with allylic amine bound at one of the two equivalent coordination sites (S = solvent). (c) Transition structure for insertion into the C–H_{Re} bond. (d) Transition structure for insertion into the C–H_{Si} bond.

The catalytic cycle shown in Scheme 6.46 has been proposed to account for the kinetics and observable intermediates in the reaction [259]. Starting from the top, the allylic amine displaces a solvent to form the *N*-coordinated rhodium species. The rate-determining step in the cycle is the substitution of the enamine ligand by a new allylic amine substrate, which probably proceeds *via* the substrate–product mixed complex shown on the left.

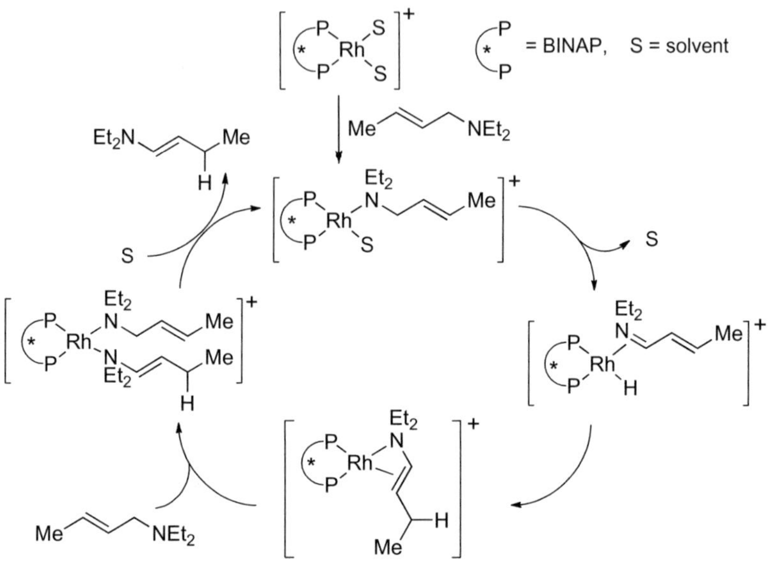

SCHEME 6.46 Catalytic cycle for the rhodium-catalyzed rearrangement of allylic amines.

Investigation of the scope of the asymmetric rearrangement of allylic amines has led to the following generalizations [259]: (i) both C_1 and C_2 should have no alkyl substituents (substitution at either position would erase the preference for the *antiperiplanar* and *s-trans* conformations, *cf.* Figure 6.13); (ii) C_3 may be substituted with an aryl group (or also be only monosubstituted, but the latter circumstance has no stereochemical consequence); (iii) the nitrogen substitutents must not be aryl (a less basic nitrogen fails to bind the rhodium).

The related isomerization of allylic alcohols is attractive to synthetic chemists because this reaction affords β,β-disubstituted ketones *via* the corresponding enol (Scheme 6.47a; for a review of early work in this field see ref. [270]). Following some pioneering but poorly selective examples published by Tani in 1985 [271], Fu found that these reactions could be carried out in steadily improving enantioselectivities using rhodium catalysts outfitted with planar-chiral ferrocenyl ligands (Schemes 6.47b and c, [272,273]). In part, difficulties in obtaining high selectivities in some of these early examples could be traced to the fact that these isomerizations are highly substrate dependent. To wit, isomerization using chiral iridium catalysts was found by Mazet to afford some product types in very high selectivities, as shown in Scheme 6.47d [274]. Extension of the concept to meso 1,4-enediol ethers also affords ketones in good-to-excellent stereoselectivities (Scheme 6.47e [275]). Moreover, subjecting chiral propargylic alcohols to the Rh system has been reported to result in kinetic resolutions [276].

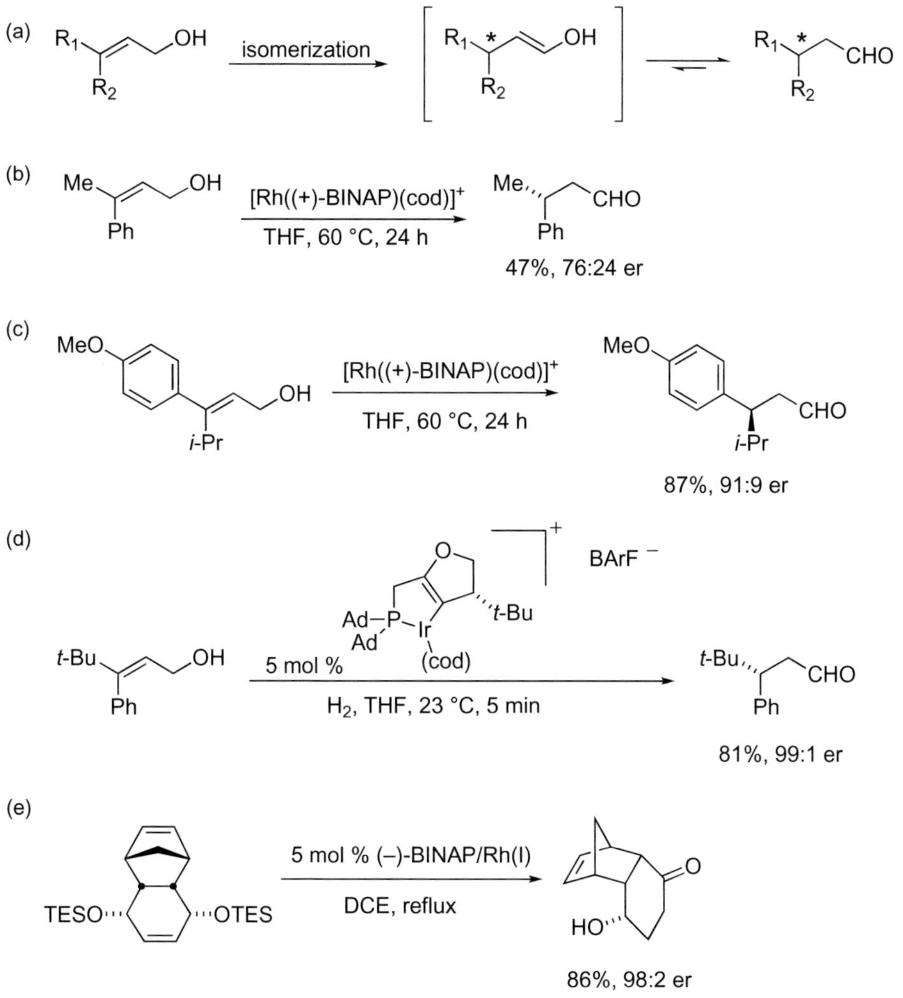

SCHEME 6.47 (a) Asymmetric isomerizations of allylic alcohols and silyl ethers: (b) Tani [271], (c) Fu [272], (d) Mazet [274], and (e) Ogasawara [275].

6.2.2 [2,3]-Wittig Rearrangements

What is now known as the [1,2]-Wittig rearrangement was apparently first observed in the 1920s by Schorigin [277,278], and by Schlenk and Bergmann [279], who reported that reductive metalation of benzyl alkyl ethers with lithium or sodium afforded rearranged products. In 1942, Wittig reported that benzyl ethers could be deprotonated with phenyl lithium, and similarly rearranged [280,281] (Scheme 6.48a). It is now agreed that the [1,2]-rearrangement involves successive deprotonation, homolysis of the opposite carbon—oxygen bond, and recombination to an alkoxide [282,283].[25] The [2,3]-variant was first observed by Wittig

---

25. Note that a concerted [1,2]-carbanion migration is symmetry forbidden [284].

(although not recognized as such) in 1949 [285] and by Hauser 2 years later (Scheme 6.48b, [286]), and was shown in subsequent studies to proceed by a concerted S_Ni mechanism [287,288]. When the [1,2]- and [2,3]-rearrangements can compete, the [2,3]-Wittig rearrangement predominates at low temperatures [289–291].[26]

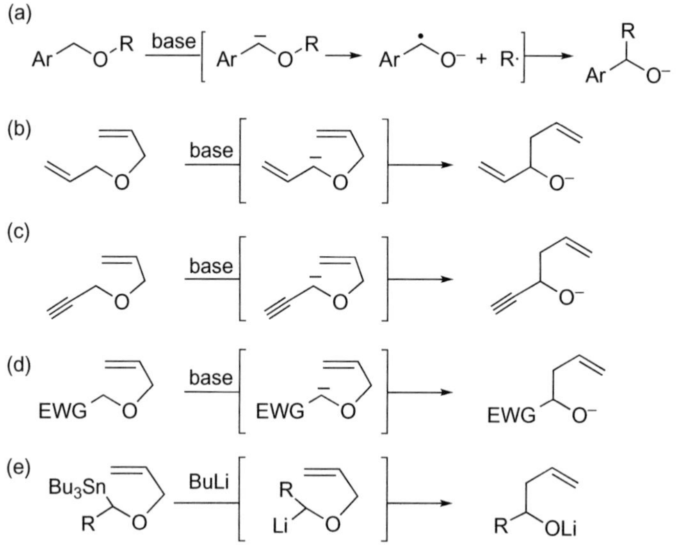

SCHEME 6.48   (a) The [1,2]-Wittig rearrangement [280,281]. (b) The [2,3]-Wittig rearrangement [286]. (c) The [2,3]-Wittig rearrangement of propargyl allyl ethers occurs by deprotonation at the propargylic position. (d) Similarly, electron-withdrawing groups (EWG) can be used to influence the site of deprotonation. (e) The Still variant of the [2,3]-Wittig, which uses a tin–lithium transmetalation to effect anion formation [301].

With unsymmetrical ethers, the problem of the regiochemistry of metalation arises. Three approaches have successfully addressed this issue. One takes advantage of the fact that propargyl allyl ethers deprotonate exclusively at the propargylic position ([302,303] Scheme 6.48c). Second, an electron-withdrawing group (EWG) that stabilizes the anion on one side of the ether can be used to control the site of deprotonation, although enolates may suffer competitive [3,3]-rearrangement ([304,305], Scheme 6.48d). Finally, the regiochemical issue can be eliminated by using tin–lithium exchange to generate the carbanion at a specific site ([301], Scheme 6.48e).

The migration across the allyl system is suprafacial [291], as illustrated by the example shown in Scheme 6.49a [306,307]. The configuration of the carbanionic carbon[27] inverts during the rearrangement, as predicted by theory in 1990 [309], and subsequently proven by three independent studies in 1992 [310–312], the simplest of which is illustrated in Scheme 6.49b. Thus, the [2,3]-Wittig rearrangement is a $[_\pi 2_s + _\sigma 2_a + _\sigma 2_a]$-rearrangement, which is symmetry allowed for a concerted six-electron process with two inversions [284].

---

26. For reviews of the Wittig rearrangements, see refs. [292–300].
27. α-Alkoxyorganolithiums are configurationally stable below about −30 °C (see Section 3.2.1 [308]).

(a)
Me
H
Me
CH2SnBu3
Me
O
BuLi
83%
OMe
Me
H
CH2OH
Me
Me
OMe
(b)
R
SnBu3
O
BuLi
95%
R
OH
R = Et
OSiR3
R =
i-Pr
CH2

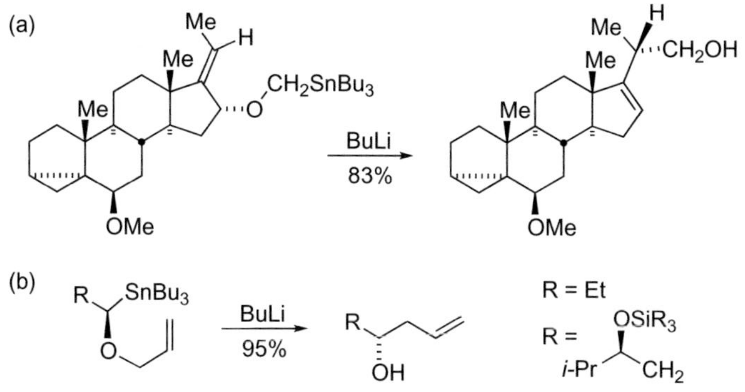

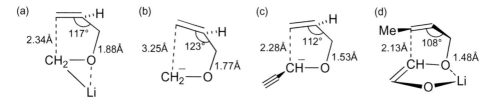

SCHEME 6.49 (a) Example illustrating the suprafacial nature of the migration across the allyl moiety [306]. (b) Examples illustrating inversion of configuration at the metalated carbon [311,312].

The approximate geometries of four calculated transition structures are shown in Figure 6.15. When the lithium is included in the calculation (Figure 6.15a), all nonhydrogen atoms except the middle carbon of the allyl system are approximately coplanar [309]. When the lithium is removed, the envelope conformation is maintained, but the bond lengths and angles change dramatically, as shown in Figure 6.15b [313]. For the naked anion, the transition structure is extremely early, with practically no bond making or breaking having occurred. When the lithium is present, the transition structure is somewhat later, which may be an artifact of the method, since the calculation requires that the lithium be unsolvated and in the gas phase. Since one cannot ignore the presence of the cation, we may assume that the real transition state geometry probably lies somewhere between these two structures. When the carbanion is stabilized by an alkynyl group (Figure 6.15c) or is an enolate (Figure 6.15d), the calculated transition structure is much more compressed [313]. Note for example, that the forming and breaking bonds are shorter than in the other two structures, and also note that the bond angle is smaller.

(a)
H
2.34Å
117°
1.88Å
CH2—O
Li
(b)
H
3.25Å
123°
1.77Å
CH2—O
(c)
H
2.28Å
112°
1.53Å
CH—O
(d)
Me
2.13Å
108°
1.48Å
CH—O
Li
O

FIGURE 6.15 *Ab initio* transition structures for the [2,3]-Wittig rearrangement. (a) Structure including a lithium, in which the metal is *antiperiplanar* to both carbons of the allyl system and bridges the carbon and oxygen [309]. (b) Calculated transition structure for [2,3]-rearrangement of the naked $ROCH_2^-$ anion [313]. (c) Calculated transition structure for the [2,3]-rearrangement of a propargyl anion. Orientation of the alkynyl moiety on the convex face is favored by 2.1 kcal/mol [313]. (d) For the rearrangement of a lithium enolate, the endo structure is favored [313]. (The authors are grateful to Professors Y. Wu and K.N. Houk, who kindly supplied the indicated bond lengths and angles in a private communication.)

6.2.2.1 Simple Diastereoselectivity

The aspects of diastereoselectivity in the [2,3]-Wittig rearrangement that we will be concerned with involve the geometry of the double bond and the configuration of the allylic and "carbanionic" carbons in the allyl ether. Figure 6.16a illustrates diastereomeric transition structures for [2,3]-rearrangements of α-allyloxy organolithiums. If there is a substituent (R_1) at the allylic position, $A^{1,2}$ and $A^{1,3}$ allylic strain will play a role. If both R_1 and R_2 are not hydrogen, $A^{1,2}$ strain will disfavor the left conformer. If the alkene has the Z configuration, $A^{1,3}$ strain is particularly severe in the structure on the right. With reference to Figure 6.15, note that $A^{1,2}$ strain will be alleviated by a small allylic bond angle and that $A^{1,3}$ strain will be enhanced by a small bond angle.

Similar transition structures having stereogenic carbanionic carbons are illustrated in Figure 6.16b. For electrostatic reasons, an electron-rich substituent such as an alkyl, vinyl, or alkynyl group will preferably occupy the convex face of the envelope conformation, while an electropositive substituent favors the concave side [309,313].

If the carbanionic carbon is trigonal, such as with enolates, the preference is to occupy the concave face, as shown in Figure 6.16c. This effect is reminiscent of the endo effect in Diels–Alder reactions (Section 6.6), and is also consistent with Seebach's topological rule suggesting a preference of *synclinal* donor/acceptor orientations in a Newman projection along the forming bond (*cf.* Figure 5.2, [314]). For $Z(O)$-enolates, additional stabilization can be had by metal chelation with the ether oxygen (*cf.* Figure 6.15d).

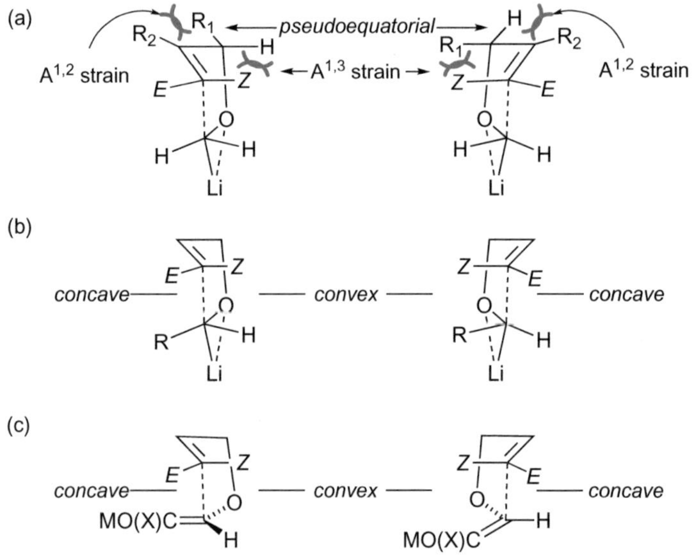

FIGURE 6.16 Factors influencing the relative configuration of the products in [2,3]-Wittig rearrangements: (a) Diastereomeric transition states illustrating the possibility of allylic strain. (b) The conformation having R on the convex face of the envelope is preferred for alkyl, vinyl, and alkynyl substituents. (c) For enolates, the concave orientation (*synclinal* double bonds) is preferred.

Each one of the effects illustrated in Figure 6.16 is attributable to a stereogenic element in the starting material (olefin geometry or absolute configuration at the allylic or carbanionic carbon), and is an example of single asymmetric induction. When more than one element is present these effects can operate as matched or mismatched pairs of double asymmetric induction, and very high selectivities can be achieved when they operate in concert. Additionally, it is possible to introduce a stereogenic element elsewhere, such as a chiral auxiliary (X of Figure 6.16c). Conversely, when two elements are dissonant, lower selectivity may be expected. The reader should recognize that these 5-membered-ring transition states are considerably more flexible than, for example, a chair structure such as the Zimmerman–Traxler transition state in aldol additions (*cf.* Scheme 5.1),[28] which complicates the analysis of the various effects.

Scheme 6.50 illustrates the influence of allylic strain between alkyl substituents on the double bond and allylic positions, uncomplicated by substitution at the carbanionic carbon. As shown in Scheme 6.50a, tin–lithium exchange affords an anion that rearranges (*cf.* Figure 6.16a, R_1 = Bu, R_2 = Me, $E = Z = H$) to give a near quantitative yield of alkene with 96–97% diastereoselectivity [301]. In this example, $A^{1,2}$ strain is relieved when the butyl group adopts the pseudoaxial orientation. Scheme 6.50b illustrates the influence of $A^{1,3}$ strain between two alkyl groups (*cf.* Figure 6.16a, R_1 = *n*-heptyl, R_2 = H, E = H, Z = Me), this time favoring the pseudoequatorial conformation for the allylic substituent, so as to avoid the Z-methyl. The [2,3]-Wittig rearrangement is 100% stereoselective for the *E*-alkene [301]. In contrast, when the alkene is unsubstituted in the "Z-position," the selectivity for a particular olefin geometry is severely diminished. Scheme 6.50c lists two such examples having *E* or unsubstituted alkene as educt, which are only 60–65% selective for the Z-product. It was noted that propargylic anions rearrange *via* a transition structure that has significantly shorter bond lengths, and also a compressed allylic bond angle (*cf.* Figure 6.15c, [313]). The latter effect amplifies $A^{1,3}$ strain, and *E* selectivity is restored when the carbanionic carbon is propargylic (Scheme 6.50d, [302]).

As seen in Scheme 6.50d, if both carbons involved in bond formation have heterotopic faces, two adjacent stereocenters are formed in the rearrangement. The topicity of these examples can be analyzed by reference to Figure 6.17, which defines the facial topicity for the components of the bond-forming reaction, and also shows how these heterotopic faces are combined to form either *syn* or *anti* relative configurations in the product. Figure 6.17a and c

---

28. Indeed, transition state models having slightly different envelope or half-chair conformations have been proposed (*cf.* refs. [294–296,315]).

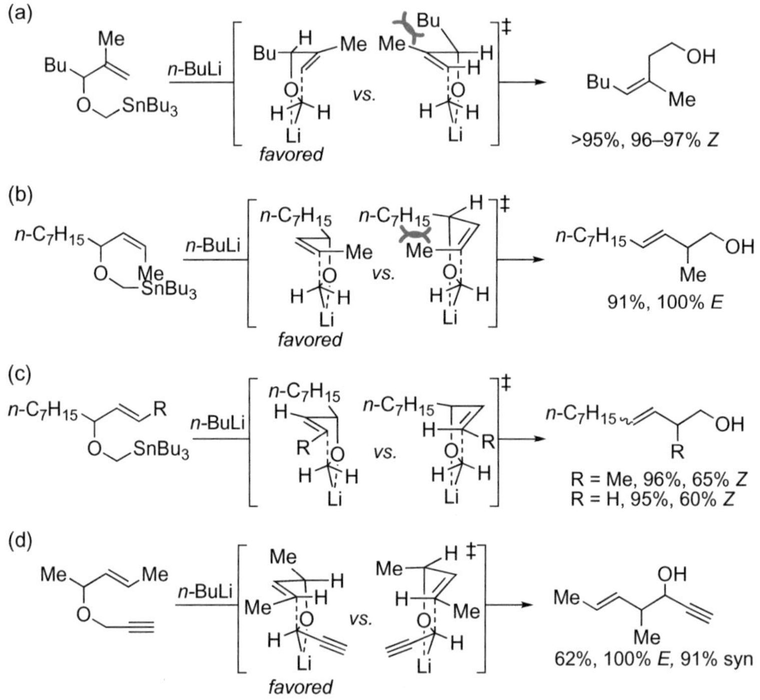

SCHEME 6.50 The effects of allylic strain on the stereoselectivity of alkene formation [301]. (a) $A^{1,2}$ strain and the selective formation of Z-alkenes. (b) $A^{1,3}$ strain causes selective formation of E-alkenes. (c) If one or both of the "partners" (*cf.* Figure 6.16a, R_1, R_2, or Z) is hydrogen, the selectivity is diminished. (d) 100% E selectivity is observed when the carbanionic carbon is propargylic and the alkene is E [302].

shows the topicities for Z-alkenes, while Figure 6.17b and d illustrates similar transition structures for E-alkenes. Note that the preceding discussion analyzed the combined effects of substitution on the double bond and at the allylic position. The structures in Figure 6.17 are unsubstituted at the allylic position, so that the factors affecting relative configuration can be analyzed independent of the effects of an allyl substituent.

Many examples of this type of reaction have been reported in the literature, but only with a few alkyl substituents on the metalated carbon are high selectivities consistently achieved. Table 6.7 lists several such examples, which can be rationalized by the indicated structures in Figure 6.17. Recall (Figure 6.16b and accompanying discussion) that theory predicts that electron-rich alkyl substituents will prefer the convex face of the transition structures (*i.e.*, Figure 6.16a and d), for electrostatic reasons [309].

Entry 1 was the first example, reported in 1970 [290], of a highly stereoselective [2,3]-Wittig rearrangement, but comparison with entry 5 shows that only the Z isomer is selective. Entries 2–4 illustrate substituted propargyl Z allyl ethers, which show a consistently high Z → *syn* selectivity, consistent with the transition structure in Figure 6.17a being favored over Figure 6.17c [302]. Entry 4 (trimethylsilylalkyne) is particularly striking because the product has a higher *syn/anti* ratio than the *Z/E* ratio in the starting material! This is not experimental error, as shown by Entry 8, which is also highly *syn* selective even though the starting material is 93% *E*, and *anti* product was expected [302]. Entries 6 and 7 show a more predictable tendency for very high *E* → *anti* stereoselectivity (Figure 6.17d favored over Figure 6.17c), underscoring the anomalous nature of Entry 8. Entry 9 demonstrates that carbanions that are not resonance stabilized are also highly selective. In this case, the organolithium was generated by transmetalation of an organostannane, and again high *E* → *anti* stereoselectivity is observed [312].

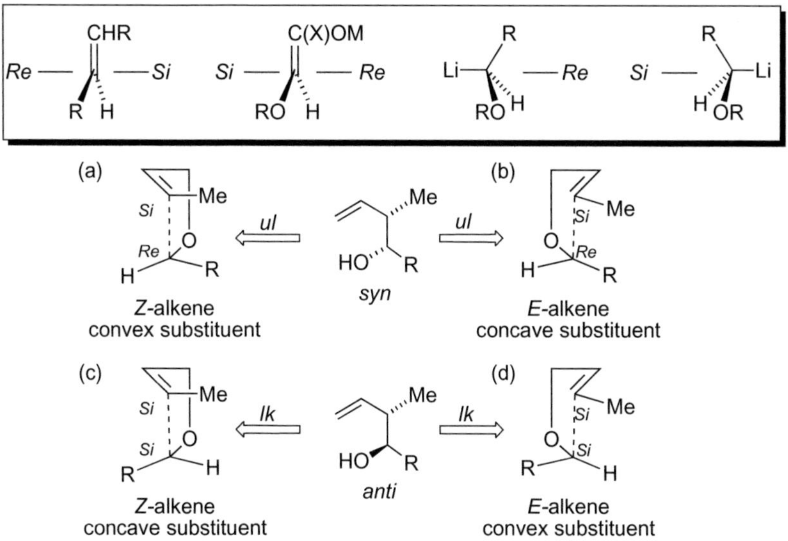

FIGURE 6.17 *Inset:* Heterotopic faces for determining relative topicity (note inversion at the stereogenic RLi). (a) and (b) *Syn* product is formed by two combinations of *ul* topicity. (c) and (d) *Anti* product is formed by two combinations of *lk* topicity. In transition states (a)–(d), the metal is omitted. When R is an alkyl group, it would be bridged to the C—O bond, *antiperiplanar* to the allyl group (*cf.* Figure 6.15a and, b). If R is a carbonyl, the metal will be attached to the enolate oxygen (*cf.* Figure 6.16c).

TABLE 6.7 Selective [2,3]-Wittig Rearrangements of α-Phenyl, α-Propargyl, and α-Alkyl Organolithiums, Showing a High $Z \to syn/E \to anti$ Correlation

Entry	R	E/Z	Path[a]	Config.	% ds	% Yield	Ref.
1	Ph	Z	a	100% syn	100	–	[290]
2	HC≡C	98% Z	a	88% syn	90	56	[302]
3	MeC≡C	98% Z	a	98% syn	100	55	[302]
4	TMSC≡C	93% Z	a (and b)	98% syn	105 (!)	74	[302]
5	Ph	E	b and d	50:50	50	–	[290]
6	HC≡C	93% E	d	93% anti	100	72	[302]
7	MeC≡C	93% E	d	92% anti	99	65	[302]
8	TMSC≡C	93% E	b	75% syn	73	72	[302]
9	Et	E	d	99% anti	99	95	[312]

[a]The "Path" column refers to the transition structures in Figure 6.17.

When the alkenyl component is an *O-tert*-butyldimethylsilyl (TBDMS) enol ether, another anomaly occurs: independent of enol ether geometry, the *anti* product is favored (Scheme 6.51) [316]. With trimethylsilylpropargyl ethers, the *anti* selectivity is 95–98%, making this reaction an excellent route for the preparation of *anti* 1,2-diols. In these cases, transition structures similar to Figures 6.17c and d are operative, the dominant influence being mutual repulsion between the carbanion substituent, R, and the *O*-silyl group.

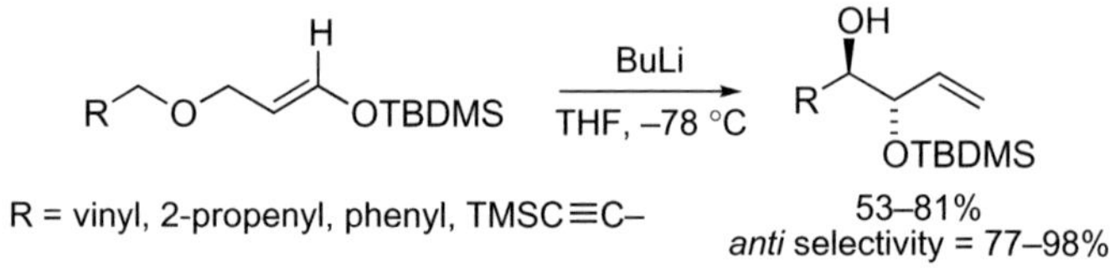

R = vinyl, 2-propenyl, phenyl, TMSC≡C–

53–81%
anti selectivity = 77–98%

SCHEME 6.51 The [2,3]-Wittig rearrangement of silyl enol ethers is *anti* selective independent of carbanion substituent and double bond geometry [316].

Lithium enolates tend to occupy the concave face of the transition structure (*cf.* Figures 6.15d and 6.16c) and therefore prefer transition structures such as those illustrated in Figure 6.17b and c.[29] Table 6.7 lists several examples of simple acyclic diastereoselection, which show a tendency for $Z \to syn$ and $E \to anti$ selectivity, in contrast to the tendency observed for hydrocarbon substituted carbanions. Entries 1 and 2 involve dianions of crotyloxy acetates, and show $Z \to syn$ and $E \to anti$ selectivity. A more complex example involving extension of a steroid side chain (similar to Scheme 6.50a) is 100% *anti* selective from an "*E*"-alkene, however [307].

---

29. Enolates may also suffer competitive [3,3]-sigmatropic rearrangement [294,304,305].

The ester enolates illustrated in Entries 3 and 4 are considerably more selective when the lithium cation is exchanged for dicyclopentadienyl zirconium [317]. It is suggested that the zirconium chelates the α-alkoxy oxygen in these examples, and that the cyclopentadienyl ligands influence the topicity in the transition state [317]. Scheme 6.52 illustrates how the *lk* topicity may be disfavored by a steric interaction between a pseudoaxial allylic hydrogen and a cyclopentadienyl ligand. The Z-alkene isomer (Entry 4) is also *syn*-selective, although less so than the *E* isomer, and the yield is not encouraging. The rationale illustrated in Scheme 6.52 [317] implies that deprotonation of the ester affords the *Z(O)*-enolate, in contrast to the expected (Section 3.1.1) tendency of esters to afford *E(O)*-enolates. In his review of enolate formation [318], Wilcox notes that *Z(O)*-enolate formation by deprotonation of α-alkoxy esters would be expected if chelation were the dominant influence, but that the results reported in the literature show no consistent trend.[30]

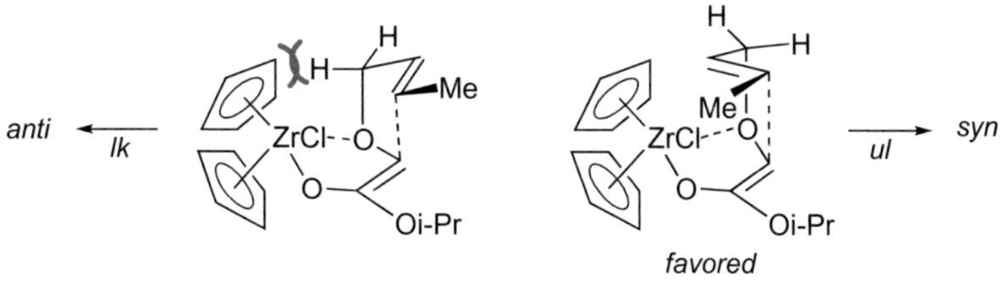

SCHEME 6.52 Rationale for the *ul* selectivity of dicyclopentadienylzirconium ester enolates in [2,3]-Wittig rearrangements.

Pyrrolidinyl amides undoubtedly form *Z(O)*-enolates, and the [2,3]-Wittig rearrangement of the *E*-alkene (Entry 5) is highly selective [323]. The Z-alkene was not tested, and propargylic amide enolates do not rearrange [324]. Entry 5 also shows the highest yield in Table 6.8. As will be seen, amides of C_2-symmetric amines can be excellent chiral auxiliaries in this process.

---

30. Based on the relative configuration of the products of Ireland-Claisen rearrangements, two groups have concluded that *Z(O)*-enolate formation predominates [319,320]. On the other hand, two other groups quenched α-alkoxy ester enolates with trialkylsilyl chlorides and found mixtures of enol ether (ketene acetal) isomers [321,322].

TABLE 6.8 [2,3]-Wittig Rearrangements of α-Allyloxy Enolates

Entry	X	E/Z	Path[a]	Config.	% ds	% Yield	Ref.
1	OH	93% E	b	65% syn[b]	70	60	[325]
2	OH	95% Z	c	75% anti[b]	79	73	[325]
3	Oi-Pr, as Cp$_2$ZrCl enolate	E	b	98% syn	98	47	[317]
4	Oi-Pr, as Cp$_2$ZrCl enolate	Z	a	88% syn	88	15	[317]
5	Pyrrolidinyl	E	b	96% syn	96	97	[323]

[a]The "Path" column refers to the transition structures in Figure 6.17. All examples used LDA as base; entries 3 and 4 also have Cp$_2$ZrCl$_2$ as additive (Cp = η^5-cyclopentadienyl).
[b]The original reference [325] used ambiguous erythro/threo nomenclature without drawing a reference structure (this was a common problem in the 1970s and early 1980s, as was disco). Later, in a review by the same authors [294], the same nomenclature is used but apparently to indicate the opposite relative configurations. Additionally, the review [294] states a different selectivity for the E-alkene than is given in the original article [325]. The table lists the selectivities from the original article with the relative configurations as drawn in the review.

6.2.2.2 Chirality Transfer in Enantiopure Educts

Scheme 6.53 illustrates generic schemes for the asymmetric synthesis of homoallylic alcohols using a [2,3]-Wittig reaction as a key step. In these sequences, the absolute configuration and enantiomeric purity of the starting materials are determined by their method of preparation (or commercial source), and the following examples will show that the chirality sense of the starting material controls the absolute configuration of the product *via* the principles of simple diastereoselectivity outlined in the preceding sections. The absolute configuration of a stereocenter at the allylic (or propargylic) position may be set by asymmetric reduction of an allylic or propargylic ketone (Chapter 7) or asymmetric addition to an aldehyde (Chapter 4). The absolute configuration at the tin-bearing carbon can be set by asymmetric reduction of acyl stannanes [326–328], kinetic resolution using a lipase [329], or oxidation of α-stannyl-borates [330]. In certain cases, the carbanion configuration can be controlled by enantioselective deprotonation.

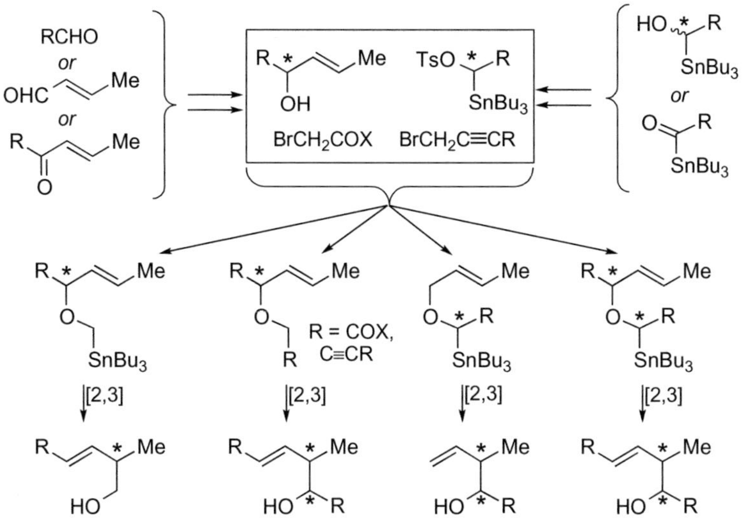

SCHEME 6.53 *Top:* Some routes for preparing chiral substrates for a [2,3]-Wittig rearrangement. *Bottom:* Intramolecular asymmetric induction in [2,3]-Wittig rearrangements.

Qualitative evidence that the [2,3]-Wittig rearrangement of nonracemic substrates might have high enantioselectivities was reported in the early 1970s (*e.g.*, see refs. [291,331]), but it was some years before this aspect of the reaction was quantitated. For example, in 1984 Midland [332] and Nakai [333] showed that nonracemic ethers with stereocenters at the allylic position *and* having the *Z* configuration at the double bond were highly selective for the product having the *E*-configured double bond and *syn* relative configuration at the two new stereocenters. In addition, the chirality transfer was quantitative, as illustrated in Scheme 6.54a. Substituents at the allylic position and the *Z*-olefinic site are susceptible to severe $A^{1,3}$ strain in one of the conformers of the transition state (*cf.* Figure 6.16a, Scheme 6.50b), and this effect determines the absolute configuration at one of the two new stereocenters. Additionally, the two faces of the carbanionic carbon in these examples are heterotopic; the topicity is determined by the greater preference of the carbanionic substituent to occupy the convex face of the envelope transition structure (*cf.* Figure 6.17a). When R_1 is isopropyl, the *E*-alkene isomers show only 60−62% selectivity for the *anti* isomer [332]. In these cases, the isopropyl probably favors a pseudoequatorial conformation and there is only a slight preference for the phenyl or vinyl carbanion substituent to occupy the convex face of the transition structure. Marshall has reported examples that show how chirality transfer may be adversely affected by untoward steric interactions, as in the case where an additional methyl group was introduced into the transannular case shown in Scheme 6.54b. [334].

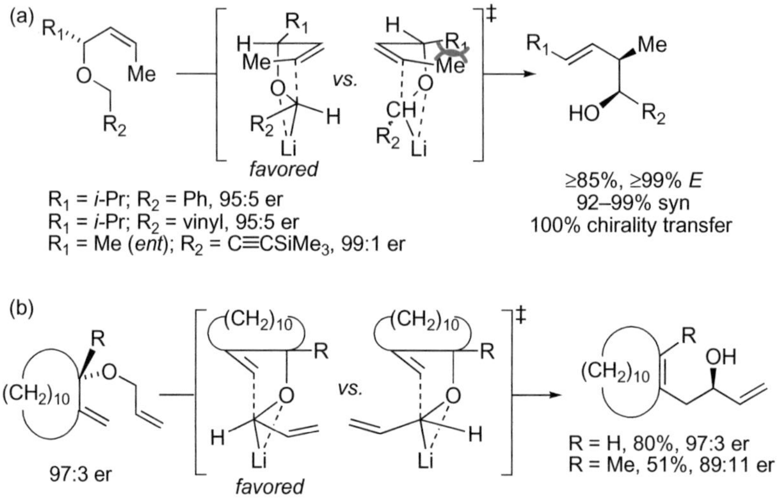

SCHEME 6.54 (a) Asymmetric induction and chirality transfer in [2,3]-Wittig rearrangements of allylic benzyl [332], allyl [332], and trimethylsilylpropargyl [333] ethers. (b) A low enantioselectivity may ensue in some instances, for example, when a transannular interaction destabilizes the favored carbanion configuration [334].

In 1986 [317], Katsuki showed that the dicyclopentadienylzirconium ester enolates shown in Scheme 6.55 afforded products where three stereochemical elements in the product were controlled with a high degree of selectivity: the double bond geometry, the relative configuration, and the absolute configuration. Only one double bond isomer was observed, the *syn/anti* diastereoselectivity was 98–99%, and the enantioselectivity was >98%.

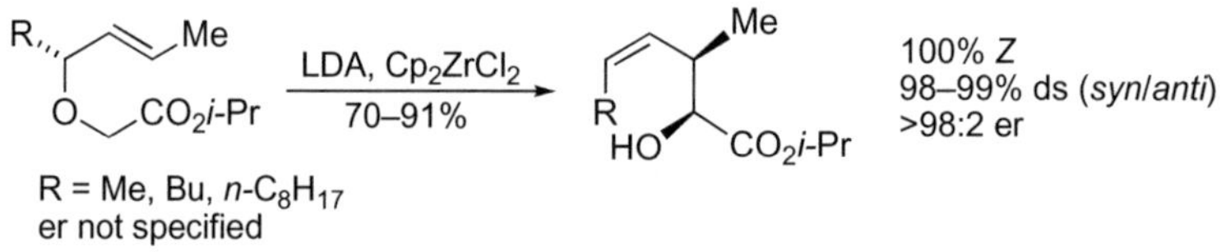

SCHEME 6.55 Asymmetric induction in [2,3]-Wittig rearrangements of chiral α-alkoxy esters [317].

With three stereogenic elements in the product, there are a total of eight possible stereoisomers. However, if it is assumed that the possible transition structures are similar to those shown in Scheme 6.53, there are only four possibilities for the [2,3]-rearrangement, as shown in Scheme 6.56. (Recall from Figure 6.15d that enolate transition structures have shorter bond lengths and smaller allylic bond angles than the other transition structures.) The two having *lk* topicity, Schemes 6.56a and c, are disfavored by having the pseudoaxial allylic substituent in close proximity to the cyclopentadiene ligand. Of the two possible *ul* transition structures, the R group is on the less crowded convex face of the bicyclic structure in Scheme 6.56b, but on the concave face in d, where it encounters the cyclopentadiene ligand.

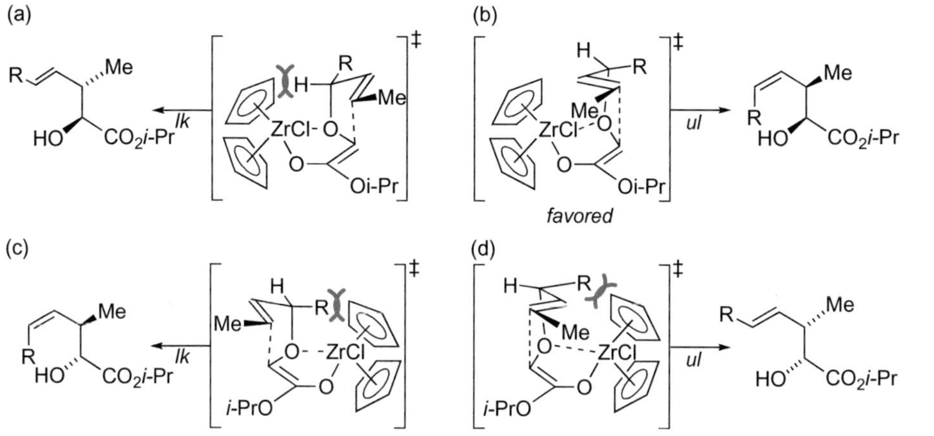

SCHEME 6.56 Proposed transition structures for the [2,3]-Wittig rearrangement of the *R*-allylic ester enolates shown in Scheme 6.55.

Marshall has similarly shown that propargyloxy esters undergo highly selective [2,3]-Wittig rearrangements with 100% chirality transfer (Scheme 6.57a [335–337]), presumably resulting from a transition state assembly having the propargylic substituent on the convex face of the transition state assembly (*cf.* Scheme 6.56b). In addition, this group utilized trimethylsilyl triflate (previously reported by Nakai [338]) to trigger the rearrangement, this time affording the different relative configurations between the allenyl group and the C-2 stereocenter [337]. Nakai suggested an "oxygen ylide" as the intermediate in these silicon-mediated [2,3]-rearrangements, with the silicon and the enolate moieties *trans* to each other in the 5-membered-ring transition structure [338], as shown for the propargyl ether in Scheme 6.57b [337].

Chirality transfer in the rearrangement of allyloxymethyl stannanes is complete, even in cases where the rearrangement itself is not selective for one product, as shown by the

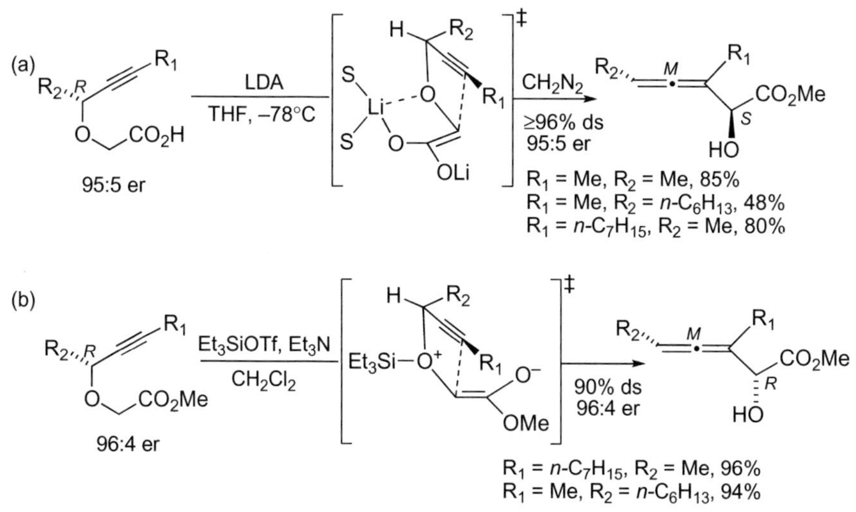

SCHEME 6.57 (a) [2,3]-Wittig rearrangement of a lithium enediolate (S = solvent [335–337]). (b) Silicon-mediated [2,3]-Wittig rearrangement of chiral propargyloxy acetates [337]. The minor diastereomer is the C-2 *S* hydroxyl.

examples in Scheme 6.58 [339]. Recall from Scheme 6.50b and c that in the Still−Wittig rearrangement, one product double bond configuration is formed selectively only when the educt has the Z configuration. This is due to severe $A^{1,3}$ strain in one of the two transition structures (*e.g.*, between the isopropyl and the methyl in Scheme 6.58a). In 1985, Midland reported that rearrangement of the Z-olefin illustrated in Scheme 6.58a was 100% selective for the E-double bond geometry and that the enantiomeric purity of the product matched the enantiomeric purity of the starting material. In contrast, the isomeric E-educt affords a 53:47 mixture of E and Z products, as shown in Scheme 6.58b. However, chirality transfer for the formation of each of these products is 100%, even though the absolute configurations of the newly created stereocenter in the two products are opposite. This result may be explained by examining the two transition structures illustrated. The conformation that presents the *Si* face of the olefin to the metalated carbon (Scheme 6.58b, top) is destabilized by $A^{1,2}$ strain (between the isopropyl and the neighboring vinyl proton) while the conformer that presents the *Re* face is destabilized by $A^{1,3}$ strain (between the isopropyl and the other vinylic hydrogen). These two effects are approximately equal in this relatively "loose" transition

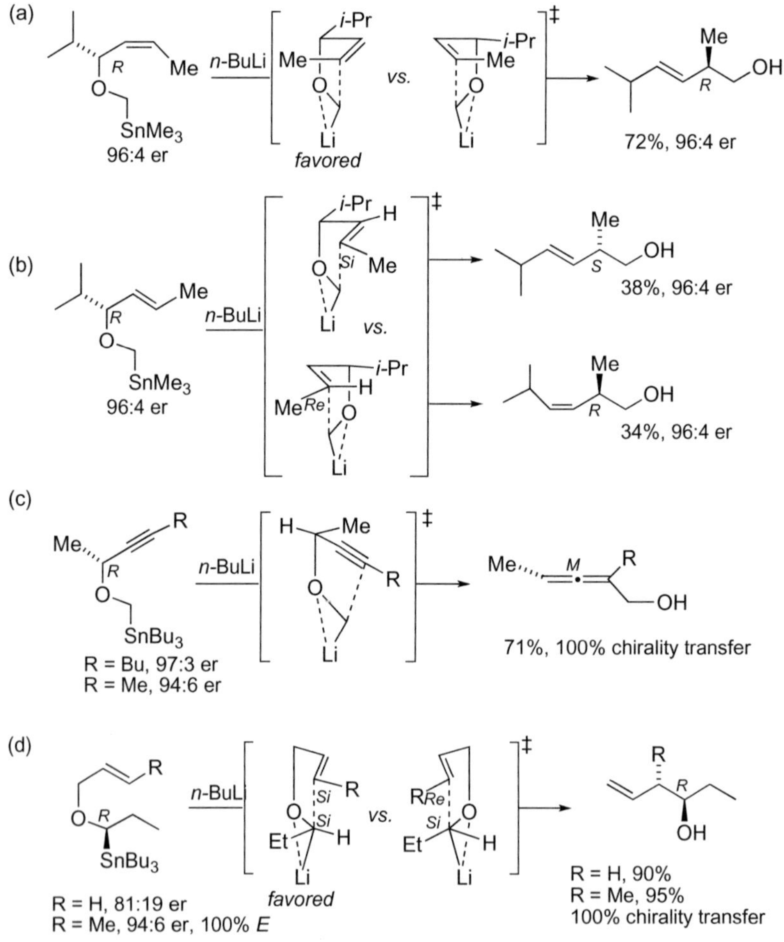

SCHEME 6.58 [2,3]-Still−Wittig rearrangements of (a) and (b) double bond isomers of allyl ethers [339], (c) a propargylic ether [335], ands (d) an allyl ether containing stereogenic carbon [312].

structure (*cf.* Figure 6.15a and b), so the product ratio is nearly equal. The rearrangement of propargyloxy stannanes is highly selective, as shown by Marshall in 1989 [335]. The two examples illustrated in Scheme 6.58 show 100% chirality transfer. In this case, there is no conformational ambiguity, since neither of the carbons involved in bond formation are heterotopic. Chirality transfer is also quantitative when the metalated carbon is stereogenic, as shown by the examples in Scheme 6.58 [312]. When R is hydrogen, the two faces of the terminal allylic carbon are homotopic and it does not matter which of the illustrated transition structures is involved. The only important point is that the metal-bearing carbon undergoes inversion of configuration (see also Scheme 6.50b). When R is methyl, the metal-bearing carbon still undergoes inversion, but the configuration at the second stereocenter is determined by consideration of the two illustrated transition structures. Here, the *ul* topicity is favored (the reaction is 99% diastereoselective for the *anti* relative configuration) because of the preference for the ethyl group to occupy the convex face of the transition structure (see Table 6.7, entry 9).

6.2.2.3 Asymmetric Induction

Chiral auxiliaries, in the form of esters of chiral alcohols and amides of C_2-symmetric chiral amines have been evaluated in these rearrangements. For example, Nakai showed that the lithium enolates of 8-phenylmenthol esters afford good simple diastereoselectivity with good asymmetric induction as well (Scheme 6.59, [340]. As before, the rationale invokes an α-alkoxyenolate that chelates the lithium metal. The inset illustrates the most stable conformation of the chelated enolate and shows the rationale for preferential attack on the *Si* face of the enolate. While monosubstituted pyrrolidine amides were poor auxiliaries for this reaction ($\leq$76% ds) [323],

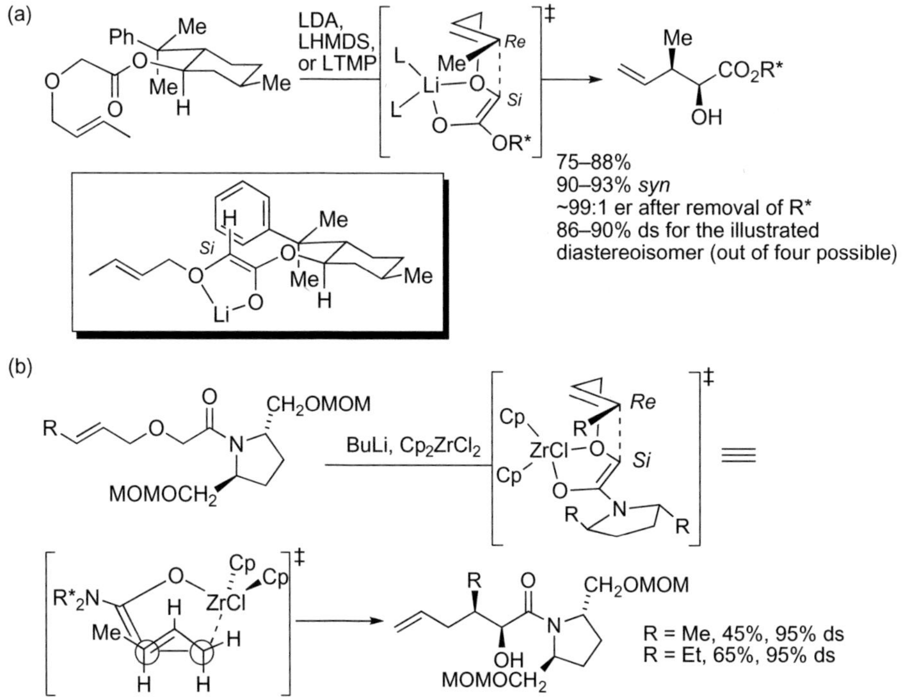

SCHEME 6.59   (a) Ester [340] and (b) amide [341] chiral auxiliaries in the [2,3]-Wittig rearrangement. *Inset:* Rationale for the *Si*-face selectivity of the enolate in (a).

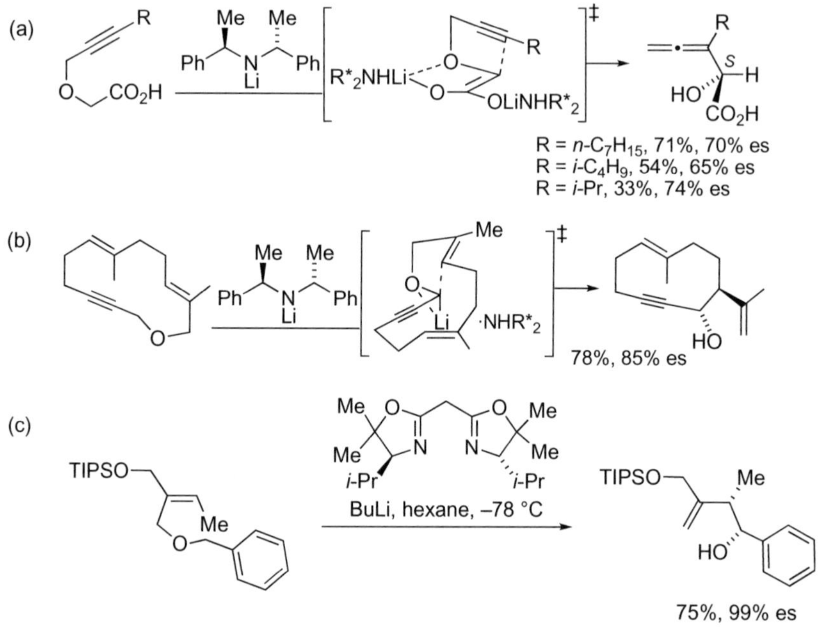

SCHEME 6.60 Asymmetric [2,3]-Wittig rearrangements using (a) and (b) a chiral lithium amide base [324,342,343,345] or (c) chiral ligand [344].

C_2-symmetric pyrrolidines were more successful when applied to zirconium enolates, as shown in Scheme 6.59 [341]. In this reaction, the enolate *Si* face was favored because the closer of the two pyrrolidine stereocenters blocks the *Re* face. In addition, *ul* topicity was favored because when the enolate moiety is on the concave face of the cyclopentane envelope, a severe interaction between a pseudoaxial hydrogen and a cyclopentadiene ligand is avoided.

One can exploit the close association of the amine with the lithium ion to effect interligand asymmetric induction in [2,3]-Wittig reactions of achiral substrates. In Scheme 6.60a [324], Overberger's base was used to doubly deprotonate a propargyloxy acetic acid; presumably, the enolate is chelated by the α oxygen, as shown in the illustrated transition structure. Higher enantioselectivity was obtained with the 13-membered propargyl ether shown in Scheme 6.60b [342,343]. In a related approach, a highly enantioselective reaction was accomplished using an added Box ligand to coordinate the lithium counterion [344]. Interestingly, carrying out the analogous conditions on the corresponding *Z* isomer of the substrate shown in Scheme 6.60c was essentially nonselective.

6.2.3 Other Rearrangements

The concepts introduced in this chapter can be applied to other rearrangements, of which we will introduce a few examples. One of the most important rearrangements in all of stereoselective synthesis is the Claisen rearrangement, which exists in numerous variations and is a mainstay of complex molecule synthesis. As noted in the introduction to this chapter, however, the majority of this work has focused on diastereoselective reactions and has been expertly reviewed (one might start with a comprehensive monograph published in 2007 [346]). Here, we limit our discussion to a brief summary of asymmetric Claisen reactions that begin with achiral starting materials and in which the rearrangement is the stereochemically relevant step (Scheme 6.61).

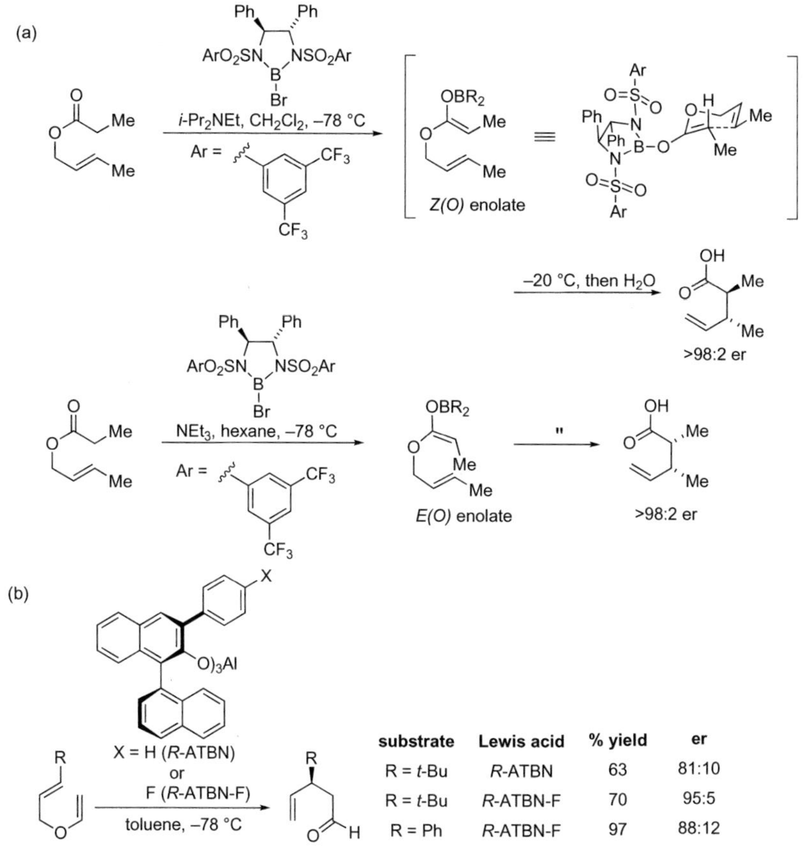

substrate	Lewis acid	% yield	er
R = *t*-Bu	*R*-ATBN	63	81:10
R = *t*-Bu	*R*-ATBN-F	70	95:5
R = Ph	*R*-ATBN-F	97	88:12

SCHEME 6.61 Early examples of asymmetric Claisen rearrangements (a) Corey [347] and (b) Yamamoto [348].

The pioneering contributions shown in Schemes 6.61a and b were reported by Corey in 1991 [347] and Yamamoto in 1995 [348] for the Ireland-type enolate Claisen and the reaction of a simple vinyl ether, respectively. Note the effect of adding an electron-withdrawing group to the aryl ligand on the stereoselectivity for the latter example. The bracketed structures in Scheme 6.61a show how the chair-like transition state normally associated with these reactions can be folded into a particular enantiomeric sense in response to the configuration of the boronate unit. In addition, it features one of the most useful aspects of all stereoselective Claisen rearrangements: the ability to control the *syn/anti* relative configuration of the product through generating a specific enolate (as pioneered by Ireland [349]).

Despite this early promise, there followed some years of fits and starts (for some examples, see ref. [350]) before additional catalyst systems were brought to bear. To this end, MacMillan demonstrated useful selectivities in an interesting Claisen variant using an *N*-acylation event to trigger the reaction; the rather unusual ArBox ligand shown proved superior in this case (Scheme 6.62a [351]). Around the same time, Hiersemann applied more common Box-based Lewis acid systems to carry out the reaction shown in Scheme 6.62b [352,353].

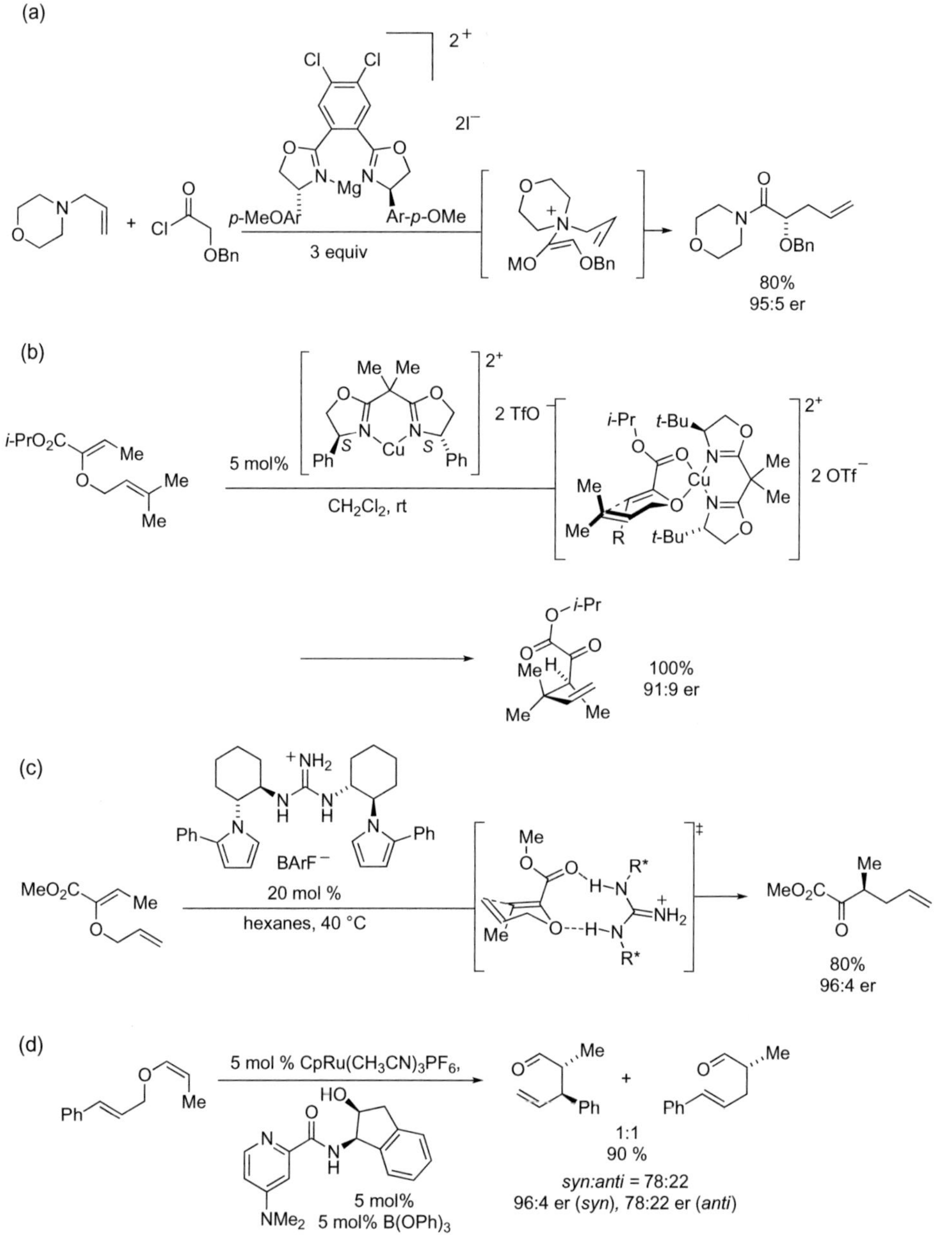

(a)

80%
95:5 er

(b)

100%
91:9 er

(c)

80%
96:4 er

(d)

1:1
90 %

syn:anti = 78:22
96:4 er (syn), 78:22 er (anti)

SCHEME 6.62 Asymmetric Claisen rearrangements by (a) MacMillan [351], (b) Hiersemann, (c) Jacobsen [354], and (d) Nelson [355].

He proposed that bidentate coordination of the copper atom with the "extra" carbonyl group installed on the substrate permitted the high stereoselectivity observed in this case. Seven years later, Jacobsen explored a similarly equipped substrate using hydrogen bond catalysis [354]. Note the similar conformations adopted by the two rather different catalysts used in Schemes 6.62b and c. In 2010, Nelson reported that a cationic Ru complex could, in the presence of a suitable chiral ligand provide Claisen rearrangement in high selectivity even in the absence of additional ligating substituents (Scheme 6.62d [355]; the free hydroxy group in the catalyst shown is essential). Nelson proposed that this reaction occurs stepwise *via* an ionic intermediate based in part on the observation of a formal [1,3] rearrangement product resulting from recombination of the two pieces with the opposite regiochemistry to the product given (1:1 product ratio in the example shown).

Another [3,3] rearrangement is the *Overman rearrangement*, of which two asymmetric examples are provided in Scheme 6.63. This reaction is of obvious utility in alkaloid work because it represents a convenient way of introducing an allylic nitrogen substituent into a molecule. Two data points in the development of this reaction are provided for the reader's interest, along with the outline of Overman's original mechanistic hypothesis for this reaction, termed "cyclization induced rearrangement catalysis" [356,357]. In this scenario, the main stereochemically relevant step is presumably the *anti* addition of the nitrogen nucleophile to the olefin activated by the metal catalyst, with face selectivity arising from interactions encountered along the way [356].

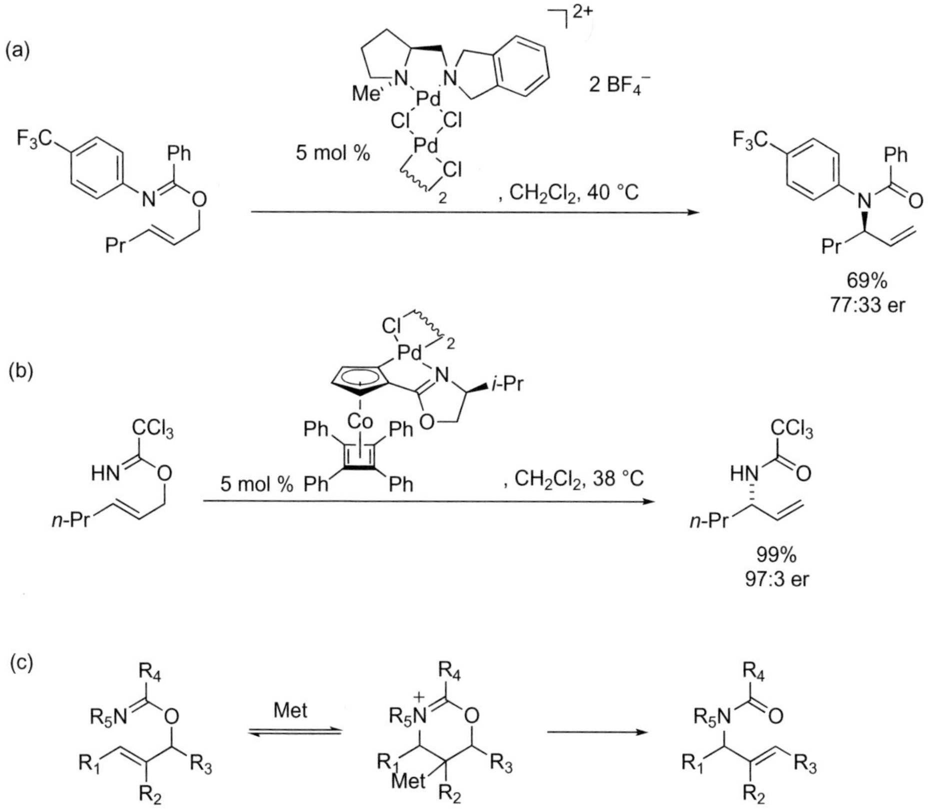

SCHEME 6.63 (a) and (b) Asymmetric examples of the Overman rearrangement [356,357]. (c) The basic concept of "cyclization-induced rearrangement catalysis" as mediated by a metallic catalyst [356].

The *Steglich rearrangement* is a decidedly ionic affair, involving the migration of an acyl group from oxygen to carbon and, in so doing, providing an opportunity for the establishment of a new stereocenter (for some recent reviews that cover this as well as other asymmetric rearrangements see refs. [358,359]). Scheme 6.64 provides some examples, beginning with an

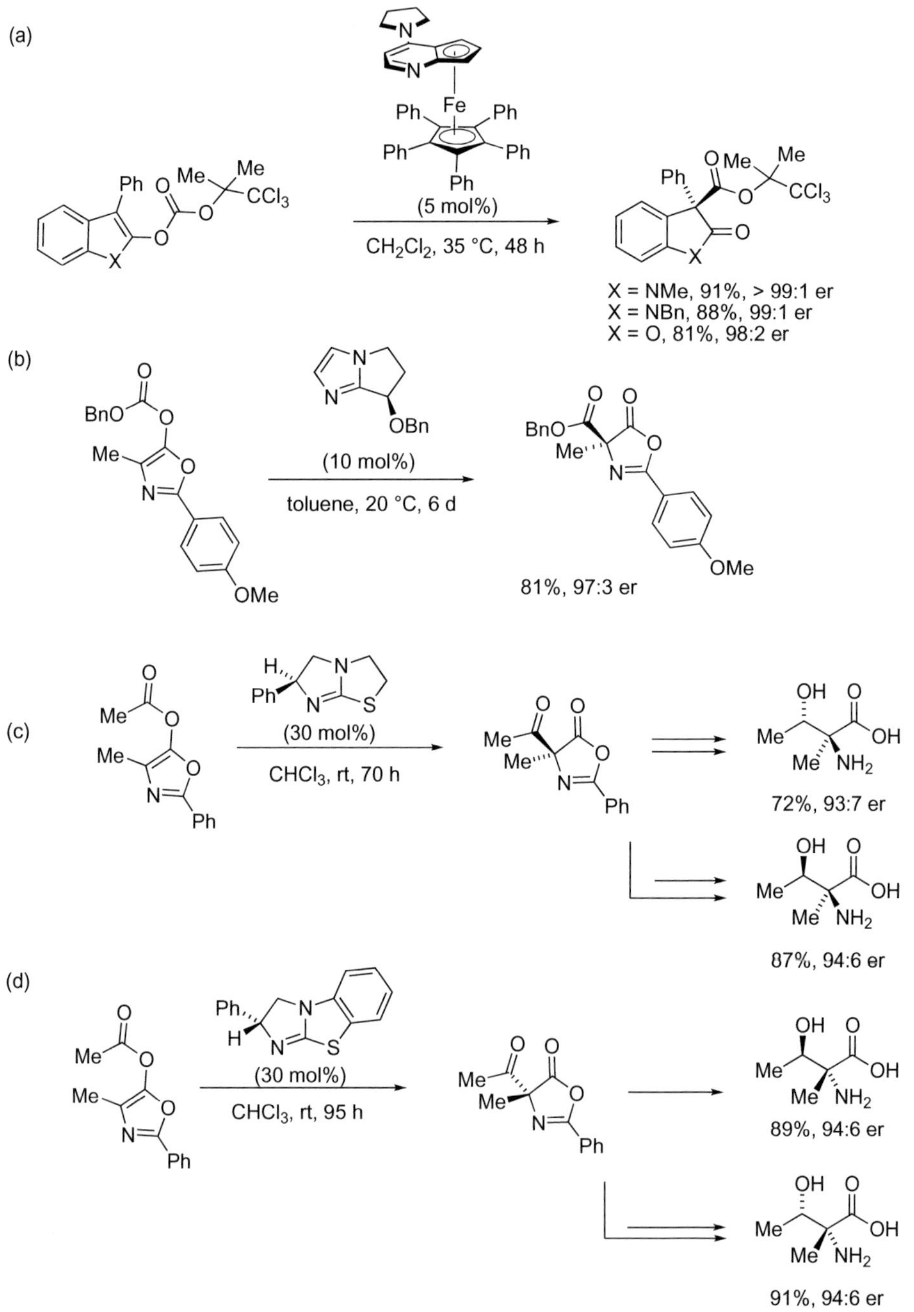

SCHEME 6.64   Steglich rearrangements reported by (a) Fu [360], (b) Zhang [362], and (c) and (d) Gröger [363].

apt use of Fu's catalyst, shown in Scheme 6.64a [360]. This can be viewed as a hybrid between the highly nucleophilic dimethylaminopyridine (the original catalyst for the reaction and a reagent championed by Steglich himself [361]) and an ortho-substituted ferrocene, which generates planar chirality in the Fu catalyst. The mechanism likely involves the nucleophilic attack of the catalyst onto substrate, which then rearranges to the more thermodynamically stable C-acylated product. This mechanism is strongly supported by an X-ray structure of the ion pair intermediate [360]. Other versions that use entirely organocatalytic means of effecting the reaction are shown by Zhang (Scheme 6.63b [362]) and Gröger (Schemes 6.64c and d [363]). The latter two examples show how similar catalysts having opposite configurations combine with different reagent-controlled carbonyl reduction reactions to afford all four stereoisomers of α-methylthreonine. Note that each of the examples shown provides products containing challenging fully-substituted stereogenic centers.

We conclude this chapter by showing a couple of C-to-C migration reactions that provide lovely carbocylic frameworks containing quaternary stereocenters *via* group-selective rearrangement reactions. The first example, by Zhang (Scheme 6.65a), involves an acid-promoted semipinacol rearrangement [364] and the second is Tu's vinylogous α-ketol rearrangement (Scheme 6.65b [365]). The former reaction is triggered by an asymmetric protonation of a nucleophilic enol ether whereas the latter involves iminium ion catalysis on an electron-deficient double bond.

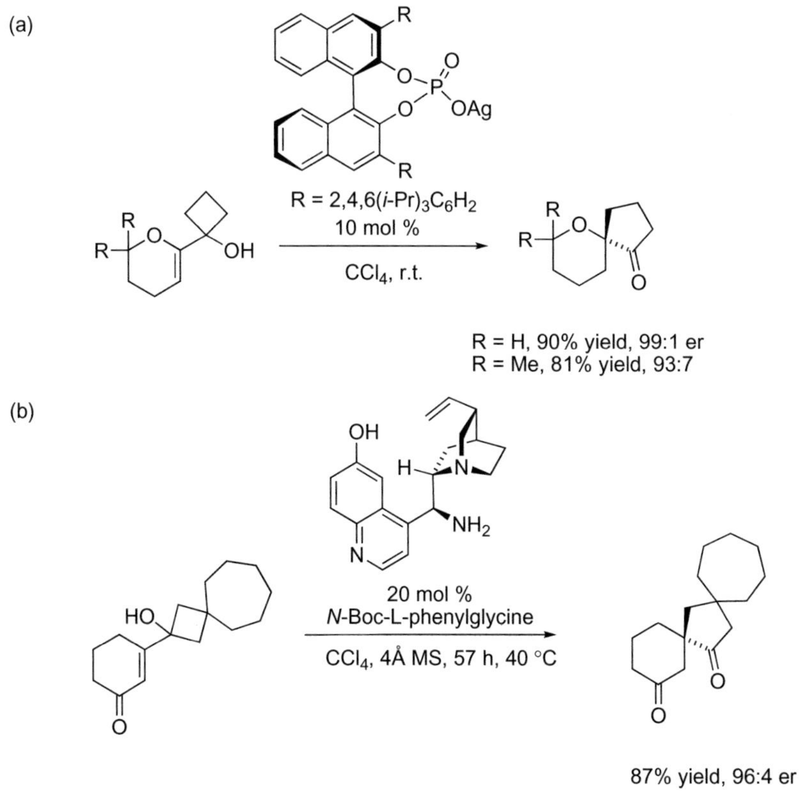

SCHEME 6.65 Examples of carbon-to-carbon migration reactions that proceed with high stereoselectivity by (a) Zhang [364] and (b) Tu [365].

REFERENCES

[1] Oppolzer, W. *Angew. Chem., Int. Ed. Engl.* **1984**, *23*, 876−889.
[2] Helmchen, G.; Karge, R.; Weetman, J. In *Modern Synthetic Methods*; Scheffold, R., Ed.; Springer: Berlin, 1986; Vol. 4, pp. 262−306.
[3] Taschner, M. In *Organic Synthesis. Theory and Applications*; Hudlicky, T., Ed.; JAI: Greenwich, CT, 1989; Vol. 1, pp. 1−101.
[4] Kagan, H. B.; Riant, O. *Chem. Rev.* **1992**, *92*, 1007−1019.
[5] Ma, S. *Handbook of Cyclization Reactions*; Wiley-VCH: Weinheim, 2010.
[6] Ishihara, K.; Sakakura, A. In *Science of Synthesis: Stereoselective Synthesis*; De Vries, J.G., Molander, G. A., Evans, P. A., Eds.; Georg Thieme Verlag: Stuttgart, 2011, Vol. 3, pp. 67−123.
[7] Sauer, J.; Kredel, J. *Tetrahedron Lett.* **1966**, *7*, 6359−6364.
[8] Corey, E. J.; Ensley, H. E. *J. Am. Chem. Soc.* **1975**, *97*, 6908−6909.
[9] Oppolzer, W.; Kurth, M.; Reichlin, D.; Moffatt, F. *Tetrahedron Lett.* **1981**, *22*, 2545−2548.
[10] Boeckman, R. K., Jr; Naegley, P. C.; Arthur, S. D. *J. Org. Chem.* **1980**, *45*, 752−754.
[11] LeDrain, C.; Greene, A. E. *J. Am. Chem. Soc.* **1982**, *104*, 5473−5483.
[12] Corey, E. J.; Cheng, X.-M.; Crimprich, K. A. *Tetrahedron Lett.* **1991**, *32*, 6839−6842.
[13] Oppolzer, W.; Chapuis, C.; Dao, G. M.; Reichlin, D.; Godel, T. *Tetrahedron Lett.* **1982**, *23*, 4781−4784.
[14] Oppolzer, W.; Chapuis, C.; Bernardinelli, G. *Tetrahedron Lett.* **1984**, *25*, 5885−5888.
[15] Wipf, P. In *Comprehensive Organic Synthesis. Selectivity, Strategy, and Efficiency in Modern Organic Chemistry*; Trost, B. M., Fleming, I., Eds.; Pergamon: Oxford, 1991; Vol. 5, pp. 827−873.
[16] Poll, T.; Sobczak, A.; Hartmann, H.; Helmchen, G. *Tetrahedron Lett.* **1985**, *26*, 3095−3098.
[17] Poll, T.; Hady, A. F. A.; Karge, R.; Linz, G.; Weetman, J.; Helmchen, G. *Tetrahedron Lett.* **1989**, *30*, 5595−5598.
[18] Gras, J.-L.; Pellissier, H. *Tetrahedron Lett.* **1991**, *32*, 7043−7046.
[19] Oppolzer, W.; Chapuis, C.; Bernardinelli, G. *Helv. Chim. Acta* **1984**, *67*, 1397−1401.
[20] Evans, D. A.; Chapman, K. T.; Bisaha, J. *J. Am. Chem. Soc.* **1988**, *110*, 1238−1256.
[21] Wiberg, K. B.; Laidig, K. E. *J. Am. Chem. Soc.* **1987**, *109*, 5935−5943.
[22] Loncharich, R. J.; Schwarz, T. R.; Houk, K. N. *J. Am. Chem. Soc.* **1987**, *109*, 14−23.
[23] Castellino, S.; Dwight, W. J. *J. Am. Chem. Soc.* **1993**, *115*, 2986−2987.
[24] Poll, T.; Metter, J. O.; Helmchen, G. *Angew. Chem., Int. Ed. Engl.* **1985**, *24*, 112−114.
[25] Oppolzer, W.; Chapuis, C. *Tetrahedron Lett.* **1984**, *25*, 5383−5386.
[26] Taber, D. F. *Intramolecular Diels-Alder and Alder Ene Reactions*; Springer: Berlin, 1984.
[27] Ciganek, E. *Org. React.* **1984**, *32*, 1−374.
[28] Roush, W. R. In *Comprehensive Organic Synthesis. Selectivity, Strategy, and Efficiency in Modern Organic Chemistry*; Trost, B. M., Fleming, I., Eds.; Pergamon: Oxford, 1991; Vol. 5, pp. 513−550.
[29] Deslongchamps, P. *Pure Appl. Chem.* **1992**, *64*, 31−47.
[30] Oppolzer, W.; Dupuis, D. *Tetrahedron Lett.* **1985**, *26*, 5437−5440.
[31] Roush, W. R.; Gillis, H. R.; Ko, A. I. *J. Am. Chem. Soc.* **1982**, *104*, 2269−2283.
[32] Walborsky, H. M.; Barash, L.; Davis, T. C. *J. Org. Chem.* **1961**, *26*, 4778−4779.
[33] Walborsky, H. M.; Barash, L.; Davis, T. C. *Tetrahedron* **1963**, *19*, 2333−2351.
[34] Furuta, K.; Iwanaga, K.; Yamamoto, H. *Tetrahedron Lett.* **1986**, *27*, 4507−4510.
[35] Ito, Y. N.; Beck, A. K.; Bohá., A.; Ganter, C.; Gawley, R. E.; Kühnle, F. N. M.; Piquer, J. A.; Tuleja, J.; Wang, Y. M.; Seebach, D. *Helv. Chim. Acta* **1994**, *77*, 2071−2110.
[36] Maruoka, K.; Saito, S.; Yamamoto, H. *J. Am. Chem. Soc.* **1992**, *114*, 1089−1090.
[37] Tomioka, K.; Hamada, N.; Suenaga, T.; Koga, K. *J. Chem. Soc., Perkin Trans. 1* **1990**, 426−428.
[38] Helmchen, G.; Schmierer, R. *Angew. Chem., Int. Ed. Engl.* **1981**, *20*, 205−207.
[39] Baldwin, S. W.; Greenspan, P.; Alaimo, C.; McPhail, A. T. *Tetrahedron Lett.* **1991**, *32*, 5877−5880.
[40] Maruoka, K.; Akakura, M.; Saito, S.; Ooi, T.; Yamamoto, H. *J. Am. Chem. Soc.* **1994**, *116*, 6153−6158.
[41] Enders, D.; Meyer, O.; Raabe, G. *Synthesis* **1992**, 1242−1244.
[42] Krohn, K. *Angew. Chem., Int. Ed. Engl.* **1993**, *32*, 1582−1584.
[43] Kozmin, S. A.; Rawal, V. H. *J. Am. Chem. Soc.* **1997**, *119*, 7165−7166.
[44] Trost, B. M.; O'Krongly, D.; Belletire, J. L. *J. Am. Chem. Soc.* **1980**, *102*, 7595−7596.
[45] Siegel, C.; Thornton, E. R. *Tetrahedron Lett.* **1988**, *29*, 5225−5228.
[46] Raw, S. A.; Taylor, R. J. K. *Adv. Heterocycl. Chem.* *100*:75−100.

[47] Evans, D. A.; Lectka, T.; Miller, S. J. *Tetrahedron Lett.* **1993**, *34*, 7027−7030.

[48] Evans, D. A.; Miller, S. J.; Lectka, T. *J. Am. Chem. Soc.* **1993**, *115*, 6460−6461.

[49] Corey, E. J.; Ishihara, K. *Tetrahedron Lett.* **1992**, *33*, 6807−6810.

[50] Kobayashi, S.; Hachiya, I.; Ishitani, H.; Araki, M. *Tetrahedron Lett.* **1993**, *34*, 4535−4538.

[51] Kobayashi, S.; Ishitani, H. *J. Am. Chem. Soc.* **1994**, *116*, 4083−4084.

[52] Mamedov, E.; Klabunovskii, E. *Russ. J. Org. Chem.* **2008**, *44*, 1097−1120.

[53] Oppolzer, W. In *Comprehensive Organic Synthesis. Selectivity, Strategy, and Efficiency in Modern Organic Chemistry*; Trost, B. M., Fleming, I., Eds.; Pergamon: Oxford, 1991; Vol. 5, pp. 315−399.

[54] Tomioka, K. *Synthesis* **1990**, 541−549.

[55] Narasaka, K. *Synthesis* **1991**, 1−11.

[56] Corey, E. J. *Angew. Chem., Int. Ed.* **2002**, *41*, 1650−1667.

[57] Hashimoto, S.; Komeshima, N.; Koga, K. *J. Chem. Soc. Chem. Commun.* **1979**, 437−438.

[58] Hawkins, J. M.; Loren, S. *J. Am. Chem. Soc.* **1991**, *113*, 7794−7795.

[59] Hawkins, J. M.; Loren, S.; Nambu, M. *J. Am. Chem. Soc.* **1994**, *116*, 1657−1660.

[60] Furuta, K.; Shimizu, S.; Miwa, Y.; Yamamoto, H. *J. Org. Chem.* **1989**, *54*, 1481−1483.

[61] Furuta, K.; Kanamatsu, A.; Yamamoto, H.; Takaoka, S. *Tetrahedron Lett.* **1989**, *30*, 7231−7232.

[62] Ishihara, K.; Gao, Q.; Yamamoto, H. *J. Am. Chem. Soc.* **1993**, *115*, 10412−10413.

[63] Aggarwal, V. K.; Anderson, E.; Giles, R.; Zaparucha, A. *Tetrahedron: Asymmetry* **1995**, *6*, 1301−1306.

[64] Hayashi, Y.; Rohde, J. J.; Corey, E. J. *J. Am. Chem. Soc.* **1996**, *118*, 5502−5503.

[65] Corey, E. J.; Shibata, T.; Lee, T. W. *J. Am. Chem. Soc.* **2002**, *124*, 3808−3809.

[66] Ryu, D. H.; Lee, T. W.; Corey, E. J. *J. Am. Chem. Soc.* **2002**, *124*, 9992−9993.

[67] Ryu, D. H.; Zhou, G.; Corey, E. J. *J. Am. Chem. Soc.* **2004**, *126*, 4800−4802.

[68] Corey, E. J.; Imai, N.; Zhang, H.-Y. *J. Am. Chem. Soc.* **1991**, *113*, 728−729.

[69] Narasaka, K.; Inoue, M.; Okada, N. *Chem. Lett.* **1986**, *7*, 1109−1112.

[70] Seebach, D.; Beck, A. K.; Imwinkelried, R.; Roggo, S.; Wonnacott, A. *Helv. Chim. Acta* **1987**, *70*, 954−974.

[71] Seebach, D.; Dahinden, R.; Marti, R. E.; Beck, A. K.; Plattner, D. A.; Kühnle, F. N. M. *J. Org. Chem.* **1995**, *60*, 1788−1799.

[72] Engler, T. A.; Letavic, M. A.; Lynch, K. O., Jr.; Takusagawa, F. *J. Org. Chem.* **1994**, *59*, 1179−1183.

[73] Mikami, K.; Motoyama, Y.; Terada, M. *J. Am. Chem. Soc.* **1994**, *116*, 2812−2820.

[74] Kelly, T. R.; Whiting, A.; Chandrakumar, N. S. *J. Am. Chem. Soc.* **1986**, *108*, 3510−3512.

[75] Chapuis, C.; Jurczak, J. *Helv. Chim. Acta* **1987**, *70*, 436−440.

[76] Ishihara, K.; Yamamoto, H. *J. Am. Chem. Soc.* **1994**, *116*, 1561−1562.

[77] Ishihara, K.; Kurihara, H.; Yamamoto, H. *J. Am. Chem. Soc.* **1996**, *118*, 3049−3050.

[78] Hawkins, J. M.; Nambu, M.; Loren, S. *Org. Lett.* **2003**, *5*, 4293−4295.

[79] Maruoka, K.; Murase, N.; Yamamoto, H. *J. Org. Chem.* **1993**, *58*, 2938−2939.

[80] Corey, E. J.; Loh, T.-P. *J. Am. Chem. Soc.* **1991**, *113*, 8966−8967.

[81] Corey, E. J.; Loh, T.-P.; Roper, T. D.; Azimioara, M. D.; Noe, M. C. *J. Am. Chem. Soc.* **1992**, *114*, 8290−8292.

[82] Seerden, J.-P. G.; Scheeren, H. W. *Tetrahedron Lett.* **1993**, *34*, 2669−2672.

[83] Narasaka, K.; Iwasawa, N.; Inoue, M.; Yamada, T.; Nakashima, M.; Sugimori, J. *J. Am. Chem. Soc.* **1989**, *111*, 5340−5345.

[84] Haase, C.; Sarko, C. R.; DiMare, M. *J. Org. Chem.* **1995**, *60*, 1777−1787.

[85] Gothelf, K. V.; Hazell, R. G.; Jørgensen, K. A. *J. Am. Chem. Soc.* **1995**, *117*, 4435−4436.

[86] Boyle, T. J.; Eilerts, N. W.; Heppert, J. A.; Takusagawa, F. *Organometallics* **1994**, *13*, 2218−2229.

[87] Ahrendt, K. A.; Borths, C. J.; MacMillan, D. W. C. *J. Am. Chem. Soc.* **2000**, *122*, 4243−4244.

[88] List, B. *Chem. Rev.* **2007**, *107*, 5413−5415.

[89] Pellissier, H. In *Recent Developments in Asymmetric Organocatalysis*; RSC Publishing: Cambridge, 2010, pp. 1−241.

[90] *Asymmetric Organocatalysis*; List, B., Ed.; Springer: Heidelberg, 2010.

[91] Berkessel, A.; Groerger, H. In *Asymmetric Organocatalysis*; Wiley-VCH: Weinheim, 2005, pp. 1−454.

[92] Gordillo, R.; Houk, K. N. *J. Am. Chem. Soc.* **2006**, *128*, 3543−3553.

[93] List, B. *Tetrahedron* **2002**, *58*, 5573−5590.

[94] Northrup, A. B.; MacMillan, D. W. C. *J. Am. Chem. Soc.* **2002**, *124*, 2458−2460.

[95] Gotoh, H.; Hayashi, Y. *Org. Lett.* **2007**, *9*, 2859−2862.

[96] Kano, T.; Tanaka, Y.; Maruoka, K. *Org. Lett.* **2006**, *8*, 2687−2689.

[97] Ishihara, K.; Nakano, K. *J. Am. Chem. Soc.* **2005**, *127*, 10504−10505.

[98] Sakakura, A.; Suzuki, K.; Ishihara, K. *Adv. Synth. Catal.* **2006**, *348*, 2457−2465.

[99] Wilson, R. M.; Jen, W. S.; MacMillan, D. W. C. *J. Am. Chem. Soc.* **2005**, *127*, 11616−11617.

[100] Zheng, C.; Lu, Y.; Zhang, J.; Chen, X.; Chai, Z.; Ma, W.; Zhao, G. *Chem. Eur. J.* **2010**, *16*, 5853−5857.

[101] Doyle, A. G.; Jacobsen, E. N. *Chem. Rev.* **2007**, *107*, 5713−5743.

[102] Yu, X.; Wang, W. *Chem. Asian J.* **2008**, *3*, 516−532.

[103] Wassermann, A. *J. Chem. Soc.* **1942**, 618−621.

[104] Kelly, T. R.; Meghani, P.; Ekkundi, V. S. *Tetrahedron Lett.* **1990**, *31*, 3381−3384.

[105] Thadani, A. N.; Stankovic, A. R.; Rawal, V. H. *Proc. Natl. Acad. Sci. U.S.A.* **2004**, *101*, 5846−5850.

[106] Nakashima, D.; Yamamoto, H. *J. Am. Chem. Soc.* **2006**, *128*, 9626−9627.

[107] Stork, G.; van Tamelen, E. E.; Friedman, L. J.; Burgstahler, A. W. *J. Am. Chem. Soc.* **1953**, *75*, 384−392.

[108] Masamune, S.; Reed, L. A.; Davis, J. T.; Choy, W. *J. Org. Chem.* **1983**, *48*, 4441−4444.

[109] Wrobleski, A.; Sahasrabudhe, K.; Aubé, J. *J. Am. Chem. Soc.* **2002**, *124*, 9974−9975.

[110] Kong, K.; Moussa, Z.; Romo, D. *Org. Lett.* **2005**, *7*, 5127−5130.

[111] Hu, Q.-Y.; Zhou, G.; Corey, E. J. *J. Am. Chem. Soc.* **2004**, *126*, 13708−13713.

[112] Yamatsugu, K.; Yin, L.; Kamijo, S.; Kimura, Y.; Kanai, M.; Shibasaki, M. *Angew. Chem., Int. Ed.* **2009**, *48*, 1070−1076.

[113] Jones, S. B.; Simmons, B.; MacMillan, D. W. C. *J. Am. Chem. Soc.* **2009**, *131*, 13606−13607.

[114] Gioia, C.; Hauville, A.; Bernardi, L.; Fini, F.; Ricci, A. *Angew. Chem., Int. Ed.* **2008**, *47*, 9236−9239.

[115] Boger, D. L.; Weinreb, S. M. *Hetero Diels-Alder Methodology in Organic Synthesis*; Academic: New York, 1987.

[116] Weinreb, S. M. In *Comprehensive Organic Synthesis. Selectivity, Strategy, and Efficiency in Modern Organic Chemistry*; Trost, B. M., Fleming, I., Eds.; Pergamon: Oxford, 1991; Vol. 5, pp. 401−449.

[117] Boger, D. L. In *Comprehensive Organic Synthesis. Selectivity, Strategy, and Efficiency in Modern Organic Chemistry*; Trost, B. M., Fleming, I., Eds.; Pergamon: Oxford, 1991; Vol. 5, pp. 451−512.

[118] Waldmann, H. *Synthesis* **1994**, *6*, 535−551.

[119] Pellissier, H. *Tetrahedron* **2009**, *65*, 2839−2877.

[120] Jorgensen, K. A. *Angew. Chem., Int. Ed.* **2000**, *39*, 3558−3588.

[121] Danishefsky, S.; Kerwin, J. F.; Kobayashi, S. *J. Am. Chem. Soc.* **1982**, *104*, 358−360.

[122] Larson, E. R.; Danishefsky, S. *J. Am. Chem. Soc.* **1982**, *104*, 6458−6460.

[123] Danishefsky, S. J. *Aldrichim. Acta* **1986**, *19*, 59−69.

[124] Bednarski, M.; Danishefsky, S. *J. Am. Chem. Soc.* **1983**, *105*, 3716−3717.

[125] Maruoka, K.; Itoh, T.; Shirasaka, T.; Yamamoto, H. *J. Am. Chem. Soc.* **1988**, *110*, 310−312.

[126] Schaus, S. E.; Brnalt, J.; Jacobsen, E. N. *J. Org. Chem.* **1998**, *63*, 403−405.

[127] Yamamoto, Y.; Yamamoto, H. *Angew. Chem., Int. Ed.* **2005**, *44*, 7082−7085.

[128] Dossetter, A. G.; Jamison, T. F.; Jacobsen, E. N. *Angew. Chem., Int. Ed.* **1999**, *38*, 2398−2400.

[129] Huang, Y.; Unni, A. K.; Thadani, A. N.; Rawal, V. H. *Nature* **2003**, *424*, 146.

[130] Unni, A. K.; Takenaka, N.; Yamamoto, H.; Rawal, V. H. *J. Am. Chem. Soc.* **2005**, *127*, 1336−1337.

[131] Kozmin, S. A.; Rawal, V. H. *J. Org. Chem.* **1997**, *62*, 5252−5253.

[132] Huang, Y.; Rawal, V. H. *Org. Lett.* **2000**, *2*, 3321−3323.

[133] Huang, Y.; Rawal, V. H. *J. Am. Chem. Soc.* **2002**, *124*, 9662−9663.

[134] Zhuang, W.; Poulsen, T. B.; Jorgensen, K. A. *Org. Biomol. Chem.* **2005**, *3*, 3284−3289.

[135] Buonora, P.; Olsen, J. C.; Oh, T. *Tetrahedron* **2001**, *57*, 6099−6138.

[136] Devine, P. N.; Reilly, M.; Oh, T. *Tetrahedron Lett.* **1993**, *34*, 5827−5830.

[137] Barluenga, J.; Aznar, F.; Ribas, C.; Valdes, C.; Fernandez, M.; Cabal, M.-P.; Trujillo, J. *Chem.−Eur. J.* **1996**, *2*, 805−811.

[138] Cravotto, G.; Nano, G. M.; Palmisano, G.; Tagliapietra, S. *Tetrahedron: Asymmetry* **2001**, *12*, 707−709.

[139] Dujardin, G.; Leconte, S.; Coutable, L.; Brown, E. *Tetrahedron Lett.* **2001**, *42*, 8849−8852.

[140] Audrain, H.; Thorhauge, J.; Hazell, R. G.; Jorgensen, K. A. *J. Org. Chem.* **2000**, *65*, 4487−4497.

[141] Schmidt, R. R. *Acc. Chem. Res.* **1986**, *19*, 250−259.

[142] Tietze, L.; Kettschau, G. In *Stereoselective Heterocyclic Synthesis I*, Vol. 189. Metz, P. Ed.; Springer: Berlin/Heidelberg, 1997. pp. 1−120.

[143] Gademann, K.; Chavez, D. E.; Jacobsen, E. N. *Angew. Chem., Int. Ed.* **2002**, *41*, 3059−3061.

[144] Juhl, K.; Jorgensen, K. A. *Angew. Chem., Int. Ed.* **2003**, *42*, 1498−1501.

[145] Xu, H.; Zuend, S. J.; Woll, M. G.; Tao, Y.; Jacobsen, E. N. *Science* **2010**, *327*, 986−990.

[146] Kouznetsov, V. V. *Tetrahedron* **2009**, *65*, 2721−2750.

[147] Huisgen, R. *Angew. Chem., Int. Ed. Engl.* **1963**, *2*, 565−598.

[148] Hein, C. D.; Liu, X.-M.; Wang, D. *Pharm. Res* **2008**, *25*, 2216−2230.

[149] Mamidyala, S. K.; Finn, M. G. *Chem Soc Rev* **2010**, *39*, 1252−1261.

[150] Ramón, D. J.; Guillena, G.; Seebach, D. *Helv. Chim. Acta* **1996**, *79*, 875−894.

[151] Gothelf, K. V.; Jørgensen, K. A. *Chem. Rev.* **1998**, *98*, 863−910.

[152] Pellissier, H. *Tetrahedron* **2007**, *63*, 3235−3285.

[153] Stanley, L. M.; Sibi, M. P. *Chem. Rev.* **2008**, *108*, 2887−2902.

[154] Houk, K. N.; Sims, J.; Duke, R. E.; Strozier, R. W.; George, J. K. *J. Am. Chem. Soc.* **1973**, *95*, 7287−7301.

[155] Houk, K. N.; Sims, J.; Watts, C. R.; Luskus, L. J. *J. Am. Chem. Soc.* **1973**, *95*, 7301−7315.

[156] Belzecki, C.; Panfil, I. *J. Org. Chem.* **1979**, *44*, 1212−1218.

[157] Baldwin, S. W.; Aubé, J.; McPhail, A. T. *J. Org. Chem.* **1991**, *56*, 6546−6550.

[158] Romeo, G.; Iannazzo, D.; Piperno, A.; Romeo, R.; Corsaro, A.; Rescifina, A.; Chiacchio, U. *Mini-Rev. Org. Chem.* **2005**, *2*, 59−77.

[159] Hyrosova, E.; Fisera, L.; Medvecky, M.; Reissig, H.-U.; Al-Harrasi, A.; Koos, M. *ARKIVOC* **2008**, 122−142.

[160] Vasella, A. *Helv. Chim. Acta* **1977**, *60*, 1273−1295.

[161] Kanemasa, S. *Synlett* **2002**, *9*, 1371−1387.

[162] Hanselmann, R.; Zhou, J.; Ma, P.; Confalone, P. N. *J. Org. Chem.* **2003**, *68*, 8739−8741.

[163] Gothelf, K. V.; Hazell, R. G.; Jørgensen, K. A. *J. Org. Chem.* **1996**, *61*, 346−355.

[164] Gothelf, K. V.; Jørgensen, K. A. *Acta Chem. Scand.* **1996**, *50*, 652−660.

[165] Saito, T.; Yamada, T.; Miyazaki, S.; Otani, T. *Tetrahedron Lett.* **2004**, *45*, 9581−9584.

[166] Yoon, T. P.; Jacobsen, E. N. *Science* **2003**, *299*, 1691−1693.

[167] Kano, T.; Hashimoto, T.; Maruoka, K. *J. Am. Chem. Soc.* **2005**, *127*, 11926−11927.

[168] Jen, W. S.; Wiener, J. J. M.; MacMillan, D. W. C. *J. Am. Chem. Soc.* **2000**, *122*, 9874−9875.

[169] Sibi, M. P.; Ma, Z.; Jasperse, C. P. *J. Am. Chem. Soc.* **2005**, *127*, 5764−5765.

[170] Onishi, T.; Sebahar, P. R.; Williams, R. M. *Org. Lett.* **2003**, *5*, 3135−3137.

[171] Chen, C.; Li, X.; Schreiber, S. L. *J. Am. Chem. Soc.* **2003**, *125*, 10174−10175.

[172] Kobayashi, S.; Shimizu, H.; Yamashita, Y.; Ishitani, H.; Kobayashi, J. *J. Am. Chem. Soc.* **2002**, *124*, 13678−13679.

[173] Shirakawa, S.; Lombardi, P. J.; Leighton, J. L. *J. Am. Chem. Soc.* **2005**, *127*, 9974−9975.

[174] Pohlhaus, P. D.; Johnson, J. S. *J. Am. Chem. Soc.* **2005**, *127*, 16014−16015.

[175] Helquist, P. In *Comprehensive Organic Synthesis. Selectivity, Strategy, and Efficiency in Modern Organic Chemistry*; Trost, B. M., Fleming, I., Eds.; Pergamon: Oxford, 1991; Vol. 4, pp. 951−997.

[176] Davies, H. M. L. In *Comprehensive Organic Synthesis. Selectivity, Strategy, and Efficiency in Modern Organic Chemistry*; Trost, B. M., Fleming, I., Eds.; Pergamon: Oxford, 1991; Vol. 4, pp. 1031−1067.

[177] Simmons, H. E., Jr.; Cairns, T. L.; Vladuchick, S. A.; Hoiness, C. M. *Org. React.* **1973**, *20*, 1−131.

[178] Doyle, M. P.; Dorow, R. L.; Buhro, W. E.; Griffin, J. H.; Tamblyn, W. H.; Trudell, M. L. *Organometallics* **1984**, *3*, 44−52.

[179] Doyle, M. P.; Dorow, R. L.; Tamblyn, W. H.; Buhro, W. E. *Tetrahedron Lett.* **1982**, *23*, 2261−2264.

[180] Doyle, M. P.; Bagheri, V.; Wandless, T. J.; Harn, N. K.; Brinker, D. A.; Eagle, C. T., et al. *J. Am. Chem. Soc.* **1990**, *112*, 1906−1912.

[181] Davies, H. M. L.; Clark, T. J.; Church, L. A. *Tetrahedron Lett.* **1989**, *30*, 5057−5060.

[182] Lebel, H.; Marcoux, J.-F.; Molinaro, C.; Charette, A. B. *Chem. Rev.* **2003**, *103*, 977−1050.

[183] Pellissier, H. *Tetrahedron* **2008**, *64*, 7041−7095.

[184] Abdallah, H.; Grée, R.; Carrié, R. *Tetrahedron Lett.* **1982**, *23*, 503−506.

[185] Arai, I.; Mori, A.; Yamamoto, H. *J. Am. Chem. Soc.* **1985**, *107*, 8254−8256.

[186] Mori, A.; Arai, I.; Yamamoto, H.; Nakai, H.; Arai, Y. *Tetrahedron* **1986**, *42*, 6447−6458.

[187] Mash, E. A.; Nelson, K. A. *J. Am. Chem. Soc.* **1985**, *107*, 8256−8258.

[188] Mash, E. A.; Nelson, K. A. *Tetrahedron* **1987**, *43*, 679−692.

[189] Mash, E. A.; Torok, D. S. *J. Org. Chem.* **1989**, *54*, 250−253.

[190] Mash, E. A.; Hemperly, S. B.; Nelson, K. A.; Heidt, P. C.; Deusen, S. V. *J. Org. Chem.* **1990**, *55*, 2045−2055.

[191] Sugimura, T.; Futugawa, T.; Tai, A. *Tetrahedron Lett.* **1988**, *29*, 5775−5778.

[192] Sugimura, T.; Yoshikawa, M.; Futugawa, T.; Tai, A. *Tetrahedron* **1990**, *46*, 5955−5966.

[193] Charette, A. B.; Côté, B.; Marcoux, J.-F. *J. Am. Chem. Soc.* **1991**, *113*, 8166−8167.

[194] Charette, A. B.; Marcoux, J.-F. *Tetrahedron Lett.* **1993**, *34*, 7157−7160.

[195] Charette, A. B.; Côté, B. *J. Org. Chem.* **1993**, *58*, 933−936.

[196] Aggarwal, V. K.; Fang, G. Y.; Meek, G. *Org. Lett.* **2003**, *5*, 4417−4420.

[197] Doyle, M. P. *Rec. Trav. Chim. Pays-Bas* **1991**, *110*, 305−316.

[198] Doyle, M. P.; Winchester, W. R.; Hoorn, J. A. A.; Lynch, V.; Simonsen, S. H.; Ghosh, R. *J. Am. Chem. Soc.* **1993**, *115*, 9968−9978.

[199] Krieger, P. A.; Landgrebe, J. A. *J. Org. Chem.* **1978**, *43*, 4447−4452.

[200] Doyle, M. P.; Protopopova, M. N.; Brandes, B. D.; Davies, H. M. L.; Hruby, N. J. S.; Whitesell, J. K. *Synlett* **1993**, *2*, 151−153.

[201] Davies, H. M. L.; Huby, N. J. S.; Cantrell, W. R., Jr.; Olive, J. L. *J. Am. Chem. Soc.* **1993**, *115*, 9468−9479.

[202] Davies, H. M. L. *Tetrahedron Lett.* **1991**, *32*, 6509−6512.

[203] Davies, H. M. L.; Huby, N. J. S. *Tetrahedron Lett.* **1992**, *33*, 6935−6938.

[204] Davies, H. M. L. *Tetrahedron* **1993**, *49*, 5203−5223.

[205] Charette, A. B.; Juteau, H. *J. Am. Chem. Soc.* **1994**, *116*, 2651−2652.

[206] Charette, A. B. *Chem. Eng. News* **1995**, *February 6*, p.2.

[207] Charette, A. B.; Prescott, S.; Brochu, C. *J. Org. Chem.* **1995**, *60*, 1081−1083.

[208] Lacasse, M.-C.; Poulard, C.; Charette, A. B. *J. Am. Chem. Soc.* **2005**, *127*, 12440−12441.

[209] Voituriez, A.; Charette, A. B. *Adv. Synth. Catal.* **2006**, *348*, 2363−2370.

[210] Long, J.; Du, H.; Li, K.; Shi, Y. *Tetrahedron Lett.* **2005**, *46*, 2737−2740.

[211] Miura, T.; Murakami, Y.; Imai, N. *Tetrahedron: Asymmetry* **2006**, *17*, 3067−3069.

[212] Lorenz, J. C.; Long, J.; Yang, Z.; Xue, S.; Xie, Y.; Shi, Y. *J. Org. Chem.* **2003**, *69*, 327−334.

[213] Takayashi, H.; Yoshioka, M.; Ohno, M.; Kobayashi, S. *Tetrahedron Lett.* **1992**, *33*, 2575−2578.

[214] Ukaji, Y.; Nishimura, M.; Fujisawa, T. *Chem. Lett.* **1992**, *1*, 61−64.

[215] Nozaki, H.; Moriuti, S.; Takaya, H.; Noyori, R. *Tetrahedron Lett.* **1966**, *7*, 5239−5242.

[216] Aratani, T.; Yoneyoshi, Y.; Nagase, T. *Tetrahedron Lett.* **1977**, *18*, 2599−2602.

[217] Aratani, T.; Yoneyoshi, Y.; Nagase, T. *Tetrahedron Lett.* **1982**, *23*, 685−688.

[218] Aratani, T. *Pure Appl. Chem.* **1985**, *57*, 1839−1844.

[219] Fritschi, H.; Leutenegger, U.; Pfaltz, A. *Helv. Chim. Acta* **1988**, *71*, 1553−1565.

[220] Leutenegger, U.; Umbricht, G.; Fahrni, C.; von Matt, P.; Pfaltz, A. *Tetrahedron* **1992**, *48*, 2143−2156.

[221] Lowenthal, R. E.; Abiko, A.; Masamune, S. *Tetrahedron Lett.* **1990**, *31*, 6005−6008.

[222] Lowenthal, R. E.; Masamune, S. *Tetrahedron Lett.* **1991**, *32*, 7373−7376.

[223] Evans, D. A.; Woerpel, K. A.; Hinman, M. M.; Faul, M. M. *J. Am. Chem. Soc.* **1991**, *113*, 726−728.

[224] Ito, K.; Katsuki, T. *Tetrahedron Lett.* **1993**, *34*, 2662−2664.

[225] Nishiyama, H.; Itoh, Y.; Matsumoto, H.; Park, S.-B.; Itoh, K. *J. Am. Chem. Soc.* **1994**, *116*, 2223−2224.

[226] Klyne, W.; Prelog, V. *Experientia* **1960**, *16*, 521−523.

[227] Doyle, M. P.; Brandes, B. D.; Kazala, A. P.; Pieters, R. J.; Jartsfer, M. B.; Watkins, L. M.; Eagle, C. T. *Tetrahedron Lett.* **1990**, *31*, 6613−6616.

[228] Davies, H. M. L.; Hutcheson, D. K. *Tetrahedron Lett.* **1993**, *34*, 7243−7246.

[229] Davies, H. M. L.; Venkataramani, C. *Org. Lett.* **2003**, *5*, 1403−1406.

[230] Davies, H. M. L. *Eur. J. Org. Chem.* **1999**, *1999*, 2459−2469.

[231] Müller, P.; Bernardinelli, G.; Allenbach, Y. F.; Ferri, M.; Flack, H. D. *Org. Lett.* **2004**, *6*, 1725−1728.

[232] Müller, P.; Bernardinelli, G.; Allenbach, Y. F.; Ferri, M.; Grass, S. *Synlett* **2005**, 1397−1400.

[233] Doyle, M. P.; Pieters, R. J.; Martin, S. F.; Austin, R. E.; Oalmann, C. J.; Müller, P. *J. Am. Chem. Soc.* **1991**, *113*, 1423−1424.

[234] Martin, S. F.; Oalmann, C. J.; Liras, S. *Tetrahedron Lett.* **1992**, *33*, 6727−6730.

[235] Martin, S. F.; Spaller, M. R.; Liras, S.; Hartmann, B. *J. Am. Chem. Soc.* **1994**, *116*, 4493−4494.

[236] Bruneau, C.; Dixneuf, P. H.; Nishiyama, H. *Ruthenium Catalysts and Fine Chemistry*, Vol. 11. Springer: Berlin/Heidelberg, 2004, pp. 81−92.

[237] Bonaccorsi, C.; Mezzetti, A. *Organometallics* **2005**, *24*, 4953−4960.

[238] Berkessel, A.; Kaiser, P.; Lex, J. *Chem.−Eur. J* **2003**, *9*, 4746−4756.

[239] Quinkert, G.; Schwartz, U.; Stark, H.; Weber, W.-D.; Baier, H.; Adam, F.; Dürner, G. *Angew. Chem., Int. Ed. Engl.* **1980**, *19*, 1029−1030.

[240] Quinkert, G.; Schwartz, U.; Stark, H.; Weber, W.-D.; Adam, F.; Baier, A.; Frank, G.; Dürner, G. *Liebigs Ann. Chem* **1982**, 1999−2040.

[241] Romo, D.; Romine, J. L.; Midura, W.; Meyers, A. I. *Tetrahedron* **1990**, *46*, 4951−4994.

[242] Romo, D.; Meyers, A. I. *J. Org. Chem.* **1992**, *57*, 6265−6270.

[243] Corey, E. J.; Chaykovsky, M. J. *J. Am. Chem. Soc.* **1965**, *87*, 1353−1364.

[244] Johnson, C. R. *Aldrichim. Acta* **1985**, *18*, 3−11.

[245] Williams, R. M.; Fegley, G. J. *J. Am. Chem. Soc.* **1991**, *113*, 8796−8806.

[246] Aggarwal, V. K.; Richardson, J. *Chem. Commun.* **2003**, *21*, 2644−2651.

[247] Riches, S. L.; Saha, C.; Filgueira, N. F.; Grange, E.; McGarrigle, E. M.; Aggarwal, V. K. *J. Am. Chem. Soc.* **2010**, *132*, 7626−7630.

[248] Papageorgiou, C. D.; de Dios, M. A. C.; Ley, S. V.; Gaunt, M. J. *Angew. Chem., Int. Ed.* **2004**, *43*, 4641−4644.

[249] Kunz, R. K.; MacMillan, D. W. C. *J. Am. Chem. Soc.* **2005**, *127*, 3240−3241.

[250] Kakei, H.; Sone, T.; Sohtome, Y.; Matsunaga, S.; Shibasaki, M. *J. Am. Chem. Soc.* **2007**, *129*, 13410−13411.

[251] Rhoads, S. J.; Raulins, N. R. *Org. React.* **1975**, *22*, 1−252.

[252] Ziegler, F. E. *Acc. Chem. Res.* **1977**, *10*, 227−232.

[253] Mori, K.; Nomi, H.; Chuman, T.; Kohno, M.; Kato, K.; Noguchi, M. *Tetrahedron* **1982**, *38*, 3705−3711.

[254] Ireland, R. E. *Aldrichim. Acta* **1988**, *21*, 59−69.

[255] Blechert, S. *Synthesis* **1989**, *2*, 71−82.

[256] Hill, R. K. In *Comprehensive Organic Synthesis. Selectivity, Strategy, and Efficiency in Modern Organic Chemistry*; Trost, B. M., Fleming, I., Eds.; Pergamon: Oxford, 1991; Vol. 5, pp. 785−826.

[257] Pereira, S.; Srebnik, M. *Aldrichim. Acta* **1993**, *26*, 17−29.

[258] Davies, S. G. *Organotransition Metal Chemistry Applications to Organic Synthesis*; Pergamon: Oxford, 1982. pp. 266−303.

[259] Otsuka, S.; Tani, K. *Synthesis* **1991**, *9*, 665−680.

[260] Birch, A. J.; Jenkins, I. D. In *Transition Metal Organometallics in Organic Synthesis*; Alper, H., Ed.; Academic: New York, 1976; Vol. 1, pp. 1−82.

[261] Colquhoun, H. M.; Holton, J.; Thompson, D. J.; Twigg, M. V. In *New Pathways for Organic Synthesis. Practical Applications of Transition Metals*; Plenum: New York, 1984, pp. 173−193.

[262] Kumobayashi, H.; Akutagawa, S.; Otsuka, S. *J. Am. Chem. Soc.* **1978**, *100*, 3949−3950.

[263] Tani, K.; Yamagata, T.; Otsuka, S.; Akutagawa, S.; Kumobayashi, H.; Taketomi, T.; Takaya, H.; Miyashita, A.; Noyori, R. In *Asymmetric Reactions and Processes in Organic Chemistry. ACS Symposium Series 185*; Eliel, E. L., Otsuka, S., Eds.; American Chemical Society: Washington, 1982, pp. 187−193.

[264] Akutagawa, S.; Tani, K. In *Catalytic Asymmetric Synthesis*; Ojima, I., Ed.; VCH: New York, 1993, pp. 41−113.

[265] Sheldon, R. A. *Chirotechnology. Industrial Synthesis of Optically Active Compounds*; Marcel Dekker: New York, 1993. p. 304.

[266] Prelog, V.; Helmchen, G. *Angew. Chem., Int. Ed. Engl.* **1982**, *21*, 567−583.

[267] Tani, K.; Yamagata, T.; Akutagawa, S.; Kumobayashi, H.; Taketomi, T.; Takaya, H., et al. *J. Am. Chem. Soc.* **1984**, *106*, 5208−5217.

[268] Ohta, T.; Takaya, H.; Noyori, R. *Inorg. Chem.* **1988**, *27*, 566−569.

[269] Mashima, K.; Kusano, K.; Ohta, T.; Noyori, R.; Takaya, H. *J. Chem. Soc., Chem. Commun.* **1989**, 1208−1210.

[270] Uma, R.; Crévisy, C.; Grée, R. *Chem. Rev.* **2003**, *103*, 27−51.

[271] Tani, K. *Pure Appl. Chem.* **1985**, *57*, 1845−1854.

[272] Tanaka, K.; Qiao, S.; Tobisu, M.; Lo, M. M. C.; Fu, G. C. *J. Am. Chem. Soc.* **2000**, *122*, 9870−9871.

[273] Tanaka, K.; Fu, G. C. *J. Org. Chem.* **2001**, *66*, 8177−8186.

[274] Mantilli, L.; Gerard, D.; Torche, S.; Besnard, C.; Mazet, C. *Angew. Chem., Int. Ed.* **2009**, *48*, 5143−5147.

[275] Hiroya, K.; Kurihara Y, Ogasawara K. *Angew. Chem., Int. Ed. Engl.* **1995**, 34, 2289.

[276] Tanaka, K.; Shoji, T. *Org. Lett.* **2005**, *7*, 3561−3563.

[277] Schorigin, P. *Chem. Ber.* **1924**, *57*, 1627−1634.

[278] Shorigin, P. *Chem. Ber.* **1925**, *58*, 2028−2036.

[279] Schlenk, W.; Bergmann, E. *Liebigs Ann. Chem.* **1928**, *464*, 35−42.

[280] Wittig, G.; Löhmann, L. *Liebigs Ann. Chem.* **1942**, *550*, 260−268.
[281] Wittig, G.; Löhmann, L. *Liebigs Ann. Chem.* **1947**, *557*, 205−220.
[282] Lansbury, P. T.; Pattison, V. A.; Sidler, J. D.; Bieber, J. B. *J. Am. Chem. Soc.* **1966**, *88*, 78−84.
[283] Tomooka, K.; Igrashi, T.; Nakai, T. *Tetrahedron Lett.* **1993**, *34*, 8139−8142.
[284] Woodward, R. B.; Hoffmann, R. *The Conservation of Orbital Symmetry*; Academic: New York, 1970.
[285] Wittig, G.; Döser, H.; Lorenz, I. *Liebigs Ann. Chem.* **1949**, *562*, 192−205.
[286] Hauser, C. R.; Kantor, S. W. *J. Am. Chem. Soc.* **1951**, *73*, 1437−1441.
[287] Schöllkopf, U.; Feldenberger, K. *Liebigs Ann. Chem* **1966**, *698*, 80−85.
[288] Makisumi, Y.; Notzumoto, S. *Tetrahedron Lett.* **1966**, *7*, 6393−6397.
[289] Baldwin, J. E.; DeBernardis, J.; Patrick, J. E. *Tetrahedron Lett.* **1970**, *11*, 353−356.
[290] Rautenstrauch, V. *J. Chem. Soc., Chem. Commun.* **1970**, 4−6.
[291] Baldwin, J. E.; Patrick, J. E. *J. Am. Chem. Soc.* **1971**, *93*, 3556−3558.
[292] Hoffmann, R. *Angew. Chem., Int., Ed. Engl.* **1979**, *18*, 563−640.
[293] Hill, R. K. In *Asymmetric Synthesis*; Morrison, J. D., Ed.; Academic: Orlando, 1984; Vol. 3, pp. 503−572.
[294] Nakai, T.; Mikami, K. *Chem. Rev.* **1986**, *86*, 885−902.
[295] Mikami, K.; Nakai, T. *Synthesis* **1991**, *8*, 594−604.
[296] Marshall, J. A. In *Comprehensive Organic Synthesis. Selectivity, Strategy, and Efficiency in Modern Organic Chemistry*; Trost, B. M., Fleming, I., Eds.; Pergamon: Oxford, 1991; Vol. 3, pp. 975−1014.
[297] Brückner, R. In *Comprehensive Organic Synthesis. Selectivity, Strategy, and Efficiency in Modern Organic Chemistry*; Trost, B. M., Fleming, I., Eds.; Pergamon: Oxford, 1991; Vol. 6, pp. 873−908.
[298] Nakai, T.; Mikami, K. *Org. React.* **1994**, *46*, 105−209.
[299] Ahmad, N. M. In *Name Reactions for Homologations* Pt. 2 ed.; Li, J J., Ed.; John Wiley & Sons, Inc: New York, 2009, pp. 241−256.
[300] Wolfe, J. P.; Guthrie, N. J. In *Name Reactions for Homologations Part. 2*; Li, J J., Ed.; John Wiley & Sons, Inc: New York, 2009, pp. 226−240.
[301] Still, W. C.; Mitra, A. *J. Am. Chem. Soc.* **1978**, *100*, 1927−1928.
[302] Mikami, K.; Azuma, K.; Nakai, T. *Tetrahedron* **1984**, *40*, 2303−2308.
[303] Mikami, K.; Maeda, T.; Nakai, T. *Tetrahedron Lett.* **1986**, *27*, 4189−4190.
[304] Koreeda, M.; Luengo, J. I. *J. Am. Chem. Soc.* **1985**, *107*, 5572−5573.
[305] Luengo, J. I.; Koreeda, M. *J. Org. Chem.* **1989**, *54*, 5415−5417.
[306] Castedo, L.; Granja, J. R.; Mouriño, A. *Tetrahedron Lett.* **1985**, *26*, 4959−4960.
[307] Koreeda, M.; Ricca, D. J. *J. Org. Chem.* **1986**, *51*, 4090−4092.
[308] Still, W. C.; Sreekumar, C. *J. Am. Chem. Soc.* **1980**, *102*, 1201−1202.
[309] Wu, Y.-D.; Houk, K. N.; Marshall, J. A. *J. Org. Chem.* **1990**, *55*, 1421−1423.
[310] Verner, E. J.; Cohen, T. *J. Am. Chem. Soc.* **1992**, *114*, 375−377.
[311] Hoffmann, R.; Brückner, R. *Angew. Chem., Int. Ed. Engl.* **1992**, *31*, 647−649.
[312] Tomooka, K.; Igarashi, T.; Watanabe, M.; Nakai, T. *Tetrahedron Lett.* **1992**, *33*, 5795−5798.
[313] Eichenauer, H.; Friedrich, E.; Lutz, W.; Enders, D. *Angew. Chem., Int. Ed. Engl.* **1978**, *17*, 206−208.
[314] Seebach, D.; Golinski, J. *Helv. Chim. Acta* **1981**, *64*, 1413−1423.
[315] Mikami, K.; Kimura, Y.; Kishi, N.; Nakai, T. *J. Org. Chem.* **1983**, *48*, 279−281.
[316] Nakai, E.; Nakai, T. *Tetrahedron Lett.* **1988**, *29*, 5409−5412.
[317] Uchikawa, M.; Katsuki, T.; Yamaguchi, M. *Tetrahedron Lett.* **1986**, *27*, 4581−4582.
[318] Mekelburger, H. B.; Wilcox, C. S. In *Comprehensive Organic Synthesis. Selectivity, Strategy, and Efficiency in Modern Organic Chemistry*; Trost, B. M., Fleming, I., Eds.; Pergamon: Oxford, 1991; Vol. 2, pp. 99−131.
[319] Burke, S. D.; Fobare, W. F.; Pacofsky, G. J. *J. Org. Chem.* **1983**, *48*, 5221−5228.
[320] Kallmerten, J.; Gould, T. J. *Tetrahedron Lett.* **1983**, *24*, 5177−5180.
[321] Whitesell, J. K.; Helbing, A. M. *J. Org. Chem.* **1980**, *45*, 4135−4139.
[322] Ireland, R. E.; Thaisrivongs, S.; Vanier, N.; Wilcox, C. S. *J. Org. Chem.* **1980**, *45*, 48−61.
[323] Mikami, K.; Takahashi, O.; Kasuga, T.; Nakai, T. *Chem. Lett.* **1985**, 1729−1732.
[324] Marshall, J. A.; Wang, X. *J. Org. Chem.* **1992**, *57*, 2747−2750.
[325] Nakai, T.; Mikami, K.; Taya, S.; Kimura, Y.; Mimura, T. *Tetrahedron Lett.* **1981**, *22*, 69−72.
[326] Marshall, J. A.; Gung, W. Y. *Tetrahedron Lett.* **1988**, *29*, 1657−1660.
[327] Chan, P. C.-M.; Chong, J. M. *J. Org. Chem.* **1988**, *53*, 5584−5586.

[328] Marshall, J. A.; Welmaker, G. S.; Gung, B. W. *J. Am. Chem. Soc.* **1991**, *113*, 647−656.

[329] Chong, J. M.; Mar, E. K. *Tetrahedron Lett.* **1991**, *32*, 5683−5686.

[330] Matteson, D. S.; Tripathy, P. B.; Sarkur, A.; Sadhu, K. N. *J. Am. Chem. Soc.* **1989**, *111*, 4399−4402.

[331] Thomas, A. F.; Dubini, R. *Helv. Chim. Acta* **1974**, *57*, 2084−2087.

[332] Tsai, D. J.-S.; Midland, M. M. *J. Org. Chem.* **1984**, *49*, 1842−1843.

[333] Sayo, N.; Azuma, K.; Mikami, K.; Nakai, T. *Tetrahedron Lett.* **1984**, *25*, 565−568.

[334] Marshall, J. A.; Jenson, T. M. *J. Org. Chem.* **1984**, *49*, 1707−1712.

[335] Marshall, J. A.; Robinson, E. D.; Zapata, A. *J. Org. Chem.* **1989**, *54*, 5854−5855.

[336] Marshall, J. A.; Wang, X. *J. Org. Chem.* **1990**, *55*, 2995−2996.

[337] Marshall, J. A.; Wang, X. *J. Org. Chem.* **1991**, *56*, 4913−4918.

[338] Mikami, K.; Takahashi, O.; Tabei, T.; Nakai, T. *Tetrahedron Lett.* **1986**, *27*, 4511−4514.

[339] Midland, M. M.; Kwon, Y. C. *Tetrahedron Lett.* **1985**, *26*, 5013−5016.

[340] Takahashi, O.; Mikami, K.; Nakai, T. *Chem. Lett.* **1987**, 69−72.

[341] Uchikawa, M.; Hanemoto, T.; Katsuki, T.; Yamaguchi, M. *Tetrahedron Lett.* **1986**, *27*, 4577−4580.

[342] Marshall, J. A.; Lebreton, J. *Tetrahedron Lett.* **1987**, *28*, 3323−3326.

[343] Marshall, J. A.; Lebreton, J. *J. Am. Chem. Soc.* **1988**, *110*, 2925−2931.

[344] Kitamura, M.; Hirokawa, Y.; Maezaki, N. *Chem.−Eur. J.* **2009**, *15*, 9911−9917.

[345] Marshall, J. A.; Lebreton, J. *J. Org. Chem.* **1988**, *53*, 4108−4112.

[346] *The Claisen Rearrangement*; Hiersemann, M.; Nubbemeyer, U., Eds.; Wiley-VCH: New York, 2007.

[347] Corey, E. J.; Lee, D. H. *J. Am. Chem. Soc.* **1991**, *113*, 4026−4028.

[348] Maruoka, K.; Saito, S.; Yamamoto, H. *J. Am. Chem. Soc.* **1995**, *117*, 1165−1166.

[349] Ireland, R. E.; Mueller, R. H.; Willard, A. K. *J. Am. Chem. Soc.* **1976**, *98*, 2868−2877.

[350] Kazmaier, U.; Mues, H.; Krebs, A. *Chem.−Eur. J.* **2002**, *8*, 1850−1855.

[351] Yoon, T. P.; MacMillan, D. W. C. *J. Am. Chem. Soc.* **2001**, *123*, 2911−2912.

[352] Abraham, L.; Czerwonka, R.; Hiersemann, M. *Angew. Chem., Int. Ed.* **2001**, *40*, 4700−4703.

[353] Abraham, L.; Körner, M.; Schwab, P.; Hiersemann, M. *Adv. Synth. Catal.* **2004**, *346*, 1281−1294.

[354] Uyeda, C.; Jacobsen, E. N. *J. Am. Chem. Soc.* **2008**, *130*, 9228−9229.

[355] Geherty, M. E.; Dura, R. D.; Nelson, S. G. *J. Am. Chem. Soc.* **2010**, *132*, 11875−11877.

[356] Calter, M.; Hollis, T. K.; Overman, L. E.; Ziller, J.; Zipp, G. G. *J. Org. Chem.* **1997**, *62*, 1449−1456.

[357] Anderson, C. E.; Overman, L. E. *J. Am. Chem. Soc.* **2003**, *125*, 12412−12413.

[358] Clayden, J.; Donnard, M.; Lefranc, J.; Tetlow, D. *J. Chem. Commun.* **2011**, 4624−4639.

[359] Alba, A.-N. R.; Rios, R. *Chem.−Asian J.* **2011**, *6*, 720−734.

[360] Hills, I. D.; Fu, G. C. *Angew. Chem., Int. Ed.* **2003**, *42*, 3921−3924.

[361] Höfle, G.; Steglich, W.; Vorbrüggen, H. *Angew. Chem., Int. Ed. Engl.* **1978**, *17*, 569−583.

[362] Zhang, Z.; Xie, F.; Jia, J.; Zhang, W. *J. Am. Chem. Soc.* **2010**, *132*, 15939−15941.

[363] Dietz, F. R.; Gröger, H. *Synthesis* **2009**, *2009*, 4208−4218.

[364] Zhang, Q.-W.; Fan, C.-A.; Zhang, H.-J.; Tu, Y.-Q.; Zhao, Y.-M.; Gu, P.; Chen, Z.-M. *Angew. Chem., Int. Ed.* **2009**, *48*, 8572−8574.

[365] Zhang, E.; Fan, C.-A.; Tu, Y.-Q.; Zhang, F.-M.; Song, Y.-L. *J. Am. Chem. Soc.* **2009**, *131*, 14626−14627.

Chapter 7

Reductions and Hydroborations

Nothing captures the attention of the scientific community quite like the bestowal of a Nobel prize, and while a number of Nobel prizes have been awarded to individuals who have made important contributions to our understanding of stereochemistry or stereoselective synthesis (Prelog, Woodward, Barton, Brown, and Corey, to name a few) only one award to date has been exclusively dedicated to asymmetric catalysis. In 2001, the Royal Swedish Academy of Sciences awarded one half of the prize to William S. Knowles and Ryoji Noyori "for their work on chirally catalyzed hydrogenation reactions" [1,2] and the other half to K. Barry Sharpless "for his work on chirally catalyzed oxidation reactions" [3]. This chapter addresses asymmetric reduction chemistry and the next chapter covers oxidations. Within the scope of the present chapter are additions of dihydrogen, hydrides, and hydroborations. For the latter, the product boranes may be converted to a number of useful functional groups, but this chemistry is not covered here (reviews: refs. [4,5]). The chapter is divided into three parts: reduction of carbon—heteroatom double bonds, reduction of carbon—carbon double bonds, and hydroborations. As in previous chapters, the selective coverage of this topic is intended to highlight particularly important and selective reagents, with an emphasis on understanding the factors that influence stereoselectivity.

7.1 REDUCTION OF CARBON—HETEROATOM DOUBLE BONDS

Numerous reagents are used to effect the asymmetric reduction of ketones (for one comprehensive review, see ref. [6]). The nonenzymatic versions can be divided into several categories based on reagent type and/or mechanism: lithium aluminum hydrides modified with chiral ligands, borohydrides modified (sometimes catalytically) with chiral ligands, chiral boranes that reduce carbonyls in a self-immolative chirality transfer process, and chiral transition metal complexes that catalyze hydrogenation or transfer hydrogenation. Each of these involves interligand asymmetric induction (Section 1.3). Only selected examples from each category will be presented in detail, the objective being to analyze the factors that determine enantioselectivity for each reaction. In contrast to the numerous approaches available for reducing ketones, highly selective reductions of the carbon—nitrogen double bond were slower to develop although there are now a number of practical approaches available [7].

Principles of Asymmetric Synthesis.

7.1.1 Modified Lithium Aluminum Hydride

The first efforts to modify lithium aluminum hydride (LAH) with a chiral ligand were reported by Bothner-by in 1951 [8]. Although the result was later challenged, the seed was planted and many attempts have been made to produce an efficient chiral reducing agent using this strategy (reviews: refs. [2,9−12]). Of these, we will examine the binaphthol-LAH-ROH reagent (BINAL-H) introduced by Noyori in 1979 [13−16]. As has been amply demonstrated throughout this book, binaphthol and its cousin BINAP are popular ligands for asymmetric synthesis because these have a pleasing C_2 symmetry which, when bound in a bidentate fashion to a metal, often affords excellent differentiation between heterotopic faces of a bound ligand.

Noyori's reagent is prepared by addition of binaphthol to a *solution* of LAH in THF, then adding another equivalent of an alcohol such as ethanol or methanol to form the reagent (Scheme 7.1).[1] The ethanol or methanol is a pragmatic necessity, as the reagent having two (presumed) active hydrides shows poor enantioselectivity in asymmetric reductions [15]. The exact nature of the reagent is not known, since aluminum hydrides may disproportionate and/or aggregate, processes that may continue as the product of the reduction (an alkoxide) accumulates during the reaction. Perhaps because of such processes, optimal selectivity is achieved with a 3-fold excess of the hydride reagent. Under these conditions, the reagent is highly enantioselective in reductions of certain classes of ketones. Some examples are listed in Table 7.1.

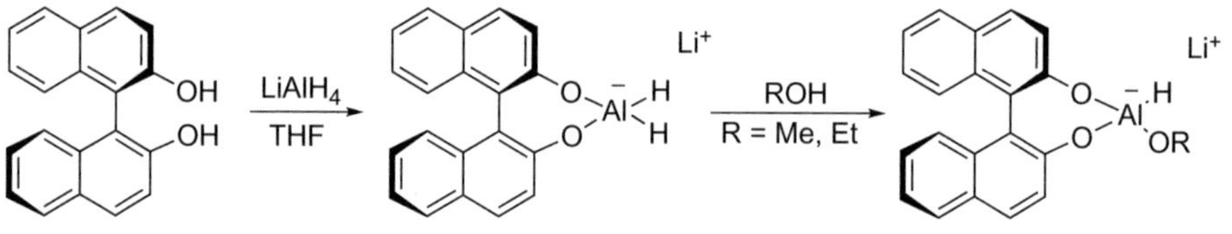

M-(+)-binaphthol   *presumed reagent*

SCHEME 7.1   Preparation of Noyori's BINAL-H reagents [15]. The aluminum complexes shown are postulated structures that likely represent "time averages" of several equilibrating species.

Entries 1 and 2 show the reagent's ability to reduce deuterated aldehydes to afford primary alcohols that are chiral by virtue of isotopic substitution. Note that the rest of the examples showing high selectivity (entry 13 being the exception) have one substituent on the ketone that is unsaturated and one that is not. Note also that in the saturated substituent, branching at the α-position lowers enantioselectivity significantly (compare entries 4/5 and 7/8). The effect of unsaturation on selectivity is exemplified by the observation that 3-octyn-2-one (entry 9) is reduced with 92% enantioselectivity whereas 2-octanone (entry 13) is reduced with only 62% enantioselectivity.

---

1. In using this reagent, care should be taken to follow the Noyori experimental procedure exactly. Precipitous drops in enantioselectivity result from very minor changes in protocol. Note that a "milk-white" or "cloudy" reagent solution is OK; but when there is "extensive precipitation," the reagent should not be expected to perform as advertised [15] (see also refs. [17,18]).

TABLE 7.1 Asymmetric Reductions Using BINAL-H[a]

Entry	R_1	R_2	RO	% Yield	% es	Ref.
1	Ph	D	EtO	59	93	[16]
2	(see structure)	D	EtO	91	92	[16]
3	Ph	Me	EtO	61	97	[15]
4	Ph	Pr	EtO	92	>99	[15]
5	Ph	i-Pr	EtO	68	85	[15]
6	α-Tetralone		EtO	91	87	[15]
7	HC≡C	n-C_5H_{11}	MeO	87	92	[16]
8	HC≡C	i-Pr	MeO	84	79	[16]
9	n-C_4H_9C≡C	Me	MeO	79	92	[16]
10	n-C_4H_9C≡C	n-C_5H_{11}	MeO	85	95	[16]
11	E-n-C_4H_9CH=CH	Me	EtO	47	89	[16]
12	E-n-C_4H_9CH=CH	n-C_5H_{11}	EtO	91	95	[16]
13	n-C_6H_{13}	Me	EtO	67	62	[15]
14	(see structure)		EtO	87	>99	[16]

[a]*The reactions were conducted by initial reaction at −100 °C for 3 h, followed by several hours at −78 °C. All examples favor* ul *relative topicity (see Figure 7.1a). Thus, the M reagent adds to the Si face to give the R product, and vice versa for the P reagent.*

The rationale offered by the Noyori group to explain the chirality sense of the observed products is predicated on the 6-membered ring transition structures shown in Figures 7.1a and b. These structures differ only in the orientation of the two ketone substituents. Another pair, in which the 6-membered ring is flipped, is destabilized by a steric repulsion between the alkoxy methyl (or ethyl) and the C-3 position of the binaphthol. Figure 7.1c shows this interaction, which is (note the bold lines) a "gauche pentane-like" conformation (*cf.* Figure 5.5 and accompanying discussion).[2] With respect to the 6-membered ring in Figures 7.1a and b, note that one

---

2. Note that in Figures 7.1a–c, the alkoxy "R group" is always axial. The authors point out that structures in which the R group occupies an equatorial position would be further destabilized by repulsive interactions between R and the BINOL moieties [15]. It may be useful to note that the configuration of the alkoxy oxygen in the favored chairs (Figures 7.1a and b) having the *P*-BINOL ligand is *R*. The configuration of the oxygen in the disfavored chair (Figure 7.1c, *P*-BINOL ligand) is *S*.

of the ketone substituents is equatorial and one is axial. The interaction of the latter with the axial naphthyloxy ligand is postulated to account for the enantioselectivity. This interaction is suggested to be one of two types: steric interactions, which are repulsive, and electronic, which may (in principle) be either repulsive or attractive, but which are repulsive for all the examples in Table 7.1 (other substrates are suggested to have dominant *attractive* electronic interactions [15]). For the examples in Table 7.1, it is observed that the *P*-BINAL-H reagent selectively adds hydride to the *Re* face (*ul* relative topicity) as shown in the transition structure of Figure 7.1a. In this structure, the saturated ligand (R_{sat}) bears a 1,3-diaxial relationship to the naphthyloxy ligand on aluminum. Since an alkene or alkyne ligand is generally considered to be "smaller" than an *n*-alkyl ligand,[3] this is somewhat counterintuitive. Noyori suggests that the reason for this topicity has to do with an unfavorable repulsive electronic interaction between the unpaired electrons on the axial naphthyl oxygen and the π orbital of the unsaturated ligand (R_{un}) in the transition structure having *lk* topicity, shown in Figure 7.1b, and that this interaction causes greater repulsion than that of an axial saturated ligand.

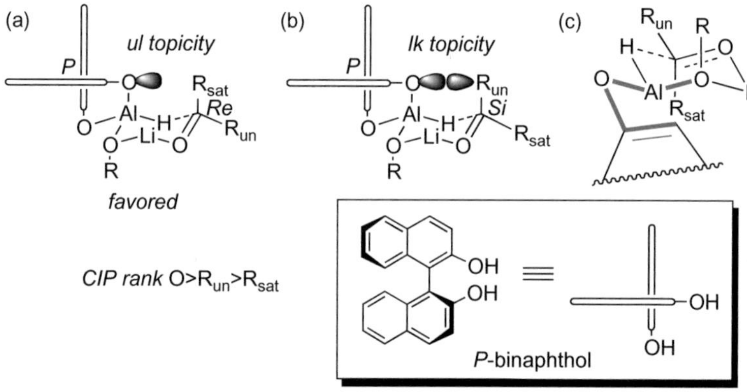

FIGURE 7.1 Postulated transition structures for the asymmetric reduction of unsaturated ketones by BINAL-H [15]. Structures (a) and (b) differ in the orientation of R_{sat} and R_{un}, the saturated and unsaturated ketone ligands, respectively. (a) Topicity *ul*: *P* reagent adds hydride to the *Re* face of the ketone. (b) Topicity *lk*: *P* reagent adds hydride to the *Si* face of the ketone. (c) Alternate chair that is destabilized by the "gauche-pentane" conformation accentuated by the bold lines (*cf.* Figure 5.5). Transition structures containing this conformation were considered by Noyori to be unimportant [15].

7.1.2 Modified Borane

In 1969, Kagan sought to carry out the enantioselective reduction of acetophenone using amphetamine−desoxyephedrine−borane complexes, but only obtained 1-phenyl ethanol in <5% es [20]. Subsequently, a breakthrough was realized in the introduction of the oxazaborolidines, which were introduced by Hirao in 1981, developed by Itsuno, and further developed by Corey several years later (selected reviews: refs. [21−26]). These have become the

---

3. The *A* values of $-CH_2CH_3$, $-CH=CH_2$, and $-C\equiv CH$ are ~ 1.75, 1.7, and 0.41 kcal/mol, respectively [19].

most successful boranes used for asymmetric carbonyl reduction and provide outstanding examples of reagent design.

Figure 7.2 illustrates several of the Hirao-Itsuno and Corey oxazaborolidines that have been evaluated to date. All these examples are derived from amino acids by reduction or Grignard addition. Hirao originally investigated the reagent derived from condensation of amino alcohols such as valinol and prolinol with borane (Figures 7.2a–c, e), and found enantioselectivities in the neighborhood of 70–80% [27]. Optimization studies revealed that enantioselectivities of ∼85% (for the reduction of acetophenone) could be obtained in THF solvent at 30 °C, using amino alcohol:borane ratios of 1:2 [28]. In 1983, Itsuno found that the reagent was much more selective (96–100% es with acetophenone) if tertiary alcohols derived from addition of phenyl magnesium bromide to valine (Figure 7.2d) were used [29,30]. Additionally, Itsuno found that a polymer-bound amino alcohol could be used for the process with equal facility [31]. Reduction of aliphatic ketones was not quite as selective, affording reduction products in 77–87% es [30,32]. Itsuno [30] and Corey [33] demonstrated the synthesis of oxiranes by asymmetric reduction of α-halo ketones followed by cyclization. In 1985, Itsuno showed that oxime ethers (but not oximes) could be enantioselectively reduced to primary amines (84–99% es) using the valinol-derived reagent (Figure 7.2d [30]), and in 1987 showed that this process could be catalytic in oxazaborolidine: acetophenone O-methyloxime was reduced to α-methylbenzyl amine in 90% yield and 100% es [34]. In 1987, Corey characterized the Itsuno reagent (Figure 7.2d) and showed that the diphenyl derivative (Figure 7.2f) of the Hirao reagent (Figure 7.2e) afforded excellent enantioselectivities (≥95%) when used in catalytic amounts [35]. In the same year, the Corey group reported that B-alkyl oxazaborolidines (Figure 7.2g and h) were easier to prepare, could be stored at room temperature, could be weighed and transferred in air, and afforded enantioselectivities comparable to the B-H reagents [33,36]. In 1989, Corey found that the β-naphthyl (β-Np) derivatives of prolinol afforded reagents with still higher enantioselectivities (Figures 7.2i and j, e.g., 99% es with acetophenone [37]).

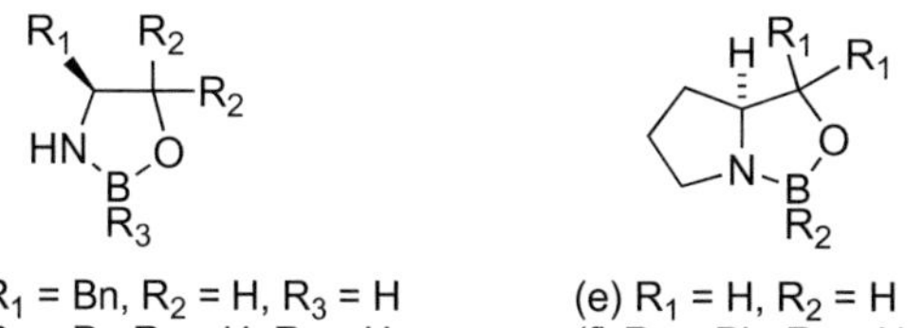

(a) R$_1$ = Bn, R$_2$ = H, R$_3$ = H
(b) R$_1$ = Pr, R$_2$ = H, R$_3$ = H
(c) R$_1$ = i-Pr, R$_2$ = H, R$_3$ = H
(d) R$_1$ = i-Pr, R$_2$ = Ph, R$_3$ = H

(e) R$_1$ = H, R$_2$ = H
(f) R$_1$ = Ph, R$_2$ = H
(g) R$_1$ = Ph, R$_2$ = Me
(h) R$_1$ = Ph, R$_2$ = Bu
(i) R$_1$ = β-Np, R$_2$ = H
(j) R$_1$ = β-Np, R$_2$ = Me

FIGURE 7.2   Oxazaborolidines for the asymmetric reduction of ketones: (a–c) [27,28], (d) [29–32,34], (e) [27], (f) [35], (g) [36], (h) [49], and (i) and (j) [37].

X-ray crystal structures of the oxazaborolidine reagent [38] and a derivative [39] have been published, and a mechanistic hypothesis has been formulated [35]. Heterocycles such as the boranes shown in Figures 7.2a–f and i do not, by themselves, reduce carbonyls. This is an important feature because it minimizes the liability of background reductions via nonopti-mal reaction ensembles. In contrast, the presence of excess borane leads to reduction by the mechanism shown in Scheme 7.2 for the *B*-methyl catalyst of Figure 7.2g. In the first step, borane coordinates to the nitrogen of the oxazaborolidine on the less hindered convex face of the fused bicyclic system; the ketone then coordinates to the convex face. From the perspective of the ketone, the Lewis acid (boron atom) is *trans* to the larger ketone substituent [40]. Hydride transfer occurs *via* a 6-membered chair transition structure [41,42] having *lk* relative topicity (the *R*-enantiomer of the catalyst favoring the *Re* face of the carbonyl carbon). Elimination of the alkoxy borane completes the catalytic cycle [43]. Table 7.2 lists representative examples of oxazaborolidine reductions. Entry 4 is one example (among several) of the asymmetric reduction [44] of trichloromethyl ketones [45]. Corey's group has shown that the resulting carbinols are versatile intermediates for the preparation of α-amino acids [44], α-hydroxy and α-aryloxy acids [46], and terminal epoxides [47].

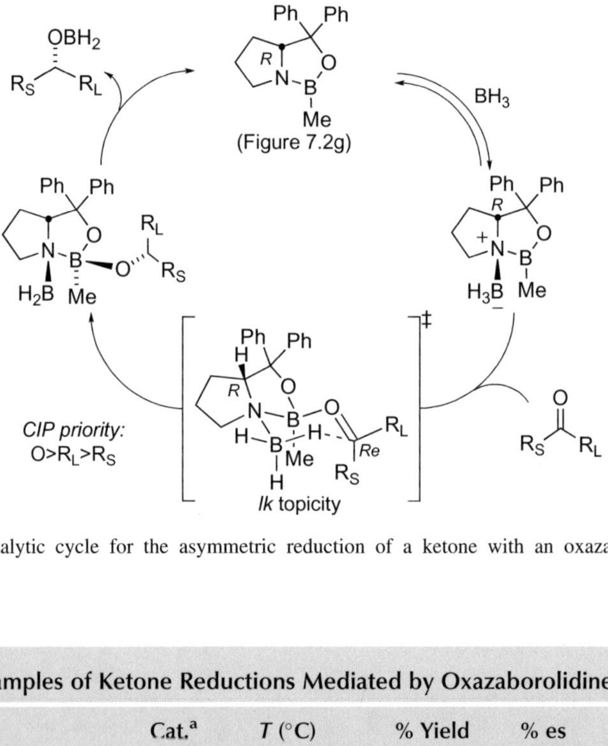

SCHEME 7.2 Catalytic cycle for the asymmetric reduction of a ketone with an oxazaborolidine catalyst [35,41,42].

TABLE 7.2 Examples of Ketone Reductions Mediated by Oxazaborolidines

Entry	Ketone	Cat.[a]	T (°C)	% Yield	% es	Ref.
1	PhCOMe	c, f–j	2–30	95–100	≥97	[29,30,35–37,48]
2	PhCOEt	c, f, g, j	−10 to 30	100	≥94	[29,30,35–37]
3	PhCOCH₂Cl	c, f, g	25–32	100	98	[30,33,35,36]
4	t-BuCOCCl₃	h	−20	96	99	[44]
5	α-Tetralone	f, g, i, j	−10 to 31	100	≥93	[35–37]
6	t-BuCOMe	f, g, j	−10 to 25	100	≥96	[35–37]
7	CyCOMe	g, j	−10 to 0	100	91–92	[36,37]

(*Continued*)

TABLE 7.2 (Continued)

Entry	Ketone	Cat.[a]	T (°C)	% Yield	% es	Ref.
8	i-PrCOMe	c	30	100	80	[30]
9	n-C$_6$H$_{13}$COMe	c	30	100	79	[30,32]
10	PhCO(CH$_2$)$_n$CO$_2$Me, n = 2, 3	g, j	0	100	97–98	[36,37]
11	(cyclohexenone with Br)	g, i	23–36	100	95	[36,37]
12	(cyclohexenyl COMe)	h	−78[b]	>95	90	[49]
13	E-PhCH=CHCOMe	h	−78[b]	>95	96	[49]
14	(cyclohexenone with Me)	h	−78[b]	>95	96	[49]
15	(PhC(=NOMe)Me)	c	30	90–100	99–100	[30,34]
16	(tetralone NOMe oxime)	c	30	100	84	[30]

[a]The "Cat." column refers to the catalysts in Figure 7.2. The reductant is borane, unless otherwise noted. For entries 3 and 10, the product may spontaneously cyclize. The products of entries 15 and 16 are primary amines.
[b]Catechol borane as reductant.

Operationally, these reagents are effective at or near room temperature, which may be of significant benefit to large-scale employment. The preparation of the S-diphenylprolinol ligand (cf. Figures 7.2f–h) is most easily accomplished by addition of a phenyl Grignard reagent to L-proline N-carboxyanhydride (73% yield, >99:1 er [39]). The R enantiomer of the amino alcohol may be made by a similar addition to D-proline, but may also be made by enantioselective lithiation of Boc-pyrrolidine and addition to benzophenone (70% yield, >99% es, as illustrated in Scheme 3.56 [50,51]). The catalysts may be made by condensation of the amino alcohol with methyl boronic acid [36,37,39] or trimethylboroxine [39] with simultaneous water removal. B-Methyl or B-butyl catalysts can be made by condensation of the amino alcohol with bis(trifluoroethyl) alkylboronate and removal of trifluoroethanol in vacuo [48]. The catalysts may be used in 5–10 mol% loadings, with either borane or catechol borane [49] as the stoichiometric reductant. Use of the more reactive catechol borane allows one to conduct the reduction at lower temperature, a feature that may be advantageous in cases where selectivity at room temperature is not high enough. The reductions are sensitive to moisture: Jones et al. [52] found that the presence of 1 mg water/g ketone lowered the enantioselectivity from 97% to 75% es.

The above reductions distinguish the enantiotopic faces of aldehydes and ketones. An interesting extension is the group-selective asymmetric reduction of *meso* anhydrides or imides to afford lactones or lactams. An early example of the process, along with yields and enantioselectivities of several substrates, was published in 1993 by Takeda, who used BINAL-H as the reagent (Scheme 7.3a [53]). Other reducing agents are also effective in this chemistry. Accordingly, reduction of achiral imides with an oxazaborolidine as shown in Scheme 7.3b followed by equilibration of the initially formed product in acidic ethanol afforded a series of amides bearing the thermodynamically more stable ethoxy group [54,55]. Alternatively, one example is shown where the initially formed kinetic product is trapped by acetylation. In Scheme 7.3c, reduction using the zinc reagent shown affords lactone products [56]. This reaction provided good results in the few monocyclic examples examined as well.

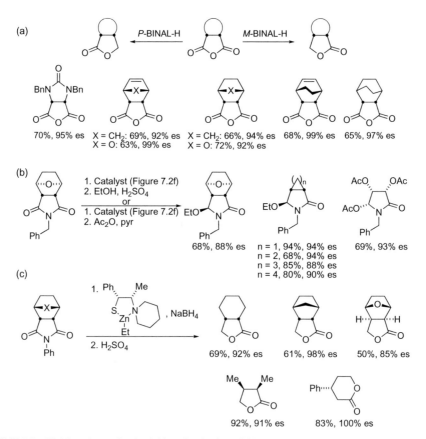

SCHEME 7.3 Yields and enantioselectivities of reduction of (a) *meso* anhydrides [53] or (b) and (c) imides [54–57].

In Praise of Enzymes

If there were a single subject to which this book has given the short end of the stick it has to be the use of enzymes in asymmetric synthesis. Not only is this a very old topic—going back to Marckwald's original definition of an "asymmetric synthesis" (Section 1.2)—but it is also one that continues to find new uses, especially in industrial settings. Although outside the scope of this book, any reader who wants to be a "compleat asymmetric synthesizer" should study important enzyme promoted reactions as well. Of course, *any* biocatalytic process that produces a natural product or a bioengineered variant of one will typically produce it in enantiomerically pure form, but there is also an enormous utility in asymmetric reactions where enzymes act more or less like chemical reagents. In fact, nearly every reaction in this book has a useful enzymatic equivalent, so besides using one's favorite database or search engine, one might seek an introduction to the topic with some excellent monographs [1–3]. For just a taste, here are a couple of nice examples.

Acylation and hydrolysis. Many enzymes work by differentiating enantiotopic groups, and ester hydrolases provide typical examples of this [4]. Consider the achiral starting material *cis-1,3-cyclopent-4-ene diacetate*; all one needs to do is differentiate the two ester groups in this molecule to imagine an effective route to the protected silyl ether-containing enone, an intermediate useful in the preparation of many prostaglandin congeners (and other things; see ref. [5]). Although a number of hydrolases carry out this conversion, the use of electric eel acetylcholinesterase (EEAC) stands out as a useful method [6]. As is often the case, the success of enzymes in carrying out stereoselective hydrolysis reactions has inspired chemists to develop biomimetic versions of related reactions (e.g., see refs. [7,8]).

OAc / OAc →(EEAC)→ OH / OAc 87% 98% es ----→ OTMS (enone, O) prostaglandin intermediate

Since enzymes naturally exert their action in a dizzyingly complex biological milieu, they are beautifully equipped to work in biosynthetic pathways involving other enzymes. This can be done *in vitro* or even extended to reactions in tandem with synthetic catalysts, such as the asymmetric transformation of the second kind reported by Bäckvall [9]. Here, a Ru catalyst is used to interconvert the two enantiomers present in a racemic mixture of 2-indanol, with the R isomer being selectively converted to the corresponding acetate in the presence of an acyl donor.

OH ⇌ (Ru catalyst (2 mol%), racemization) OH ; slow / fast ; enzyme-mediated acylation ; OAc / OAc 81% >99% es

Dihydroxylation of aromatic rings. One doesn't often consider aromatic rings as the starting point for organic synthesis, except for making other aromatic compounds or the occasional

Birch reduction. But perhaps this should change, as evidenced by the use of toluene dioxygenase for the oxidative de-aromatization of substituted benzene rings (the two faces of C-2 of bromobenzene are enantiotopic). Pioneered by Gibson [10] and utilized in a microbial setting, these enzymes provide access to numerous synthetic intermediates. Hudlicky, in particular, has demonstrated these methods in numerous syntheses, including a number of routes to the Amaryllidaceae alkaloid pancratistatin (reviews: refs. [11,12]).

Desymmetrization of meso pyrrolidines. Oxidative reactions can be promoted by numerous enzymes, many of which mediate metabolism of both natural compounds and xenobiotics (such as drugs). The monoamine oxidase shown allows the use of molecular oxygen as the terminal oxidant and carries out the group-selective oxidation of *cis*-3,4-dimethylpyrrolidine [13]. Trapping of the imine intermediate with cyanide is shown but numerous other applications can be readily imagined (for a nice example, see ref. [14]).

References

[1] Wong, C.-H.; Whitesides, G. M. *Enzymes in Synthetic Organic Chemistry*; Pergamon: Oxford, 1994.

[2] Patel, R. N.; Ed.; *Stereoselective Biocatalysis*; Dekker, 2000.

[3] Gotor, V.; Alfonso, I.; Garcia-Urdiales, E.; Eds.; *Asymmetric Organic Synthesis with Enzymes*; Wiley-VCH: Weinheim, 2008.

[4] Ohno, M.; Otsuka, M. *Org. React.* **1989**, *37*, 1–55.

[5] Roche, S. P.; Aitken, D. J. *Eur. J. Org. Chem.* **2010**, 5339–5358.

[6] Deardorff, D. R.; Windham, C. Q.; Craney, C. L. *Org. Synth.* **1998**, *Coll. Vol. 9*, 487–493.

[7] Fu, G. C. *Acc. Chem. Res.* **2000**, *33*, 412–420.

[8] Davie, E. A. C.; Mennen, S. M.; Xu, Y.; Miller, S. J. *Chem. Rev.* **2007**, *107*, 5759–5812.

[9] Larsson, A. L. E.; Persson, B. A.; Bäckvall, J.-E. *Angew. Chem., Int. Ed. Engl.* **1997**, *36*, 1211–1212.

[10] Zylstra, G. J.; Gibson, D. T. *J. Biol. Chem.* **1989**, *264*, 14940–14946.

[11] Boyd, D. R.; Bugg, T. D. H. *Org. Biomol. Chem.* **2006**, *4*, 181–192.

[12] Hudlicky, T. *Pure Appl. Chem.* **2010**, *82*, 1785–1796.

[13] Köhler, V.; Bailey, K. R.; Znabet, A.; Raftery, J.; Helliwell, M.; Turner, N. J. *Angew. Chem., Int. Ed.* **2010**, *49*, 2182–2184.

[14] Znabet, A.; Zonneveld, J.; Janssen, E.; De Kanter, F. J. J.; Helliwell, M.; Turner, N. J.; Ruijter, E.; Orru, R. V. A. *Chem. Commun.* **2010**, *46*, 7706–7708.

A related reaction has been used in the total synthesis of biotin [57] as shown in Scheme 7.4a. In this, reduction with the unusual reagent LiH in the presence of the chiral ligand shown afforded a partially reduced imide in high enantioselectivity; subsequent reduction using the stronger reagent, KBH_4, finished the job by reducing this carbon to an alcohol that then attacked the amide carbonyl to afford the lactone shown. Finally, Spivey has reported a particularly interesting example whereby a centrosymmetric, achiral bis-N-Boc lactam lacking an internal mirror plane is selectively converted to a hydroxylactam with high enantioselectivity (Scheme 7.4b [58]). Inspection of the proposed reaction ensemble for this reaction demonstrates a feature typical of group-selective reactions, which is that they typically require both reagent-based enantioselectivity and substrate-based diastereoselectivity to afford the product in high selectivity (this concept has been eloquently espoused by Schreiber in the context of group-selective oxidations (*cf.* Scheme 1.8c and Section 8.14 [59])). Thus, for the example in Scheme 7.4b, attack by the *Si*-selective oxazaborolidine reagent occurs selectively at the carbonyl shown because the *Si* face of the alternative, nonreacting carbonyl is blocked by the second ring. A similar argument can be made for the reactions of other facially biased dicarbonyl compounds, such as the tricyclic anhydride shown in Scheme 7.4c, for which preferential selectivity for attack by a nucleophile occurs *syn* to the smaller oxygen bridge instead of the two-carbon bridge.

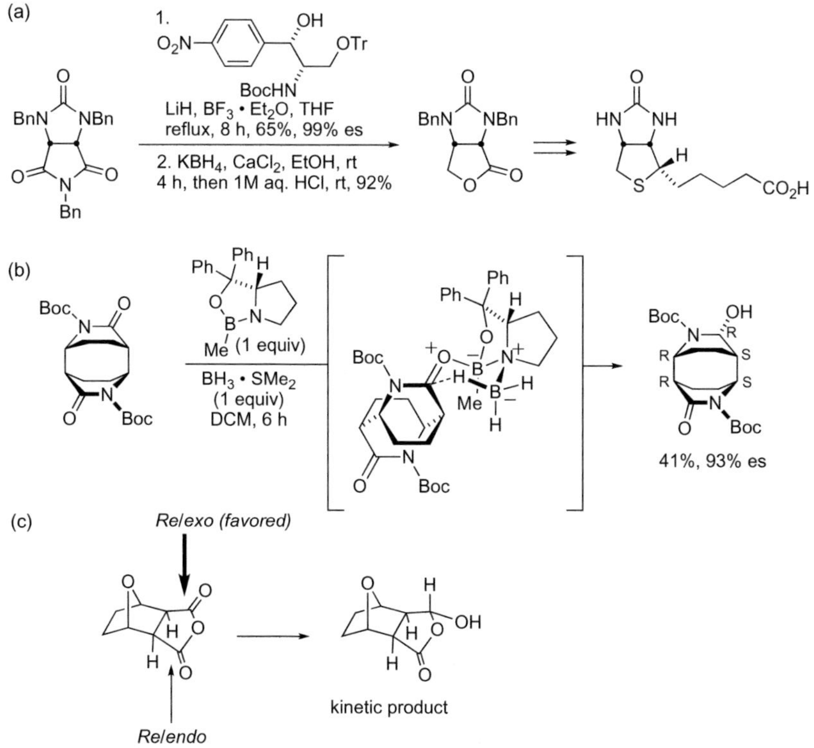

SCHEME 7.4 Yields and enantioselectivities of reduction of (a) imides [54–57]. (b) Reduction of a centrosymmetric bis-N-Boc lactam [58]. (c) The combination of a *Re*-selective reagent with substrate-directed exo selectivity leads to a group-selective reduction.

7.1.3 Chiral Organoboranes[4]

The reaction of a chiral alkene with borane may afford an alkyl or dialkyl borane, depending on stoichiometry. Attempts to achieve highly selective reductions of ketones using such reagents have met with little success, except for the particular case of alkynyl ketones.[5] Trialkyl boranes R_3B were first reported to reduce aldehydes and ketones (under forcing conditions) in 1966 by Mikhailov [64]. Mechanistic studies (summarized in ref. [60]) showed that there are two limiting mechanisms for the reduction of a carbonyl compound by a trialkylborane, as shown in Scheme 7.5: a pericyclic process reminiscent of the Meerwein−Pondorf−Verley reaction (Scheme 7.5a), and a two-step process that involves dehydroboration to a dialkylborane plus olefin, followed by carbonyl reduction by the dialkylborane (Scheme 7.5b). With unhindered carbonyl compounds such as aldehydes, the reaction is bimolecular and appears to proceed by the pericyclic pathway [65]. With ketones, the rate is independent of ketone concentration, indicating a switch to the dehydroboration−reduction pathway.

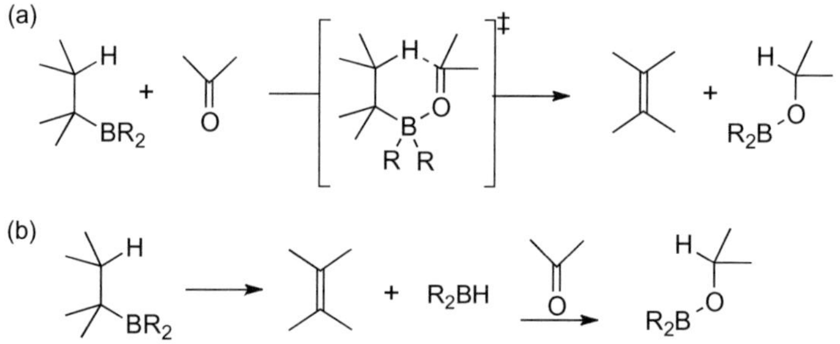

(a)

(b)

SCHEME 7.5 Limiting mechanisms, generically depicted, for reduction of carbonyls by a trialkylborane: (a) pericyclic mechanism. (b) Two-step mechanism involving dehydroboration of a trialkylborane followed by carbonyl reduction by the resultant dialkylborane.

In 1979−1980, Midland showed that the trialkyl borane formed by hydroboration of α-pinene by 9-borabicyclononane (9-BBN), known as B-isopinocampheyl-9-borabicyclo [3.3.1]nonane or Alpine-Borane®, efficiently reduces aldehydes [66,67] and propargyl ketones [68,69] with a high degree of enantioselectivity, as shown in Scheme 7.6. The mechanism was shown to be a self-immolative chirality transfer process (Scheme 7.5a), proceeding through the 6-membered ring boat transition structure shown in Schemes 7.6b and c [60]. This reduction is probably the method of choice for the production of enantiomerically enriched primary alcohols that are chiral by virtue of isotopic substitution, provided enantiopure α-pinene is used [70]. Most ketones other than propargyl ketones are not readily reduced by trialkylboranes, making this process highly chemoselective for aldehydes and propargyl ketones in the presence of other ketones, esters, acid chlorides, alkyl halides, alkenes, and alkynes. Under forcing conditions, Alpine-borane dehydroborates (the reverse of Scheme 7.6a) with a half-life of 500 min in refluxing THF [60], and nonselective reduction by 9-BBN becomes competitive (*cf.* Scheme 7.5b).

---

4. Reviews: refs. [60−62].
5. For a notable exception, see ref. [63].

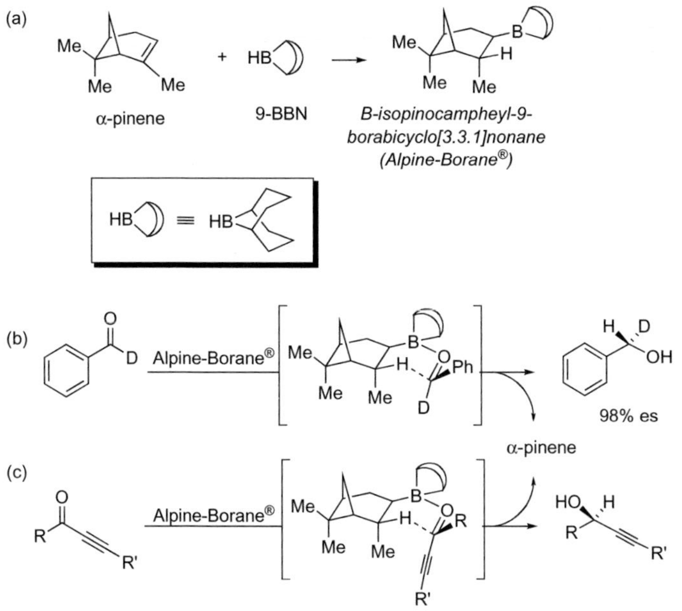

SCHEME 7.6 Alpine-borane method of asymmetric reduction (Midland reduction). (a) Preparation of Alpine-Borane®. (b) Reduction of deuterio benzaldehyde [66]. (c) Reduction of propargyl ketones [68,69].

To circumvent the problem of competitive dehydroboration with ketones, the Alpine-borane reductions can be conducted in neat (excess) reagent [71] or at high pressure (6000 atm [72]). Experiments done in neat reagent take several days to go to completion, and afford enantioselectivities of 70–98% [71]. At pressures of 6000 atm, the reactions are faster and dehydroboration is completely suppressed. Ketones are reduced with slightly higher enantioselectivities (75–100% es) under these conditions [72].

A better solution to asymmetric ketone reduction is to make a more reactive borane. Brown showed that hindered dialkylchloroboranes (R_2BCl) are less prone to dehydroboration than hindered trialkylboranes (R_3B) such as Alpine-borane and are excellent reagents for the reduction of aldehydes and ketones. Inductive electron withdrawal by the chlorine also increases the Lewis acidity of the boron. B-Chlorodiisopinocampheylborane (Ipc_2Cl, DIP-chloride™) is such a reagent, and is an excellent reagent for the asymmetric reduction of aryl–alkyl ketones [73,74]. Scheme 7.7 shows the preparation of Ipc_2Cl and the postulated transition structure to rationalize the chirality sense of the products. Table 7.3 lists several examples. Note that dialkyl ketones and alkynyl–alkyl ketones are reduced with low selectivity unless one of the substituents is tertiary. For a summary of other pinene-based self-immolative reducing agents (reviews: refs. [61,62]).

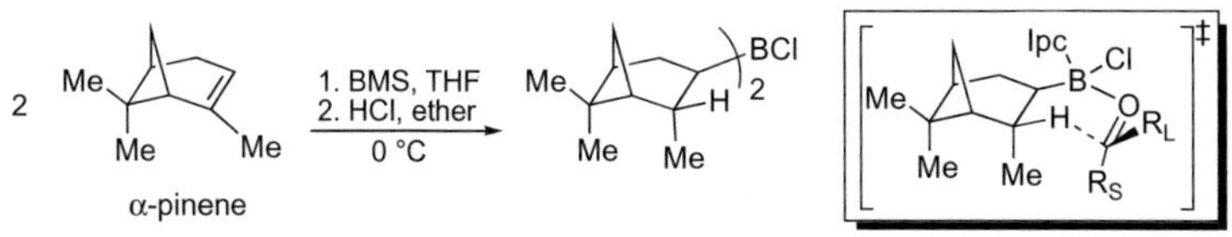

SCHEME 7.7 Preparation of Ipc_2Cl. *Inset*: Proposed transition structure for asymmetric reductions using Ipc_2Cl [73].

TABLE 7.3 Asymmetric Reduction of Ketones, $R_1C(=O)R_2$, with Ipc_2Cl

Entry	R_1	R_2	% Yield	% es	Ref.
1	Me	Et	—	52	[73]
2	Me	i-Pr	—	66	[73]
3	Me	t-Bu	50	93	[73]
4	2,2-Dimethylcyclopentanone		71	98	[73]
5	2,2-Dimethylcyclohexanone		60	91	[73]
6	1-Indanone		62	97	[73]
7	α-Tetralone		70	86	[73]
8	$HC\equiv C$	Me	83	58	[74]
9	$PhC\equiv C$	Me	92	60	[74]
10	$PhC\equiv C$	i-Pr	85	92	[74]
11	$PhC\equiv C$	t-Bu	80	>99	[74]
12	$cyclo\text{-}C_5H_{11}C\equiv C$	i-Pr	81	69	[74]
13	$cyclo\text{-}C_5H_{11}C\equiv C$	t-Bu	76	98	[74]
14	$n\text{-}C_8H_{17}C\equiv C$	i-Pr	86	63	[74]
15	$n\text{-}C_8H_{17}C\equiv C$	t-Bu	77	99	[74]

7.1.4 Chiral Transition Metal Catalysts

Early attempts to discover catalysts that deliver consistently high selectivities in reactions of simple ketones were only partially successful [75–78], with more success being recorded in the asymmetric reduction of functionalized ketones and imines (reviews: refs. [79,80]). Breakthroughs in the mid 1990s, largely led by Noyori, led to considerably greater scope of transition metal-promoted ketone and imine reductions. The field can be divided into reactions that use dihydrogen as the stoichiometric reductant and those that do not, such as dihydrosilanes (reviews: refs. [81,82]) or transfer catalytic processes [83–90]. The latter class of reactions are those in which the elements of dihydrogen are derived from donors such as isopropanol, the formic acid/triethylamine azeotrope, or Hantzsch esters.

Ketone reductions using hydrogen as the primary reductant. For the asymmetric hydrogenation of functionalized ketones, a team led by Noyori in Nagoya and Akutagawa in Tokyo introduced ruthenium(II) BINAP catalysts that produce excellent enantioselectivities for a number of functionalized ketones [91–97] (review: ref. [98]). The topicity of the reduction is illustrated in Scheme 7.8a, and suggests a mechanism in which the heteroatom X and the carbonyl oxygen chelate the metal (*vide infra*). The catalyst is thought to be a monomeric BINAP ruthenium(II) dichloride, which was originally prepared by a tedious process using Schlenk techniques [91]; however, improved procedures were subsequently introduced [93–95]. In 1995, Burk introduced a diphosphine ligand useful in this sort of catalytic hydrogenation (Scheme 7.8b [99]). The same year (in retrospect, sort of an *annus mirabilis* for catalytic hydrogenation — see the next section), Noyori reported that the addition of a chiral amine to the (BINAP)RuCl$_2$ system (or versions where the phenyl groups on BINAP were replaced by tolyl groups) resulted in a catalyst particularly useful for the highly enantioselective reductions of aryl ketones (Scheme 7.8c [100]). An important bonus of the Noyori procedure is that, in addition to enantioselectivity, it permits the selective catalytic reduction of ketones in the presence of alkenes and alkynes [101].

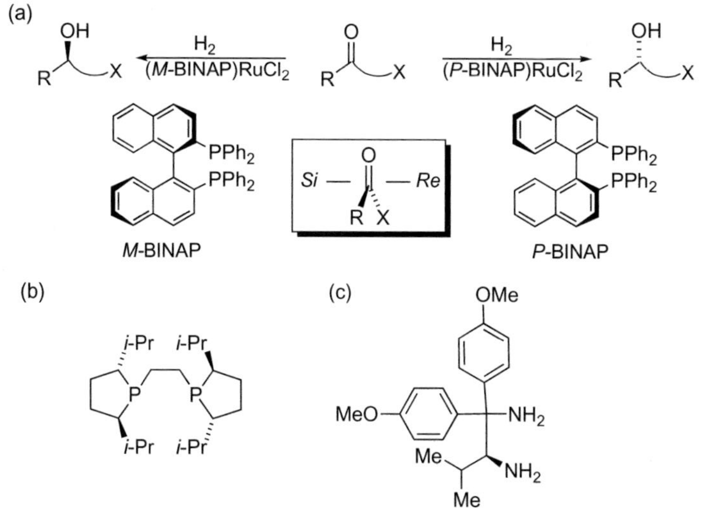

SCHEME 7.8   (a) Relative topicity *ul* (*e.g.*, *P*-BINAP/*Re* face) is uniformly observed for ruthenium BINAP catalyzed asymmetric reduction of functionalized ketones [92]. (b) *R,R-i*-Pr-BPE (1,2-bis(*trans*-2,5-diisopropylphospholano)ethane) [99]. (c) A diamine additive used by Noyori along with BINAP in ketone reductions [100].

Selected examples that proceed in high selectivity are listed in Table 7.4. Several β-keto esters are reduced with excellent enantioselectivity (entries 1, 4−7); however, α-keto esters were reduced with somewhat diminished enantioselectivity [92]. A variety of substituted ketones were revealed to be good substrates. Particularly striking was the chemoselectivity observed when the reductions were conducted at low pressures: isolated double bonds were left intact (entries 6 and 7). Bifunctional ketones may be problematic, since chelation might occur by more than one functional group. In cases like this, protection of hydroxyl as a relatively nonchelatable group like a triisopropylsilyl (TIPS) ether could help; note the excellent enantioselectivity in the example shown in entry 5.

TABLE 7.4 Selected Examples of Asymmetric Ketone Reductions Using the Catalyst Systems Noted in Scheme 7.8[a]

Entry	Ketone	Catalyst	% Yield	% es	Ref.
1	Me–C(=O)–CH$_2$–CO$_2$R; R = Me, Et, i-Pr, t-Bu	a	≥97	≥99	[91–93]
2	Me–C(=O)–CH$_2$–CO$_2$Me	b	100	>99	[99]
3	Me–C(=O)–CH$_2$–C(=O)–X; X = NMe$_2$	a	100	98	[92]
	X = SEt		42[b]	96	
4	Ph–C(=O)–CH$_2$–CO$_2$Et	a	99	92	[91]
5	RO–(CH$_2$)$_n$–C(=O)–CH$_2$–CO$_2$Et; R = TIPS, n = 1	a	100	97	[92]
	R = Bn, n = 2		94	99	
6	Me–CH=CH–CH$_2$CH$_2$–C(=O)–CH$_2$–CO$_2$Me[c]; R = H	a	73	99	[94]
	R = Me		96	99	
7	Ph–C(=O)–CH$_2$CH$_2$–CH=CH$_2$	c	>99	95	[101]
8	Me–C(=O)–CH$_2$–X; X = NMe$_2$	a	72	98	[92]
	X = OH		100	96	
	X = CH$_2$OH		100	99	
9	R–C(=O)–CH$_2$–NMe$_2$; R = i-Pr	a	83	97	[92]
	R = Ph		85	97	
10	(4-Me-C$_6$H$_4$)–C(=O)–Me	c	99	96	[101]

[a]Reactions were run at room temperature and 50–100 atm unless otherwise noted. Yields were determined by GC or spectroscopically unless noted.
[b]Isolated yield.
[c]50 psi, 80 °C.

These catalytic reductions are relatively slow, requiring high pressures or high temperatures, and chiral β-ketoesters racemize faster than they can be reduced, a case of Curtin–Hammett kinetics [102,103]. As it happens, asymmetric reduction of one enantiomer is considerably faster than reduction of the other. Since the enantiomers racemize rapidly, the ruthenium BINAP catalyst can be used to effect a dynamic kinetic resolution (Section 1.9 and ref. [104], as shown in Scheme 7.9a [105]. In this example, the *racemic* β-keto ester is completely converted to the *syn* amino alcohol with a diastereoselectivity (*syn:anti*) of 99:1. The *syn* product is obtained in 97:3 er, indicating that of the four possible stereoisomeric products (*syn* and *anti* enantiomers), the major product is 96% of the mixture. The simple explanation for this beautiful result is shown in Scheme 7.9b: racemization under the reaction conditions is fast compared to reduction of either enantiomer, but reduction of the *S*-enantiomer by the *M*-BINAP catalyst (double asymmetric induction matched pair, addition to the ketone *Si* face) is faster than reduction of the *R*-enantiomer (mismatched pair, not shown), so the net result is a draining of the fast racemization equilibrium.

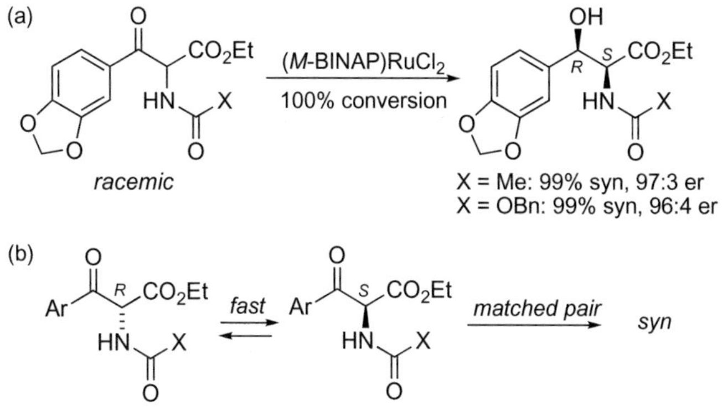

SCHEME 7.9   (a) Asymmetric reduction of chiral β-keto esters may be used in an asymmetric transformation of the first kind (dynamic kinetic resolution; see also Scheme 1.12) [105]; (b) Rationale for the selectivity.

The proposed catalytic cycle for these reductions is shown in Scheme 7.10 [98]. In this scheme, it is assumed that the polymeric catalyst precursor [(BINAP)RuCl$_2$]$_n$ is dissociated to monomer by the methanolic solvent. Reduction and loss of hydrogen afford the putative catalyst, (BINAP)RuHCl(MeOH)$_2$. Displacement of the two methanols by the bidentate substrate then sets the stage for the hydrogen transfer step (*vide infra*). Exchange of the alkoxide product with the methanolic solvent and reaction with hydrogen to regenerate the catalyst completes the catalytic cycle. Deuterium labeling experiments showed that the mechanism involves C=O reduction, and not the alternative C=C reduction of an enol tautomer [105].

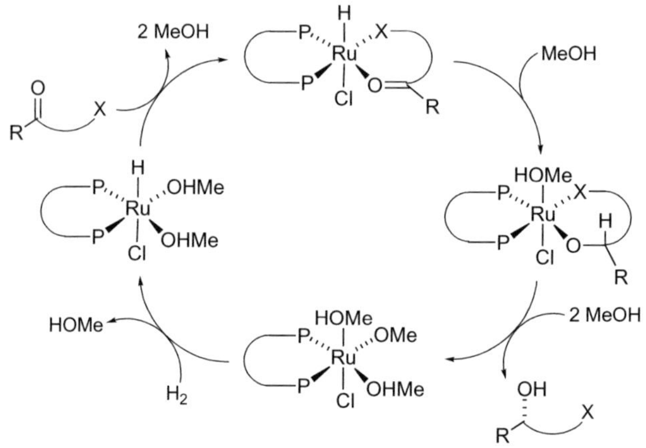

SCHEME 7.10 Catalytic cycle proposed for the asymmetric reduction of functionalized ketones by ruthenium BINAP catalyst [98].

The X-ray crystal structures of two ruthenium BINAP complexes have been determined [96,106]. Figure 7.3a illustrates the structural features that are thought to influence stereoselectivity. In both crystal structures, the 7-membered chelate ring formed by the *P*-enantiomer of the BINAP ligand and the metal adopt similar conformations and have the pseudoequatorial phenyl groups occupying the lower left and upper right quadrants, as viewed from the P−Ru−P plane with the BINAP to the rear. The pseudoaxial phenyls are slanted to the rear and would not significantly interact with a ligand bound *trans* to either phosphorous atom. For reduction of methyl acetoacetate, *ul* relative topicity is observed (*P*-BINAP catalyst preferentially attacking the *Re* face of the ketone). Assuming that the catalyst is a mononuclear monohydride complex having the hydrogen and chlorine *trans*, with the substrate chelated to the ruthenium (each carbonyl oxygen being *trans* to a phosphorous), the chirality sense may be rationalized by the two transition structures illustrated in Figures 7.3b and c. A 4-membered transition structure having *lk* topicity (Figure 7.3b) would force the C-4 methyl into the crowded lower left quadrant, while the transition structure with *ul* topicity (Figure 7.3c) is less hindered [98].

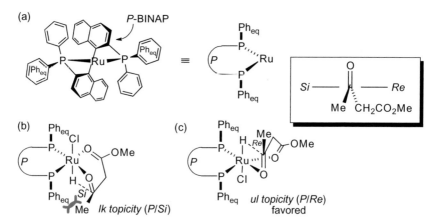

FIGURE 7.3 (a) Conformation of *P*-BINAP in two crystal structures [96,106]; (b) *lk* topicity transition structure for asymmetric reduction of methyl acetoacetate; (c) *ul* topicity transition structure [98].

Catalytic transfer hydrogenation. The benefits of replacing dihydrogen with another hydrogen source for reduction reactions primarily accrue from avoiding highly flammable H_2 gas and the attendant experimental setups needed to use it.[6] The general moniker given to stereoselective hydrogenation procedures that eschew H_2 is asymmetric transfer hydrogenation (ATH; reviews: refs. [84,85,87,88,107]). A classic example of an organic reaction that transfers hydride from an organic donor is the aluminum-promoted Meerwein−Pondorf−Verley reduction, which was invented in the 1920s. It is therefore fitting that the first attempt (1950) to carry out an asymmetric hydrogen transfer reaction was reported by Doering and Young, who used the aluminum oxide of (+)-methyl isopropyl carbinol to reduce methyl cyclohexyl ketone to the corresponding alcohol in 61:39 er [108]. Early attempts to carry out analogous reduction reactions using chiral ruthenium reagents were published independently by Sinou and Ohkubo in 1976 [109,110], but the breakthrough paper in the area was published by Noyori in 1995 [111]. In that work, isopropanol was used as the hydrogen source and a chiral Ru complex was generated *in situ* from a $[RuCl_2(arene)]_2$ precursor and a sulfonylated chiral diamine in the presence of base (Scheme 7.11a). An immediate advantage of the Noyori method over related ones (*e.g.*, a Sm-based MPV reaction reported by Evans [112]) was the low catalyst loadings (substrate:catalyst = *ca.* 200:1). The overall mechanism of the reaction using isopropanol, supported in part by some beautiful X-ray crystallography carried out by Noyori [113], is summarized in Scheme 7.11b (which depicts the *in situ* generation of the active catalyst) and Scheme 7.12a (for a review of the mechanism of the reaction, see ref. [107]).

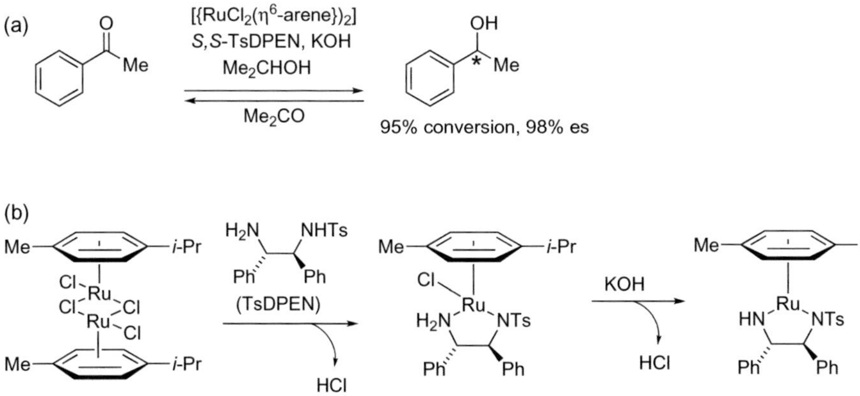

SCHEME 7.11 (a) Noyori's asymmetric transfer hydrogenation of acetophenone [111]; (b) mechanisms for the *in situ* formation of the active catalyst from [Ru(cymene)Cl₂]₂.

6.

The key series of events are the transfer of hydride from isopropanol to the Ru complex, which occurs concomitantly with loss of acetone byproduct, followed by the analogous binding and subsequent hydride transfer from complex to the substrate ketone. The stereoselectivity depends on the nature of the arene ligand used, and although benzene was initially used by Noyori, cymene gives better results and is shown in the figure [114]. As is often (almost always?) the case, the actual mechanism is more complex and likely involves solvent in a more explicit way than is typically depicted [115]. Noyori invoked the transition state assembly for the reduction shown in Scheme 7.12b [116]. Note the proposed edge−face interaction (a favorable C−H/aryl interaction) in the favored transition state leading to the observed *Re* addition.

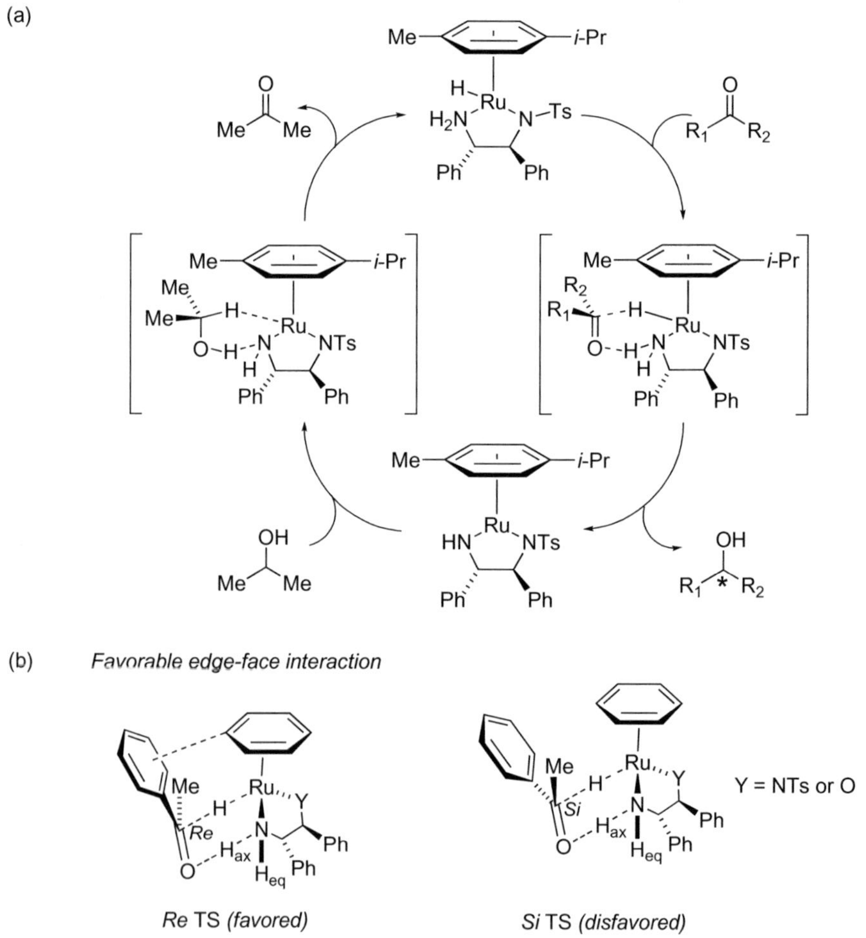

SCHEME 7.12   (a) The asymmetric transfer hydrogenation using *i*-PrOH as hydride donor and (b) a proposed model to explain the sense of enantioselectivity [116].

A challenge unique to asymmetric transfer hydrogenations using an alcohol as the hydrogen source is that the reverse reaction (*i.e.*, transfer of hydride from the chiral alcohol product back to acetone formed in the course of the reaction) can lead to racemization. Initially, the problem was addressed by minimizing the concentration of substrate, running the reaction in isopropanol, and by keeping the reaction time to a minimum. An alternative is to use an irreversible hydride donor instead, such as formic acid, which is usually introduced as an ammonia−formic acid mixture. In this case, the Ru catalyst undergoes oxidative addition of formic acid that then expels CO_2 to form the same hydride intermediate as obtained from isopropanol (Scheme 7.13, ref. [114]). Trialkylsilanes such as $HSiMe_3$ are another alternative hydrogen equivalent that have received substantial interest but are not covered here in detail (for reviews, see refs. [88,117−119]). Thus, the hydrosilylation of a ketone affords a silylated ether that can be easily converted to the parent alcohol by treatment with fluoride or acid (this reaction has also been applied to imine reductions).[7]

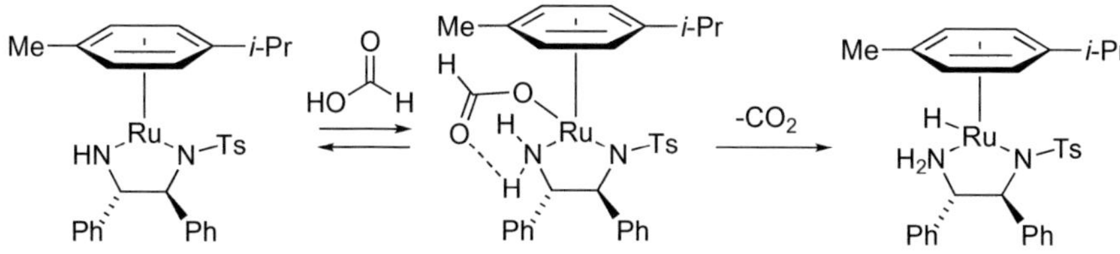

SCHEME 7.13 Alternative formation of the active catalytic species from formic acid (generally introduced as ammonium formate).

Numerous other researchers have followed up on Noyori's initial work to outfit the Ru metal with a variety of ligands such as the representative sampling shown in Figure 7.4. As indicated in Table 7.5, these catalysts and catalyst systems (comprising ligand and pre-catalyst that combine to generate a catalyst *in situ*) have helped to establish ATH as an impactful method in asymmetric synthesis of considerable scope, particularly involving aromatic ketones. Besides the Noyori systems, represented here in Figures 7.4a and b, a number of noteworthy examples are listed, such as the comparison between the BINAP-derived catalyst shown in Figure 7.4c and the closely related ketone in Figure 7.4d. In the latter case, the interpolation of a carbonyl between the two phenyl groups means the ligand lacks the configurationally stable axial chirality typical of binaphthyl systems in the free ligand, but coordination to Ru stabilizes the illustrated diastereomer in response to the chiral diamine ligand and leads to superior selectivity (see entries 8 and 9) [120]. Adolfsson has done a great deal of creative work in the area of ATH utilizing peptide-derived ligands [121], with three interesting examples represented in entries 10−12 of Table 7.5. Combining the observations that (1) Ru dipeptide complexes were excellent catalysts for ATH and (2) that the Ru pre-catalyst was capable of promoting the formation of peptide bonds between two ligands [121], he was able to carry out the modular assembly of a highly selective hydrogenation catalyst from its component elements *in situ* [122]. Another interesting feature was that it proved possible to use a single enantiomeric series of amino acids in preparing catalysts able to switch from one sense of enantioselectivity to another by slight modification of the other functional group

---

7. The reductive hydrosilylation of electron-deficient C=C bonds will be briefly discussed in the following section.

elements in the catalyst (*e.g.*, amide to thioamide, addition of a hydroxy group; *cf.* entries 11 and 12 of Table 7.5 [121,123]). Other recent work has focused on designing catalysts for use in different solvents, such as water (entries 13 and 16 [124,125]) or mixtures involving ionic liquids (entry 14 [126]), use of host—guest chemistry principles (entry 15 [127]), or a version in which the arene fragment is tethered to the chiral diamine ligand [125].

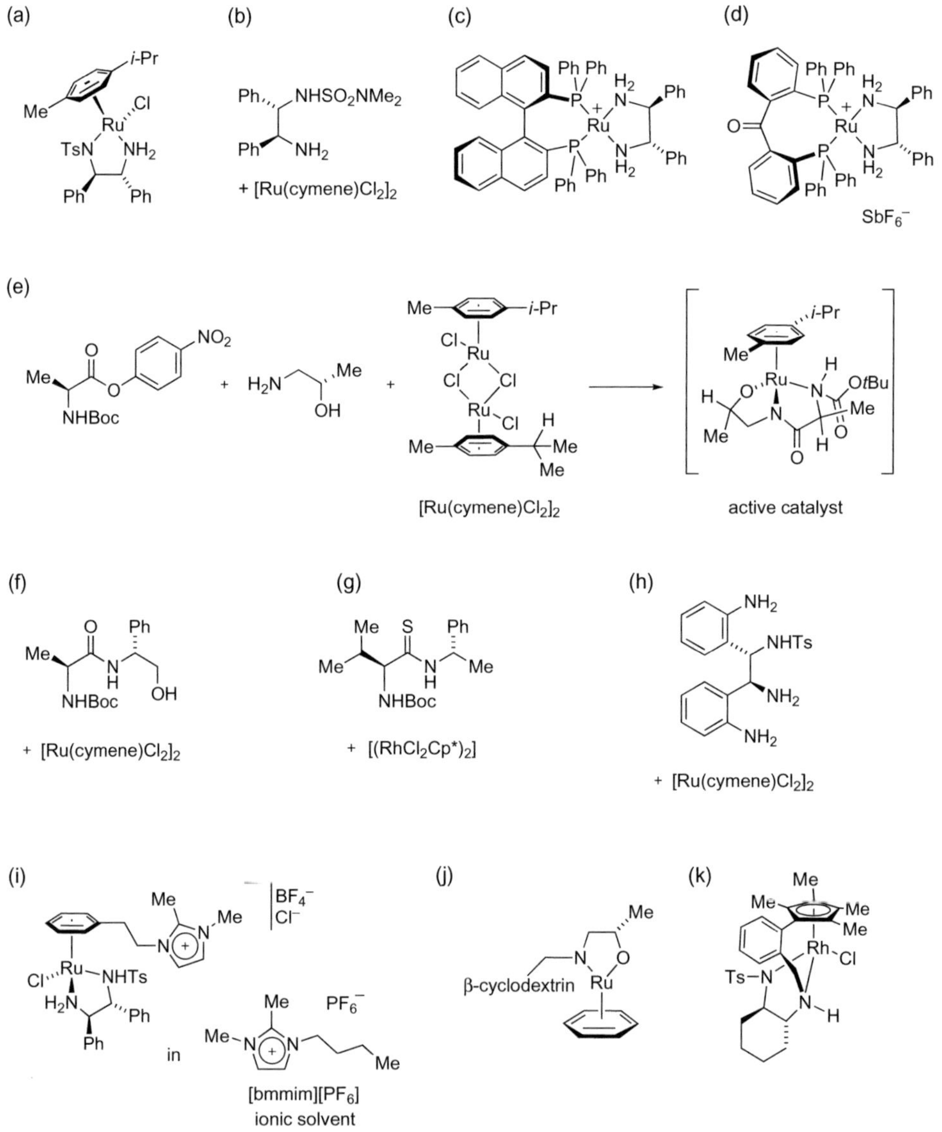

FIGURE 7.4 Catalysts and catalyst systems used in asymmetric transfer hydrogenation: (a) [128], (b) [129], (c) and (d) [120], (e) [122], (f) [121], (g) [123], (h) [124], (i) [126], (j) [127], and (k) [125].

TABLE 7.5 Catalytic Asymmetric Transfer Hydrogenation Reactions

Entry	Starting Material	Catalyst (Other Details)[a]	[H] Source	% Conversion	% es (Major Isomer)/(Abs Config)	Ref.
1	O / Me (acetophenone)	a	HCO$_2$Na/ CTAB[b]	99	98 (R)	[128]
2	O / Me (propiophenone)	a	HCO$_2$Na/ CTAB[b]	97	95 (R)	[128]
3	O / Me Me (isobutyrophenone)	a	HCO$_2$Na/ CTAB[a]/SDS[c]	78	69 (R)	[128]
4	O (indanone)	a	HCO$_2$Na/ CTAB[b]	98	99 (R)	[128]
5	Ph / O / O / Ph	a (S,S)	HCO$_2$H/Et$_3$N	97–100	>99 (R,R-diol) (syn:anti = 98.6:1.4)	[130]
6	O / Me Ph	a (S,S)	i-PrOH	>99	99 (S)	[131]
7	O O / Ph OEt	b	HCO$_2$H/Et$_3$N	100	99 (S)	[129]
8	Me O / Me	c	i-PrOH	61	79 (R)	[120]
9	"	d	i-PrOH	99	>99 (R)	[120]
10	O / Me	e	i-PrOH	85	99 (S)	[122]
11	"	f	i-PrOH	95	97 (S)	[121]
12	"	g	i-PrOH	88	98 (R)	[123]

(Continued)

TABLE 7.5 (Continued)

Entry	Starting Material	Catalyst (Other Details)[a]	[H] Source	% Conversion	% es (Major Isomer)/(Abs Config)	Ref.
13	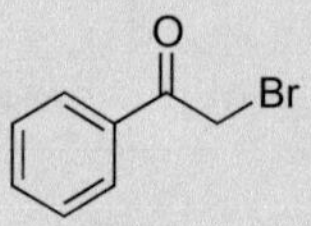	h (H_2O solvent)	HCO_2Na	91	99 (S)	[124]
14	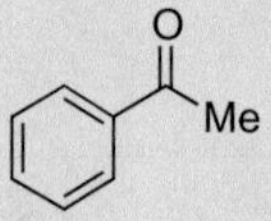	i	HCO_2H/Et_3N	>99	>99	[126]
15	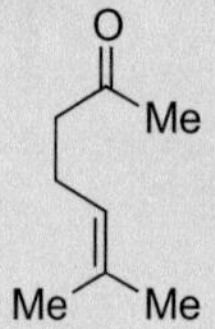	j	HCO_2Na (3:1 water/DMP)	95	93 (S)	[127]
16	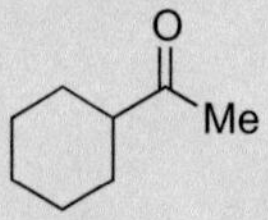	k	HCO_2H/Et_3N	96	94 (S)	[125]

[a] See Figure 7.4 for Catalyst Structures.
[b] CTAB = cetyl trimethyl ammonium bromide.
[c] SDS = sodium dodecyl sulfate.

Figure 7.5 illustrates a few natural products that have been synthesized using ruthenium (II)·BINAP-mediated ketone reduction or catalytic transfer hyrdogenation as the key step. In all but one of the cases shown, the secondary carbinol is retained; but for indolizidine 223AB, the Mitsunobu reaction was used to convert the alcohol to an amine [95].

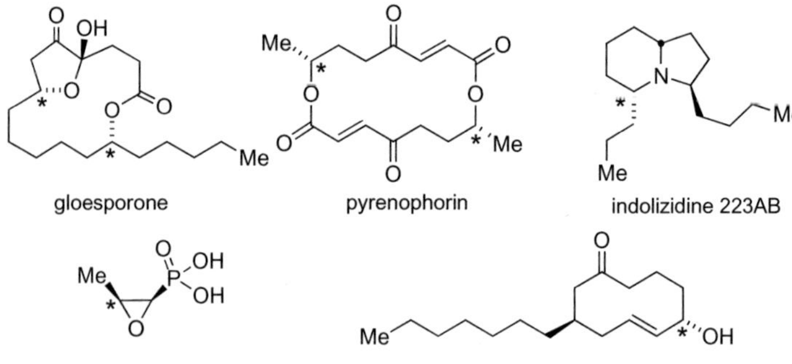

gloesporone   pyrenophorin   indolizidine 223AB

fosfomycin   hypocreolide A

FIGURE 7.5 Ruthenium(II)·BINAP catalysts have been used as a key step in the asymmetric synthesis of gloesporone [132], pyrenophorin [133], indolizidine 223AB [95], fosfomycin [134] and hypocreolide A [135]. Stereocenters formed by asymmetric reduction are indicated (*).

Imine reductions and related reactions. A medicinal chemist might well argue that the reduction of imines or analogs such as hydrazones and oximes to amines (whether carried out as a discrete reaction or folded into a reductive amination process) is the single most important reaction in the world, simply because nitrogen is present in so many drugs. In that context, it is interesting that the detailed discussion of this topic in the first edition of this book (published in 1995) was limited to a single case: Buchwald's reduction of imines, especially cyclic ones, using a titanocene catalyst. The late 1990s and early 2000s saw tremendous advances in this field, with many useful catalysts based on palladium, rhodium, ruthenium, and especially iridium making the scene in that period. This work, along with highlights of other catalysts will be covered below, but the reader is, as always, directed to some excellent review articles for more in-depth information and particularly, to see the wide array of catalysts available for such transformations [7,89,136].

Compounds containing C=N bonds often require additional activation for smooth reduction. Imines, in particular have the additional complications of imine/enamine equilibria and *E/Z* isomerization (Scheme 7.14a). For this reason, chemists have often resorted to imines bearing electron-withdrawing groups on the nitrogen atom, which both increase the stability of the C=N species relative to other tautomers and additionally render the compounds more electrophilic. The complementary reduction of enamine derivatives — typically enamides — will be addressed in the following section on C=C bond reduction. Some of the most common examples are shown in Scheme 7.14b. Finally, the "old-fashioned" approach of reducing chiral imines (Scheme 7.14c) still has practical importance, especially in industrial settings where these approaches are used on production scale. A couple of interesting examples are shown in Scheme 7.14d and e. In the former case, the relatively modest diastereoselectivity provided by the workhorse α-methylbenzylamine group could be improved by taking advantage of the kinetic preference for hydrogenolysis of the *R,R*-amine shown relative to its *R,S*-diastereomer [137]. Thus, by stopping the hydrogenolysis reaction of the amine at 55% conversion, it was possible to obtain the desired amine in 95:5 er. The sulfinylimines popularized by Davis [138] and Ellman [139] are also useful for amine synthesis *via* stereoselective reduction. Some particularly nice examples reported by Anderson showed that a single enantiomeric starting imine could be diverted to either enantiomeric product by changing the reducing agent [140]. As we will restrict our discussion below to imines (the principles of which are similar to oximes and other "activated" imines[8]), the reader is directed to the review by Nugent and El-Shazly, which provides a particularly good critical overview of these various approaches, along with many examples [7].

---

8. One issue that arises in C=N derivatives containing an additional N—X bond is that the latter needs to be reduced if one wishes to make an amine. For some examples, see ref. [141].

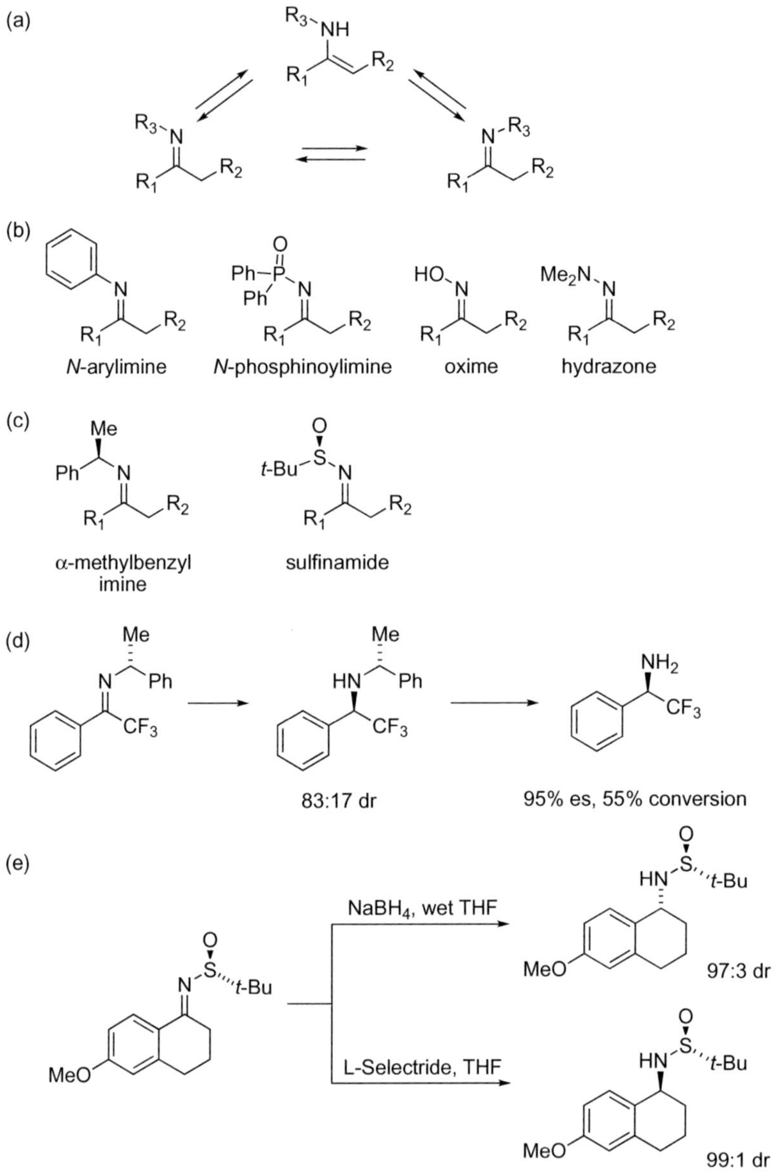

SCHEME 7.14 (a) Stereochemical and tautomeric equilibria of simple imines. (b) Some examples of N-activated imine derivatives often used in asymmetric reduction chemistry. (c) Chiral groups commonly used in imine reduction chemistry. (d) An example of asymmetric reduction combined with kinetic resolution in a downstream N-dealkylation step [137]. (e) Stereodivergent reduction of a chiral imine [139].

Buchwald's work was important because it was among the first approaches to provide good to excellent enantioselectivity in the catalytic reduction of imines [142–144] (review: ref. [145]). The reaction can be highly stereoselective for both acyclic and cyclic imines, but since acyclic imines are usually a mixture of *E* and *Z* isomers, and since the imine isomerization is catalyzed by the titanocene, the reaction is not always preparatively useful for acyclic substrates. Examples are listed in Table 7.6 (Scheme 7.15).

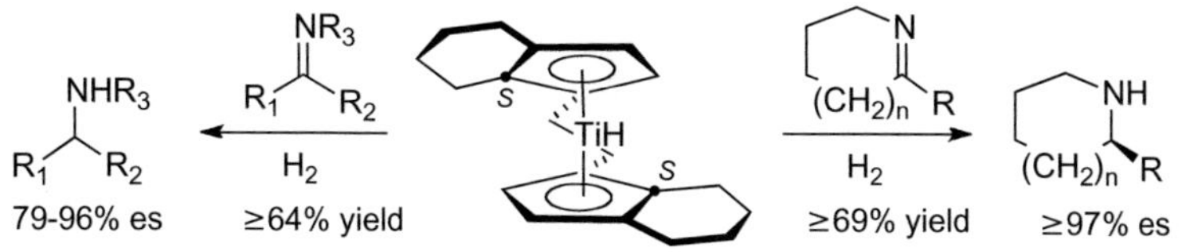

SCHEME 7.15 Titanocene-catalyzed asymmetric reduction of imines [143]. In the accompanying discussion, the catalyst shown is designated the *S,S*-enantiomer, in accord with the CIP rules for describing metal arenes [146]. This is a different designation than that used by Buchwald, however.[9]

TABLE 7.6 Examples of Asymmetric Imine Reduction using Buchwald's Chiral Titanocene Catalyst[a]

Entry	Imine	Amine	% Yield	% es	Ref.
1	 Ph—N(CH₂)n n = 1–3	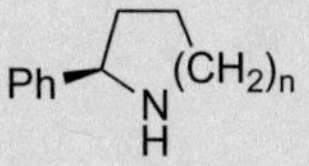 Ph—N(H)(CH₂)n	71–83	≥99	[142,143,153]
2	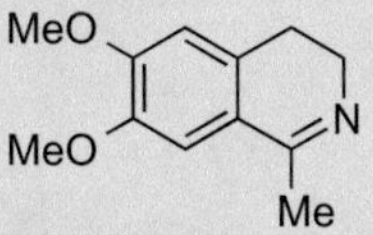 MeO, MeO isoquinoline with Me	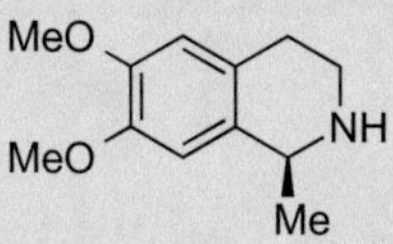 MeO, MeO with NH, Me	79	97	[142,143,153]
3	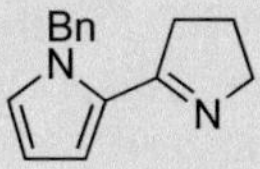 Bn pyrrole with N	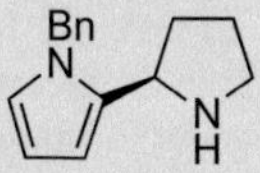 Bn pyrrole with NH	72	>99	[143,153]
4	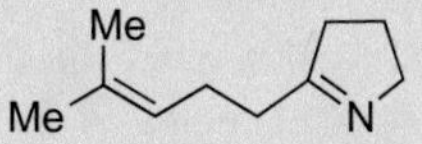 Me, Me with N	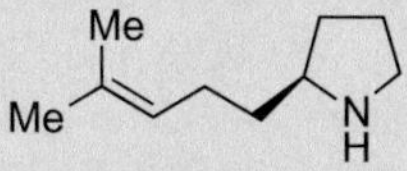 Me, Me with NH	79	>99	[143,153]
5	R (CH₂)₄ N R = H R = TMS	R (CH₂)₄ NH	72 73[b]	>99 >99	[143,153]

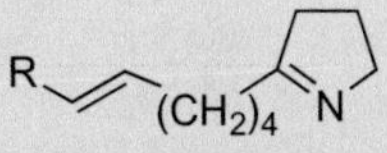

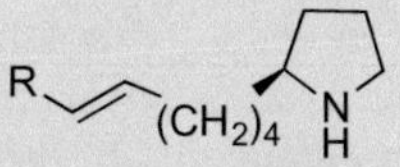

(Continued)

9. *For nomenclature buffs*: In the original paper describing the preparation of the titanocene catalyst precursor [147], Brintzinger specified the chirality sense of the ansa-metallocene by referring to the absolute configuration at C-1 of the indene (the carbon bearing the ethylene bridge), and Buchwald adopted this usage. However, the *CIP* system states that the chirality sense of the *complex* should be assigned with reference to *the arene ring atom* (or in general, the π-complexed atom of any ligand) *having the highest CIP rank* [146]. In this case, the highest-ranking atom is the C7a (indicated by the ● in Scheme 7.15), which has the opposite *CIP* designation of C-1. For rules on assigning a *CIP* descriptor to π-complexes, see refs. [148−151]. For another method (Ω+/Ω−), see ref. [152].

TABLE 7.6 (Continued)

Entry	Imine	Amine	% Yield	% es	Ref.
6	NR ⟍⟋ Me (cyclohexane) R = Me (92% E) R = Bn (92% E)	NHR ⟍⟋ Me (cyclohexane)	85^c 85^c	96^c 71^c	[142,143]
7	Me ⟍ NBn Me⟍⟍⟍Me (75% E)	Me ⟍ NHBn Me⟍⟍⟍Me	64^d	81^d	[143]
8	NBn ‖ R⟍Me R = i-Pr (93% E) R = Ph (94% E) R = 2-naphthyl (98% E)	NHBn ⋮ R⟍Me	66^d 81^d 82^d	88^d 88^d 85^d	[142,143]

^{a}Reactions were run at 45 °C and 80 psi, with 5 mol% S,S catalyst, unless noted otherwise.
^{b}Yield includes 5–8% of product having a saturated side chain.
^{c}500 psi H_2.
^{d}2000 psi H_2.

Examination of the enantioselectivities in Table 7.6 indicates a striking difference in selectivity achieved in the reduction of cyclic (entries 1−5) *vs.* acyclic imines (entries 6−8). The former is very nearly 100% stereoselective. The simple reason for this is that the acyclic imines are mixtures of E and Z stereoisomers, which reduce to enantiomeric amines (*vide infra*). The mechanism proposed for this reduction is shown in Scheme 7.16 [144]. The putative titanium(III) hydride catalyst is formed *in situ* by sequential treatment of the titanocene BINOL complex with butyllithium and phenylsilane. The latter reagent serves to stabilize the catalyst. Kinetic studies show that the reduction of cyclic imines is first order in hydrogen and first order in titanium but zero order in imine. This (and other evidence) is consistent with a fast 1,2-insertion followed by a slow hydrogenolysis (σ-bond metathesis), as indicated in Scheme 7.16 [144]. Although β-hydride elimination of the titanium amide intermediate is possible, it appears to be slow relative to the hydrogenolysis.

Note the η^2 bonding of the imine to the titanium at the transition state for insertion. The geometry of this complex is critical to the stereoselectivity of the reaction, since it is in this step that the stereocenter in the product is created. A dichlorotitanocene is tetrahedral around titanium, as verified by X-ray crystallography. Note the C_2 symmetry of the complex in Figure 7.6a, the orientation of the two cyclohexane moieties in the upper left and lower right quadrants, and the placement of the two chlorines with respect to the cyclohexanes.

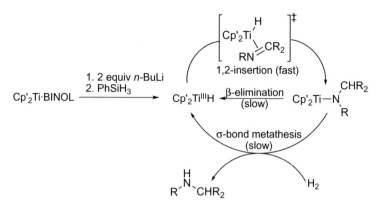

SCHEME 7.16 Proposed catalytic cycle for the titanocene-catalyzed reduction of imines [144].

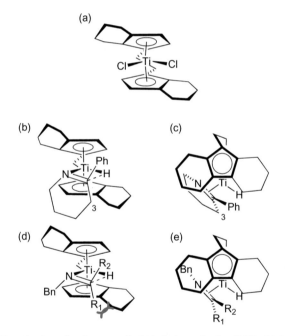

FIGURE 7.6 Transition structures for titanocene hydride imine reduction [144]: (a) Perspective drawing of catalyst. (b) Front view of heterocycle reduction. (c) Top view of heterocycle reduction. (d) Front view of acyclic imine reduction. (e) Top view of imine reduction.

Buchwald proposed that the configuration of the titanium in the transition state is similar to that of the dichlorotitanium complex, with one chloride being replaced by a hydride, the other by the η^2 imine ligand, as shown in Figure 7.6b [144]. In a tetrahedral geometry, the imine can only coordinate to the titanium as shown, with the *N*-methylene oriented to the lower left quadrant of the drawing in Figure 7.6b. That this can be the only possible orientation is shown clearly by the top view in Figure 7.6c. This view also illustrates positioning of the phenyl in the vacant upper right quadrant, with a minor interaction taking place between C-3 of the heterocycle and the cyclohexyl in the lower right quadrant. This aspect of binding in the transition structure is important in the analysis of the reduction of acyclic imines, as shown in Figures 7.6d and e.

For acyclic imines, note that interchange of R_1 and R_2 in the transition structure is equivalent to an E/Z isomerization of the educt. Reduction of cyclohexyl methyl N-benzyl imine, using a stoichiometric amount of catalyst affords a 92:8 R/S enantiomer ratio that is identical to the 92:8 E/Z ratio of the educt (*i.e.*, the reaction is stereospecific). This is interpreted as follows: the major imine isomer is E (R_2 = cyclohexyl, R_1 = methyl). Addition to the Si face gives the R-enantiomer of the amine. With the Z imine, R_2 is methyl and R_1 is cyclohexyl. Addition to the Re face gives the S amine. Entry 6 of Table 7.6 (R = Bn) is the same reaction, but using only 5 mol% of catalyst. Under catalytic conditions, the reaction is no longer stereospecific for two reasons: first, the E and Z isomers interconvert slowly under the reaction conditions (probably catalyzed by the titanium), and second, the Z isomer is reduced faster than the E isomer. If the hydrogen pressure is reduced from 2000 to 500 psi, the enantioselectivity drops to 71%, consistent with a slower rate of reduction relative to E/Z isomerization [144].[10]

The titanocene catalyzed asymmetric imine reduction may be used in kinetic resolutions of racemic pyrrolines [154]. The most efficient kinetic resolution was observed for 5-substituted pyrrolines, and the mechanistic postulate outlined above readily accommodates the experimental results, as shown by the matched pair transition structure in Scheme 7.17 [154].[11] Pyrrolines substituted at the 3- and 4-positions were reduced with excellent enantioselectivity, but kinetic resolution of the starting material was only modest [154]. Finally, Buchwald has also applied titanocenes to the problem of ketone hydrosilylation [155].

In the following decade, progress in catalytic imine reduction accelerated, providing the chemist with a number of options for catalysts (Figure 7.7) and substrates (see representative examples in Table 7.7). Although a variety of metals can be used in this context, iridium has proven especially useful and efficient, either by generating a complex *in situ*

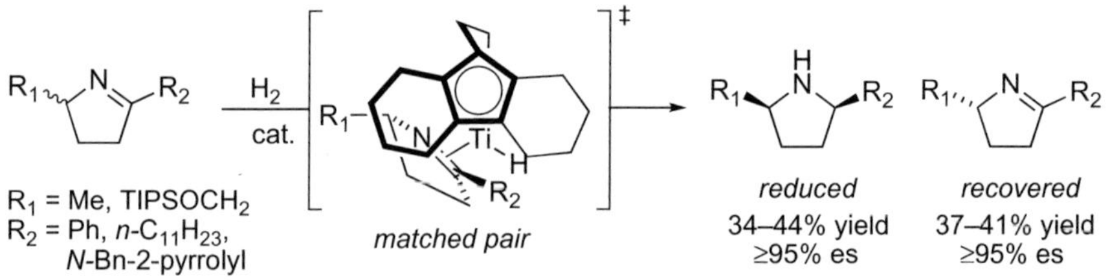

R_1 = Me, TIPSOCH$_2$
R_2 = Ph, n-C$_{11}$H$_{23}$,
        N-Bn-2-pyrrolyl

matched pair

reduced
34–44% yield
≥95% es

recovered
37–41% yield
≥95% es

SCHEME 7.17 Kinetic resolution of 5-substituted 1-pyrrolines by asymmetric reduction using the S,S titanocene catalyst [154].

---

10. In contrast, the enantioselectivity of cyclic imine reduction is independent of hydrogen pressure.
11. In ref. [154] the R,R-enantiomer of the catalyst (*cf.* Scheme 7.15) was employed. To maintain consistency with Scheme 7.15 and Figure 7.6, we illustrate the S,S catalyst.

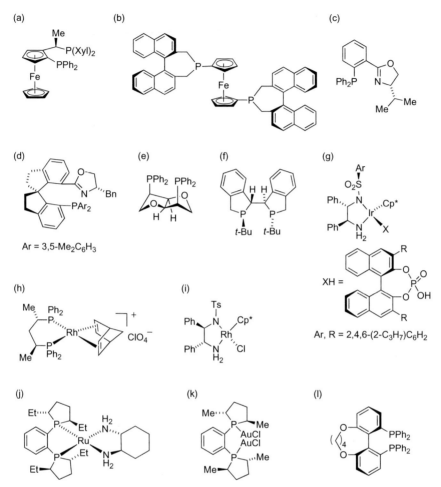

(a) (b) (c)

(d) (e) (f) (g)

Ar = 3,5-Me₂C₆H₃

(h) (i)

XH =

Ar, R = 2,4,6-(2-C₃H₇)C₆H₂

(j) (k) (l)

FIGURE 7.7 Representative ligands and catalysts used in imine hydrogenations: (a) [156], (b) [157], (c) [158], (d) [159], (e) [160], (f) [161], (g) [162], (h) [163], (i) [164], (j) [165], (k) [166], and (l) [167].

(*e.g.*, Figures 7.7a−d) or by preforming the reactive catalyst. Anyone who has read other parts of this book dealing with catalysts would by now notice many familiar ligand motifs: planar chiral ferrocenes, binaphthyl-based axial chirality, and oxazolines, to name a few. Some of the more interesting ligands used here include spirocyclic groups (Figure 7.7d) and phosphines that are either forced into a particular orientation by a bicyclic ring system (e) or that represent a stereogenic center (f). Besides iridium, catalysts abetted by Brønsted acid co-catalysts (g), rhodium (h-i), ruthenium (j), gold (k) or palladium (l) have also been reported. With respect to substrates, imines activated by aryl or benzyl groups or cyclic imines still predominate, with benzyl imines having the obvious attribute that the *N*-benzyl group is generally removable under reductive conditions to afford a primary or secondary amine product.

An important type of imine reduction occurs in reductive aminations, in which the imine is formed and directly reduced *in situ* (a common variant of this reaction is the Borch reduction). Such reactions are especially important when one is laboring under the constraints of process chemistry and step count becomes critical, as does the necessity to handle sometimes large quantities of imines (which can be a problem due to the borderline stability of many C=N containing compounds). In principle and in practice, many of the same procedures

TABLE 7.7 Catalytic Asymmetric Reduction of Imines

Entry	Starting Material	Catalyst	% Conversion	% es (Major Isomer)	Ref.
1		a + Ir(cod)$_2$BARF	>99	89 (S)	[156]
2		b + [Ir(cod)Cl]$_2$	>99	92[b]	[157]
3	"	c + Ir(cod)$_2$BARF	>99	92 (R)	[158]
4	"	d + Ir(cod)$_2$BARF	>99	96 (R)	[159]
5	"	e + Ir(cod)$_2$PF$_6$	>99	92 (R)	[160]
6	"	f + Ir(cod)$_2$BARF	>99	96 (R)	
5	"	g	94[a]	97 (S)	[162]
6		h +0.1 M sodium bis(2-ethyl-hexyl) sulfosuccinate	>99[a]	93 (R)	[163]
7		i	94[a]	99 (S)	[164]
8		"	90[a]	40 (S)	"
9		j	92	96 (S)	[165]
10		"	98	70 (S)	"

(*Continued*)

TABLE 7.7 (Continued)

Entry	Starting Material	Catalyst	% Conversion	% es (Major Isomer)	Ref.
11	$\underset{Ph}{\overset{N\text{-}Bn}{\bigg\|}}Me$	k	>99	87[b]	[166]
12	$\underset{Ph}{\overset{N\text{-}Ph}{\bigg\|}}Me$	l + Pd(OCOCF$_3$)$_2$	92[a]	85 (R)	[167]
13	(3,4-dihydronaphthalen-1(2H)-ylidene)-N-Ph	"	95[a]	95[b]	"

[a]Isolated yields.
[b]Absolute configuration not determined.

used to reduce pre-formed imines are applied to reductive aminations (reviews: refs. [7,168]). A few representative examples using chiral metal complexes are provided in Scheme 7.18a and b. The iridium/xyliphos example shown in Scheme 7.18a was adapted to an industrial scale synthesis of the herbicide agent metolachlor [156]. An organocatalytic approach published by MacMillan (Scheme 7.18c) is interesting because it utilizes both a chiral Brønsted acid promoter and a Hantzsch dihydropyridine ester as a biologically inspired transfer hydrogen source [169]). The Hantzsch ester has a more-than-passing resemblance to the common reduction cofactor NADH; for early applications of this agent to pre-formed imine reduction by Reuping and List, see refs. [170,171]. The generalized mechanism of this reaction is that the bulky Brønsted acid forms a complex with the transiently generated imine, both blocking one face and activating the imine for hydride transfer from the Hantzsch ester. For a review of applications and related approaches, see ref. [172].

7.2 REDUCTION OF CARBON–CARBON BONDS

Reduction of a carbon–carbon double bond will produce a chiral product if the olefin is unsymmetrically and at least geminally disubstituted. The potential inherent in this transformation is such that hundreds of catalysts having chiral ligands have been examined and the field has been exhaustively reviewed [79,175–195]. To provide the reader with a toehold into this vast literature while providing a sense of the key issues involved, we will first present one of the most historically important hydrogenation reactions of C=C bonds, namely the reduction of acetamido cinnamates using soluble rhodium catalysts (reviews: refs. [175,178,196–198]). Following this discussion, we will briefly consider more recent work in two areas: the conjugate reduction of olefins bearing electron-withdrawing groups (such as α,β-unsaturated ketones or nitroolefins) and finally summarize progress made in the asymmetric reduction of "nonfunctionalized" double bonds.

Acetamido cinnamates are a subset of enamides, that is, enamines that have been "trapped" as their *N*-acylated cousins. These reductions would be worth discussing just for the role that this reaction played in developing the field of asymmetric catalysis and the importance of some

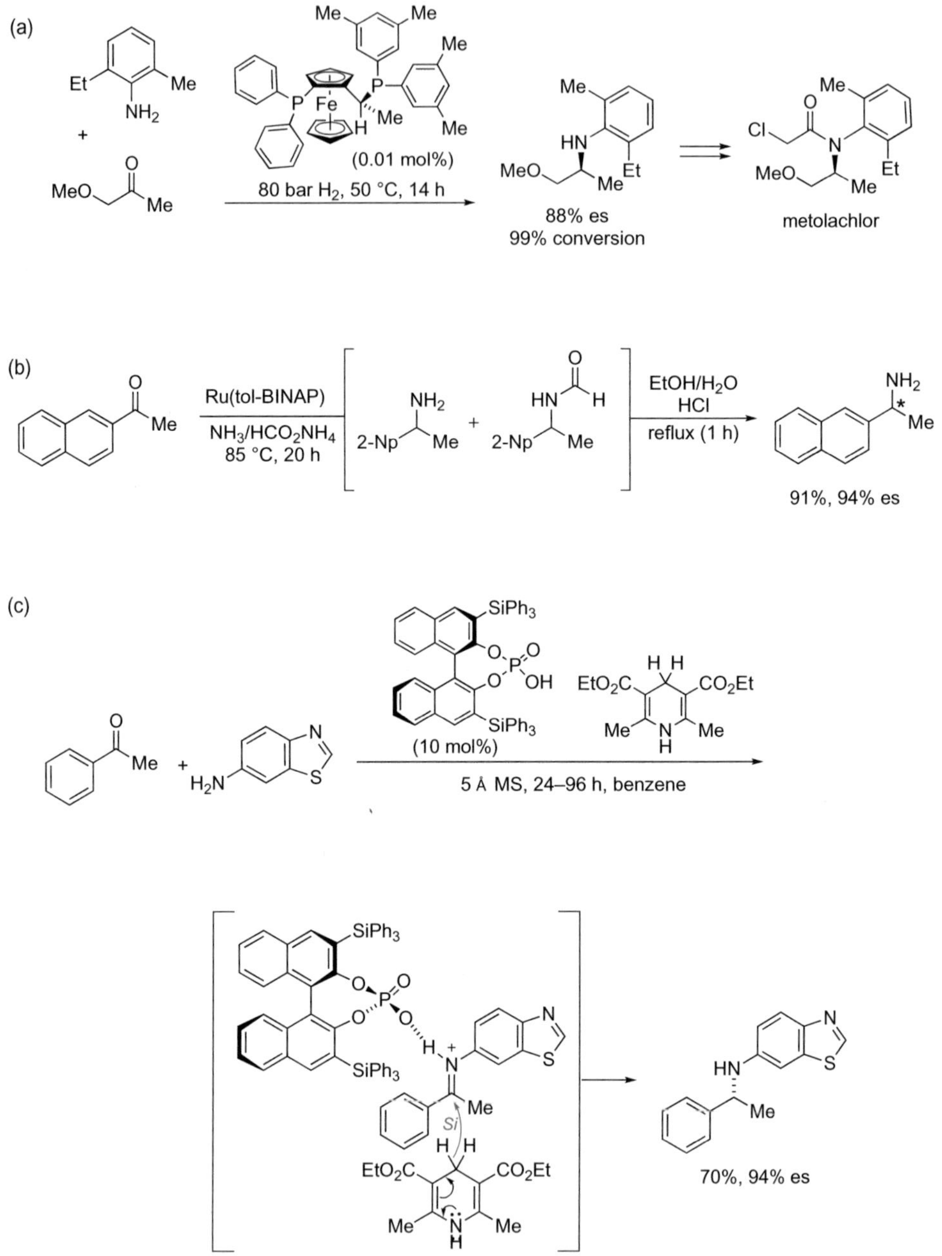

SCHEME 7.18 Catalytic and asymmetric reductive aminations: (a) [173], (b) [174], and (c) [169].

products that have been made using it (e.g., L-DOPA). However, there is also much to learn about asymmetric synthesis by studying this reaction in detail. The development of chiral bisphosphine ligands and the Herculean effort that led to the elucidation of the mechanism of this reaction make it an important example for study, since we now know that the major enantiomer of the product arises from a minor (often invisible) component of a pre-equilibrium [197,199]. This aspect of chemical reactivity is an important lesson whose importance cannot be overemphasized: when we strive to understand the forces that govern reactivities and selectivities, we must never overlook the fact that an observable intermediate in a chemical process may not be the one responsible for the observed products.

Following Wilkinson's detailed studies of tris-triphenylphosphine rhodium chloride as a soluble catalyst for hydrogenations [200], it did not take long for chemists to realize that chiral phosphines could be substituted for triphenylphosphine so as to effect an asymmetric reduction [201]. Following Mislow's development of a synthesis of chiral phosphine oxides [202], the groups of Knowles [203] and Horner [201] tested methyl phenyl n-propyl phosphine in the Wilkinson catalyst system, but found only low selectivities in the reduction of substrates such as α-phenylacrylic acid. These efforts were predicated on the very reasonable assumption that the chiral rhodium complex should contain chirality centers at phosphorous (since these are close to the metal). However, this assumption was proven wrong in 1971 when Morrison [204] and Kagan [205] independently showed that ligands such as R*$-$PPh$_2$ (Morrison) and Ph$_2$P$-$R*$-$PPh$_2$ (Kagan) (where R* contains a chirality center) were capable of reducing substituted cinnamic acids with enantioselectivities in the 80% range, as shown by the "record-setting" examples in Scheme 7.19. The Kagan ligand, derived from tartaric acid, later became known as "DIOP," and served as the prototype for many more chiral, chelating diphosphine ligands. The Kagan example also demonstrates the utility of an asymmetric reduction protocol for the synthesis of α-amino acids. A similar reaction is now used industrially for the enantioselective production of $(-)$-dihydroxyphenylalanine (L-DOPA, a drug for treating Parkinson's disease) and aspartame (an artificial sweetener). It can be fairly stated that these spectacular early successes served to heighten optimism for the prospects of asymmetric synthesis in general, and asymmetric catalysis in particular — hopes that have been well rewarded.

These examples were followed with a continuous stream of ligands [80,183,185,196,206$-$208] that were tested with rhodium and other metals in asymmetric reductions and other reactions catalyzed by transition metals [180$-$182,209]. Simultaneously, studies of the mechanism of the asymmetric hydrogenation were pursued, most aggressively in the labs of Halpern and Brown. The currently accepted mechanism is shown in Scheme 7.20 [199]. The substrate (methyl Z-acetamidocinnamate, middle left) displaces two solvent molecules from the cationic rhodium catalyst (center) in an equilibrium that favors the diastereomer in which the rhodium is bound to the Si face (at C-2) of the alkene [210]. This

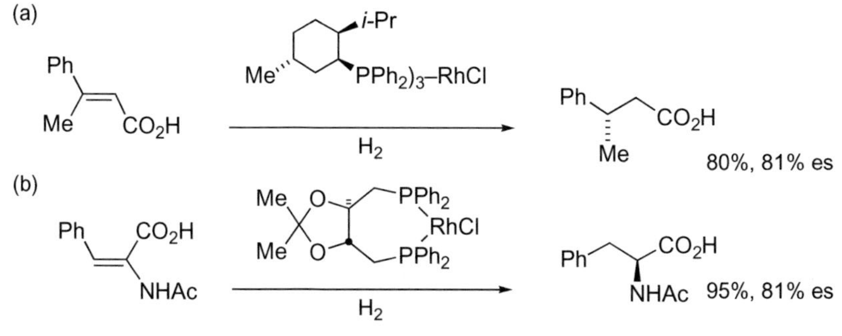

SCHEME 7.19   (a) Morrison's asymmetric reduction of β-methyl cinnamic acid [204]. (b) Kagan's asymmetric reduction of N-acetyl dehydrophenylalanine and the debut of the DIOP ligand [205].

equilibrium defines the "major" and "minor" manifolds of the reaction. The sequence of oxidative addition of dihydrogen, migratory insertion, and reductive elimination, completes the cycle in both manifolds. With some bisphosphines, both of the initially formed diastereomeric complexes are visible by NMR; with others, signals from the minor diastereomer are lost in the noise. Each subsequent reaction is irreversible, but at low temperature the rhodium alkyl hydride product of migratory insertion can be intercepted and characterized spectroscopically [211,212]. Surprisingly, the intercepted complex (leading to the S product) has the metal on the Re face of C-2! *Thus, the major product of the reduction is produced by oxidative addition of dihydrogen to the minor diastereomer of the catalyst–substrate complex* [199,211].

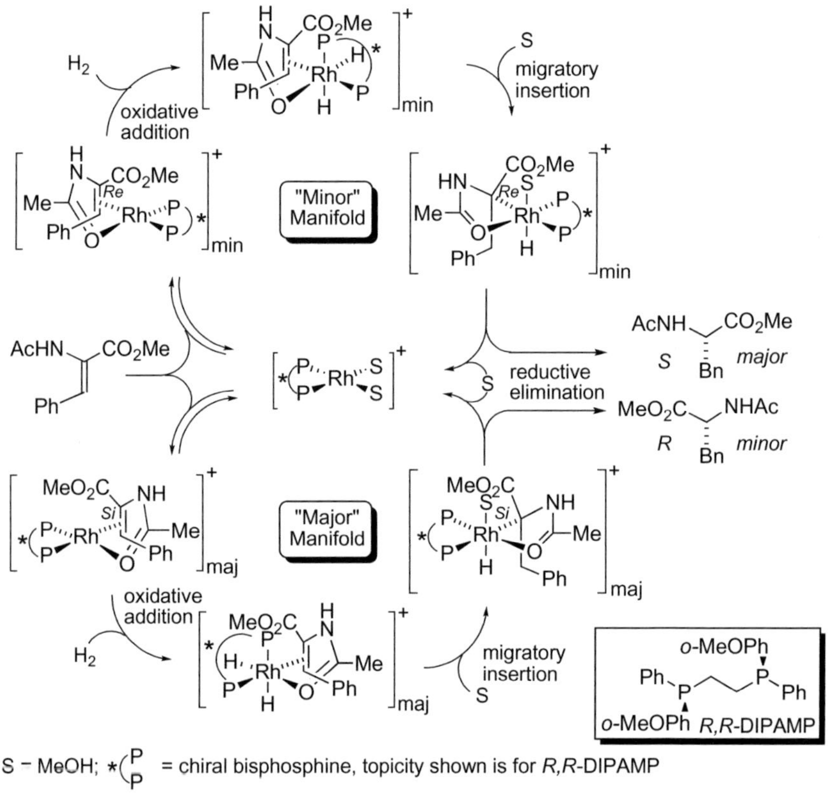

S – MeOH; *$\left(\begin{smallmatrix}P\\P\end{smallmatrix}\right.$ = chiral bisphosphine, topicity shown is for *R,R*-DIPAMP

SCHEME 7.20 Mechanism of asymmetric hydrogenation of *N*-acetyl dehydrophenylalanine [199].

At temperatures above −40 °C, the rate-determining step of the reaction is the oxidative addition of hydrogen to the catalyst–substrate complex. Because the interconversion of the two diastereomeric catalyst–substrate complexes is fast relative to the rate of oxidative addition [199], and because the migratory insertion and reductive elimination steps are kinetically invisible, the complex equilibria in Scheme 7.18 reduce to a classic case of Curtin–Hammett kinetics (Section 1.8 [103]), whereby the relative rate of the formation of the two enantiomers is determined solely by the relative energy of the two transition states, as illustrated in Scheme 7.21. This energy difference is 2.3 kcal/mol, corresponding to a relative rate of about

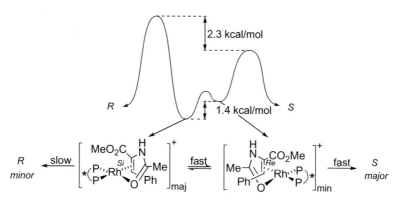

SCHEME 7.21 The asymmetric hydrogenation of *N*-acetyl dehydrophenylalanine ester as an example of Curtin–Hammett kinetics. Energy values taken from ref. [199].

50:1 for the formation of the *S* product over the *R* (from *R,R*-DIPAMP catalyst at 25 °C), or 98% es (*cf.* Figure 1.3).

Still, the question remains: why is the one diastereomer of the catalyst–substrate complex so much more reactive than the other? In 1977, Knowles suggested, based on examination of the crystal structures of several metal complexes having chiral bisphosphine ligands, that the orientation of the *P*-phenyl groups could be the source of the enantioselectivity of these reactions [213]. The common feature Knowles observed (Figure 7.8) is that the two equatorial phenyls are oriented such that — with the bisphosphine in the horizontal plane and the metal in front — their "faces" are exposed to a ligand coordinated at a site toward the viewer, while the two axial phenyls expose "edges." This conformation produces two crowded quadrants (those having the axial phenyls) and two vacant quadrants, as shown in Figure 7.8.

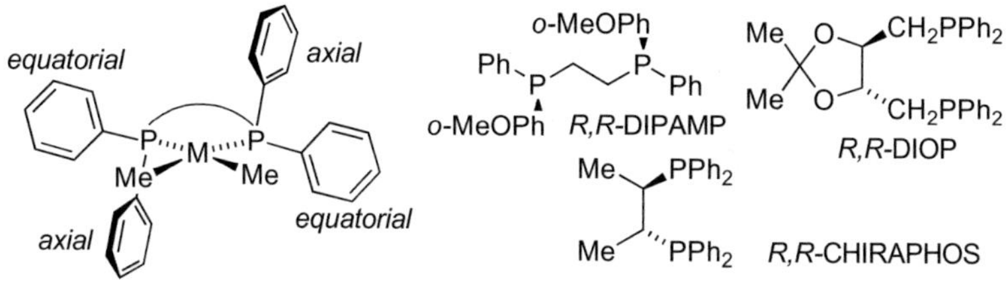

FIGURE 7.8 Bisphosphine ligands and the common structural feature that affects steric crowding of ligands bound *trans* to phosphorus [175,213]. Note vacant upper left and lower right quadrants in the generalized structure on the left.

For the asymmetric hydrogenation, both substrate–catalyst complexes are square–planar, and hydrogen could, in principle, add from either the top or the bottom of either complex, as illustrated in Scheme 7.22. From a molecular orbital standpoint, there is no reason to expect any one of these four possibilities to dominate. Indeed, it is likely that the oxidative addition of dihydrogen occurs stepwise, and that the hydrogen binds "edgewise" initially (making a square pyramid complex), followed by H–H cleavage to form the octahedral dihydride product. Further, it is likely that the dihydrogen associates with the metal many times before

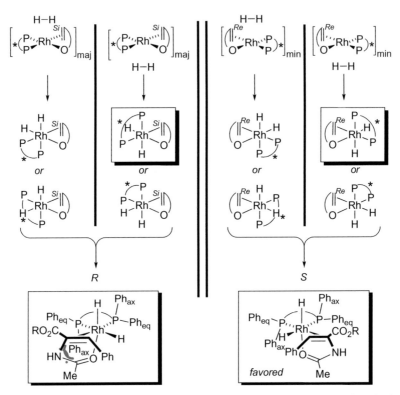

SCHEME 7.22 Possible orientations for oxidative addition of dihydrogen to the major (left) and minor (right) diastereomers of the catalyst—substrate complex (for simplicity, the linkages connecting the atoms bonded to the metal are indicated with a curved line). The boxed structures are the only octahedral structures that are not encumbered by severe nonbonded interactions; these are redrawn at the bottom with the bisphosphine to the rear and in the horizontal plane. The topicity illustrated is for ligands having the structure of Figure 7.8, such as *R,R*-DIPAMP, *R,R*-DIOP, or *R,R*-CHIRAPHOS.

H—H bond cleavage occurs.[12] Lacking an electronic rationale, the only other possible explanation is that the movement of the ligands (as the oxidative addition proceeds) is determinant.

Scheme 7.22 illustrates the eight possible octahedral complexes that could arise by addition of dihydrogen to either the top or the bottom of the two square complexes. Each is drawn so that the orientation of the substrate remains unchanged, and one of the phosphines is moved *trans* to the incoming hydrogen. A molecular mechanics investigation [215] indicates that only two of the eight structures are viable (boxed, redrawn at the bottom of Scheme 7.22); all the others suffer severe nonbonded interactions between ligands. Note the similarity of the two boxed structures: in both, a hydride is *trans* to the chelating oxygen and *cis* to both phosphorous atoms. The double bond and the second hydride are then meridonal with respect to the two phosphorous atoms. The reason for the difference in stability can be seen by examining the orientation of the substrate relative to the axial phenyls: in the favored configuration, the substrate occupies the less crowded "lower right" quadrant. It is implicit in

---

12. Hoff has shown that the rate of dissociation of a tungsten—dihydrogen complex is at least one order of magnitude faster than the rate of oxidative addition (H—H bond cleavage) to a dihydride complex [214].

this analysis that the energetic consequences of these various structural features must be felt in the competing transition states for oxidative addition.

Two factors contribute to the success of this reaction: the outstanding enantioselectivity achieved and efficiency of the catalyst (*i.e.*, high turnover). The above analysis emphasizes only the former, but the latter also varies with the nature of the chiral bisphosphine ligand and the structure of the substrate. The structural features of the substrate and the catalyst are mutually optimal in the example cited above. Perturbation of any of these features usually lowers either the enantioselectivity or the turnover rate. The range of substrates that is amenable to asymmetric hydrogenation with this catalyst system is therefore limited. Figure 7.9 illustrates the classes of substrate that can be accommodated by cationic rhodium bisphosphine catalysts [182]. For a more extensive summary, see ref. [198].

R_1 = H, alkyl, aryl
R_2 = CO_2R, Ph

FIGURE 7.9   Substrate tolerance in the asymmetric hydrogenation [182].

Work pioneered in the laboratories of Pringle [216] and Reetz [217] established that monodentate phosphines can also be used to great effect in Knowles-type hydrogenation reactions. An important catalyst of this type is the BINOL-derived phosphoramide MonoPhos, a Rh complex shown to be highly effective in reducing both dehydroamino acids and itaconic acid derivatives (Schemes 7.23a and b [218]). Since, numerous monodentate ligands have been introduced, extending the range of substrates that can be reduced by catalysts in this class, for example, the enol ether shown in Scheme 7.23c (for reviews covering much of the work that has followed these initial efforts, see refs. [219,220]). Although the mechanism of the reaction is complex and under continuing investigation [221], there is wide support for the early suggestion that two ligands may well be associated with the active rhodium catalyst [222].

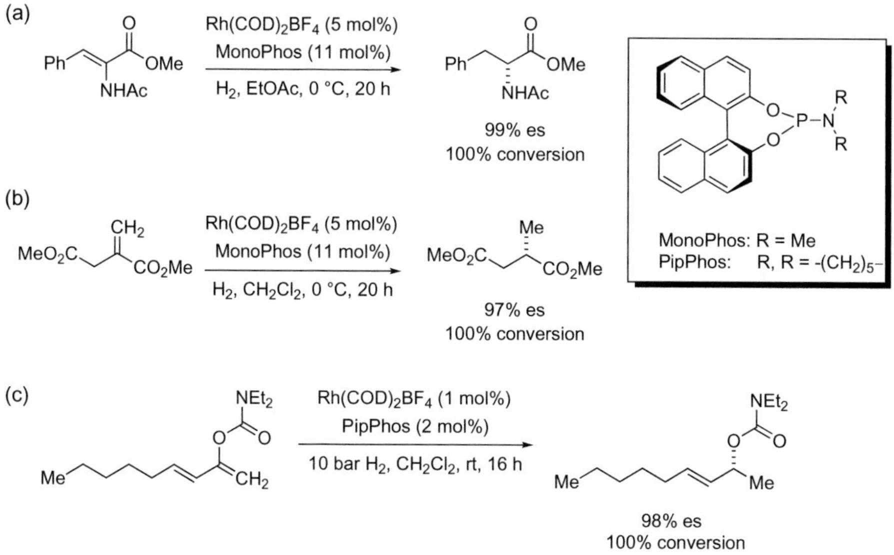

SCHEME 7.23   Asymmetric hydrogenations using monodentate phosphoramidites MonoPhos and PipPhos (inset); (a) and (b) [218]; (c) [223].

(a)

(b)

SCHEME 7.24 (a) Comparison between proposed binding pockets for a metal complex containing two phosphoramidite ligands (left) and a bidentate bisphosphine ligand (right) [225]. (b) A catalytic reduction used in a process chemistry route to aliskiren [224,225].

Scheme 7.24a shows how the quadrant formed by incorporating two monodentate phosphoramidites onto a metal (left) might be similar to the analogous complex formed with a bisphosphine ligand. Finally, consider the important practical application of this catalyst in a scalable process route to the renin inhibitor aliskiren, a potential agent for the treatment of high blood pressure (Scheme 7.24b; refs. [224,225].). In seeking an optimal hydrogenation catalyst for this purpose, de Vries and co-workers used reaction condition screening, in which various combinations of ligands were systematically surveyed. The team eventually settled on a mixed ligand system that gave the satisfactory results shown in Scheme 7.24b.

"Unfunctionalized" or "minimally functionalized" alkenes present special difficulties because by definition these lack coordinating groups and therefore rely almost exclusively on spatial differentiation to achieve selectivity.[13] Since the late 1990s, chemists have been chipping away at this problem with increasing levels of success to the point where certain classes of olefins may now be viewed as largely solved problems. Following important early breakthroughs using titanium [227], ruthenium [228], and lanthanides [229], iridium emerged as the predominant metal in this field. Figure 7.10 provides the structures of some typical catalysts used for these reductions.

Scheme 7.25a gives results of three different catalysts on a typical reduction substrate; one of the attributes of the Andersson catalyst types shown in Figure 7.10c is higher reactivity due to the presence of aryloxy groups on the phosphorous ligand and good scope. Largely led by Pfaltz and co-workers, highly successful reductions of 1,1-disubstituted, trisubstituted (Scheme 7.25b), and even fully substituted olefins (Scheme 7.26a) have been registered.

---

13. Burgess has discussed issues specific to the reduction of unsubstituted olefins, including the slippery slope of olefin descriptions ranging from "unfunctionalized" to "largely unfunctionalized" to "functionalized" [226]. For general reviews on the asymmetric hydrogenation of less-than-highly-functionalized olefins see refs. [189,190,193,195,226].

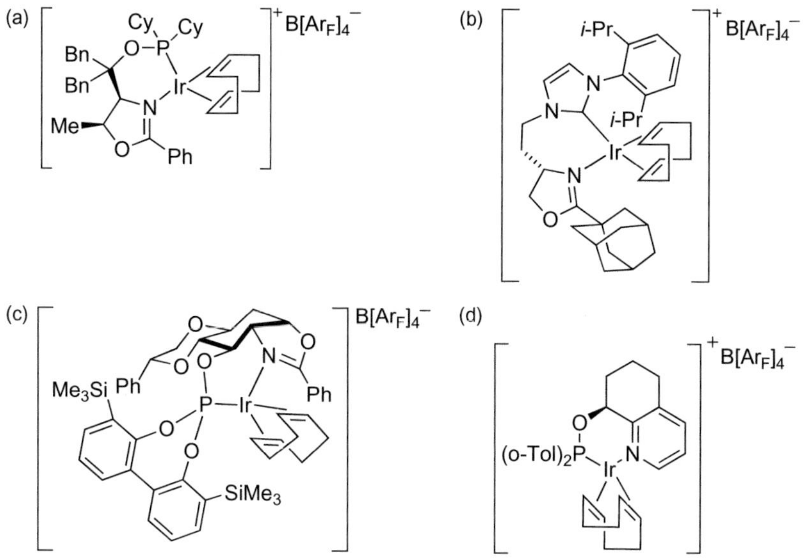

FIGURE 7.10 Iridium catalysts used for the reduction of barely functionalized C=C bonds. (a) [230], (b) [231], (c) [232,233], and (d) [234].

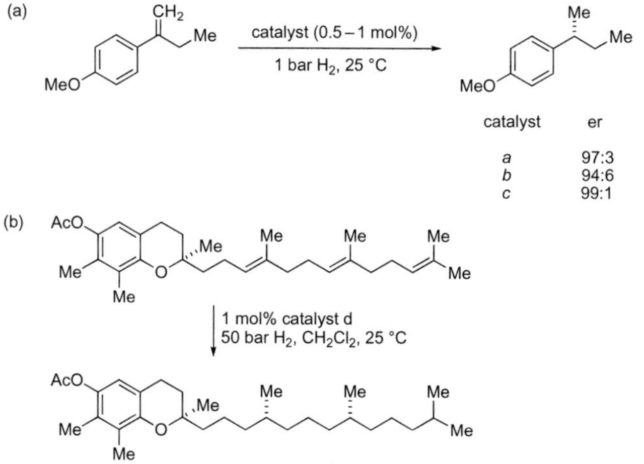

SCHEME 7.25 Reductions of modestly functionalized double bonds. (a) Comparison of various catalysts on a commonly used substrate (see Figure 7.10 for structures of catalysts a-c). (b) Serial reduction of trisubstituted double bonds leading to γ-tocopheryl acetate [234].

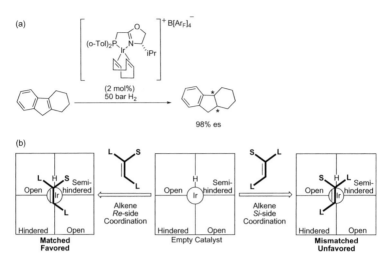

SCHEME 7.26 Reductions of modestly functionalized double bonds. (a) Reduction of a fully substituted olefin [240]. (b) Andersson's model for the steric course of reducing trisubstituted olefins [239].

As of this writing, however, a nagging restriction is that most highly selective reductions of lowly functionalized olefins have at least one aryl group attached to the reacting double bond. Although the details of the mechanism are still evolving [235−237], the enhancement of reaction rate using a weaker coordinating counterion to the positively charged catalyst, such as $B(Ar_F)_4^-$, has been recognized [238] and a useful quadrant model for predicting the enantioselectivity of trisubstituted olefins by iridium catalysts has appeared (Scheme 7.26b [239]).

Electron-poor olefins. Because of their electrophilicity, definitely functionalized olefins such as α,β-unsaturated ketones or esters can be reduced either by catalytic hydrogenation or through the addition of a hydride equivalent in a conjugate fashion to the β carbon. In the latter case, the resulting enolate or enolate equivalent (*e.g.*, a silyl enol ether in the event of hydrosilylation) can be further manipulated to link a C−C bond formation with the reducing event (early examples of which were reviewed in refs. [119,241]).

Buchwald published some copper hydride-based approaches to catalytic conjugate reduction based on Stryker's reagent, [(PPh₃)CuH]₆. Using *p*-tol-BINAP as the chiral ligand and PMHS, a polymeric silane reductant, it proved possible to reduce unsaturated esters [242] with high enantioselectivity as shown in Scheme 7.27a; the reaction could also be extended to lactams and lactones [243]. Two useful types of selectivity accrue from this procedure. First, the use of a nucleophilic hydride species (see the proposed catalytic cycle in Scheme 7.27b) permitted the chemoselective reduction of unsaturated esters in the presence of unconjugated or unfunctionalized olefins. Second, the absolute configuration in the product of the reaction differed according to the geometry of the starting alkene. The methodology could be adapted to a conjugate reduction/alkylation sequence to afford *trans*-2,3-disubstituted cyclopentanones (Scheme 7.27c [244].

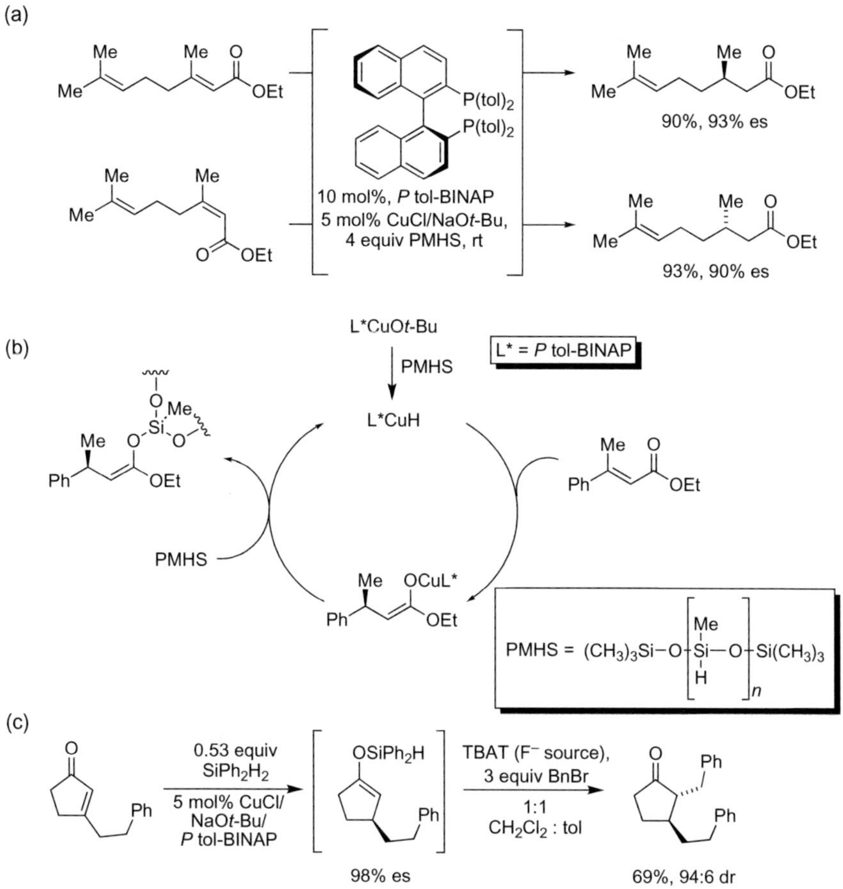

(a)

(b)

(c)

SCHEME 7.27 (a) Reductive hydrosilylation of α,β-unsaturated esters and (b) the proposed catalytic cycle [242]. (c) Adaptation of the general reaction to a reductive alkylation sequence [244].

In a neat twist, if racemic 3,5-disubstituted 2-cyclopentenes were used, the high suscepti-bility of C-5 to undergo epimerization under basic conditions could be leveraged in a dynamic kinetic resolution, resulting in highly enantio- and diastereoselective formation of 2,4-disubstiuted cyclopentanes (Scheme 7.28a [245]). Nishiyama has pursued conceptually related hydrosilylation reactions using rhodium Phebox catalysts [246]. Scheme 7.28b shows a typical result of the standard procedure, whereby the reduced ester was obtained after acid hydrolysis of the silyl ketene acetal. For added value, the ketene acetal intermediate could be utilized in a subsequent aldol reaction [247].

Organocatalytic reactions utilizing Hantzsch esters as the stoichiometric reductant have been applied to α,β-unsaturated aldehydes [248,249] and ketones [250,251] that have been activated activated *via* iminium catalysis (for reviews, see ref. [87,252]). As in Diels−Alder reactions

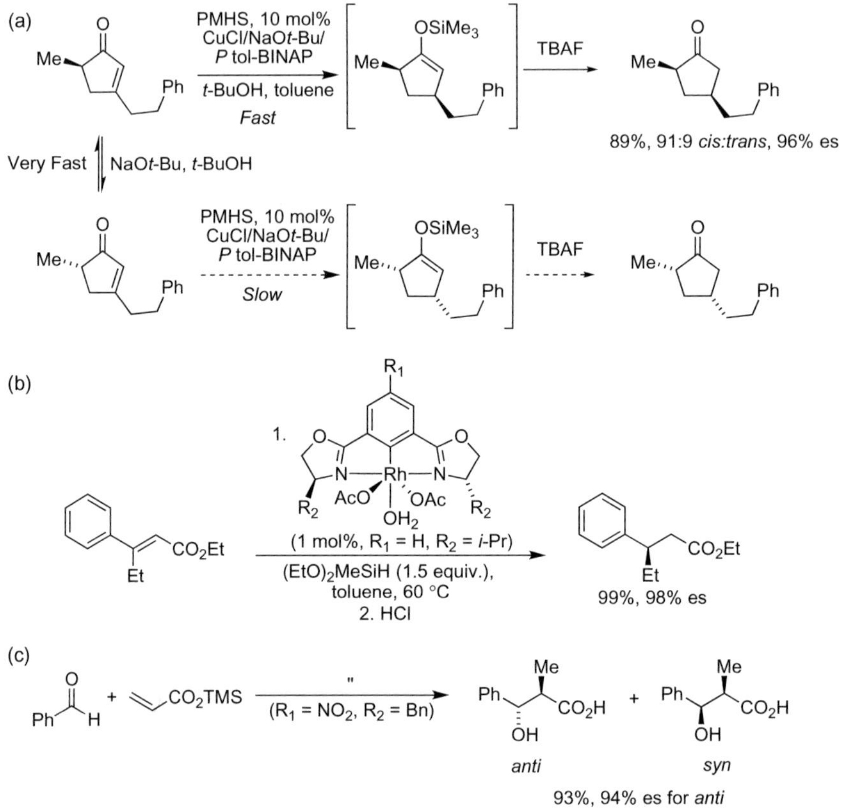

SCHEME 7.28 (a) Dynamic kinetic resolution [245]). Rh-promoted (b) asymmetric hydrosilylation [246] and (c) a related reductive aldol [247].

(*cf.* Section 6.1.1.3), MacMillan found different catalysts to be preferable for aldehydes (Scheme 7.29a) as opposed to ketones (Scheme 7.29b).

List employed a chiral Brønsted catalyst/amino ester salt to carry out similar transformations (Scheme 7.30a). A particularly instructive sequence, depicted in Scheme 7.30b, demonstrates the potential inherent in combining multiple asymmetric catalytic reactions in a cascade [253]. In this case, initial conjugate reduction provides a saturated aldehyde intermediate in enantiomerically enriched form. Instead of isolating this material and subjecting it to a separate reaction, a second catalyst is added, thus generating a reactive enamine that reacts with a fluorinating reagent to accomplish the formal stereoselective addition of "HF" across the double bond. Not only is the second reaction essentially independent of the first, meaning that either relative configuration can be obtained depending on the chirality of the second catalyst added, but also the net effect of carrying out the two reactions in tandem is that the overall enantiomeric purity of the product can be higher than would have been obtained from each individual step.[14]

---

14. This is true because reacting the minor enantiomer of the initially formed aldehyde along the preferred facial attack of the second halogenation step affords a different diastereomer than the major isomer does; this "mismatched" product is thus readily removed by simple chromatography. See Scheme 7.27 and Section 8.14 for other examples.

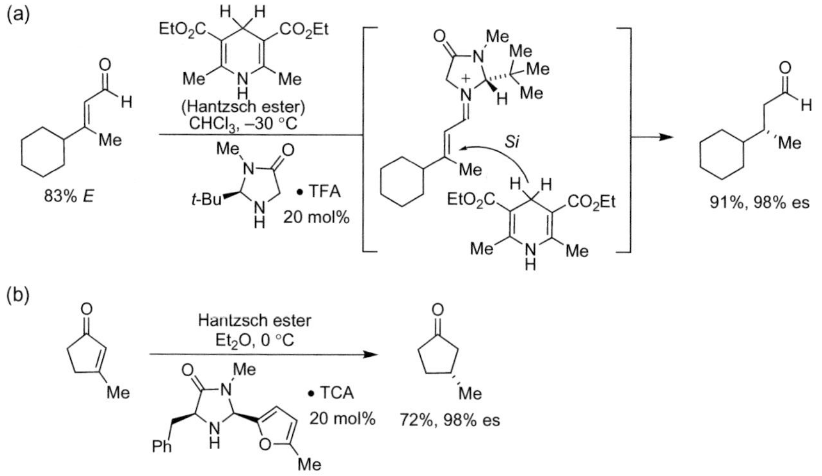

SCHEME 7.29 Organocatalytic reductions of an α,β-unsaturated (a) aldehyde [248] and (b) ketone [250] using MacMillan imidazolones.

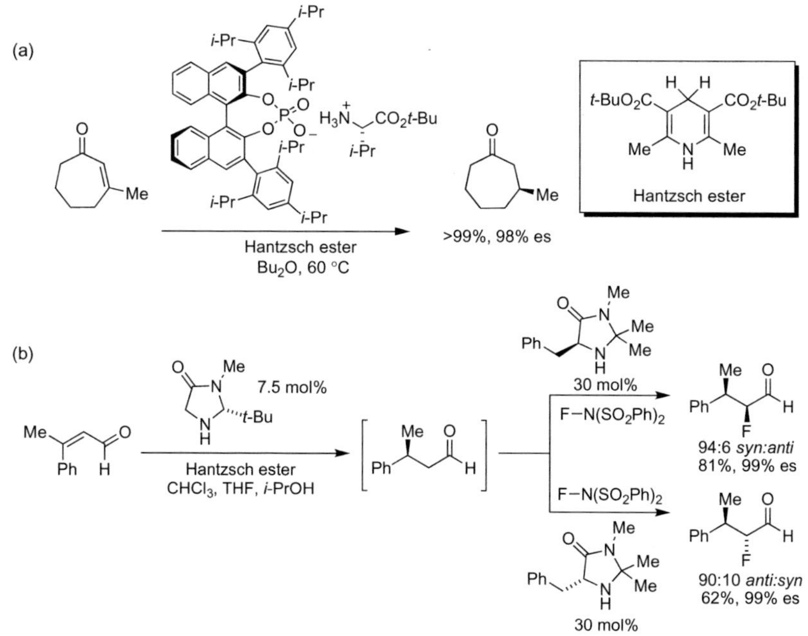

SCHEME 7.30 (a) List's complex salt approach [251]. (b) Cascade reduction/fluorination reported by MacMillan [253].

7.3 HYDROBORATIONS

The first asymmetric synthesis to achieve >90% optical yield was Brown's hydroboration of *cis* alkenes with diisopinocampheylborane (Ipc$_2$BH, Figure 7.11) in 1961 [254,255].[15] The 2-butanol obtained from hydroboration/oxidation of *cis*-2-butene was 87% optically pure, indicating an optical yield of 90%. *cis*-3-Hexene was hydroborated in ~100% optical yield. Since then, simple methods for the enantiomer enrichment of Ipc$_2$BH (and IpcBH$_2$) have been developed [256−258], and enantioselectivities have been evaluated more carefully with the purified material. For example, Ipc$_2$BH of >99% optical purity[16] affords 2-butanol (from *cis*-2-butene) in 99% optical purity and 3-hexanol (from *cis*-3-hexene) in 96% optical purity, both determined by rotation (see Table 7.8, entries 1 and 6) [256].[17]

Today, Ipc$_2$BH is still as good a reagent as any for achieving enantioselective hydroboration of *cis* alkenes (Table 7.8, entries 1, 6, 10, 12−17; reviews: refs. [5,262−264]). However, it does not afford good enantioselectivities with *trans*-disubstituted or trisubstituted alkenes. For these classes of compounds, monoisopinocampheylborane, IpcBH$_2$ (Figure 7.11), gives good selectivities, as indicated by the examples in Table 7.8, entries 2, 4, 8, 11, 18, 20, 22, 24, and 26 [259]; recrystallization of the intermediate borane may be used to purify the major borane diastereomer in some cases, and serves to improve the overall enantioselectivity of the process [257].

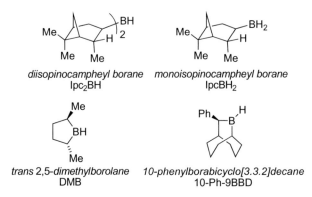

diisopinocampheyl borane
Ipc$_2$BH

monoisopinocampheyl borane
IpcBH$_2$

trans 2,5-dimethylborolane
DMB

10-phenylborabicyclo[3.3.2]decane
10-Ph-9BBD

FIGURE 7.11 Chiral hydroborating reagents: Ipc$_2$BH [254−258], IpcBH$_2$ [257,259], DMB [260], and 10-Ph-9BBD [261].

15. At that time, selectivity was described in terms of "optical yield," which is a measure of the ratio of two specific rotations. In these early examples, the α-pinene had an er of only 95:5, or 90% optical purity. Due to the inaccuracies of specific rotations, these optical yields only approximate the actual enantioselectivity of the reactions described.

16. This is the enantiomeric purity of isopinocampheol produced by oxidation of the purified Ipc$_2$BH, measured by rotation [256].

17. The discrepancy between the optical yields using enantiopure α-pinene and that of ~90% optical purity is probably due to experimental error in the measurement of rotations. See Chapter 2 for a discussion of the relative merits of enantiomeric purity measurements using rotation *vs.* other methods.

TABLE 7.8 Examples of Enantioselective Hydroborations. The "Reagent" Column Refers to the Structures in Figure 7.11

Entry	Alkene	Reagent	% Yield	% es[a]	Ref.
1	cis-2-Butene	Ipc$_2$BH	74	99	[254–256]
2	"	IpcBH$_2$	78	99	[257]
3	"	DMB	75	97	[260]
4	trans-2-Butene	IpcBH$_2$	73	86	[259]
5	"	DMB	71	100	[260]
6	cis-3-Hexene	Ipc$_2$BH	68–81	95–96	[254–256]
7	"	DMB	83	100	[260]
8	trans-3-Hexene	IpcBH$_2$	83	87	[259]
9	"	DMB	83	>99	[260]
10	Norbornene	Ipc$_2$BH	62	91	[254–256]
11	trans-Stilbene	IpcBH$_2$	69	82	[259]
12		Ipc$_2$BH X = O	74	>99	[266]
13		X = NCO$_2$Bn	85	>99	[266]
14		Ipc$_2$BH X = O	87	>99	[266]
15		X = S	80	>99	[266]
16		Ipc$_2$BH X = O	81–85	91–93	[266]
17		X = S	68	83	[266]
18	Me Me Me	IpcBH$_2$	78	99	[257]
19		DMB	90	94	[260]
20		n = 1: IpcBH$_2$	65	100	[257]
21	Me	n = 1: DMB	79	98	[260]
22		n = 2: IpcBH$_2$	75	99	[257]
23		n = 2: DMB	60	96	[260]
24		IpcBH$_2$ n = 1	72–92	100	[257,259]
25	Ph	n = 2	79	94	[259]

(Continued)

TABLE 7.8 (Continued)

Entry	Alkene	Reagent	% Yield	% es[a]	Ref.
26	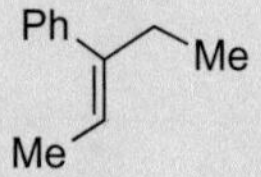 Ph⎯Me / Me (see structure)	IpcBH$_2$	77	100	[257]
27	 Ph⎯Me (see structure)	IpcBH$_2$	77	52	[261,267]
28		10-Ph-9DBB	95	89	
29	 t-Bu⎯Me (see structure)	10-Ph-9DBB	84	96	[261]

[a] *The "% es" column reflects the overall enantioselectivity of the process, including any diastereomeric enrichment, and is corrected for the enantiomeric purity of the borane.*

In 1985 [260], Masamune introduced *trans*-2,5-dimethylborolane (Figure 7.11) as a chiral hydroborating agent that works well for *cis* and *trans*-disubstituted and trisubstituted alkenes (Table 7.8, entries 3, 5, 7, 9, 19, 21, and 23). Although this reagent is quite versatile, its preparation is sufficiently cumbersome that its synthetic utility is limited. On the other hand, the conformational rigidity of this reagent allows one to postulate a reasonable transition structure to account for the topicity of the hydroboration (Scheme 7.31). Specifically, when $R_1 \neq H$, and either R_2 or $R_3 = H$, the boron of the *R,R*-borolane adds preferentially to the *Si* face of the alkene carbon. Good stereoselectivity will result when either R_2 or R_3 (or both) are $\neq H$, since the carbon that is attacked by boron determines the stereoselectivity. Conversely, if $R_1 = H$, there is little difference in energy between the illustrated transition structure and an alternative one in which R_2 and R_3 are interchanged. In fact, 1,1-disubstituted alkenes constitute one of the most challenging chemotypes in all of asymmetric synthesis [265]. To date,

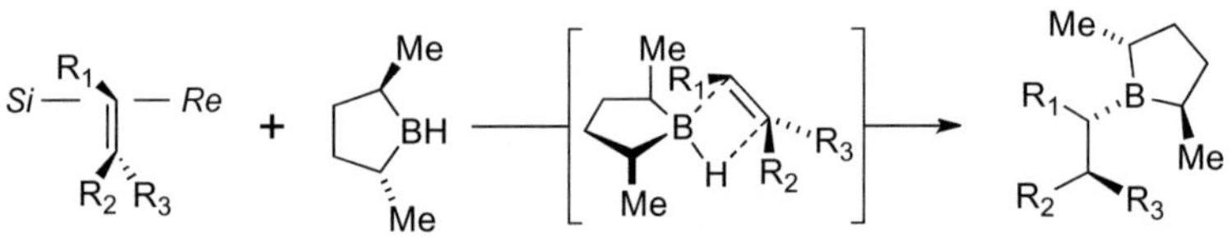

SCHEME 7.31 Favored transition structure for asymmetric hydroboration by Masamune's borolane [260].

the only hydroboration reagents able to carry out reasonably selective reactions on this type of substrate are Soderquist's 9BBDs (Figure 7.11 and Table 7.8, entries 27 and 28 [261]).

The conformational mobility around the B−C bond(s) in IpcBH$_2$ and Ipc$_2$BH complicates the analysis for these terpene-derived boranes, but Figure 7.12 gives a simplified picture.

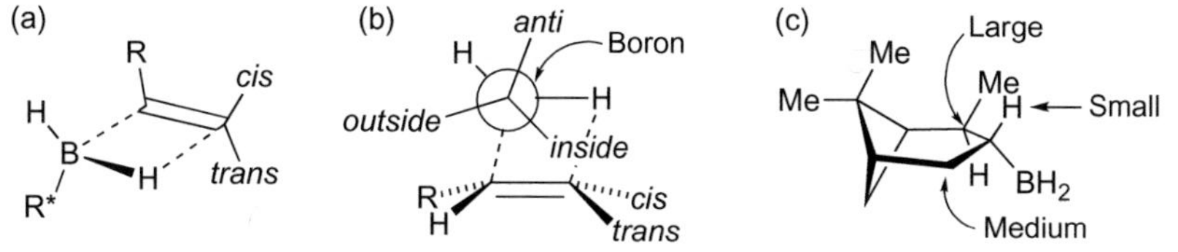

FIGURE 7.12   Terminology definitions for hydroboration transition structures [268]: (a) The auxiliary may be either *syn* or *anti* to the alkene substituents, but *anti* to the substituent (R) on the nearest carbon. (b) A stereocenter attached to boron, in a staggered conformation with respect to the forming C−B bond, has substituents in *anti*, *inside*, and *outside* positions. (c) Definition of the Large, Medium, and Small substituents of IpcBH$_2$.

Using *ab initio* techniques, Houk and co-workers [268] located the transition structures for the hydroboration of simple alkenes, and found that the most consistent feature of the most stable transition structures has the auxiliary (R*) and the substituent on carbon (R) *anti* to each other, as shown in Figure 7.12a. Analysis of the conformational motion of the B−R* bond revealed that the substituents on boron prefer to be staggered with respect to the forming C−B bond. Furthermore, the most stable position is *anti* to this bond. The so-called "*outside*" position is less encumbered sterically than the "*inside*" position, and the difference in energy between these two is affected by whether the alkene is *cis* or *trans* (Figure 7.12b). In Figure 7.12c the pinene substituent is reduced to a shorthand notation of **S**mall (H), **M**edium, and **L**arge substituents on the carbon bearing the boron.

With these generalizations in mind, it is possible to qualitatively[18] rationalize the results with IpcBH$_2$ and Ipc$_2$BH. The more easily understood example is IpcBH$_2$, since there is only one pinene moiety involved. This reagent is most selective with *trans* alkenes, so this olefin type is illustrated first. The lowest energy (molecular mechanics) transition structure for the addition of boron to the *Si* face of the alkene (Figure 7.13a) has the pinene *anti* to methyl group and has the small, medium, and large ligands in the most stable positions relative to the newly forming C−B bond: **L**arge-*anti*, **M**edium-*outside*, and **S**mall (H)-*inside*. In contrast, the transition structure for addition to the *Re* face (Figure 7.13b) has **L** in the less favorable *outside* position [268]. Note that in both of these structures the *inside* position is in close proximity to the second methyl group, which increases the destabilization of any conformer in which either **M** or **L** occupy the *inside* position. IpcBH$_2$ is also fairly selective with trisubstituted alkenes (Table 7.8), and the transition structures of Figure 7.12 show why this should be the case: an additional substituent in the *cis* position (*cf.* Figures 7.12a and b) imposes no additional crowding on the transition structure. On the other hand, IpcBH$_2$ is

---

18. Houk notes that the magnitude of the experimentally observed selectivities do not correspond to the energy differences their molecular mechanics calculations indicate, so this analysis and the calculated transition structures may only be taken as a first approximation [268].

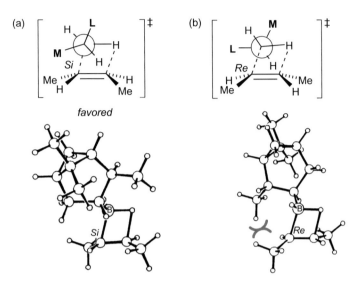

FIGURE 7.13 Transition structures for the asymmetric hydroboration of *trans*-2-butene with IpcBH$_2$. *Reprinted with permission from ref. [268], ©1984 Elsevier Science, Ltd.*

much less selective with *cis* alkenes. Here, the position of the alkyl group is moved away from its close proximity to the *inside* position, and a number of other transition structures become feasible [268].

For hydroborations with Ipc$_2$BH, there are two pinene moieties to consider. Ipc$_2$BH is only selective for *cis* alkenes, and the alkene substituents (R in Figure 7.14a) must be near one of them (R* in Figure 7.14a). Houk et al. found that there is only one conformation that the two pinenes may adopt relative to each other, and that is shown schematically in Figure 7.14b [268]. In the conformation shown, the olefin can align itself with the B–H bond and have the two R groups oriented either toward the proximal or distal pinene. Note that the proximal pinene has the Small hydrogen in the *inside* position, whereas the Large substituent of the distal pinene is in the crowded *inside* position. The alkene is least hindered in the orientation shown in Figure 7.14b. Figure 7.14c shows the transition structure with the distal pinene deleted for clarity [268]. Note that in this structure, the alkene

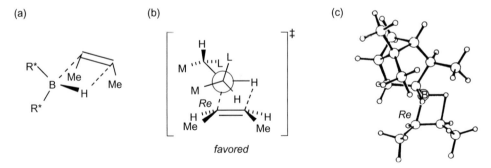

FIGURE 7.14 Transition structure for hydroboration of a *cis* alkene with Ipc$_2$BH. (a) The alkene substituents *must* be *syn* to one of the pinenes (R*). (b) Schematic representation of the lowest energy conformation. (c) Molecular mechanics—derived structure, with the rear (distal) pinene deleted for clarity. *Reprinted with permission from ref. [268], ©1984, Elsevier Science, Ltd.*

substituent is oriented toward the proximal pinene (with respect to the 4-membered ring), and it is therefore clear why Ipc$_2$BH preferentially attacks the *Re* face (Figure 7.14). This is in contrast to IpcBH$_2$, which prefers the *Si* face (Figure 7.13), because the alkene substituent is *anti* to the pinene.

The most important post-H.C. Brown innovation in the hydroboration field has been the development of catalytic hydroboration reactions and related hydrosilylations (reviews: refs. [269−273]). Männig and Nöth first demonstrated the promotion of hydroborations with catechol borane with Wilkinson's catalyst [274], and numerous workers have subsequently examined the chemoselectivity, reactivity, and most relevant in the present context, regioselectivity differences between the catalyzed and the thermal reactions. The breakthrough experiment for asymmetric synthesis was reported by Hayashi and Ito as shown in Scheme 7.32a. First, note that the catalyzed reaction affords a secondary alcohol after oxidation, which stands in contrast to the usual expectation of hydroborations to afford *anti*-Markovnikov addition to the terminal carbon of a single-substituted alkene. Now that the

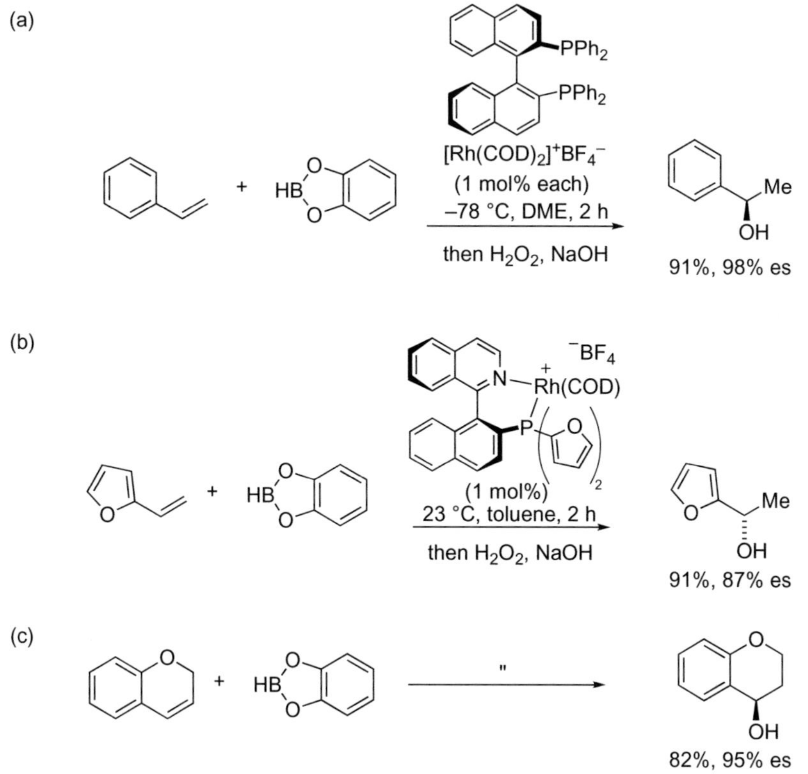

SCHEME 7.32   Examples of catalytic asymmetric hydroboration reactions: (a) [275]; (b) and (c) [277].

formation of a stereocenter is even possible for this reaction, the BINAP system of Hayashi and Ito provides the benzylic alcohol in excellent yield and enantioselectivity [275]. Despite efforts to use chiral boranes in tandem with achiral catalysts [276], the greatest success has been achieved through the screening and development of other rhodium-attached ligands for the process (for two examples out of many, see Schemes 7.32b and c [277]). Here, the use of the quinoline-containing catalyst derived from QUINAP demonstrates the utility of *P,N*-bound rhodium in promoting this reaction.

The mechanistic aspects of this reaction are complex, but most authors invoke some version of the mechanism shown in Scheme 7.33a (for an informative discussion, see ref. [270]).

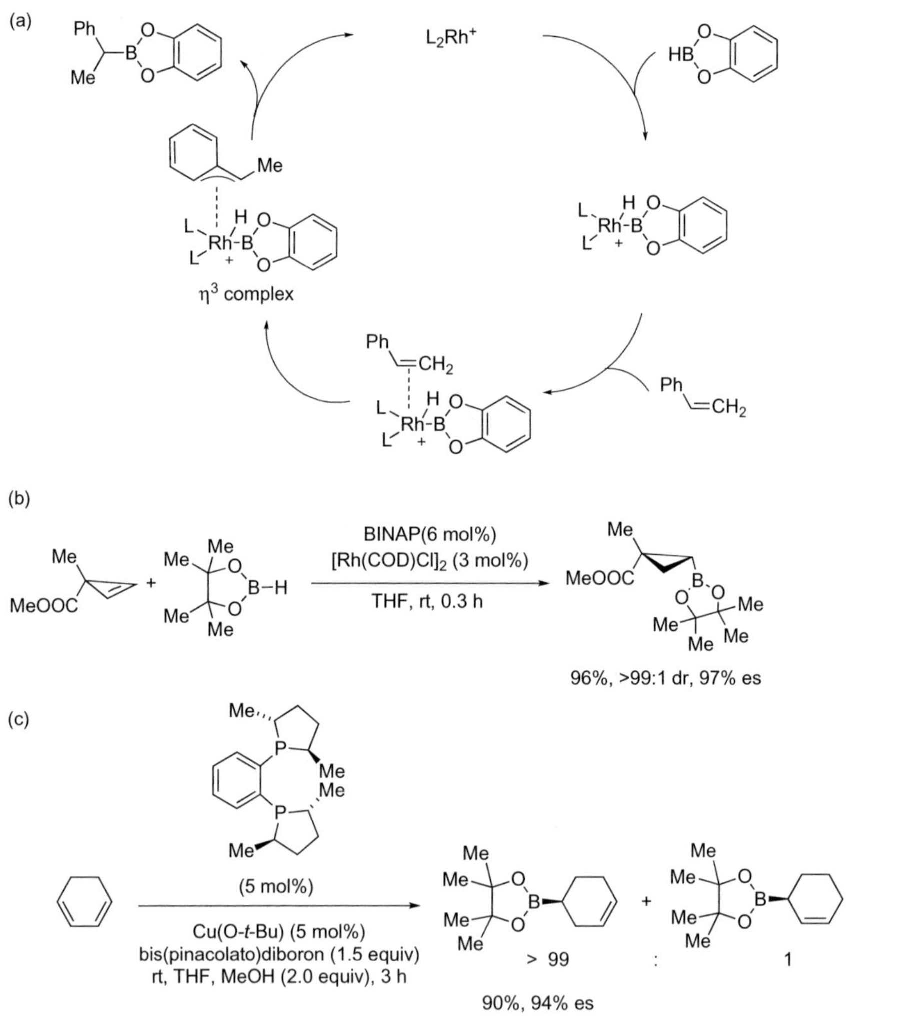

SCHEME 7.33 (a) Generalized mechanism of the reaction. Reactions of nonstyrenyl substrates: (b) [278] and (c) [279].

It begins with a fairly standard oxidative addition of borane and subsequent alkene coordination, but then leads to the quite unusual η^3 complex initially proposed by Hayashi and Ito [278], which is consistent with the nonstandard regiochemistry observed in these reactions. Issues of mechanism aside, it is certainly true that the vast majority of catalytic asymmetric hydroborations have been accomplished using aromatic olefins. Two of the handful of exceptions to this generalization are the interesting cyclopropene reaction reported by Gevorgyan [278] and the copper(I) promoted reaction of 1,3-dienes [279] shown in Schemes 7.33b and c, respectively.

When paired with the Fleming—Tamao oxidation of electron-poor alkylsilanes (known to convert C−Si bonds to C−OH bonds with retention of configuration), the palladium-catalyzed hydrosilylation of double bonds represents a useful alternative to hydroboration/oxidation (reviewed in ref. [273]). We close with a single impressive example of this reaction, in which norbornadiene was allowed to react with HSiCl$_3$ in the presence of Me-mop ligand and Pd and then oxidized to afford a C_2 symmetric diol with practically complete stereoselectivity (Scheme 7.34a [280]). This reaction leverages a high level of *exo* substrate selectivity along with high *Re* selectivity imparted by the reagent; together these result in the >99% enantioselective formation of the C_2 symmetric product, which is also enantiomerically enriched relative to the monosilylation product (obtained in an already-high 97% es) because the leakages in *Re*/*Si* selectivity predominantly lead to the meso/bis−exo product. The diol obtained in this way has been pressed into service in the context of natural product total synthesis (Schemes 7.34b [281]) or chiral ligand preparation (Schemes 7.34c and d [282,283]).

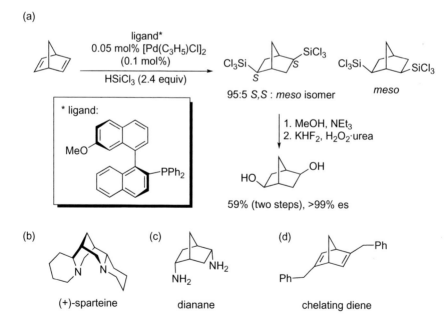

SCHEME 7.34   (a) A double asymmetric hydrosilylation reaction [280,282]. Applications to (b) natural product [281], and (c) and (d) chiral ligand synthesis [282,283].

REFERENCES

[1] Knowles, W. S. *Angew. Chem., Int. Ed.* **2002**, *41*, 1998−2007.
[2] Noyori, R. *Angew. Chem., Int. Ed.* **2002**, *41*, 2008−2022.
[3] Sharpless, K. B. *Angew. Chem., Int. Ed.* **2002**, *41*, 2024−2032.
[4] Brown, H. C.; Jadhav, P. K.; Mandal, A. K. *Tetrahedron* **1981**, *37*, 3574−3587.
[5] Brown, H. C.; Singaram, B. *Pure Appl. Chem.* **1987**, *59*, 879−894.
[6] Itsuno, S. *Org. React.* **2004**, *52*, 395−576.
[7] Nugent, T. C.; El-Shazly, M. *Adv. Synth. Catal.* **2010**, *352*, 753−819.
[8] Bothner-by, A. A. *J. Am. Chem. Soc.* **1951**, *73*, 846.
[9] Morrison, J. D.; Mosher, H. S. *Asymmetric Organic Reactions*; Prentice-Hall: Englewood Cliffs, NJ, 1971, pp. 202−215.
[10] Grandbois, E. R.; Howard, S. I.; Morrison, J. D. In *Asymmetric Synthesis*; Morrison, J. D., Ed.; Academic Press: Orlando, FL, 1983; Vol. 2, pp. 71−79.
[11] Nishizawa, M.; Noyori, R. In *Comprehensive Organic Synthesis. Selectivity, Strategy, and Efficiency in Modern Organic Chemistry*; Trost, B. M., Fleming, I., Eds.; Pergamon: Oxford, 1991; Vol. 8, pp. 159−182.
[12] Seyden-Penne, J. *Reductions by the Alumino- and Borohydrides in Organic Synthesis*; VCH: New York, 1991.
[13] Noyori, R.; Tomino, R.; Tomino, I.; Tanimoto, Y. *J. Am. Chem. Soc.* **1979**, *101*, 3129−3131.
[14] Nishiwaza, M.; Yamada, M.; Noyori, R. *Tetrahedron Lett.* **1981**, *22*, 247−250.
[15] Noyori, R.; Tomino, I.; Tanimoto, Y.; Nishizawa, M. *J. Am. Chem. Soc.* **1984**, *106*, 6709−6716.
[16] Noyori, R.; Tomino, I.; Yamada, M.; Nishizawa, M. *J. Am. Chem. Soc.* **1984**, *106*, 6717−6725.
[17] Seebach, D.; Beck, A. K.; Dahinden, R.; Hoffmann, M.; Kühnle, F. N. M. *Croatia Chem. Acta* **1996**, *69*, 459−484.
[18] Beck, A. K.; Dahinden, R.; Kühnle, F. N. M. In *ACS Symposium Series. Reduction in Organic Chemistry*; Abdel-Magid, A. F., Ed.; American Chemical Society: Washington, D.C, 1996, pp. 52−69.
[19] March, J. *Advanced Organic Chemistry, 4th Ed.*; Wiley-Interscience: New York, 1992. p. 145.
[20] Fiaud, J. C.; Kagan, H. B. *Bull. Soc. Chim. Fr.* **1969**, 2742−2743.
[21] Wallbaum, S.; Martens, J. *Tetrahedron: Asymmetry* **1992**, *3*, 1475−1504.
[22] Deloux, L.; Srebnik, M. *Chem. Rev.* **1993**, *93*, 763−784.
[23] Wills, M.; Hannedouche, J. *Curr. Opin. Drug Discov. Devel.* **2002**, *5*, 881−891.
[24] Cho, B. T. *Tetrahedron* **2006**, *62*, 7621−7643.
[25] Kim, J.; Suri, J. T.; Cordes, D. B.; Singaram, B. *Org. Process Res. Dev.* **2006**, *10*, 949−958.
[26] Cho, B. T. *Chem. Soc. Rev.* **2009**, *38*, 443−452.
[27] Hirao, A.; Itsuno, S.; Nakahama, S.; Yamazaki, N. *J. Chem. Soc., Chem. Commun.* **1981**, 315−317.
[28] Itsuno, S.; Hirao, A.; Nakahama, S.; Yamazaki, N. *J. Chem. Soc., Perkin Trans. 1* **1983**, 1673−1676.
[29] Itsuno, S.; Ito, K.; Hirao, A.; Nakahama, S. *J. Chem. Soc., Chem. Commun.* **1983**, 469−470.
[30] Itsuno, S.; Nakano, M.; Miyazaki, K.; Masuda, H.; Ito, K. *J. Chem. Soc., Perkin Trans. 1* **1985**, 2039−2044.
[31] Itsuno, S.; Ito, K.; Hirao, A.; Nakahama, S. *J. Chem. Soc., Perkin Trans. 1* **1984**, 2887−2893.
[32] Itsuno, S.; Ito, K.; Hirao, A.; Nakahama, S. *J. Org. Chem.* **1984**, *49*, 555−557.
[33] Corey, E. J.; Bakshi, R. K.; Shibata, S. *J. Org. Chem.* **1988**, *53*, 2861−2863.
[34] Itsuno, S.; Sakurai, Y.; Ito, K.; Hirao, A.; Nakahama, S. *Bull. Chem. Soc. Jpn.* **1987**, *60*, 395−396.
[35] Corey, E. J.; Bakshi, R. K.; Shibata, S. *J. Am. Chem. Soc.* **1987**, *109*, 5551−5553.
[36] Corey, E. J.; Bakshi, R. K.; Shibata, S.; Chen, C.-P.; Singh, V. K. *J. Am. Chem. Soc.* **1987**, *109*, 7925−7926.
[37] Corey, E. J.; Link, J. O. *Tetrahedron Lett.* **1989**, *30*, 6275−6278.
[38] Corey, E. J.; Azomioara, M.; Sarshar, S. *Tetrahedron Lett.* **1992**, *33*, 3429−3430.
[39] Mathre, D. J.; Jones, T. K.; Xavier, L. C.; Blacklock, T. J.; Reamer, R. A.; Mohan, J. J.; Jones, E. T. T.; Hoogsteen, K.; Baum, M. W.; Grabowski, E. J. J. *J. Org. Chem.* **1991**, *56*, 751−762.
[40] Nevalainen, V. *Tetrahedron: Asymmetry* **1991**, *2*, 429−435.
[41] Jones, D. K.; Liotta, D. C.; Shinkai, I.; Mathre, D. J. *J. Org. Chem.* **1993**, *58*, 799−801.
[42] Quallich, G. J.; Blake, J. F.; Woodall, T. M. *J. Am. Chem. Soc.* **1994**, *116*, 8516−8525.

[43] Nevalainen, V. *Tetrahedron: Asymmetry* **1992**, *3*, 921−932.

[44] Corey, E. J.; Link, J. O. *J. Am. Chem. Soc.* **1992**, *114*, 1906−1908.

[45] Corey, E. J.; Link, J. O.; Shao, Y. *Tetrahedron Lett.* **1992**, *33*, 3435−3438.

[46] Corey, E. J.; Link, J. O. *Tetrahedron Lett.* **1992**, *33*, 3431−3434.

[47] Corey, E. J.; Helal, C. J. *Tetrahedron Lett.* **1993**, *34*, 5227−5230.

[48] Corey, E. J.; Link, J. O. *Tetrahedron Lett.* **1992**, *33*, 4141−4144.

[49] Corey, E. J.; Bakshi, R. K. *Tetrahedron Lett.* **1990**, *31*, 611−614.

[50] Beak, P.; Kerrick, S. T.; Wu, S.; Chu, J. *J. Am. Chem. Soc.* **1994**, *116*, 3231−3239.

[51] Nikolic, N. A.; Beak, P. *Org. Synth.* **1998**, *Coll Vol. 9*, 391−396.

[52] Jones, T. K.; Mohan, J. J.; Xavier, L. C.; Blacklock, T. J.; Mathre, D. J.; Sohar, P.; Jones, E. T. T.; Reamer, R. A.; Roberts, F. E.; Grabowski, E. J. J. *J. Org. Chem.* **1991**, *56*, 763−769.

[53] Matsuki, K.; Inoue, H.; Takeda, M. *Tetrahedron Lett.* **1993**, *34*, 1167−1170.

[54] Romagnoli, R.; Roos, E. C.; Hiemstra, H.; Moolenaar, M. J.; Speckamp, W. N.; Kaptein, B.; Schoemaker, H. E. *Tetrahedron Lett.* **1994**, *35*, 1087−1090.

[55] Ostendorf, M.; Romagnoli, R.; Pereiro, I. C.; Roos, E. C.; Moolenaar, M. J.; Speckamp, W. N.; Hiemstra, H. *Tetrahedron: Asymmetry* **1997**, *8*, 1773−1789.

[56] Kang, J.; Lee, J. W.; Kim, J. I.; Pyun, C. *Tetrahedron Lett.* **1995**, *36*, 4265−4268.

[57] Chen, F.-E.; Dai, H.-F.; Kuang, Y.-Y.; Jia, H.-Q. *Tetrahedron: Asymmetry* **2003**, *14*, 3667−3672.

[58] Spivey, A. C.; Andrews, B. I.; Brown, A. D.; Frampton, C. S. *Chem. Commun.* **1999**, 2523−2524.

[59] Schreiber, S. L.; Schreiber, T. S.; Smith, D. B. *J. Am. Chem. Soc.* **1987**, *109*, 1525−1529.

[60] Midland, M. M. *Chem. Rev.* **1989**, *89*, 1553−1561.

[61] Brown, H. C.; Ramachandran, P. V. *Pure Appl. Chem.* **1991**, *63*, 307−316.

[62] Brown, H. C.; Ramachandran, P. V. *Acc. Chem. Res.* **1992**, *25*, 16−24.

[63] Imai, T.; Tamura, T.; Yamamuro, A.; Sato, T.; Wollmann, T. A.; Kennedy, R. M.; Masamune, S. *J. Am. Chem. Soc.* **1986**, *108*, 7402−7404.

[64] Mikhailov, B. M.; Bubnov, Y. N.; Kiselev, V. G. *J. Gen. Chem. USSR (Engl. Transl.)* **1966**, *36*, 65−69.

[65] Midland, M. M.; Zderic, S. A. *J. Am. Chem. Soc.* **1982**, *104*, 525−528.

[66] Midland, M. M.; Greer, S.; Tramontano, A.; Zderic, S. A. *J. Am. Chem. Soc.* **1979**, *101*, 2352−2355.

[67] Midland, M. M.; Greer, S. *Synthesis* **1978**, *11*, 845−846.

[68] Midland, M. M.; McDowell, D. C.; Hatch, R. L.; Tramontano, A. *J. Am. Chem. Soc.* **1980**, *102*, 867−869.

[69] Midland, M. M.; Graham, R. S. *Org. Synth.* **1990**, *Coll. Vol. 7*, 402−406.

[70] Midland, M. M.; Tramontano, A.; Kazubski, A.; Graham, R.; Tsai, D. J.-S.; Cardin, D. B. *Tetrahedron* **1984**, *40*, 1371−1380.

[71] Brown, H. C.; Pai, G. G. *J. Org. Chem.* **1985**, *50*, 1384−1394.

[72] Midland, M. M.; McLoughlin, J. I.; Gabriel, J. *J. Org. Chem.* **1989**, *54*, 159−165.

[73] Brown, H. C.; Chandrasekharan, J.; Ramachandran, P. V. *J. Am. Chem. Soc.* **1988**, *110*, 1539−1546.

[74] Ramachandran, P. V.; Teodorovic, A. V.; Rangaishenvi, M. V.; Brown, H. C. *J. Org. Chem.* **1992**, *57*, 2379−2386.

[75] Zassinovich, G.; Betella, R.; Mestroni, G.; Bresciani-Pahor, N.; Geremia, S.; Randaccio, L. *J. Organomet. Chem.* **1989**, *370*, 187−202.

[76] Gladiali, S.; Pinna, L.; Deloga, G.; Martin, S. D.; Zassinovich, G.; Mestroni, G. *Tetrahedron: Asymmetry* **1990**, *1*, 635−648.

[77] Bolm, C. *Angew. Chem., Int. Ed. Engl.* **1991**, *30*, 542−543.

[78] Zhang, X.; Taketomi, T.; Yoshizumi, T.; Kumobayashi, H.; Akutagawa, S.; Mashima, K.; Takaya, H. *J. Am. Chem. Soc.* **1993**, *115*, 3318−3319.

[79] Takaya, H.; Ohta, T.; Noyori, R. In *Catalytic Asymmetric Synthesis*; Ojima, I., Ed.; VCH: New York, 1993, pp. 1−39.

[80] Noyori, R. *Asymmetric Catalysis in Organic Synthesis*; Wiley-Interscience: New York, 1994.

[81] Brunner, H.; Nishiyama, H.; Itoh, K. In *Catalytic Asymmetric Synthesis*; Ojima, I., Ed.; VCH: New York, 1993, pp. 303−322.

[82] Noyori, R. *Asymmetric Catalysis in Organic Synthesis*; Wiley-Interscience: New York, 1994, pp. 124−131.

[83] Noyori, R.; Hashiguchi, S.; Iwasawa, Y. *Acc. Chem. Res.* **1997**, *30*, 97−102.

[84] Gladiali, S.; Alberico, E. *Chem. Soc. Rev.* **2006**, *35*, 226−236.

[85] Ikariya, T.; Blacker, A. J. *Acc. Chem. Res.* **2007**, *40*, 1300–1308.

[86] You, S.-L. *Chem.–Asian J.* **2007**, *2*, 820–827.

[87] Wang, C.; Wu, X.; Xiao, J. *Chem.–Asian J.* **2008**, *3*, 1750–1770.

[88] Morris, R. H. *Chem. Soc. Rev.* **2009**, *38*, 2282–2291.

[89] Fleury-Bregeot, N.; de la Fuente, V.; Castillon, S.; Claver, C. *ChemCatChem* **2010**, *2*, 1346–1371.

[90] Zaidlewicz, M.; Pakulski, M. M. *Reduction of Carbonyl Groups: Transfer Hydrogenation, Hydrosilylation, Catalytic Hydroboration, and Reduction with Borohydrides, Aluminum Hydrides, or Boranes*; Georg Thieme Verlag: Stuttgart, 2011, Vol 2.

[91] Noyori, R.; Ohkuma, T.; Kitamura, M.; Takaya, H.; Sayo, N.; Kumobayashi, H.; Akutagawa, S. *J. Am. Chem. Soc.* **1987**, *109*, 5856–5858.

[92] Kitamura, M.; Ohkuma, T.; Inoue, S.; Sayo, N.; Kumobayashi, H.; Akutagawa, S.; Ohta, T.; Takaya, H.; Noyori, R. *J. Am. Chem. Soc.* **1988**, *110*, 629–631.

[93] Kitamura, M.; Tokunaga, M.; Ohkuma, T.; Noyori, R. *Tetrahedron Lett.* **1991**, *32*, 4163–4166.

[94] Taber, D. F.; Silverberg, L. J. *Tetrahedron Lett.* **1991**, *32*, 4227–4230.

[95] Taber, D. F.; Deker, P. B.; Silverberg, L. J. *J. Org. Chem.* **1992**, *57*, 5990–5994.

[96] Mashima, K.; Kusano, K.; Ohta, T.; Noyori, R.; Takaya, H. *J. Chem. Soc., Chem. Commun.* **1989**, 1208–1210.

[97] Mashima, K.; Kusano, K.; Sato, N.; Matsumura, Y.; Nozaki, K.; Kumobayashi, H.; Sayo, N.; Hori, Y.; Ishikazi, T.; Akutagawa, S.; Takaya, H. *J. Org. Chem.* **1994**, *59*, 3064–3076.

[98] Noyori, R. In *Asymmetric Catalysis in Organic Synthesis*; Wiley-Interscience: New York, 1994, pp. 63–66.

[99] Burk, M. J.; Harper, T. G. P.; Kalberg, C. S. *J. Am. Chem. Soc.* **1995**, *117*, 4423–4424.

[100] Ohkuma, T.; Ooka, H.; Hashiguchi, S.; Ikariya, T.; Noyori, R. *J. Am. Chem. Soc.* **1995**, *117*, 2675–2676.

[101] Ohkuma, T.; Ooka, H.; Ikariya, T.; Noyori, R. *J. Am. Chem. Soc.* **1995**, *117*, 10417–10418.

[102] Curtin, D. Y. *Rec. Chem. Progr.* **1954**, *15*, 111–128.

[103] Seeman, J. I. *Chem. Rev.* **1983**, *83*, 83–134.

[104] Coldham, I.; Sheikh, N. S. *Top. Stereochem.* **2010**, *26*, 253–293.

[105] Noyori, R.; Ikeda, T.; Ohkuma, T.; Widhalm, M.; Kitamura, M.; Takaya, H.; Akutagawa, S.; Sayo, N.; Saito, T.; Taketomi, T.; Kumobayashi, H. *J. Am. Chem. Soc.* **1989**, *111*, 9134–9135.

[106] Ohta, T.; Takaya, H.; Noyori, R. *Inorg. Chem.* **1988**, *27*, 566–569.

[107] Samec, J. S. M.; Backvall, J.-E.; Andersson, P. G.; Brandt, P. *Chem. Soc. Rev.* **2006**, *35*, 237–248.

[108] Doering, W. v. E.; Young, R. W. *J. Am. Chem. Soc.* **1950**, *72*, 631.

[109] Ohkubo, K.; Hirata, K.; Yoshinaga, K.; Okada, M. *Chem. Lett.* **1976**, *5*, 183–184.

[110] Descotes, G.; Sinou, D. *Tetrahedron Lett.* **1976**, *17*, 4083–4086.

[111] Hashiguchi, S.; Fujii, A.; Takehara, J.; Ikariya, T.; Noyori, R. *J. Am. Chem. Soc.* **1995**, *117*, 7562–7563.

[112] Evans, D. A.; Nelson, S. G.; Gagne, M. R.; Muci, A. R. *J. Am. Chem. Soc.* **1993**, *115*, 9800–9801.

[113] Haack, K.-J.; Hashiguchi, S.; Fujii, A.; Ikariya, T.; Noyori, R. *Angew. Chem., Int. Ed. Engl.* **1997**, *36*, 285–288.

[114] Fujii, A.; Hashiguchi, S.; Uematsu, N.; Ikariya, T.; Noyori, R. *J. Am. Chem. Soc.* **1996**, *118*, 2521–2522.

[115] Handgraaf, J.-W.; Meijer, E. J. *J. Am. Chem. Soc.* **2007**, *129*, 3099–3103.

[116] Noyori, R.; Yamakawa, M.; Hashiguchi, S. *J. Org. Chem.* **2001**, *66*, 7931–7944.

[117] Malacea, R.; Poli, R.; Manoury, E. *Coord. Chem. Rev.* **2010**, *254*, 729–752.

[118] Arena, C. G. *Mini-Rev. Org. Chem.* **2009**, *6*, 159–167.

[119] Riant, O.; Mostefai, N.; Courmarcel, J. *Synthesis* **2004**, *18*, 2943–2958.

[120] Mikami, K.; Wakabayashi, K.; Yusa, Y.; Aikawa, K. *Chem. Commun.* **2006**, *22*, 2365–2367.

[121] Pastor, I. M.; Västilä, P.; Adolfsson, H. *Chem.–Eur. J.* **2003**, *9*, 4031–4045.

[122] Vastila, P.; Wettergren, J.; Adolfsson, H. *Chem. Commun.* **2005**, *32*, 4039–4041.

[123] Zaitsev, A. B.; Adolfsson, H. *Org. Lett.* **2006**, *8*, 5129–5132.

[124] Li, L.; Wu, J.; Wang, F.; Liao, J.; Zhang, H.; Lian, C.; Zhu, J.; Deng, J. *Green Chem.* **2007**, *9*, 23–25.

[125] Matharu, D. S.; Morris, D. J.; Clarkson, G. J.; Wills, M. *Chem. Commun.* **2006**, *30*, 3232–3234.

[126] Geldbach, T. J.; Dyson, P. J. *J. Am. Chem. Soc.* **2004**, *126*, 8114–8115.

[127] Schlatter, A.; Kundu, M. K.; Woggon, W.-D. *Angew. Chem., Int. Ed.* **2004**, *43*, 6731–6734.

[128] Wang, F.; Liu, H.; Cun, L.; Zhu, J.; Deng, J.; Jiang, Y. *J. Org. Chem.* **2005**, *70*, 9424–9429.

[129] Šterk, D.; Stephan, M. S.; Mohar, B. *Tetrahedron: Asymmetry* **2002**, *13*, 2605–2608.

[130] Murata, K.; Okano, K.; Miyagi, M.; Iwane, H.; Noyori, R.; Ikariya, T. *Org. Lett.* **1999**, *1*, 1119–1121.

[131] Matsumura, K.; Hashiguchi, S.; Ikariya, T.; Noyori, R. *J. Am. Chem. Soc.* **1997**, *119*, 8738–8739.
[132] Schreiber, S. L.; Kelly, S. E.; Porco, J. A., Jr; Sammakia, T.; Suh, E. M. *J. Am. Chem. Soc.* **1988**, *110*, 6210–6218.
[133] Baldwin, J. E.; Adlington, R. M.; Ramcharitar, S. H. *Synlett* **1992**, 11, 875–877.
[134] Kitamura, M.; Tokunaga, M.; Noyori, R. *J. Am. Chem. Soc.* **1995**, *117*, 2931–2932.
[135] Gotz, K.; Liermann, J. C.; Thines, E.; Anke, H.; Opatz, T. *Org. Biomol. Chem.*, **2010**, *8*, 2123–2130.
[136] Xie, J.-H.; Zhu, S.-F.; Zhou, Q.-L. *Chem. Rev.* **2011**, *111*, 1713–1760.
[137] Török, B.; Surya Prakash, G. K. *Adv. Synth. Catal.* **2003**, *345*, 165–168.
[138] Zhou, P.; Chen, B.-C.; Davis, F. A. *Tetrahedron* **2004**, *60*, 8003–8030.
[139] Liu, G.; Cogan, D. A.; Ellman, J. A. *J. Am. Chem. Soc.* **1997**, *119*, 9913–9914.
[140] Colyer, J. T.; Andersen, N. G.; Tedrow, J. S.; Soukup, T. S.; Faul, M. M. *J. Org. Chem.* **2006**, *71*, 6859–6862.
[141] Laczkowski, K. Z.; Pakulski, M. M.; Krzeminski, M. P.; Jaisankar, P.; Zaidlewicz, M. *Tetrahedron: Asymmetry* **2008**, *19*, 788–795.
[142] Willoughby, C. A.; Buchwald, S. L. *J. Am. Chem. Soc.* **1992**, *114*, 7562–7564.
[143] Willoughby, C. A.; Buchwald, S. L. *J. Am. Chem. Soc.* **1994**, *116*, 8952–8965.
[144] Willoughby, C. A.; Buchwald, S. L. *J. Am. Chem. Soc.* **1994**, *116*, 11703–11714.
[145] Bolm, C. *Angew. Chem., Int. Ed. Engl.* **1993**, *32*, 232–233.
[146] Cahn, R. S.; Ingold, C. K.; Prelog, V. *Angew. Chem., Int. Ed. Engl.* **1966**, *5*, 385–415, 511.
[147] Wild, F. R. W. P.; Zsolnai, L.; Huttner, G.; Brintzinger, H. H. *J. Organomet. Chem.* **1982**, *232*, 233–247.
[148] Schlögl, K. *Top. Stereochem.* **1967**, *1*, 39–91.
[149] Sloan, T. E. *Top. Stereochem.* **1981**, *12*, 1–36.
[150] Eliel, E. L.; Wilen, S. H.; Mander, L. N. *Stereochemistry of Organic Compounds*; Wiley-Interscience: New York, 1994; Chapter 14, pp. 1121–1122.
[151] Helmchen, G. In *Stereoselective Synthesis*; Helmchen, G., Hoffmann, R. W., Mulzer, J., Schaumann, E., Eds.; Georg Thieme: Stuttgart, 1995; Vol. E21a, pp. 1–74.
[152] Hortmann, K.; Brintzinger, H.-H. *New J. Chem.* **1992**, *16*, 51–55.
[153] Willoughby, C. A.; Buchwald, S. L. *J. Org. Chem.* **1993**, *58*, 7627–7629.
[154] Viso, A.; Lee, N. E.; Buchwald, S. L. *J. Am. Chem. Soc.* **1994**, *116*, 9373–9374.
[155] Yun, J.; Buchwald, S. L. *J. Am. Chem. Soc.* **1999**, *121*, 5640–5644.
[156] Blaser, H.-U. *Adv. Synth. Catal.* **2002**, *344*, 17–31.
[157] Xiao, D.; Zhang, X. *Angew. Chem., Int. Ed.* **2001**, *40*, 3425–3428.
[158] Baeza, A.; Pfaltz, A. *Chem.–Eur. J.* **2010**, *16*, 4003–4009.
[159] Zhu, S.-F.; Xie, J.-B.; Zhang, Y.-Z.; Li, S.; Zhou, Q.-L. *J. Am. Chem. Soc.* **2006**, *128*, 12886–12891.
[160] Dervisi, A.; Carcedo, C.; Ooi, L.-l *Adv. Synth. Catal.* **2006**, *348*, 175–183.
[161] Li, W.; Hou, G.; Chang, M.; Zhang, X. *Adv. Synth. Catal.* **2009**, *351*, 3123–3127.
[162] Li, C.; Wang, C.; Villa-Marcos, B.; Xiao, J. *J. Am. Chem. Soc.* **2008**, *130*, 14450–14451.
[163] Buriak, J. M.; Osborn, J. A. *Organometallics* **1996**, *15*, 3161–3169.
[164] Li, C.; Xiao, J. *J. Am. Chem. Soc.* **2008**, *130*, 13208–13209.
[165] Cobley, C. J.; Henschke, J. P. *Adv. Synth. Catal.* **2003**, *345*, 195–201.
[166] Gonzalez-Arellano, C.; Corma, A.; Iglesias, M.; Sanchez, F. *Chem. Commun.* **2005**, *27*, 3451–3453.
[167] Zhou, X.-Y.; Bao, M.; Zhou, Y.-G. *Adv. Synth. Catal.* **2011**, *353*, 84–88.
[168] Nugent, T. C. *Chiral Amine Synthesis*; Wiley-VCH: Weinheim, 2010, pp. 225–245.
[169] Storer, R. I.; Carrera, D. E.; Ni, Y.; MacMillan, D. W. C. *J. Am. Chem. Soc.* **2005**, *128*, 84–86.
[170] Rueping, M.; Sugiono, E.; Azap, C.; Theissmann, T.; Bolte, M. *Org. Lett.* **2005**, *7*, 3781–3783.
[171] Hoffmann, S.; Seayad, A. M.; List, B. *Angew. Chem., Int. Ed.* **2005**, *44*, 7424–7427.
[172] de Vries, J. G.; Mrsic, N. *Catal.: Sci. Technol.* **2011**, *1*, 727–735.
[173] Blaser, H.-U.; Buser, H.-P.; Jalett, H.-P.; Pugin, B.; Spindler, F. *Synlett* **1999**, *S1*, 867–868.
[174] Kadyrov, R.; Riermeier, T. H. *Angew. Chem., Int. Ed.* **2003**, *42*, 5472–5474.
[175] Kagan, H. B. *Pure Appl. Chem.* **1975**, *43*, 401–421.
[176] Morrison, J. D.; Masler, W. F.; Neuberg, M. K. *Adv. Catal.* **1976**, *25*, 81–124.
[177] Caplar, V.; Comisso, G.; Sunjic, V. *Synthesis* **1981**, *2*, 85–116.
[178] Knowles, W. S. *Acc. Chem. Res.* **1983**, *16*, 106–112.
[179] Takaya, H.; Noyori, R. In *Comprehensive Organic Synthesis. Selectivity, Strategy, and Efficiency in Modern Organic Chemistry*; Trost, B. M., Fleming, I., Eds.; Pergamon: Oxford, 1991; Vol. 8, pp. 443–469.

[180] Brunner, H. *Synthesis* **1988**, *9*, 645–654.

[181] Blystone, S. L. *Chem. Rev.* **1989**, *89*, 1663–1679.

[182] Kagan, H. B. *Bull. Soc. Chim. Fr.* **1988**, *5*, 846–853.

[183] Noyori, R.; Takaya, H. *Acc. Chem. Res.* **1990**, *23*, 345–350.

[184] Noyori, R. In *Asymmetric Catalysis in Organic Synthesis*; Wiley-Interscience: New York, 1994, pp. 16–94.

[185] Inoguchi, K.; Sakuraba, S.; Achiwa, K. *Synlett* **1992**, *3*, 169–178.

[186] Halterman, R. L. In *Comprehensive Asymmetric Catalysis*, Jacobsen, E. N.; Pfaltz, A.; Yamamoto, H., Eds., Springer: Berlin, 1999; Vol. 1, pp. 183–195.

[187] Ohkuma, T.; Kitamura, M.; Noyori, R., In *Catalytic Asymmetric Synthesis, 2nd Ed.*, Ojima, I., Ed., Wiley-VCH: New York, 2000; pp. 1–110.

[188] Genet, J.-P. *Acc. Chem. Res.* **2003**, *36*, 908–918.

[189] Kaellstroem, K.; Munslow, I.; Andersson, P. G. *Chem. Eur. J.* **2006**, *12*, 3194–3200.

[190] Roseblade, S. J.; Pfaltz, A. *Acc. Chem. Res.* **2007**, *40*, 1402–1411.

[191] Genet, J.-P. In *Modern Reduction Methods*, Andersson, P. G.; and Munslow, I. J., Eds., Wiley-VCH: Weinheim 2008, pp. 3–38.

[192] Noyori, R.; Ohkuma, T.; Sandoval, C. A.; Muniz, K. In *Asymmetric Synthesis – The Essentials*, Christmann, M.; Bräse, S., Eds., Wiley-VCH: Weinheim 2006, pp. 337–341.

[193] Pamies, O.; Andersson, P. G.; Dieguez, M. *Chem. Eur. J.* **2010**, *16*, 14232–14240.

[194] Ager, D., In *Science of Synthesis: Stereoselective Synthesis*, De Vries, J. G.; Molander, G. A.; Evans, P. A., Eds. Georg Thieme Verlag: Stuttgart, 2011; Vol. 1, pp. 185–256.

[195] Woodmansee, D. H.; Pfaltz, A. *Chem. Commun.* **2011**, *47*, 7912–7916.

[196] Kagan, H. B. In *Asymmetric Synthesis*; Morrison, J. D., Ed.; Academic Press: Orlando, FL, 1985; Vol. 5, pp. 1–39.

[197] Halpern, J. In *Asymmetric Synthesis*; Morrison, J. D., Ed.; Academic Press: Orlando, FL, 1985; Vol. 5, pp. 41–69.

[198] Koenig, K. E. In *Asymmetric Synthesis*; Morrison, J. D., Ed.; Academic Press: Orlando, FL, 1985; Vol. 5, pp. 71–101.

[199] Landis, C. R.; Halpern, J. *J. Am. Chem. Soc.* **1987**, *109*, 1746–1754.

[200] Osborn, J. A.; Jardine, F. H.; Young, J. F.; Wilkinson, G. *J. Chem. Soc. (A)* **1966**, *88*, 1711–1732.

[201] Horner, L.; Büthe, H.; Siegel, H. *Tetrahedron Lett.* **1968**, *37*, 4023–4026.

[202] Korpium, O.; Lewis, R. A.; Chickos, J.; Mislow, K. *J. Am. Chem. Soc.* **1968**, *90*, 4842–4846.

[203] Knowles, W. S.; Sabacky, M. J. *J. Chem. Soc. Chem. Commun.* **1968**, 1445–1446.

[204] Morrison, J. D.; Burnett, R. E.; Aguiar, A. M.; Morrow, C. J.; Phillips, C. *J. Am. Chem. Soc.* **1971**, *93*, 1301–1303.

[205] Dang, T. P.; Kagan, H. B. *Chem. Commun.* **1971**, 481.

[206] Burk, M. J.; Feaster, J. E.; Harlow, R. L. *Tetrahedron: Asymmetry* **1991**, *2*, 569–592.

[207] *Catalytic Asymmetric Synthesis*; Ojima, I., Ed.; VCH: New York, 1993.

[208] Burk, M. J.; Gross, M. F.; Martinez, J. P. *J. Am. Chem. Soc.* **1995**, *117*, 9375–9376.

[209] Noyori, R. *Science* **1990**, *248*, 1194–1199.

[210] Chan, A. S. C.; Pluth, J. J.; Halpern, J. *J. Am. Chem. Soc.* **1980**, *102*, 5952–5954.

[211] Brown, J. M.; Chaloner, P. A. *J. Chem. Soc. Chem. Commun.* **1980**, *76*, 344–346.

[212] Chan, A. S. C.; Halpern, J. *J. Am. Chem. Soc.* **1980**, *102*, 838–840.

[213] Vineyard, B. D.; Knowles, W. S.; Sabacky, M. J.; Bachman, G. L.; Weinkauff, D. J. *J. Am. Chem. Soc.* **1977**, *99*, 5946–5952.

[214] Zhang, K.; Gonzalez, A. A.; Hoff, C. D. *J. Am. Chem. Soc.* **1989**, *111*, 3627–3632.

[215] Brown, J. M.; Evans, P. L. *Tetrahedron* **1988**, *44*, 4905–4916.

[216] Claver, C.; Fernandez, E.; Gillon, A.; Heslop, K.; Hyett, D. J.; Martorell, A.; Orpen, A. G.; Pringle, P. G. *Chem. Commun.* **2000**, *11*, 961–962.

[217] Reetz, M. T.; Sell, T. *Tetrahedron Lett.* **2000**, *41*, 6333–6336.

[218] van den Berg, M.; Minnaard, A. J.; Schudde, E. P.; van Esch, J.; de Vries, A. H. M.; de Vries, J. G.; Feringa, B. L. *J. Am. Chem. Soc.* **2000**, *122*, 11539–11540.

[219] Eberhardt, L.; Armspach, D.; Harrowfield, J.; Matt, D. *Chem. Soc. Rev.* **2008**, *37*, 839–864.

[220] Minnaard, A. J.; Feringa, B. L.; Lefort, L.; de, V. J. G. *Acc. Chem. Res.* **2007**, *40*, 1267–1277.

[221] Alberico, E.; Baumann, W.; de Vries, J. G.; Drexler, H.-J.; Gladiali, S.; Heller, D.; Henderickx, H. J. W.; Lefort, L. *Chem.–Eur. J.* **2011**, *17*, 12683–12695.

[222] Komarov, I. V.; Börner, A. *Angew. Chem., Int. Ed.* **2001**, *40*, 1197–1200.
[223] Panella, L.; Feringa, B. L.; de Vries, J. G.; Minnaard, A. J. *Org. Lett.* **2005**, *7*, 4177–4180.
[224] Boogers, J. A. F.; Felfer, U.; Kotthaus, M.; Lefort, L.; Steinbauer, G.; de Vries, A. H. M.; de Vries, J. G. *Org. Process Res. Dev.* **2007**, *11*, 585–591.
[225] Boogers, J. A. F.; Sartor, D.; Felfer, U.; Kotthaus, M.; Steinbauer, G.; Dielemans, B.; Lefort, L.; de Vries, A. H. M.; de Vries, J. G. In *Asymmetric Catalysis on Industrial Scale*; Blaser, H.-U., Federsel, H.-J., Eds.; Wiley-VCH: Weinheim, 2010, pp. 127–150.
[226] Cui, X.; Burgess, K. *Chem. Rev.* **2005**, *105*, 3272–3296.
[227] Broene, R. D.; Buchwald, S. L. *J. Am. Chem. Soc.* **1993**, *115*, 12569–12570.
[228] Forman, G. S.; Ohkuma, T.; Hems, W. P.; Noyori, R. *Tetrahedron Lett.* **2000**, *41*, 9471–9475.
[229] Conticello, V. P.; Brard, L.; Giardello, M. A.; Tsuji, Y.; Sabat, M.; Stern, C. L.; Marks, T. J. *J. Am. Chem. Soc.* **1992**, *114*, 2761–2762.
[230] McIntyre, S.; Hörmann, E.; Menges, F.; Smidt, S. P.; Pfaltz, A. *Adv. Synth. Catal.* **2005**, *347*, 282–288.
[231] Perry, M. C.; Cui, X.; Powell, M. T.; Hou, D.-R.; Reibenspies, J. H.; Burgess, K. *J. Am. Chem. Soc.* **2002**, *125*, 113–123.
[232] Diéguez, M.; Mazuela, J.; Pàmies, O.; Verendel, J. J.; Andersson, P. G. *Chem. Commun.* **2008**, *33*, 3888–3890.
[233] Diéguez, M.; Mazuela, J.; Pàmies, O.; Verendel, J. J.; Andersson, P. G. *J. Am. Chem. Soc.* **2008**, *130*, 7208–7209.
[234] Bell, S.; Wüstenberg, B.; Kaiser, S.; Menges, F.; Netscher, T.; Pfaltz, A. *Science* **2006**, *311*, 642–644.
[235] Cui, X.; Fan, Y.; Hall, M. B.; Burgess, K. *Chem.–Eur. J.* **2005**, *11*, 6859–6868.
[236] Mazet, C. m.; Smidt, S. P.; Meuwly, M.; Pfaltz, A. *J. Am. Chem. Soc.* **2004**, *126*, 14176–14181.
[237] Church, T. L.; Rasmussen, T.; Andersson, P. G. *Organometallics* **2011**, *29*, 6769–6781.
[238] Lightfoot, A.; Schnider, P.; Pfaltz, A. *Angew. Chem., Int. Ed.* **1998**, *37*, 2897–2899.
[239] Hedberg, C.; Källström, K.; Brandt, P.; Hansen, L. K.; Andersson, P. G. *J. Am. Chem. Soc.* **2006**, *128*, 2995–3001.
[240] Schrems, M. G.; Neumann, E.; Pfaltz, A. *Angew. Chem., Int. Ed.* **2007**, *46*, 8274–8276.
[241] Nishiyama, H.; Ito, J.-i.; Shiomi, T.; Hashimoto, T.; Miyakawa, T.; Kitase, M. *Pure Appl. Chem.* **2008**, *80*, 743–749.
[242] Appella, D. H.; Moritani, Y.; Shintani, R.; Ferreira, E. M.; Buchwald, S. L. *J. Am. Chem. Soc.* **1999**, *121*, 9473–9474.
[243] Hughes, G.; Kimura, M.; Buchwald, S. L. *J. Am. Chem. Soc.* **2003**, *125*, 11253–11258.
[244] Yun, J.; Buchwald, S. L. *Org. Lett.* **2001**, *3*, 1129–1131.
[245] Jurkauskas, V.; Buchwald, S. L. *J. Am. Chem. Soc.* **2002**, *124*, 2892–2893.
[246] Tsuchiya, Y.; Kanazawa, Y.; Shiomi, T.; Kobayashi, K.; Nishiyama, H. *Synlett* **2004**, *14*, 2493–2496.
[247] Shiomi, T.; Ito, J.-i.; Yamamoto, Y.; Nishiyama, H. *Eur. J. Org. Chem.* **2006**, *24*, 5594–5600.
[248] Ouellet, S. p. G.; Tuttle, J. B.; MacMillan, D. W. C. *J. Am. Chem. Soc.* **2004**, *127*, 32–33.
[249] Yang, J. W.; Hechavarria Fonseca, M. T.; List, B. *Angew. Chem., Int. Ed.* **2004**, *43*, 6660–6662.
[250] Tuttle, J. B.; Ouellet, S. p. G.; MacMillan, D. W. C. *J. Am. Chem. Soc.* **2006**, *128*, 12662–12663.
[251] Martin, N. J. A.; List, B. *J. Am. Chem. Soc.* **2006**, *128*, 13368–13369.
[252] Ouellet, S. p. G.; Walji, A. M.; Macmillan, D. W. C. *Acc. Chem. Res.* **2007**, *40*, 1327–1339.
[253] Huang, Y.; Walji, A. M.; Larsen, C. H.; MacMillan, D. W. C. *J. Am. Chem. Soc.* **2005**, *127*, 15051–15053.
[254] Brown, H. C.; Zweifel, G. *J. Am. Chem. Soc.* **1961**, *83*, 486–487.
[255] Brown, H. C.; Ayyangar, N. R.; Zweifel, G. *J. Am. Chem. Soc.* **1964**, *86*, 397–403.
[256] Brown, H. C.; Desai, M. C.; Jadhav, P. K. *J. Org. Chem.* **1982**, *47*, 5065–5069.
[257] Brown, H. C.; Singaram, B. *J. Am. Chem. Soc.* **1984**, *106*, 1797–1800.
[258] Brown, H. C.; Dhotke, U. P. *J. Org. Chem.* **1994**, *59*, 2365–2369.
[259] Brown, H. C.; Jadhav, P. K.; Mandal, A. K. *J. Org. Chem.* **1982**, *47*, 5074–5083.
[260] Masamune, S.; Kim, B. M.; Peterson, J. S.; Sato, T.; Veenstra, S. J.; Imai, T. *J. Am. Chem. Soc.* **1985**, *107*, 4549–4551.
[261] Gonzalez, A. Z.; Román, J. G.; Gonzalez, E.; Martinez, J.; Medina, J. R.; Matos, K.; Soderquist, J. A. *J. Am. Chem. Soc.* **2008**, *130*, 9218–9219.
[262] Brown, H. C.; Jadhav, P. K. In *Asymmetric Synthesis*; Morrison, J. D., Ed.; Academic Press: Orlando, FL, 1983; Vol. 5, pp. 1–43.

[263] Brown, H. C.; Jadhav, P. K.; Desai, M. C. *Tetrahedron* **1984**, *40*, 1325−1332.

[264] Brown, H. C.; Singaram, B. *Acc. Chem. Res.* **1988**, *21*, 287−293.

[265] Thomas, S. P.; Aggarwal, V. K. *Angew. Chem., Int. Ed.* **2009**, *48*, 1896−1898.

[266] Brown, H. C.; Prasad, J. V. N. V. *J. Am. Chem. Soc.* **1986**, *108*, 2049−2054.

[267] Zweifel, G.; Ayyangar, N. R.; Munekata, T.; Brown, H. C. *J. Am. Chem. Soc.* **1964**, *86*, 1076−1079.

[268] Houk, K. N.; Rondan, N. G.; Wu, Y.-D.; Metz, J. T.; Paddon-Row, M. N. *Tetrahedron* **1984**, *40*, 2257−2274.

[269] Beletskaya, I.; Pelter, A. *Tetrahedron* **1997**, *53*, 4957−5026.

[270] Crudden, C. M.; Edwards, D. *Eur. J. Org. Chem.* **2003**, *24*, 4695−4712.

[271] Carroll, A.-M.; O'Sullivan, T. P.; Guiry, P. J. *Adv. Synth. Catal.* **2005**, *347*, 609−631.

[272] Vogels, C. M.; Westcott, S. A. *Current Organic Chemistry* **2005**, *9*, 687−699.

[273] Han, J. W.; Hayashi, T. In *Catalytic Asymmetric Synthesis*; *3rd Ed.*; Ojima, I., Ed. John Wiley & Sons, Inc.: Hoboken, NJ, 2010, pp. 771−798.

[274] Männig, D.; Nöth, H. *Angew. Chem., Int. Ed. Engl.* **1985**, *24*, 878−879.

[275] Hayashi, T.; Matsumoto, Y.; Ito, Y. *J. Am. Chem. Soc.* **1989**, *111*, 3426−3428.

[276] Brown, J. M.; Lloyd-Jones, G. C. *Tetrahedron: Asymmetry* **1990**, *1*, 869−872.

[277] Doucet, H.; Fernandez, E.; Layzell, T. P.; Brown, J. M. *Chem. Eur. J.* **1999**, *5*, 1320−1330.

[278] Hayashi, T.; Matsumoto, Y.; Ito, Y. *Tetrahedron: Asymmetry* **1991**, *2*, 601−612.

[279] Sasaki, Y.; Zhong, C.; Sawamura, M.; Ito, H. *J. Am. Chem. Soc.* **2010**, *132*, 1226−1227.

[280] Uozumi, Y.; Lee, S.-Y.; Hayashi, T. *Tetrahedron Lett.* **1992**, *33*, 7185−7188.

[281] Smith, B. T.; Wendt, J. A.; Aubé, J. *Org. Lett.* **2002**, *4*, 2577−2579.

[282] Berkessel, A.; Schroeder, M.; Sklorz, C. A.; Tabanella, S.; Vogl, N.; Lex, J.; Neudoerfl, J. M. *J. Org. Chem.* **2004**, *69*, 3050−3056.

[283] Hayashi, T.; Ueyama, K.; Tokunaga, N.; Yoshida, K. *J. Am. Chem. Soc.* **2003**, *125*, 11508−11509.

Chapter 8

Oxidations

Some of the most commonly-used asymmetric transformations are oxidation reactions. The reason for this is the widespread existence of heteroatoms in interesting organic molecules and the high utility of oxygenated compounds as synthetic intermediates. For example, ring strain means that epoxides are excellent partners for substitution reactions by a wide variety of nucleophiles; they can also be readily converted to allylic alcohols by elimination or to ketones by rearrangement. Given all of this, it is no wonder that K. Barry Sharpless's contributions to oxidation chemistry were celebrated by his recognition, along with William S. Knowles and Ryoji Noyori (both for reduction reactions) by the 2001 Nobel Prize in Chemistry [1].

Some of the most pertinent virtues of asymmetric epoxidations and dihydroxylations were already present in their classical versions. Both reactions are highly chemoselective and can be carried out in the presence of many other functional groups. More important with respect to stereochemistry, each reaction is stereospecific in that the product faithfully reflects the *E* or *Z* configuration of the starting olefin (the nucleophilic epoxidation of α,β-unsaturated carbonyl compounds is an important exception). And one should not underestimate the importance of experimental simplicity: in most cases, one can carry out these reactions by simply adding the often commercially available reagents to a substrate in solvent, without extravagant precautions to avoid moisture or air.

This chapter summarizes a variety of important asymmetric oxidation reactions, concentrating mainly on the most commonly-used examples such as epoxidation and dihydroxylation reactions but also glancing at some paths less traveled (*e.g.*, asymmetric ring-expansion reactions). Throughout our book, since most of these reactions have been thoroughly reviewed, coverage is selective. Once again, the emphasis is on utility and rationales of stereoselectivity.

8.1 EPOXIDATIONS AND RELATED REACTIONS

8.1.1 Early Approaches

Most early approaches to the incorporation of enantioselectivity into oxidation chemistry utilized straightforward chiral variants of the peracids so popular in standard epoxidation reactions; the essential aspects of this work have been summarized [2]. The main difficulties arose from the nature of the transition state in peracid-mediated epoxidations, as illustrated for a simple *trans*-alkene (Scheme 8.1). Regardless of the size differential of the ligands in a chiral peracid R*CO_3H, any stereogenic center on R* is too far away from the developing

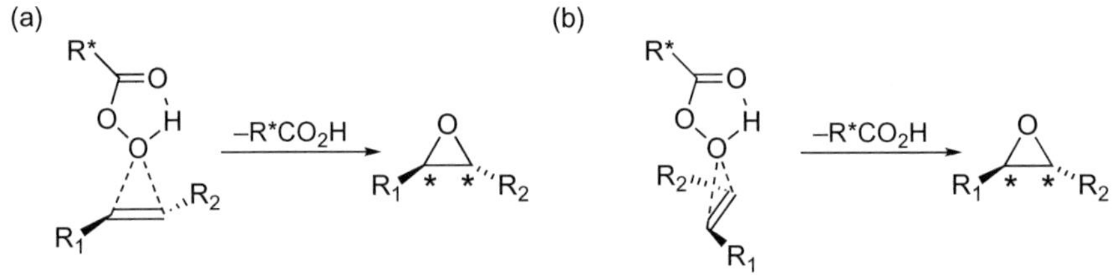

SCHEME 8.1 Generalized illustration of epoxidation of a *trans*-alkene using a chiral peracid; R* = a generic chiral substituent (in early work, monoperoxycamphoric acid was often used). (a) Butterfly and (b) spiro arrangements.

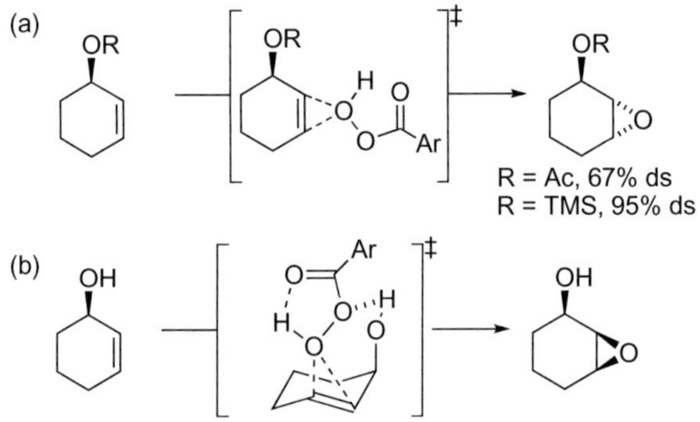

SCHEME 8.2 (a) Addition of *m*-CPBA from the face opposite to the allylic acyloxy or trimethylsilyloxy ligand. (b) Proposed delivery of peracid to the β-face of the substrate mediated by the allylic alcohol group. Other modes of hydrogen bonding have been proposed for this type of reagent delivery [7,8].

stereogenic centers in the epoxide to exert much influence between the two possible transition structures shown in Scheme 8.1. This is true whether the transition structure has the peracid functional group and the developing epoxide in a plane (the butterfly arrangement, Scheme 8.1a) or within planes perpendicular to each other (the spiro arrangement, Scheme 8.1b). Clearly, a transition state in which the chirality in the reagent is closer to the reacting olefin was required.

An important clue as to how this could be attained was provided by Henbest and co-workers [3]. This group compared the diastereoselectivity of peracid oxidation reactions of 3-hydroxy and 3-acyloxycyclohex-2-enes (Scheme 8.2). When the alcohol was capped by an acetate group, the *trans* addition product predominated. Better selectivity was later obtained by placing a larger trimethylsilyl group on the allylic alcohol [4]. In both cases, the source of the selectivity could be ascribed to the approach of the reagent from the least hindered side of the molecule (*anti* to OR; Scheme 8.2a). In contrast, attack was found to occur *syn* to an allylic hydroxy group; obviously, simple steric effects do not account for this result. Instead, it appears that the alcohol is hydrogen bonded to the peracid in the transition state. One possible transition structure for this is shown in Scheme 8.2b; note that the allylic alcohol must occupy a pseudoaxial position to "deliver" the reagent to the olefin. In addition to this stereochemical feature, reagent delivery might be expected to lower the activation barrier of the reaction due to favorable entropy. Thus, rather than achieving selectivity by blocking an unfavorable path relative to an achiral model system, one might affect facial selectivity by enhancing the rate of attack from one face relative to the other. Similar directing effects have been observed in a wide variety of oxidations [5] and other reactions [6].

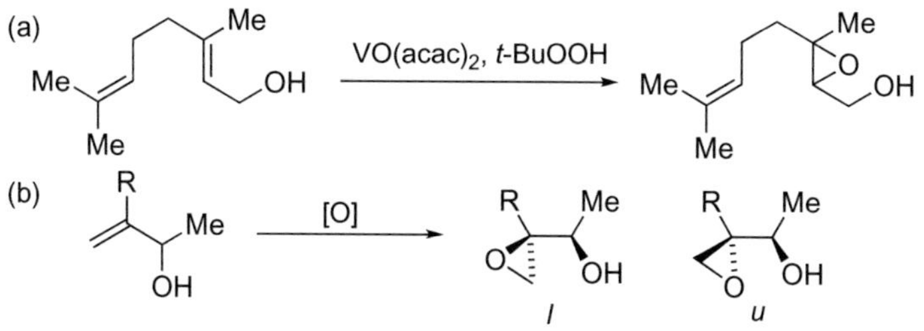

(a) ... VO(acac)₂, t-BuOOH ...

(b) ... [O] ... *l* ... *u*

R	[O]	*l:u* ratio
H	m-CPBA	60 : 40
H	VO(acac)₂, t-BuOOH	20 : 80
CH₃	m-CPBA	45 : 55
CH₃	VO(acac)₂, t-BuOOH	5 : 95

SCHEME 8.3 Some examples of V^{+5}-mediated reactions of allylic alcohols with *t*-BuOOH. (a) A chemoselective reaction [10]. (b) Stereoselective reactions of acyclic allylic alcohols, compared to results obtained using *m*-CPBA [11]. Note that better selectivity is usually obtained using the metal-based oxidation system, but not always with the same relative topicity as observed using a peracid.

This idea was later extended by Sharpless to include epoxidation reactions mediated by transition metals, notably those based on vanadium [8] (for a general review of transition metal mediated epoxidations, see ref. [9]). These diastereoselective epoxidation reactions laid the groundwork for the development of the catalytic asymmetric epoxidation reactions. Thus, soluble metal complexes such as VO(acac)₂ react with simple organic peroxides, such as *tert*-butylhydroperoxide, to form a potent oxidizing system *in situ*. However, an allylic alcohol is *essential* for the oxidation reaction to proceed: other alkenes do not react under similar conditions. Accordingly, a mechanism involving intimate contact between all three components of the reaction around the transition metal was proposed. The various components of the oxidizing system seemed to be close to the reacting olefin in the transition state, as reflected in higher diastereoselectivities relative to peracid oxidations. Some outstanding results were obtained; several chemo- and stereoselective examples are depicted in Scheme 8.3 [8]. The requirement for coordination of an allylic alcohol to the metal and the lack of epoxidation by *t*-BuOOH in the absence of metal guaranteed a substantial rate acceleration for suitable substrates. In addition, this phenomenon allowed the useful chemoselective differentiation between allylic alcohols and other olefins. These experiments set the stage for the development of an efficient asymmetric epoxidation reaction.

8.1.2 Epoxidations

Katsuki—Sharpless asymmetric epoxidation. Upon its introduction in 1980 [12], the Katsuki—Sharpless asymmetric epoxidation (AE) reaction of allylic alcohols quickly became one of the most popular methods in asymmetric synthesis [13—16]. In this work, the metal-catalyzed epoxidation of allylic alcohols described in the previous section was rendered asymmetric by switching from vanadium catalysts to titanium ones and by the addition of various tartrate esters as chiral ligands. Although subject to some technical improvements (most notably the addition of molecular sieves, which allowed the use of low substoichiometric amounts of the titanium—tartrate complex), this recipe has persisted to this writing.

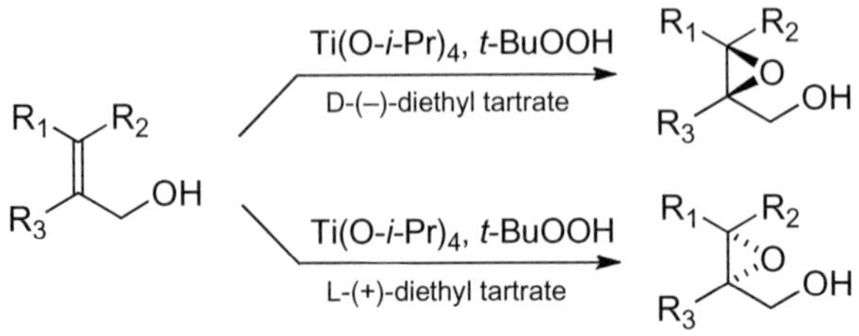

SCHEME 8.4 The asymmetric epoxidation reaction of allylic alcohols generally affords the product epoxides in excellent yields ($>70\%$) and enantioselectivities ($>95\%$). In addition, the reaction is predictable with respect to the predominant enantiomer obtained as shown.

In general, the reaction accomplishes the efficient asymmetric synthesis of hydroxymethyl epoxides from allylic alcohols (Scheme 8.4). Operationally, the catalyst is prepared by combining titanium isopropoxide, diethyl or diisopropyltartrate (DET or DIPT, respectively), and molecular sieves in CH_2Cl_2 at $-20\,^\circ C$, followed by addition of the allylic alcohol or t-BuOOH. After a brief waiting period (presumably to allow the ligand equilibration to occur on titanium), the final component of the reaction is added. The virtues of the AE are obvious. In each case, the components are commercially available at reasonable cost. The availability of tartrate esters in both enantiomeric forms is especially pertinent, allowing the synthesis of either enantiomer of a desired product. A key feature in this regard is the predictability of the enantioselectivity as shown in Scheme 8.4. And the experimental simplicity of standard epoxidation reactions has been effectively retained, since the chiral catalyst system is prepared *in situ*.

A simplified version of the mechanism proposed by Sharpless is given in Scheme 8.5. Early work on the mechanism of the reaction has been reviewed [13]. Evidence in support of this mechanism has included extensive kinetic studies, spectroscopy, molecular weight determinations, and theoretical investigations [15,17−22].[1] A very important aspect of this mechanism, not shown in the scheme, is the formation of the titanium−tartrate species from its commercially available precursors, Ti(O-i-Pr)$_4$ and the dialkyl tartrate. The equilibrium in this step lies far toward the formation of the chiral complex formed; this is critical because the enantioselectivity of the process depends on the absence of any active *achiral* catalyst. Note that the complex as drawn (in the upper left of Scheme 8.5) is dimeric and has a C_2 axis of symmetry. This structure has not been isolated in the solid state, but is similar to that observed in the X-ray structure of a related tartramide complex [24].

Without specifying the order of events, two isopropoxide ligands must be replaced by one molecule of peroxide and one molecule of allylic alcohol to give the species shown in the upper right of Scheme 8.5 (recall that, in reality, the peroxide and allylic alcohol are added at different times). The ease of such ligand exchange reactions in these titanium complexes largely accounts for their utility here. At this point (lower right of Scheme 8.5), the complex is fully loaded and ready for oxygen transfer to the alkene. In this mechanism, the allylic alcohol occupies a position *cis* to the reactive peroxide oxygen. In the AE reaction ($R_{Si} = R_{Re} = H$), the diastereofacial selectivity of the olefin in the complex results from the avoidance of the allylic carbon and a carboxylic ester (Figure 8.1b). After oxygen transfer, the final step is the exchange of the reaction products, epoxy alcohol and t-BuOH, with other

---

1. An alternative mechanism involving a monomeric complex has also appeared [23].

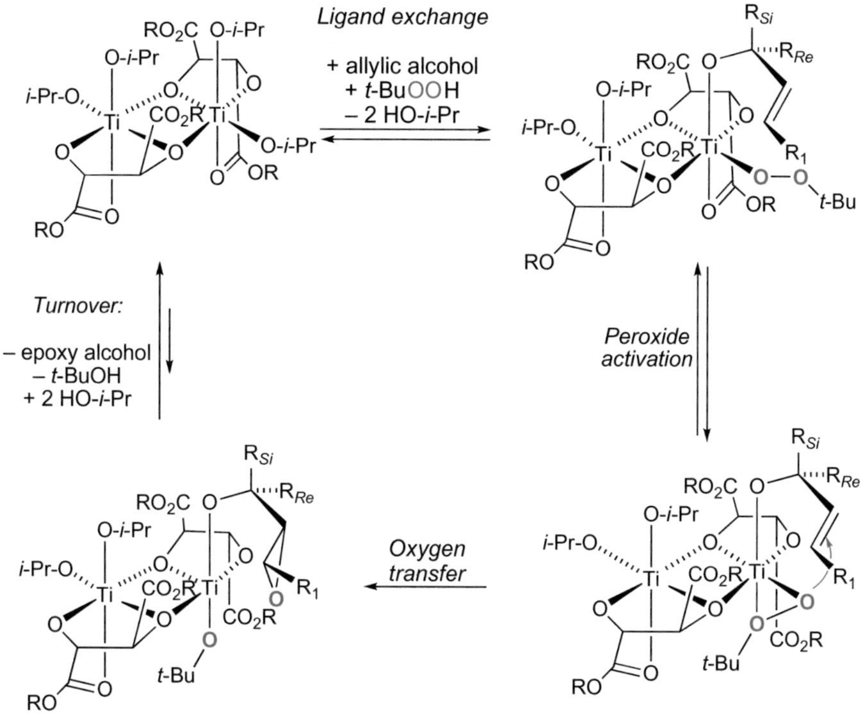

SCHEME 8.5 Proposed mechanism for the Sharpless asymmetric epoxidation reaction of allylic alcohols, shown here for a simple *trans*-allylic alcohol and L-(+)-tartrate. For the AE reaction, $R_{Si} = R_{Re} = H$. When one (or occasionally both) of these substituents are alkyl groups, the Scheme pertains to the kinetic resolution sequence described in the next section.

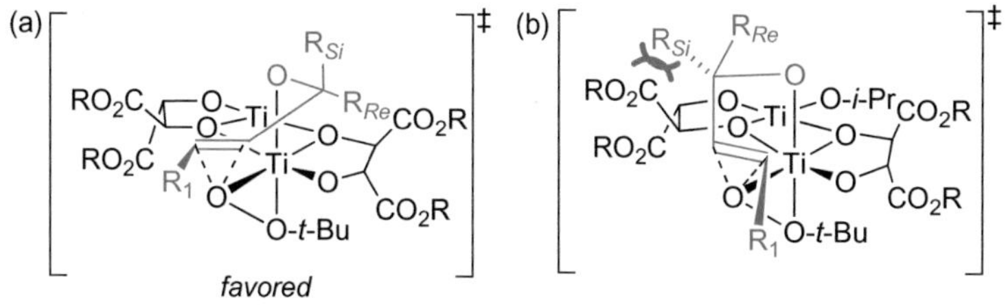

FIGURE 8.1 Proposed steric interactions leading to enantioselectivity in the Sharpless AE reaction.

ligands to give either the starting complex or some other species on the way to the loaded catalyst. The importance of turnover must not be underappreciated, for without it one may have a reagent but never a catalyst.

This model is consistent with much that is known about the scope of the Sharpless AE. The most common and best-behaved substrates are simple *trans*-allylic alcohols; their reactions are generally fast and reliably give products with very good enantioselectivity (>95% es). Inspection of the loaded complex in Figure 8.1 might suggest that substrates with an alkyl group *cis* to the hydroxymethyl substituent (*i.e.*, where $R_2 \neq H$ in Scheme 8.4) may be less stable due to steric interactions with the main portion of the catalyst. Indeed, such compounds are the slowest reacting and subject to the most variation in levels of enantioselectivity. However, there are examples of excellent results using alkenes of every conceivable type, although some work may need to be invested in optimizing reaction conditions (Table 8.1). The epoxidation of the allylic alcohol in the presence of another olefin in entry 9 exemplifies the chemoselectivity that is inherent in this reaction.

TABLE 8.1 Examples of Sharpless AE Reactions[a]

Entry	Product	Tartrate	% Yield	% es	Ref.
1		(−)-DIPT	50−60	94−96	[25]
2		(+)-DET	85	97	[25]
3		(+)-DET	54	83	[26]
4		(+)-DIPT	63	>90	[25]
5		(+)-DET	88	97	[25]
6		(−)-DIPT	87	95	[27]
7		(+)-DET	77	96	[25]
8		(+)-DET	80	94	[12]
9		(+)-DET	95	95	[25]
10		(+)-DET	Not reported	>95	[28]

[a]These reactions were carried out under catalytic conditions (<10 mol% of Ti(OR)₄ and tartrate), except for entry 8 (done using stoichiometric catalyst).

Asymmetric epoxidation reactions of simple olefins. Since the discovery of the Katsuki-Sharpless AE reaction, a major goal became to obviate the need for an allylic alcohol. Attempts to carry out asymmetric epoxidation reactions on simple olefins began with the ill-fated chiral peracid approaches noted in Section 8.1.1. In addition, numerous attempts using transition-metal-containing catalysts such as porphyrins as stoichiometric chiral reagents have been reported, with various degrees of success (peroxides, dioxiranes, and oxaziridines). These approaches have been summarized [29]. We concentrate here on the two catalytic methods that have found the most use by practicing synthetic organic chemists: the salen-catalyzed reactions of activated oxygen donors and reactions of chiral dioxiranes.

The introduction of salen catalysts for epoxidation reactions was a major breakthrough in the field, with the key papers being published by Jacobsen [30,31] and Katsuki in 1990 and 1991 [32–34]; for reviews, see refs. [29,35–37]. The Jacobsen–Katsuki epoxidation uses chiral and typically C_2-symmetric (salen)Mn complexes, such as those shown in Scheme 8.6 (the initial discovery of salen-based catalysis in oxidation reactions is generally attributed to Kochi [38]). The chiral salen catalysts are very easily prepared by the condensation of a chiral diamine with a substituted salicylaldehyde, followed by coordination of the metal. The ready availability of both components and the swift synthesis of the target complexes permit

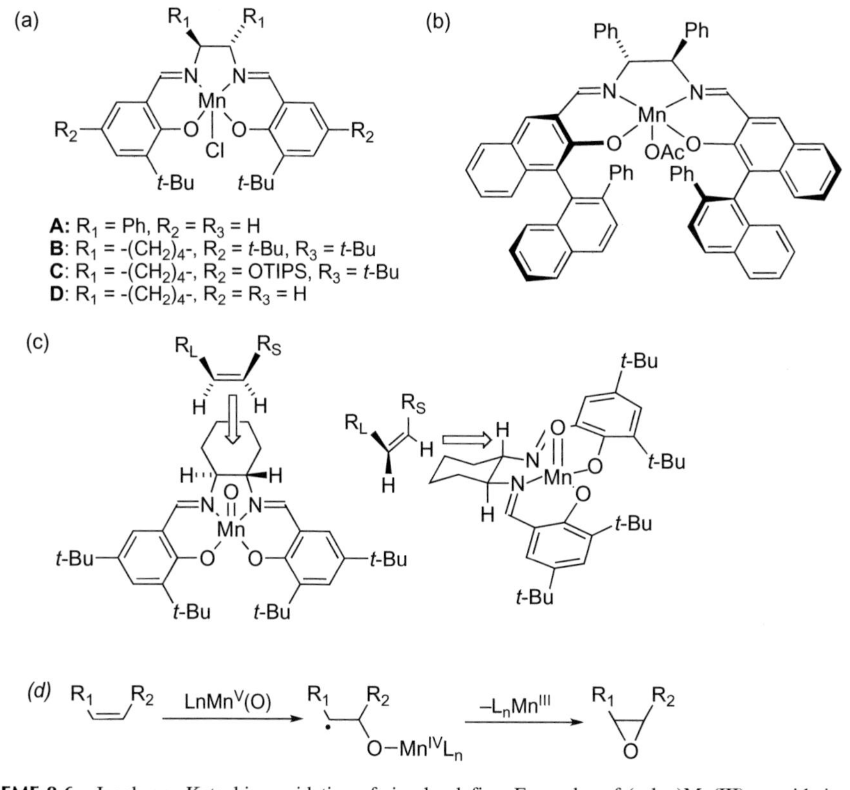

SCHEME 8.6 Jacobsen–Katsuki epoxidation of simple olefins. Examples of (salen)Mn(III) epoxidation catalysts prior to reaction with NaOCl by (a) Jacobsen [29–31,41,44] and (b) Katsuki [32–34]. (c) Two views of the proposed side-on approach of a generic *cis* olefin to the loaded catalyst. (d) Proposed stepwise mechanism of the reaction [44].

easy access to a great many catalyst variations, which facilitates reaction optimization. The starting Mn(III) complex is subjected to *in situ* oxidation with the stoichiometric oxidant, usually NaOCl (bleach!). The use of this inexpensive and relatively safe oxidant is another virtue of this system.

Although many outstanding results have been obtained, there are some limitations to the scope of this process (Table 8.2). The reaction works best with *cis*-olefins and affords the highest selectivities with conjugated, preferentially cyclic olefins. Further improvements resulted in a substantial broadening of this profile, obtaining some good-to-excellent selectivities from styrene [39] and certain tri- and tetra-substituted olefins (especially those that are not subject to isomerization due to symmetry or by constraining the double bond in a ring) [40,41].

TABLE 8.2 Examples of Jacobsen AE Reactions[a]

Entry	Olefin	Catalyst	% Yield (*cis/trans*)	% es	Ref.
1	Ph / Me	A	71 (*trans* only)	60	[30]
2	Ph Me	B	84 (92:8)	96 (*cis*), 92 (*trans*)	[29,31]
3	Ph Me	C[b]	Not reported (5:95)	90 (*trans*)	[44]
4	NC—chromene, O, Me, Me	B	96	98	[29,31]
5	Me₃Si—alkyne—cyclohexyl	B	65 (16:84)	82 (*cis*), 99 (*trans*)	[45]
6	cyclooctene ring	B	73	82	[46]
7	Ph, Me / Ph	B	87	94	[40]
8	Br—chromene, O, Me, Me, Me, Ph	D	72	90	[41]
9	Ph, Ph / Ph, Me	A	12	72	[41]

[a]See Scheme 8.6a for catalyst structures.
[b]A cinchona alkaloid additive was used.

Jacobsen proposed the approach vector for substrate shown in Scheme 8.6c, with selectivity arising from the minimization of steric interactions when the smaller alkene substituent R_S is closer to the *syn* axial hydrogen atom of the catalyst than R_L [31]. While Katsuki proposed a different approach from the side of the catalyst (not shown, ref. [32]), both models are consistent with the observed sense of stereoinduction (this issue is extensively considered in ref. [42]). Some acyclic *cis*-olefins were found to afford various amounts of *trans* epoxides, one observation that led to a radical mechanism being proposed for some substrates and supported through a radical probe study (Scheme 8.6d [43]). This isomerization could be facilitated by the addition of chiral quaternary ammonium salts, leading to synthetically useful ($> 10:1$ *trans:cis*, $> 90\%$ es) conversions of *cis*-olefins to *trans*-dialkyl epoxides (*cf.* entries 2 and 3 in Table 8.2) [44].

The 1990s also saw significant efforts toward the development of chiral dioxiranes for epoxidations (reviews: refs. [47−49]). Dioxiranes are highly active oxidizing agents that can be generated *in situ* by reacting a ketone precursor with inorganic Oxone, a commercially available form of potassium peroxomonosulfate ($KHSO_5$) formulated as $2KHSO_5 \cdot KHSO_4 \cdot K_2SO_4$. In contrast to simple peracids, the 3-membered ring of a dioxirane places the elements of the chiral ketone precursor closer to the reacting olefin, and in principle high enantioselectivity can result (Scheme 8.7a). In practice, the development of appropriate catalysts proved challenging, with the carbohydrate-derived system developed by Shi being the predominant version used in laboratories today [50−52]. The design features that led to this particular ketone, which is prepared from fructose [50], include highly differentiated α and β faces of the ring system, the presence of σ electron-withdrawing groups adjacent to the carbonyl to facilitate the formation of the dioxirane group, and the confinement of the latter in fused rings or upon a quaternary center to avoid epimerization of the ketone during the course of the reaction. In practice, the Shi ketone is mixed with Oxone under buffered conditions of pH *ca.* 7−10, found to minimize decomposition of the Oxone on one hand and to avoid the destruction of the catalyst *via* a Baeyer−Villiger reaction on the other [51]. Table 8.3 shows that the Shi epoxidation works well with a variety of olefin types, notably providing needed complementarity to the Jacobsen epoxidation due to its utility with *trans*-substituted olefins. Although the reaction does succeed with electron-deficient olefins (Table 8.3, entry 10 − more about those shortly), it preferentially reacts with a more electron-rich one when there is a choice (entry 8). While the enantiomer of the catalyst shown is most easily prepared from D-fructose, L-fructose is available from sorbose and can be used to prepare the opposite enantiomer of the Shi catalyst [53]. Additionally, other ketones and means of generating the active species have been studied [48].

The reaction is generally considered to proceed with a more or less concerted transfer of the active oxygen from the dioxirane to the substrate *via* one of the transition states shown in Scheme 8.7c. Shi has rationalized the preferred enantioselectivity of the reaction in the context of the spiro transition structure in which nonbonded interactions are minimized. One of several disfavored spiro structures shows an unfavorable steric interaction between one of the alkene substituents and a nearby oxygen atom on the catalyst, while one of the four analogous planar transition states is also depicted. This planar structure was proposed to be on the predominant minor pathway and presumably disfavored for subtle electronic and structural reasons. Kinetic isotope effects and calculations have been carried out by Singleton, who has emphasized the importance of reaction asynchronicity in rationalizing the results [60].

Epoxidation of electron-deficient olefins. In contrast to the oxidation reactions of electron-rich olefins just described, highly selective nucleophilic epoxidation reactions of α,β-unsaturated carbonyl compounds were slower to appear on the scene (Scheme 8.8, for reviews, see refs. [61−65]).

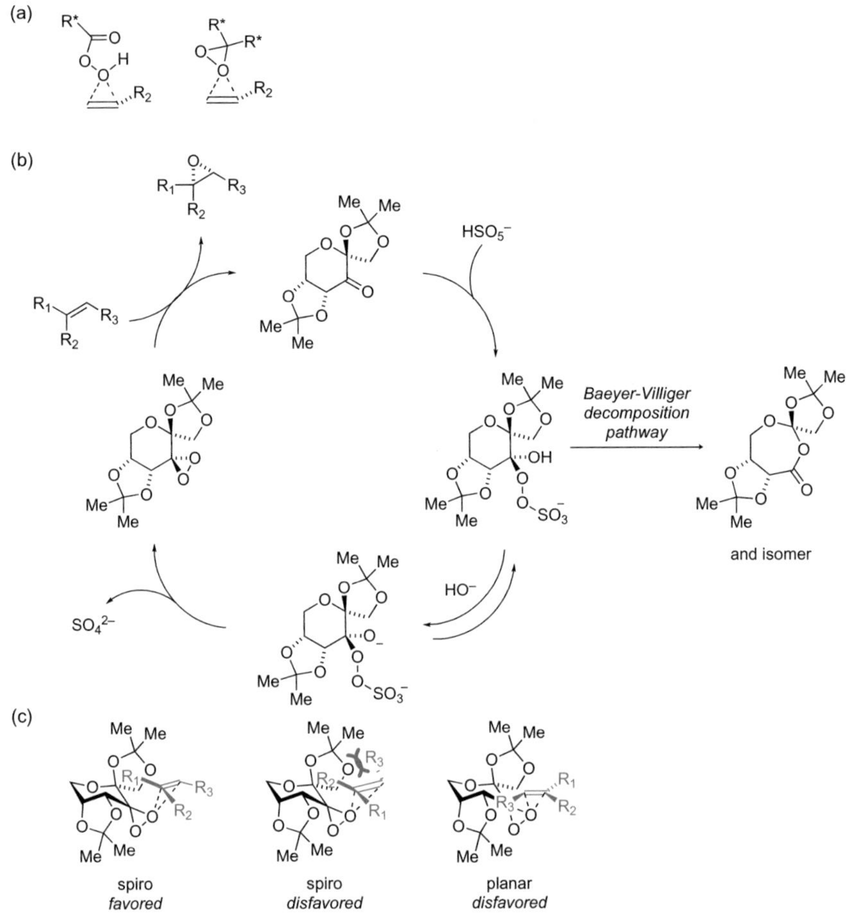

SCHEME 8.7 (a) Comparison of generalized transition structures for peracid-mediated *vs.* dioxirane-mediated epoxidations. (b) Shi catalyst structure and catalytic cycle, including the Baeyer–Villiger pathway for catalyst decomposition. (c) Comparison of three possible orientations leading to epoxidation. Two spiro pathways are shown (out of four), with the leftmost one being the most favorable. The one planar orientation shown (again, out of four possible) is cited as the second-most favored overall pathway; it leads to the opposite enantiomer of that resulting from the most favored spiro pathway. For a full discussion, see ref. [54].

The most successful early attempts were carried out on chalcones, using standard basic peroxidation conditions with additives such as a quinine-derived phase-transfer catalyst first reported by Wynberg in 1976 [66] and explored by others (*e.g.*, Scheme 8.8a [67]). This constituted an early example of organocatalysis, as were reactions promoted by the unusual catalyst poly-L-leucine, first reported in 1980 and now known as the Juliá–Colonna epoxidation [68]. Mechanistically, epoxidations of most electron-poor olefins do not take place by concerted (if highly asynchronous) bond formation across both olefin carbons typical for other epoxidations. Instead, most occur by asymmetric conjugate addition of a peroxide anion followed by nucleophilic ring closure onto oxygen, which is now acting as an electrophile (Scheme 8.8b). When this happens, the configuration of the double bond is lost and *trans* epoxide is normally formed; in such cases the reaction is stereoselective but *not* stereospecific. Although exceptions exist

TABLE 8.3 Shi Epoxidation Reaction

Entry	Substrate	% Yield	% es	Ref.
1[a]	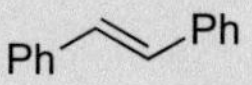 Ph—CH=CH—Ph	85	99 (R,R)	[52]
2[a]	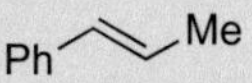 Ph—CH=CH—Me	94	98 (R,R)	[52]
3[a]	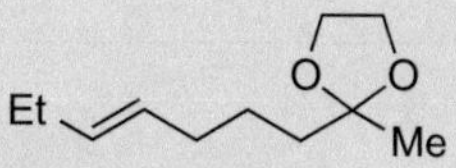 Et—CH=CH—CH₂CH₂—C(O—CH₂CH₂—O)—Me (cyclic acetal)	92	96 (R,R)	[52]
4[a]	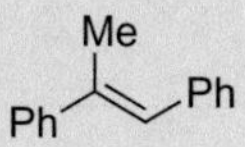 Ph—C(Me)=CH—Ph	89	98 (R,R)	[52]
5[a]	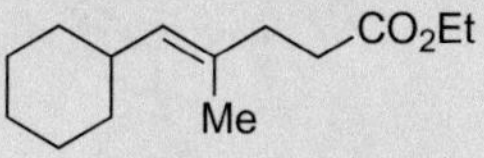 cyclohexyl—CH=C(Me)—CH₂CH₂—CO₂Et	89	97 (R,R)	[52]
6[b]	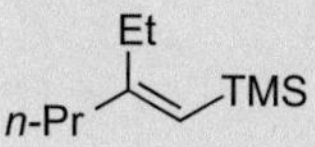 n-Pr—C(Et)=CH—TMS	51	95 (R,R)	[55]
7[a]	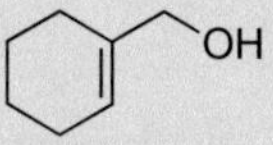 cyclohexene—CH₂OH	93	97 (R)	[56]
8[a,c]	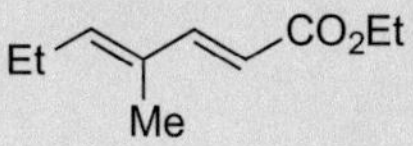 Et—CH=C(Me)—CH=CH—CO₂Et	82	97 (R,R)	[57]
9[a]	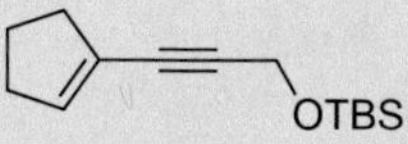 cyclopentene—C≡C—CH₂OTBS	97	88 (R,R)	[58]
10	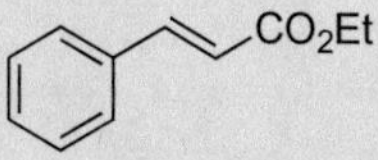 phenyl—CH=CH—CH₂—CO₂Et	62[d]	91 (R,R)	[59]

[a]Conditions: Ketone (0.3 equiv), Oxone (1.38 equiv), K₂CO₃ (5.8 equiv), MeCN–DMM–0.05 M Na₂B₄O₇·10H₂O of aq. Na₂EDTA (1:2:2, v/v).
[b]Conditions: Ketone (0.65 equiv), Oxone (1.38 equiv), K₂CO₃ (5.8 equiv), MeCN–DMM–0.05 M Na₂B₄O₇·10H₂O of aq. Na₂EDTA (1:2:2, v/v).
[c]Epoxidizes across the γ, δ-unsaturated double bond.
[d]Conditions: Ketone (0.3 equiv), Oxone (3 equiv), K₂CO₃ (6 equiv), dioxane/H₂O, 6 h, rt.

(obviously for cyclic substrates but also in a few acyclic examples [69,70]), the stereospecificity or lack thereof of a given reaction is generally taken as evidence for a concerted or a two-step mechanism, respectively. In a two-stage mechanism, the absolute configuration can in principle be set in the first step *via* a kinetically controlled enantiofacially selective conjugate addition reaction. However, in his extensive explorations of the Juliá–Colonna epoxidation [62,71], Roberts' analysis showed that this first step was reversible and that the high stereoselectivity could be best explained by a kinetic resolution of the initially formed β adduct isomers through a stereoselective ring closure of the substrate bound to the leucine oligomer, which adopts a highly helical structure in solution (Scheme 8.8c). Although

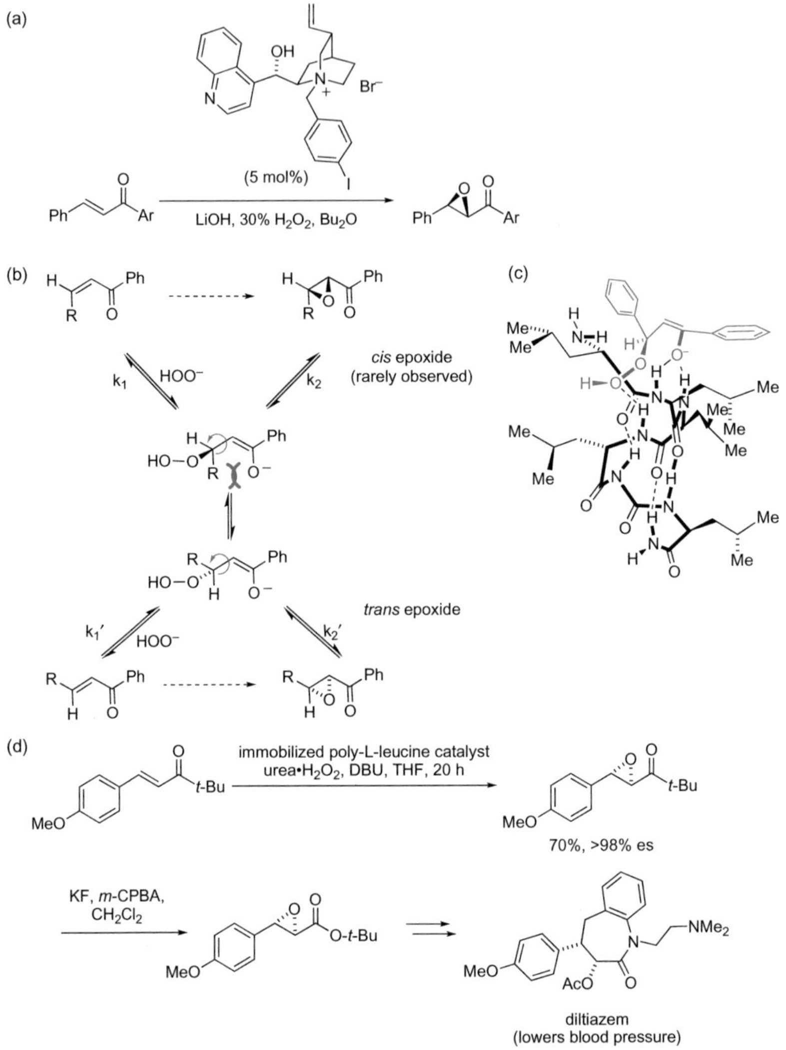

SCHEME 8.8 Nucleophilic epoxidation reactions of enones using (a) phase transfer catalysis [67] or the Juliá–Colonna epoxidation, for which is shown (b) the kinetic profile, (c) Roberts' proposed model for the helical peptide complexed to a β-peroxy adduct (shown in red), and (d) an example used in the synthesis of diltiazem [73].

limited to α,β-unsaturated ketones, the products can often be converted to the corresponding α,β-epoxy esters *via* a regioselective Baeyer–Villiger oxidation, a reaction that was central to the use of the Juliá–Colonna epoxidation in the synthesis of the blood pressure regulator diltiazem (Scheme 8.8d [72]).

More recent work has centered on catalysts that activate the electrophilic alkene to nucleophilic attack; although little mechanistic information is available, it is reasonable to expect that the stereoselectivity of these reactions occurs *via* a kinetically controlled addition of the peroxide to the α,β-unsaturated ketone. An early selective example was published by Jackson using a tartrate ligand (Scheme 8.9a [74]) who later proposed the

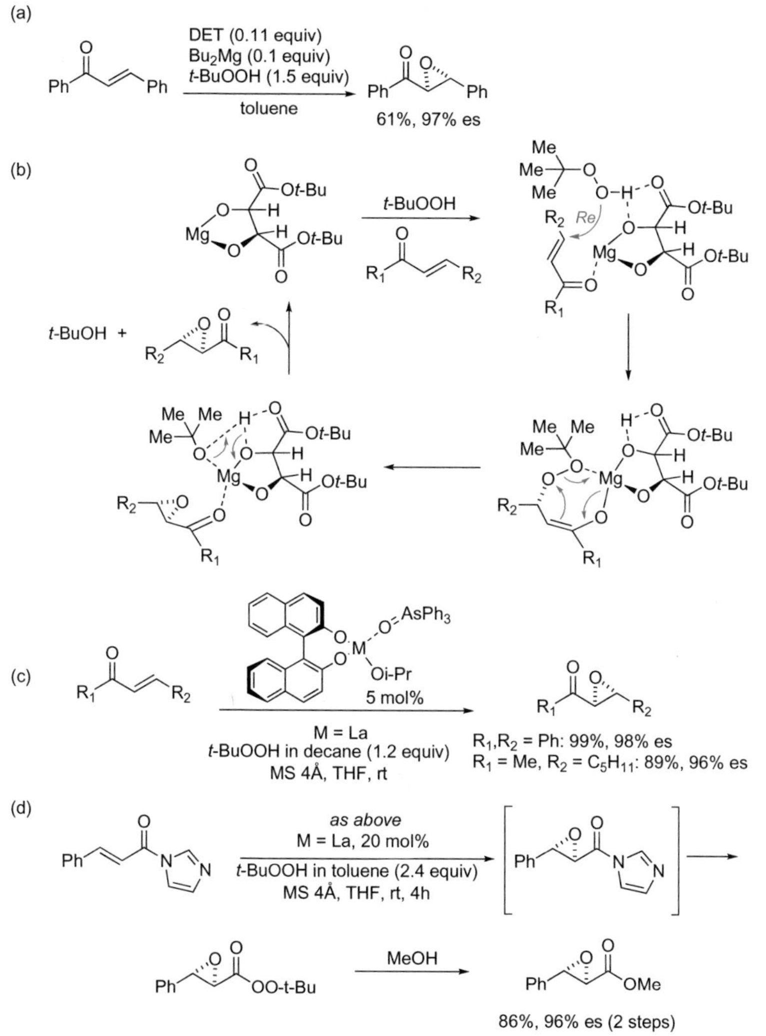

SCHEME 8.9 Metal-promoted epoxidations of electron-poor olefins by (a) [74], (b) [75], (c) [76], and (d) [77].

magnesium-coordinated delivery of *tert*-butylhydroperoxide as shown in Scheme 8.9b [75]. Probably the most general approach to this problem is Shibasaki's BINOL coordinated lanthanides, which were first applied to chalcone epoxidation (Scheme 8.9c, ref. [76]) and later extended to other substrate types [77,78]. Note the unusual use of a triarylarsine as a ligand; this proved more effective than other additives. This catalyst did not initially give good results with α,β-unsaturated esters. The apparent need for steric bulk on either side of the carbonyl group led to the ingenious work-around shown in Scheme 8.9d [77]. Here, a imidazoyl-substituted substrate was converted under the reaction conditions by *tert*-butyl hydroperoxide to a mixed perester that could be subsequently converted to the methyl ester. Later papers described the versions that could be used for the direct and selective epoxidation of α,β-unsaturated amides [78] and esters [79].

8.1.3 Sharpless Kinetic Resolution

Inspection of the mechanism in Scheme 8.5 suggests that the Sharpless epoxidation should be relatively insensitive to configuration of any stereocenter in an alkene substituent with one very important exception: the allylic carbon bearing the alcohol. Indeed, good diastereoselectivity was often obtained in reactions of various chiral allylic alcohols with achiral epoxidizing agents (Scheme 8.3). Substitution at this particular position is important because of its proximity to the bulk of the catalyst. Thus, one might expect substitution at R_{Si} to be well-tolerated because this group points away from the catalyst, whereas R_{Re} should be much more sterically encumbered (Figure 8.2). Some experimental observations that address this issue and permit the application of the Katsuki−Sharpless catalyst to kinetic resolution reactions are shown in Scheme 8.10 [80].

Like the vanadium-based catalysts, the Sharpless AE system intrinsically favors 1,2-*anti* products; this is because the cyclohexyl group in Scheme 8.10a occupies the position denoted by group R_{Si} in Figure 8.2, away from the catalyst. In fact, this diastereoselectivity is somewhat amplified relative to achiral titanium catalysts. When the *S* allylic alcohol is used with (+)-DIPT, a matched pair results (Scheme 8.10a). The strong enantiofacial selectivity of the L-(+)-DIPT catalyst clashes with the *R* substrate's resident chirality (this is the case shown in Figure 8.2 with R_{Re} = cyclohexyl). In this mismatched pair, the preference of the chiral catalyst for α attack moderately exceeds that of the allylic alcohol for 1,2-*anti* product (Scheme 8.10b). The most important consequence is that *the mismatched reaction is 140 times slower than the matched one*. That is, the selectivity factor, s = 140 (see Section 1.7).

Using a racemic allylic alcohol, one can take advantage of this rate differential to selectively epoxidize the more reactive *S* isomer in the presence of its enantiomer. This procedure is known as a Sharpless kinetic resolution (KR) [80]. The KR has very wide applicability for the preparation of both 1,2-*anti* epoxy alcohols and the unreacted allylic alcohol, often with very high enantioselectivities (note that the diastereomeric 1,2-*syn* series is not generally available by this technique). In general terms, carrying out the reaction to lower conversions will maximize the enantiomeric purity of the epoxide, while greater conversions sometimes lead to very high (>99%) enantiomeric purities of the allylic alcohol, albeit in a reduced yield.[2] Scheme 8.10c shows an example of what is possible under optimized conditions with a favorable substrate.

The KR procedure is not limited to making simple epoxides bearing an adjacent stereogenic center. Figure 8.3 depicts several interesting classes of molecules that have been

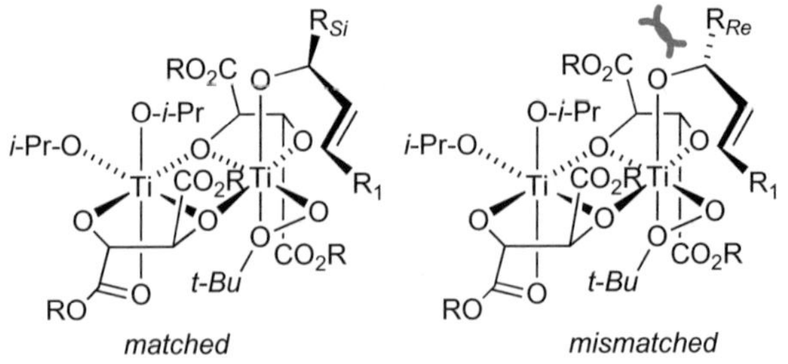

FIGURE 8.2 Origins of selectivity in the Sharpless kinetic resolution.

---

2. Interested readers are directed to the original literature for a quantitative treatment [15,80].

resolved using KR procedures. Although results have been spotty, alternative sites of oxidation have included attempts with alkynes, furans, and β-amino alcohols. Of particular interest to stereochemistry buffs are procedures that result in different classes of enantiomerically pure compounds, such as those with axial chirality (cycloalkylidenes or allenes) or planar

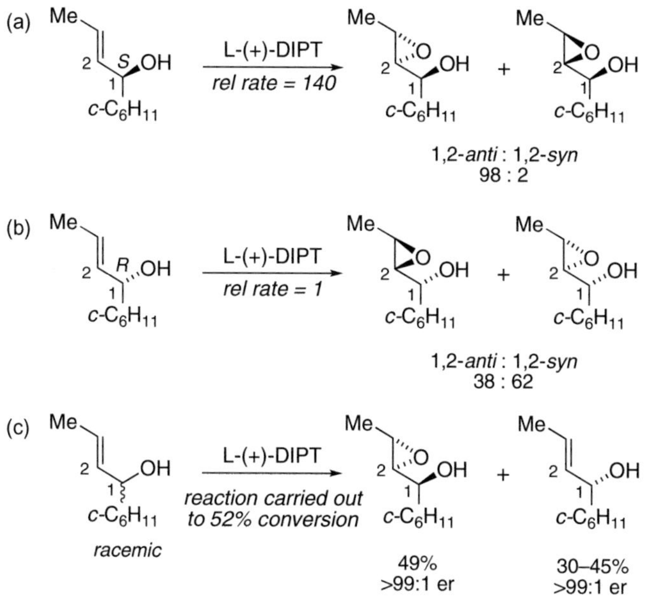

(a) Me S OH / L-(+)-DIPT, rel rate = 140 → 2 ""O OH + 2 O OH (c-C₆H₁₁)

1,2-anti : 1,2-syn
98 : 2

(b) Me R "OH / L-(+)-DIPT, rel rate = 1 → 2 O "OH + 2 "O OH (c-C₆H₁₁)

1,2-anti : 1,2-syn
38 : 62

(c) Me OH (c-C₆H₁₁) racemic / L-(+)-DIPT, reaction carried out to 52% conversion → 2 ""O OH (49% >99:1 er) + 2 "OH (30–45% >99:1 er)

SCHEME 8.10 Reactions of a chiral allylic alcohol under Sharpless epoxidation conditions (Ti(O-i-Pr)₄, t-BuOOH) using the chiral tartrates given (DIPT = diisopropyltartrate, ref. [80]). (a) The "matched" case, in which the preferred approach of the asymmetric catalyst and the diastereoselectivity of the substrate are the same. (b) The "mismatched" case. (c) An example of a Sharpless kinetic resolution (KR).

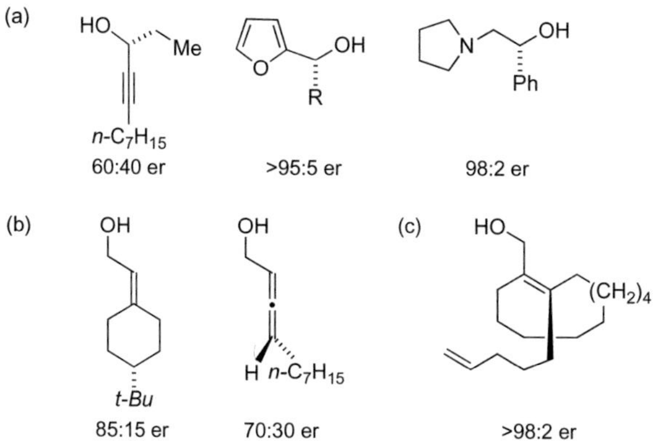

(a) HO "Me / n-C₇H₁₅ / 60:40 er — O R OH / >95:5 er — N OH Ph / 98:2 er

(b) OH t-Bu / 85:15 er — OH H n-C₇H₁₅ / 70:30 er — (c) HO (CH₂)₄ / >98:2 er

FIGURE 8.3 Examples of molecules prepared in enantiomerically enriched form using Sharpless KR procedure. (a) Compounds having alternative sites of oxidation; the enantioenriched (unreacted isomers) products of kinetic resolutions of an acetylene [82], a furan [83], and an amine [84] are shown. (b) Compounds bearing axial chirality [82]. (c) An alkene with planar chirality following KR enrichment [85].

chirality. Finally, some success has been seen in extending both the AE and KR procedures to homoallylic alcohols [81].

8.1.4 Some Applications of Asymmetric Epoxidation and Kinetic Resolution Procedures

The influence of asymmetric epoxidations has been far too pervasive to allow even a partial representative listing here. However, it's worth recalling a few illustrations of the power of epoxidation chemistry in organic synthesis.

Carbohydrate synthesis. Save the all-important hydroxymethyl group that the titanium reagent uses as its handle, the Sharpless AE is remarkably insensitive to stereogenic centers extant in the substrate. This has led to the wide use of this system for *reagent-based stereocontrol*, wherein the chirality of a new stereocenter is determined simply by pulling the appropriate reagent off the shelf (as opposed to *substrate control*, in which a new element of chirality is installed under the influence of those already in the reactant; see Section 1.6). This strategy was beautifully illustrated by the synthesis of all eight isomeric hexoses in their unnaturally-occurring L-series, summarized for L-allose in Scheme 8.11 [86,87]. This iterative sequence prepares the target carbohydrate in the C-6→C-1 direction and starts with a readily prepared *trans*-allylic alcohol. The first AE directly sets the C-5 stereogenic center (carbohydrate numbering), now requiring that the epoxide be opened in a regio- and stereoselective manner and that the primary alcohol be converted to the oxidation state of an aldehyde. Both tasks were accomplished by a Payne rearrangement using base, which isomerized the epoxy alcohol with inversion at the C-4 center. The new epoxide thus formed is monosubstituted and therefore suffers a kinetically favored attack by an external nucleophile, in this case the thiophenolate anion.[3]

Next, the researchers took advantage of the acetonide protecting group to control the relative configuration between C-4 and C-5 (the use of a protecting group for this kind of stereochemical finesse has been termed *ancillary stereocontrol* [89]). A mild, nonbasic unraveling of the aldehyde by reduction at the acetate carbonyl group was accomplished with diisobutyl aluminum hydride, which left the target in its initial *cis* configuration about the 5-membered ring. Alternatively, basic deprotection led to epimerization to the *trans* isomer. This was possible in this case because of poor overlap between the aldehyde enolate and the σ* orbital associated with the C-5 carbon−oxygen bond − otherwise β-elimination of one of the acetal oxygens might have occurred. Also, using an epimerization step obviated the necessity of preparing and working with the less reactive *cis*-olefins to switch between diastereomeric series. Overall, the conversions in Schemes 8.11a and b constitute a single iteration of the synthesis.

Scheme 8.11c shows how the aldehyde could be homologated to a new allylic alcohol and how simple choice of tartrate ligand afforded the diastereomeric epoxides shown, since the AE process effectively ignores the resident stereocenter in the new substrate. This is the essence of reagent-controlled synthesis: the utilization of a tool for enantioselective elaboration to permit the selective synthesis of diastereomeric compounds. Once prepared, the utilization of the diisobutyl aluminum hydride variant of the iterative sequence followed by final deprotection steps led to the synthesis of L-allose. A useful exercise is to arbitrarily draw an

---

3. Although known prior to the discovery of the Sharpless AE, this use of the Payne rearrangement is a good example of how the availability of a particular functional array by asymmetric synthesis provoked a reaction's further development [88]. In this case, the product sulfide allows the chemoselective conversion of this carbon to the oxidation state of the aldehyde, in the guise of an acetoxy sulfide.

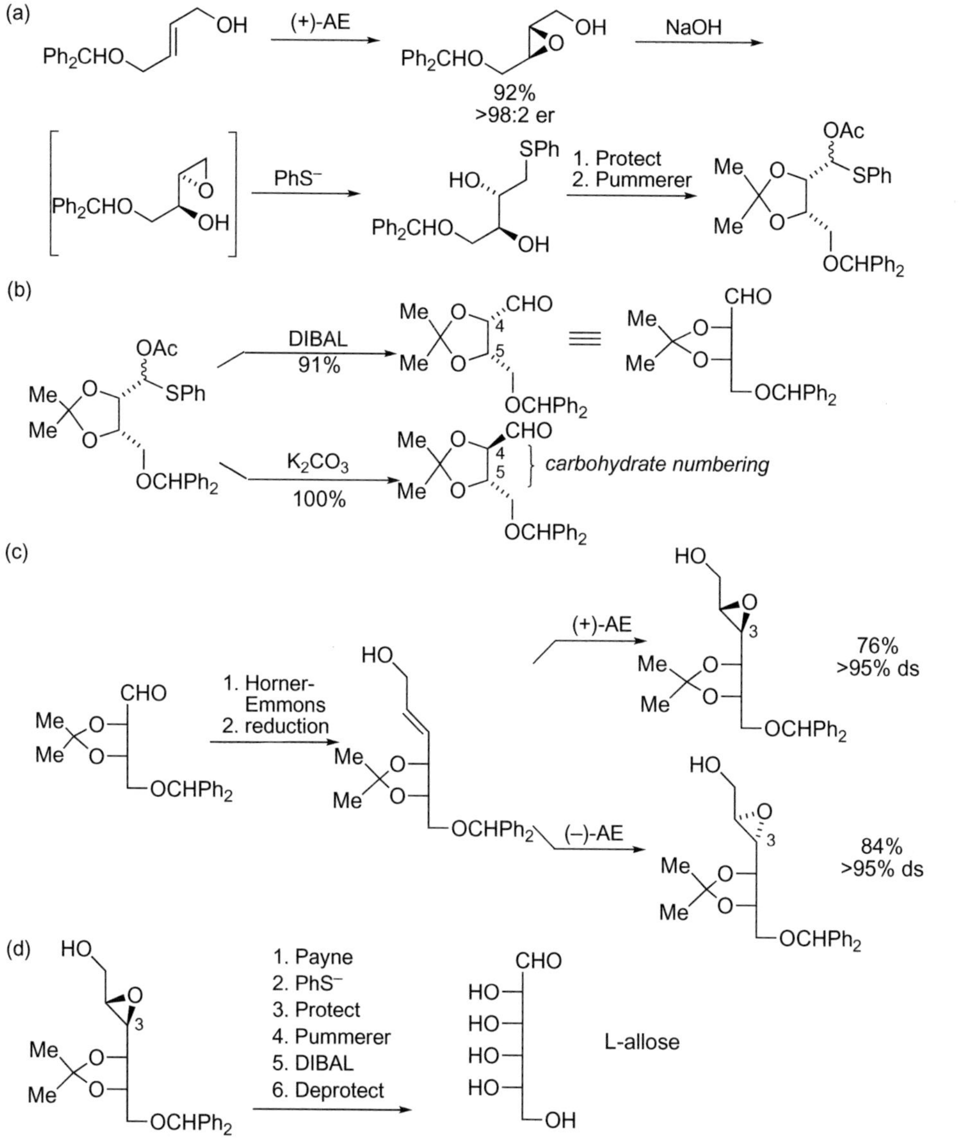

SCHEME 8.11   Reagent-controlled synthesis of L-allose ((+)-AE = Sharpless AE using L-(+)-DIPT; (−)-AE = Sharpless AE using D-(−)-DIPT). (a) A Sharpless AE followed by Payne rearrangement and oxidation. (b) Stereodifferentiation of the C-4 and C-5 stereocenters. (c) Chain extension followed by reagent-controlled oxidation of the olefin. (d) Completion of the synthesis.

isomer of allose and synthesize it using this technique (on paper, of course), or to imagine a modification that would lead to the corresponding pentoses [90].

   Group-selective reactions of divinyl carbinols. Recall that the reagent control strategy is inapplicable to situations where the resident chirality is on the allylic position bearing the

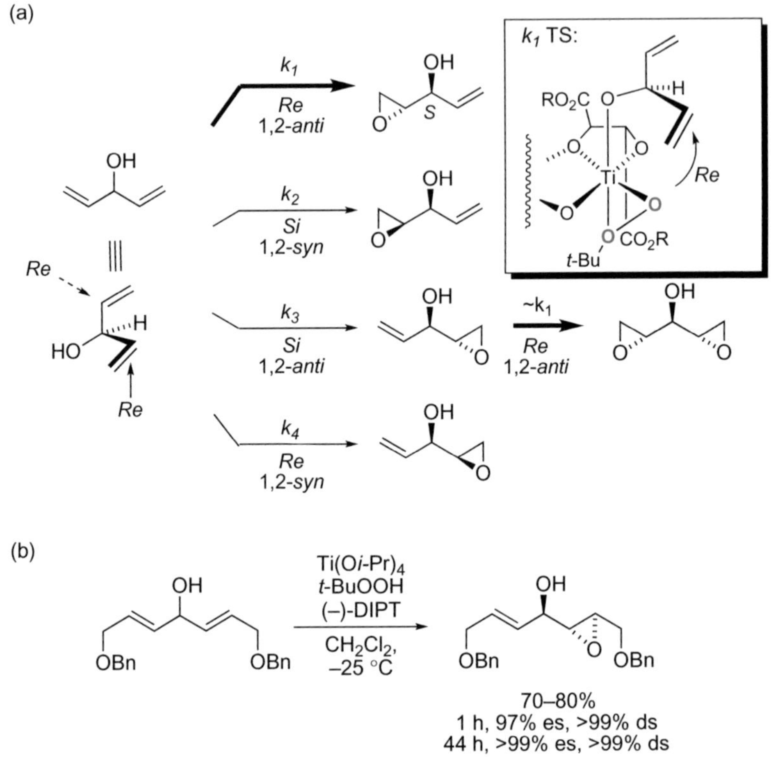

(a)

(b)

70–80%
1 h, 97% es, >99% ds
44 h, >99% es, >99% ds

SCHEME 8.12 Reaction of divinyl carbinol under (+)-AE conditions as an example of enantiotopic group selectivity in epoxidation chemistry. Matched cases of enantiofacial selectivity are shown with bold arrows, *via* the k_1 partial transition structure shown in the inset (see also Scheme 8.5). Qualitative rate differences are on the order $k_1 \gg k_2$, $k_3 \gg k_4$ (without specifying an order for k_2 vs. k_3). Note that the products arising from the pairs k_1/k_3 and k_2/k_4 are enantiomers. (a) Theoretical treatment and (b) an experimental example [94].

hydroxyl "handle" for the catalyst. However, the preference for 1,2-*anti* product has been cleverly applied to a problem in diastereotopic group selectivity (Scheme 8.12) [91–95]. The two olefins carry a total of two enantiomeric pairs of diastereotopic faces. When a tartrate–titanium epoxidation system is allowed to react with this substrate, approach to only one of these four faces simultaneously satisfies the requirements of both the catalyst (which prefers the *Re* face) and the substrate (which prefers 1,2-*anti* addition). To the extent that the rate of epoxidation at this face exceeds that of the others (k_1 in Scheme 8.12), one product predominates. Minor diastereomers result from pathways k_2 and k_4. However, note that the pathway with a rate of k_3 (mismatched: 1,2-*anti* diastereoselectivity combined with disfavored *Si* enantiofacial attack) affords the enantiomer of the major isomer.

Schreiber has noted that this group-selective process is expected to provide products with very high enantioselectivity because the disfavored enantiomer resulting from pathway k_3 *still has the most favorable face available for reaction* [94]. Thus, to the extent that product from pathway k_3 accumulates, it is rapidly siphoned off at a rate comparable to k_1. Thus, the problem of enantiomer separation at the end of the reaction can largely be replaced by the problem of diastereomer and side product separation (although itself never an issue to be taken lightly!). The availability of a path for selective destruction of the unwanted enantiomer means that the desired product can be obtained with very high enantioselectivity when the reaction is pushed to higher conversions. However, this will come at the expense of overall yield because some of the desired product will also react further under such conditions [94,96]. Provided that one is able to distinguish either end of a developing chain, such reactions have promise in applications involving two-dimensional chain elongation strategies [97].

Epoxide-opening reactions. The most common use of epoxides is in S_N2 ring-opening reactions leading to 1,2-difunctionalized compounds.[4] As just discussed, the availability of enantiomerically enriched epoxides, when combined with appropriate regiochemical control of their opening, has enhanced the applicability of this approach to the preparation of enantiomerically pure compounds. An alternative approach is to begin with a *meso* epoxide, and then follow this reaction with a sequence able to distinguish between the enantiotopic carbons of the epoxide (Scheme 8.13a). One can open an epoxide with a chiral amine and separate the products [99] but catalytic means of carrying out the reaction enantioselectively have been reported by Nugent [100] and Jacobsen [101]. The salen-promoted epoxide openings by Jacobsen are also portable to the kinetic resolution of terminal epoxides by a variety of useful nucleophiles such as

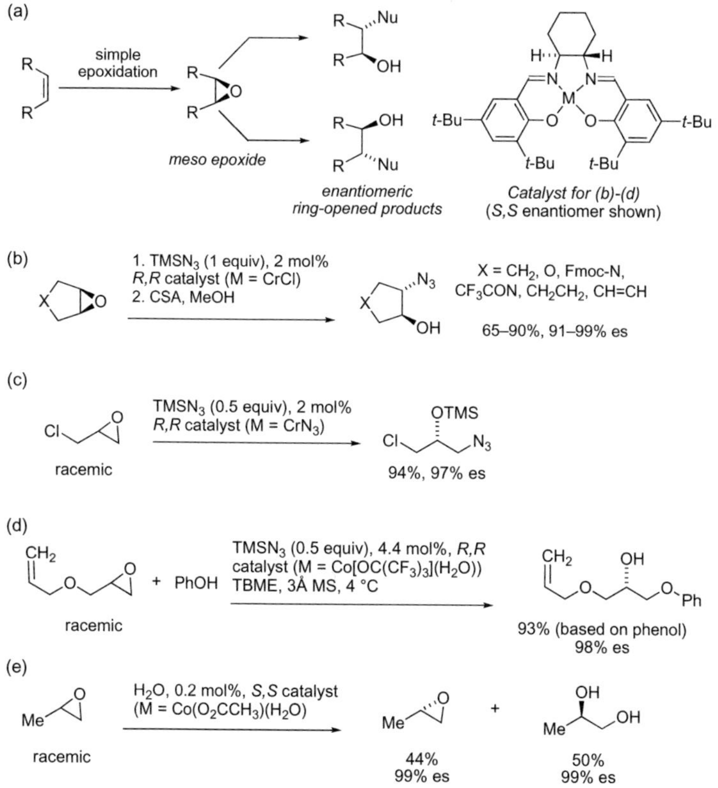

SCHEME 8.13 (a) Group-selective ring-opening of meso epoxides by nucleophiles leads to enantioselective syntheses of 1,2-difunctionalized compounds. (b) Azido alcohol synthesis from epoxides and trimethylsilyl azide as catalyzed by the salen–metal complexes shown in (a) [101]. Kinetic epoxide resolutions with (c) azide [102,103], (d) phenol [104], and (e) water [105].

---

4. For a review of such reactions in asymmetric synthesis, see ref. [98].

azide [102,103], phenol [104], and water [105]. The value of these procedures lies both in obtaining the ring-opened products in high enantiopurity and also in the isolation of the terminal epoxides, which are not generally accessible by direct asymmetric epoxidation.

8.1.5 Aziridinations

The asymmetric synthesis of aziridines has not received the same level of attention when compared to epoxides, not only because these reactions have posed particular challenges but also because their scope as synthetic intermediates is more narrow (various approaches have been reviewed in ref. [106]). Nonetheless, some catalytic, enantioselective aziridination reactions have been developed. In influential work, Evans showed that a copper bisoxazoline complex catalyzed the addition of *N*-(*p*-toluenesulfonylimino)phenyliodinane across an olefin (Scheme 8.14a [107]). Numerous investigators have examined similar ligands in aziridination reactions, *e.g.*, the intramolecular reaction shown in Scheme 8.14b [108]. In addition, chiral salen complexes of Mn [109], Cu [110], or Ru have been found to catalyze similar reactions with moderate to excellent enantioselectivities (Scheme 8.14c [108]). In all of these approaches, a metal-bound nitrene is generally assumed and has been supported by some mechanistic work [111]. Alternative approaches include the use of a cinchona alkaloid derivative to promote the conjugate addition of a hydroxamic acid as nitrogen source across an electron-deficient olefin followed by cyclization to aziridine product (Scheme 8.14d [112]).

8.2 ASYMMETRIC DIHYDROXYLATION (AD) REACTION

8.2.1 Reaction Development

The synthesis of vicinal diols from olefins using OsO_4 complements epoxidation/hydrolysis as a route to 1,2-diols (Scheme 8.15a, for reviews see refs. [115−120]). Both reagents effect *cis* bis-functionalization of an olefin, but since the epoxide-opening step involves an inversion of configuration, the two routes afford opposite diastereomers beginning with a single olefin geometry. The development of an efficient asymmetric dihydroxylation process, again pioneered by the Sharpless laboratory, was the first general method for the asymmetric oxygenation of unactivated olefins (*i.e.*, those without an allylic alcohol) and remains a valuable tool in synthesis. A general overview of the reaction sequence is outlined in Scheme 8.15b. All osmylation reactions ultimately afford an osmate ester[5] with concomitant reduction of osmium. Unlike most of the epoxidation reactions discussed earlier in this chapter, which require the formation of the active species in a chiral environment (either as a peroxo ligand or metal oxo species), the achiral precursor to most dihydroxylation catalysts, OsO_4, is itself a competent dihydroxylation reagent. An important antecedent for the development of the asymmetric osmylation was therefore the introduction of *achiral* means of accelerating the reaction rate and increasing turnover of osmium, with the goal of minimizing the amount of toxic metal necessary. Thus, catalytic methods using a variety of stoichiometric reoxidants for osmium were introduced, such as $KClO_4$ in 1912 [121] and the convenient Upjohn process, which uses *N*-methylmorpholine-*N*-oxide for this purpose [122]. Finally, a critical clue

---

5. Both [3 + 2] concerted mechanisms and a stepwise process involving the formation of a metallooxetane intermediate by [2 + 2] cycloaddition followed by rearrangement were proposed, the latter most vigorously by Sharpless in his early papers introducing this reaction. On the strength of both theory and experiment, most now accept the simpler [3 + 2] mechanism.

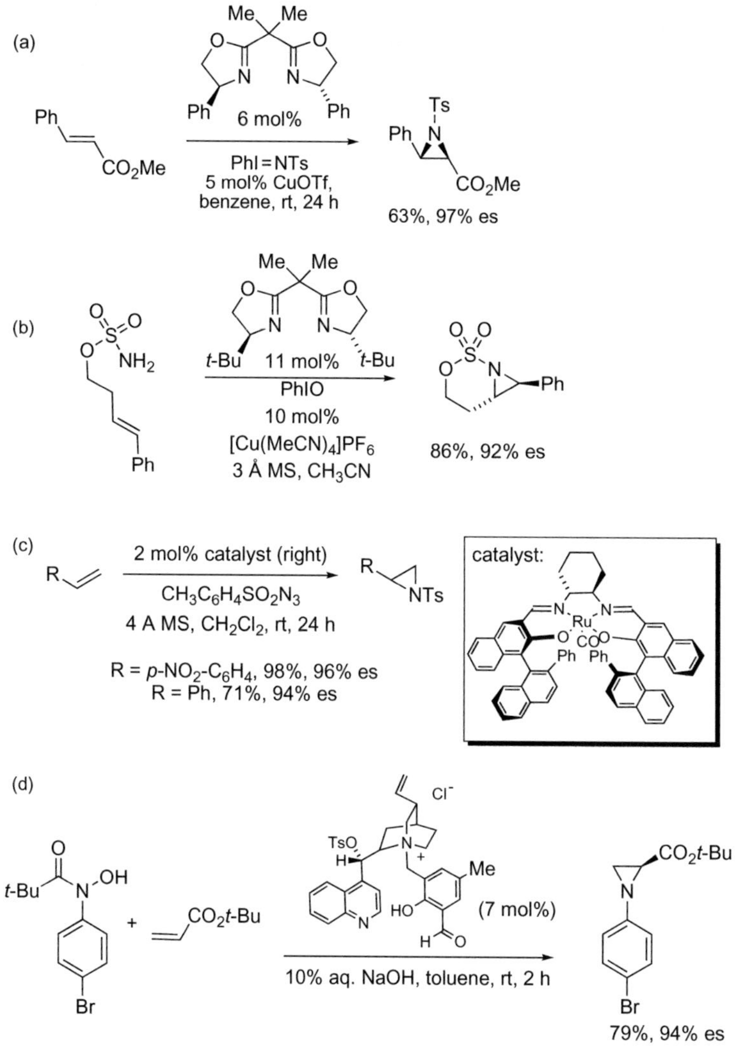

SCHEME 8.14 Asymmetric aziridination reactions: (a) [113], (b) [108], (c) [114], and (d) [112].

was found in the work of Criegee, who had shown in 1942 that the reaction was accelerated by the addition of pyridine [123].

Taking this cue from Criegee, Sharpless reported that enantioselectivities of up to 97% could be realized when stoichiometric amounts of chiral amines were added to OsO_4-mediated oxidation reactions [124]. Sharpless used the cinchona alkaloids dihydroquinidine (DHQD) and dihydroquinine (DHQ) (Figures 8.4a and b). Note that although these ligands are nearly enantiomeric, their mirror symmetry is spoiled by the placement of the ethyl group on the bicyclic portion of the molecules. Nonetheless, they act as if they were enantiomers when applied to various asymmetric transformations. Other workers also reported a variety of

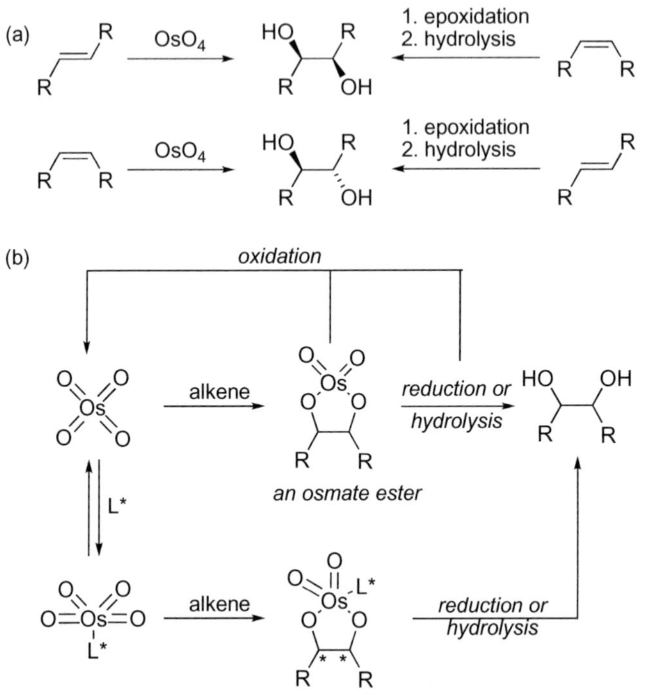

SCHEME 8.15 (a) 1,2-Diol synthesis from alkenes *via* direct osmylation or epoxidation followed by hydrolysis. (b) Generalized mechanism of the osmylation reaction.

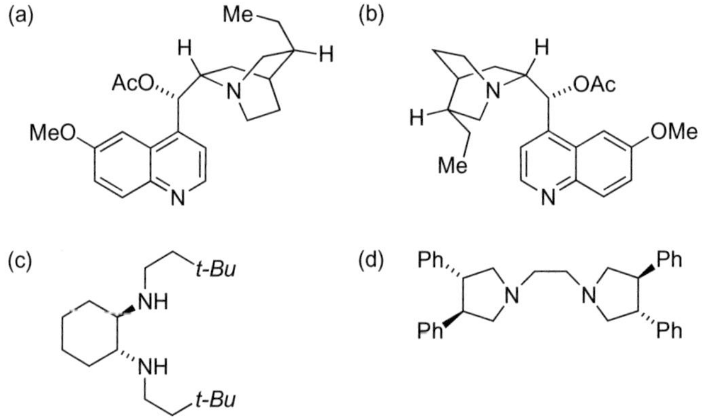

FIGURE 8.4 Representative ligands used in stoichiometric, asymmetric dihydroxylation reactions. (a) Dihydroquinidine (DHQD) and (b) dihydroquinine (DHQ) used by Sharpless [124]. Examples of C_2-symmetrical ligands used in stoichiometric osmylation reactions: (c) [125] and (d) [126]. These catalysts gave the following es values in dihydroxylation reaction of styrene: (a) 82%, (b) 80%, (c) >99%, and (d) 95%.

additional ligands able to effect highly selective stoichiometric dihydroxylation reactions; these ligands very often incorporate C_2 symmetry into their design (*e.g.*, Figures 8.4c and d).

Sharpless reported the first generally useful catalytic version of the reaction in 1987 [127]. This landmark paper showed that the reaction could be rendered catalytic by

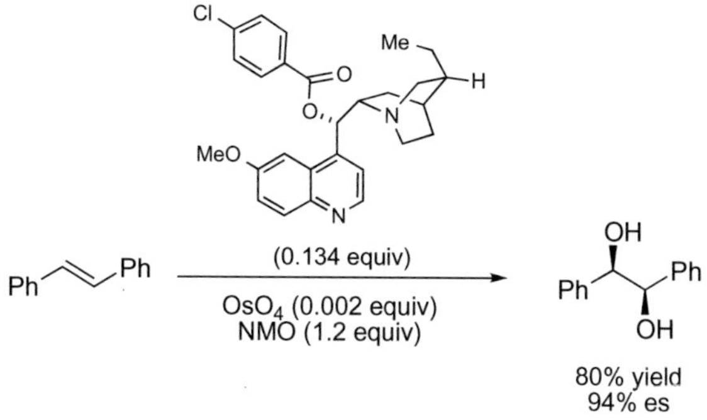

SCHEME 8.16 An early example of a catalytic asymmetric dihydroxylation reaction [127].

combining modified cinchona ligands with the Upjohn process (Scheme 8.16). This use of *ligand-accelerated catalysis* was critical to the success of this catalytic AD reaction because of the pre-equilibrium present between OsO_4 and OsO_4L^* in solution. Unless the equilibrium lies so far to the latter species as to effectively lower the concentration of OsO_4 to zero, the nonselective background reaction of OsO_4 with the olefin could compete with that of OsO_4L^*, lowering the enantioselectivity of the overall process. The ligand acceleration effect provided by the chiral amine ensured that the asymmetric pathway involving OsO_4L^* was also the most kinetically competent.

An interesting contrast exists between the development of the Sharpless asymmetric epoxidation reaction and the asymmetric dihydroxylation process. In the former case, the original reagents and protocol for carrying out the reaction have basically survived in their original form. However, the AD has been subjected to a great deal of optimization since its introduction, both in terms of ligand design and modification of conditions. In particular, protocols that cut down on interference by nonselective pathways have helped raise the utility of the overall procedure to its current high level. For example, the intrusion of a second catalytic cycle was proposed to lower overall stereoselectivity of the AD (Scheme 8.17). In this second cycle, the osmate ester formed by the reaction of one olefin with the chiral Os–cinchona complex was proposed to undergo oxidation and become itself a reactive dihydroxylation reagent, albeit one that had little enantiofacial selection. This pathway could be minimized by

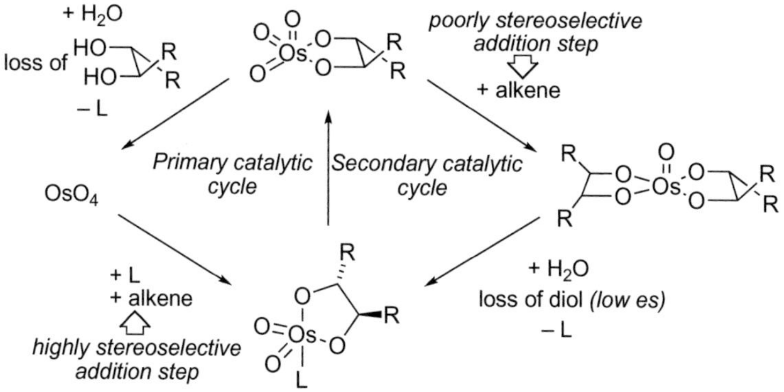

SCHEME 8.17 The two catalytic cycles proposed for the Sharpless AD reaction [128].

slow addition of the alkene (allowing the osmate ester time to undergo hydrolysis and reoxidation [128]), through the use of $K_3Fe(CN)_6$ as the reoxidant in place of NMO [129], or by increasing the rate of hydrolysis by adding $MeSO_2NH_2$ to the reaction mixture [130]. In particular, the use of the iron-based reoxidant remands the job of Os reoxidation to the aqueous portion of a biphasic reaction mixture, thus "protecting" the organic osmate ester from inopportune oxidation prior to hydrolysis. The addition of sulfonamide is doubly useful because it increases the turnover rate of the reaction and facilitates the dihydroxylation of otherwise sluggish substrates.

A mind-boggling number of AD catalysts have been examined. Some of the most commonly-used ligands are given in Figure 8.5 with a generalized correlation of ligand type with olefin class provided in Table 8.4. Examples of representative reactions are given in

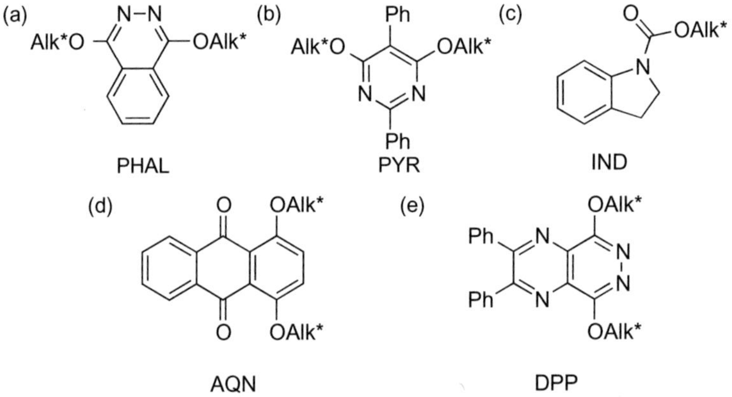

FIGURE 8.5 Ligands for the Sharpless AD process. (a) PHAL [130,131], (b) PYR [132,133], (c) IND [134], (d) AQN [135], and (e) DPP [136]. For each, the Alk* bound to each position is a cinchona alkaloid derivative (see Figures 8.4a and b).

TABLE 8.4 Generalized Utility of Ligand Types for AD Reactions According to Olefin Type [120]

	PHAL	PYR	IND	AQN	DPP
R ⌒					
aromatic	X				X
aliphatic				X	
branched		X			
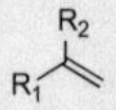 R₂ / R₁ ⌒					
aromatic	X				X
aliphatic				X	
branched		X			

(Continued)

TABLE 8.4 (Continued)

	1	2	3	4	5
R₁–R₂ (terminal alkene)					
acylic			X		
cyclic		X		X	X
R₁=R₂ (1,2-disubstituted)					
aromatic	X				X
aliphatic				X	
R₁(R₂)=R₃ (trisubstituted)	X			X	X
R₁(R₂)=R₃(R₄) (tetrasubstituted)	X	X			
R₁–R₂					
aromatic	X			X	X
aliphatic				X	
R₁(R₂)=R₃	X			X	X
R₁(R₂)=R₃(R₄)	X	X			

Table 8.5. The most striking advance was the use of dimeric species such as PHAL and PYR, which are still among the most general of the catalysts (Figures 8.5a and b). The former ligand has been formulated along with $K_2OsO_2(OH)_4$ (a nonvolatile source of Os), $K_3Fe(CN)_6$, and either DHQ or DHQD, respectively; these stable, storable powders contain all of the necessary ingredients for AD reactions and are marketed as AD-mix-α or AD-mix-β. Interestingly, although the hydroxyl substituent on the cinchona alkaloid platforms for these catalysts tolerates and benefits from a great many variations, the rest of the alkaloid has proven much less flexible [118], and this portion of the catalytic system can usually be left alone. Numerous catalyst types have been devised to test mechanistic hypotheses or to provide various conveniences (such as immobilization for easy removal); the review literature may be consulted for leading references [120].

TABLE 8.5 Examples of Asymmetric Dihydroxylation Reactions

Entry	Diol product	Ligand[a]	% es	Ref.
1		(DHQD)₂-PHAL	95	[137]
2		(DHQ)₂-PHAL	98	[130,132]
3		(DHQD)₂-PHAL	89	[138]
4	(from *trans* olefin)	(DHQD)₂-PHAL	98	[130]
5		(DHQD)₂-PHAL	79 (R = Me), > 99 (R = phenyl)	[118,130]
6	(from *cis* olefin)	DHQD-IND	78 (R = c-C₆H₁₁), 86 (R = Ph)	[134]
7		(DHQD)₂-PYR	61	[133]
8		(DHQD)₂-PYR	92	[133]
9		(DHQD)₂-AQN	96	[135]
10		(DHQD)₂-AQN	99	[135]

(*Continued*)

TABLE 8.5 (Continued)

Entry	Diol product	Ligand[a]	% es	Ref.
11	MeO₂C ... OH / N * OH / (phenyl)	(DHQD)₂-AQN	99	[139]
12	OH / OH Me (phenyl) (from *cis* olefin)	(DHQD)₂-DPP	84	[136]
13	HO OH / BzO OH (from *cis* olefin)	(DHQD)₂-DPP	91	[136]

[a]See Figure 8.5 for ligand structures.

The results of many dihydroxylation reactions have led to mnemonic devices for the prediction of enantioselectivity (Figure 8.6a). The first such model was introduced by Sharpless [140] and was later modified by Norrby, following computational studies based on results obtained with an expanded set of ligands introduced in the late 1990s [140]. Although this model is useful, there can be some ambiguity as to which group is the large one and which is the medium (especially with *trans*-disubstituted olefins) and electronic characteristics cannot be ignored [117]. Some useful generalizations are that the "large" group is very often aromatic, and indeed, aromatic olefins are some of the best substrates for this reaction. Experimental results have led to the suggestion that the loaded catalyst is very forgiving for *trans* olefins (the best substrates), but that it begins to experience some interference at the R_S position. That the binding site is even less favorable toward substituents *cis* to the R_L position (H in Figure 8.6a) is surmised by the difficulty of carrying out AD reactions with fully substituted [133] and *cis*-olefins [134]. However, very good-to-excellent results have been wrested from all alkene types. In many cases, the enantiomeric purity of the diol products can be increased by simple recrystallization, which increases the practicality of the method.

The structural basis for enantioselectivity in the AD reaction has been the subject of vigorous debate over the years (for a review, see ref. [143]). One view of an intermediate proposed by the Sharpless school is given in Figure 8.6b, depicting the osmate ester of the reaction between styrene and a (DHQD)₂-PHAL-derived complex. In the transition state leading to this product, one DHQD binds the active osmium species and the other is a bystander ligand. Overall, the binding pocket is proposed to have an "L" shape, with the floor made up by the flat PHAL heterocycle (the better to accommodate aromatic alkenes) and one wall coming from the bystander DHQD ligand. Favorable aromatic stacking interactions along with the minimization of steric interactions between the alkene group and the methoxyquinone wall result in the observed diastereoselectivity. Corey has offered mechanistic proposals and

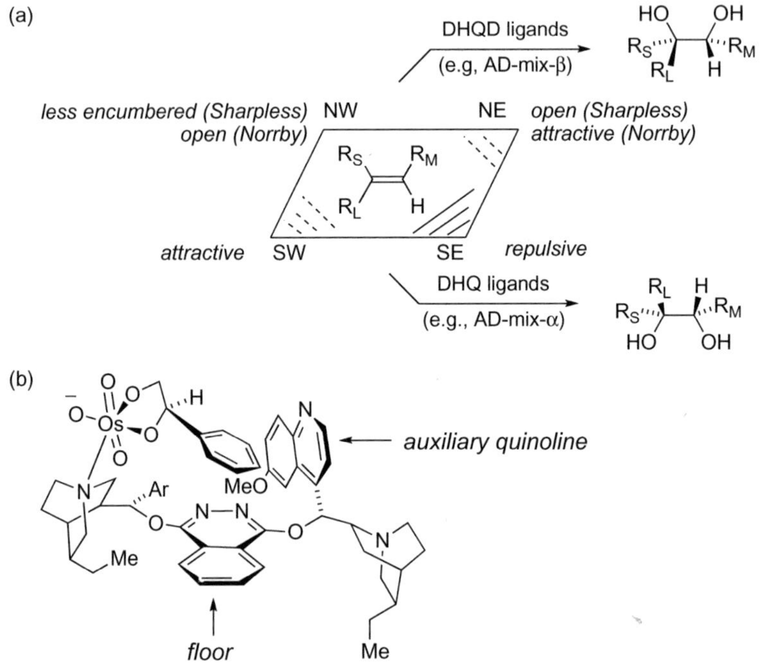

(a)

(b)

FIGURE 8.6 (a) The predicted enantioselectivity of Sharpless AD reactions using DHQD or DHQ ligands, as originally formulated by Sharpless [140] and modified by Norrby [141]. This model is used by orienting the substrate so that the large (often aromatic), medium, and small substituents match up best with the R_L, R_M, and R_S positions. Application of this model to some alkenes requires some compromises in placing the groups. (b) One view of a proposed intermediate in the Sharpless AD reaction of the (DHQD)$_2$-PHAL-derived osmium species reacting with styrene [142].

transition structures [144,145] that are also consistent with the observed course of the reaction, differing mostly in the relative orientations of the two cinchona moieties and in the orientation of the substrate prior to oxidation. A molecular mechanics study has determined the Sharpless and Corey transition states to be close in energy in numerous individual reactions [146].

8.2.2 Applications of Enantioselective Dihydroxylations

The diol products of an AD reaction are not intrinsically activated for further chemistry as are epoxides. Accordingly, it has proven necessary to develop schemes for the incorporation of the AD into synthetic programs. A few examples confer some of the flavor of this work (Scheme 8.18). For one, glyceraldehyde and its acetonides have found very wide acceptance as chiral starting materials in asymmetric synthesis [147]. The straightforward conversion of a diol to the epoxide afforded a building block that nicely complements the use of the naturally occurring material (Scheme 8.18a) [148]. An alternative to converting diols to reactive epoxides is to activate the diols themselves to nucleophilic attack; this has been accomplished by converting them into cyclic sulfates (Scheme 8.18b) [118,149]. These species are subject to substitution by many nucleophiles, including halides, azides, reducing agents, and sulfur and carbon nucleophiles. Scheme 8.19b depicts a strategy involving irreversible epoxide formation (cf. the Payne rearrangement; Section 8.1.4) [150]. Examples of using the reaction for

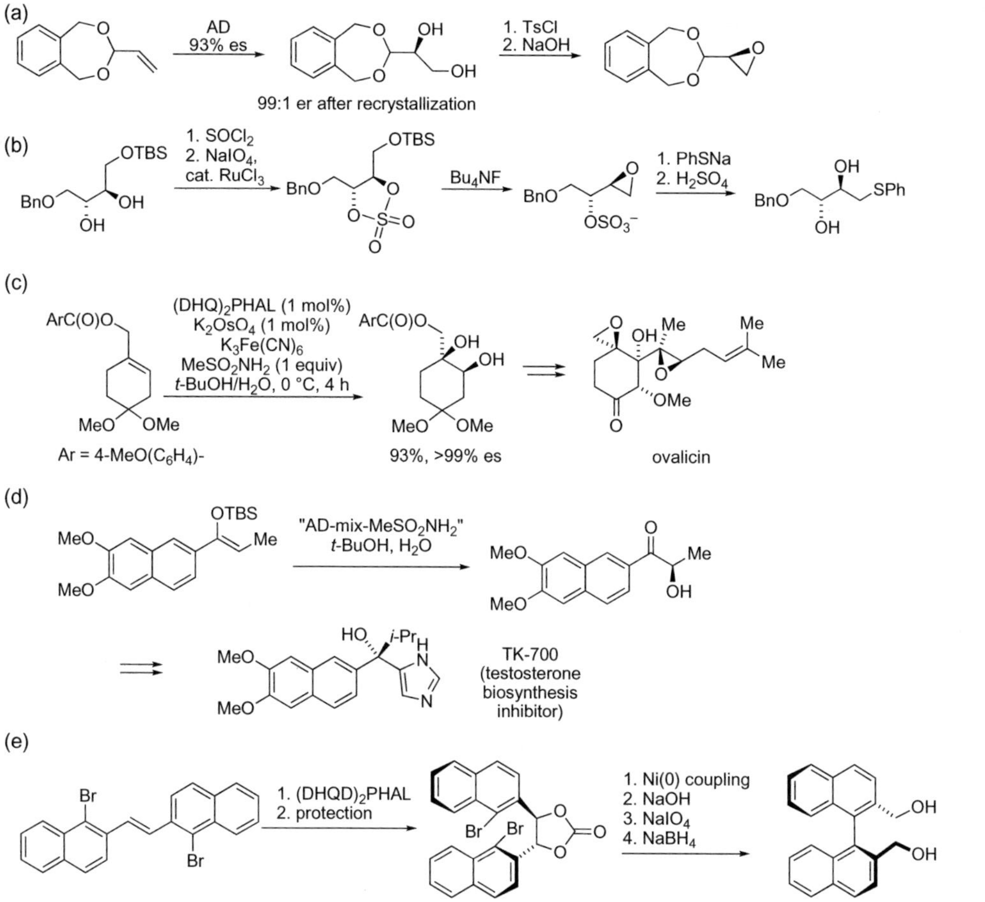

SCHEME 8.18 Synthetic applications of diols obtained by AD chemistry. (a) Synthesis of a glyceraldehyde equivalent and conversion to an epoxide [148]. (b) Formation and reactivity of cyclic sulfates. (b) Application of cyclic sulfates to the synthesis of erythrose [150]. Here, the epoxide formation is irreversible because the sulfate leaving group is no longer nucleophilic. (c) Synthesis of ovalicin [151]. (d) Synthesis of TK-700 [152]. (e) Synthesis of an enantiomerically enriched (>99% es) biaryl diol [153].

natural product (Scheme 8.19c, ref. [151]) or drug candidate synthesis (Scheme 8.19d, ref. [152]) are also provided — in neither case do both of the stereocenters generated during the AD reaction make it to the end product of the synthesis. In the latter example, directed toward a candidate for the treatment of prostate cancer, note the good results obtained for the oxidation of a silyl enol ether to afford an α-hydroxy ketone. Finally, a diastereofacially selective AD reaction affords a nice starting material for the preparation of an axially chiral binaphthyl derivative [153]. As shown in Scheme 8.18e, cyclization of the diol restricts the motion of the aryl groups so that they can only undergo intramolecular biaryl coupling to give one configuration about the newly-formed single bond. Diol oxidation removed the original stereocenters installed by the AD reaction.

   Issues of double asymmetric induction [154] and the potential for kinetic resolution arise in the reactions of chiral alkenes. Prior to the development of this asymmetric

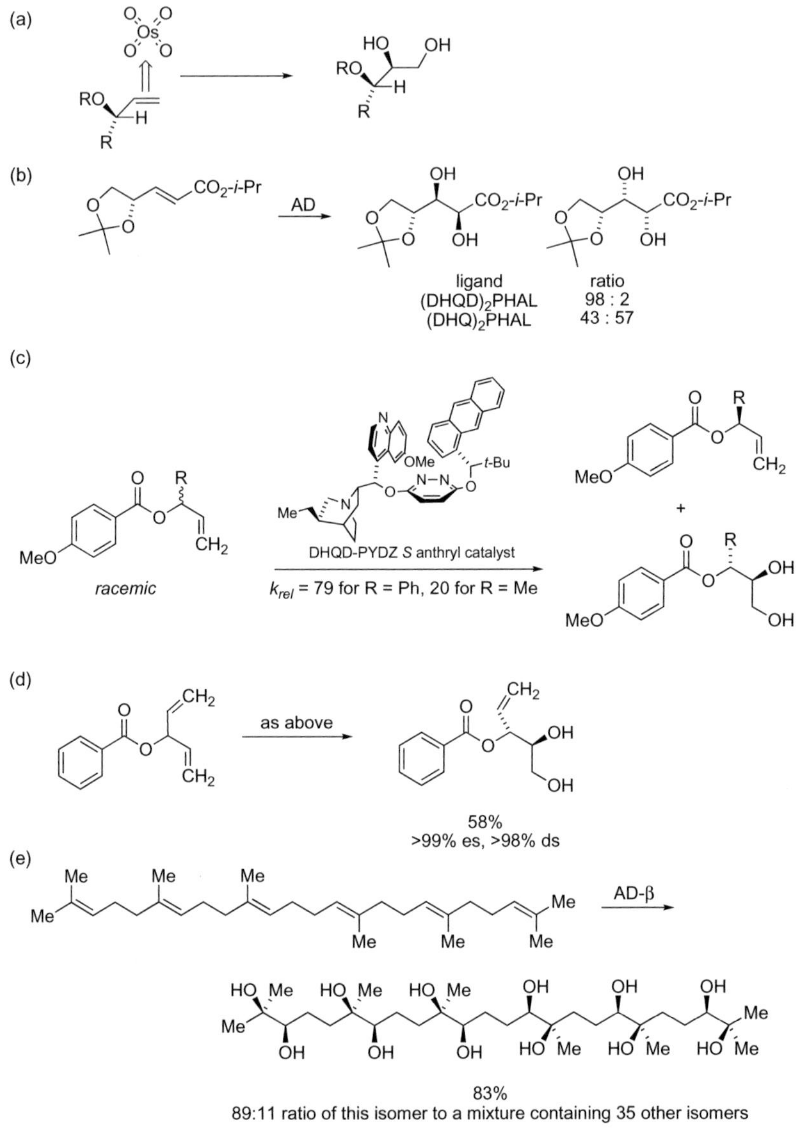

SCHEME 8.19 (a) Kishi model for acyclic control in osmylation reactions [155]. (b) Double diastereoselectivity in an AD reaction [160]. (c) Kinetic resolution or a chiral alkene and (d) group-selective conversion of an achiral substrate [157]. (e) Multiple AD reactions carried out on squalene [158].

dihydroxylation, the dependence of diastereofacial selection in alkenes bearing allylic substitution had been cataloged by Kishi (Scheme 8.19) [155,156]. When AD reactions were carried out on substrates already bearing stereogenic centers, matched *vs.* mismatched situations developed (Section 1.6), with the former affording very high selectivity. However, the ability of the AD system to induce enantiofacial selectivity is often high enough that varying levels of selectivity in either direction can be obtained. Although the high tolerance of the standard

AD ligands for a wide variety of substrates translates to mediocre results in kinetic resolutions using those ligands [118], Corey has designed a more demanding version and showed it to be useful in both kinetic resolution and the selective reaction of one of the enantiotopic alkenes in a symmetrical starting material (Schemes 8.19c and d, ref. [157]). The high selectivity possible with trisubstituted alkenes was manifested in the multiple AD of a hexalkenyl substrate, squalene, which gave a high yield of 1 out of the 36 possible isomers for this substrate ([158]; for a thematically related example using the Shi epoxidation, see ref. [159]).[6]

8.3 α-FUNCTIONALIZATION OF CARBONYL GROUPS AND THEIR EQUIVALENTS

Most methods to formally oxidize the carbon adjacent to a carbonyl group by converting a C–H bond to a C–X group where X = OR, NR_2, or halogen leverage the rich chemistry developed for the asymmetric alkylation of carbonyl groups (Chapter 3) with the involvement of an appropriately electrophilic agent. Given the numerous ways of differentiating the faces of a ketone or aldehyde that have been developed (from chiral auxiliaries and coordinating ligands to catalysts, with or without metals), much of the innovation required in this field has been to identify appropriate reagents for adding the heteroatom of interest under the right reaction conditions ("X[+]" in Scheme 8.20). Since most such routes involve the intermediacy of an enolate or its equivalent, another approach is to functionalize an electron-rich double bond present in a discretely trapped enol ether — another process having considerable precedent in epoxidation or dihydroxylation chemistry (see Scheme 8.18d for an example). This discussion will concentrate on those methods that have been developed particularly for these types of conversions.[7]

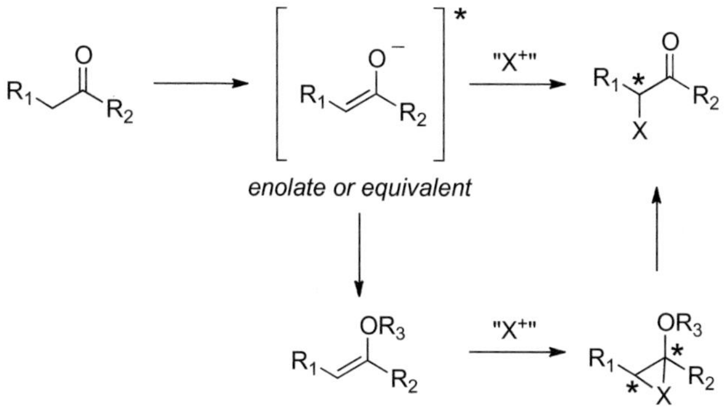

enolate or equivalent

SCHEME 8.20 General approaches for α-functionalization of a ketone rendered chiral in a so far unspecified way.

---

6. Figuring out exactly why there are 36 possible isomers of this polyol is an entertaining way to spend a rainy day; see the discussion in ref. [158].

7. For general reviews, see refs. [161–163].

8.3.1 Hydroxylations

In general, chiral auxiliaries that are effective for alkylation reactions work well when combined with electrophilic sources of oxygen, and some of the usual suspects are shown in Scheme 8.21. All presumably involve mechanisms for diastereofacial discrimination similar to those involved in carbon–carbon bond-forming reactions (Chapters 3 and 5). These

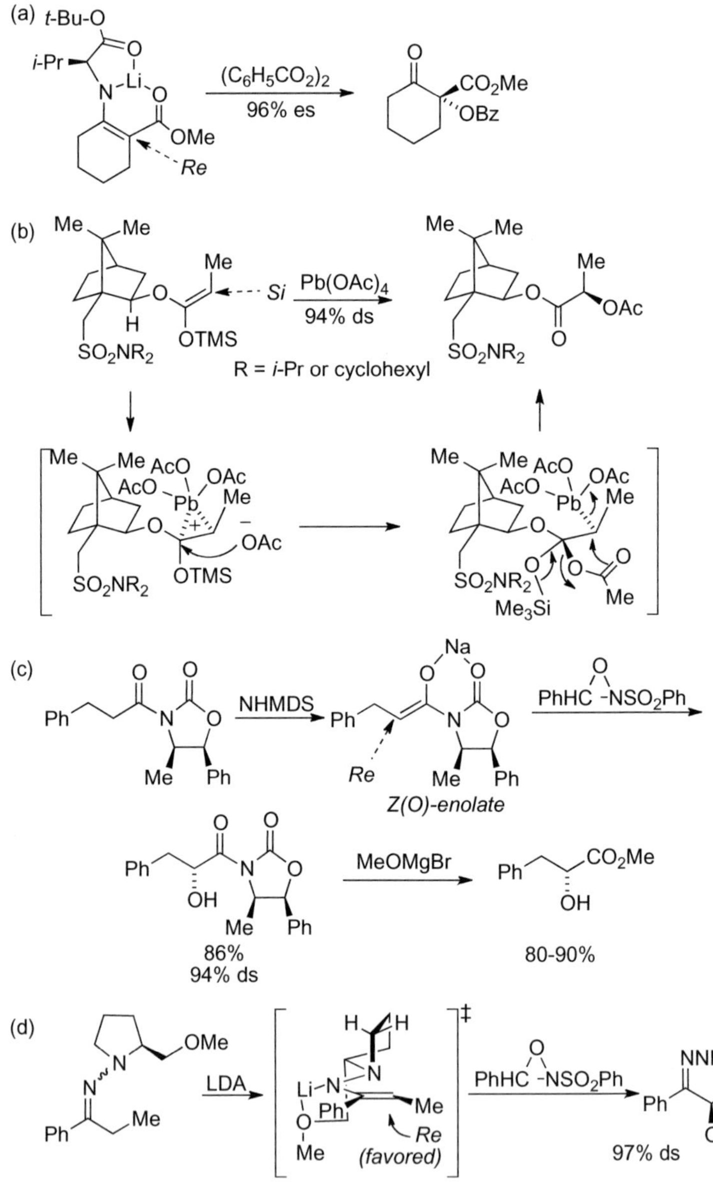

SCHEME 8.21 Representative hydroxylation reactions of chiral enolates, using (a) metalloenamine [166], (b) sulfonamide [167] (*cf.* Scheme 3.19), (c) oxazolidinone [168] (*cf.* Scheme 3.21), and (d) hydrazone [169] (*cf.* Scheme 3.27) auxiliaries.

examples show some of the reagents that have been used to transfer oxygen. In early years, chemists commonly used for this purpose reagents like MoOPh (the complex of oxodiperoxy-molybdenum with pyridine and HMPA – not shown, but see ref. [164]), benzoyl peroxide (Scheme 8.21a), or Pb(OAc)$_4$ (demonstrated for the oxidation of a pre-formed silyl enol ether in Scheme 8.21b). Since their popularization by Davis in the 1990s, N-sulfonyl oxaziridines have become the reagents of choice (Scheme 8.21c and d; for a review, see ref. [165]).

Chiral N-sulfonyl oxaziridines can react with enolates to afford α-hydroxy carbonyl compounds in excellent yield and enantioselectivity. An application of a highly selective sulfony-loxaziridine derived from camphor to the synthesis of daunomycin is shown in Scheme 8.22. Attack of the oxaziridine presumably occurs such that the enolate ester avoids nonbonded interactions with the *exo* methoxy group on the bicyclic ring system. This is a reaction of wide scope, and can be carried out on both stabilized enolates derived from keto esters as shown and simple ketone enolates [170].

Catalytic approaches to α-carbonyl oxygenation include epoxidation and dihydroxylation reactions, as discussed in the relevant sections earlier in this chapter (for a review, see ref. [162]). In addition, the catalytic generation of enolates and selective reaction through interli-gand chirality transfer has been reported using oxidants such as Davis' oxaziridines (Scheme 8.23a [173]), dimethyldioxirane (Scheme 8.23b [174]), or nitrosobenzene (Scheme 8.23c [175]). Note the use of either 1,3-dicarbonyl-containing compounds (which have high enol content) or the pre-formed tin enolate in Scheme 8.23b. Nitrosobenzene is an interesting oxidant that can in principle react with a nucleophile at either oxygen or nitrogen. Which one occurs depends on the particular conditions employed, although the majority of its use in asymmetric catalysts is for α-hydroxylation rather than α-amination.

Nitrosobenzene has also found considerable use in organocatalytic α-oxygenation reactions (reviewed in ref. [162]), such as the examples shown in Schemes 8.24a and b [176,177]. Both reactions use enamine catalysts to α-oxygenate an aldehyde or ketone, respectively; they also demonstrate how the nitroso adducts may be converted to more standard functional group arrays useful in synthesis. Mechanistically, these reactions involve reactive ensembles

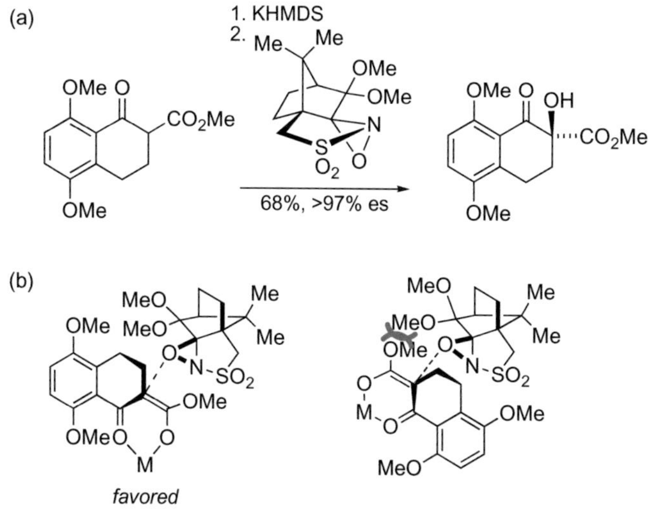

SCHEME 8.22   (a) Application of enolate oxidation reactions of a chiral oxaziridine to the synthesis of an AB ring synthon of daunomycin [171]. (b) Proposed competing transition structures [172].

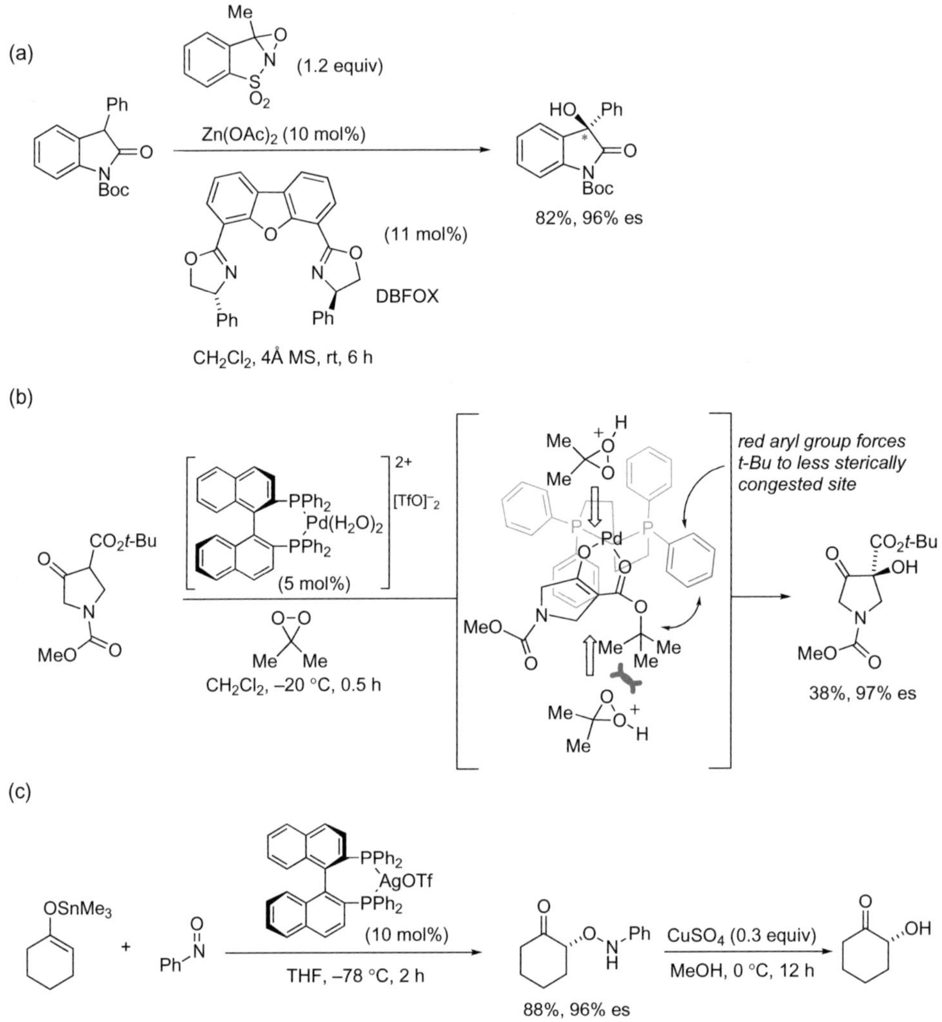

SCHEME 8.23 Catalytic methods for α-hydroxylation using a (a) an oxaziridine [173], (b) dimethyldioxirane [174], and (c) nitrosobenzene [175]. In (b), note the blockage of the bottom face due to steric interactions between catalyst and the *tert*-butyl ester.

that resemble those discussed for enamine-promoted aldol reactions (Section 5.2.3). In 2009, three groups reported a comeback for benzoyl peroxide as an α-hydroxylating agent in the context of organocatalysis (Scheme 8.24c, ref. [178]; also see refs. [179,180]). Mechanistically, these reactions might involve either direct nucleophilic attack at one of the oxygens of benzoyl peroxide or possible N-attack followed by a rearrangement in the resulting intermediate [178].

8.3.2 Aminations and Halogenations

Amination reactions of carbonyl compounds provide access to useful building blocks for nitrogen-containing compounds, with the conversion of esters to amino acid derivatives being

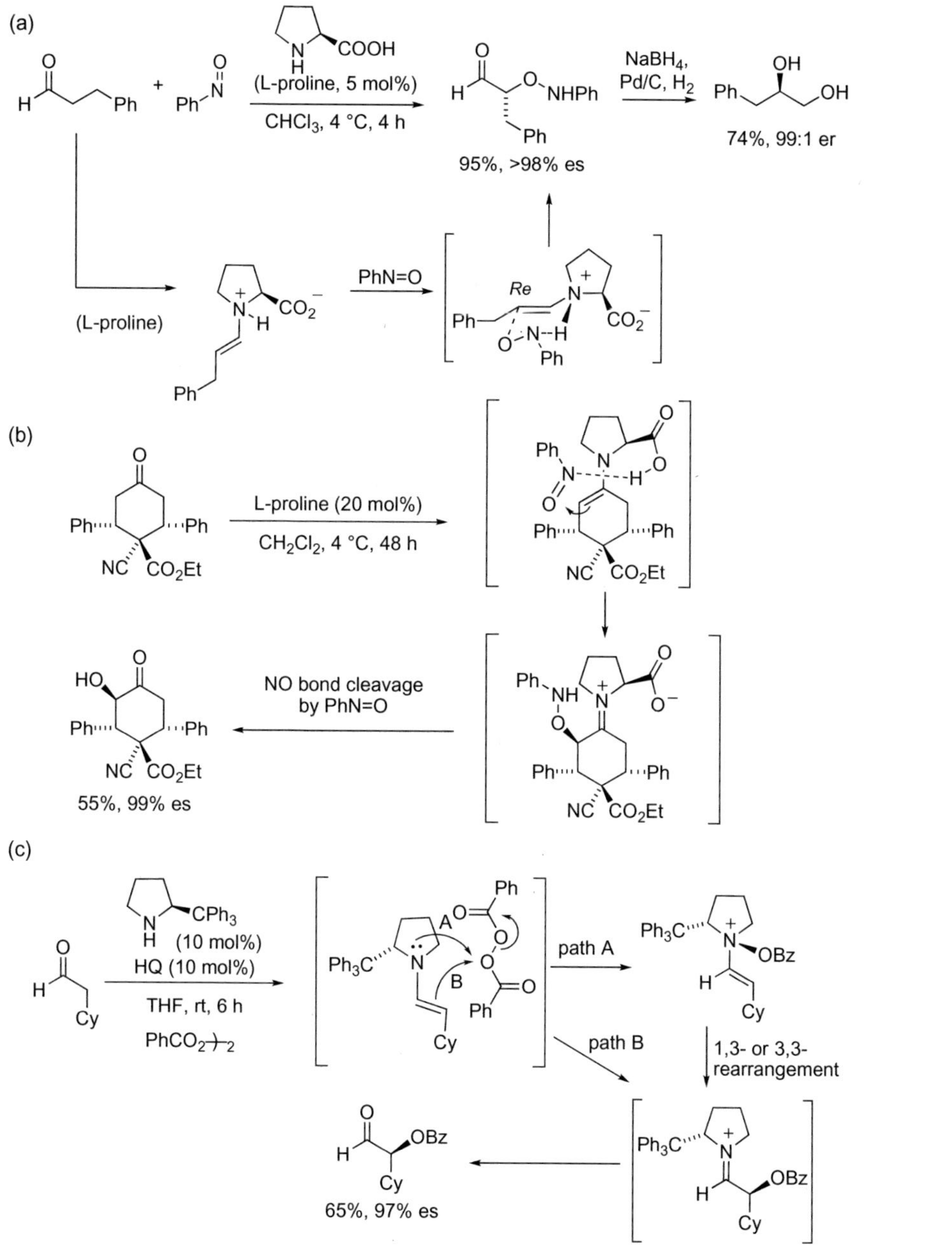

SCHEME 8.24 Organocatalytic α-oxygenations: (a) [176], (b) [177], and (c) [178]. Although not shown, the cleavage of the NO bond in (b) is caused by excess nitrosobenzene; see ref. [177] for a full mechanistic proposal.

particularly important. Likewise, the α-halogenation of carbonyl groups is a reaction of potentially enormous utility given the existence of the S_N2 reaction. Obviously, one useful way of installing a nitrogen into an organic molecule is through the displacement of a halide and for that reason we discuss these two methods together, although asymmetric halogenation chemistry was introduced much later than amination approaches.

In 1986, the groups of Gennari, Evans, and Vederas simultaneously published routes to α-hydrazino ester derivatives by the addition of the electrophilic reagent di(*tert*-butyl)azodi-carboxylate (DBAD) to enolates or trimethylsilyl ketene acetals (Scheme 8.25) [181–184]. Excellent yields were obtained, and the products were formed in accord with the diastereofa-cial selectivity of the nucleophiles in alkylation or aldol reactions (Chapters 3 and 5).

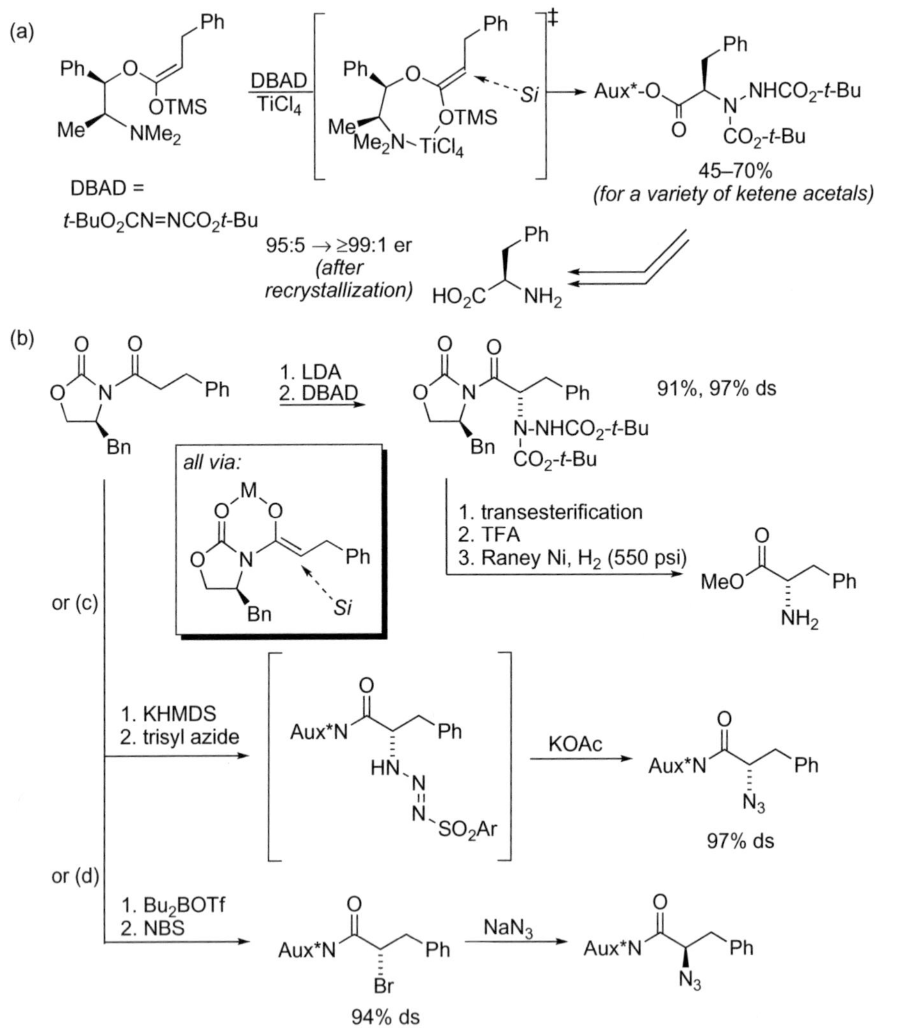

SCHEME 8.25 α-Amidation of chiral ester enolates using di(*tert*-butyl)azodicarboxylate and (a) *N*-methyle-phedrine [181] or (b) oxazolidinone chiral auxiliaries [182]. Azidation of a chiral enolate (c) directly or (d) via bromination/azidation [185].

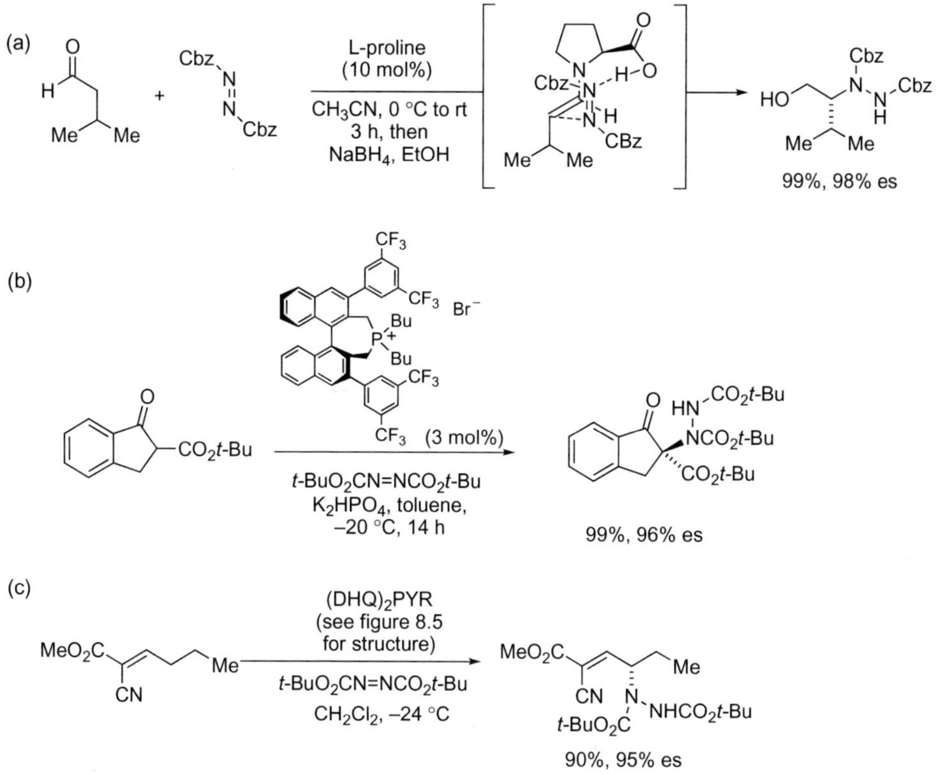

SCHEME 8.26 Asymmetric α-amination promoted by (a) enamine [186] and (b) phase transfer [188] catalysis. (c) An approach to formal allylic amination involving formation of an extended dienolate [189].

Unfortunately, the hydrazino esters or amides required inconveniently high pressures for their hydrogenolysis (500 psi). An improvement involved the direct azidation of the same enolates using arylazide derivatives, that were found to undergo reactions with enolate nucleophiles to provide a *C*-sulfonyltriazene intermediate that could be decomposed to the α-azido ester (Scheme 8.25c) [185]. Alternatively, azides may be obtained by enolate bromination followed by S_N2 azide displacement; this allows access to both enantiomers from the same auxiliary.

Azodicarboxylate esters are similarly useful electrophiles for organocatalytic α-aminations, two of the many examples reported are depicted in Schemes 8.26a and b (for an extensive review of this literature, see ref. [162]). Scheme 8.26a shows an early example of enamine catalysis in this context, which was discovered essentially simultaneously by List and Jørgensen [186,187]. Once again, the proposed transition structure resembles the cyclic intermediates suggested in the pioneering aldol organocatalysis literature (Section 5.2.3). Scheme 8.26b demonstrates a selective example of phase transfer catalysis in this context [188]. The concept may also be extended to dienolates for a formal allylic amination reaction (Scheme 8.26c, ref. [189]).

The extension of these concepts to halogenation chemistry requires the development of appropriate electrophilic donors of "X^+". Perhaps the most familiar of these are *N*-chloro- and *N*-bromosuccinimide (see Scheme 8.25 for an example of enolate bromination using the latter). Table 8.6 provides an introduction to the types of catalysts and reagents used for this

TABLE 8.6 Examples of Asymmetric α-Halogenation

Entry	Product	Catalyst type	Halogenating Agent	% Yield	% es	Ref.
1		Ti-TADDOL	(SelectFluor™)	80–95	81	[191]
2		Ti-TADDOL	NCS	85	79	[190]
3		Pd-xylylBINAP	PhO₂S–N–SO₂Ph (NFSI)	91	97	[194]
4		Ni-BINAP	NFSI	99	94	[195]
5		Cu-t-Bu-BOX	NCS	99	88	[196]
6		Ni-DBFOX	NFSI	86	99	[197]
7				93	94	[192]

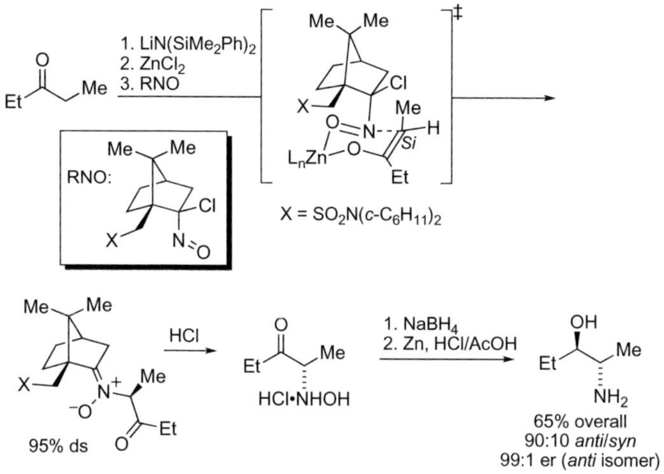

SCHEME 8.27   Reaction of an achiral enolate with a chiral α-chloro-α-nitroso reagent [198].

purpose, beginning with Togni's early innovations in the field using TADDOL derivatives (entries 1 and 2, refs. [190,191]) and ending with an organocatalytic approach toward direct chlorination published by MacMillan (entry 7, ref. [192]). The increasing attention noted toward selective introduction of fluorine reflects not so much synthetic utility but rather the importance of fluorine in drugs and imaging agents (reviews: refs. [161,193]).

Finally, a different approach toward amination utilizes achiral enolates and a chiral amination reagent (interligand asymmetric induction) [198]. α-Chloro-α-nitrosocyclohexane had previously been used as an aminating reagent with chiral enolates, providing nitrones as the primary product [199]. The adaptation of this chemistry to chiral aminating agents gave the nitrones with high diastereoselectivity (Scheme 8.27), which could be hydrolyzed to give α-hydroxylamino ketones. These were further reduced to the amino alcohols using borohydride reagents and zinc/HCl. The reactions were proposed to proceed through a Zimmerman–Traxler-type transition structure in which the Z(O)-enolate of the ketone was coordinated *via* zinc to the nitroso group and the whole ensemble oriented to avoid steric interactions between the incoming nucleophile and the sulfonamide group.

8.4 MISCELLANEOUS OXIDATIONS THAT NECESSITATE DIFFERENTIATION OF ENANTIOTOPIC GROUPS

8.4.1 Oxidation of Sulfides[8]

A great deal of effort has been expended in the development of ways to carry out the asymmetric oxidation of sulfides to sulfoxides (for reviews, see refs. [200–205]). This is interesting from both a theoretical viewpoint and from the utility of certain chiral sulfoxides as reagents in asymmetric synthesis [206]. Some natural products and drugs also contain sulfoxide stereogenic centers.

---

8. Here the "enantiotopic groups" are the lone pairs on sulfur, which we acknowledge is stretching the section title a little bit.

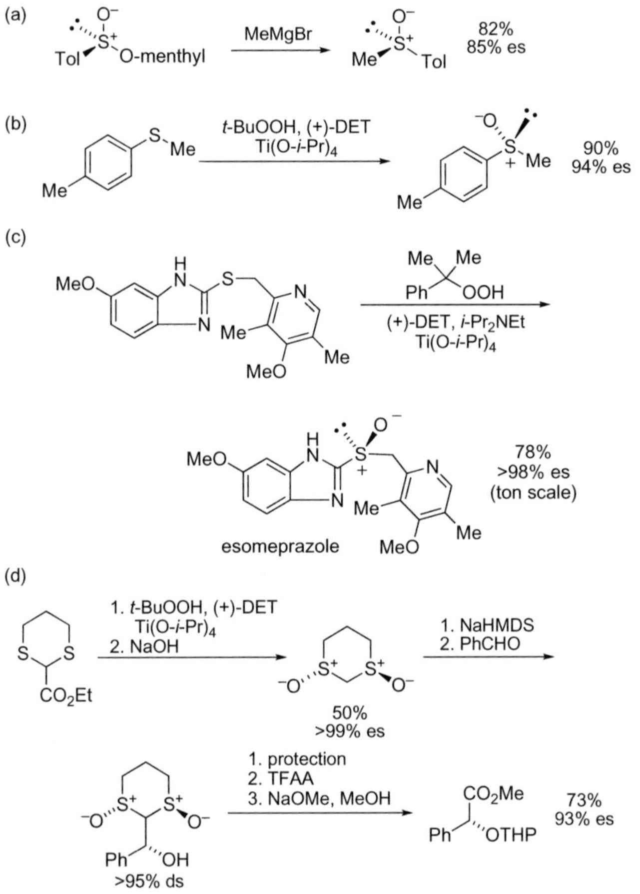

SCHEME 8.28 (a) An example of the Andersen synthesis of chiral sulfoxides [216]. (b) Catalytic oxidation of an aromatic sulfide using a chiral titanium complex [209]. (c) Industrial-scale synthesis of esomeprazole, a proton pump inhibitor [211]. (d) Synthesis of a C_2-symmetrical *trans*-1,3-dithiane-1,3-dioxide and its use as an asymmetric acyl anion equivalent [214,215].

An important classical method for obtaining chiral sulfur compounds is the Andersen synthesis, which utilizes a chiral sulfinate ester such as that derived from menthol by diastereo-selective oxidation and isomeric enrichment *via* epimerization and recrystallization [206,207]. Scheme 8.28a shows a simple example; note that the Grignard reaction occurs with inversion of configuration at the sulfur atom. Additional approaches use chiral reagents (*i.e.*, chiral *N*-sulfonyloxaziridines [208]) and catalytic systems to address this problem, such as the titanium-diethyl tartrate system (Scheme 8.28b); this reaction can proceed with high enantioselectivity when appropriate substrates such as aryl sulfides are used [200,209]. An important application is the industrial-scale preparation of esomeprazole, the active ingredient in Nexium®, a stomach acid reducing agent [210,211]. An interesting application of chiral sulfoxide chemistry has been reported by the Aggarwal group (reviewed in refs. [212,213]) and exemplified in Scheme 8.28d. These workers prepared the *trans* isomer of 1,3-dithiane-1,3-dioxide in high enantioselectivity;

note the use of a temporary carboethoxy group, which proved necessary for high enantio-selectivity (Scheme 8.28d, ref. [214]). Deprotonation gave an acyl anion equivalent which reacted with aromatic aldehydes with high diastereoselectivity [215]. Pummerer removal of the heterocycle followed by basic transesterification met with some isomerization and loss of enantiomeric purity, although this problem could be mitigated by a multistep procedure involving intermediate thioesters.

8.4.2 Group-Selective Oxidation of C—H Bonds

The functionalization of C—H bonds became a major area of expansion in the early twenty-first century due to the attractiveness of functionalizing both very basic building blocks (such as methanol [217]) or very complex ones (such as natural products [218]). Although the field is far beyond the scope of this book, it is worth providing the reader with a taste of how chemists have attempted the oxidative conversion of enantiotopic hydrogen atoms to provide chiral molecules (Scheme 8.29; for a review, see ref. [219]). It is worth noting that the oxidation of an apparently unactivated C—H bond is "business as usual" for the enzymes of metabolism, particularly those in the family of cytochrome P450. Since such reactions often occur with high levels of enantioselectivity, engineered versions of these enzymes can be attractive tools for particular applications (e.g., ref. [220]).

A few nonenzymatic methods that permit the formal group-selective oxidation of enantiotopic C—H bonds are shown in Scheme 8.29 (for a review, see ref. [219]). The asymmetric Kharasch reaction in Scheme 8.29a uses a chiral bisoxazoline ligand and a Cu(I) salt [221,222]. The reaction involves generation of an allylic radical that reacts with an in situ-generated Cu(II) complex with the chiral ligand (Scheme 8.29b). Note that although the overall reaction selectively replaces one of the enantiotopic hydrogen atoms with a benzyloxy group, the actual enantioselective step is the face-selective recombination of the allylic radical with the copper(II)-benzoate. The low reactivities and yields reported for some substrates suggest room for improvement (e.g., see related work by Singh [223]). Katsuki has used the salen catalyst of his design to oxidize positions adjacent to heteroatoms in appropriate ring systems (Scheme 8.29c, refs. [224,225]). While formally another C—H oxidation, one cannot rule out the intermediacy of an sp^2 hybridized acyliminium ion that is attacked from the exo face of the bicyclic ring system to afford the product shown.

A truer sense of the promise inherent in direct C—H oxidation can be seen in the allylic/benzylic amination shown in Scheme 8.30a reported by Du Bois (Scheme 8.30 [226]).[9] Although these reactions require relatively weak C—H bonds, it is not thought that they proceed through radical intermediates. Rather, evidence suggests that this reaction proceeds via a rhodium nitrene complex that directly inserts into the C—H$_{Si}$ bond. Impressive results were also registered in the diastereomeric manifold by Dodd and Dauban as shown in Scheme 8.30b [228].

8.4.3 Group-Selective Oxidative Ring Expansions

In a symmetrical ketone like 4-tert-butylcyclohexanone, the two methylene groups adjacent to the ketone are enantiotopic. The formal "oxidation" of these bonds through asymmetric

9. For a review of this field of insertion chemistry see ref. [227].

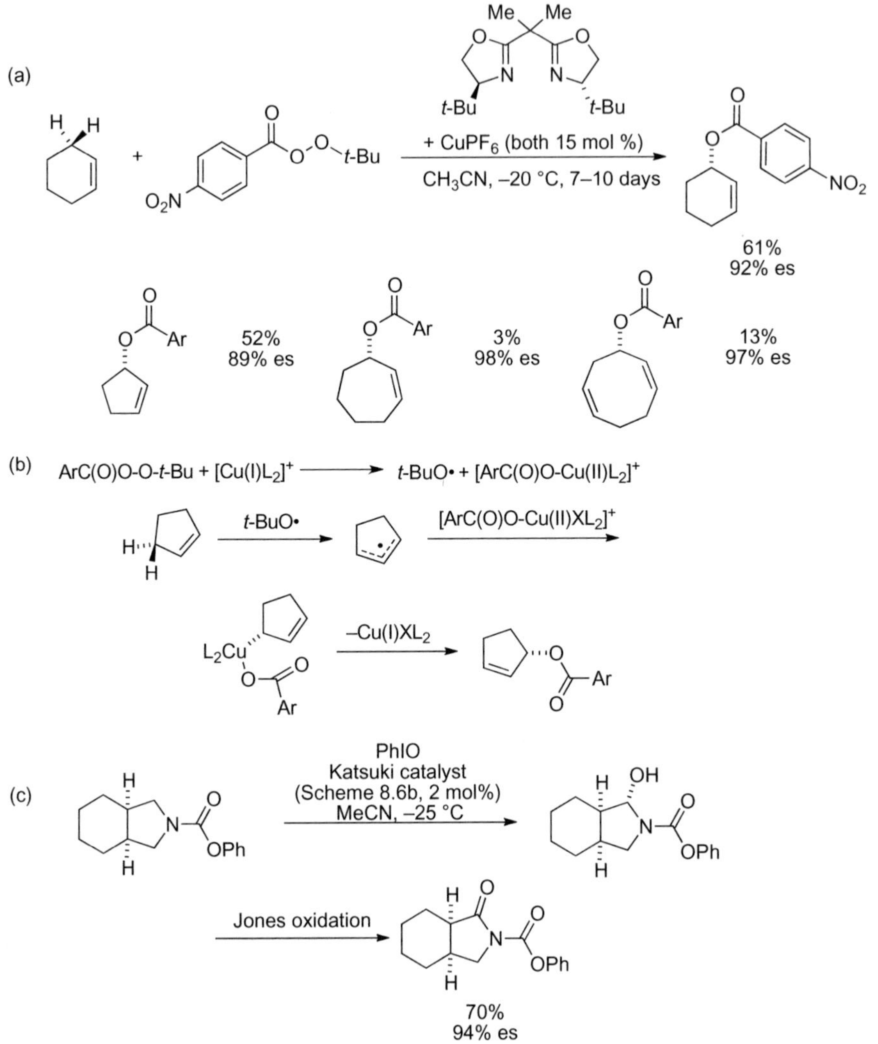

SCHEME 8.29 Formally group-selection insertion of oxygen into enantiotopic C–H bonds. (a) An asymmetric Kharasch reaction [221] and (b) its proposed mechanism [221,222]. (c) Oxidation of an isoindoline derivative [224,225].

versions of the classical Baeyer–Villiger and Schmidt reactions results in useful routes to lactones and lactams, respectively (Scheme 8.31a).

Although highly efficient enzymatic Baeyer–Villiger reactions have been reported [229–232], results of attempts using abiotic catalysts have been much less successful (reviews: [230,233,234]). In landmark work, an early kinetic resolution published by Bolm used a chiral copper complex to activate molecular oxygen for addition to racemic 2-substituted

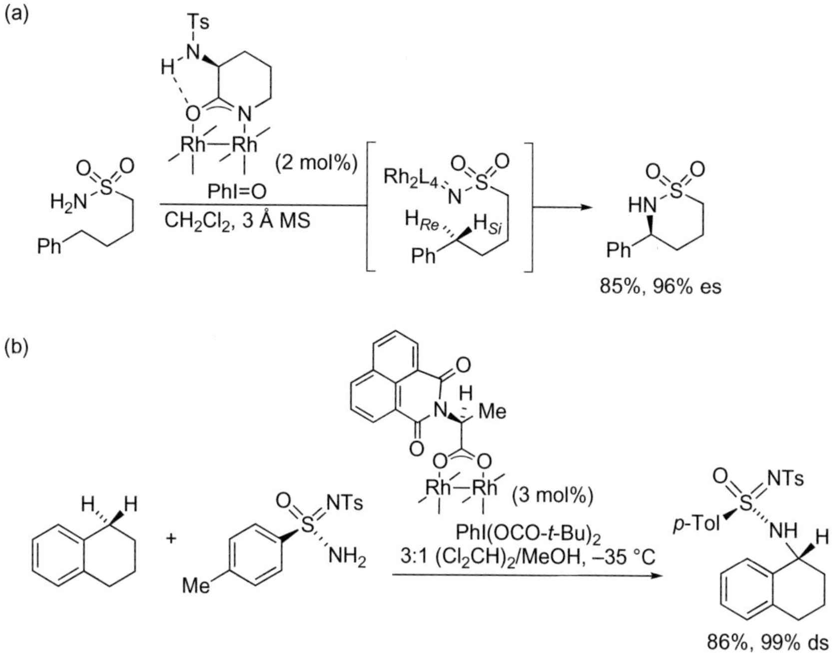

(a)

85%, 96% es

(b)

86%, 99% ds

SCHEME 8.30 (a) Enantioselective benzylic amination by Du Bois [226] and (b) a diastereoselective version by Dodd and Dauban [228]. The catalysts are formulated as Rh₂L₄, with only one of the four ligands explicitly shown.

cyclohexanone (Scheme 8.31b [235]). In this way, the corresponding lactone was prepared and enriched in the *R* isomer in good enantioselectivity. Numerous attempts to improve on these results were made over the next two decades but only rarely approached the selectivities routinely obtained by enzymes. One exception is shown in Scheme 8.31c. Katsuki and co-workers used a salen catalyst to promote the ring-expansion of 4-phenylcyclohexanone to the valerolactam analog with solid enantioselectivity [236]. The inset to Scheme 8.31c summarizes the issues that are necessary to obtain good selectivity: (1) facial selectivity for the addition of the hydroperoxide to the ketone and (2) selective orientation of one of the two potential migrating carbons *antiperiplanar* to the O—O being cleaved in the course of the reaction. The latter condition depends on the chiral environment formed by the large ligands of the catalyst, whereas the former is a function of the substrate's tendency to direct any reagent from a particular direction (which might be one factor leading to the highly selective reactions obtained on the niche ketone shown in Scheme 8.31d).

Similar systems have also been employed for the kinetic resolution of unsymmetrical ketones (Scheme 8.32 [237]). The particular results depicted were obtained using the

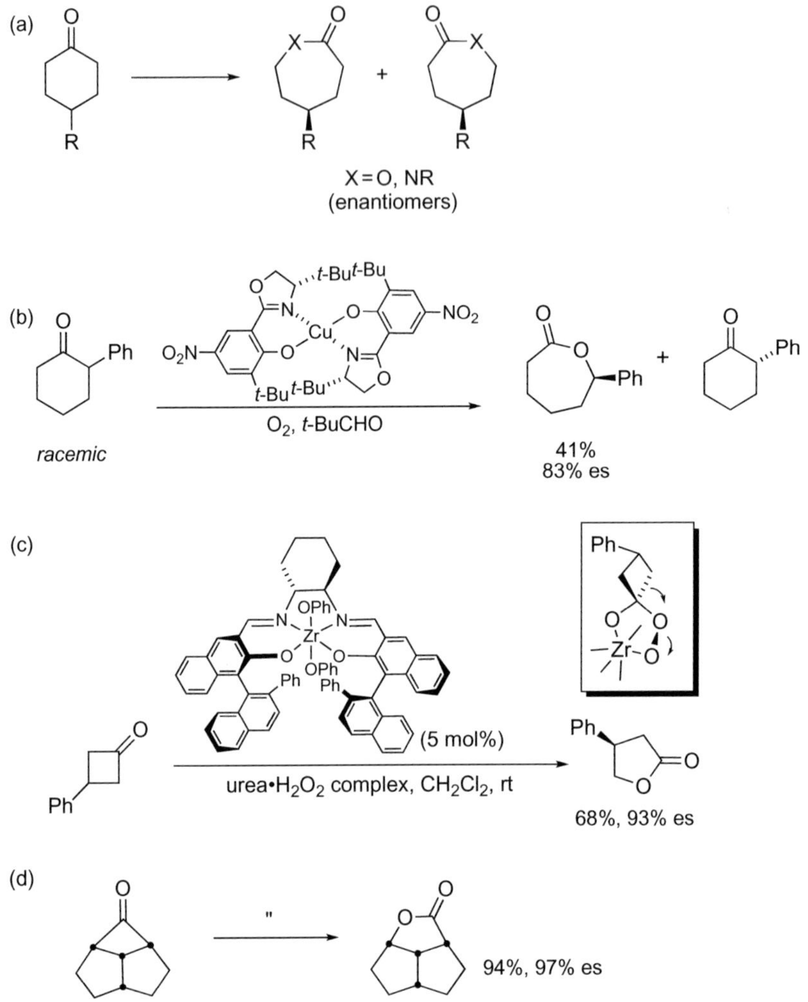

SCHEME 8.31   (a) A generic asymmetric ring-expansion reaction. (b) A kinetic resolution using an asymmetric Baeyer–Villiger catalyst [235]. (c) and (d) Asymmetric Baeyer–Villiger reactions [233,236].

above-noted Katsuki protocol. In this example, although the difference in reactivity between the two isomeric ketones was modest ($s \sim 4.2$), both isomers reacted with good-to-high stereoselectivity at 76% conversion. Note that in this chiral ketone, which group migrates is now viewed as a matter of regioselectivity (or product selectivity), with the preference of the catalyst edging out that of the substrate in determining which products are preferred (recall that standard Baeyer–Villiger oxidations generally occur with migration of the more substituted carbon, hence the designation of these products in Scheme 8.32 as "normal").

In contrast, there is no known enzymatic version of an asymmetric nitrogen insertion process, although two methods that utilize variations on known ring-expansion processes have been reported. The first utilizes oxaziridines as the first isolated intermediate in a three-step

Racemic mixture used: "Normal lactones"   "Abnormal lactones"

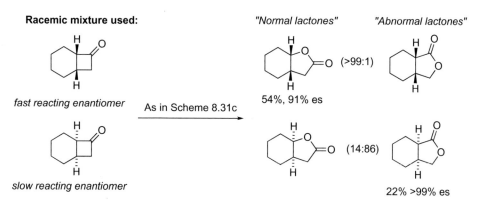

fast reacting enantiomer   As in Scheme 8.31c   54%, 91% es

slow reacting enantiomer   22% >99% es

SCHEME 8.32 Kinetic resolution using enantioselective Baeyer—Villiger reactions [237].

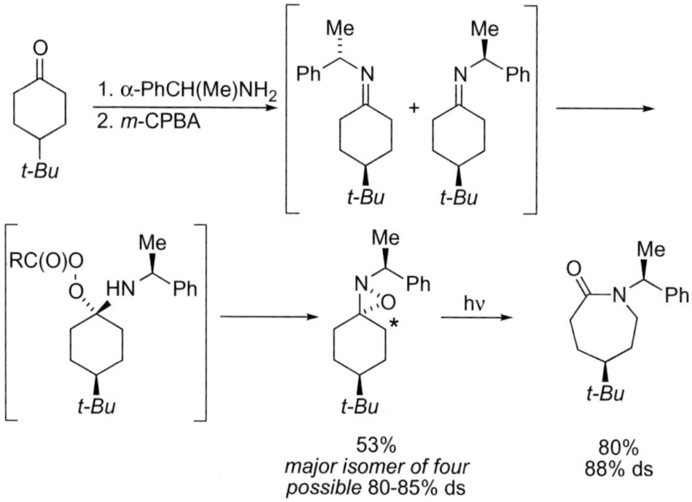

SCHEME 8.33 Asymmetric nitrogen ring-expansion reactions of ketones utilizing oxaziridine synthesis and photolysis [238].

overall sequence (Scheme 8.33 [238]). Axially dissymmetric spirocyclic oxaziridines are available by the oxidation of imines derived from the starting ketone and α-methylbenzyla-mine. The reaction utilizes one element of diastereofacial selectivity (interpreted here as equatorial attack of the peracid oxidizing agent) and an interesting kind of selectivity whereby intramolecular attack of the now-secondary nitrogen causes ejection of a carboxylic acid; in this latter reaction, the stereogenic nitrogen atom of the oxaziridine (which is not epimerizable at room temperature; [239]) is formed with good diastereoselectivity. The oxaziridine is then photolyzed, which causes the molecule to undergo bond reorganization to give the lactam. This reaction takes advantage of the known (but not well-understood) tendency of the oxaziridine to react with regioselective migration of the bond *antiperiplanar* to the

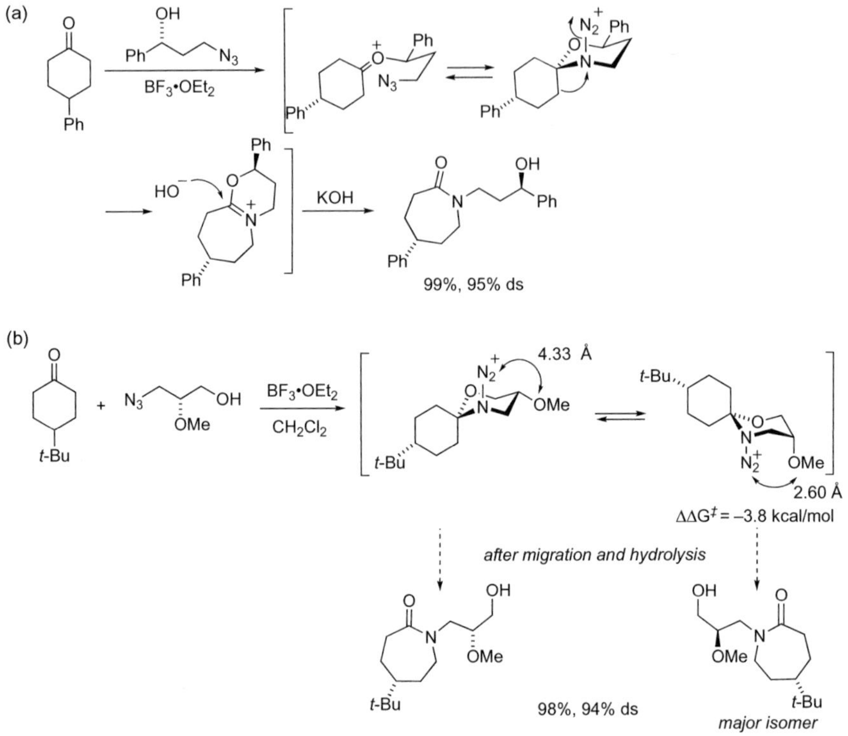

SCHEME 8.34 (a) Asymmetric ring-expansion azide-Schmidt reaction [241,242]. (b) Evidence for an attractive electrostatic interaction resulting from an n-cation interaction [243]. The energy differences noted were calculated using *ab initio* methods.

lone pair on the nitrogen atom (marked with an asterisk in the scheme) [240]. Reductive removal of the chiral substituent on nitrogen then finishes off this overall ring-expansion protocol.

A few examples of a similar conversion utilizing an azide-based variant of the Schmidt reaction have also appeared (Scheme 8.34 [241,242]). The reaction is thought to entail the formation of a hemiacetal between the hydroxyl group of the reagent and the ketone; dehydration of the hemiacetal leads to an oxonium ion that is subject to attack by the now-tethered azido group (Scheme 8.34a). Formation of the more stable spirocyclic intermediate followed by migration of the bond *antiperiplanar* to the departing N_2 substituent was proposed to lead to the observed lactam. Again, removal of the chiral substituent on nitrogen afforded the formal asymmetric Schmidt reaction product. A provocative element of this chemistry is that attractive interactions might have an effect on particular examples of this reaction specifically those in which the substituent on the newly-formed heterocyclic ring and the positively charged N_2^+ group have a 1,3-diaxial relationship in the intermediate.[10] In such cases,

---

10. We can conservatively estimate that $>95\%$ of the asymmetric transformations described in this book depend ultimately on the relief of repelling through-space (steric) interactions of one type or another.

nonbonded stabilization of the intermediate that bears an electron-rich group in the axial orientation occurs, leading to the observed product (Scheme 8.34b); this can either be the interactions shown involving the oxygen nonbonded electrons and the cationic nitrogen [243], or analogous (and more common) π-cation interactions where the methoxy group in the scheme is replaced by an aromatic group [244,245].

C—C Bonds from C—H Bonds

Forming C—C bonds is very often cited as an activity at the core of organic chemistry and the subject has accordingly received much attention throughout this book. Functionalizing "unactivated" C—H bonds is attractive from a synthetic planning perspective because it is desirable to inexorably increase complexity as one moves along a synthetic pathway and especially desirable to avoid introducing functional groups for the express purpose of converting them into something else.[1] The challenges of doing so in complex organic molecules, which may contain *numerous* unactivated C—H bonds, entail both obtaining appropriately chemoselective reactions and, if one is doing asymmetric synthesis, requiring that one differentiate between enantiotopic C—H bonds as well (the general issues involved in C—H functionalization have been thoughtfully reviewed by Baran [1]).

An early (2005!) example published by Ellman and Bergman utilized a chiral auxiliary approach that entailed the *in situ* generation of an imine and subsequent activation/cyclization using a rhodium catalyst system. The imine substituent plays a directing role in this reaction, which was able to afford high stereoselectivities in a variety of cases [2]. It proved necessary to employ a chiral auxiliary approach in the application of this reaction to a total synthesis of lithospermic acid [3]. Here the C—H bond substituted was the only one available for intramolecular cyclization.

---

1. What actually constitutes an "unactivated" functional group is up for some debate. For example, are C—H bonds of an aromatic ring "unactivated" in the context of the Friedel—Crafts reaction? Here, we take a "we know it when we see it" stance in which obviously activated C—H bonds (such as those on a methylene next to a carbonyl) are excluded from discussion.

No one would accuse cyclohexane as being functionalized, which is one reason that the conversion of C_6H_{12} to the ester shown is so impressive [4]. Davies has utilized rhodium carbene complexes of the type shown to tremendous effect not only in insertion chemistry but also for cyclopropanations and other transformations (see Section 6.1.5.2 of the present book and also an instructive tutorial review [5]). In this case, one might reasonably consider the roles of "reagent" and "substrate" reversed from their usual understanding since the stereocenter ends up on the carbenoid precursor and not on the cyclohexane, but this is not a general limitation of the technology (cf. the corresponding diastereoselective reaction of a silyl enol ether [6]).

Finally, consider the differentiation of enantiotopic ethyl groups accomplished by Sames en route to (−)-rhazinilam [7]. Once again, the expedience of attaching a chiral auxiliary that controls the delivery of a metal to a particular group was needed for high selectivity; the auxiliary also allowed the separation of diastereomers prior to transamination, which afforded an aniline product. Carbonylation followed by amide bond formation finished off the total synthesis. In this case, the differentiation of the two ethyl groups took place via the equilibration of the heterocyclic ring prior to CH activation under strong acid promotion.

There is little doubt that a third edition of this book, 15 years after this writing, would be able to track the maturation of this field from dependence on chiral auxiliaries to the wide availability of truly effective and general catalysts, just as has been the case for numerous other reactions contained in this edition.

References
[1] Gutekunst, W. R.; Baran, P. S. *Chem. Soc. Rev.* **2011**, *40*, 1976–1991.
[2] Thalji, R. K.; Ellman, J. A.; Bergman, R. G. *J. Am. Chem. Soc.* **2004**, *126*, 7192–7193.
[3] O'Malley, S. J.; Tan, K. L.; Watzke, A.; Bergman, R. G.; Ellman, J. A. *J. Am. Chem. Soc.* **2005**, *127*, 13496–13497.
[4] Davies, H. M. L.; Hansen, T.; Churchill, M. R. *J. Am. Chem. Soc.* **2000**, *122*, 3063–3070.
[5] Davies, H. M. L.; Morton, D. *Chem. Soc. Rev.* **2011**, *40*, 1857–1869.
[6] Davies, H. M. L.; Antoulinakis, E. G.; Hansen, T. *Org. Lett.* **1999**, *1*, 383–386.
[7] Johnson, J. A.; Li, N.; Sames, D. *J. Am. Chem. Soc.* **2002**, *124*, 6900–6903.

REFERENCES

[1] Sharpless, K. B. *Angew. Chem., Int. Ed.* **2002**, *41*, 2024–2032.
[2] Morrison, J. D.; Mosher, H. S. In *Asymmetric Organic Reactions*; American Chemical Society: Washington, D.C., 1976. pp. 258–262.
[3] Henbest, S. W.; Wilson, R. A. L. *J. Chem. Soc.* **1957**, 1965–1968.
[4] Chavdarian, C. G.; Heathcock, C. H. *Synth. Commun.* **1976**, *6*, 277–280.
[5] Kocovsky, P.; Stary, I. *J. Org. Chem.* **1990**, *55*, 3236–3243.
[6] Hoveyda, A. H.; Evans, D. A.; Fu, G. C. *Chem. Rev.* **1993**, *93*, 1307–1370.
[7] Chamberlain, P.; Roberts, M. L.; Whitham, G. H. *J. Chem. Soc. B* **1970**, 1374–1381.
[8] Sharpless, K. B.; Verhoeven, T. R. *Aldrichim. Acta* **1979**, *12*, 63–74.
[9] Jørgensen, K. A. *Chem. Rev.* **1989**, *89*, 431–458.
[10] Sharpless, K. B.; Michaelson, R. C. *J. Am. Chem. Soc.* **1973**, *95*, 6136–6137.
[11] Rossiter, B. E.; Verhoeven, T. R.; Sharpless, K. B. *Tetrahedron Lett.* **1979**, *20*, 4733–4736.
[12] Katsuki, T.; Sharpless, K. B. *J. Am. Chem. Soc.* **1980**, *102*, 5974–5976.
[13] Finn, M. G.; Sharpless, K. B. In *Asymmetric Synthesis*; Morrison, J. D., Ed.; Academic: Orlando, 1985; Vol. 5, pp. 247–308.
[14] Rossiter, B. E. In *Asymmetric Synthesis*; Morrison, J. D., Ed.; Academic: Orlando, 1985; Vol. 5, pp. 194–246.
[15] Johnson, R. A.; Sharpless, K. B. In *Catalytic Asymmetric Synthesis*; Ojima, I., Ed.; VCH: New York, 1993, pp. 103–158.
[16] Katsuki, T.; Martin, V. S. *Org. React.* **1996**, *48*, 1–299.
[17] Woodard, S. S.; Finn, M. G.; Sharpless, K. B. *J. Am. Chem. Soc.* **1991**, *113*, 106–113.
[18] Finn, M. G.; Sharpless, K. B. *J. Am. Chem. Soc.* **1991**, *113*, 113–126.
[19] Wu, Y.-D.; Lai, D. K. W. *J. Am. Chem. Soc.* **1995**, *117*, 11327–11336.
[20] Wu, Y.-D.; Lai, D. K. W. *J. Org. Chem.* **1995**, *60*, 673–680.
[21] Cui, M.; Adam, W.; Shen, J. H.; Luo, X. M.; Tan, X. J.; Chen, K. X.; Ji, R. Y.; Jiang, H. L. *J. Org. Chem.* **2002**, *67*, 1427–1435.
[22] Kuznetsov, M. L.; Pessoa, J. C. *Dalton Trans.* **2009**, 5460–5468.
[23] Corey, E. J. *J. Org. Chem.* **1990**, *55*, 1693–1694.
[24] Williams, I. D.; Pederson, S. F.; Sharpless, K. B.; Lippard, S. J. *J. Am. Chem. Soc.* **1984**, *106*, 6430–6431.
[25] Gao, Y.; Hanson, R.; Klunder, J. M.; Ko, S. Y.; Masamune, H.; Sharpless, K. B. *J. Am. Chem. Soc.* **1987**, *109*, 5765–5780.
[26] Wood, R. D.; Ganem, B. *Tetrahedron Lett.* **1982**, *23*, 707–710.
[27] Garner, P.; Park, J. M.; Rotello, V. *Tetrahedron Lett.* **1985**, *26*, 3299–3302.
[28] Hamon, D. P. G.; Shirley, N. J. *J. Chem. Soc. Chem. Commun.* **1988**, 425–426.
[29] Jacobsen, E. N. In *Catalytic Asymmetric Synthesis*; Ojima, I., Ed.; VCH: New York, 1993, pp. 159–202.

[30] Zhang, W.; Loebach, J. L.; Wilson, S. R.; Jacobsen, E. N. *J. Am. Chem. Soc.* **1990**, *112*, 2801−2803.

[31] Jacobsen, E. N.; Zhang, W.; Muci, A.; Ecker, J. R.; Deng, L. *J. Am. Chem. Soc.* **1991**, *113*, 7063−7064.

[32] Irie, R.; Noda, K.; Ito, Y.; Matsumoto, N.; Katsuki, T. *Tetrahedron Lett.* **1990**, *31*, 7345−7348.

[33] Irie, R.; Noda, K.; Ito, Y.; Matsumoto, N.; Katsuki, T. *Tetrahedron: Asymmetry* **1991**, *2*, 481−494.

[34] Irie, R.; Noda, K.; Ito, Y.; Katsuki, T. *Tetrahedron Lett.* **1991**, *32*, 1055−1058.

[35] Katsuki, T. *Synlett* **2003**, 281−297.

[36] McGarrigle, E. M.; Gilheany, D. G. *Chem. Rev.* **2005**, *105*, 1563−1602.

[37] Matsumoto, K.; Katsuki, T. In *Catalytic Asymmetric Synthesis, Third Edition*; Ojima, I., Ed.; John Wiley & Sons, Inc.: Hoboken, NJ, USA, 2010, pp. 839−890.

[38] Samsel, E. G.; Srinivasan, K.; Kochi, J. K. *J. Am. Chem. Soc.* **1985**, *107*, 7606−7617.

[39] Palucki, M.; Pospisil, P. J.; Zhang, W.; Jacobsen, E. N. *J. Am. Chem. Soc.* **1994**, *116*, 9333−9334.

[40] Brandes, B. D.; Jacobsen, E. N. *J. Org. Chem.* **1994**, *59*, 4378−4380.

[41] Brandes, B.; Jacobsen, E. N. *Tetrahedron Lett.* **1995**, *36*, 5123−5126.

[42] Jacobsen, H.; Cavallo, L. *Chem.−Eur. J.* **2001**, *7*, 800−807.

[43] Fu, H.; Look, G. C.; Zhang, W.; Jacobsen, E. N.; Wong, C. H. *J. Org. Chem.* **1991**, *56*, 6497−6500.

[44] Chang, S.; Galvin, J. M.; Jacobsen, E. N. *J. Am. Chem. Soc.* **1994**, *116*, 6937−6938.

[45] Lee, N. H.; Jacobsen, E. N. *Tetrahedron Lett.* **1991**, *32*, 6533−6536.

[46] Chang, S.; Heid, R. M.; Jacobsen, E. N. *Tetrahedron Lett.* **1994**, *1994*, 669−672.

[47] Frohn, M.; Shi, Y. *Synthesis* **2000**, 1979−2000.

[48] Wong, O. A.; Shi, Y. *Chem. Rev.* **2008**, *108*, 3958−3987.

[49] Shi, Y. In *Modern Oxidation Methods, Second Edition*; Bäckvall, J.-E., Ed.; Wiley-VCH: Weinheim, Germany, 2010, pp. 85−115.

[50] Tu, Y.; Wang, Z.-X.; Shi, Y. *J. Am. Chem. Soc.* **1996**, *118*, 9806−9807.

[51] Wang, Z.-X.; Tu, Y.; Frohn, M.; Shi, Y. *J. Org. Chem.* **1997**, *62*, 2328−2329.

[52] Wang, Z.-X.; Tu, Y.; Frohn, M.; Zhang, J.-R.; Shi, Y. *J. Am. Chem. Soc.* **1997**, *119*, 11224−11235.

[53] Zhao, M.-X.; Shi, Y. *J. Org. Chem.* **2006**, *71*, 5377−5379.

[54] Shi, Y. *Acc. Chem. Res.* **2004**, *37*, 488−496.

[55] Warren, J. D.; Shi, Y. *J. Org. Chem.* **1999**, *64*, 7675−7677.

[56] Wang, Z.-X.; Shi, Y. *J. Org. Chem.* **1998**, *63*, 3099−3104.

[57] Frohn, M.; Dalkiewicz, M.; Tu, Y.; Wang, Z.-X.; Shi, Y. *J. Org. Chem.* **1998**, *63*, 2948−2953.

[58] Wang, Z.-X.; Cao, G.-A.; Shi, Y. *J. Org. Chem.* **1999**, *64*, 7646−7650.

[59] Solladié-Cavallo, A.; Jierry, L.; Klein, A. *C. R. Chim.* **2003**, *6*, 603−606.

[60] Singleton, D. A.; Wang, Z. *J. Am. Chem. Soc.* **2005**, *127*, 6679−6685.

[61] Porter, M. J.; Skidmore, J. *Chem. Commun.* **2000**, 1215−1225.

[62] Lauret, C.; Roberts, S. M. *Aldrichim. Acta* **2002**, *35*, 47−51.

[63] Diez, D.; Nunez, M. G.; Anton, A. B.; Garcia, P.; Moro, R. F.; Garrido, N. M.; Marcos, I. S.; Basabe, P.; Urones, J. G. *Curr. Org. Synth.* **2008**, *5*, 186−216.

[64] Lattanzi, A. *Curr. Org. Synth.* **2008**, *5*, 117−133.

[65] Porter, M. J.; Skidmore, J. *Org. React.* **2009**, *74*, 425−672.

[66] Helder, R.; Hummelen, J. C.; Laane, R. W. P. M.; Wiering, J. S.; Wynberg, H. *Tetrahedron Lett.* **1976**, 1831−1834.

[67] Arai, S.; Tsuge, H.; Oku, M.; Miura, M.; Shioiri, T. *Tetrahedron* **2002**, *58*, 1623−1630.

[68] Colonna, S.; Molinari, H.; Banfi, S.; Juliá, S.; Masana, J.; Alvarez, A. *Tetrahedron* **1983**, *39*, 1635−1641.

[69] Dorow, R. L.; Tymonko, S. A. *Tetrahedron Lett.* **2006**, *47*, 2493−2495.

[70] Lygo, B.; Gardiner, S. D.; McLeod, M. C.; To, D. C. M. *Org. Biomol. Chem.* **2007**, *5*, 2283−2290.

[71] Geller, T.; Roberts, S. M. *J. Chem. Soc., Perkin Trans. 1* **1999**, 1397−1398.

[72] Flisak, J. R.; Gombatz, K. J.; Holmes, M. M.; Jarmas, A. A.; Lantos, I.; Mendelson, W. L.; Novack, V. J.; Remich, J. J.; Snyder, L. *J. Org. Chem.* **1993**, *58*, 6247−6254.

[73] Adger, B. M.; Barkley, J. V.; Bergeron, S.; Cappi, M. W.; Flowerdew, B. E.; Jackson, M. P.; McCague, R.; Nugent, T. C.; Roberts, S. M. *J. Chem. Soc., Perkin Trans. 1* **1997**, 3501−3508.

[74] Elston, C. L.; Jackson, R. F. W.; MacDonald, S. J. F.; Murray, P. J. *Angew. Chem., Int. Ed. Engl.* **1997**, *36*, 410−412.

[75] Hinch, M.; Jacques, O.; Drago, C.; Caggiano, L.; Jackson, R. F. W.; Dexter, C.; Anson, M. S.; Macdonald, S. J. F. *J. Mol. Catal. A: Chem.* **2006**, *251*, 123−128.

[76] Nemoto, T.; Ohshima, T.; Yamaguchi, K.; Shibasaki, M. *J. Am. Chem. Soc.* **2001**, *123*, 2725−2732.

[77] Nemoto, T.; Ohshima, T.; Shibasaki, M. *J. Am. Chem. Soc.* **2001**, *123*, 9474−9475.

[78] Nemoto, T.; Kakei, H.; Gnanadesikan, V.; Tosaki, S.-Y.; Ohshima, T.; Shibasaki, M. *J. Am. Chem. Soc.* **2002**, *124*, 14544−14545.

[79] Kakei, H.; Tsuji, R.; Ohshima, T.; Shibasaki, M. *J. Am. Chem. Soc.* **2005**, *127*, 8962−8963.

[80] Martin, V. S.; Woodard, S. S.; Katsuki, T.; Yamada, Y.; Ikeda, M.; Sharpless, K. B. *J. Am. Chem. Soc.* **1981**, *103*, 6237−6240.

[81] Rossiter, B. E.; Sharpless, K. B. *J. Org. Chem.* **1984**, *49*, 3707−3711.

[82] Sharpless, K. B.; Behrens, C. H.; Katsuki, T.; Lee, A. W. M.; Martin, V. S.; Takatani, M.; Viti, S. M.; Walker, F. J.; Woodard, S. S. *Pure Appl. Chem.* **1983**, *55*, 589−604.

[83] Kusakabe, M.; Sato, F. *J. Org. Chem.* **1989**, *54*, 3486−3487.

[84] Miyano, S.; Lu, L. D.-L.; Viti, S. M.; Sharpless, K. B. *J. Org. Chem.* **1985**, *40*, 4350−4360.

[85] Marshall, J. A.; Flynn, K. E. *J. Am. Chem. Soc.* **1983**, *105*, 3360−3362.

[86] Ko, S. Y.; Lee, A. W. M.; Masamune, S.; Reed, L. A., III; Sharpless, K. B.; Walker, F. J. *Science* **1983**, *220*, 949−951.

[87] Ko, S. Y.; Lee, A. W. M.; Masamune, S.; Reed, I. L. A.; Sharpless, K. B.; Walker, F. J. *Tetrahedron* **1990**, *46*, 245−264.

[88] Behrens, C. H.; Ko, S. Y.; Sharpless, K. B.; Walker, F. J. *J. Org. Chem.* **1985**, *50*, 5687−5696.

[89] Stork, G.; Rychnovsky, S. D. *Pure Appl. Chem.* **1987**, *59*, 345.

[90] Masamune, S.; Choy, W. *Aldrichim. Acta* **1982**, *15*, 47−63.

[91] Takano, S.; Sakurai, K.; Hatakeyama, S. *J. Chem. Soc., Chem. Commun.* **1985**, 1759−1761.

[92] Babine, R. E. *Tetrahedron Lett.* **1986**, *27*, 5791−5794.

[93] Häfele, B.; Schröter, D.; Jäger, V. *Angew. Chem., Int. Ed. Engl.* **1986**, *25*, 87−89.

[94] Schreiber, S. L.; Schreiber, T. S.; Smith, D. B. *J. Am. Chem. Soc.* **1987**, *109*, 1525−1529.

[95] Smith, D. B.; Wang, Z.; Schreiber, S. L. *Tetrahedron* **1990**, *46*, 4793−4808.

[96] Wang, Y.-F.; Chen, C.-S.; Girdaukas, G.; Sih, C. J. *J. Am. Chem. Soc.* **1984**, *106*, 3695−3696.

[97] Schreiber, S. L. *Chem. Scr.* **1987**, *27*, 563−566.

[98] Johnson, J. B. In *Science of Synthesis, Stereoselective Synthesis*; De Vries, J. G., Molander, G. A., Evans, P. A., Eds.; Georg Thieme Verlag: Stuttgart, 2011; Vol. 3, pp. 759−827.

[99] Overman, L. E.; Sugai, S. *J. Org. Chem.* **1985**, *50*, 4154−4155.

[100] Nugent, W. A. *J. Am. Chem. Soc.* **1992**, *114*, 2768−2769.

[101] Martinez, L. E.; Leighton, J. L.; Carsten, D. H.; Jacobsen, E. N. *J. Am. Chem. Soc.* **1995**, *117*, 5897−5898 (see this paper for references to earlier approaches to this problem).

[102] Larrow, J. F.; Schaus, S. E.; Jacobsen, E. N. *J. Am. Chem. Soc.* **1996**, *118*, 7420−7421.

[103] Lebel, H.; Jacobsen, E. N. *Tetrahedron Lett.* **1999**, *40*, 7303−7306.

[104] Ready, J. M.; Jacobsen, E. N. *J. Am. Chem. Soc.* **1999**, *121*, 6086−6087.

[105] Tokunaga, M.; Larrow, J. F.; Kakiuchi, F.; Jacobsen, E. N. *Science* **1997**, *277*, 936−938.

[106] Pellissier, H. *Tetrahedron* **2010**, *66*, 1509−1555.

[107] Evans, D. A.; Woerpel, K. A.; Hinman, M. M.; Faul, M. M. *J. Am. Chem. Soc.* **1991**, *113*, 726−728.

[108] Estéoule, A.; Durán, F.; Retailleau, P.; Dodd, R. H.; Dauban, P. *Synthesis* **2007**, *2007*, 1251,, 1260.

[109] O'Connor, K. J.; Wey, S.-J.; Burrows, C. J. *Tetrahedron Lett.* **1992**, *33*, 1001−1004.

[110] Li, Z.; Conser, K. R.; Jacobsen, E. N. *J. Am. Chem. Soc.* **1993**, *115*, 5326−5327.

[111] Li, Z.; Quan, R. W.; Jacobsen, E. N. *J. Am. Chem. Soc.* **1995**, *117*, 5889−5890.

[112] Murugan, E.; Siva, A. *Synthesis* **2005**, 2022−2028.

[113] Evans, D. A.; Miller, S. J.; Lectka, T. *J. Am. Chem. Soc.* **1993**, *115*, 6460−6461.

[114] Omura, K.; Murakami, M.; Uchida, T.; Irie, R.; Katsuki, T. *Chem. Lett.* **2003**, *32*, 354−355.

[115] Schröder, M. *Chem. Rev.* **1980**, *80*, 187−213.

[116] Lohray, B. B. *Tetrahedron: Asymmetry* **1992**, *3*, 1317−1349.

[117] Johnson, R. A.; Sharpless, K. B. In *Catalytic Asymmetric Synthesis*; Ojima, I., Ed.; VCH: New York, 1993, pp. 227−272.

[118] Kolb, H. C.; VanNieuwenhze, M. S.; Sharpless, K. B. *Chem. Rev.* **1994**, *94*, 2483−2547.

[119] Noe, M. C.; Letavic, M. A.; Snow, S. L. *Org. React.* **2005**, *66*, 109−625.

[120] Zaitsev, A. B.; Adolfsson, H. *Synthesis* **2006**, 1725−1756.

[121] Hofmann, K. *Chem. Ber.* **1912**, *45*, 3329−3336.

[122] Van Rheenen, V.; Kelly, R. C.; Cha, D. Y. *Tetrahedron Lett.* **1976**, 1973−1976.

[123] Criegee, R.; Marchand, B.; Wannowius, H. *Liebigs Ann. Chem.* **1942**, *550*, 99−133.

[124] Hentges, S. G.; Sharpless, K. B. *J. Am. Chem. Soc.* **1980**, *102*, 4263−4265.
[125] Hanessian, S.; Meffre, P.; Girard, M.; Beaudoin, S.; Sancéau, J.-Y.; Bennani, Y. *J. Org. Chem.* **1993**, *58*, 1991−1993.
[126] Tomioka, K.; Nakajima, M.; Koga, K. *J. Am. Chem. Soc.* **1987**, *109*, 6213−6215.
[127] Jacobsen, E. N.; Markó, I.; Mungall, W. S.; Schröder, G.; Sharpless, K. B. *J. Am. Chem. Soc.* **1988**, *110*, 1968−1970.
[128] Wai, J. S. M.; Markó, I.; France, M. B.; Svendsen, J. S.; Sharpless, K. B. *J. Am. Chem. Soc.* **1989**, *111*, 1123−1125.
[129] Kwong, H.; Sorato, C.; Ogino, Y.; Chen, H.; Sharpless, K. B. *Tetrahedron Lett.* **1990**, *31*, 2999−3002.
[130] Sharpless, K. B.; Amberg, W.; Bennani, Y. L.; Crispino, G. A.; Hartung, J.; Jeong, Kyu-Sung; Kwong, H.-L.; Morikawa, K.; Wang, Z.-M.; Xu, D.; Zhang, X.-L. *J. Org. Chem.* **1992**, *57*, 2768−2771.
[131] Amberg, W.; Bennani, Y. L.; Chadha, R. K.; Crispino, G. A.; Davis, W. D.; Hartung, J.; Jeong, K.-S.; Ogino, Y.; Shibata, T.; Sharpless, K. B. *J. Org. Chem.* **1993**, *58*, 844−849.
[132] Crispino, G.; Jeong, K.-S.; Kolb, H. C.; Wang, Z.-M.; Xu, D.; Sharpless, K. B. *J. Org. Chem.* **1993**, *58*, 3785−3786.
[133] Morikawa, K.; Park, J.; Andersson, P. G.; Hashiyama, T.; Sharpless, K. B. *J. Am. Chem. Soc.* **1993**, *115*, 8463−8464.
[134] Wang, L.; Sharpless, K. B. *J. Am. Chem. Soc.* **1992**, *114*, 7568−7570.
[135] Becker, H.; Sharpless, K. B. *Angew. Chem., Int. Ed. Engl.* **1996**, *35*, 448−451.
[136] Becker, H.; King, S. B.; Taniguchi, M.; Vanhessche, K. P. M.; Sharpless, K. B. *J. Org. Chem.* **1995**, *60*, 3940−3941.
[137] Wang, Z.-M.; Zhang, X.-L.; Sharpless, K. B. *Tetrahedron Lett.* **1993**, *34*, 2267−2270.
[138] Wang, Z. M.; Sharpless, K. B. *Synlett* **1993**, 603−604.
[139] Wang, X.; Zak, M.; Maddess, M.; O'Shea, P.; Tillyer, R.; Grabowski, E. J. J.; Reider, P. J. *Tetrahedron Lett.* **2000**, *41*, 4865−4869.
[140] Vanhessche, K. P. M.; Sharpless, K. B. *J. Org. Chem.* **1996**, *61*, 7978−7979.
[141] Fristrup, P.; Tanner, D.; Norrby, P.-O. *Chirality* **2003**, *15*, 360−368.
[142] Norrby, P.-O.; Kolb, H. C.; Sharpless, K. B. *J. Am. Chem. Soc.* **1994**, *116*, 8470−8478.
[143] Drudis-Solé, G.; Ujaque, G.; Maseras, F.; Lledós, A. In *Topics in Organometallic Chemistry*; Frenking, G., Ed.; Springer: Heidelberg, 2005; Vol. 12, pp. 311−327.
[144] Corey, E. J.; Noe, M. C. *J. Am. Chem. Soc.* **1993**, *115*, 12579−12580.
[145] Corey, E. J.; Noe, M. C. *J. Am. Chem. Soc.* **1996**, *118*, 319−329.
[146] Moitessier, N.; Henry, C.; Len, C.; Chapleur, Y. *J. Org. Chem.* **2002**, *67*, 7275−7282.
[147] Jurczak, J.; Pikul, S.; Bauer, T. *Tetrahedron* **1986**, *42*, 447−488.
[148] Oi, R.; Sharpless, K. B. *Tetrahedron Lett.* **1992**, *33*, 2095−2098.
[149] Gao, Y.; Sharpless, K. B. *J. Am. Chem. Soc.* **1988**, *110*, 7538−7539.
[150] Ko, S. Y.; Malik, M. *Tetrahedron Lett.* **1993**, *34*, 4675−4678.
[151] Corey, E. J.; Guzman-Perez, A.; Noe, M. C. *J. Am. Chem. Soc.* **1994**, *116*, 12109−12110.
[152] Matsunaga, N.; Kaku, T.; Ojida, A.; Tasaka, A. *Tetrahedron: Asymmetry* **2004**, *15*, 2021−2028.
[153] Rawal, V. J.; Florjancic, A. S.; Singh, S. P. *Tetrahedron Lett.* **1994**, *35*, 8985−8988.
[154] Masamune, S.; Choy, W.; Petersen, J. S.; Sita, L. R. *Angew. Chem., Int. Ed. Engl.* **1985**, *24*, 1−30.
[155] Cha, J. K.; Christ, W. C.; Kishi, Y. *Tetrahedron* **1984**, *40*, 2247−2255.
[156] Stork, G.; Kahn, M. *Tetrahedron Lett.* **1983**, *24*, 3951−3954.
[157] Corey, E. J.; Noe, M. C.; Guzman-Perez, A. *J. Am. Chem. Soc.* **1995**, *117*, 10817−10824.
[158] Crispino, G. A.; Ho, P. T.; Sharpless, K. B. *Science* **1993**, *259*, 64−66.
[159] Vilotijevic, I.; Jamison, T. F. *Science* **2007**, *317*, 1189−1192.
[160] Morikawa, K.; Sharpless, K. B. *Tetrahedron Lett.* **1993**, *34*, 5575−5578.
[161] Ma, J.-A.; Cahard, D. *Chem. Rev.* **2004**, *104*, 6119−6146.
[162] Vilaivan, T.; Bhanthumnavin, W. *Molecules* **2010**, *15*, 917−958.
[163] Smith, A. M. R.; Hii, K. K. *Chem. Rev.* **2011**, *111*, 1637−1656.
[164] Vedejs, E.; Engler, D. A.; Telschow, J. E. *J. Org. Chem.* **1978**, *43*, 188−196.
[165] Davis, F. A.; Chen, B.-C. *Chem. Rev.* **1992**, *92*, 919−934.
[166] Lee, J.; Oya, S.; Snyder, J. K. *Tetrahedron Lett.* **1991**, *32*, 5899−5902.
[167] Oppolzer, W.; Dudfield, P. *Helv. Chim. Acta* **1985**, *68*, 216−219.
[168] Evans, D. A.; Morrissey, M. M.; Dorow, R. L. *J. Am. Chem. Soc.* **1985**, *107*, 4346−4348.

[169] Enders, D.; Bhushan, V. *Tetrahedron Lett.* **1988**, *29*, 2437−2440.

[170] Davis, F. A.; Chen, B.-C. *J. Org. Chem.* **1993**, *58*, 1751−1753.

[171] Davis, F. A.; Clark, C.; Kumar, A.; Chen, B.-C. *J. Org. Chem.* **1994**, *59*, 1184−1190.

[172] Davis, F. A.; Weismiller, M. C.; Murphy, C. K.; Reddy, R. T.; Chen, B.-C. *J. Org. Chem.* **1992**, *57*, 7274−7285.

[173] Ishimaru, T.; Shibata, N.; Nagai, J.; Nakamura, S.; Toru, T.; Kanemasa, S. *J. Am. Chem. Soc.* **2006**, *128*, 16488−16489.

[174] Smith, A. M. R.; Billen, D.; Hii, K. K. *Chem. Commun.* **2009**, 3925−3927.

[175] Momiyama, N.; Yamamoto, H. *J. Am. Chem. Soc.* **2003**, *125*, 6038−6039.

[176] Brown, S. P.; Brochu, M. P.; Sinz, C. J.; MacMillan, D. W. C. *J. Am. Chem. Soc.* **2003**, *125*, 10808−10809.

[177] Ramachary, D. B.; Barbas, C. F. *Org. Lett.* **2005**, *7*, 1577−1580.

[178] Kano, T.; Mii, H.; Maruoka, K. *J. Am. Chem. Soc.* **2009**, *131*, 3450−3451.

[179] Gotoh, H.; Hayashi, Y. *Chem. Commun.* **2009**, 3083−3085.

[180] Vaismaa, M. J. P.; Yau, S. C.; Tomkinson, N. C. O. *Tetrahedron Lett.* **2009**, *50*, 3625−3627.

[181] Gennari, C.; Colombo, L.; Bertolini, G. *J. Am. Chem. Soc.* **1986**, *108*, 6394−6395.

[182] Evans, D. A.; Britton, T. C.; Ellman, J. A.; Dorow, R. L. *J. Am. Chem. Soc.* **1986**, *108*, 6395−6397.

[183] Trimble, L. A.; Vederas, J. A. *J. Am. Chem. Soc.* **1986**, *108*, 6397−6399.

[184] Evans, D. A.; Britton, T. C.; Dorow, R. L.; Dellaria, J. F. *Tetrahedron* **1988**, *44*, 5525−5540.

[185] Evans, D. A.; Britton, T. C.; Ellman, J. A.; Dorow, R. L. *J. Am. Chem. Soc.* **1990**, *112*, 4011−4030.

[186] List, B. *J. Am. Chem. Soc.* **2002**, *124*, 5656−5657.

[187] Bøgevig, A.; Juhl, K.; Kumaragurubaran, N.; Zhuang, W.; Jørgensen, K. A. *Angew. Chem., Int. Ed.* **2002**, *41*, 1790−1793.

[188] He, R.; Wang, X.; Hashimoto, T.; Maruoka, K. *Angew. Chem., Int. Ed.* **2008**, *47*, 9466−9468.

[189] Poulsen, T. B.; Alemparte, C.; Jørgensen, K. A. *J. Am. Chem. Soc.* **2005**, *127*, 11614−11615.

[190] Hintermann, L.; Togni, A. *Helv. Chim. Acta* **2000**, *83*, 2425−2435.

[191] Hintermann, L.; Togni, A. *Angew. Chem., Int. Ed.* **2000**, *39*, 4359−4362.

[192] Brochu, M. P.; Brown, S. P.; MacMillan, D. W. C. *J. Am. Chem. Soc.* **2004**, *126*, 4108−4109.

[193] Furuya, T.; Kamlet, A. S.; Ritter, T. *Nature* **2011**, *473*, 470−477.

[194] Hamashima, Y.; Suzuki, T.; Takano, H.; Shimura, Y.; Tsuchiya, Y.; Moriya, K.-i.; Goto, T.; Sodeoka, M. *Tetrahedron* **2006**, *62*, 7168−7179.

[195] Suzuki, T.; Hamashima, Y.; Sodeoka, M. *Angew. Chem., Int. Ed.* **2007**, *46*, 5435−5439.

[196] Marigo, M.; Kumaragurubaran, N.; Jørgensen, K. A. *Chem.−Eur. J.* **2004**, *10*, 2133−2137.

[197] Shibata, N.; Kohno, J.; Takai, K.; Ishimaru, T.; Nakamura, S.; Toru, T.; Kanemasa, S. *Angew. Chem., Int. Ed.* **2005**, *44*, 4204−4207.

[198] Oppolzer, W.; Tamura, O.; Sundarababu, H.; Signer, M. *J. Am. Chem. Soc.* **1992**, *114*, 5900−5902.

[199] Oppolzer, W.; Tamura, O. *Tetrahedron Lett.* **1990**, *31*, 991−994.

[200] Kagan, H. B. In *Catalytic Asymmetric Synthesis*; Ojima, I., Ed.; VCH: New York, 1993, pp. 203−226.

[201] Bolm, C. *Med. Res. Rev.* **1999**, *19*, 348−356.

[202] Colonna, S.; Gaggero, N.; Carrea, G.; Pasta, P. In *Asymmetric Oxidation Reactions*; Katsuki, T., Ed.; Oxford University Press: Oxford, 2001, pp. 227−235.

[203] Kagan, H. B.; Luukas, T. O. In *Transition Metals for Organic Synthesis*; 2nd ed., Beller, M., Bolm, C., Eds.; Wiley-VCH: Weinheim, 2004; Vol. 2, pp. 479−495.

[204] Bryliakov, K. P.; Talsi, E. P. *Curr. Org. Chem.* **2008**, *12*, 386−404.

[205] O'Mahony, G. E.; Kelly, P.; Lawrence, S. E.; Maguire, A. R. *ARKIVOC* **2011**, 1−110.

[206] Barbachyn, M. R.; Johnson, C. R. In *Asymmetric Synthesis*; Morrison, J. D., Scott, J. W., Eds.; Academic: Orlando, 1984; Vol. 4, pp. 227−261.

[207] Andersen, K. K. *Tetrahedron Lett.* **1962**, 93−95.

[208] Davis, F. A.; Sheppard, A. C. *Tetrahedron* **1989**, *45*, 5703−5742.

[209] Pritchen, P.; Deshmukh, M.; Dunach, E.; Kagan, H. B. *J. Am. Chem. Soc.* **1984**, *106*, 8188−8193.

[210] Federsel, H.-J.; Larsson, M. In *Asymmetric Catalysis on Industrial Scale*; Blaser, H.-U., Schmidt, E., Eds.; Wiley-VCH: Weinheim, 2004, pp. 413−436.

[211] Seenivasaperumal, M.; Federsel, H.-J.; Szabo, K. J. *Adv. Synth. Catal.* **2009**, *351*, 903−919.

[212] Aggarwal, V. K.; Richardson, J. *Chem. Commun.* **2003**, 2644−2651.

[213] Aggarwal, V. K.; Winn, C. L. *Acc. Chem. Res.* **2004**, *37*, 611−620.

[214] Aggarwal, V. K.; Evans, G.; Moya, E.; Dowden, J. *J. Org. Chem.* **1992**, *57*, 6390−6391.

[215] Aggarwal, V. K.; Thomas, A.; Franklin, R. J. *J. Chem. Soc., Chem. Commun.* **1994**, 1653−1655.

[216] Drabowicz, J.; Buknicki, B.; Mikolajczyk, M. *J. Org. Chem.* **1982**, *47*, 3325−3327.

[217] Conley, B. L.; Tenn, W. J.; Young, K. J. H.; Ganesh, S. K.; Meier, S. K.; Ziatdinov, V. R.; Mironov, O.; Oxgaard, J.; Gonzales, J.; Goddard, W. A.; Periana, R. A. *J. Mol. Catal. A: Chem.* **2006**, *251*, 8−23.

[218] Chen, M. S.; White, M. C. *Science* **2010**, *327*, 566−571.

[219] Giri, R.; Shi, B.-F.; Engle, K. M.; Maugel, N.; Yu, J.-Q. *Chem. Soc. Rev.* **2009**, *38*, 3242−3272.

[220] Peters, M. W.; Meinhold, P.; Glieder, A.; Arnold, F. H. *J. Am. Chem. Soc.* **2003**, *125*, 13442−13450.

[221] Andrus, M. B.; Argade, A. B.; Chen, X.; Pamment, M. G. *Tetrahedron Lett.* **1995**, *36*, 2945−2948.

[222] Andrus, M. B.; Zhou, Z. *J. Am. Chem. Soc.* **2002**, *124*, 8806−8807.

[223] Ginotra, S. K.; Singh, V. K. *Org. Biomol. Chem.* **2006**, *4*, 4370−4374.

[224] Punniyamurthy, T.; Miyafuji, A.; Katsuki, T. *Tetrahedron Lett.* **1998**, *39*, 8295−8298.

[225] Punniyamurthy, T.; Katsuki, T. *Tetrahedron* **1999**, *55*, 9439−9454.

[226] Zalatan, D. N.; Du Bois, J. *J. Am. Chem. Soc.* **2008**, *130*, 9220−9221.

[227] Du Bois, J. *Org. Proc. Res. Dev.* **2011**, *15*, 758−762.

[228] Liang, C.; Collet, F.; Robert-Peillard, F.; Müller, P.; Dodd, R. H.; Dauban, P. *J. Am. Chem. Soc.* **2007**, *130*, 343−350.

[229] Alphand, V.; Furstoss, R. *Asymm. Oxid. React.* **2001**, 214−227.

[230] Mihovilovic, M. D. *Curr. Org. Chem.* **2006**, *10*, 1265−1287.

[231] Reetz, M. T.; Wu, S. *J. Am. Chem. Soc.* **2009**, *131*, 15424−15432.

[232] Leisch, H.; Morley, K.; Lau, P. C. K. *Chem. Rev.* **2011**, *111*, 4165−4222.

[233] Katsuki, T. *Russ. Chem. Bull.* **2004**, *53*, 1859−1870.

[234] ten Brink, G. J.; Arends, I. W. C. E.; Sheldon, R. A. *Chem. Rev.* **2004**, *104*, 4105−4124.

[235] Bolm, C.; Schlingloff, G.; Weickhardt, K. *Angew. Chem., Int. Ed. Engl.* **1994**, *33*, 1848−1849.

[236] Watanabe, A.; Uchida, T.; Ito, K.; Katsuki, T. *Tetrahedron Lett.* **2002**, *43*, 4481−4485.

[237] Watanabe, A.; Uchida, T.; Irie, R.; Katsuki, T.; Trost, B. M. *Proc. Natl. Acad. Sci. U.S.A.* **2004**, *101*, 5737−5742.

[238] Aubé, J.; Wang, Y.; Hammond, M.; Tanol, M.; Takusagawa, F.; Vander Velde, D. *J. Am. Chem. Soc.* **1990**, *112*, 4879−4891.

[239] Davis, F. A.; Jenkins, R. H., Jr. In *Asymmetric Synthesis*; Morrison, J. D., Scott, J. W., Eds.; Academic: Orlando, 1984; Vol. 4, pp. 313−353.

[240] Lattes, A.; Oliveros, E.; Rivière, M.; Belzecki, C.; Mostowicz, D.; Abramskj, W.; Piccinni-Leopardi, C.; Germain, G.; Van Meerssche, M. *J. Am. Chem. Soc.* **1982**, *104*, 3929−3934.

[241] Gracias, V.; Milligan, G. L.; Aubé, J. *J. Am. Chem. Soc.* **1995**, *117*, 8047−8048.

[242] Sahasrabudhe, K.; Gracias, V.; Furness, K.; Smith, B. T.; Katz, C. E.; Reddy, D. S.; Aubé, J. *J. Am. Chem. Soc.* **2003**, *125*, 7914−7922.

[243] Ribelin, T.; Katz, C. E.; Winthrow, D.; Smith, S.; Manukyan, A.; Day, V. W.; Neuenswander, B.; Poutsma, J. L.; Aubé, J. *Angew. Chem., Int. Ed.* **2008**, *47*, 6233−6235.

[244] Katz, C. E.; Aubé, J. *J. Am. Chem. Soc.* **2003**, *125*, 13948−13949.

[245] Katz, C. E.; Ribelin, T.; Withrow, D.; Basseri, Y.; Manukyan, A. K.; Bermudez, A.; Nuera, C. G.; Day, V. W.; Powell, D. R.; Poutsma, J. L.; Aubé, J. *J. Org. Chem.* **2008**, *73*, 3318−3327.

Index

Note: Page numbers followed by "f" and "t" refer to Figures and Tables, respectively. Page numbers in italics refer to glossary definitions.

A

$A^{1,3}$ strain, *34*, 102, 102f, 106−109, 112, 117, 123−125, 139, 139f, 207, 213−214, 231, 311−312, 312f

AAAs. *See* Asymmetric allylic allylations

abscisic acid, 127f

Absolute configuration, *34*, 73−91, 74f, 76−80f, 82f, 84f, 86−87f

Acetamido cinnamates, reduction of, 463−470, 465−472f

Acetate enolates, 263−264, 269−270, 271f

N-Acetyl dehydrophenylalanine, 465−467, 466−467f

Achiral derivatizing agents, 80, 80f

Acrolein, 307, 312f, 342, 348, 357

Acrylates, 304, 336f, 337−339, 337f, 338t, 339f

N-Acryloyloxazolidinone, 348−349, 349f

Acylation, 19, 20f, 212, 265, 439

Acylhydrazones, 373−375, 374f

N-Acylpyridinium, 212−213, 213f

AD reactions. *See* Asymmetric dihydroxylation reactions

1,2-Additions, 179, 179f, 212, 225, 236. *See also* Aldol additions; 1,2-Allylations

chiral catalysts and auxiliaries for, 196−197, 197f

Cram's rule

cyclic model, *40*, 191−196, 191f, 192t, 193−196f

open chain model, *40−41*, 180−191, 180−187f, 181t, 188t, 189−191f

hydrocyanations of azomethines, 219−221, 220−221f

hydrocyanations of carbonyls, 215−218, 216−219f

organometallic additions to aldehydes, 197−202, 198−202f, 202t

organometallic additions to azomethines, 203−211, 206t, 207−210f, 210−211t, 212f

organometallic additions to pyridinium ions, 212−214, 212−213f, 214t, 215f

1,4-Additions. *See* Conjugate additions

Aldehyde azaenolates, 306, 306−307f

Aldehydes. *See also* Aldol additions

hetero Diels−Alder reaction with, 361−365, 362−363f, 365f

hydrocyanations of, 215−218, 216−219f

organocatalytic reactions with, 473−474, 475f

organometallic additions to, 197−202, 198−202f, 202t

Aldol additions, 245, 246f, 258

double asymmetric induction, 295−300, 296−300f

organocatalysis of, 245, 256−257, 285−294, 286−288f, 289t, 292−293f, 294t

simple diastereoselectivity of, 259−263, 259f, 260−263f

single asymmetric induction, 263−284, 264−265f, 266t, 267f, 268t, 270−273f, 274t, 275−278f, 276t, 281t, 282−284f, 284t

Alkali metal enolates, 262

Alkenes. *See also* Asymmetric dihydroxylations; epoxidations

asymmetric reactions with, 393−394

chiral, 519−521

α-Alkoxyaldehydes, 251

Alkylations. *See* Azaenolate alkylations; Enolate alkylations; Organolithiums

Alkylhydrazone, 321f

N-Alkylhydroxylamine, 321f

Allenylboranes, 247f, 253

Allyl aluminums, 254

Allyl anions, 312, 314, 313−314f, 313t

Allyl organometallics, 254

Allyl tin reagents, 254−256, 255f

1,2-Allylations, 245−246, 246f

double asymmetric induction, 251−253, 251t, 253f

simple diastereoselectivity, 248−249, 249f

simple enantioselectivity, 246−247, 247−248f

single asymmetric induction, 249−250, 249t

Allylic alcohols, 492–496, 505f
Allylic amines, 395–400, 396–401f
Allylic boron compounds, 246
    double asymmetric induction, 251–253,
        251t, 253f
    simple diastereoselectivity, 248–249, 249f
    simple enantioselectivity, 246–247, 247–248f
    single asymmetric induction, 249–250, 249t
Allylic strain. See A^{1,3} strain
Allylic sulfoxide addition, 312, 313–314f, 313t, 314
Allylsilanes, 254–257, 255f, 258f
Allylstannanes, 254–255, 255f, 256
Allyltrichlorosilane, 256–257
Aluminum catalysts, 236, 236t, 254
Amide enolates, 302f, 303–304, 305f
Amides, conjugate additions to, 229–232, 230f,
        231–232t, 232f
Amidinium, 212–214, 213f
Aminations, 524–529, 526f, 527f
Amino acid synthesis, 117–119, 118–119t, 119f,
        144f, 219–221, 220–221f
Amino alcohols, 199, 200, 264, 370–373
Anh model. See Felkin–Anh model
Anomeric effect, 35, 113–114, 113f
N-Anthracylmethyl cinchona alkaloids, 137, 138f, 138t
Antiperiplanar, 36–37, 56–57
    allylations, 255–257, 256f, 258f
Antiperiplanar effect, 109, 110f
Aphidicolin, 315–316, 316f
Aratani's copper catalyst, 382–384, 384–385f
Aristeromycin, 146–147, 147f
Aromatic rings, dihydroxylation of, 439–440
Asymmetric allylic alkylations (AAAs), 145–151,
        145f, 147–150f, 151t
Asymmetric amplification, 31, 32f
Asymmetric catalysis, 28–33, 29–32f, 64t, 68–71.
        See also Chiral catalysts
Asymmetric deprotonations, 161–165, 162–166f
Asymmetric destruction. See Kinetic resolution
Asymmetric dihydroxylation (AD) reactions, 510–521
    application of, 518–521, 519–520f
    development of, 510–518, 512–513f, 516t, 518f
    enantioselectivity of, 517–518, 518f
    ligands for, 514–515, 514f, 514t
Asymmetric hydrogenation
    of N-acetyl dehydrophenylalanine, 465–467,
        466–467f
    substrate tolerance in, 469, 469f
    substrate–catalyst complexes of, 467–468, 468f
Asymmetric induction, 36. See also specific types
Asymmetric reduction, 431–432, 433t, 444t
Asymmetric synthesis (general), 2–4, 36. See also
        specific reactions
    criteria for, 3
    getting started with, 69–71

method selection for, 63–69, 64t, 67f, 69f
    reasons for, 1–2
Asymmetric transformations, 23–28, 24–28f, 36
    of the first kind, 23–28, 24f, 36
    of the second kind, 23–28, 24f, 36
Autocatalysis, 200, 201f
Aza-ene, 311–312, 312f
Azaenolates, 97–99, 97–100f. See also Michael
        additions
    carbonyl deprotonation, 100–105, 101–105f
Azaenolate alkylations
    with chiral electrophiles, 139–151,
        140–150f, 151t
    with chiral nucleophiles, 110–139, 110–113f,
        111t, 114t, 115–116f, 117–119t,
        118–132f, 122–123t, 125–126t, 129t,
        133t, 134–139f, 136t, 138t
Azidation, 526–527, 526f
Aziridinations, 510, 511f
Azomethines
    aldol additions to, 275
    conjugate additions to, 207–208, 208–209f
    hydrocyanations of, 219–221, 220–221f
    organometallic additions to, 203–211, 206t,
        207–210f, 210–211t, 212f

B

Baeyer–Villiger reaction, 499, 500f, 501t, 531–533,
        534–535f
Bakuchiol, 232, 232f
9-BBN. See 9-Borabicyclononane
Benzaldehyde, 267
N-Benzylcinchonium halide catalyst, 136–137
Bicyclic lactams, 125–128, 126–127f, 126t, 391,
        391f
Bidentate dienophiles, 348–351, 349–350f, 350t
Bienz model, 188–190, 190f
BINAL-H, 431–434, 432f, 433t, 434f
BINAP, 148–149, 150f, 222f, 223, 444–445, 445f,
        446t, 447–448, 447–448f
BINOL (binaphthol), 32–34, 32f, 198f, 201, 201f,
        226–227, 228f, 228t, 279–280, 321–322,
        431–434, 432f, 433t, 434f
Bismuth catalysts, 234–235, 234t, 235f
Bisoxazoline (Box), 157–158, 158f, 278f, 281–282,
        282f, 320–321, 321f
Bisphosphine ligands, 467, 467f
Boat conformation, 37, 46, 259, 260f, 261, 262f
9-Borabicyclononane (9-BBN), 442–443, 443f
Borane
    chiral organoborane reductions, 442–443,
        442–443f, 444t
    reductions with modified, 434–441, 435–436f,
        436t, 438f, 441f
Boron aldolates, 262

Boron catalysts, 234–235, 234t, 235f, 276–277, 276–277f, 276t. *See also* Allylic boron compounds
Boron enolates, 260, 261f, 264–265, 267–269, 271–275, 272f, 274t, 275f
Box. *See* Bisoxazoline
BozPHOS, 210–211, 210f, 211t, 212f
N-(Bromomagnesio)-2,2,6,6-tetramethylpiperidide, 267
Brønsted acid catalysis, in Diels–Alder reactions, 335–360, 358f
Bürgi-Dunitz trajectory, *37*, 182–186, 183–186f, 301, 301f
Butadiene, 340
2-Butanone, 283, 283f
tert-Butyl cinnamic ester, 319, 320f
tert-Butyl PHOX, 151, 151t
tert-Butylglycine, 264, 264f

C

Cahn-Ingold-Prelog *(CIP)* method, *38–39*
Callipeltoside C, 292, 292f
Camphor, 119–120, 120f, 130–131, 131f, 230, 230f, 264, 264f, 270
Capacity ratio. *See* Chromatographic capacity ratio
Carbanions, 97. *See also* Azaenolates; Enolates; Organolithiums
Carbenes, 375, 377f
Carbenoid cyclopropanations. *See* Cyclopropanations
Carbohydrate synthesis, 506–507, 507f
Carbon–carbon bonds
  from carbon–hydrogen bonds, 537–538
  reduction of, 463–475
Carbon–hydrogen bonds
  carbon-carbon bonds from, 537–538
  oxidation of, 531, 532–533f
Carbon-to-carbon migration reactions, 421, 421f
Carbonyls. *See also* 1,2-Additions; Aldol additions; Conjugate additions; Michael additions
  aminations and halogenations of, 524–529, 526–527f, 528t, 529f
  deprotonation of, 100–105, 101–105f
  α-functionalization of, 521–529, 521f
  hydroxylations of, 522–524, 522–525f
Carbovir, 146–147, 147f
Catalysis. *See* Asymmetric catalysis
Catalysts. *See* Chiral catalysts
Catalytic asymmetric hydroboration reactions, 481–482, 481f
Catalytic transfer hydrogenation, 449–452, 449–452f, 453t, 454, 454f
CD. *See* Circular dichroism
CDA. *See* Chiral derivatizing agent
Chair conformation, *37*, 259, 260f, 261, 262f, 312
Chelate model. *See* Cram's rule

Chelate-enforced intraannular asymmetric induction, 110, 110f, 120–123, 121–123f, 122–123t
Chelation-controlled addition, 196, 196f, 361
Chimonanthine, 142–143, 143f
Chiral auxiliaries, 3, *38. See also specific auxiliaries; specific reactions*
  acrylates, 336f, 337–339, 337f, 338t, 339f
  for carbenoid cyclopropanations, 377–380, 378–381f
  for Diels–Alder reactions, 336–343, 336–337f, 358, 359f
  for diene component, 342–343, 343f
  for dienophiles, 336, 336f
  1,3-dipolar cycloadditions with, 371–373, 372f
  fumarates and maleates, 340–341, 341f, 343f
  in hetero Diels–Alder reactions, 367–368, 367f
  for hydroxylations, 522–524, 522–525f
  intramolecular cycloadditions, 337f, 339, 340f
  for inverse demand hetero Diels–Alder reaction, 367–368, 367f
Chiral catalysts, 3, 15, 28–33, 29–32f. *See also specific catalysts; specific reactions*
  for acetamido cinnamate reduction, 463–470, 472, 465–472f
  for catalytic transfer hydrogenation, 449–452, 449–452f, 453t, 454, 454f
  for cyclopentadiene cycloadditions, 344–345, 346t
  for cyclopropanations, 381–390, 382–385f, 386t, 387–390f
  for Diels–Alder reactions, 358, 359f
  for 1,3-dipolar cycloadditions, 371–373, 372f
  for epoxidations, 493–503
  for hetero Diels–Alder reaction, 362–363, 363f, 364t
  for [1,3]-hydrogen shifts, 395–400, 396–401f
  hydrogen-bonding activation and Brønsted acid catalysis, 335–360, 358f
  for ketone reductions, 444–445, 445f, 446t, 447–448, 447–448f
  kinetic resolution with, 16–17, 16–17f
  Lewis acids, 344–351, 344–345f, 346t
  organocatalysis for, 344, 344f, 351–357, 352f, 353t, 354–356f
  synthesis using, 64t, 68–71
Chiral derivatizing agent (CDA), *37*, 80–85, 82f, 84f
Chiral donors, 303–314, 304–310f, 311t, 312f–314f, 313t
Chiral Michael acceptors, 315–316, 315–316f
Chiral phase-transfer catalysts, 136–137, 137–138f, 138t
Chiral solvating agent (CSA), *41*, 80–81, 85–86, 86f
Chiral stationary phase (CSP), *41*, 75–79, 77–79f
Chiral sulfur ylides, 392, 392f

Chirality centers, *48*, 116, 465
  self-regeneration of, 111−113, 111−112f, 111t,
    232−233
Chirality transfer, 4, *38*, 255, 344, 431, 442, 523
Chiraphos, 222f, 225, 234−235, 235f, 468f
Chiroptical methods, 86−91, 86f−87f
B-Chlorodiisopinocampheylborane (Ipc$_2$Cl), 443,
  443f
Chromatographic capacity ratio, 74−75, 74f
Chromatographic separability factor, 74f,
  75, 77−79
Chromatography, 64t, 69, 71, 73−80
  achiral derivatizing agents in, 80, 80f
  basics of, 73−76, 74f, 76f
  CSP, 75−79, 77−79f
  self-disproportionation of enantiomers during, 71,
    72f, 73
Cilistatin, 382−383, 383f
Cinchona alkaloid, 137, 138f, 498, 510−511,
  513−515, 514f
trans-Cinnamyl alcohol, 381, 382f
CIP method. *See* Cahn-Ingold-Prelog method
Circular dichroism (CD), 90−91
Claisen rearrangement, 395, 409, 416−419, 417f,
  418f
Closed transition structures, 246−247, 247f, 259,
  260f, 261, 267−268, 267f
Column efficiency, 74−75
Conducted tour mechanism, 99
Conjugate additions, 179, 179f, 221, 222f, 371−373.
    See also Michael additions
  to acyclic amides and imides, 229−232, 230f,
    231−232t, 232f
  to acyclic esters and ketones, 222−228, 222f,
    223−226t, 227−229f, 228t
  to azomethines, 207−208, 208−209f
  to cyclic ketones and lactones, 232−236, 233f,
    234t, 235f, 236t
  of nitrogen nucleophiles, 319−326, 320−322f,
    323t, 325−326f
  of oxygen nucleophiles, 325, 325f
  of sulfur nucleophiles, 325−326, 326f
Convolutamydine A, 292, 292f
Cope rearrangement, 314, 380, 381f
Copper bisoxazoline, 510, 511f
Cornforth model, *40*, 180−182, 180f, 186
Cotton effect, 90−91
Cp$_2$ZrCl$_2$, 261, 262f
Cram-controlled addition, 361
Cram's rule
  cyclic model, *40*, 191−196, 191f, 192t, 193−196f
  open chain model, *40−41*, 180−191, 180−187f,
    181t, 188t, 189−191f
Crimmins thiazolidinethione, 264

Crotonaldehyde, 302
Crotyl compounds, 248−249, 249f, 250−252t, 252,
  254, 257, 299
Crotyloxazolidinones, 349−350, 350t
Crown ethers, 77, 262, 316, 317f
CSA. *See* Chiral solvating agent
CSP. *See* Chiral stationary phase
Curtin−Hammett kinetics, 21−23, 21f, 23f, 25, 26f,
  153−154, 158−159, 158f, 180−181, 348,
  447, 466, 467f
Cyclic model. *See* Cram's rule
Cycloadditions, 335−394. *See also*
    Cyclopropanations
  Diels−Alder reaction, 335−360
  of diene, 341−342, 342f
  1,3-dipolar, 368−375, 370f
  hetero Diels−Alder reaction, 360−368
  Lewis acids for, 336, 344−351, 344−345f, 346t
Cyclohexanone lithium enolates, 317, 317f
Cyclopropanations, 375−394, 375−377f
  chiral auxiliaries for, 377−380, 378−381f
  chiral catalysts for, 381−390, 382−385f, 386t,
    387−390f
  intramolecular, 389, 389f
  stepwise, 390−394, 391−392f
  strategies for, 376, 376f
Cyclopropane, 373, 373f, 375, 375f, 381, 384, 386,
  390, 392, 392f
Cyclopropyl amino acids, 391, 391f
Cytovaricin, 296−297, 297f

D

Danishefsky's diene, 361, 362f
Daunomycin, 523, 523f
DBAD. *See* Di(tert-butyl)azodicarboxylate
Dehydroboration, 9-BBN and, 443
6-Deoxyerythronolide-B, 295−296, 297f
Deprotonation
  asymmetric, 161−165, 162f−166f
  of carbonyls, 97, 100−105, 101−105f
Dialkylboron triflate, 260
Diamines, 146, 161, 163−165, 162−165f
Diastereomer analysis. *See* Stereoisomer analysis
Diastereomer excess, *41*, 90
Diastereomer ratio (dr), 9, 9f, *41*, 90
1,3-Diaxial interactions, 102−104, 102f, 104f,
  246−249, 273
Diazaborolidines, 274−275, 275f
Di(*tert*-butyl)azodicarboxylate (DBAD), 526−527,
  526f
Dibutylboron enolates, 265, 267−268
Dibutylboron triflate, 267−268
Dichloroaluminum alkoxide, 345f, 347
Dicyclopentadienyl zirconium, 408t, 409, 412, 412f

Diels–Alder reactions, 335–360. *See also* Hetero
  Diels–Alder reaction
  chiral auxiliaries for, 336–343, 336–337f
  dienophile modification, 336–337, 336–337f
  hydrogen-bonding activation and Brønsted acid
    catalysis, 335–360, 358f
  Lewis acids for, 344–351, 344–345f, 346t
  organocatalysis for, 344, 344f, 351–357, 352f,
    353t, 354–356f
  regiochemistry of, 370
  total synthesis with, 358–360, 359–360f
Dienes, 341–343, 342–343f, 357, 358f, 361,
  383, 483
Dienophiles, 336–337, 336–337f
Diethylaluminum chloride, 268–269
Diethylzinc reaction, 197–200, 198–199f, 202, 212f
Dihydroxylation, 439–440. *See also* Asymmetric
  dihydroxylation
Diiodobinaphthol, 226–227, 228f, 228t
Diisopinocampheyl boron chloride, 273
Diisopinocampheyl boron triflate, 272, 272f
Diisopinocampheylborane (Ipc$_2$BH), 476, 476f, 477t,
  478–481, 480f
trans-2,5-Dimethylborolane (DMB), 476f, 477t, 478
Dipamp, 222f, 234–235, 235f
Diphenyloxazinone, 117
Diphenylphosphinobenzoic acid (DPPBA), 147–150,
  147–149f, 151, 151t
Diphenylphosphinoyl imines, 210, 210f, 211, 211t, 212f
1,3-Dipolar cycloadditions, 368–375, 370f
  with chiral auxiliaries, 371–373, 372f
  with chiral catalysts, 371–373, 372f
  Kinugasa reaction, 373, 373f
  with Lewis acids, 371–373, 372f
  of nitrones, 370–373, 371–373f
  regiochemistry of, 370
Direct aldol addition, 283, 283f
Directing effects, 492
Discodermolide, 299, 300f
Disopke acetals, alkylation of, 113–114, 113f, 114t
DMB. *See trans*-2,5-Dimethylborolane
Double asymmetric induction, 4, 4f, 12–15, 12–15f,
  36, 224, 377, 519–521
  aldol additions, 295–300, 296–300f
  1,2-allylations, 251–253, 251t, 253f
  dynamic resolution with, 25
  kinetic resolution with, 16–17, 17f
  matched, 4, 4f, 14, 13–15f, *36*, 224
  mismatched, 4, 4f, 14, 13–15f, *36*, 224
DPPBA. *See* Diphenylphosphinobenzoic acid
dr. *See* Diastereomer ratio
ds. *See* Diastereoselectivity
Dunitz angle, *37*, 186, 195. *See also* Bürgi-Dunitz
  trajectory

Dynamic resolutions, 23–28, 24–28f
  kinetic, 25–26, 25–26f
  thermodynamic, 25–28, 27–28f, 157–158, 158f

E

ec. *See* Enantiomer composition
ee. *See* Enantiomer excess
Electrophilic attack, in enolate alkylations, 105–109,
  106f, 109–110f, 139–151, 140–150f, 151t
Electrophilic auxiliaries, 391, 391f
Electrophilic substitution
  inversion competition with, 155, 155f
  organolithium, 151–167, 152–166f, 167t, 168f
Enamine intermediates, 307–312, 307–310f,
  311t, 312f
Enantiomers, *43*
  self-disproportionation of, 71, 72f, 73
Enantiomer analysis. *See* Stereoisomer analysis
Enantiomer composition (ec), 43
Enantiomer excess (ee), *43*, 90
Enantiomer ratio (er), 9, 9f, 18–20, 31, 43, *44*, 90
Enantiomerically enriched molecules, *44*, 71, 72f, 73
Enantiomerization, 25–27, 25f, *44*, *51*, 114–116,
  115–116f, 117t
Enantiopure educts, chirality transfer in, 410–415,
  411–414f
Enantioselectivity (es), 6f, 8, 29, 34, *43*, *44*
  of 1,2-allylations, 246–247, 247–248f
Enolates, 97–99, 97–100f. *See also* Aldol additions;
    Michael additions; *specific enolates*
  carbonyl deprotonation, 100–105, 101–105f
  E(O) and *Z(O)* isomer formation, 102, 102f,
    103–105, 103–104f
Enolate alkylations
  with chiral electrophiles, 139–151, 140–150f,
    151t
  with chiral nucleophiles, 110–139, 110–113f,
    111t, 114t, 115–116f, 117–119t,
    118–132f, 122–123t, 125–126t, 129t,
    133t, 134–139f, 136t, 138t
  transition state of, 105–109, 106f, 109–110f
Enthalpy, 10–12, 11f, 180, 194, 220, 336
Entropy, 10–12, 11f, 180, 194, 492
Enzymes, 65t, 285, 439–440, 531, 533
Ephedrine, 229–230, 230f, 247, 378
Epoxidations, 491–510
  applications of, 506–510, 507–509f
  aziridinations, 510, 511f
  chiral catalysts in, 493–503
  diastereoselectivity of, 492, 492f
  early approaches to, 491–493, 492–493f
  of electron-deficient olefins, 499–503, 502–503f
  Katsuki–Sharpless asymmetric, 493–496,
    494–495f, 496t

Epoxidations (*Continued*)
  Sharpless kinetic resolution, 504–506, 504–505f
  of simple olefins, 497–498, 497f, 498t, 499, 500f
  with vanadium catalysts, 493, 493f
Epoxide-opening reactions, 509–510, 509f
er. *See* Enantiomer ratio
Eremantholide A, 141–142, 142f
Eremophilenolide, 130, 130f
Erythronolide B, 71, 297f
es. *See* Enantioselectivity
Esermethole, 140, 141f
Esomeprazole, 530f
Ester enolates, 260, 302–303, 302f, 304f
Esters, conjugate additions to, 222–228, 222f,
      223–226t, 227–229f, 228t
Estradiol, 315–316, 316f
Estrone, 315–316, 316f
cis-2-Ethyl-1-isopropylcyclohexane, 273, 273f
Evans imide enolates, 267–268, 267f, 268t, 269,
      296–297, 296f
Evolution, homochirality and, 107
Extraannular intraligand asymmetric induction, 110,
      110f, 117–119, 119f, 119t

F

Felkin control, 257, 258f
Felkin model, 181–182, 181t, 182f
Felkin–Anh model, *40–41*, 180, 186, 187f, 188,
      188t, 189–190, 190f
Felkin–Anh–Heathcock model, 251
Fleming–Tamao oxidation, 483, 483f
Flippin–Lodge angle, 187, 187f
Formamicinone, 196, 196f
Formamidines, 159–160, 159–160f
Free energy, chromatographic separation based on,
      71, 75, 78
Fumarates, 340–341, 341f, 343f
Functionalized organolithiums, 157, 157f

G

Gas chromatography (GC), 73–74, 76–77
Gauche pentane interactions, 273, 298, 298f
GC. *See* Gas chromatography
Geminal dimethyl substitution, 356–357, 356f
Group-selective reactions, 134–135, 135f, 161

H

Hajos–Parrish–Eder–Sauer–Weichert reactions,
      285–288, 286–287f
Half-life, racemization, 114–116, 116f
Halogenations, 524–529, 528t, 529f
Halohydrins, 269, 275
Hammond postulate, 106, 180
Hawkins' (2-aryl)cyclohexylboron dichloride catalyst,
      345f, 347, 347f

Heathcock model, 187, 187f, 188–189, 189f, 190, 191f
Helmchen acetate, 264
Hetero Diels–Alder reaction, 360–368
  with aldehydes, 361–365, 362–363f, 365f
  chiral catalyst for, 362–363, 364t
  with imines, 365–368, 366f–367f, 369f
  inverse electron demand, 367–368, 367f
  organocatalysis of, 368, 369f
  TADDOL catalysis of, 364–365, 365f
Heterochiral dimers, 73, 89
Hexoses, 506
High performance liquid chromatography (HPLC),
      73–74, 76–77
Hoffmann test, 155, 156f
Homochiral, 11, 30–32, *44*, *46*, 107
Homochiral dimers, *53*, 73, 89, 200
Horeau effect, 89
Host–guest chemistry principles, 451–452
Houk model, 188–190, 190f
HMPA. *See* hexamethylphosphoramide
HPLC. *See* High performance liquid chromatography
Hydrazones, 128–131, 128–131f, 129t, 203, 206t,
      207, 207f, 264, 306, 306f
Hydroborations, 476–483, 476f, 477t, 478–483f
Hydrocyanations, 215–221, 216–221f
Hydrogen bond catalysis, 364–365, 365f
[1,3]-Hydrogen shifts, 395–400, 396–401f
Hydrogenation. *See* Asymmetric hydrogenation
Hydrogen-bonding activation, in Diels–Alder
      reactions, 335–360, 358f
Hydrolysis, enzymes for, 439
α-Hydroxy ketone, 518–519, 520f
2-Hydroxyacetone, 288
3-Hydroxypipecolic acid, 215f
Hydroxylamines, 320–321, 321f
Hydroxylations, 522–524, 522–525f

I

Imide enolates, 304, 305f
Imides
  conjugate additions to, 229–232, 230f, 231–232t,
      232f
  reduction of, 438, 438f, 441, 441f
Imines. *See also* Azomethines
  hetero Diels–Alder reaction with, 365–368,
      366–367f, 369f
  reduction of, 455–463, 456–457f, 457t,
      459–461f, 462t, 464f
Iminium ion catalysis, 351–357, 352f, 353t,
      354–356f
Interligand asymmetric induction, 5, 6f, 110, 110f,
      135–137, 136f, 136t
  aldol additions, 263, 271–275, 272–273f, 274t,
      275f
  Michael additions, 316–319, 317–318f

Interligand chirality transfer, 523, 524f
Intermolecular aldol reactions, 286, 286f
Internal solvation effect, 308, 310f
Intraannular asymmetric induction, 110, 110−111f, 125−126. *See also* Chelate-enforced intraannular asymmetric induction
Intraligand asymmetric induction, 5, 6f, 111, 110−111f, 117−123, 119f, 119t, 121−122f, 122t, 123f, 123t, 125−126
  aldol additions, 263−270, 264−265f, 266t, 267f, 268t, 270f, 271f
Intramolecular aldol reactions, 285−286
Inversion, *57*
  of organolithiums, 153−156, 153−156f
  substitution competition with, 155, 155f
Inversion temperature, 10−11, 11f, 12
Ionomycin, 121−122, 122f
Ipc₂BH. *See* Diisopinocampheylborane
Ipc₂Cl. *See* B-Chlorodiisopinocampheylborane
Ireland model, 102−104, 102−104f, 128−130, 260
Iridium catalysts, 470, 471f
Isobutyraldehyde, 270
Isoinversion principle, 10−12, 11f
Isopinocampheyl ligands, 272−273, 272f
Isosteres, 287−288
Itsuno reagent, 435, 435f
Ivanov reaction, 245, 246f, 258

J
Jacobsen catalysts, 362−363, 363f, 367−368
Jacobsen−Katsuki epoxidation, 497−498, 497f, 498t, 499
Josiphos, 222f, 223−226, 227f
Juliá−Colonna epoxidation, 499−502, 502f

K
Karabatsos model, 180−181, 181f
Katsuki−Sharpless asymmetric epoxidation, 493−496, 494−495f, 496t, 504−506, 504−505f
Ketone azaenolates, 306, 306−307f
Ketone enolates, 302−303, 302f
Ketones. *See also* Aldol additions
  asymmetric reduction of, 431, 444t
  conjugate additions to, 222−228, 222f, 223−226t, 227−229f, 228t, 232−236, 233f, 234t, 235f, 236t
  hydrocyanations of, 215−218, 216−219f
  hydrogen reductions of, 444−445, 445f, 446t, 447−448, 447−448f
  organocatalytic reactions with, 473−474, 475f
  oxidative ring expansions of, 531−534, 534−535f
Kinetic control, *7−8*, 7−9, 8−9f
  of aldol additions, 259−263, 259−263f

Kinetic resolution, 25−26, 25−26f, *47*
  dynamic, 16−20, 16−17f, 19−20f, 473
  Hoffmann test with, 155, 156f
  of imine reduction, 455, 456f, 460, 460f
  of Juliá−Colonna epoxidation, 499−502
  of Katsuki-Sharpless asymmetric epoxidation, 504−506, 504−505f
Kinugasa reaction, 373, 373f
Kiyooka catalyst, 276−277, 276f
Kobayashi catalysts, 278f
Kobayashi−Mukaiyama aldol addition, 278f

L
Lactams, 125−126, 126−127f, 303, 534, 535−538
Lactones, 534, 535−538
  conjugate additions to, 232−236, 233f, 234t, 235f, 236t
LAH. *See* Lithium aluminum hydride
Lanthanide catalysts, 277
Lasalocid, 196, 196f
LDA. *See* Lithium diisopropyl amide
tert-Leucine, 264, 264f
Lewis acids
  in aldol additions, 268
  for Diels−Alder reactions, 336, 344−351, 344−345f, 346t, 358, 359f
  for 1,3-dipolar cycloadditions, 371−373, 372f
  for hetero Diels−Alder reaction, 362−363, 363f, 364t
  hydrogen bond catalysis and, 364−365, 365f
  monodentate dienophiles, 345f, 347−348, 347−348f
Ligancy complementation, *38*
Ligand-accelerated catalysis, 512−513
Lithium. *See* Organolithiums
Lithium aluminum hydride (LAH), modified, 431−434, 432f, 433t, 434f
Lithium bromide, 104−105, 105f, 135−136, 136f, 136t
Lithium chloride, 104−105, 105f, 123−125, 148−149, 150f
Lithium diisopropyl amide (LDA), 99, 99f, 101−105, 101f, 104f, 117−119, 119f, 119t, 123−125, 124f, 128, 128f, 265−266
Lithium enolates, 260−261, 265−266, 317, 317f, 403−404f, 408
Lithium salts, 104−105, 105f, 156

M
MA. *See* Mandelic acid
MacMillan catalyst, 287−288, 288f
Magnesium catalysts, 236, 236t
Maleates, in Diels−Alder reactions, 340−341, 341f, 343f
Mandelic acid (MA), 82f, 264f, 342

Masamune catalyst, 276, 276f, 277
Matched double asymmetric induction, 4, 4f, 14, 13−15f, *36*, 224
MBH reaction. *See* Morita−Baylis−Hillman reaction
Meerwein−Pondorf−Verley reaction, 2, 442, 449
MeMn ligand. *See* Methylmenthyl ligand
Menthone, 273, 273f
α-Methoxy-α-phenylacetic acid (MPA), 82f, 84, 84f
α-Methoxy-α-phenylpropionic acid (MPPA), 82, 82f
α-Methoxy-α-trifluoromethylphenylacetic acid (MTPA), 82−85, 82f, 84f
Methyl DuPHOS, 210−211, 210f, 211t, 212f
Methyl jasmonate, 315−316, 316f
Methyl vinyl ketone, 283, 283f
α-Methylbenzylamine, 279, 319, 320f, 535−536
2-Methylcinnamic ester, 319, 320f
Methylmenthyl (MeMn) ligand, 273, 273f
Michael additions, 301. *See also* Conjugate additions
  chiral acceptors for, 315−316, 315−316f
  chiral donors in, 303−314, 304−310f, 311t, 312−314f, 313t
  interligand asymmetric induction, 316−319, 317−318f
  simple diastereoselectivity, 301−303, 301−302f
Mismatched double asymmetric induction, 4, 4f, 13−15f, 14, *36*, 224
Modified Eyring plot, 10−12, 11f
Monensin, 196, 196f
Monodentate dienophiles, 345f, 347−348, 347−348f
Monodentate ligands, 336−337
Monodentate phosphines, 469, 469f
Monodentate phosphoramidites, 470, 470f
Morita−Baylis−Hillman (MBH) reaction, 293, 293f, 294t
Mosher's acid. *See* α-Methoxy-α-trifluoromethylphenylacetic acid
MPA. *See* α-Methoxy-α-phenylacetic acid
MPPA. *See* α-Methoxy-α-phenylpropionic acid
MTPA. *See* α-Methoxy-α-trifluoromethylphenylacetic acid
Mukaiyama aldol addition, 275−277, 276−278f, 276t, 281t
Mutarotation, 23, 24f

N

Nakamura−Morokuma modeling, 184−186, 185−186f, 195, 196f
Naphthyloxazolines, 208, 209f
Naphthyloxy ligand, 433−434
Newman projection, *47*, 47f, 56, 187, 256, 387, 404
Nickel Box complex, 278f, 281−282, 282f
Nitriles, 198, 205, 301, 304
Nitrogen insertion, 535−537, 535−536f
Nitrogen nucleophiles, 319−326, 320−322f, 323t, 325−326f

Nitrones, 370−373, 371−373f
Nitrosobenzene, 523−524, 524−525f
NMR. *See* Nuclear magnetic resonance
Nonbonded interactions, *47*, 468−469, 468f
Nonlinear effects, 28−33, 29−32f, 200
Norpectinatone, 228, 229f
Noyori catalyst, 199−200, 200f
Noyori's reagent. *See* BINAL-H
Nuclear magnetic resonance (NMR), 80−86
  CDAs in, 80−85, 82f, 84f
  CSAs in, 80−81, 85−86, 86f
Nucleophilic addition. *See also specific nucleophilic additions*
  Cram's rule, cyclic model of, *40*, 191−196, 191f, 192t, 193f−196f
  Cram's rule, open chain model of, *40−41*, 180−191, 180−187f, 181t, 188t, 189−191f
Nucleophilic epoxidation reactions, 499−502, 502f

O

Ocoteine, 161f
Olefins, electron-poor
  epoxidation of, 499−503, 502−503f
  reduction of, 472−473, 473−474f
Oleandolide, 300
Olefins, simple, epoxidations of, 497−498, 497f, 498t, 499, 500f
Open chain model. *See* Cram's rule
Open transition structures, 254−255, 256f, 257, 259, 260f, 261, 262f, 267−268, 267f
Oppolzer's sultams, 123t
Optical purity, 43, *47*, 89−90
Optical rotatory dispersion (ORD), 89−91
Organoboranes, chiral, reductions with, 442−443, 442−443f, 444t
Organocatalysis
  for aldehyde and ketone reduction, 473−474, 475f
  for aldol reactions, 245, 256−257, 285−294, 286−288f, 289t, 292−293f, 294t
  for conjugate additions of nitrogen nucleophiles, 321−322
  for Diels−Alder reactions, 344, 344f, 351−357, 352f, 353t, 354−356f, 360, 360f
  for hetero Diels−Alder reaction, 368, 369f
  for hydroxylations, 523−524, 525f
  for imine reduction, 462−463, 464f
  for Michael additions, 318−319, 318f
Organolithiums, 97, 97f, 151−167, 152f
  asymmetric deprotonations, 161−165, 162−166f
  functionalized, 157, 157f
  inversion dynamics of, 153−156, 153−156f
  stereochemically defining step in reactions of, 157−160, 158−161f
  unstabilized, 165−167, 167t, 168f

Organometallic compounds
    for additions to aldehydes, 197–202, 198–199f, 201–202f, 202t
    for additions to azomethines, 203–211, 206t, 207–210f, 210–211t, 212f
    for additions to pyridinium ions, 212–214, 212–213f, 214t, 215f
    allyl, 254
    for conjugate additions to acyclic esters and ketones, 223–226, 224–225t, 228, 229f
Organozinc compounds, 197–202, 198–202f, 202t, 210–211, 210f, 211t, 212f, 236, 236t
Osmylation reactions, 510–511, 512f, 520
Overberger's base, 416
Overhauser effect, 348
Overman rearrangement, 419, 419f
Oxazaborolidines, 218, 219f, 276–277, 276–277f, 276t
    catalytic cycle with, 435–436, 436f
    in Diels–Alder reactions, 348, 348f
    examples of, 435, 435f
    in reductions, 434–441, 435–436f, 436t, 438f, 441f
Oxazinones, 117, 118t, 144f
Oxazolidinethiones, 230f, 231, 264
Oxazolidinone imides, 267–269, 267f, 268t
Oxazolidinones, 121–122, 121–122f, 122t, 130–131, 131f, 230f, 231, 287
Oxazolines, 120–122, 121f, 158–159, 158f, 207–208, 208–209f
Oxidations, 491–544. *See also* Epoxidations
    aminations and halogenations, 524–529, 526–527f, 528t, 529f
    asymmetric dihydroxylation reactions, 510–521
    of carbon–hydrogen bonds, 531, 532–533f
    carbonyl groups, 521–529, 521f
    hydroxylations, 522–524, 522–525f
    ring expansions, 531–537, 534f, 536f
    of sulfides, 529–531, 530f
Oximes, *35*, 204, 322, 325, 325f, 435, 455
Oxone, 499, 500f, 501t
Oxygen nucleophiles, 325, 325f
Oxygen ylide, 413, 413f

P

Palladium complexes
    in allylic alkylations, 145–150, 145f, 147–149f, 151
    in conjugate additions to cyclic ketones and lactones, 234–235, 234t, 235f
Parallel kinetic resolution, 16–17, 17f
Peak coalescence, 77, 77f
Pectinatone, 130, 130f
Peracids, 491–492, 492f, 499

Percent diastereoselectivity (% ds), 9, 9f, *41*
Percent enantioselectivity (% es), 9, 9f, *44*
Phenserine, 140, 141f
Phenylalaninol, 264, 264f
Phosphaimidazolidine, 314, 314f
Phosphaoxazolidine, 314, 314f
Phosphoramides, 256–257, 357, 358f
Phosphoramidite ligand, 224, 232, 236
Phosphorous-stabilized allyllithiums, 312, 313f, 313t, 314, 314f
Phthioceranic acid, 228, 229f
Physostigmine, 140, 141f
Pictet–Spengler cyclization, 204–205
α-Pinene, 117–119, 119f, 119t
Pirkle columns, 77–79, 79f
PNNP ligand, 390
Polarimetry, 86–91, 86–87f
Polypropionate, 245
Porphyrin, 390, 390f, 497
Povarov reaction, 368, 369f
Prelog–Djerassi lactone, 121–122, 122f, 295–296
Prochiral molecule, *49*
    acceptor, 301
    donor, 301
Proline, 279, 285–288, 286–288f
Prolinol amides, 120–122, 121f
Prolinol methyl ether, 308, 309–310f
Propargylboranes, 247f, 253
ProPhenol ligand, 278f, 282, 282f
Pseudoaxialphenyls, 448
Pseudoephedrine amides, 123–125, 124–125f, 125t, 229–230, 230f, 304, 305f
PyBox. *See* Pyridyl Box
Pyridinium ions, 212–214, 212–213f, 214t, 215f
Pyridyl Box (PyBox), 278f, 282
Pyrrolidinoindoline alkaloids, 140–142, 141–142f
Pyrrolidinone auxiliary, 341, 341f

Q

Quasi-enantiomers, 16–17, 17f
Quinine, 318, 318f
Quinine-derived phase-transfer catalyst, 499–502

R

Racemization, *51*, 114–116, 115–116f, 117t, 451, 451f
*R*AMP hydrazones. *See* SAMP/RAMP hydrazones
Rate acceleration, 124, 493
Rate-determining step, 160, 217–219, 226, 399, 466
re, 33, *52*
Reagent-based stereocontrol, 15, 15f, 225t, 295–297, 296f, 506
Reagents. *See also* Chiral reagents
    stereospecific, 254

Rearrangements, 395–421
  [1,3]-hydrogen shifts, 395–400, 396–401f
  [2,3]-Wittig, 401–416, 402–403f
  Claisen, 416, 417–419, 417–418f
  C-to-C migration, 421, 421f
  Overman, 419, 419f
  Steglich, 420–421, 420f
Reductions
  of acetamido cinnamates, 463–470, 465–472f, 472
  of carbon–carbon bonds, 463–475
  of carbon–heteroatom double bonds, 431–463
  catalytic transfer hydrogenation, 449–452,
    449–452f, 453t, 454, 454f
  with chiral organoboranes, 442–443, 442–443f,
    444t
  chiral transition metal catalysts, 444–463
  of electron-poor olefins, 472–473, 473–474f
  of imines, 455–463, 456–457f, 457t, 459–461f,
    462t, 464f
  of ketones with hydrogen, 444–445, 445f, 446t,
    447–448, 447–448f
  with modified borane, 434–441, 435–436f, 436t,
    438f, 441f
  with modified LAH, 431–434, 432f, 433t, 434f
Reductive aldol reactions, 283, 284f, 284t
Reductive aminations, 462–463, 464f
Reductive elimination, 146, 146f, 226, 235, 466, 466f
Reframoline, 161f
Relative topicity, of Trost auxiliary, 342–343, 343f
Resolution, 53. See also Dynamic resolutions; Kinetic
    resolution
  of chromatographic peaks, 75–76, 76f
  classical, 2, 16, 64–65t, 68, 69f
Retention time, 74–75, 74f
Retro aldol addition, 262
Rhodium carbenes, 379–380
Rhodium ligands, 235, 235f
Rigid model. See Cram's rule
Roush's tartrate ligand, 252

S

Salen catalysts, 217–218, 217f, 278f, 279–280,
    497–498, 497f, 509–510, 509f
SAMP/RAMP hydrazones, 128–131, 128f–130f,
    129t, 203, 206t, 207, 207f, 264, 306, 306f
Schmidt reaction, 531–532, 534f, 536–537, 536f
S_E2inv, 53, 152, 152f, 154–156, 154f, 159f, 162f,
    168f, 161–162, 167
S_E2ret, 53, 152, 152f, 154–156, 154f, 159f,
    161–162, 162f, 167, 168f
Selectivity factor, 16, 18–20, 19f, 53, 155–156, 156f
Self-disproportionation of enantiomers, 71, 72f, 73
Self-immolative chirality transfer process, 442, 442f

Self-regeneration of stereocenters (SRS), 111–114,
    111–112f, 111t, 232–233
Semipinacol rearrangement, 421
Separability factor. See Chromatographic separability
    factor
SFC. See Supercritical fluid chromatography
Sharpless asymmetric epoxidation, 15, 15f, 19–20,
    20f. See also Katsuki–Sharpless
    asymmetric epoxidation
Shi catalyst, 499, 500f, 501t
Si, 33, 52
Sigmatropic rearrangements, 395
Silanes, 252, 254–257, 255f, 258f
Silyl ethers, 259, 518–519, 520f
Simmons–Smith reaction, 375, 377–379, 378f, 381,
    382f
Single asymmetric induction, 12–15, 12–14f
  aldol additions, 263–284, 264–265f, 266t, 267f,
    268t, 270–273f, 274t, 275–278f, 276t,
    281t, 282–284f, 284t
  1,2-allylations, 249–250, 249t
Soderquist reagents, 247f, 253
SOMO activation, 131–132, 132f, 133t
Soraphen $A_{1\alpha}$, 196, 196f
sp^2-hybridized electrophiles, 139, 140f
Sparteine, 161, 162–165f, 163–165, 269, 279–280
Specific rotation, 54, 86, 88–90
SRS. See Self-regeneration of stereocenters
Staudinger reaction, 205
Steglich rearrangement, 420–421, 420f
Stereoisomer analysis, 71–73, 72f
  chiroptical methods in, 86–91, 86–87f
  chromatography in, 73–80, 74f, 76–80f
  NMR in, 80–86, 82f, 84f, 86f
Stereoselectivity, 55
  achievement of, 4–7, 4–6f
  entropy and, 10–12, 11f
  isoinversion principle and, 10–12, 11f
  kinetic and thermodynamic control in, 7–9, 8–9f
  temperature effect on, 10–12, 11f
Still–Wittig rearrangement, 413–415, 414f
Stoichiometric inducers, in aldol additions, 271–275,
    272–273f, 274t, 275f
Strecker reaction, 219–221, 220–221f
Substrate control, 506
Sulfides, oxidations of, 529–531, 530f
Sulfinimines, 209, 209f, 210t
Sulfur nucleophiles, 325–326, 326f
Sulfur-stabilized allyllithiums, 312, 313–314f, 313t, 314
Sultams, 122–123, 123f, 123t, 230, 230f, 264, 270
Supercritical fluid chromatography (SFC), 73–74
Synclinal, 35, 56–57
  allylations, 255–257, 256f, 258f

T

TADDOL, 15, 15f, 198f, 201, 202f, 349–351, 350f, 350t, 357, 358f, 364–365, 365f
Takasago process, 396, 397f, 398
Tartrate–titanium epoxidation system, 507–508
β-Tetralone, 309, 310f
Tetraphenyl dimethyldioxolane ligand, 349–350
Theoretical plates, 74–76
Thermodynamic control, 7, 7–9, 8–9f
  of aldol additions, 262
Thermodynamic dynamic resolutions, 25–28, 27–28f, 157–158, 158f
Thiazolidincthione, 269, 270f, 281
Thorpe–Ingold effect, 56, 277
Tin catalysts, allyl, 254–255, 255f, 256
Tin triflates, 277, 278f
Titanium complexes, 216, 216f, 217–218, 217f
Titanium enolates, 267–268, 269
Titanocene
  asymmetric imine reduction with, 456–457, 457f, 457t, 460, 460f
  catalytic cycle for imine reduction with, 458, 459f
  transition structures in imine reduction with, 458–460, 459f
TMEDA, 99, 103, 162f, 166–167f, 269, 270f, 305f
α-Tocopherol, 120, 120f
π-Transfer additions, 245
Transition metal catalysts, for aldol additions, 277
Transition metal enolates, 258, 261
Transition state theory, 7–8, 155
Triethyl amine, 274–275
Trimethylsilyl cyanide, 215–216, 216–219f
Trimethylsilyl enol ethers, 270
Trimethylsilyl ketene acetals, 526–527, 526f
Trocade, 121–122, 122f
Trost auxiliary, 278f, 282, 282f, 342–343, 343f
Turnover number, 3, 28–29

U

Umpolung, 131
Unstabilized organolithiums, 165–167, 167t, 168f
α,β-Unsaturated acid chlorides, 347
α,β-Unsaturated aldehydes, 301, 473
α,β-Unsaturated carbonyls, 258
α,β-Unsaturated esters, 321, 473f, 503

V

Valine, 264, 264f, 279, 306, 307f
Valinol, 126, 212, 264, 279, 435
van der Waals interactions, 57, 260–261, 263, 272
Vinyl bromides, 252
Vanadium, 493, 504

W

Whelk-O, 78, 79f
Wieland–Meischer ketone, 285–286
[1,2]-Wittig rearrangements, 401–402, 402f
[2,3]-Wittig rearrangements, 401–416, 402–403f
  asymmetric induction, 415–416, 415–416f
  chirality transfer in enantiopure educts, 410–415, 411f–414f
  simple diastereoselectivity, 404–409

X

X-14547A, 130, 130f

Y

Yohimbone, 161f
Ytterbium/BINOL, 322

Z

Z-crotyl, 248–249, 249f, 250t
Zimmerman–Traxler transition structure, 245, 246f, 259, 405, 529
Zinc catalysts, 236, 236t. See also Organozinc compounds
Zirconium enolates, 261, 262f, 264